OXFORD MASTER SERIES IN ATOMIC, OPTICAL, AND LASER PHYSIC

OXFORD MASTER SERIES IN PHYSICS

The Oxford Master Series is designed for final year undergraduate and beginning graduate students in physics and related disciplines. It has been driven by a perceived gap in the literature today. While basic undergraduate physics texts often show little or no connection with the huge explosion of research over the last two decades, more advanced and specialized texts tend to be rather daunting for students. In this series, all topics and their consequences are treated at a simple level, while pointers to recent developments are provided at various stages. The emphasis in on clear physical principles like symmetry, quantum mechanics, and electromagnetism which underlie the whole of physics. At the same time, the subjects are related to real measurements and to the experimental techniques and devices currently used by physicists in academe and industry. Books in this series are written as course books, and include ample tutorial material, examples, illustrations, revision points, and problem sets. They can likewise be used as preparation for students starting a doctorate in physics and related fields, or for recent graduates starting research in one of these fields in industry.

CONDENSED MATTER PHYSICS

1. M.T. Dove: *Structure and dynamics: an atomic view of materials*
2. J. Singleton: *Band theory and electronic properties of solids*
3. A.M. Fox: *Optical properties of solids, second edition*
4. S.J. Blundell: *Magnetism in condensed matter*
5. J.F. Annett: *Superconductivity, superfluids, and condensates*
6. R.A.L. Jones: *Soft condensed matter*
17. S. Tautz: *Surfaces of condensed matter*
18. H. Bruus: *Theoretical microfluidics*
19. C.L. Dennis, J.F. Gregg: *The art of spintronics: an introduction*

ATOMIC, OPTICAL, AND LASER PHYSICS

7. C.J. Foot: *Atomic physics*
8. G.A. Brooker: *Modern classical optics*
9. S.M. Hooker, C.E. Webb: *Laser physics*
15. A.M. Fox: *Quantum optics: an introduction*
16. S.M. Barnett: *Quantum information*

PARTICLE PHYSICS, ASTROPHYSICS, AND COSMOLOGY

10. D.H. Perkins: *Particle astrophysics, second edition*
11. Ta-Pei Cheng: *Relativity, gravitation and cosmology, second edition*

STATISTICAL, COMPUTATIONAL, AND THEORETICAL PHYSICS

12. M. Maggiore: *A modern introduction to quantum field theory*
13. W. Krauth: *Statistical mechanics: algorithms and computations*
14. J.P. Sethna: *Statistical mechanics: entropy, order parameters, and complexity*
20. S.N. Dorogovtsev: *Lectures on complex networks*

Laser Physics

SIMON HOOKER
and
COLIN WEBB

Department of Physics, University of Oxford

OXFORD
UNIVERSITY PRESS

Great Clarendon Street, Oxford OX2 6DP

Oxford University Press is a department of the University of Oxford.
It furthers the University's objective of excellence in research, scholarship,
and education by publishing worldwide in

Oxford New York

Auckland Cape Town Dar es Salaam Hong Kong Karachi
Kuala Lumpur Madrid Melbourne Mexico City Nairobi
New Delhi Shanghai Taipei Toronto

With offices in

Argentina Austria Brazil Chile Czech Republic France Greece
Guatemala Hungary Italy Japan Poland Portugal Singapore
South Korea Switzerland Thailand Turkey Ukraine Vietnam

Oxford is a registered trade mark of Oxford University Press
in the UK and in certain other countries

Published in the United States
by Oxford University Press Inc., New York

© Oxford University Press 2010

The moral rights of the authors have been asserted
Database right Oxford University Press (maker)

First published 2010
Reprinted 2011, 2012

All rights reserved. No part of this publication may be reproduced,
stored in a retrieval system, or transmitted, in any form or by any means,
without the prior permission in writing of Oxford University Press,
or as expressly permitted by law, or under terms agreed with the appropriate
reprographics rights organization. Enquiries concerning reproduction
outside the scope of the above should be sent to the Rights Department,
Oxford University Press, at the address above

You must not circulate this book in any other binding or cover
and you must impose the same condition on any acquirer

British Library Cataloguing in Publication Data

Data available

Library of Congress Cataloging in Publication Data

Hooker, Simon, 1965–
Laser physics / Simon Hooker and Colin Webb.
    p. cm.
Includes bibliographical references and index.
ISBN 978-0-19-850692-8—ISBN 978-0-19-850691-1   1. Lasers.
2. Optics.   3. Light.   I. Webb, Colin E., 1937- II. Title.
QC688.H66 2010
621.36'6—dc22   2010019282

Typeset by SPI Publisher Services, Pondicherry, India
Printed in Great Britain
on acid-free paper by
CPI Group (UK) Ltd, Croydon, CR0 4YY

ISBN 978–0–19–850691–1 (hbk)
      978–0–19–850692–8 (pbk)

3 5 7 9 10 8 6 4

Preface

This book aims to give a comprehensive description of the physics underlying the operation of lasers. It is intended for undergraduates in the third or final year of a four-year physics course, and for graduate students—either those engaged in research into new laser systems, or those who wish to know more about the 'black box' in the middle of their optical table.

Much of the material included stems from a final-year option course on laser physics developed and taught by CEW for many years. This formed the basis of a short introductory course for third-year physicists, and an optional final-year course, both presently taught by SMH. We assume that the reader is familiar with the core material of the first two years of an undergraduate degree in physics, and in particular electromagnetism, quantum mechanics, and atomic physics.

The book roughly divides into: (i) chapters describing the physics of laser operation, usually without reference to any particular system; and (ii) chapters giving a detailed description of some representative types of laser. Laser oscillation has been demonstrated on thousands of transitions in atoms, ions, and molecules and in the gas, liquid, solid, and plasma states. A book of this type can only include a tiny fraction of these; the laser systems included here have been selected because they illustrate one or more of the key concepts described in the introductory chapters, or because they are of particular technical or scientific importance. If the laser system you are working with is not included, we hope that nevertheless the material covered here will help you understand its operation.

Most topics are dealt with in a single chapter, rather than being covered in several places at an increasingly advanced level. Although this approach helps reduce repetition, and allows a comprehensive understanding of each subject, we recognize that a newcomer to the field cannot be expected to master all the material in one go. We have indicated the more advanced sections by a dagger†; these may be safely skipped on a first reading. We would encourage the reader to revisit these as their expertise develops.

We have included problems at the end of each chapter. These are designed to help test understanding, or to lead the reader through proofs that would otherwise be distracting. The problems range from simple numerical calculations, designed to give an appreciation of the orders of magnitude involved, to more advanced questions (also denoted by a dagger†).

We hope that the book will be used in many ways: as an introduction to the subject; as a text for intermediate-level courses; and as a reference for those using lasers in their work. A suggested introductory course is given in the table

Table 1 Suggested sections for an introductory course on laser physics.

Chapter	Topics	Sections[a]	Suggested problems
1	Electromagnetic radiation	all	1.1, 1.2, 1.3, 1.4
2	Einstein treatment, conditions for optical gain	all	2.1, 2.2, 2.3, 2.4
3	Line broadening	all	3.1, 3.2, 3.3, 3.4, 3.5
4	The optical gain cross-section, gain narrowing	all	4.1, 4.2, 4.3, 4.4, 4.5
5	Gain saturation	all	5.1, 5.3, 5.4, 5.5, 5.11
6	Optical cavities, Gaussian beams, laser oscillation, output power	all	6.1, 6.2, 6.3, 6.4, 6.5, 6.12, 6.13
7	Solid-state lasers, the ruby and Nd:YAG lasers	7.1, 7.2.1–7.2.5, 7.3.2	7.1, 7.2, 7.3, 7.7

[a] Advanced[†] sections should be omitted.

above; we trust that more advanced readers will be able to identify the sections relevant to their needs.

Many students and colleagues have contributed to the development of the text. We would like to record our thanks to the generations of undergraduate and graduate students who road-tested many of the problems and who, by constant questioning, helped reduce our confusion to its current level. We would like to express particular thanks to Geoffrey Brooker for detailed and constructive criticism of several chapters, for help with the technicalities of producing a textbook, and for Fig. 6.12. We are grateful to several colleagues for careful reading of draft sections of the book: Patrick Baird and Ian Walmsley of the Department of Physics at Oxford; Roy Taylor and Gareth Parry of Imperial College, London; and Mark Fox of Sheffield University. We thank Rodney Loudon of the University of Essex, for many informative discussions. CEW would also like to record his thanks to Marty Fejer of Stanford University for very informative discussions on the teaching of semiconductor laser physics; and to Johan Nilsson of the Optoelectonics Research Centre of the University of Southampton and Cyril Renaud of University College, London for providing the references for the data of Table 10.5.

Finally, we are sincerely grateful to our friends and family, who provided a healthy mixture of tolerance, encouragement and distraction.

Website:
www.physics.ox.ac.uk/users/hooker. This site contains supplementary information and corrections.

Contents

1 Introduction — 1
  1.1 The laser — 1
  1.2 Electromagnetic radiation in a closed cavity — 3
    1.2.1 The density of modes — 7
  1.3 Planck's law — 7
    1.3.1 The energy density of blackbody radiation — 8
  Further reading — 9
  Exercises — 9

2 The interaction of radiation and matter — 12
  2.1 The Einstein treatment — 12
    2.1.1 Relations between the Einstein coefficients — 14
  2.2 Conditions for optical gain — 16
    2.2.1 Conditions for steady-state inversion — 16
    2.2.2 Necessary, but not sufficient condition — 18
  2.3 The semi-classical treatment† — 19
    2.3.1 Outline — 19
    2.3.2 Selection rules for electric dipole transitions — 20
  2.4 Atomic population kinetics† — 21
    2.4.1 Rate equations — 22
    2.4.2 Semi-classical equations — 22
    2.4.3 Validity of the rate-equation approach — 23
  Further reading — 24
  Exercises — 25

3 Broadening mechanisms and lineshapes — 27
  3.1 Homogeneous broadening mechanisms — 27
    3.1.1 Natural broadening — 27
    3.1.2 Pressure broadening — 32
    3.1.3 Phonon broadening — 35
  3.2 Inhomogeneous broadening mechanisms — 35
    3.2.1 Doppler broadening — 36
    3.2.2 Broadening in amorphous solids — 38
  3.3 The interaction of radiation and matter in the presence of spectral broadening — 38
    3.3.1 Homogeneously broadened transitions — 38
    3.3.2 Inhomogeneously broadened atoms† — 39
  3.4 The formation of spectral lines: The Voigt profile† — 40

		3.5 Other broadening effects	42
		3.5.1 Self-absorption	42
	Further reading		43
	Exercises		43

4 Light amplification by the stimulated emission of radiation — 46

4.1 The optical gain cross-section — 46
    4.1.1 Condition for optical gain — 48
    4.1.2 Frequency dependence of the gain cross-section — 48
    4.1.3 The gain coefficient — 49
    4.1.4 Gain narrowing — 49
4.2 Narrowband radiation — 50
    4.2.1 Amplification of narrowband radiation — 50
    4.2.2 Form of rate equations — 51
4.3 Gain cross-section for inhomogeneous broadening† — 52
4.4 Orders of magnitude — 53
4.5 Absorption — 54
    4.5.1 The absorption cross-section — 54
    4.5.2 Self-absorption — 55
    4.5.3 Radiation trapping — 56
Further reading — 56
Exercises — 57

5 Gain saturation — 60

5.1 Saturation in a steady-state amplifier — 60
    5.1.1 Homogeneous broadening — 60
    5.1.2 Inhomogeneous broadening† — 67
5.2 Saturation in a homogeneously broadened pulsed amplifier† — 73
5.3 Design of laser amplifiers — 77
Exercises — 78

6 The laser oscillator — 83

6.1 Introduction — 83
6.2 Amplified spontaneous emission (ASE) lasers — 83
6.3 Optical cavities — 85
    6.3.1 General considerations — 85
    6.3.2 Low-loss (or 'stable') optical cavities — 89
    6.3.3 High-loss (or 'unstable') optical cavities† — 97
6.4 Beam quality† — 103
    6.4.1 The M^2 beam-propagation factor — 103
6.5 The approach to laser oscillation — 106
    6.5.1 The 'cold' cavity — 106
    6.5.2 The laser threshold condition — 110
6.6 Laser oscillation above threshold — 111
    6.6.1 Condition for steady-state laser oscillation — 112
    6.6.2 Homogeneously broadened systems — 113

	6.6.3 Inhomogeneously broadened systems[†]	115
6.7	Output power	117
	6.7.1 Low-gain lasers	117
	6.7.2 High-gain lasers: the Rigrod analysis[†]	120
	6.7.3 Output power in other cases	123
	Further reading	123
	Exercises	123

7 Solid-state lasers — 132

7.1	General considerations	132
	7.1.1 Energy levels of ions doped in solid hosts[†]	132
	7.1.2 Radiative transitions[†]	137
	7.1.3 Non-radiative transitions[†]	138
	7.1.4 Line broadening[†]	142
	7.1.5 Three- and four-level systems	142
	7.1.6 Host materials	146
	7.1.7 Techniques for optical pumping	149
7.2	Nd^{3+}:YAG and other trivalent rare-earth systems	157
	7.2.1 Energy-level structure	157
	7.2.2 Transition linewidth	157
	7.2.3 Nd:YAG laser	158
	7.2.4 Other crystalline hosts	163
	7.2.5 Nd:glass laser	164
	7.2.6 Erbium lasers	165
	7.2.7 Praseodymium ions	169
7.3	Ruby and other trivalent iron-group systems	169
	7.3.1 Energy-level structure[†]	169
	7.3.2 The ruby laser	174
	7.3.3 Alexandrite laser	177
	7.3.4 Cr:LiSAF and Cr:LiCAF	180
	7.3.5 Ti:sapphire	180
	Further reading	184
	Exercises	184

8 Dynamic cavity effects — 188

8.1	Laser spiking and relaxation oscillations	188
	8.1.1 Rate-equation analysis	190
	8.1.2 Analysis of relaxation oscillations	190
	8.1.3 Numerical analysis of laser spiking	192
8.2	Q-switching	193
	8.2.1 Techniques for Q-switching	194
	8.2.2 Rate-equation analysis of Q-switching	198
	8.2.3 Comparison with numerical simulations	203
8.3	Modelocking	203
	8.3.1 General ideas	204
	8.3.2 Simple treatment of modelocking	206
	8.3.3 Active modelocking techniques	208
	8.3.4 Passive modelocking techniques	214

	8.4	Other forms of pulsed output	221
	Further reading		222
	Exercises		222

9 Semiconductor lasers — 226

- 9.1 Basic features of a typical semiconductor diode laser — 226
- 9.2 Review of semiconductor physics — 228
  - 9.2.1 Band structure — 228
  - 9.2.2 Density of states and the Fermi energy ($T = 0K$) — 231
  - 9.2.3 The Fermi–Dirac distribution ($T \neq 0\,K$) — 232
  - 9.2.4 Doped semiconductors — 233
- 9.3 Radiative transitions in semiconductors — 235
- 9.4 Gain at a p-i-n junction — 236
- 9.5 Gain in diode lasers — 238
- 9.6 Carrier and photon confinement: the double heterostructure — 241
- 9.7 Laser materials — 243
- 9.8 Quantum-well lasers[†] — 244
- 9.9 Laser threshold — 247
- 9.10 Diode laser beam properties — 250
  - 9.10.1 Beam shape — 250
  - 9.10.2 Transverse modes of edge-emitting lasers — 250
  - 9.10.3 Longitudinal modes of diode lasers — 251
  - 9.10.4 Single longitudinal mode diode lasers — 253
  - 9.10.5 Diode laser linewidth — 254
  - 9.10.6 Tunable diode laser cavities[†] — 255
- 9.11 Diode laser output power[†] — 257
- 9.12 VCSEL lasers[†] — 259
- 9.13 Strained-layer lasers — 261
- 9.14 Quantum cascade lasers[†] — 262
- Further reading — 264
- Exercises — 264

10 Fibre lasers — 267

- 10.1 Optical fibres — 267
  - 10.1.1 The importance of optical-fibre technology — 267
  - 10.1.2 Optical-fibre properties: Ray optics — 268
  - 10.1.3 Optical-fibre properties: Wave optics — 271
  - 10.1.4 Dispersion in optical fibres — 274
  - 10.1.5 Fabrication of optical fibres — 276
  - 10.1.6 Fibre-optic components — 277
- 10.2 Wavelength bands for fibre-optic telecommunications — 280
- 10.3 Erbium-doped fibre amplifiers — 282
  - 10.3.1 Energy levels and pumping schemes — 282
  - 10.3.2 Gain spectra — 282
  - 10.3.3 EDFA design and layout — 284
  - 10.3.4 Fabrication of erbium-doped fibre amplifiers — 285
- 10.4 Fibre Raman amplifiers — 285
  - 10.4.1 Introduction — 285

		10.4.2 Raman scattering	285

        10.4.2 Raman scattering   285
        10.4.3 Fibre Raman amplifiers   286
        10.4.4 Long-haul optical transmission systems   287
   10.5 High-power fibre lasers   289
        10.5.1 The revolution in fibre-laser performance   289
        10.5.2 Cladding-pumped fibre-laser design   290
        10.5.3 Materials and mechanisms of cladding-pumped fibre-laser systems   291
        10.5.4 High-power fibre lasers: Linewidth considerations   291
   10.6 High-power pulsed fibre lasers   293
        10.6.1 Large mode area (LMA) fibres   293
        10.6.2 Q-switched fibre lasers   294
        10.6.3 Oscillator–amplifier pulsed fibre lasers   294
   10.7 Applications of high-power fibre lasers   295
   Further reading   296
   Exercises   296

11 Atomic gas lasers   **298**
   11.1 Discharge physics interlude   298
        11.1.1 Low-pressure and high-pressure discharges   298
        11.1.2 Low-pressure glow discharge   299
        11.1.3 Temperatures   300
        11.1.4 The steady-state positive column   303
        11.1.5 Ionization rates   306
        11.1.6 Excitation rates   307
        11.1.7 Second-kind or superelastic collisions   310
        11.1.8 Excited-state populations in low-pressure discharges   311
   11.2 The helium-neon laser   314
        11.2.1 Introduction   314
        11.2.2 Energy levels, transitions and excitation mechanisms   316
        11.2.3 Laser construction and operating parameters   318
        11.2.4 Output-power limitations of the He-Ne laser   319
        11.2.5 Applications of He-Ne lasers   321
   11.3 The argon-ion laser   321
        11.3.1 Introduction   321
        11.3.2 Energy levels, transitions and excitation mechanisms   322
        11.3.3 Laser construction and operating parameters   325
        11.3.4 Argon-ion laser: Power limitations   327
        11.3.5 Krypton-ion lasers   328
        11.3.6 Applications of ion lasers   329
   Further reading   329
   Exercises   329

12 Infra-red molecular gas lasers   **332**
   12.1 Efficiency considerations   332
        12.1.1 Energy levels of atoms and molecules   332
        12.1.2 Quantum ratio   333

12.2 Partial population inversion between vibrational energy levels of molecules   335
12.3 Physics of the CO_2 laser   338
  12.3.1 Levels and lifetimes   338
  12.3.2 The effect of adding N_2 341
  12.3.3 Effect of adding He   342
12.4 CO_2 laser parameters   343
12.5 Low-pressure c.w. CO_2 lasers   344
12.6 High-pressure pulsed CO_2 lasers   346
12.7 Other types of CO_2 laser   349
  12.7.1 Gas-dynamic CO_2 lasers   349
  12.7.2 Waveguide CO_2 lasers   351
12.8 Applications of CO_2 lasers   351
Further reading   352
Exercises   352

13 Ultraviolet molecular gas lasers   355

13.1 The UV and VUV spectral regions   355
13.2 Energy levels of diatomic molecules   356
  13.2.1 Separation of the overall wave function   356
  13.2.2 Vibrational eigenfunctions   357
13.3 Electronic transitions in diatomic molecules: The Franck–Condon principle   358
  13.3.1 Absorption transitions   358
  13.3.2 The 'Franck–Condon loop'   360
13.4 The VUV hydrogen laser   361
13.5 The UV nitrogen laser   364
13.6 Excimer molecules   364
13.7 Rare-gas excimer lasers   367
13.8 Rare-gas halide excimer lasers   370
  13.8.1 Spectroscopy of the rare-gas halides   370
  13.8.2 Rare-gas halide laser design   371
  13.8.3 Pulse-length limitations of discharge-excited RGH lasers   373
  13.8.4 Cavity design and beam properties of RHG lasers   373
  13.8.5 Performance and applications of RGH excimer laser   375
Further reading   377
Exercises   378

14 Dye lasers   380

14.1 Introduction   380
14.2 Dye molecules   380
14.3 Energy levels and spectra of dye molecules in solution   382
  14.3.1 Energy-level scheme   382
  14.3.2 Singlet–singlet absorption   382
  14.3.3 Singlet–singlet emission spectra   385
  14.3.4 Triplet–triplet absorption   387
14.4 Rate-equation models of dye laser kinetics   387

14.5	Pulsed dye lasers	388
	14.5.1 Flashlamp-pumped systems	388
	14.5.2 Dye lasers pumped by pulsed lasers	389
14.6	Continuous-wave dye lasers	391
	14.6.1 Population kinetics	391
	14.6.2 Continuous waves dye laser design	393
14.7	Solid-state dye lasers	395
14.8	Applications of dye lasers	396
	Further reading	398
	Exercises	398

15 Non-linear frequency conversion — 400

15.1	Introduction	400
15.2	Linear optics of crystals	400
	15.2.1 Classes of anisotropic crystals	400
	15.2.2 Vectors	402
	15.2.3 Field directions for o- and e-rays in a uniaxial crystal	403
15.3	Basics of non-linear optics	405
	15.3.1 Maxwell's equations for non-linear media	405
	15.3.2 Second-harmonic generation in anisotropic crystals	406
	15.3.3 The requirement for phase matching	408
15.4	Phase-matching techniques	409
	15.4.1 Birefringent phase matching in uniaxial crystals	409
	15.4.2 Critical and non-critical phase matching	412
	15.4.3 Poynting vector walk-off in birefringent phase matching	414
	15.4.4 Other factors affecting SHG conversion efficiency	414
	15.4.5 Phase-matched SHG in biaxial crystals	415
	15.4.6 Birefringent materials for SHG	416
	15.4.7 Quasi-phase matching techniques	418
15.5	SHG: practical aspects	420
15.6	Three-wave mixing and third-harmonic generation (THG)	421
	15.6.1 Three-wave mixing processes in general	421
	15.6.2 Third-harmonic generation (THG)	423
15.7	Optical parametric oscillators (OPOs)	424
	15.7.1 Parametric interactions	424
	15.7.2 Optical parametric oscillators (OPOs)	425
	15.7.3 Practical parametric devices	426
	Further reading	428
	Exercises	428

16 Precision frequency control of lasers[†] — 431

16.1	Frequency pulling	431
16.2	Single longitudinal mode operation	433
	16.2.1 Short cavity	434
	16.2.2 Intra-cavity etalons	435
	16.2.3 Ring resonators	437
	16.2.4 Other techniques	440

	16.3 Output linewidth	440
		16.3.1 The Schawlow–Townes limit	441
		16.3.2 Practical limitations	444
		16.3.3 Intensity noise	446
	16.4 Frequency locking	448
		16.4.1 Locking to atomic or molecular transitions	450
		16.4.2 Locking to an external cavity	452
	16.5 Frequency combs	453
	Further reading	456
	Exercises	456

17 Ultrafast lasers **462**

17.1 Propagation of ultrafast laser pulses in dispersive media — 462
 17.1.1 The time–bandwidth product — 462
 17.1.2 General considerations — 463
 17.1.3 Propagation through a dispersive system — 466
 17.1.4 Propagation of Gaussian pulses — 469
 17.1.5 Non-linear effects: self-phase modulation and the B-integral — 472
17.2 Dispersion control — 474
 17.2.1 Geometric dispersion control — 474
 17.2.2 Chirped mirrors — 478
 17.2.3 Pulse shaping — 480
17.3 Sources of ultrafast optical pulses — 482
 17.3.1 Modelocked lasers — 482
 17.3.2 Oscillators — 483
 17.3.3 Chirped-pulse amplification (CPA) — 483
17.4 Measurement of ultrafast pulses — 489
 17.4.1 Autocorrelators — 489
 17.4.2 Methods for exact reconstruction of the pulse — 492
Further reading — 495
Exercises — 495

18 Short-wavelength lasers **502**

18.1 Definition of wavelength ranges — 503
18.2 Difficulties in achieving optical gain at short wavelengths — 503
 18.2.1 Pump-power scaling — 503
18.3 General properties of short-wavelength lasers — 505
 18.3.1 Travelling-wave pumping — 505
 18.3.2 Threshold and saturation behaviour in an ASE laser — 506
 18.3.3 Spectral width of the output — 508
 18.3.4 Coherence properties of ASE lasers — 509
18.4 Laser-generated plasmas[†] — 510
 18.4.1 Inverse bremsstrahlung heating — 510
 18.4.2 Generation of highly ionized plasmas from laser-solid interactions — 511
 18.4.3 Optical field ionization — 514
18.5 Collisionally excited lasers — 517

	18.5.1 Ne-like ions†	518
	18.5.2 Ni-like ions†	520
	18.5.3 Methods of pumping	520
	18.5.4 Collisionally excited OFI lasers	528
18.6	Recombination lasers	530
	18.6.1 H-like carbon	532
	18.6.2 OFI recombination lasers	533
18.7	Other sources	535
	18.7.1 High-harmonic generation	535
	18.7.2 Free-electron lasers	537
Further reading		541
Exercises		541

Appendix A: The semi-classical theory of the interaction of radiation and matter — **548**

A.1 The amplitude equations — 548
    A.1.1 Derivation of the amplitude equations — 548
    A.1.2 Solution of the amplitude equations — 550
A.2 Calculation of the Einstein B coefficient — 551
    A.2.1 Polarized atoms and radiation — 551
    A.2.2 Unpolarized atoms and/or radiation — 553
    A.2.3 Treatment of degeneracy — 554
A.3 Relations between the Einstein coefficients — 555
A.4 Validity of rate equations — 555

Appendix B: The spectral Einstein coefficients — **557**

Appendix C: Kleinman's conjecture — **560**

Bibliography — **563**

Index — **579**

Introduction

1.1 The laser

1.1 The laser	1
1.2 Electromagnetic radiation in a closed cavity	3
1.3 Planck's law	7
Further reading	9
Exercises	9

In the 1960s the laser was rather disparagingly characterized as a solution in search of a problem. Today lasers are everywhere. The enormous information-carrying capacity of the optical-fibre networks upon which internet, telephone, and television communications depend, and the information-storing capacity of CDs have been made possible by lasers. Most people are familiar with the use of lasers in eye surgery and other medical procedures, but few know the vital role that lasers play in the marking and machining of precision parts in industry, as well as cutting and welding of components in automobile manufacture. To list just a few of their many other applications, lasers are used in the detection of trace chemical species, such as atmospheric pollutants; to image fast events; to guide and aim weaponry; and in entertainment displays.

The invention of the laser is truly one of the towering achievements of the twentieth century. Many scientists and engineers played a part in its evolution. Foremost among these is Albert Einstein. In a famous paper published in 1917, he postulated the existence of the process of **stimulated emission** by which a beam of radiation passing through a medium can gain energy by inducing excited atoms to emit photons identical to those in the beam. However, stimulated emission has to compete with the inverse process of absorption by which energy is removed from the beam by promoting atoms in lower levels back to the excited state. In ordinary situations there are more atoms in lower energy levels than in excited levels and hence absorption wins over stimulated emission.

For many years it was thought that absorption must *always* dominate, but in 1954 the research group headed by Charles Townes at Columbia University showed otherwise.[1] They built a device in which a beam of excited molecules of ammonia gas (NH_3) was separated from their unexcited counterparts and directed into a rectangular metal box with highly reflecting sides.[2] The dimensions of the box were precisely a few half-wavelengths of the microwave frequency emitted by the excited molecules. As a result, standing electromagnetic waves were set up at this frequency, with the box acting as a resonant cavity, much as an organ pipe resonates to sound waves at a particular frequency. The name given by the Columbia group to the device, which acted as a very low noise amplifier for the microwaves, was **maser**—the acronym being formed from the initials of microwave amplification by stimulated emission of radiation see Fig. 1.1.

[1] See Gordon et al. (1954) and (1955).

[2] The first maser operated on the 'inversion transition' in NH_3 that occurs between the ground state and a level lying only 24 GHz above it. At room temperature the populations of these two levels are essentially equal. The population inversion was established by passing a beam of room-temperature NH_3 through a quadrupolar electrostatic field. This field provided a focusing force for those molecules in the upper level, but defocused molecules in the lower level. As a consequence, only molecules in the upper level were able to pass through a small hole and into the resonant microwave cavity. Hence the population inversion was formed simply by selecting those molecules in the upper level; this is quite different from the techniques discussed later in this book for establishing population inversions on visible and other high-energy transitions. For further details of the first maser see (Gordon et al., 1954; Gordon et al., 1955).

Fig. 1.1: Schematic diagram of the first maser. This operated on the 'inversion transition' in NH_3 that occurs between the ground state and a level lying only 24 GHz above it. At room temperature the populations of these two levels are essentially equal. The population inversion was established by passing a beam of room-temperature NH_3 through a quadrupolar electrostatic field, formed by an array of electrodes in the 'focuser.' This field provided a focusing force for those molecules in the upper level, but defocused molecules in the lower level. As a consequence, only molecules in the upper level were able to pass through a small hole and into the resonant microwave cavity. Hence, within the cavity a 'population inversion' was formed, i.e. there were more molecules in the upper level than in the lower level. This method of forming a population inversion—which is necessary for maser or laser action to occur—is quite different from the techniques used to generation population inversions on higher-energy transitions. Reprinted figure with permission from J. P. Gordon, H. J. Zeiger and C. H. Townes, *Physical Review* **99** 1264 (1955). ©(1955) by the American Physical Society. http://link.aps.org/doi/10.1103/PhysRev.99.1264

[3] See Schawlow and Townes (1958).

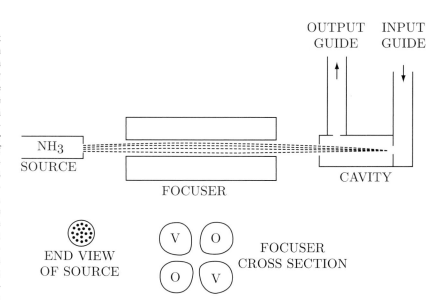

The extension of the maser principle to the visible and infra-red regions of the spectrum was envisioned in a paper by Arthur Schawlow and Charles Townes submitted to Physical Review in July 1958.[3] In that publication they predicted the properties of such a device, giving it the name the **optical maser**. One key feature was the design of the resonant cavity—no longer a six-sided box, but simply two parallel high-reflectance mirrors facing one another and separated now by many thousands of half-wavelengths (see Fig. 1.2). This feature was also described in an historic notebook entry of November 1957 by Gordon Gould. He called the device a **laser** substituting the word 'light' for 'microwave' in the maser acronym, and it is this name that has stuck.

Spurred by the Schawlow and Townes publication, researchers in many laboratories set out to demonstrate a working laser. The race was won by

Fig. 1.2: Schematic diagram of the first laser, the ruby laser demonstrated by Maiman et al. (1961). In this laser intense visible radiation from the flashlamp excited Cr^{3+} ions in the ruby crystal to the upper laser level. A temporary population inversion was generated with respect to transitions to the ground state, with a wavelength of 694 nm. The cavity was formed by the silvered ends of the ruby rod, one of which was only partially silvered to allow the laser radiation to be coupled out of the laser rod. Reprinted figure with permission from T. H. Maiman, R. H. Hoskins, I. J. D'Haenens, C. K. Asawa and V. Evtuhov, *Physical Review* **123** 1151 (1961). ©(1961) by the American Physical Society. http://link.aps.org/doi/10.1103/PhysRev.123.1151

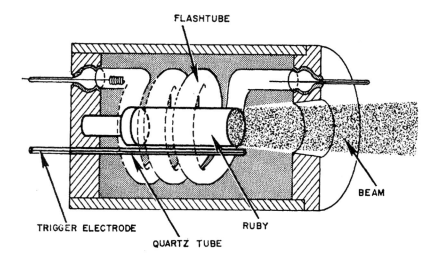

Theodore Maiman of Hughes Laboratories who, in May 1960, was able to obtain pulses of red laser light from a ruby crystal rod excited by light from a flash-tube. Later the same year, a continuously operating gas laser, with output at 1.15 μm in the near-infra-red region, was demonstrated at Bell Labs by Javan, Bennett, and Herriott.[4]

[4]See Maiman et al. (1961) and Javan et al. (1961). For an account of the race to build the first laser see Hecht (2005b; 2005a).

That was the just the beginning. From 1960 onwards, the number of different types of laser has proliferated and their uses have multiplied. The advances in wavelength coverage, power capabilities and miniaturization have been spectacular, and many new industries have grown up as a result.

Our aim in this book is to provide an understanding of the basic physics underlying laser operation. We shall not endeavour to review the almost endless list of laser systems. Instead, we shall only discuss those lasers that either illustrate a key principle of laser physics, or which are so important in terms of their applications that they could not be left out.

In order to set the foundations, we first investigate electromagnetic radiation from the classical viewpoint of Maxwell's equations.

1.2 Electromagnetic radiation in a closed cavity

In many situations involving the interaction of radiation and matter no cavity is involved, and the boundaries of the region under consideration may be ill-defined, or non-existent. However, from a theoretical standpoint it is convenient to consider a finite region of space within a well-defined cavity. Free space can then be thought of as the limit of the cavity becoming extremely large.

It will turn out that the details of the construction and shape of the cavity will not affect our conclusions, and hence for simplicity we consider a cubic cavity as illustrated schematically in Fig. 1.3. Again, for simplicity we will take the

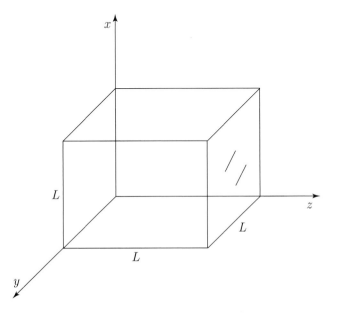

Fig. 1.3: A cubic cavity of side L with perfectly conducting walls.

[5] For a full discussion of Maxwell's equations consult any undergraduate textbook on electromagnetism, such as Bleaney and Bleaney (1989).

walls of the cavity to be perfect conductors, and the space within the cavity to be vacuum.

The electric field $\boldsymbol{E}$ must satisfy Maxwell's equations,[5] and in particular

$$\nabla \cdot \boldsymbol{E} = 0, \tag{1.1}$$

where we have assumed that for the region within the cavity the charge density is zero. The electric field will also obey the wave equation,

$$\nabla^2 \boldsymbol{E} = \frac{1}{c^2} \frac{\partial^2 \boldsymbol{E}}{\partial t^2}, \tag{1.2}$$

in which c is the speed of light in vacuo.

The previous two equations are satisfied in any region of vacuum with zero charge density; the particular solutions to the problem under consideration follow from the additional requirement that the field must satisfy the boundary conditions imposed by the properties of the cavity walls.[6] The electric field is always zero within a perfect conductor, and since the tangential components of the electric field are constant across any boundary, the tangential components of the electric field at the boundary must also be zero. Furthermore, since the tangential components of the electric field are zero at the walls, eqn (1.1) then implies that $\partial E_n / \partial n = 0$, where E_n is the normal component of the field, and $\partial/\partial n$ indicates that the derivative is evaluated along the direction of the normal to the wall.

[6] For further details see Exercise 1.1.

Suppose now that the field is harmonic with angular frequency ω, that is to say that we can write the field in the form

$$\boldsymbol{E}(\boldsymbol{r}, t) = \frac{1}{2} \left[\boldsymbol{U}(\boldsymbol{r}) e^{-i\omega t} + \text{c.c.} \right], \tag{1.3}$$

where $\boldsymbol{U}(\boldsymbol{r})$ describes the spatial variation of the field, and we use the usual convention that c.c. means that we add the complex conjugate of the preceding terms. Substitution into the wave equation then yields the well-known **Helmholtz equation** for the spatially dependent part of the electric field:

$$\nabla^2 \boldsymbol{U} = -\frac{\omega^2}{c^2} \boldsymbol{U}. \tag{1.4}$$

It is clear that we can solve for each component of $\boldsymbol{U}(\boldsymbol{r})$ separately, and so we now concentrate on one such component, which we choose to be the x-component, $U_x(\boldsymbol{r})$. The symmetry of the cavity suggests that we write $U_x(\boldsymbol{r})$ as a product of functions that depend on x, y, and z only:

$$U_x(\boldsymbol{r}) = X(x) Y(y) Z(z). \tag{1.5}$$

Upon substitution into eqn (1.3) we find,

$$\frac{1}{X} \frac{d^2 X}{dx^2} + \frac{1}{Y} \frac{d^2 Y}{dy^2} + \frac{1}{Z} \frac{d^2 Z}{dz^2} = -\frac{\omega^2}{c^2}. \tag{1.6}$$

We may now use the method of separation of variables to write,

$$\frac{d^2 X}{dx^2} = -k_x^2 X \tag{1.7}$$

$$\frac{d^2 Y}{dy^2} = -k_y^2 Y \tag{1.8}$$

$$\frac{d^2 Z}{dz^2} = -k_z^2 Z, \tag{1.9}$$

in which the separation constants $-k_x^2$, $-k_y^2$, and $-k_z^2$ must obey the relation,

$$k_x^2 + k_y^2 + k_z^2 = \frac{\omega^2}{c^2}. \tag{1.10}$$

The solutions to eqns (1.7), (1.8), and (1.9) are straightforward, and application of the boundary conditions then yields,

$$U_x(\mathbf{r}) = E_{0x} \cos\left(\frac{\pi}{L} l_x x\right) \sin\left(\frac{\pi}{L} m_x y\right) \sin\left(\frac{\pi}{L} p_x z\right), \tag{1.11}$$

where l_x, m_x, and p_x are integers, and E_{0x} is the amplitude of the field.

The other components of $\mathbf{U}(\mathbf{r})$ may be found by the same method, and we find,

$$U_y(\mathbf{r}) = E_{0y} \sin\left(\frac{\pi}{L} l_y x\right) \cos\left(\frac{\pi}{L} m_y y\right) \sin\left(\frac{\pi}{L} p_y z\right), \tag{1.12}$$

$$U_z(\mathbf{r}) = E_{0z} \sin\left(\frac{\pi}{L} l_z x\right) \sin\left(\frac{\pi}{L} m_z y\right) \cos\left(\frac{\pi}{L} p_z z\right). \tag{1.13}$$

At this point the l_i, m_i, and p_i are apparently unrelated. However, substitution of our solution into the Maxwell equation (1.1) yields,

$$\frac{\pi}{L}\left[-l_x E_{0x} \sin\left(\frac{\pi}{L} l_x x\right) \sin\left(\frac{\pi}{L} m_x y\right) \sin\left(\frac{\pi}{L} p_x z\right)\right]$$
$$+ \frac{\pi}{L}\left[-m_y E_{0y} \sin\left(\frac{\pi}{L} l_y x\right) \sin\left(\frac{\pi}{L} m_y y\right) \sin\left(\frac{\pi}{L} p_y z\right)\right]$$
$$+ \frac{\pi}{L}\left[-p_z E_{0z} \sin\left(\frac{\pi}{L} l_z x\right) \sin\left(\frac{\pi}{L} m_z y\right) \sin\left(\frac{\pi}{L} p_z z\right)\right]$$
$$= 0. \tag{1.14}$$

Apart from the trivial solution $E_{0x} = E_{0y} = E_{0z} = 0$, eqn (1.14) can only be satisfied *at all points in the cavity* if,

$$l_x = l_y = l_z = l \tag{1.15}$$

$$m_x = m_y = m_z = m \tag{1.16}$$

$$p_x = p_y = p_z = p, \tag{1.17}$$

and,

$$l E_{0x} + m E_{0y} + p E_{0z} = 0. \tag{1.18}$$

Each set of integers (l, m, p) is said to define a **mode** of the cavity. It is worth noting here that if more than one of l, m, or p is zero then $\mathbf{E}(\mathbf{r}, t)$ is also

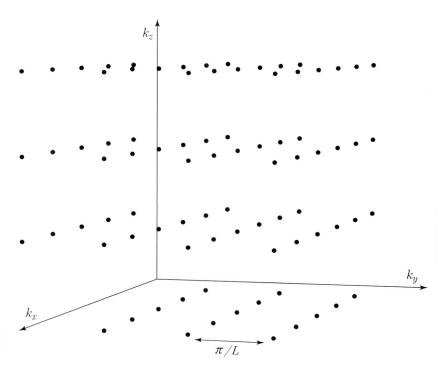

Fig. 1.4: The distribution of allowed electromagnetic modes for a cubic cavity. Note that no modes appear on the axis, since not more than one of l, m, or p may be zero.

zero and hence there is no electromagnetic field in the cavity. Furthermore, as discussed below, the integers l, m, p must all be positive.

The **wave vector** of the mode is defined as,

$$\boldsymbol{k} = k_x \boldsymbol{i} + k_y \boldsymbol{j} + k_z \boldsymbol{k} \quad (1.19)$$

$$= \frac{\pi}{L} [l\boldsymbol{i} + m\boldsymbol{j} + p\boldsymbol{k}]. \quad (1.20)$$

The restrictions on the set of integers (l, m, p) therefore restricts the allowed wave vectors, as shown in Fig. 1.4, which plots in so-called k-space the allowed values of k_x, k_y, and k_z.

If we define the vector,

$$\boldsymbol{E}_0 = E_{0x}\boldsymbol{i} + E_{0y}\boldsymbol{j} + E_{0z}\boldsymbol{k}, \quad (1.21)$$

eqn (1.18) then gives the condition,

$$\boldsymbol{k} \cdot \boldsymbol{E}_0 = 0. \quad (1.22)$$

We see that the vector $\boldsymbol{E}_0$, which gives the amplitude of the radiation field, must be perpendicular to $\boldsymbol{k}$, and hence that there are *two independent solutions*, or 'polarizations,' for each mode.[7]

We note also that according to eqn (1.10) only certain values of $|\boldsymbol{k}|$ are possible, and hence that only certain frequencies, ω_{lmp}, are allowed. Those modes with the same value of $l^2 + m^2 + p^2$ will have the same frequency and are said to be degenerate. This degeneracy, and that due to the two independent

---

[7]We can now see why l, m, p must be positive. Consider a mode described by the integers (l, m, p) and the polarization vector $\boldsymbol{E}_0 = E_{0x}\boldsymbol{i} + E_{0y}\boldsymbol{j} + E_{0z}\boldsymbol{k}$. Now imagine letting $l \to -l$. The condition (1.22) would be satisfied if we let $E_{0x} \to E_{0x}, E_{0y} \to -E_{0y}, E_{0z} \to -E_{0z}$. However, from eqns (1.11), (1.12), and (1.13) we see that the radiation field would be *unchanged* from that of the original mode. Hence, the 'new' mode is not a new mode after all.

polarizations, is typical of electromagnetic modes in any cavity including, for example, the cavity of a laser oscillator.

1.2.1 The density of modes

In the sections below we will be interested in the density of modes, that is the number $p(k)\mathrm{d}k$ of modes with a wave vector of magnitude lying in the range k to $k + \mathrm{d}k$. This number will be the number of points in Fig. 1.4 lying in the positive octant of a spherical shell of radius k and thickness $\mathrm{d}k$. Along the axes of Fig. 1.4 the modes are separated in k-space by a distance π/L, and hence the volume of k-space occupied by each mode is $(\pi/L)^3$. The volume of the spherical shell in the positive octant is $\frac{1}{8}4\pi k^2 \mathrm{d}k$, and hence, provided the radius of the shell is sufficiently large, the number of modes in the shell is,

$$p(k)\mathrm{d}k = \frac{\frac{1}{8}4\pi k^2 \mathrm{d}k}{(\pi/L)^3} \times 2, \tag{1.23}$$

where the final factor of 2 takes into account the two independent polarizations allowed for each set of (l, m, p). Tidying, we find,

$$p(k)\mathrm{d}k = \frac{L^3 k^2}{\pi^2} \mathrm{d}k. \tag{1.24}$$

Free-space modes

If we allow the size L of the cavity to increase, the frequency spacing of the allowed modes decreases, and the number of modes in the interval $k, k + \mathrm{d}k$ increases. However, the number of modes per unit volume of space, the **spatial density of modes**, $g(k)\mathrm{d}k$, remains constant:[8]

$$g(k)\mathrm{d}k = \frac{k^2}{\pi^2}\mathrm{d}k. \quad \textbf{Density of free-space modes} \tag{1.25}$$

[8] This is our first example of a spectral quantity, that is one that is *per unit wave vector or per unit frequency interval*; in this case the number density of radiation modes per unit wave vector.

Of great importance is the fact that, provided the cavity is large enough, this density of modes is independent of the size or shape of the cavity, and, indeed, the material of the cavity walls. Free space may be thought of as the limit $L \to \infty$, such that the allowed modes are distributed continuously in wave vector (and frequency) space with a mode density given by eqn (1.25).

Finally, we note that we will often find it more useful to express the mode density in terms of the number of modes in the angular frequency interval ω to $\omega + \mathrm{d}\omega$. Using the fact that $\omega = ck$, we find straightforwardly,

$$g(\omega)\mathrm{d}\omega = \frac{\omega^2}{\pi^2 c^3}\mathrm{d}\omega. \quad \textbf{Density of free-space modes} \tag{1.26}$$

1.3 Planck's law

We now consider the energy stored in the electromagnetic field. According to classical electromagnetism, for each mode this energy is given by,[9]

[9] See, for example, Bleaney and Bleaney (1989).

$$\int_V \left(\frac{1}{2} \boldsymbol{E} \cdot \boldsymbol{D} + \frac{1}{2} \boldsymbol{B} \cdot \boldsymbol{H}\right) \mathrm{d}\tau, \tag{1.27}$$

where the integral is taken over the region in which the fields exist, and $\boldsymbol{E}$ is the electric field, $\boldsymbol{D}$ the electric displacement, $\boldsymbol{B}$ the magnetic flux density, and $\boldsymbol{H}$ the magnetic field strength of the mode. Taking an average over the period of oscillation of the field, and making use of the fact that the mean energies stored in the electric and magnetic fields are equal, we find the mean energy in the mode is given by[10]

$$E_{\mathrm{mode}} = \frac{1}{2} \int_V \epsilon_0 \boldsymbol{E}_0^2 \mathrm{d}\tau, \tag{1.28}$$

[10] Note that $\boldsymbol{E} \cdot \boldsymbol{D} = \boldsymbol{E}_0 \cos^2(\omega t)$, and the time-average of the term in $\cos^2(\omega t)$ is equal to 1/2.

where ϵ_0 is the permittivity of free space, and we have used the fact that for vacuo $\boldsymbol{D} = \epsilon_0 \boldsymbol{E}$. We see that this classical expression for the energy of the mode can take *any* positive value, since the magnitude of $\boldsymbol{E}_0$ may take any value.

These purely classical considerations, however, do not provide a complete description and we should move to a quantum-mechanical picture in which the electromagnetic fields are replaced by operators. Such an approach is beyond the scope of this book,[11] and instead we adopt the hypothesis of Planck: we assume that since the electromagnetic fields of the modes oscillate harmonically, the energy stored in the fields are quantized as they would be for any other quantum-mechanical oscillator.[12] In other words, we must have

[11] A detailed discussion of the quantization of the radiation field may be found in the book by Loudon (2000).

[12] In fact, Planck assumed that the *walls* of the cavity were composed of oscillators that could only absorb or emit discrete quanta of energy. He also took the energy of the oscillators to be of the form $n\hbar\omega$, and hence did not include the zero-point energy.

$$E_{\mathrm{mode}} = \left(n + \frac{1}{2}\right)\hbar\omega, \quad n = 0, 1, 2, 3, ..., \tag{1.29}$$

where ω is the angular frequency of the mode.

Note that the lowest energy of the mode—known as the **zero-point energy**—is not zero, but equal to $\frac{1}{2}\hbar\omega$. For $n > 0$ the mode is said

- to be in the nth excited state; or
- to contain n quanta of energy $\hbar\omega$; or
- to contain n photons.

1.3.1 The energy density of blackbody radiation

Under thermal equilibrium at temperature T, the probability P_n that the mode contains n photons is given by the usual Boltzmann expression:

$$P_n = \frac{\exp(-E_n/k_\mathrm{B}T)}{\sum_{j=0}^{\infty} \exp(-E_j/k_\mathrm{B}T)}, \tag{1.30}$$

where $E_j = (j + \frac{1}{2})\hbar\omega$. The mean number of photons in the mode, $\bar{n}$, is then given by

$$\bar{n} = \sum_{n=0}^{\infty} n P_n. \tag{1.31}$$

In Exercise 1.2 this is shown to be

$$\bar{n} = \frac{1}{\exp(\hbar\omega/k_B T) - 1}. \quad (1.32)$$

We now know the density of modes, the mean number of photons per mode, and the energy of each photon. Hence, we may write the energy density $\varrho(\omega)d\omega$ of the radiation field with frequencies lying between ω and $\omega + d\omega$ as

$$\varrho(\omega)d\omega = \begin{pmatrix} \text{Density} \\ \text{of} \\ \text{modes} \end{pmatrix} \times \begin{pmatrix} \text{Number of} \\ \text{photons} \\ \text{per mode} \end{pmatrix} \times \begin{pmatrix} \text{Energy} \\ \text{of} \\ \text{photon} \end{pmatrix}$$

$$= g(\omega)d\omega \times \bar{n} \times \hbar\omega. \quad (1.33)$$

It is worth realizing that eqn (1.33) is true for *any* radiation field.[13] For the particular case of radiation that is in thermal equilibrium, with a well-defined temperature T, the number of photons per mode is given by eqn (1.32). Using this and eqns (1.29) and (1.26) we find, for the special case of radiation in thermal equilibrium —known as **blackbody radiation**:

$$\varrho_B(\omega)d\omega = \frac{\hbar\omega^3}{\pi^2 c^3} \frac{d\omega}{\exp(\hbar\omega/k_B T) - 1}. \quad \textbf{Blackbody radiation}$$

$$(1.34)$$

This result is known as **Planck's law**.[14]

[13] For an example of the use of this result, see Exercise 1.3.

[14] This is another example of a spectral quantity. $\varrho_B(\omega)$ is the energy density of the radiation field per unit frequency interval. We should be aware that often we wish to deal with non-spectral quantities. For example, we may wish to consider the total energy density U of the radiation field, i.e. $U = \int \varrho_B(\omega)d\omega$. Where confusion might arise between spectral and non-spectral quantities we will write spectral quantities in the form $a(\omega)d\omega$ to indicate the presence of a frequency interval.

Further reading

An intriguing account of the race to build the first laser may be found in the book devoted to this subject by Hecht (2005*a*). An extract from this, describing the early work on the ruby laser, is also available Hecht (2005*b*).

A detailed treatment of the quantum theory of light may be found in the book by Loudon (2000).

Exercises

(1.1) **Boundary conditions**
Here, we consider in more detail the boundary conditions imposed by a closed cavity with perfectly conducting walls.

(a) Write down in general the boundary conditions relating the components of the electric field E and displacement vector D either side of a boundary between two media. Assume that there are no free charges or surface currents.

(b) Explain briefly why in a good conductor $E = 0$, and hence find a condition for the tangential components of the electric field on the walls of a cavity with perfectly conducting walls.

(c) Let us now suppose that the cavity is a cube. We will consider the boundary conditions at one of the walls, which for convenience we will take to lie in the plane $z = 0$. Use eqn (1.1) to show that

$$\frac{\partial E_z}{\partial z} = 0.$$

(d) Hence, argue that in general $\partial E_n/\partial n = 0$, where E_n is the normal component of the field, and $\partial/\partial n$ indicates that the derivative is evaluated along the direction of the normal to the wall.

(e) Comment on whether in general E_n is also zero.

(1.2) **Mean number of photons per mode (blackbody radiation)**

(a) We first find the probability P_n of there being n photons in a mode with frequency ω. Show that eqn (1.30) may be written as

$$P_n = \frac{\exp(-n\beta)}{\sum\limits_{j=0}^{\infty} \exp(-j\beta)},$$

where $\beta = \hbar\omega/k_B T$.

(b) Using the fact that the denominator is a geometric progression, show that

$$P_n = \exp(-n\beta)\left[1 - \exp(-\beta)\right].$$

(c) The mean number of photons in the mode is given by eqn (1.31). Show that

$$\bar{n} = \left[1 - \exp(-\beta)\right] \sum_{n=0}^{\infty} n \exp(-n\beta).$$

(d) Show that

$$\sum_{n=0}^{\infty} n \exp(-n\beta) = -\frac{\partial S}{\partial \beta},$$

where

$$S = \sum_{n=0}^{\infty} \exp(-n\beta).$$

(e) Evaluate S, and hence show that

$$\bar{n} = \frac{1}{\exp(\hbar\omega/k_B T) - 1}.$$

(1.3) **Differences between classical and laser sources** In this problem we compare the number of photons per mode $\bar{n}$ for various light sources. In order to do this we must first find, for non-blackbody sources, a relation between measurable quantities, such as the total intensity I_T and the bandwidth $\Delta\omega$, and the energy density $\varrho(\omega)$.

(a) By considering the flux of photons passing a fixed plane, or otherwise, show that for a *collimated beam* of radiation

$$I_T = \varrho(\omega)\Delta\omega\, c.$$

(b) To derive a similar relation for *isotropic radiation* we may treat the radiation as a gas of photons. Using this model, show that for isotropic radiation

$$I_T = \frac{1}{4}\varrho(\omega)\Delta\omega\, c.$$

(c) Hence, using the general relation between the energy density and the mean number of photons per mode (eqn (1.33)), show that

$$\bar{n} = \alpha \frac{\pi^2 c^2}{\hbar\omega^3} \frac{I_T}{\Delta\omega},$$

where $\alpha = 1$ for radiation in the form of a beam, and $\alpha = 4$ for isotropic radiation.

(d) Calculate $\bar{n}$ for the following sources:
  (i) Radiation with wavelengths near 500 nm from a blackbody with a temperature of 3000 K.
  (ii) As an example of one of the most spectrally intense non-laser sources consider the emission at 254 nm from a cooled, single-isotope mercury lamp. Suppose that the bandwidth of the radiation is 0.0003 nm and the radiation is focused to an intensity of 1 W cm^{-2}.
  (iii) A stabilized He-Ne laser produces an output power of 1 mW at 633 nm in a beam of diameter 1 mm. Take the bandwidth of the laser to be 1 MHz.
  (iv) A high-power Nd:glass laser system operating at a wavelength of 1.06 µm can generate peak focal intensities of more than 10^{15} W cm^{-2}. The bandwidth of the laser transition is approximately 30 GHz.
  (v) Finally, Ti:sapphire laser systems can amplify pulses only 50 fs long to energies of order 100 mJ. Calculate $\bar{n}$ assuming that the pulses are focused to a spot of diameter 10 µm, and that the bandwidth is limited by the pulse duration.

(1.4) **Stefan–Boltzmann law**

(a) By considering the photons within an enclosed cavity at temperature T to act as an ideal gas, show that the spectral intensity $\mathcal{I}(\omega)$—i.e. the intensity per unit bandwidth—of radiation leaving the cavity through a small hole in the cavity wall is given by,

$$\mathcal{I}(\omega) = \frac{1}{4}\varrho_B(\omega)\,c. \qquad (1.35)$$

(b) Hence, show that the total intensity of radiation emitted by the blackbody is given by the **Stefan–Boltzmann Law**:

$$I = \sigma T^4, \qquad (1.36)$$

where the Stefan–Boltzmann constant σ is given by,

$$\sigma = \frac{\pi^2 k_B^4}{60\hbar^3 c^2} \approx 5.67 \times 10^{-8} \,\mathrm{W\,m^{-2}\,K^{-4}}. \qquad (1.37)$$

[You may use the fact that

$$\int_0^\infty \frac{x^3 \, dx}{\exp(x) - 1} = \frac{\pi^4}{15}.\Big]$$

2 The interaction of radiation and matter

2.1 The Einstein treatment   12
2.2 Conditions for optical gain   16
2.3 The semi-classical treatment[†]   19
2.4 Atomic population kinetics[†]   21
Further reading   24
Exercises   25

[1] See Einstein (1917). An English translation of this paper is available in Knight and Allen (1983).

The amplification of radiation by stimulated emission necessarily involves the interaction of radiation and matter. Here, we discuss two treatments of this interaction: that due to Einstein, and the so-called semi-classical model.

2.1 The Einstein treatment

In 1917 Einstein introduced a model of the interaction of radiation and matter.[1] Although superseded by developments in quantum theory his approach is still used very widely owing to its simplicity and the insight that it affords. Further, the application of rate equations to calculate atomic populations, as is done extensively in this book, is linked intrinsically to the Einstein approach.

Einstein considered two levels of an atom, an upper level 2 with energy E_2, and a lower level 1 with energy E_1, and identified three processes by which radiation could interact with atoms in these levels:

1. **Spontaneous emission**—in which an atom in the upper level decays to the lower level by the emission of a photon with angular frequency ω_{21} and energy $\hbar\omega_{21} = E_2 - E_1$. The spontaneously emitted photon can be emitted in any direction.
2. **Absorption**—in which an atom in the lower level is excited to the upper level by the absorption of a photon of energy $\hbar\omega_{21}$.
3. **Stimulated emission**—in which an incident photon of energy $\hbar\omega_{21}$ stimulates an atom in the upper level to decay to the lower level by the emission of a second photon of energy $\hbar\omega_{21}$. The stimulated photon is emitted into the same radiation mode as the incident photon, and hence has exactly the same frequency, direction, and polarization as the incident photon.

This third process, stimulated emission, was postulated by Einstein and is key to laser operation. As we shall see, the inclusion of stimulated emission is necessary in order that the theory is consistent with Planck and Boltzmann's laws.[2]

It seems reasonable that the rate of spontaneous emission should be independent of the conditions of the radiation field in which the atom finds itself. It is also clear that the rates at which absorption and stimulated emission occur must depend in some way on the density of photons of energy $\hbar\omega_{21}$, or,

[2] This can be seen from eqn. (2.1). If stimulated emission were not included (i.e. setting $B_{21} = 0$) the Einstein theory would not predict a Boltzmann distribution over the energy levels of an atom in thermal equilibrium with a radiation field, as it should do.

equivalently, on the energy density of the radiation field at this photon energy. In three postulates Einstein went further and stated that the rates are *linearly* dependent on this energy density:

1. The rate per unit volume[3] at which atoms in the upper level 2 decay to the lower level 1 is equal to $N_2 A_{21}$, where N_2 is the number of atoms per unit volume in level 2. Within the Einstein picture A_{21} will depend on the characteristics of the transition under consideration—i.e. the quantum numbers of the transition, and the atomic species—but will be independent of the radiation field.
2. The rate per unit volume at which atoms in the lower level are excited to the upper level by the absorption of photons of energy $\hbar\omega_{21}$ is equal to $N_1 B_{12} \varrho(\omega_{21})$, where N_1 is the number of atoms per unit volume in level 1, $\varrho(\omega_{21})$ is the energy density of radiation of angular frequency ω_{21}, and B_{12} is a constant. Again, within the Einstein picture this Einstein B coefficient will depend on the details of the atomic transition but will be independent of the radiation field.
3. The rate per unit volume at which atoms in the upper level make transitions to the lower level by the stimulated emission of photons of energy $\hbar\omega_{21}$ is equal to $N_2 B_{21} \varrho(\omega_{21})$, where, once again, B_{21} depends on the properties of the atomic transition but not on those of the radiation field.

[3] Notice that we have decided to deal with the population densities, rather than the absolute number of atoms. As we will see, this turns out to be the most convenient way of considering the populations of atomic levels. Note also that we will often talk of the 'population in level j,' by which we will virtually always mean the number of atoms per unit volume in level j.

The coefficients A_{21}, B_{12}, and B_{21} are known as the Einstein A and B coefficients. The three fundamental interactions between atoms and radiation are shown schematically in Fig. 2.1

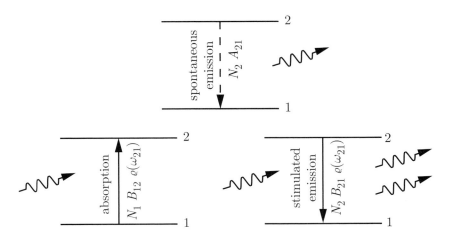

Fig. 2.1: Schematic diagram of the processes of absorption, stimulated emission, and spontaneous emission between two levels of an atom in a radiation field of energy density $\varrho(\omega_{21})$, where $\hbar\omega_{21}$ is the energy separation of the levels. For each process the transition rates are given in terms of the Einstein coefficients and the number densities N_2 and N_1 of atoms in the upper and lower levels. The photons have energy $\hbar\omega_{21}$ and the figure should be read from left to right. For example, in the case of stimulated emission an incident photon of energy $\hbar\omega_{21}$ stimulates an excited atom to make a transition to the lower level by emitting a second photon of energy $\hbar\omega_{21}$.

2.1.1 Relations between the Einstein coefficients

For a given transition the three Einstein coefficients are postulated to be constant (for a given atomic transition), that is independent of the details of the radiation field. We can use this fact to find relations between the Einstein coefficients by considering a special case in which the relative populations of the upper and lower levels are known.

To that end we consider an ensemble of stationary atoms immersed in a bath of blackbody radiation of temperature T and energy density $\varrho_B(\omega)$. We suppose the atoms to have many energy levels $\ldots E_i, E_j, E_k \ldots$ with degeneracies[4] $\ldots g_i, g_j, g_k \ldots$. Now, atoms in any level j will, in principle, undergo transitions to all higher levels k by absorbing radiation of angular frequency ω_{kj}, where $\hbar\omega_{kj} = E_k - E_j$, as well as making transitions to all lower levels i by spontaneous and stimulated emission of radiation of angular frequency ω_{ji}, where $\hbar\omega_{ji} = E_j - E_i$. The atoms will be in a dynamic equilibrium such that, whilst an individual atom will continually make transitions between its energy levels, the total number of atoms in a given level remains, on average, constant.

Clearly, atoms in any level can make transitions to a large, even infinite, number of other levels. The **Principle of Detailed Balance** states, however, that in thermal equilibrium the transitions between *any pair* of levels are also in dynamic equilibrium, as illustrated schematically in Fig. 2.2.

The reasoning behind the Principle of Detailed Balance is most easily seen by considering the radiation field, rather than the populations of the atomic levels. Since the system is in thermodynamic equilibrium, it must be that the rate at which the radiation field loses photons of angular frequency ω by absorption must be balanced by the rate at which it gains photons of this frequency by spontaneous and stimulated emission. Now, an arbitrary pair of atomic levels 1, 2 will only absorb or emit photons of a particular frequency,

---

[4] The degeneracy of a level is equal to the number of independent states with the same energy. For example, in an isolated atom each atomic level of total angular momentum J has $(2J + 1)$ magnetic sublevels (corresponding to states labelled with the magnetic quantum number $m_J = -J, -J+1, \ldots, J-1, J$) all of which have the same energy. The degeneracy of the level is therefore $g = 2J + 1$.

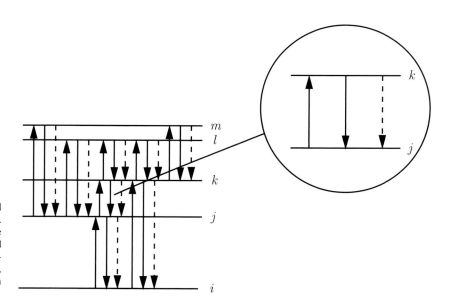

Fig. 2.2: Illustrating the dynamic thermal equilibrium between levels in an atom. Atoms in any level j may undergo radiative transitions to a large number of higher and lower levels. However, the Principle of Detailed Balance states that any *pair* of levels j, k must be in dynamic equilibrium between themselves.

ω_{21}, and this frequency is unique to that pair of levels.[5] It must be, then, that transitions between levels 1 and 2, and that pair alone, maintain the density of photons of frequency ω_{21}.

In terms of the Einstein coefficients, equating the rate of transitions from $2 \rightarrow 1$ with that for transitions from $1 \rightarrow 2$ yields:

$$N_1 B_{12} \varrho_B(\omega_{21}) = N_2 B_{21} \varrho_B(\omega_{21}) + N_2 A_{21}. \quad (2.1)$$

[5]This argument can be extended to allow for spectral broadening of the transitions, as discussed in Appendix B.

It is straightforward to rearrange this result to find an expression for the energy density of the blackbody radiation at ω_{21}:

$$\varrho_B(\omega_{21}) = \frac{A_{21}/B_{21}}{\frac{N_1}{N_2} \frac{B_{12}}{B_{21}} - 1}. \quad (2.2)$$

Since the atom is in thermal equilibrium with the radiation field, the ratio of the populations in the upper and lower level is given by the Boltzmann law:

$$\frac{N_2}{N_1} = \frac{g_2}{g_1} \exp\left(-\frac{\hbar \omega_{21}}{k_B T}\right), \quad (2.3)$$

where k_B is the Boltzmann constant. Substituting eqn (2.3) into eqn (2.2) we find

$$\varrho_B(\omega_{21}) = \frac{A_{21}/B_{21}}{\frac{g_1}{g_2} \frac{B_{12}}{B_{21}} \exp\left(\frac{\hbar \omega_{21}}{k_B T}\right) - 1}. \quad (2.4)$$

As discussed in Section 1.3, blackbody radiation of temperature T has an energy density given by Planck's law:

$$\varrho_B(\omega) = \frac{\hbar \omega^3}{\pi^2 c^3} \frac{1}{\exp\left(\frac{\hbar \omega}{k_B T}\right) - 1}. \quad (2.5)$$

Clearly, for eqns. (2.4) and (2.5) to be consistent for all temperatures T, we must have,

$$\frac{A_{21}}{B_{21}} = \frac{\hbar \omega_{21}^3}{\pi^2 c^3} \quad \textbf{Einstein relations} \quad (2.6)$$

$$g_1 B_{12} = g_2 B_{21}. \quad (2.7)$$

It should be stressed that although the relationship between the Einstein coefficients were derived under conditions of thermal equilibrium, since the A and B coefficients are postulated to be independent of the radiation field eqns (2.6) and (2.7) hold *irrespective of whether or not the atom is thermal equilibrium or the radiation field is blackbody.*

Finally, we note that the energy levels of the atom were assumed to be infinitely sharp. In Chapter 3 we describe the various processes that lead to broadening of the energy levels, and how the Einstein coefficients must be modified in the presence of such broadening.

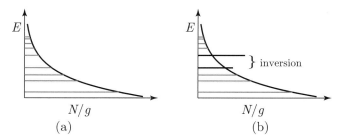

Fig. 2.3: Comparison of the populations per state for: (a) atoms in thermal equilibrium; (b) atoms with a population inversion between two of the levels.

2.2 Conditions for optical gain

As we will see more formally in Section 4.1.1, for amplification of a beam of radiation we require the rate of stimulated emission to be greater than the rate of absorption:

$$N_2 B_{21} \varrho(\omega_{21}) > N_1 B_{12} \varrho(\omega_{21}), \tag{2.8}$$

from which we find, using eqn (2.7):

$$\boxed{\frac{N_2}{g_2} > \frac{N_1}{g_1}.} \quad \textbf{Condition for optical gain} \tag{2.9}$$

In other words, for optical gain the population *per state* must be greater in the upper level than in the lower level, a situation called a **population inversion**.

It is important to realize that a population inversion is unusual. For example, if the level populations were in thermal equilibrium at temperature T the populations of the levels would be described by the Boltzmann distribution of eqn (2.3). From this we may write,

$$\frac{(N_2/g_2)}{(N_1/g_1)} = \exp\left(-\frac{E_2 - E_1}{k_\mathrm{B} T}\right) < 1. \tag{2.10}$$

from which it is clear that for a system in thermal equilibrium the population per state of higher-lying level is always lower than that for any lower-lying level. Figure 2.3 illustrates schematically the differences in the distributions over atomic levels when thermal equilibrium obtains and when a population exists.

2.2.1 Conditions for steady-state inversion

Having demonstrated that a population inversion cannot exist under conditions of thermal equilibrium, it is useful briefly to explore the conditions under which a population inversion *can* be produced. Figure 2.4 shows schematically the kinetic processes that affect the populations of the upper and lower levels of a laser transition. We suppose that atoms in the upper level are produced, or 'pumped' at a rate R_2 atoms per second per unit volume, and that the lifetime of the upper level is τ_2. Note that the pump rate includes all processes that excite the upper level such as direct optical pumping, electron collisional excitation, and radiative and non-radiative cascade from higher-lying levels. The lifetime

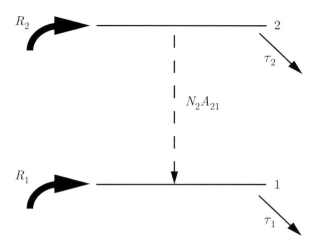

Fig. 2.4: Schematic diagram showing the kinetic processes considered in the calculations of the minimum conditions required to achieve a steady-state population inversion between levels 2 and 1. Note that absorption and stimulated emission are ignored since we are calculating the *minimum* requirements to produce a population inversion; in an operating laser these additional processes will make it more difficult to achieve and maintain an inversion.

τ_2 is the lifetime against all types of decay (radiative, collisional de-excitation, etc.) and includes radiative decay to level 1. It is known as the **fluorescence lifetime**, since it is the lifetime with which the strength of the fluorescence on any radiative transition from level 2 would be observed to decay. We define the pump rate and lifetime of the lower laser level in a similar way, but do not include spontaneous emission on the laser transition itself in R_1. This contributes population to level 1 at a rate $N_2 A_{21}$ per unit volume, and will be handled explicitly for reasons that will become clear below.

The evolution of the population densities in the two levels may then be written as,

$$\frac{dN_2}{dt} = R_2 - \frac{N_2}{\tau_2} \tag{2.11}$$

$$\frac{dN_1}{dt} = R_1 + N_2 A_{21} - \frac{N_1}{\tau_1}. \tag{2.12}$$

Note that the symmetry of the two equations is broken by the spontaneous emission term, $N_2 A_{21}$. It is straightforward to solve the above for steady-state conditions by setting $dN_2/dt = dN_1/dt = 0$. We find,

$$N_2 = R_2 \tau_2 \tag{2.13}$$

$$N_1 = R_1 \tau_1 + R_2 \tau_2 A_{21} \tau_1. \tag{2.14}$$

For optical gain we require $N_2/g_2 > N_1/g_1$, which yields the following condition for a steady-state population inversion:

$$\frac{R_2}{R_1} \frac{\tau_2}{\tau_1} \frac{g_1}{g_2} \left[1 - \frac{g_2}{g_1} A_{21} \tau_1 \right] > 1. \quad \textbf{Steady-state inversion} \tag{2.15}$$

For the moment we will ignore the term in square brackets by setting it equal to unity.[6] We conclude that for a steady-state population inversion to be achieved *at least one* of the following must be true:

[6] We will deal with this term in Section 2.2.2. Here, we merely note that, since all the parameters are positive, this term can be no bigger than unity.

Selective pumping $R_2 > R_1$: i.e. the upper laser level is pumped more rapidly than the lower laser level;

Favourable lifetime ratio $\tau_2 > \tau_1$: i.e. the lower laser level decays more rapidly than the upper level, which keeps the population of the lower level small;

Favourable degeneracy ratio $g_1 > g_2$: which ensures that the population *per state* of the lower laser level is small.

In addition, we need to satisfy the condition described in the following section.

2.2.2 Necessary, but not sufficient condition

The factor in square brackets in eqn. (2.15) is interesting in that it can be negative as well as positive; it also depends only on the parameters of the laser transition, and hence is independent of the pumping rates R_1 and R_2. Clearly, if the bracket is negative the inequality of eqn. (2.15) cannot be satisfied since the R_i, τ_i, and g_i are all positive. In other words, for some systems no matter how selective the pumping is, or how favourable are the lifetime and degeneracy ratios, it is not possible to achieve a steady-state population inversion.

The reason for this somewhat surprising result is that increasing the population of the upper laser level by pumping harder also increases the rate at which the lower level is populated by spontaneous emission on the laser transition itself.

This final factor in eqn. (2.15) therefore yields **a necessary, but not sufficient condition** for achieving a steady-state population inversion. The condition that this factor is positive can be rewritten as,

$$A_{21} < \frac{g_1}{g_2}\frac{1}{\tau_1}. \quad \textbf{Necessary, but not sufficient condition} \quad (2.16)$$

Hence, the rate of spontaneous decay from the upper to the lower laser level must be smaller than the total rate of decay from the lower laser level (multiplied by a factor of g_1/g_2). In other words, the lower level has to empty sufficiently quickly for population not to build up by spontaneous emission on the laser transition.

It should be stressed that satisfying eqn. (2.16) does not *ensure* that a steady-state population inversion *will* be achieved on a given transition—the pumping may not preferentially populate the upper laser level, or the lifetime ratio may be very unfavourable. However, *unless* eqn. (2.16) is satisfied, *no* pumping technique, no matter how selective, will be able to create a steady-state population inversion on the transition.

Of course, a *transient* inversion can always be created with suitably selective pumping, but if eqn (2.16) is not satisfied the laser output must be pulsed, usually with a pulse duration of order A_{21}^{-1}. Lasers of this type are said to be **self-terminating**.

2.3 The semi-classical treatment[†]

2.3.1 Outline

We now outline the semi-classical description of the interaction of radiation and matter.[7] In this approach the matter is treated quantum mechanically but the radiation field remains classical. The semi-classical theory leads directly to expressions for the Einstein B coefficients in terms of quantum-mechanical matrix elements. However, the theory cannot explain spontaneous emission, and the expression for the Einstein A coefficient in quantum-mechanical terms can only be obtained by using eqn (2.6) to relate A_{21} to the expression for B_{21}.[8]

In principle, provided the wave functions of the atomic levels are known, the formulae of the semi-classical model yield numerical values for the Einstein coefficients. In practice, of course, it is often the case that the wave functions are not known sufficiently accurately for reliable transition rates to be calculated. However, even then the semi-classical theory is useful in that it provides us with not only the frequency scaling of the Einstein coefficients, but also certain **selection rules** that state whether the rates of a given transition are zero (or should at least be small).

We consider a simplified atom consisting of just two non-degenerate levels 1 and 2 in the presence of a plane harmonic wave with angular frequency ω, the electric field of which is given by

$$\boldsymbol{E}(\boldsymbol{r}, t) = \boldsymbol{E}_0 \cos(\boldsymbol{k} \cdot \boldsymbol{r} - \omega t), \qquad (2.17)$$

where $\boldsymbol{E}_0$ is the peak electric field vector, and $\boldsymbol{k}$, the wave vector of the radiation, has magnitude $k = \omega/c$ and points in the direction of propagation.

The electric field will interact with the electric dipole moment $\boldsymbol{p}$ of the optically active electrons of the atom with an energy H' given by,[9]

$$H' = -\boldsymbol{p} \cdot \boldsymbol{E}. \qquad (2.18)$$

This extra energy will be treated as a small perturbation to the Hamiltonian of the isolated atom. The electric dipole moment of the atom is given by $\boldsymbol{p} = -e \sum_j \boldsymbol{r}_j$, where $\boldsymbol{r}_j$ is the position of the jth electron, taking the atomic nucleus to be at the origin, and the sum is over all the electrons of the atom. We will assume that the spatial extent of the atom is very much smaller than the wavelength of the radiation, so that the phase $\boldsymbol{k} \cdot \boldsymbol{r}$ of the wave does not vary significantly from one part of the atom to another, and hence we may write $\boldsymbol{k} \cdot \boldsymbol{r} \approx 0$. With these assumptions we then have,

$$H' = \left(e \sum_j \boldsymbol{r}_j \cdot \hat{\boldsymbol{\epsilon}} \right) E_0 \cos(\omega t), \qquad (2.19)$$

where $\hat{\boldsymbol{\epsilon}}$ is a unit vector in the direction of $\boldsymbol{E}_0$.

In the absence of the radiation field, the energy eigenvalues E_1 and E_2, and wave functions Ψ_1 and Ψ_2 of the two levels are given by solutions of the time-dependent Schrödinger equation:

[7] A fuller account is provided in Appendix A. See also standard texts on this subject, such as Loudon (2000) and Foot (2005).

[8] A direct calculation of the rate of spontaneous emission requires Dirac's treatment of the interaction of radiation and matter, in which the radiation field is also quantized. In this approach the rate of absorption is found to be proportional to n, the number of photons in the mode of the radiation field, whilst the rate of emission is found to be proportional to $(n + 1)$. The 'extra' photon appearing in the expression for the rate of emission arises from the zero-point energy of the electromagnetic radiation field. See Exercise 2.2 for an argument based on the Einstein treatment that gives the same dependencies on the photon numbers.

[9] As discussed on page 21, higher-order interactions between the atom and the radiation field will also occur. However, these are only important if the transition rate arising from the dipole interaction considered here is small.

$$H_0 \Psi_i = i\hbar \frac{\partial \Psi_i}{\partial t}, \qquad (2.20)$$

where H_0 is the Hamiltonian of the unperturbed atom. These solutions are of the form,

$$\Psi_2(\boldsymbol{r}, t) = \phi_2(\boldsymbol{r}) e^{-iE_2 t/\hbar} \qquad (2.21)$$

$$\Psi_1(\boldsymbol{r}, t) = \phi_1(\boldsymbol{r}) e^{-iE_1 t/\hbar}, \qquad (2.22)$$

where $\phi_2(\boldsymbol{r})$ and $\phi_1(\boldsymbol{r})$ are the spatially dependent part of the wave functions.[10,11]

As shown in Appendix A, by using time-dependent perturbation theory to obtain solutions for the Schrödinger equation for the atom in the presence of the perturbation H', we may obtain the following expressions for the Einstein coefficients:[12,13]

$$g_1 B_{12} = g_2 B_{21} \qquad (2.23)$$

$$g_2 A_{21} = \frac{\omega_{21}^3}{3\pi \epsilon_0 \hbar c^3} \sum_{m_1} \sum_{m_2} \left| \langle 2 m_2 | -e \sum_j \boldsymbol{r}_j | 1 m_1 \rangle \right|^2, \qquad (2.24)$$

where the m_i are the quantum labels of the degenerate states[14] forming the level i, and where the electric dipole matrix element $\boldsymbol{D}_{21}$ for the transition between two states is given by:

$$\boldsymbol{D}_{21} = \langle 2 m_2 | -e \sum_j \boldsymbol{r}_j | 1 m_1 \rangle \qquad (2.25)$$

$$\equiv \int \left(\phi_2^{m_2} \right)^* \left(-e \sum_j \boldsymbol{r}_j \right) \phi_1^{m_1} d\tau. \qquad (2.26)$$

2.3.2 Selection rules for electric dipole transitions

For our purposes the key feature of the semi-classical expressions for the Einstein coefficients is that they are proportional to the square of the electric dipole moment of the optical transition. Whilst detailed knowledge of the atomic wave functions is required to calculate values for the Einstein coefficients from eqns (2.23) or (2.24), quite general considerations can determine those cases for which the coefficients are zero—leading to so-called selection rules that must be satisfied if the transition rate is to be non-zero.

We outline here the derivation of just one of the selection rules. The parity operator $\widehat{P}$ inverts all coordinates through the origin, i.e. it corresponds to the operation $\boldsymbol{r} \to -\boldsymbol{r}$. It is straightforward to show[15] that the eigenvalues of the parity operator are ± 1, corresponding to so-called **even** or **odd** parity, respectively. In the absence of external forces the atomic Hamiltonian commutes with the parity operator,[16] and consequently the atomic wave functions may be written as eigenfunctions of the parity operator with a definite parity.[17]

---

[10] The wave functions ψ_i and ϕ_i depend on the coordinates of *all* the electrons and so should be written $\psi_i(\boldsymbol{r}_1, \boldsymbol{r}_2, \boldsymbol{r}_3 \ldots, \boldsymbol{r}_N, t)$, etc., where N is the number of electrons in the atom. To avoid clutter, in this section we will simply let $\boldsymbol{r} \equiv \boldsymbol{r}_1, \boldsymbol{r}_2, \boldsymbol{r}_3 \ldots, \boldsymbol{r}_N$.

[11] The $\phi_i(\boldsymbol{r})$ are eigenfunctions of the time-independent Schrödinger equation. The functions $|\phi_i(\boldsymbol{r})|^2 d\tau_1 d\tau_2 d\tau_3 \ldots d\tau_N$ give the probability of finding electron 1 within the volume $d\tau_1$ centred at $\boldsymbol{r} = \boldsymbol{r}_1$, electron 2 within the volume $d\tau_2$ centred at $\boldsymbol{r} = \boldsymbol{r}_2$, etc.

[12] In general, the semi-classical expressions for the Einstein B coefficients are *not* obtained by substituting eqn (2.6) into eqn (2.24), since the B coefficients depend on the orientation of the axes of the atoms and the polarization of the radiation. See Appendix A.

[13] Here, we have used Dirac's bra-ket notation: $\langle 2| \hat{O} |1 \rangle \equiv \int \phi_2^*(\boldsymbol{r}) \hat{O} \phi_1(\boldsymbol{r}) d\tau$, where $\hat{O}$ is an operator, and the integral is taken over all space.

[14] As an example, the energy of an isolated atom not subjected to any external fields is independent of the z-component of the total orbital angular momentum, labelled by m_J. Hence, each energy level is formed from several states with the same energy; these states are labelled by m_J, which are equivalent to the m_i appearing in eqn (2.24).

[15] See Exercise 2.5.

[16] This is *not* the case for molecules.

[17] For example, for a one-electron wave function (an 'orbital') the parity of the wave function is given by $(-1)^\ell$, where ℓ is the orbital angular momentum quantum number of the orbital.

From eqn (2.26) we see that the Einstein A coefficient is proportional to a sum of terms of the form,

$$\left| \int (\phi_2^{m_2})^* \, \boldsymbol{r}_j \phi_1^{m_1} \mathrm{d}\tau_j \right|^2.$$

Since $\boldsymbol{r}_j$ has odd parity, if the two wave functions $\phi_2^{m_2}$ and $\phi_1^{m_1}$ have the same parity the integrand is overall odd, and consequently the integral over all space is zero. Hence, we have the selection rule that for the Einstein A and B coefficients to be non-zero the two wave functions must have opposite parity.

This last result is an example of a selection rule for the atomic transition. It is a particularly powerful selection rule, since it is independent of the angular momentum coupling scheme of the atomic levels. Other selection rules may be derived once an angular momentum coupling scheme is assumed.[18]

[18] See, for example, Foot (2005).

Higher-order transitions

Most laser transitions in atomic and molecular gases (with the notable exception of the atomic iodine laser operating at 1.3 μm) operate on electric dipole transitions. However, if a particular transition does not obey the selection rules as an electric dipole transition it may still occur via a higher-order multipole interaction—albeit at a rate that is typically slower by several orders of magnitude.

For example, the magnetic field of the radiation may interact with the magnetic moment of the atomic electrons to cause a magnetic dipole transition —as in the case of the atomic iodine laser. The selection rules for higher multipole forms of transition, such as magnetic dipole or electric quadrupole transitions, will be found in texts on atomic and molecular physics.[19]

[19] See, for example, Foot (2005).

As discussed in Chapter 7, many solid-state lasers operate on transitions that are not allowed by the electric dipole selection rules (and in particular the change of parity rule). In these systems, radiative decay occurs as a result of configuration mixing (so that the levels do not have a well-defined parity), or as a result of vibrations of the crystal lattice. However, the radiative transition rates are much slower than for a dipole-allowed transition of the same frequency, which proves to be useful in enabling population to be stored in the upper laser level.

2.4 Atomic population kinetics[†]

As discussed in more detail in Appendix A, the semi-classical treatment can be used directly to calculate the temporal evolution of the populations of atomic levels of an atom subjected to a radiation field. However, as we saw in Section 2.2.1, we can also use the Einstein treatment to determine this behaviour. We therefore have two different methods for calculating the level populations of atoms exposed to a radiation field; in general these two approaches give different answers.

This difference might seem surprising since the semi-classical model can be used to derive expressions for the Einstein coefficients. However, those derivations make certain assumptions about the bandwidth and intensity of the radiation field, meaning that the two calculational approaches will only agree under certain conditions.

To illustrate this we will use both approaches to find the evolution of the level populations of an atom with just two energy levels 1 and 2, subjected to radiation of spectral energy density $\varrho(\omega)$. This radiation is assumed to be switched on at $t = 0$ and to remain at a constant intensity thereafter. Further, if the total number of atoms per unit volume is N, and the atoms are initially all in the ground level 1. Then we must have,

$$N = N_1(t) + N_2(t) = N_1(0), \qquad (2.27)$$

where $N_i(t)$ is equal to the number density of atoms in level i at time t.

2.4.1 Rate equations

We first use the Einstein treatment to find the populations of the two levels. In doing so we will derive and solve **rate equations** for the level populations. From the definitions of the Einstein coefficients these are:

$$\frac{dN_2}{dt} = N_1 B_{12} \varrho(\omega_{21}) - N_2 B_{21} \varrho(\omega_{21}) - N_2 A_{21} \qquad (2.28)$$

$$\frac{dN_1}{dt} = -N_1 B_{12} \varrho(\omega_{21}) + N_2 B_{21} \varrho(\omega_{21}) + N_2 A_{21} = -\frac{dN_2}{dt}. \qquad (2.29)$$

The solutions to these rate equations may be found straightforwardly,[20] and are plotted in Fig. 2.5. It is seen that at small values of t the population in level 2 increases linearly with t, but that at later times the population tends towards a steady-state value.

[20]See Exercise 2.3.

2.4.2 Semi-classical equations

We do not solve the semi-classical equations here, but simply quote the solutions; further details will be found in Section A.1.2.

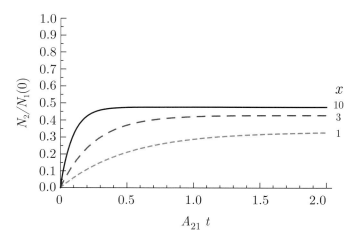

Fig. 2.5: Solutions to the rate equations (2.28) and (2.29) for the case $g_1 = g_2$ and three different values of the parameter $x = B_{21}\varrho(\omega_{21})/A_{21}$. Note that for larger values of x—corresponding to an increased radiation energy density at the frequency of the transition, and hence larger rates of stimulated emission and absorption relative to the rate of spontaneous emission—the population density in the upper level saturates at half of that of the lower level and the time taken to reach the steady-state solution decreases.

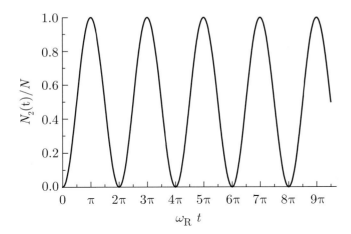

Fig. 2.6: Solutions of the semi-classical equations for a two-level atom subject to intense radiation very close to the resonance frequency.

For a two-level atom subject to an intense radiation field of constant intensity, and with a frequency close to the transition frequency ω_{21}, it is found that the entire population of atoms is periodically excited to the upper level and back again to the ground state, according to

$$N_2(t) = N \sin^2\left(\frac{1}{2}\omega_\text{R} t\right). \tag{2.30}$$

Figure 2.6 plots this behaviour, which is known as **optical nutation** or **Rabi flopping**. The characteristic frequency, the **Rabi frequency**, is given by

$$\hbar\omega_\text{R} = |\boldsymbol{D}_{12} \cdot \boldsymbol{E}_0|, \tag{2.31}$$

where $\boldsymbol{D}_{12}$, and $\boldsymbol{E}_0$ are as defined in Section 2.3.1.

The phenomenon of Rabi flopping is important in many areas of modern atomic physics in which low-density, thermally isolated samples of gas are subjected to high-intensity, narrow-bandwidth laser beams. This is in distinct contrast to the conditions typically found in the gain medium of a laser, where collisions with other atoms, electrons, or phonons perturb the atomic levels.

The effect of collisions[21] on the solution to the semi-classical equations[22] is shown in Fig. 2.7. It is seen that collisions damp the Rabi oscillations, allowing the population of the upper level to reach a steady-state value equal to that calculated by the rate equations. Indeed, if the rate of collisions is sufficiently high the solutions to the semi-classical and rate equations become indistinguishable at all times, not just once the system has reached a steady state.

[21] The collisions considered in the solution presented in Fig. 2.7 are phase-changing collisions. These are elastic collisions that perturb the phase of the atomic wave functions, but do not induce transitions between the energy levels.

[22] When extended to allow for spontaneous decay and the effect of elastic and inelastic collisions, the semi-classical equations become the optical Bloch equations.

2.4.3 Validity of the rate-equation approach

We have seen the the semi-classical and rate-equation models can give very different answers. There is no doubt that the semi-classical treatment is the more correct. However, using this approach to solve the temporal behaviour

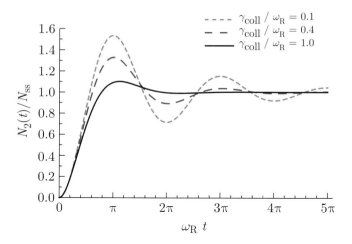

Fig. 2.7: Solutions of the semi-classical equations for various degrees of collisional damping. The plots show the calculated temporal evolution of the population density of the upper level, $N_2(t)$, normalized to the steady-state population density, N_{ss}. It can be seen that higher rates of collisions, corresponding to larger values of the parameter γ_{coll}, damp the Rabi oscillations more quickly.

of the level populations in a real laser system is usually very complicated and, more importantly, is unnecessary.

In general, the rate equations are valid provided the rate of perturbation of the phases of the atomic wave functions is large compared to the Rabi frequency ω_R. The Rabi flopping behaviour will also be washed out—and hence the temporal evolution of the level populations will follow the rate equation solution—if the bandwidth of the radiation is large compared to the Rabi frequency. In summary, we expect the rate equations to be valid provided that

$$\omega_R \ll \Delta\omega', \quad \textbf{Validity of rate equations} \tag{2.32}$$

where $\Delta\omega'$ is the *larger* of: (i) the radiation bandwidth; (ii) frequency width arising from phase-changing collisions.[23]

The phases of the atomic wave functions may be perturbed by collisions with: other atoms; electrons; or phonons. For all cases of practical interest to laser physics, the rate of these processes is sufficiently high that the rate-equation approach provides an entirely satisfactory description of the phenomena encountered. However, of course, at sufficiently high intensities the rate-equation approximation does break down; this point is explored further in Section A.4.

[23] For further information consult Loudon (2000) and Siegman (1986). The case of inhomogeneous broadening is discussed by Smith (1978b) and (1979).

Further reading

A summary of the selection rules for some of the common types of radiative transition may be found in Foot (2005).

More detailed treatments of the semi-classical theory are provided in many texts, such as those by Loudon (2000) and Foot (2005). Detailed consideration of the regime of applicability of the rate equation model is provided in a series of papers by Smith (1978a; 1978b; 1979).

Exercises

(2.1) **The relative importance of stimulated emission**
A two-level atom is placed within a cavity and is allowed to come into equilibrium with blackbody radiation of temperature T.

(a) Show that the condition for the rate of stimulated emission from the upper level being equal to the rate of spontaneous emission is
$$k_B T = \frac{\hbar\omega_{21}}{\ln 2}, \qquad (2.33)$$
where $\hbar\omega_{21}$ is the energy spacing of the levels.

(b) Find the temperature in electron volts (eV) and kelvin when this condition is met for transitions in the following regions of the electromagnetic spectrum:
  (i) radio frequencies at 50 MHz;
  (ii) microwaves at 1 GHz;
  (iii) visible light at 500 nm;
  (iv) X-rays of energy 1 keV.

(c) Under this condition, what is the ratio of the populations per state in the upper and lower level? Explain your answer in terms of the spontaneous and stimulated transition rates between the two levels.

(2.2) **Stimulated emission and the number of photons per mode**
Suppose that the two levels of Exercise 2.1 are now subjected to radiation of energy density $\varrho(\omega)$.

(a) Show that total rate of radiative transitions from the upper level can be written as:
$$R_2^{\text{rad}} = N_2 A_{21} \left[\frac{\pi^2 c^3}{\hbar \omega_{21}^3} \varrho(\omega_{21}) + 1 \right].$$

(b) Show that this may be written in the form,
$$R_2^{\text{rad}} = N_2 A_{21} [\bar{n} + 1], \qquad (2.34)$$
where $\bar{n}$ is the mean number of photons per mode.

(c) Use this result to explain the difference in the number of photons per mode for laser and non-laser sources.

(2.3) **Solution to the simple rate equations for a two-level atom**

(a) Show that the solution to the rate eqns (2.28) and (2.29) is,
$$N_2(t) = N_1(0) B_{12} \varrho(\omega) \tau' \left[1 - \exp\left(-t/\tau'\right)\right] \quad (2.35)$$
$$N_1(t) = N_1(0) - N_2(t), \qquad (2.36)$$
where,
$$\frac{1}{\tau'} = (1 + g_1/g_2) B_{12} \varrho(\omega) + A_{21}. \qquad (2.37)$$

(b) Hence, show that initially the population in level 2 grows at a rate $\dot{N}_2(t) = N_1(0) B_{12} \varrho(\omega)$.

(c) Show that the steady-state population density of the upper level is,
$$N_2(\infty) = \frac{N_1(0) B_{12} \varrho(\omega)}{A_{21} + (1 + g_1/g_2) B_{12} \varrho(\omega)}. \qquad (2.38)$$

(d) What happens to this steady-state population as the energy density of the radiation field is increased?

(e) Show that the steady-state population always obeys the condition
$$\frac{N_2(\infty)}{g_2} \leq \frac{N_1(\infty)}{g_1}. \qquad (2.39)$$

(f) †Show that at *all* times the populations obey
$$\frac{N_2(t)}{g_2} \leq \frac{N_1(t)}{g_1}. \qquad (2.40)$$

This problem demonstrates that, within the validity of the rate-equation approximation, absorption of population on a transition $1 \to 2$ cannot generate a population inversion on the same transition, no matter how intense the driving or 'pump' radiation is. Optical pumping *is* able to generate population inversions with respect to transitions from 2 to levels other than 1.

(2.4) **A proposed c.w. laser scheme**
It is proposed to construct a gas laser operating on a transition between an upper level 2 with $J = 7/2$ and a lower level 1 of $J = 5/2$. Under the operating conditions of the discharge the lower level has an effective lifetime of 30 ns and the Einstein coefficient for the laser transition $A_{21} = 5.0 \times 10^7 \, \text{s}^{-1}$. Is c.w. oscillation possible for such a laser?

(2.5) **Eigenvalues of the parity operator**
Suppose that the parity operator $\widehat{P}$ has eigenfunctions ϕ with eigenvalues a so that:

$$\widehat{P}\phi = a\phi.$$

(a) By applying the parity operator to ϕ twice in succession, show that $a^2 = 1$.

(b) Hence, show that the eigenvalues of $\widehat{P}$ are ± 1.

(2.6) **The parity selection rule for electric dipole radiation**

(a) Use eqn (2.24) to show that the Einstein A coefficient for electric dipole transitions between the states $|1m_1\rangle$ and $|2m_2\rangle$ of an atom (assumed to have only a single optically active electron) is proportional to a sum of terms of the form,

$$|\langle 1m_1| - ex |2m_2\rangle|^2 + |\langle 1m_1| - ey |2m_2\rangle|^2$$
$$+ |\langle 1m_1| - ez |2m_2\rangle|^2.$$

(b) Consider now the term involving x. What is the parity of the operator x?

(c) Write down the integral represented by

$$<1m_1| - ex|2m_2>.$$

(d) Explain why this integral is zero if the two states $|im_i\rangle$ have the same parity.

(e) Hence, explain why, for electric dipole radiation, the Einstein A and B coefficients will be zero if the two states have the same parity.

Broadening mechanisms and lineshapes

3

3.1 Homogeneous broadening mechanisms	27
3.2 Inhomogeneous broadening mechanisms	35
3.3 The interaction of radiation and matter in the presence of spectral broadening	38
3.4 The formation of spectral lines: The Voigt profile[†]	40
3.5 Other broadening effects	42
Further reading	43
Exercises	43

So far, we have assumed that the atomic levels are infinitely sharp, and consequently the radiation emitted in transitions between them is purely monochromatic. In reality, of course, this is not the case and for any given transition a number of mechanisms can lead to broadening of the radiation emitted. The broadening mechanisms may be divided into two classes: **homogeneous broadening** and **inhomogeneous broadening**. We will see that whether a laser transition is homogeneously or inhomogeneously broadened is of crucial importance in determining the details of the behaviour of the laser.

3.1 Homogeneous broadening mechanisms

Broadening mechanisms that affect all atoms in a sample equally, such that all atoms in a given excited level have the same probability of emitting a photon of a given frequency, are said to be homogeneous. We discuss below several homogeneous broadening mechanisms of importance in laser physics.

3.1.1 Natural broadening

Natural, or lifetime, broadening is intrinsic to all laser transitions since it arises as a consequence of the finite lifetimes of the upper and lower laser levels.

Classical model of natural broadening

In order to gain some understanding of natural broadening, it is useful to consider the classical electron oscillator model of the atom. In this picture the active electron[1] of an atom is imagined to be bound to the atomic nucleus by a harmonic restoring force of the form $-m_\mathrm{e}\omega_0^2 r$, where r is the displacement of the electron from its equilibrium position. The motion of the electron is assumed to be damped by a force $-m_\mathrm{e}\gamma\dot{r}$, where γ is a constant.

The great advantage of the classical electron oscillator model is that it is able to reproduce the results of the considerably more complex quantum-mechanical calculation. According to the classical electron oscillator model ω_0 is the resonant frequency of the oscillating electron, but there is no classical method for calculating its value. However, as will become clear below, a

---

[1] The selection rules for electric dipole radiation state that transitions of this type can only occur if only a single electron changes state in the transition. This is known as the active electron.

comparison with the results of quantum theory shows that ω_0 should be interpreted as the frequency of the transition under consideration.

In the absence of any applied force the equation of motion of the electron may be written as,

$$m_e \frac{d^2 \boldsymbol{r}}{dt^2} = -m_e \omega_0^2 \boldsymbol{r} - m_e \gamma \frac{d\boldsymbol{r}}{dt}. \tag{3.1}$$

By substituting a trial solution of the form $\boldsymbol{r}(t) = \boldsymbol{r}_0 e^{st}$, we find that for $\gamma \ll \omega_0$

$$\boldsymbol{r}(t) = \boldsymbol{r}_0 e^{-\frac{\gamma}{2}t} \cos \omega_0 t, \tag{3.2}$$

where we have used the boundary conditions $\boldsymbol{r}(0) = \boldsymbol{r}_0$, and $\dot{\boldsymbol{r}}(0) = 0$. Hence, we see that if the electron is initially displaced it will undergo a damped oscillation with an angular frequency ω_0.

The combination of the oscillating electron and the, assumed fixed, nucleus looks like an oscillating electric dipole moment,

$$\boldsymbol{p}(t) = -e\boldsymbol{r}(t) = -e\boldsymbol{r}_0 e^{-\frac{\gamma}{2}t} \cos \omega_0 t. \tag{3.3}$$

The oscillating dipole radiates an electromagnetic wave which, at large distances from the atom, has an amplitude proportional to $\dot{\boldsymbol{p}}(t)$. We may therefore write the amplitude of the electric field at some distant point as,

$$E(t) = \begin{cases} E_0 e^{-\frac{\gamma}{2}t} \sin \omega_0 t & t \geq 0 \\ 0 & t < 0, \end{cases} \tag{3.4}$$

where E_0 is a constant. We can then find the frequency distribution of the radiation field by taking the Fourier transform of eqn (3.4):

$$\mathcal{E}(\omega) = \sqrt{\frac{1}{2\pi}} \int_0^\infty E_0 e^{-\frac{\gamma}{2}t} \sin \omega_0 t \, e^{i\omega t} dt. \tag{3.5}$$

Providing that $\gamma \ll \omega_0$, we find that,

$$\mathcal{E}(\omega) = \frac{E_0}{2\sqrt{2\pi}} \frac{-1}{(\omega - \omega_0) + i\gamma/2}. \tag{3.6}$$

Now, the measured intensity of the wave is proportional to $|\mathcal{E}(\omega)|^2$, and hence the frequency distribution of the radiation is given by,

$$\mathcal{I}(\omega) = I_0 g_H(\omega - \omega_0), \tag{3.7}$$

where I_0 is a constant, and the **normalized lineshape** $g_H(\omega - \omega_0)$ of the emitted radiation is in this case the well-known Lorentzian distribution:[2]

$$g_H(\omega - \omega_0) = \mathcal{L}(\omega - \omega_0) \equiv \frac{1}{\pi} \frac{\gamma/2}{(\omega - \omega_0)^2 + (\gamma/2)^2}.$$

Normalized Lorentzian (3.8)

Note that the function $g_H(\omega - \omega_0)$ has been normalized such that,[3]

$$\int_0^\infty g_H(\omega - \omega_0) d\omega = 1. \tag{3.9}$$

---

[2] To be clear, we note that the constant I_0 has units of *total* intensity (i.e. W m^{-2}), which can be seen by integrating both sides of eqn (3.7) over all frequency.

[3] Strictly, $g_H(x)$ is only normalized for the region $-\infty < x < \infty$. In eqn (3.9) the lower limit has been set to the smallest physically meaningful frequency, i.e. $\omega = 0$, corresponding to $x = -\omega_0$. However, provided that $\gamma \ll \omega_0$, $g_H(x)$ will be extremely small at $x = -\omega_0$, and there is little error introduced by extending the lower limit to $-\infty$.

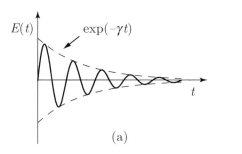 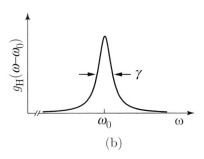

Fig. 3.1: (a) The electric field emitted by a classical atom with resonance frequency ω_0 and damping coefficient γ. (b) The frequency spectrum $g_\mathrm{H}(\omega - \omega_0)$ corresponding to the damped electromagnetic wave shown in (a).

As illustrated in Fig. 3.1, the Lorentzian distribution has a full width at half-maximum given by,

$$\Delta\omega_\mathrm{N} = \gamma. \qquad (3.10)$$

Of course, we have only considered the emission from a single atom, and we must extend our analysis to a macroscopic sample of such atoms. For such a sample, during a certain period of observation the electric field reaching a detector will consist of, say, N copies of the damped wavetrain shown in Fig. 3.1 distributed randomly in time. The electric field from the macroscopic sample can then be represented as a convolution of a single damped wavetrain with N randomly distributed Dirac δ-functions. We know from the Convolution Theorem[4] that the Fourier transform of the electric field from the macroscopic sample is proportional to the product of the transform of the single damped wavetrain with that of the array of δ-functions. Now, the Fourier transform of an array of δ-functions is simply a sum of N phases: $e^{i\phi_1} + e^{i\phi_2} + e^{i\phi_3} \ldots = A e^{i\phi}$, where, on average, $A \propto \sqrt{N}$, and ϕ is a random phase. Hence, the intensity of the detected radiation will simply be N times that of a single wavetrain, and as such will still have a Lorentzian frequency distribution given by eqn (3.8).

[4] For a detailed discussion of the Convolution Theorem and its application to problems in optics, see Brooker (2003).

Classical decay rate

We may derive a classical value for the damping coefficient. It is a well-known result from classical electromagnetism that an oscillating electric dipole will radiate an electromagnetic wave.[5] The rate at which the dipole radiates power is given by,

[5] See for example, Bleaney and Bleaney (1989).

$$W_\mathrm{dipole} = \frac{p_0^2 \omega_0^4}{12\pi\epsilon_0 c^3}, \qquad (3.11)$$

where the dipole moment is assumed to be of the form $\boldsymbol{p}(t) = \boldsymbol{p}_0 \cos\omega_0 t$. Of course, classically, this power loss is responsible for the damping of the oscillation of the electron. By relating the power loss to the rate of work done against the damping force we may derive a classical value for the damping coefficient γ:

$$\gamma_\mathrm{classical} = \frac{e^2 \omega_0^2}{6\pi\epsilon_0 m_e c^3}. \qquad \textbf{Classical decay rate} \qquad (3.12)$$

Relation of γ to the Einstein A coefficient

An atom with only two levels corresponds most closely to the classical model since in this case it is clear that the resonance frequency appearing in the classical model is equivalent to the frequency, ω_{21}, of the transition between the two levels. Suppose also that the lower level corresponds to the ground state, and hence does not decay, and that the upper level decays only by spontaneous emission to the lower level. The density of atoms in the upper level will then obey the simple rate equation,

$$\frac{dN_2}{dt} = -N_2 A_{21}. \quad \text{(two-level atom)} \quad (3.13)$$

If at $t = 0$ all the atoms in a sample are in level 2, at later times the density of population in level 2 will be given by $N_2(t) = N_2(0) \exp(-A_{21}t)$. The energy contained in the excited atoms will also decrease as $\exp(-A_{21}t)$.

According to the classical model the energy of the system is equal to the sum of its potential and kinetic energies, i.e. $\frac{1}{2}m_e\omega_0^2 r^2 + \frac{1}{2}m_e \dot{r}^2$. From eqn (3.2) we see that this energy decays as $\exp(-\gamma t)$. Hence, for this simple case we may make the identification,

$$\gamma = A_{21}. \quad (3.14)$$

Equations (3.12) and eqn (3.14) yield a classical value for the Einstein A coefficient. Frequently, however, the agreement with experiment and quantum theory is poor. This is not surprising since in real atoms there are many levels, whereas in the classical model there is only one resonance frequency. We may therefore modify eqn (3.14) and write

$$A_{21} = -3 f_{21} \gamma_{21}^{\text{classical}} = \frac{e^2 \omega_{21}^2}{2\pi \epsilon_0 m_e c^3}(-f_{21}), \quad (3.15)$$

where f_{21} is called the **emission oscillator strength**. This can be thought of as the fraction of the classical oscillator that is associated with the transition $2 \to 1$:

We can also define an **absorption oscillator strength** f_{12} by the relation,

$$g_1 f_{12} = -g_2 f_{21} = gf, \quad (3.16)$$

where gf is known as the **weighted oscillator strength**. Note that by convention the emission oscillator strength is negative.[6,7]

The radiative lifetime

Let us now consider the more usual case of an atom with many levels. Consider first the possible radiative decay transitions of an excited level k, as illustrated in Fig. 3.2. Suppose that at time $t = 0$ all the atoms in a sample are excited to level k. Subsequently, the number $N_k(t)$ of atoms per unit volume in level k will be given by the rate equation,

$$\frac{dN_k}{dt} = -N_k A_{kj} - N_k A_{ki} - N_k A_{kh} - \ldots \quad (3.17)$$

$$= -N_k \sum_{E_j < E_k} A_{kj}, \quad \text{(multilevel atom)} \quad (3.18)$$

---

[6] The form of eqn (3.16) should feel right, since it is really the same equation as that which relates the Einstein B coefficients for absorption and stimulated emission. Furthermore, the factor of 3 appearing in eqn (3.15) is related to the factor of $1/3$ that appears in the results of the semi-classical model for the case of unpolarized atoms or radiation. See Appendix A.

[7] The oscillator strength is useful in a number of ways. It represents the 'strength' of a given transition in a normalized way, which, unlike the Einstein A coefficient, is independent of the frequency of the transition. Strong, fully allowed transitions will have a weighted oscillator strength of order unity, whereas weak or disallowed transitions will have oscillator strengths that are very much smaller. Furthermore, the oscillator strengths obey a variety of **sum rules**. For a one-electron atom it may be shown that for any level k,

$$\sum_i f_{ki} = 1,$$

where the sum is over levels higher and lower than k. For more complex atoms it is found that,

$$\sum_i f_{ki} \approx n,$$

where n is the number of valence electrons. For further information see Corney (1978).

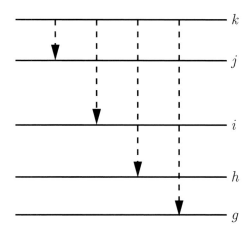

Fig. 3.2: Radiative decay of an excited level k to lower levels $j, i, h, \ldots$.

where A_{kj} is the Einstein A coefficient for spontaneous decay to the lower level j, and the sum is over all levels lying below it. It is clear that the population density of the upper level, and hence the energy stored in the sample of atoms, will decay exponentially[8] according to,

$$N_k(t) = N_k(0) e^{-t/\tau_k^{\text{rad}}}, \tag{3.19}$$

where the **radiative lifetime** of the level k is given by

$$\frac{1}{\tau_k^{\text{rad}}} = \sum_{E_j < E_k} A_{kj}. \quad \textbf{Radiative lifetime} \tag{3.20}$$

[8] If our ensemble of atoms were also subjected to collisions, the upper-level population density would still decay exponentially, but with a reduced time constant known as the fluorescence lifetime. The fluorescence lifetime takes its name from the fact that it is the characteristic time with which the intensity of fluorescence from the level would be observed to decay.

Following the arguments above, if all the atoms in a sample are initially excited to a level k, the energy of the system decays according to $\exp(-t/\tau_k^{\text{rad}})$. Hence, if we now consider the transition $k \to j$ between two excited levels we would expect that the value of γ that should be used in the classical model is equal to the sum of the decay rates of the two levels, and that—following eqn (3.10)—the natural linewidth of the transition will be given by,

$$\Delta \omega_N = \gamma = \frac{1}{\tau_k^{\text{rad}}} + \frac{1}{\tau_j^{\text{rad}}}. \tag{3.21}$$

This assertion is justified by the results of the quantum theory outlined below.

Quantum theory of natural broadening

The quantum theory of natural broadening is beyond the scope of the present text and here we merely provide a justification of the main results.

By the Uncertainty Principle, the finite radiative lifetime τ_k^{rad} of an excited level is associated with an uncertainty ΔE_k in the energy of the level, where,

$$\Delta E_k \tau_k^{\text{rad}} \approx \hbar. \tag{3.22}$$

The frequency width Γ_k associated with this is simply,

$$\Gamma_k = \frac{\Delta E_k}{\hbar} = \frac{1}{\tau_k^{\text{rad}}} = \sum_{E_j < E_k} A_{kj}. \tag{3.23}$$

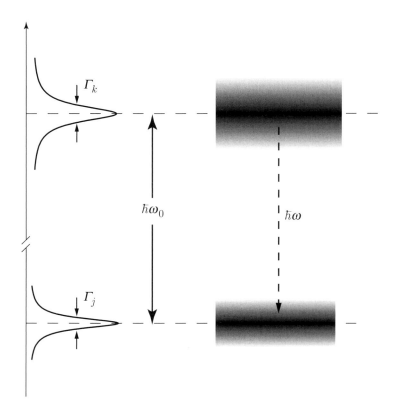

Fig. 3.3: Schematic diagram of the quantum theory of natural broadening.

As illustrated schematically in Fig. 3.3, for a given transition $k \rightarrow j$ the finite widths of *both* levels must be taken into account, and consequently the frequency width of the emitted radiation is given by,

$$\Delta \omega_N = \Gamma_k + \Gamma_j = \frac{1}{\tau_k^{\text{rad}}} + \frac{1}{\tau_j^{\text{rad}}}. \tag{3.24}$$

A more rigorous treatment shows that the spectral distribution of the emitted radiation is equal to the Lorentzian distribution of eqn (3.8) with a full width at half-maximum given by eqn (3.24).

3.1.2 Pressure broadening

Thus far, we have considered the spectral distribution of light emitted from a collection of non-interacting atoms. For atoms in a fluid, however, this is very rarely the case, and in reality the radiating atoms interact with each other, most obviously through collisions, leading to changes in the spectral content of the radiated light.

Classical model

We may extend the classical model to the case of pressure broadening in an intuitively obvious way. We suppose that a radiating atom emits a damped

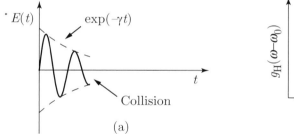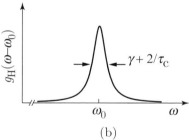

Fig. 3.4: (a) The electric field emitted by a classical atom with resonance frequency ω_0, damping constant γ, and subjected to a collision at $t = \tau_i$. (b) The frequency spectrum emitted by a macroscopic sample of such atoms with a mean collision time τ_c.

wave of the form of eqn (3.4) until such time τ_i as it experiences a collision, as illustrated schematically in Fig. 3.4. As before, we find the frequency distribution of the radiated wave by taking a Fourier transform,

$$\mathcal{E}(\omega) = \sqrt{\frac{1}{2\pi}} \int_0^{\tau_i} E_0 e^{-\frac{\gamma}{2}t} \sin \omega_0 t \, e^{i\omega t} \, dt, \tag{3.25}$$

where now the upper limit of integration is $t = \tau_i$.

It is straightforward to show that, providing $\gamma \ll \omega_0$, the frequency distribution of the detected spectral intensity $\mathcal{I}(\omega, \tau_i) \propto |\mathcal{E}(\omega)|^2$ is given by,

$$\mathcal{I}(\omega, \tau_i) \propto \frac{1 + e^{-\gamma \tau_i} - 2 \cos[(\omega - \omega_0)\tau_i] e^{-\frac{\gamma}{2}\tau_i}}{(\omega - \omega_0)^2 + (\gamma/2)^2}. \tag{3.26}$$

The spectrum emitted by a macroscopic sample of such emitters is found by averaging eqn (3.26) over the distribution $P(\tau_i)$ of times between collisions. This distribution is found from kinetic theory to be[9]

[9] See, for example, Reif (1985).

$$P(\tau_i) \, d\tau_i = e^{-\tau_i/\tau_c} \frac{d\tau_i}{\tau_c}, \tag{3.27}$$

and τ_c is the mean time between collisions.

This averaging process yields, once more, a Lorentzian distribution of frequencies, but with an increased full width at half-maximum $\Delta\omega_P$ given by,

$$\Delta\omega_p = \gamma + \frac{2}{\tau_c}. \quad \textbf{Collision broadening} \tag{3.28}$$

Note the factor of 2 in the second term!

Just as the introduction of damping led to a broadening of the radiated spectrum, collisions also lead to broadening since the radiated wavetrain from each atom is truncated. Fourier theory then tells us that a reduction in the time for which the wave is radiated must be associated with a broadening of the frequency spectrum.

Very often, the collisions of the radiating atoms with some quenching species, which may or may not be the same species as the radiators, can be described in terms of a collision cross-section σ_{coll}. The mean time between collisions is then given by,

$$\frac{1}{\tau_c} = N_q \langle v\sigma_{\text{coll}}(v) \rangle, \qquad (3.29)$$

where N_q is the number density of the quenching species, v is the velocity of the quencher relative to the radiating atom, and the brackets indicate an average over the distribution of relative velocities. In this case the pressure-broadened linewidth becomes,

$$\Delta\omega_p = \gamma + 2N_q \langle v\sigma_{\text{coll}}(v) \rangle, \qquad (3.30)$$

which shows clearly that: (i) the minimum linewidth that occurs is equal to the natural linewidth $\Delta\omega_N = \gamma$; (ii) the linewidth increases linearly with the density or partial pressure of the quenching species.

Quantum theory of pressure broadening

The quantum theory of pressure broadening considers the interaction of a radiating atom with nearby atoms or electrons in terms of a perturbation of the energy levels of the atom. The quantum theory of pressure broadening is a large and complex subject, and indeed one that remains the focus of contemporary research. It is not necessary for us to discuss the details of this theory, and we provide here only a brief outline of the subject.

In discussing pressure broadening it is useful to define the duration $\Delta\tau_{\text{coll}}$ of a collision as,

$$\Delta\tau_{\text{coll}} = \frac{d}{v}, \qquad (3.31)$$

where d is the distance of closest approach, and v is the relative velocity of the colliding partners. We may then identify two extreme regimes in which the calculations are more tractable:

Impact approximation: $\Delta\tau_{\text{coll}} \ll \tau_c$

Here, the duration of the collision is very much shorter than the mean time between collisions, and consequently the atom does not make any transitions during the collision process. During the collision the energies of the upper and lower levels are briefly shifted, and, as a result, the phase of the atomic wave function is shifted in an abrupt fashion. De-phasing of the atomic wave function in this way leads to broadening of the transition frequency.

It is clear that the impact approximation will be appropriate at low densities where the mean time between collisions is long. Typically, the impact approximation is valid for gases at pressures below 100 mbar.

Quasi-static approximation: $\Delta\tau_{\text{coll}} \gg \tau_c$

Here, the duration of a collision is much greater than the mean time between collisions, and the motion of the perturbers can be ignored. In this case the radiating atom and the perturber can be thought of as forming a diatomic molecule of internuclear separation r. The frequency of the radiation emitted by this molecule is a function of the separation r, and the frequency spectrum may be found by averaging over the statistical distribution of radiator–

perturber separations. This quasi-static approximation is particularly important in calculating the Stark broadening due to ions in a plasma, but is also used to calculate the broadening in neutral gases for pressures significantly greater than 100 mbar.

Clearly a proper quantum calculation of pressure broadening is somewhat involved. In the most general case the effect of collisions is a shift of the central frequency of the transition, and an asymmetric broadening of the radiated spectrum. However, it is quite often found that the shift in the central frequency is small, and the frequency distribution is symmetric and well approximated by a Lorentzian distribution.

3.1.3 Phonon broadening

Chapter 7 describes several important laser transitions that occur between levels of ions doped into solids. In such cases the electrons of the active ion interact with the surrounding ions of the solid via the Coulomb force, which may be conveniently described by an electric field known as the **crystal field**. The crystal field causes the energy levels of the active ion to be shifted and split from those of the free ion, leading to changes in the structure of the energy levels.

In addition to these shifts, spectral broadening occurs through dynamic changes in the crystal field arising from the thermal vibration of the host lattice. These vibrations are quantized, and hence may be described in terms of so-called lattice **phonons**. As discussed in more detail in Section 7.1.4, phonons can cause spectral broadening by two mechanisms. First, phonon collisions can cause an excited level of the active ion to decay to a lower level by emission not of radiation, but of one or more phonons. The increased rate of decay of the level results in an increased linewidth by lifetime broadening. Secondly, phonon collisions may perturb the phase of the wave function of the level, without inducing a transition. This leads to additional broadening that is proportional to the rate of such collisions.

Since the extent of thermal vibration, or equivalently the number and distribution of phonons, depends on the temperature of the lattice, phonon broadening depends strongly on temperature. At room temperature, phonon broadening usually dominates natural broadening since the radiative lifetimes of the relevant levels are usually long.

We see that phonon collisions lead to broadening of the levels in a similar way to collisional broadening in a gas. Phonon broadening is homogeneous since all active ions in the sample are affected by phonon collisions in the same way.

Finally in this section, we note that in some crystals the interaction with the crystal field is very strong leading to homogeneously broadened **vibronic transitions** with very large linewidths.[10]

[10] See Chapter 7.

3.2 Inhomogeneous broadening mechanisms

The homogeneous broadening mechanisms discussed above affect all atoms in a sample equally, so that the distribution of frequencies emitted by any atom in the sample is the same. However, other broadening mechanisms exist that cause the transition frequency of different atoms to be shifted by different amounts. Such broadening mechanisms are said to be **inhomogeneous**, the broadening being characterized by a distribution of transition frequencies $g_D(\omega - \omega_0)$.[11]

Inhomogeneous broadening will always be associated with some homogeneous broadening since natural broadening is always present, and the atoms may also be subject to other homogeneous broadening mechanisms such as phonon broadening. The transition lineshape observed from a sample subjected to both homogeneous and inhomogeneous broadening processes is discussed in more detail in Section 3.4. However, it should be clear that when, as is very often the case, the frequency width of the inhomogeneous distribution $g_D(\omega - \omega_0)$ is much greater than that of the homogeneous distribution $g_H(\omega - \omega_0)$ the observed transition lineshape will simply be the inhomogeneous lineshape.

[11] We have used the superscript 'D' rather than 'I' to distinguish the inhomogeneous lineshape from the homogeneous lineshape in order to avoid possible confusion with the use of the subscript 'I' used in later chapters to indicate saturation effects. Of course, the inhomogeneous lineshape defined here applies to any form of inhomogeneous broadening, not just Doppler broadening, as might be suggested by the use of 'D'.

3.2.1 Doppler broadening

For gas lasers the most important line-broadening mechanism is very often **Doppler broadening**, which arises from a combination of the Doppler effect and the thermal motion of the atoms.

Let us imagine using a spectrometer to measure the emission spectrum on a certain transition from a gaseous sample of atoms, as illustrated schematically in Fig. 3.5. We will ignore homogeneous broadening and assume that stationary atoms emit a single frequency ω_0. Consider now the radiation emitted by those atoms travelling with a velocity v_z towards the spectrometer. It is a well-known result of classical physics that the observed frequency ω, i.e. that measured by the spectrometer, will be shifted from the frequency ω_0 emitted by a stationary atom according to,

$$\omega - \omega_0 = \frac{v_z}{c}\omega_0, \quad (3.32)$$

where we have ignored relativistic effects, since for all cases of interest to lasers $v_z \ll c$.

We see that those atoms moving towards the spectrometer will appear to emit higher frequencies, whilst the frequencies measured for those atoms moving away from the spectrometer will be lower than ω_0. Assuming that the sample is in thermal equilibrium with temperature T, the proportion $P(v_z)dv_z$ of atoms with a z-component of velocity in the range v_z to $v_z + dv_z$ is given by the **Maxwellian distribution**:[12]

[12] See, for example, Reif (1985).

$$P(v_z)dv_z = \sqrt{\frac{M}{2\pi k_B T}} \exp\left(-\frac{Mv_z^2}{2k_B T}\right) dv_z, \quad (3.33)$$

where M is the mass of each atom, and k_B the Boltzmann constant.

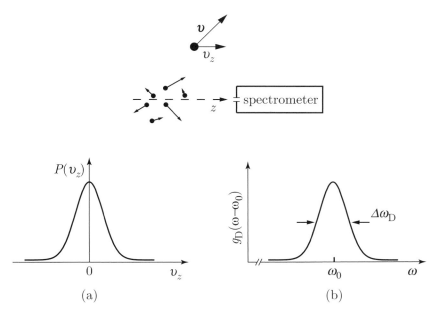

Fig. 3.5: The formation of an inhomogeneously broadened spectral line by Doppler-broadened gas atoms. Radiation emitted by the atoms in the z-direction is detected by the spectrometer. The distribution of atomic velocities in the z-direction (a) leads, via the Doppler effect, to a distribution of frequencies measured by the spectrometer, and hence broadening of the spectral line (b).

Substitution of eqn (3.32) into eqn (3.33) gives the observed lineshape $g_D(\omega - \omega_0)$:

$$g_D(\omega - \omega_0)\mathrm{d}\omega = \sqrt{\frac{M}{2\pi k_B T}} \exp\left(\frac{-Mc^2}{2k_B T}\left[\frac{\omega - \omega_0}{\omega_0}\right]^2\right) \frac{c}{\omega_0}\mathrm{d}\omega. \quad (3.34)$$

This is a Gaussian distribution with a full width at half-maximum, known as the Doppler width, given by:

$$\Delta\omega_D = 2\sqrt{2\ln 2}\frac{\omega_0}{c}\sqrt{\frac{k_B T}{M}}, \quad (3.35)$$

Doppler width

$$\frac{\Delta\omega_D}{\omega_0} = \frac{\Delta\nu_D}{\nu_0} = 7.16 \times 10^{-7}\sqrt{\frac{T}{A}}, \quad (3.36)$$

where in eqn (3.36) the temperature is in Kelvin and the atomic mass A has units of grammes per mole.

It is usually more convenient to rewrite the lineshape[13] in terms of $\Delta\omega_D$, i.e.

$$g_D(\omega - \omega_0) = \mathcal{G}(\omega - \omega_0) \equiv \frac{2}{\Delta\omega_D}\sqrt{\frac{\ln 2}{\pi}} \exp\left(-\left[\frac{\omega - \omega_0}{\Delta\omega_D/2}\right]^2 \ln 2\right),$$

Normalized Gaussian lineshape

(3.37)

[13] Whilst we have considered the lineshape of the radiation emitted from the sample, we could also have considered the absorption spectrum that would be measured and we would have found the same distribution $g_D(\omega - \omega_0)$.

[14] See Note 3 on page 28.

where the distribution is normalized[14] so that $\int_0^\infty g_D(\omega - \omega_0) d\omega = 1$.

3.2.2 Broadening in amorphous solids

As discussed above, transitions within ions doped into a solid will be broadened by phonon interactions. If the host is a good-quality single crystal, this broadening is homogeneous. However, if the solid is highly non-uniform, active ions in different locations will experience different environments. Of particular importance is the local value of the strain of the crystal lattice since this affects the local crystal field experienced by the ion, which in turn affects the energy levels of the ion through the Stark effect. Other aspects of the local environment that can affect the transition frequencies of an active ion are the presence of impurity ions, or variations in the orientation of the crystal lattice. Since such affects change the centre frequency of the active ions according to their location in the medium, the broadening of the frequency response of a macroscopic sample is inhomogeneous.

These effects are particularly important for ions doped in glasses, such as Nd^{3+} ions in the Nd:glass laser, since in a glassy material the local environment varies widely with position, leading to substantial inhomogeneous broadening. Very often, the distribution of centre frequencies is found to follow a normal, i.e. Gaussian, distribution, in which case the transition lineshape is also Gaussian.

3.3 The interaction of radiation and matter in the presence of spectral broadening

When we introduced the Einstein coefficients in Section 2.1, the transition under consideration was assumed to be infinitely sharp. As such, radiation could only interact with the atom if its angular frequency ω was exactly equal to that of the transition ω_0. In that idealized case, for atoms interacting with radiation only, we could write the rate equation for the population of the upper level as,

$$\frac{dN_2}{dt} = N_1 B_{12} \varrho(\omega_0) - N_2 B_{21} \varrho(\omega_0) - \frac{N_2}{\tau_2}, \quad (3.38)$$

where τ_2 is the fluorescence lifetime of the upper level.

In this chapter we have established that in practice all transitions are broadened to some extent. Clearly, this will mean that radiation slightly detuned from the centre transition frequency will also be able to interact with the atom. How can we modify eqn (3.38) to account for line broadening? The answer depends on the class of line broadening.

3.3.1 Homogeneously broadened transitions

For homogeneously broadened atoms, all atoms in a given level interact with radiation of angular frequency ω with the same strength. However, the strength of this interaction will depend on the extent of the frequency detuning from ω_0.

Hence, in modifying eqn (3.38) to account for homogeneous broadening, the population densities N_1 and N_2 should not be changed, but, instead the Einstein coefficients should account for the detuning from ω_0. In Appendix B we show that for homogeneously broadened transitions the Einstein postulates may be rewritten as:

1. The rate at which an atom in the upper level decays to the lower level by spontaneous emission of photons with angular frequencies lying in the range ω to $\omega + \delta\omega$ is equal to $N_2 A_{21} g_H(\omega - \omega_0) \delta\omega$, where N_2 is the density of atoms in the upper level, and where A_{21} is the Einstein A coefficient, $g_H(\omega - \omega_0)$ the normalized *homogeneous* lineshape, and ω_0 the centre frequency of the transition.
2. The rate at which atoms in the lower level are excited to the upper level by the absorption of photons with angular frequencies lying in the range ω to $\omega + \delta\omega$ is equal to $N_1 B_{12} g_H(\omega - \omega_0) \varrho(\omega) \delta\omega$, where N_1 is the density of atoms in the lower level, and $\varrho(\omega)$ is the energy density of the radiation.
3. The rate at which atoms in the upper level decay to the lower level by the stimulated emission of photons with angular frequencies lying in the range ω to $\omega + \delta\omega$ is equal to $N_2 B_{21} g_H(\omega - \omega_0) \varrho(\omega) \delta\omega$.

We may then rewrite eqn (3.38) for homogeneously broadened atoms interacting with radiation with frequencies in the range ω to $\omega + \delta\omega$ as,[15]

$$\frac{dN_2}{dt} = N_1 B_{12} g_H(\omega - \omega_0) \varrho(\omega) \delta\omega - N_2 B_{21} g_H(\omega - \omega_0) \varrho(\omega) \delta\omega - \frac{N_2}{\tau_2}. \quad (3.39)$$

[15] This modified form of the Einstein postulates and of eqn (3.38) should feel reasonable. We can think of the homogeneous lineshape as describing the reduction in strength by which radiation interacts with the atom when it is de-tuned from the centre of the transition.

3.3.2 Inhomogeneously broadened atoms[†]

We must treat the case of inhomogeneously broadened atoms somewhat differently, and rather carefully. We recall that for inhomogeneously broadened atoms the frequency emitted by an atom depends on some parameter of the atom, such as its position or velocity. The emission from each atom will also be homogeneously broadened to some extent, since at the very least the transition must exhibit natural broadening.

We can define the centre frequency ω_c emitted by each atom, and we consider all atoms with centre frequency ω_c as belonging to the same **class** of atoms. We may treat each class as being quite distinct; provided that the atoms do not change class, or at least not too frequently.[16]

In treating the interaction of inhomogeneously broadened atoms with radiation we should treat each class separately. For example, we can define the number density of atoms in the upper and lower levels with centre frequencies between ω_c and $\omega_c + \delta\omega_c$ as $\Delta N_2(\omega_c) \delta\omega_c$ and $\Delta N_1(\omega_c) \delta\omega_c$, respectively. The strength with which each class of atoms interacts with radiation of frequency ω depends on the detuning of ω from ω_c, *not* on the detuning from ω_0, the centre frequency of the inhomogeneous lineshape.

In order to clarify this point, let us write the rate equation for *each class* of atoms in the upper level interacting with radiation of spectral energy density $\varrho(\omega)$:

[16] A rapid exchange of atoms between classes (by, for example, velocity-changing collisions) would lead to an increase in the homogeneous linewidth. Describing the atomic distribution in terms of classes is still valid as long as the additional frequency width associated with the class-changing process is small compared to the inhomogeneous width.

Fig. 3.6: Steady-state distribution over centre frequencies ω_c for inhomogeneously broadened atoms in the presence of narrowband radiation at ω. Only atoms with centre frequencies close to ω are excited to the upper level, leading to very different distributions for the two levels. The dashed line gives the distribution of centre frequencies in the absence of the beam.

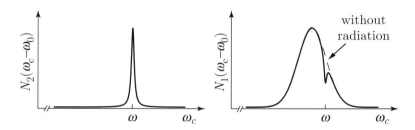

$$\frac{d\Delta N_2}{dt} = \Delta N_1 B_{12} g_H(\omega - \omega_c) \varrho(\omega) \delta\omega - \Delta N_2 B_{21} g_H(\omega - \omega_c) \varrho(\omega) \delta\omega - \frac{\Delta N_2}{\tau_2}, \quad (3.40)$$

and similarly for ΔN_1. It should be clear that this equation is analogous to eqn (3.39).

In principle, the rate equations for each class can then be solved separately, and the results integrated over ω_c to give the *total* population densities in the upper and lower levels.

The case of inhomogeneous broadening is clearly considerably more complex than that of homogeneous broadening, and considerable care must be taken when analysing such systems. In particular, it is important to realize that the distribution over frequency classes may not be the same for the upper and lower levels. Consider, for example, excitation of a previously empty upper level by absorption of a beam of narrowband radiation of frequency ω. Only ground-state atoms for which $\omega_c \approx \omega$ will interact strongly with the beam, and hence the distribution over centre frequencies of the excited atoms will be strongly peaked near ω. For the ground state, however, the distribution over centre frequencies will essentially be undistorted from that prior to the arrival of the beam, apart from the region near $\omega_c \approx \omega$, where there will be a small decrease in population, the missing atoms having been excited to the upper level. This process is illustrated schematically in Fig. 3.6.

3.4 The formation of spectral lines: The Voigt profile[†]

Finally, we determine the frequency distribution of a spectral line when both homogeneous and inhomogeneous broadening are present. This process is illustrated schematically in Fig. 3.7.

We consider the emission of radiation from atoms excited to some level 2, and suppose that the distribution over centre frequencies for atoms in level 2 is described by $g_D(\omega_c - \omega_0)$. The number of atoms in level 2 with centre frequencies lying in the range $\omega_c, \omega_c + \delta\omega_c$ is simply,

$$\Delta N_2 = N_2 g_D(\omega_c - \omega_0) \delta\omega_c, \quad (3.41)$$

where N_2 is the total population density in level 2.

This class of atoms will radiate spontaneously over a range of frequencies centred on ω_c with a distribution given by the *homogeneous*

3.4 *The formation of spectral lines: The Voigt profile*† 41

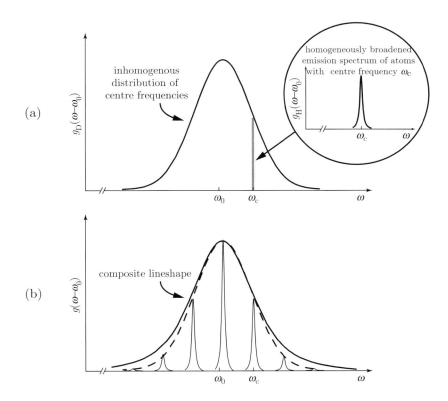

Fig. 3.7: Formation of a spectral line in the presence of both homogeneous and inhomogeneous spectral broadening. (a) The distribution $g_D(\omega - \omega_0)$ of centre frequencies arising from inhomogeneous broadening. Each class of atoms emits radiation that is homogenously broadened about the centre frequency of the class, with a strength proportional to the number of atoms in that class, as indicated by the inset. (b) The composite lineshape formed by the superposition of the spectrum emitted by each class. The thin lines show the contributions of some representative classes, the dashed line is the inhomogeneous lineshape, and the solid line is the resulting composite lineshape.

lineshape $g_H(\omega - \omega_c)$. From our discussion in Section 3.3.1, we know that the rate at which this class of atoms emits photons with frequencies in the range ω to $\omega + \delta\omega$ is,

$$\Delta N_2 A_{21} g_H(\omega - \omega_c)\delta\omega = [N_2 g_D(\omega_c - \omega_0)\delta\omega_c] \times [A_{21} g_H(\omega - \omega_c)\delta\omega]. \tag{3.42}$$

The total rate of emission of photons with frequencies in this range is given by summing over all classes, and hence the lineshape of the fluorescence is given by,

$$g(\omega - \omega_0) = \int_0^\infty g_D(\omega_c - \omega_0) g_H(\omega - \omega_c) d\omega_c. \tag{3.43}$$

By setting $\omega' = \omega_c - \omega_0$, we can recast our result as,

$$g(\omega - \omega_0) = \int_{-\infty}^\infty g_D(\omega') g_H(\omega - \omega_0 - \omega') d\omega', \tag{3.44}$$

where we have set $-\omega_0 \approx -\infty$ in the lower limit of the integral. We can now see that, as might be expected, the composite lineshape $g(\omega - \omega_0)$ is simply a convolution of the homogeneous and inhomogeneous lineshapes.

Note that if the homogeneous linewidth is extremely small compared to the inhomogeneous linewidth, $g_H(\omega - \omega_0 - \omega')$ acts rather like a Dirac δ-function, and the integrand is only significant for $\omega' \approx \omega - \omega_0$. As such, in this limit $g(\omega - \omega_0) \approx g_D(\omega - \omega_0)$ and the lineshape is essentially just the inhomogeneous lineshape. Clearly, in the other limit of the inhomogeneous

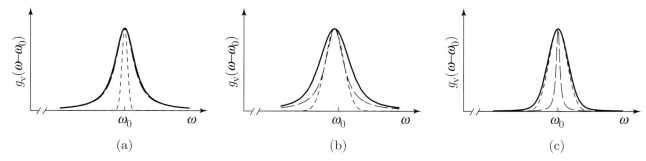

Fig. 3.8: The Voigt profile for various ratios of the homogeneous and inhomogeneous linewidths, together with the underlying Gaussian (long dash) and Lorentzian lineshapes (short dash): (a) $\Delta\omega_\mathrm{I}/\Delta\omega_\mathrm{H} = 0.2$ (b) $\Delta\omega_\mathrm{I}/\Delta\omega_\mathrm{H} = 1.0$; (c) $\Delta\omega_\mathrm{I}/\Delta\omega_\mathrm{H} = 5$. Note how in (b) the wings of the Voigt profile are well approximated by the Lorentzian lineshapes.

linewidth being very small compared to the homogeneous linewidth, the composite lineshape $g(\omega - \omega_0)$ becomes identical with the homogeneous lineshape.

A particularly important case arises when the distribution over centre frequencies for the excited atoms is Gaussian, and the homogeneous lineshape is Lorentzian. This often occurs, for example, in atoms excited in a gas discharge. Then, the composite lineshape is called a Voigt profile, and is given by,[17]

$$g_\mathrm{V}(\omega - \omega_0) = \int_{-\infty}^{\infty} \mathcal{G}(\omega')\mathcal{L}(\omega - \omega_0 - \omega')\mathrm{d}\omega'. \tag{3.45}$$

There is no analytical form for the Voigt profile, although it can be shown that the Voigt profile depends only on the ratio of the homogeneous and inhomogeneous linewidths (see Exercise 3.6). Figure 3.8 shows numerical calculations of the Voigt profile for three ratios of the homogeneous and inhomogeneous linewidths. It can be seen quite clearly that if the homogeneous linewidth is small compared to the inhomogeneous linewidth, the composite lineshape is well approximated by the inhomogeneous lineshape, and vice versa. If the two linewidths are similar in magnitude, the lineshape near the peak is approximately Gaussian, whilst in the wings it is close to Lorentzian. In fact, all Voigt profiles tend to a Lorentzian lineshape at sufficient detuning from the line centre, owing to the fact that the decrease in the lineshape with detuning is much slower for a Lorentzian than for a Gaussian.

3.5 Other broadening effects

3.5.1 Self-absorption

In some circumstances the shape of the measured emission spectrum depends on the *size and shape* of the medium. The reason for this apparently odd behaviour is that the spectrum measured is that of the photons that leave the medium; this is not necessarily the same as the lineshape of the transition since in order to be detected the photons must propagate through the medium without

---

[17] In Section 3.3.2 we warned that for inhomogeneously broadened atoms the distribution over centre frequencies might not be the same in all atomic levels. This is particularly true for levels pumped by narrow-band radiation. However, there are plenty of examples where the distribution over centre frequencies is essentially independent of the radiating atomic level. For example, in a discharge, atoms in the ground state will have a distribution over velocities v_z that is Gaussian. Electron collisional excitation to excited levels is independent of, and does not appreciably change, the velocity of the atom. Hence, in this case, the distribution of centre frequencies for atoms in excited levels will also be Gaussian.

being absorbed. For significant absorption to occur the population density of the lower level must be large, and hence the effect of **self-absorption** is most likely to be observed on transitions to the ground state.

We defer a quantitative discussion of this effect to Section 4.5.2. For the moment we note that self-absorption occurs when photons at the centre frequency of the transition have a significant probability of being absorbed before they leave the medium, and that the effect of self-absorption is a broadening of the spectral line.

Further reading

Further details on broadening mechanisms in gases may be found in the text by Corney (1978).

Exercises

(3.1) **Classical model of refractive index**

Here, we consider the classical model for the refractive index of a material. We suppose that the electron is bound to the atom by a harmonic restoring force $-m_e\omega_0^2 r$ and is damped by a force $-\gamma m_e \dot{r}$.

(a) Show that in the presence of an electric field $E(t) = E_0 \exp(-i\omega t)$ the equation of motion of the electron becomes,

$$m_e \ddot{r} = -e E_0 e^{-i\omega t} - m_e \omega_0^2 r - m_e \gamma \dot{r}.$$

(b) Show that the particular integral—i.e. steady-state solution—of this equation of motion may be written as $r(t) = r_0 \exp(-i\omega t)$ where,

$$r_0 = \frac{-(e/m_e)}{\omega_0^2 - \omega^2 - i\gamma\omega} E_0.$$

(c) Given that the dipole moment of the atom is equal to $p = -er$, show that the polarization of the medium is given by,

$$P(t) = \frac{Ne^2}{m_e} \frac{1}{\omega_0^2 - \omega^2 - i\gamma\omega} E(t),$$

where N is the density of such atoms.

(d) For an isotropic, non-magnetic medium the polarization of the medium is related to the electric field by $P = (\epsilon_r - 1)\epsilon_0 E$, where ϵ_r is the relative permittivity of the medium. Hence, show that the refractive index $n = \sqrt{\epsilon_r}$ is given by,

$$n^2 - 1 = \frac{Ne^2}{m_e \epsilon_0} \frac{1}{\omega_0^2 - \omega^2 - i\gamma\omega}. \quad (3.46)$$

We see that, in general, the refractive index is complex, with real and imaginary parts n_r and n_i, respectively.[18]

(e) Describe how the real and imaginary parts of the refractive index determine the way a plane wave propagates through this medium.

---

[18]It is worth noting here that eqn (3.46) is very close to the result of a quantum-mechanical calculation of the refractive index. In the quantum-mechanical result the parameter ω_0 corresponds to the angular frequency of the transition of the atom; if the atom has more than one transition the right-hand side of eqn (3.46) is replaced by an analogous expression comprising a weighted sum over terms in each transition frequency, where the weighting factor is the oscillator strength introduced on page 30.

(f) If the density N of the atoms of interest is very low, we may approximate eqn (3.46) as:

$$n \approx 1 + \frac{Ne^2}{2m_e\epsilon_0} \frac{1}{\omega_0^2 - \omega^2 - i\gamma\omega}.$$

We will usually be interested in the propagation of waves at frequencies ω near the resonant frequency ω_0, for which we may write $\omega_0^2 - \omega^2 = (\omega_0 + \omega)(\omega_0 - \omega) \approx 2\omega_0(\omega_0 - \omega)$. Show that within this approximation we may write

$$n_r \approx 1 + \frac{Ne^2}{4m_e\epsilon_0\omega_0} \frac{\omega_0 - \omega}{(\omega_0 - \omega)^2 + \left(\frac{\gamma}{2}\right)^2} \quad (3.47)$$

$$n_i \approx \frac{Ne^2}{4m_e\epsilon_0\omega_0} \frac{\left(\frac{\gamma}{2}\right)}{(\omega_0 - \omega)^2 + \left(\frac{\gamma}{2}\right)^2}. \quad (3.48)$$

(g) Show that the full width at half-maximum of the imaginary part of the refractive index is given by $\Delta\omega = \gamma$.

(h) Sketch the real and imaginary parts of the refractive index in the neighbourhood of ω_0, and explain this behaviour in terms of the processes that occur in a macroscopic sample of atoms.

(3.2) Classical theory of natural broadening

(a) Show that the Fourier transform of eqn (3.4) is equal to,

$$\mathcal{E}(\omega) = -\sqrt{\frac{1}{2\pi}} \frac{E_0}{2} \left[\frac{1}{(\omega - \omega_0) + i\gamma/2} - \frac{1}{(\omega + \omega_0) + i\gamma/2} \right]. \quad (3.49)$$

(b) For which frequencies will $\mathcal{E}(\omega)$ be large? Hence, by making a suitable approximation, show that the normalized lineshape is given by eqn (3.8).

(c) Derive eqn (3.12) by relating eqn (3.11) to the rate of work done against the damping force $-m_e\gamma\dot{r}$.

(3.3) Oscillator strengths

The $3p\,^2P_{3/2}$ and $3p\,^2P_{1/2}$ levels of the Na atom decay to the ground level $3s\,^2S_{1/2}$ on the so-called 'sodium-D lines' with vacuum wavelengths of 589.158 nm and 589.756 nm, respectively.

(a) Calculate the classical radiative lifetimes for these transitions, using the fact that the upper levels decay only on these transitions.

(b) The radiative lifetime of both upper levels is measured to be 16.4 ns. Deduce values for the emission and absorption oscillator strengths for the $3p\,^2P_{3/2} \to 3s\,^2S_{1/2}$ and $3p\,^2P_{1/2} \to 3s\,^2S_{1/2}$ transitions. Comment on these results.

(3.4) More oscillator strengths

In the Hg atom, the $6s6p\,^1P_1$ and $6s6p\,^3P_1$ levels decay to the $6s^2\,^1S_0$ ground level only, with vacuum wavelengths of 184.950 nm and 253.728 nm, respectively.

(a) Calculate the classical radiative lifetimes for these transitions.

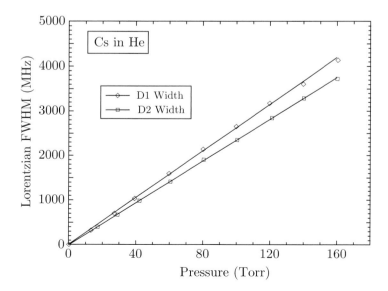

Fig. 3.9: Measured full width at half maximum of the homogeneously broadened component of the D_1 and D_2 lines of Cs as a function of the pressure of helium. Reprinted figure with permission from A. Andalkar & R.B. Warrington, *Physical Review A* **65** 032708 (2002). ©(2002) by the American Physical Society. (http://link.aps.org/abstract/PRA/v65/032708).

(b) Compare your results with the experimental values of 1.34 ns and 118.9 ns, for the 6s6p ^{1}P$_1$ and 6s6p ^{3}P$_1$ levels, respectively. Hence, deduce the emission and absorption oscillator strengths for these transitions.

(c) Comment on your results to part (b).

(3.5) Figure 3.9 shows data from measurements of the homogeneous linewidth of two transitions to the 6s ^{2}S$_{1/2}$ ground state of Cs: the D_1 (6p ^{2}P$_{1/2}$ → 6s ^{2}S$_{1/2}$) and D_2 (6p ^{2}P$_{3/2}$ → 6s ^{2}S$_{1/2}$) transitions with wavelengths of 894 and 852 nm, respectively.

(a) Calculate the Doppler width in MHz of these transitions assuming that the temperature of the Cs vapour is 21 °C, and comment on the relative magnitudes of the inhomogeneous and homogeneous linewidths.

(b) The radiative lifetimes of the 6p ^{2}P$_{1/2}$ and 6p ^{2}P$_{3/2}$ levels are 34.75 and 30.41 ns, respectively. What is the natural linewidth of the D_1 and D_2 transitions? Is your calculated value consistent with Fig. 3.9?

(c) Use the data presented to deduce the rate of increase in the homogeneous linewidth in units of MHz Torr^{-1} for each of the two transitions. Explain briefly the cause of this increase in homogenous linewidth with pressure.

(d) Use your results to calculate the mean collision time for He–Cs collisions at a He pressure of 100 Torr? [The molar mass of Cs is 132.9 g.]

(3.6) The Voigt profile

(a) By defining the normalized frequency y,

$$y = \frac{\omega}{\Delta\omega_L/2}, \qquad (3.50)$$

where $\Delta\omega_L$ is the full width at half-maximum of the Lorentzian lineshape, show that the Voigt profile may be written as,

$$g_V(y - y_0)dy = \frac{\epsilon dy}{\pi\sqrt{\pi}} \int_{-\infty}^{\infty} \frac{\exp(-[\epsilon y']^2)}{(y - y_0 - y')^2 + 1} dy', \qquad (3.51)$$

where y_0 is the value of y at $\omega = \omega_0$, and ϵ is defined by,

$$\epsilon = \frac{\Delta\omega_L}{\Delta\omega_G}\sqrt{\ln 2}, \qquad (3.52)$$

where $\Delta\omega_G$ is the full width at half-maximum of the Gaussian lineshape. We can now see that in terms of the normalized frequency y, the shape of the Voigt profile is a function of the ratio of the homogeneous and inhomogeneous widths only.

4 Light amplification by the stimulated emission of radiation

4.1 The optical gain cross-section   46
4.2 Narrowband radiation   50
4.3 Gain cross-section for inhomogeneous broadening[†]   52
4.4 Orders of magnitude   53
4.5 Absorption   54
Further reading   56
Exercises   57

In the previous chapters we considered the interaction of radiation and matter, and in particular the process of stimulated emission. Here, we develop those ideas and derive a function, the optical gain cross-section, which characterizes the strength of that interaction.

4.1 The optical gain cross-section

We consider the amplification of a beam of radiation propagating along the z-axis through a medium with population densities N_2 and N_1 in the upper and lower laser levels, respectively. In general, the radiation will have a finite spectral width, and is described by a spectral intensity $\mathcal{I}(\omega, z)$ and spectral energy density $\varrho(\omega, z)$.

We shall take the laser transition to be homogeneously broadened so that each frequency ω interacts with all the atoms equally—the strength of that interaction depending on the frequency detuning from the central angular frequency ω_0 of the transition.

We now consider the amplification of the beam as it passes through the small region lying between the planes $z = z$ and $z = z + \delta z$, as illustrated schematically in Fig. 4.1. As the beam passes through the medium it will lose energy due to absorption by atoms in the lower laser level, but gain energy due to stimulated emission from atoms in the upper laser level. From Section 3.3.1 the net rate at which atoms are transferred from the upper to the lower laser level by the stimulated emission or absorption of photons with angular frequencies lying between ω and $\omega + \delta\omega$ is

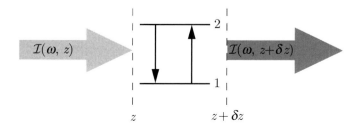

Fig. 4.1: Amplification by stimulated emission on the transition $2 \to 1$ of a beam of radiation of spectral intensity $\mathcal{I}(\omega, z)$ in passing between the planes $z = z$ and $z = z + \delta z$.

$$\left[N_2 B_{21} g_H(\omega - \omega_0)\varrho(\omega,z)\delta\omega - N_1 B_{12} g_H(\omega - \omega_0)\varrho(\omega,z)\delta\omega\right] A\delta z, \quad (4.1)$$

where A is the area of the beam. Each such transfer releases an energy of $\hbar\omega$ to the beam, and hence, within this frequency range, the power gained by the beam is:

$$\left[N_2 B_{21} g_H(\omega - \omega_0)\varrho(\omega,z)\delta\omega - N_1 B_{12} g_H(\omega - \omega_0)\varrho(\omega,z)\delta\omega\right] A\delta z \times \hbar\omega. \quad (4.2)$$

Remembering that $\mathcal{I}(\omega,z)A$ is the power per unit frequency interval carried by the beam across the plane $z = z$, this power gain may be written as

$$[\mathcal{I}(z+\delta z, \omega) - \mathcal{I}(z,\omega)] A\delta\omega. \quad (4.3)$$

Equating eqns (4.2) and (4.3) we find,

$$\left[\mathcal{I}(z,\omega) + \frac{\partial \mathcal{I}(z,\omega)}{\partial z}\delta z + \ldots - \mathcal{I}(z,\omega)\right] A\delta\omega$$
$$= [N_2 B_{21} - N_1 B_{12}] g_H(\omega - \omega_0)\varrho(\omega,z)\hbar\omega\delta\omega A\delta z, \quad (4.4)$$

and hence in the limit $\delta z \to 0$,

$$\frac{\partial \mathcal{I}(z,\omega)}{\partial z} = [N_2 B_{21} - N_1 B_{12}] g_H(\omega - \omega_0)\frac{\hbar\omega}{c}\mathcal{I}(\omega,z), \quad (4.5)$$

where we have used the fact that for a beam of radiation $\mathcal{I}(\omega,z) = \varrho(\omega,z)c$.

We can tidy this result and separate the right-hand side of eqn (4.5) into terms relating to the population densities of the upper and lower levels, the atomic physics of the laser transition, and the intensity of the radiation beam. To do this we use the relation between the Einstein B coefficients, and find[1]

$$\frac{\partial \mathcal{I}(z,\omega)}{\partial z} = N^* \sigma_{21}(\omega - \omega_0)\mathcal{I}(\omega,z), \quad (4.6)$$

$$N^* = N_2 - \frac{g_2}{g_1}N_1, \quad (4.7)$$

$$\sigma_{21}(\omega - \omega_0) = \frac{\hbar\omega}{c}B_{21}g_H(\omega - \omega_0), \quad (4.8)$$

where N^* is known as the **population inversion density** and $\sigma_{21}(\omega - \omega_0)$ is the **optical gain cross-section** of the homogeneously broadened laser transition. It will sometimes be useful to re-express eqn (4.8) in terms of the Einstein A coefficient:

$$\sigma_{21}(\omega - \omega_0) = \frac{\pi^2 c^2}{\omega_0^2} A_{21} g_H(\omega - \omega_0). \quad (4.9)$$

The optical gain cross-section $\sigma_{21}(\omega - \omega_0)$—often just called the 'gain cross-section' or 'cross-section'—parameterizes the strength with which radiation of frequency ω interacts with atoms via the transition (which has a central frequency of ω_0): a large value of $\sigma_{21}(\omega - \omega_0)$ corresponds to a strong interaction, and hence, from eqn (4.6), a rapid growth in its spectral intensity. The frequency dependence of the gain cross-section reflects the frequency

[1] We note that other definitions of the cross-section describing the interaction of atoms with radiation are possible. The definition used in eqn (4.8) is convenient for describing laser operation since it is used together with the population inversion density N^*, defined by eqn (4.7), which emphasizes the importance of the population in the upper laser level. In other contexts different definitions of the cross-section are more appropriate. For example, in Section 4.5.1 we define a different cross-section to describe the absorption of radiation; this is used with a population density that emphasizes the importance of the lower level, from which absorption occurs.

dependence of the strength of the interaction: radiation with a frequency close to the centre frequency of the transition will interact more strongly with the atoms than radiation with frequencies in the wings of the transition.

4.1.1 Condition for optical gain

From eqn (4.6) we see immediately that for the spectral intensity of the beam to grow we require the right-hand side to be positive. Since the gain cross-section is inherently positive, for optical gain we must have

$$N^* > 0 \tag{4.10}$$

$$\Rightarrow N_2 > \frac{g_2}{g_1} N_1. \quad \text{\textbf{Condition for optical gain}} \tag{4.11}$$

This is exactly the same condition we deduced in Section 2.2.

4.1.2 Frequency dependence of the gain cross-section

The frequency dependence of the optical gain cross-section is dominated by that of the lineshape function $g_H(\omega - \omega_0)$ describing the homogeneous broadening, since in almost all circumstances the variation of the term in ω in eqn (4.8), and in ω^2 in eqn (4.9), is slow on the scale of the homogeneous linewidth of the transition.

As seen in Chapter 3, two lineshapes frequently arise: a Lorentzian lineshape $\mathcal{L}(\omega - \omega_0)$, and a Gaussian lineshape $\mathcal{G}(\omega - \omega_0)$. For these cases we have:

Lorentzian lineshape

$$\sigma_{21}(\omega - \omega_0) = \frac{\hbar \omega_0}{c} B_{21} \frac{1}{\pi} \frac{\Delta\omega/2}{(\omega - \omega_0)^2 + (\Delta\omega/2)^2} \tag{4.12}$$

$$= \frac{\pi^2 c^2}{\omega_0^2} A_{21} \frac{1}{\pi} \frac{\Delta\omega/2}{(\omega - \omega_0)^2 + (\Delta\omega/2)^2}, \tag{4.13}$$

$$\sigma_{21}(0) = \frac{\hbar \omega_0}{c} \frac{2}{\pi \Delta\omega} B_{21} = 2\pi c^2 \frac{A_{21}}{\omega_0^2 \Delta\omega}. \tag{4.14}$$

Gaussian lineshape

$$\sigma_{21}(\omega - \omega_0) = \frac{\hbar \omega_0}{c} B_{21} \frac{2}{\Delta\omega} \sqrt{\frac{\ln 2}{\pi}} \exp\left(-\left[\frac{\omega - \omega_0}{\Delta\omega/2}\right]^2 \ln 2\right) \tag{4.15}$$

$$= \frac{\pi^2 c^2}{\omega_0^2} A_{21} \frac{2}{\Delta\omega} \sqrt{\frac{\ln 2}{\pi}} \exp\left(-\left[\frac{\omega - \omega_0}{\Delta\omega/2}\right]^2 \ln 2\right) \tag{4.16}$$

$$\sigma_{21}(0) = \frac{\hbar \omega_0}{c} \frac{2}{\Delta\omega} \sqrt{\frac{\ln 2}{\pi}} B_{21} = 2\sqrt{\pi^3 \ln 2} \, c^2 \frac{A_{21}}{\omega_0^2 \Delta\omega} \tag{4.17}$$

where $\Delta\omega$ is the linewidth[2] of the transition and $\sigma_{21}(0)$ is the peak cross-section, which occurs when $\omega = \omega_0$. An important point to note is that the peak gain cross-sections have the same functional dependence on: ω_0, the Einstein coefficients, and the linewidth for the two lineshapes. In fact, we will find that for *any* lineshape the peak optical gain cross-section will be of the form $\sigma_{21}(0) \propto A_{21}/\omega_0^2 \Delta\omega$.

4.1.3 The gain coefficient

Equation (4.6) describing the growth of the spectral intensity of a beam can be further simplified to

$$\frac{\partial \mathcal{I}(z,\omega)}{\partial z} = \alpha(\omega - \omega_0)\mathcal{I}(\omega, z), \quad (4.18)$$

where

$$\alpha(\omega - \omega_0) = N^* \sigma_{21}(\omega - \omega_0) \quad \textbf{gain coefficient} \quad (4.19)$$

is known as the **gain coefficient**.

We will see in Chapter 5 that it is very important to recognize that the gain coefficient is in general a *function of the beam intensity*. The reason for this is that at high intensities the increased rate of stimulated emission will reduce the population inversion, a process known as gain saturation. For the moment, however, we will neglect that complication and assume that the population inversion is positive and independent of intensity or position. Then, integrating eqn. (4.18) we find,[3]

$$\mathcal{I}(\omega, z) = \mathcal{I}(\omega, 0) \exp[\alpha_0(\omega - \omega_0)z]. \quad (4.20)$$

We see that the spectral intensity grows exponentially with propagation distance. The reason for the energy gain of the beam is straightforward: the rate of stimulated emission from the upper level is greater than the rate of absorption from the lower level.

The gain coefficient is important in determining the performance of a laser system. However, since it is proportional to the population inversion density, it is strongly dependent upon the operating conditions of the laser and in particular on the rate of pumping of the upper laser level. The optical gain cross-section is therefore a more fundamental parameter of the laser transition since it depends only on the physics of the laser transition and therefore is largely independent of the pumping conditions.[4]

4.1.4 Gain narrowing

Since the optical gain cross-section, and hence the gain coefficient, is strongly peaked about the centre frequency of the transition, it is clear from eqn (4.20) that the spectral intensity of frequencies close to ω_0 will grow much faster than that of frequencies in the wings of the lineshape. As a consequence, the spectral width of the amplified beam will decrease as the beam propagates

---

[2] For the case of homogeneous broadening considered so far $\Delta\omega$ is just the homogeneous linewidth $\Delta\omega_\mathrm{H}$. As shown in Section 4.3, in some circumstances it is also possible to define an inhomogeneous cross-section, in which case these equations also describe the frequency dependence of the cross-section of inhomogeneously broadened transitions with Lorentzian or Gaussian lineshapes of inhomogeneous width $\Delta\omega$.

[3] We have added a subscript '0' to the gain coefficient α to indicate that we have assumed that the population inversion is independent of the intensity of the radiation. In other words, we have assumed that the radiation does not saturate the population inversion. See Section 5.1.

[4] Although the conditions within the laser may affect the linewidth of the transition, and hence the gain cross-section.

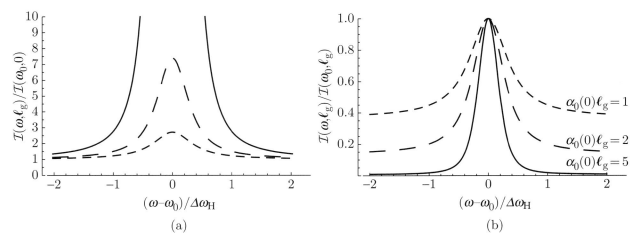

Fig. 4.2: Gain narrowing of a beam of radiation after a single pass through an amplifier of length ℓ_g and small-signal gain coefficient $\alpha_0(\omega - \omega_0)$ plotted for three different values of $\alpha_0(0)\ell_g$. (a) The spectral intensity of the beam leaving the amplifier normalized to the peak spectral intensity of the *input* beam: $\mathcal{I}(\omega, \ell_g)/\mathcal{I}(\omega_0, 0)$; this plot illustrates the enormous increase in intensity resulting from the amplification. (b) The spectral intensity of the beam leaving the amplifier, normalized to the peak spectral intensity of the *amplified* beam: $\mathcal{I}(\omega, \ell_g)/\mathcal{I}(\omega_0, \ell_g)$; this plot illustrates clearly the effect of gain narrowing. For all plots the laser amplifier is assumed to be homogeneously broadened with a linewidth of $\Delta\omega_H$; the input input beam is assumed to have a flat spectrum, i.e. $\mathcal{I}(\omega, 0)$ is independent of ω; and saturation of the gain medium (see Chapter 5) has been neglected.

through the inverted medium, a process known as **gain narrowing**. This process is illustrated in Fig. 4.2.[5]

[5] We emphasize that our treatment here is highly simplified since we have not taken saturation effects into account. For an example of gain narrowing when saturation *is* taken into account, see Section 18.3.3.

4.2 Narrowband radiation

We will often want to describe the interaction of very narrowband radiation with a population inversion. For example, the spectral width of a single mode oscillating within a laser cavity will almost always be much narrower than the linewidth of the laser transition. Here, we consider how to treat the amplification of such beams and how they interact with the atomic populations. The results will be used extensively in the remainder of this book.

4.2.1 Amplification of narrowband radiation

The *total* intensity of any beam $I_T(z)$ is related to the spectral intensity $\mathcal{I}(\omega, z)$ by,[6]

[6] Note that to the *total* intensity has units of W cm^{-2}; whereas the *spectral* intensity has units of W cm^{-2} (rad s^{-1})$^{-1}$.

$$I_T(z) = \int_0^\infty \mathcal{I}(\omega, z) d\omega. \quad (4.21)$$

Equation (4.6) gives the rate of growth of the spectral intensity with propagation distance z. Integrating both sides of this equation over frequency gives,

$$\int_0^\infty \left[\frac{\partial \mathcal{I}(z, \omega)}{\partial z}\right] d\omega = \int_0^\infty \left[N^* \sigma_{21}(\omega - \omega_0) \mathcal{I}(\omega, z)\right] d\omega$$

$$\Rightarrow \frac{d}{dz}\left[\int_0^\infty \mathcal{I}(z, \omega) d\omega\right] = N^* \sigma_{21}(\omega_L - \omega_0) \int_0^\infty \mathcal{I}(\omega, z) d\omega,$$

where, in evaluating the right-hand side we have assumed that the radiation lies in a very narrow band of frequencies about ω_L, so that $\sigma_{21}(\omega_L - \omega_0)$ can be taken outside the integral. Thus, the equation describing the growth of the total intensity is

$$\frac{dI_T}{dz} = N^* \sigma_{21}(\omega_L - \omega_0) I_T(z). \qquad \textbf{Narrowband radiation} \qquad (4.22)$$

4.2.2 Form of rate equations

Equation (3.39) gives the form of the rate equation describing the interaction of atoms in the upper level interacting with radiation with frequencies lying between ω and $\omega + \delta\omega$:

$$\frac{dN_2}{dt} = N_1 B_{12} g_H(\omega - \omega_0) \varrho(\omega) \delta\omega - N_2 B_{21} g_H(\omega - \omega_0) \varrho(\omega) \delta\omega - \frac{N_2}{\tau_2}.$$

Using the relation $g_1 B_{12} = g_2 B_{21}$ we may rewrite this in terms of the gain cross-section:

$$\frac{dN_2}{dt} = -\left(N_2 - \frac{g_2}{g_1} N_1\right) \frac{\hbar\omega}{c} B_{21} g_H(\omega - \omega_0) \frac{\varrho(\omega) c \delta\omega}{\hbar\omega} - \frac{N_2}{\tau_2}$$

$$\Rightarrow \frac{dN_2}{dt} = -N^* \sigma_{21}(\omega - \omega_0) \frac{\mathcal{I}(\omega) \delta\omega}{\hbar\omega} - \frac{N_2}{\tau_2}. \qquad (4.23)$$

To describe the interaction with all the frequencies in the beam we must integrate over frequency:

$$\frac{dN_2}{dt} = -N^* \int_0^\infty \sigma_{21}(\omega - \omega_0) \frac{\mathcal{I}(\omega)}{\hbar\omega} d\omega - \frac{N_2}{\tau_2}. \qquad (4.24)$$

Hence, for very narrowband radiation centred at ω_L we find,

$$\frac{dN_2}{dt} = -N^* \sigma_{21}(\omega_L - \omega_0) \left[\frac{I_T}{\hbar\omega_L}\right] - \frac{N_2}{\tau_2}. \qquad \textbf{Narrowband radiation}$$

$$(4.25)$$

The term on the right-hand side in square brackets is the photon flux, i.e. the number of photons crossing unit area of the beam per unit time. Hence, the net rate of transfer of atoms from the upper to the lower level by stimulated emission and absorption is equal to the product of the population inversion density, the cross-section, and the incident flux of photons. This is exactly the form we would expect for the rate of any process described in terms of a cross-section; we may regard N^* as the effective density of inverted atoms, and $\sigma_{21}(\omega_L - \omega_0)$ as their effective cross-sectional area, so that any incident photon passing within this area will transfer one atom from the upper to the lower level.

4.3 Gain cross-section for inhomogeneous broadening†

For completeness, we outline how the derivation of the optical cross-section would differ for the case of an inhomogeneously broadened laser transition. In this case we must consider the interaction of *each class* of atoms with the beam of radiation. Consider first the interaction of the class of atoms with centre frequencies lying in the range ω_c, $\omega_c + \delta\omega_c$ with radiation in the frequency range ω, $\omega + \delta\omega$. Following the approach of Section 4.1, the net rate at which atoms in this class are transferred from the upper to the lower level is:

$$\left[\Delta N_2(\omega_c)\delta\omega_c B_{21} g_H(\omega - \omega_c)\varrho(\omega, z)\delta\omega \right.$$
$$\left. - \Delta N_1(\omega_c)\delta\omega_c B_{12} g_H(\omega - \omega_c)\varrho(\omega, z)\delta\omega \right] A\delta z,$$

where $\Delta N_i(\omega_c)\delta\omega_c$ is the number density of atoms in level i within the class under consideration. The total net rate at which all atoms are transferred from the upper to the lower level is given by integrating over the frequency classes:

$$\left[\int_0^\infty B_{21} \Delta N_2(\omega_c) g_H(\omega - \omega_c) d\omega_c \right.$$
$$\left. - \int_0^\infty B_{12} \Delta N_1(\omega_c) g_H(\omega - \omega_c) d\omega_c \right] \varrho(\omega, z)\delta\omega A\delta z.$$

Hence, following the argument of Section 4.1, the rate of growth of the spectral intensity of the beam with distance is given by,

$$\frac{\partial \mathcal{I}(z, \omega)}{\partial z} = \left\{ \int_0^\infty \left[\Delta N_2(\omega_c) - \frac{g_2}{g_1} \Delta N_1(\omega_c) \right] g_H(\omega - \omega_c) d\omega_c \right\}$$
$$\times \frac{\hbar\omega}{c} B_{21} \mathcal{I}(\omega, z), \tag{4.26}$$

where we have used the relation between the Einstein B coefficients.

We may define the inhomogeneous lineshape for the population inversion by

$$\Delta N_2(\omega_c) - \frac{g_2}{g_1} \Delta N_1(\omega_c) = N^* g_D(\omega_c - \omega_0). \tag{4.27}$$

Our result then becomes,

$$\frac{\partial \mathcal{I}(z, \omega)}{\partial z} = N^* \sigma_{21}^D(\omega - \omega_0) \mathcal{I}(\omega, z), \tag{4.28}$$

where the **inhomogeneous gain cross-section** $\sigma_{21}^D(\omega - \omega_0)$ is given by,[7]

$$\sigma_{21}^D(\omega - \omega_0) = \frac{\hbar\omega}{c} B_{21} \int_0^\infty g_D(\omega_c - \omega_0) g_H(\omega - \omega_c) d\omega_c. \tag{4.29}$$

[7] We use the superscript 'D' to distinguish the inhomogeneous cross-section from the homogeneous cross-section for the reasons explained in Note 11 on page 27.

The inhomogeneous gain cross-section has the same form as the homogeneous gain cross-section, although now the frequency dependence is given by a convolution of the homogeneous and inhomogeneous lineshapes. Very often, the homogeneous linewidth is very much narrower than the inhomogeneous width, in which case eqn (4.29) reduces to,

$$\sigma_{21}^{D}(\omega - \omega_0) = \frac{\hbar \omega}{c} B_{21} g_D(\omega - \omega_0), \quad (4.30)$$

which is exactly analogous to the result we obtained for homogeneously broadened atoms.

Great care should be taken when using the inhomogeneous gain cross-section to describe the interaction of radiation with inhomogeneously broadened atoms. As we shall see in later chapters, and as is hinted in Fig. 3.6, in an operating laser system the laser radiation is always sufficiently intense to significantly distort the level populations produced by the laser-pumping mechanisms, so that the distribution over frequency classes for the upper and lower levels will generally: (i) be different; (ii) depend on the intensity of the radiation. In such cases the inhomogeneous lineshape defined in eqn (4.27)—and hence the inhomogeneous cross-section itself—will depend on the intensity of the radiation, making the inhomogeneous cross-section of little use. Instead, one should solve simultaneously eqn (4.26) and the rate equations for the populations $\Delta N_1(\omega_c)$ and $\Delta N_2(\omega_c)$ of each frequency class.

4.4 Orders of magnitude

Table 4.1 gives approximate values for the linewidth and gain cross-section for a range of laser systems. Note that we have expressed the cross-section in square centimetres, as is the convention in laser physics. We see that the gain cross-section varies over a wide range, between approximately 10^{-20} cm^2 and 10^{-12} cm^2.

Table 4.1 The primary laser wavelength λ_0, homogeneous linewidth $\Delta \omega_H$, inhomogeneous linewidth $\Delta \omega_D$, Einstein A coefficient A_{21}, and the peak gain cross-section $\sigma_{21}(0)$ for some important laser transitions.

Laser	λ_0 (nm)	$\Delta \omega_H$ (rad s^{-1})	$\Delta \omega_D$ (rad s^{-1})	A_{21} (s^{-1})	$\sigma_{21}(0)$ (cm^2)
He–Ne	633	1.9×10^8	8.0×10^9	1.0×10^7	1.2×10^{-12}
Argon-ion	488	4.0×10^9	3.1×10^{10}	1.0×10^8	1.8×10^{-12}
KrF	249	8.8×10^{13}	1.5×10^9	1.0×10^8	2.2×10^{-16}
Ruby	694	3.3×10^{11}	-	3.3×10^2	2.5×10^{-20}
Nd:YAG	1064	1.6×10^{11}	-	8.0×10^2	3.7×10^{-19}
Nd:glass	1060	1.6×10^{11}	5×10^{12}	8.0×10^2	4.1×10^{-20}
Ti:sapphire	800	1×10^{14}	-	3×10^5	2.8×10^{-19}
CO$_2$	10 600	1.5×10^8	5.5×10^7	0.2	3.9×10^{-17}

4.5 Absorption

We end this chapter with a brief discussion of absorption, how it may be described in terms of an absorption cross-section, and some important effects arising from it.

4.5.1 The absorption cross-section

Equation (4.6) describes the growth of the spectral intensity of a beam owing to stimulated emission and absorption on the transition $2 \to 1$. Of course, the same equation will also describe the attenuation of the beam in the case when $N^* < 0$, but in such cases it is useful to rewrite it as follows:

$$\frac{\partial \mathcal{I}(z, \omega)}{\partial z} = N^* \sigma_{21}(\omega - \omega_0) \mathcal{I}(\omega, z)$$

$$= -\left(N_1 - \frac{g_1}{g_2} N_2\right) \frac{g_2}{g_1} \sigma_{21}(\omega - \omega_0) \mathcal{I}(\omega, z)$$

$$= -N^{**} \sigma_{12}(\omega - \omega_0) \mathcal{I}(\omega, z), \tag{4.31}$$

where

$$N^{**} = N_1 - \frac{g_1}{g_2} N_2, \tag{4.32}$$

and the **optical absorption cross-section** $\sigma_{12}(\omega - \omega_0)$ is related to the optical gain cross-section by

$$g_1 \sigma_{12}(\omega - \omega_0) = g_2 \sigma_{21}(\omega - \omega_0). \tag{4.33}$$

The reason for rewriting the equations in this form is simply that when absorption dominates $N_1 > (g_1/g_2)N_2$ and hence N^{**} is positive.

By analogy with Section 4.1.3 we may define the **absorption coefficient** by

$$\boxed{\kappa(\omega - \omega_0) = N^{**} \sigma_{12}(\omega - \omega_0).} \quad \textbf{Absorption coefficient} \tag{4.34}$$

Very frequently, such as when the absorption occurs from the ground state of the atom, the population density of the upper level is much less than that of the lower level.[8] Then $N^{**} \approx N_1$, in which case we have simply

$$\kappa(\omega - \omega_0) \approx N_1 \sigma_{12}(\omega - \omega_0). \tag{4.35}$$

Following the approach of Sections 4.1.3 and 4.2, for the simple case when the population densities in the two levels are everywhere the same the spectral and total intensities of a beam as it propagates through the medium are given by:

$$\mathcal{I}(\omega, z) = \mathcal{I}(\omega, 0) \exp\left[-\kappa(\omega - \omega_0) z\right] \tag{4.36}$$

$$I(z) = I(0) \exp\left[-\kappa(\omega_L - \omega_0) z\right]. \tag{4.37}$$

[8] However, we should be aware that for intense beams the strong absorption will transfer a significant proportion of atoms to the upper level, which will have the effect of decreasing N^{**}. This effect is known as saturation of the absorption, and is analogous to the gain saturation discussed in Chapter 5. For further details see Exercise 5.3 and Section 8.3.4.

This exponential attenuation of the beam as it propagates through the medium is known as **Beer's law**.

4.5.2 Self-absorption

We are now in a position to discuss more quantitatively the process of self-absorption outlined in Section 3.5.1. As stated there, self-absorption will be important when the probability of photons being absorbed on their route out of the sample is high. Since the attenuation of a beam as it propagates through an absorbing medium is described by eqn (4.36), we can expect self-absorption to become important when $\tau(\omega - \omega_0) \gtrsim 1$, where the **optical thickness** is given by

$$\tau(\omega - \omega_0) = \kappa(\omega - \omega_0)R, \quad \textbf{Optical thickness} \quad (4.38)$$

and R is a characteristic dimension of the sample.[9,10] When $\tau(\omega - \omega_0) \ll 1$ the source is said to be 'optically thin'; when $\tau(\omega - \omega_0) \gg 1$ it is 'optically thick'.

The optical thickness—and hence the probability of a photon being absorbed—depends strongly on its frequency: photons with frequencies near the centre frequency ω_0 have a much larger optical thickness, and hence a higher probability of being absorbed, than those with frequencies that are significantly detuned from ω_0. As a result, for an optically thick source the measured spectral intensity at frequencies close to the line centre is suppressed by self-absorption, whereas there is no such suppression in the wings of the emission profile. Thus, self-absorption leads to a broadening of the spectral line, as shown in Fig. 4.3.

The self-absorption in a uniform slab of material is analysed quantitatively in Exercises 4.6 and 4.7. There, it is shown that the emission spectrum of an optically thin source is proportional to the lineshape of the transition, as expected. However, for an optically thick source it is found that the spectrum of radiation leaving the source is broadened significantly. In particular, it is

[9] For a cylindrical medium, for example, R would be the radius of the cylinder.

[10] For a non-uniform sample, eqn (4.38) may be generalized to

$$\tau(\omega - \omega_0) = \int \kappa(\omega - \omega_0)\mathrm{d}z,$$

where the integral is along a ray trajectory.

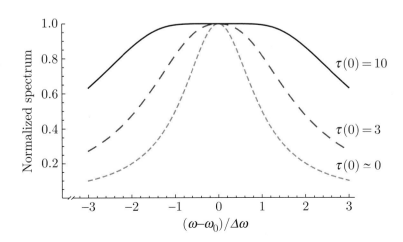

Fig. 4.3: Emission spectra of a source comprising a uniform slab of thickness ℓ for various values of the line-centre optical thickness $\tau(0) = \kappa(0)\ell$. The broadening arising from self-absorption is clear for large values of $\tau(0)$. In this plot the spectra have been normalized to unity at the centre frequency ω_0.

found that when $\tau(0) \gg 1$, and the populations of the upper and lower levels of the transition are in thermal equilibrium with a temperature T, the spectral intensity per unit solid angle emitted by the source is equal to that of a blackbody of temperature T.

4.5.3 Radiation trapping

In an optically thick sample, a photon emitted by one atom has a high probability of being absorbed by another atom before it leaves the medium, in which case there is no net change in the number of atoms in the upper level. Self-absorption therefore reduces the net rate at which atoms in the upper level decay radiatively, leading to an apparent increase in the radiative lifetime. This phenomenon is known as **radiation trapping**. It is important in the analysis of laser systems since there one is usually interested in the density of atoms in the laser levels, and it is unimportant whether the excitation has been transferred between two atoms.

From our discussion above, we would expect radiation trapping to become important when the sample is optically thick, i.e. when $\tau(0) = \kappa(0)R > 1$, where R is a length characteristic of the size of the source. Thus, as for self-absorption, radiation trapping occurs primarily on strong transitions to the ground state, for which the density of atoms in the lower level and the absorption cross-section are high.

Holstein (1951) has analysed the radiation trapping for a variety of geometries and transition lineshapes. He finds that the trapped radiative lifetime, τ_2^{trap}, is given by[11]

$$\frac{1}{\tau_2^{\text{trap}}} = \frac{1}{\tau_2^{\text{rad}}} + (g_{\text{trap}} - 1)A_{21}, \qquad \textbf{Trapped lifetime} \qquad (4.39)$$

where τ_2^{rad} is the radiative lifetime in the absence of trapping. The value of the trapping factor g_{trap} depends on the geometry of the medium and the lineshape and optical thickness of the transition. To give an example, for an infinite cylinder of radius a, and for a Doppler-broadened transition, Holstein finds in the limit $\kappa(0)a \gg 1$,

$$g_{\text{trap}} \approx \frac{1.60}{\kappa(0)a\sqrt{\pi \ln[\kappa(0)a]}}. \qquad (4.40)$$

---

[11] In the absence of collisions the fluorescence lifetime τ_2 is equal to the trapped radiative lifetime τ_2^{trap}. However, as discussed in Note 8 on page 31, collisions can increase the rate at which atoms are lost from the level, in which case the fluorescence lifetime will be smaller than the trapped lifetime given by eqn (4.39).

Further reading

A more detailed discussion of self-absorption and examples of radiation trapping may be found in Corney (1978).

Exercises

(4.1) **Maximum possible gain cross-section**
An upper limit for the gain cross-section arises when the upper laser level decays only on the laser transition itself, and the transition is purely lifetime broadened.

(a) Show that in this case the transition has a Lorentzian lineshape with a peak gain cross-section given by,

$$\sigma_{21}^{L}(0) = \frac{\lambda_0^2}{2\pi}, \qquad (4.41)$$

where λ_0 is the central vacuum wavelength of the transition.

(b) Evaluate the purely lifetime-broadened peak cross-section for a transition in the visible region of the spectrum (take $\lambda_0 = 500$ nm), and compare your answer with the cross-sections presented in Table 4.1.

(c) How does the size of this maximum possible cross-section compare with the typical dimensions of an atom?

(4.2) **Transitions in gases**
Here, we compare the inhomogeneous cross-section typical of a transition in an atom in the gaseous state with the maximum possible cross-section calculated in Exercise 4.1.

(a) For a reasonably strongly allowed transition the emission oscillator strength might be of order $f_{21} \approx -0.1$. Assuming this value, calculate the Einstein A coefficient for a transition with $\lambda_0 = 500$ nm.

(b) Assuming a molar mass of $A = 50$ g and a temperature $T = 300$ K, find the Doppler width $\Delta\omega_D$ of the transition.

(c) Discuss whether the transition is likely to be predominantly homogeneously or inhomogeneously broadened, and calculate the appropriate peak gain cross-section.

(d) How does your value compare with that calculated in Exercise 4.1?

(4.3) **Transitions in the solid-state**
Consider the case of Nd^{3+} ions doped in a crystal of YAG.[12] The laser transition at 1064 nm occurs on a dipole-forbidden transition, with upper- and lower-level fluorescence lifetimes of approximately 230 µs and 100 ps, respectively. Throughout this question you may ignore the effects of the refractive index of the host material.

(a) Estimate the natural linewidth of the Nd:YAG laser transition (for this purpose ignore the difference between the radiative and fluorescence lifetime) and estimate the peak optical gain cross-section if this were the only cause of broadening.

(b) In fact, the laser transition is homogeneously broadened by phonon collisions with a width of 190 GHz. Revise your estimate of the peak optical gain cross-section.

(c) Laser oscillation may also be observed on the same transition in Nd^{3+} ions doped into a variety of glass hosts. Under these conditions the laser transition is inhomogeneously broadened with a full width at half-maximum of order 5000 GHz. Estimate the peak optical gain cross-section in this case, assuming that the wavelength, lifetimes, and Einstein A coefficient of the transition are unchanged from those in Nd:YAG.

(4.4) **Required population inversion**
Here, we estimate of the population inversion density N^* necessary to achieve a reasonable gain on a single pass through an amplifier.

(a) Use eqn (4.22) to show that the intensity of a continuous beam of narrowband radiation centred at a frequency of ω_L is increased by a factor of $\exp(\alpha_{21}\ell_g)$, where $\alpha_{21} = N^*\sigma_{21}(\omega - \omega_L)\ell_g$, when it is passed through a length ℓ_g of the gain medium. You may assume that the population inversion density N^* is not changed by the radiation.

(b) For the laser transitions listed in Table 4.1, suggest reasonable values for the gain length ℓ_g in each case, and hence calculate the population inversion density N^* required to give a single-pass amplification by a factor of 1.1. In each case comment on the value of N^* that you estimate.

(4.5) Here, we consider a prospective laser transition operating in the X-ray region of the spectrum. Since most solid materials are strongly absorbing in this spectral region,

---

[12]See Section 7.2.3.

we will assume the active medium to be an ionized gas with a temperature of 1 keV.

(a) Use the approach of Exercise 4.2 to estimate the Einstein A coefficient in this case. Assume a laser wavelength of $\lambda_0 = 1$ nm.

(b) Estimate the Doppler and natural widths for this transition, and state which is the dominant broadening mechanism in this case.

(c) Hence, calculate the peak gain cross-section for the X-ray laser.

(d) As discussed in Chapter 18, it is difficult to make optical cavities in the X-ray spectral region, and consequently they usually operate without a cavity and output only amplified spontaneous emission. Furthermore, it is very difficult to generate a population inversion over more than a few millimetres. As such, the small-signal gain coefficient has to be of order $10 \, \text{cm}^{-1}$ if the output is to be significantly greater than just the spontaneous emission. Given this value for α_{21}, use your value for the gain cross-section to deduce the threshold population inversion density.

(e) What is the minimum rate at which atoms in the upper laser level will decay due to spontaneous emission? Hence, deduce the minimum power per unit volume that must be applied to the gain medium to maintain the threshold population inversion density calculated above. Comment on the value that you find. [Hint: see Section 18.2.1.]

(4.6) †The radiation-transfer equation

Here, we find the equation describing the variation of the spectral intensity of radiation owing to both stimulated *and* spontaneous transitions. This radiation-transfer equation is useful for a quantitative description of self-absorption, as discussed in Section 4.5.2 and in Exercise 4.7.

We find the transfer equation in terms of the properly defined spectral intensity $\mathcal{I}_\Omega$: i.e. power per unit area per unit frequency interval *per unit solid angle*. To do this we apply the method of Section 4.1 to the situation shown in Fig. 4.4.

(a) Show that the net rate at which photons with frequencies in the interval $\omega, \omega + \delta\omega$ leave the volume between z and $z + \delta z$ is given by

$$\frac{\mathcal{I}_\Omega(\omega, z + \delta z) - \mathcal{I}_\Omega(\omega, z)}{\hbar\omega} \delta A \delta\omega \delta\Omega.$$

(b) Explain why this must equal

$$\left\{ N_2 A_{21} g(\omega - \omega_0) \frac{\delta\Omega}{4\pi} \right.$$
$$\left. + [N_2 B_{21} - N_1 B_{12}] \frac{\mathcal{I}_\Omega(\omega,z)\delta\Omega}{c} \right\} \delta A \delta\omega \delta z.$$

(c) Hence, show that

$$\frac{\partial \mathcal{I}_\Omega}{\partial z} = \frac{\hbar\omega}{4\pi} N_2 A_{21} g(\omega - \omega_0) - N^{**}\sigma_{12}(\omega - \omega_0)\mathcal{I}_\Omega, \tag{4.42}$$

where N^{**} and $\sigma_{12}(\omega - \omega_0)$ are as defined in Section 4.5.1.

(4.7) Self-absorption†

Here, we solve the radiation-transfer equation derived in Exercise 4.6 for the simple case of a uniform slab of material.

(a) Solve eqn (4.42) to show that for the case of a uniform slab of thickness ℓ

$$\mathcal{I}_\Omega(\omega, \ell) = \mathcal{I}_\Omega(\omega, 0) \exp\left[-\kappa(\omega - \omega_0)\ell\right]$$
$$+ \frac{N_2 A_{21} \hbar\omega g(\omega - \omega_0)}{4\pi\kappa(\omega - \omega_0)} \left[1 - \exp\left(-\kappa(\omega - \omega_0)\ell\right)\right],$$

where $\kappa(\omega - \omega_0) = N^{**}\sigma_{12}(\omega - \omega_0)$.

(b) Consider first an *optically thin* source, that is one for which the line-centre optical depth $\tau(0) = \kappa(0)\ell$ is

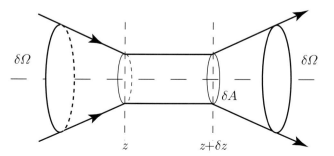

Fig. 4.4: Geometry used in Exercise 4.6.

small. Show that when there is no incident beam, so that $\mathcal{I}_\Omega(\omega, 0) = 0$, the intensity on the surface of the slab is given by

$$\mathcal{I}_\Omega(\omega, \ell) = \frac{1}{4\pi} N_2 A_{21} \hbar \omega g(\omega - \omega_0) \ell,$$

and discuss the form of this result.

(c) Consider now an optically thick source, i.e. one for which $\tau(0) \gg 1$.
  (i) Show that in this case

$$\mathcal{I}_\Omega(\omega, \ell) = \frac{\hbar \omega^3}{4\pi^3 c^2} \frac{1}{\frac{g_2 N_1}{g_1 N_2} - 1}.$$

(ii) Suppose that the atoms in the medium are in thermal equilibrium at temperature T. Show that in this case

$$\mathcal{I}_\Omega(\omega, \ell) = \frac{1}{4\pi} \varrho_B(\omega) c,$$

where $\varrho_B(\omega)$ is the spectral energy density of blackbody radiation given by eqn (1.34).

(iii) Comment on this last result.

5

Gain saturation

5.1 Saturation in a steady-state amplifier   60
5.2 Saturation in a homogeneously broadened pulsed amplifier[†]   73
5.3 Design of laser amplifiers   77
Exercises   78

In the previous chapters we discussed how a population inversion led to the amplification of a beam of light propagating through a medium. We showed there that the intensity of the propagating beam grew exponentially with propagation distance, although it was recognized that such exponential growth could not be maintained indefinitely. In deriving eqn (4.20) we assumed that the population inversion was constant and, in particular, was not affected by the presence of the beam. Clearly, this is never strictly so, since the beam gains its energy by stimulating population from the upper to the lower laser levels. The net rate at which the beam stimulates population out of the upper laser level depends on the intensity of the beam: for low intensities the rate is small and the atomic populations are essentially unperturbed by the presence of the beam; for high intensities the rate is high and the populations of the upper and lower laser level are strongly perturbed, leading to a decrease in the population inversion, reduced gain, and a departure from exponential growth of the beam intensity.

The reduction of the population inversion—and hence gain—by an intense beam of radiation is known as **saturation**. In this chapter we explore this topic fully and, amongst other things, develop an idea of when the intensity of a beam of radiation can be described as 'high' or 'low'.

5.1 Saturation in a steady-state amplifier

In order to develop some idea of the processes by which saturation occurs, we first consider the simple case of saturation in an amplifier operating under steady-state conditions. As in previous chapters, we will consider separately the cases of homogeneously and inhomogeneously broadened gain media.

5.1.1 Homogeneous broadening

In Fig. 5.1 we show schematically the processes by which the upper and lower laser levels gain or lose population when subjected to a beam of narrow-bandwidth radiation with angular frequencies close to ω_L and with a *total* intensity I. For each level i these are:

1. Pumping at a rate of R_i atoms per unit volume per second. The pumping rate R_i includes all processes that populate the level i, *except* radiative transitions on the laser transition itself, which are included explicitly. For example, R_i will include population by collisional processes, optical

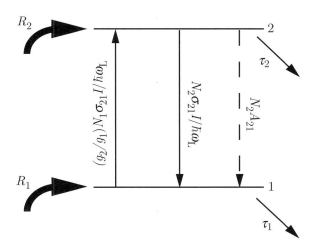

Fig. 5.1: The processes determining the populations of the upper and lower laser levels in a steady-state, homogeneously broadened amplifier in the presence of narrowband radiation with total intensity I.

pumping, and radiative cascade, etc. Note that we will assume that the pumping rates are constant; in particular that they are independent of N_1 and N_2.

2. Radiative decay to *all* levels lower in energy than i, with a fluorescence lifetime of τ_i. It is important to note that the rate of decay of the upper laser level, $1/\tau_2$, *includes* spontaneous emission on the laser transition $2 \rightarrow 1$.
3. Stimulated transitions (absorption or emission) on the laser transition arising from interaction with the beam of radiation.
4. Finally, the lower level can also be populated by spontaneous emission on the laser transition itself.

With these definitions, the rate equations for the populations of the upper and lower laser levels are:[1]

$$\frac{dN_2}{dt} = R_2 - N^* \sigma_{21}(\omega_L - \omega_0)\frac{I}{\hbar \omega_L} - \frac{N_2}{\tau_2} \tag{5.1}$$

$$\frac{dN_1}{dt} = R_1 + N^* \sigma_{21}(\omega_L - \omega_0)\frac{I}{\hbar \omega_L} - \frac{N_1}{\tau_1} + N_2 A_{21}. \tag{5.2}$$

[1] Note that by allowing the lower laser level to decay with a fluorescence lifetime τ_1 we have excluded three-level lasers systems, discussed in Section 7.1.5, in which the lower laser level is the ground state. The saturation behaviour of three-level laser systems is considered in Exercise 5.2.

It is straightforward to find the steady-state populations of the laser levels from eqns (5.1) and (5.2) by setting $dN_2/dt = dN_1/dt = 0$:

$$N_2 = R_2 \tau_2 - N^* \sigma_{21}\frac{I}{\hbar \omega_L}\tau_2 \tag{5.3}$$

$$N_1 = R_1 \tau_1 + N^* \sigma_{21}\frac{I}{\hbar \omega_L}\tau_1 + N_2 A_{21} \tau_1, \tag{5.4}$$

where, to avoid clutter, we have dropped the frequency dependence of the gain cross-section σ_{21}.

We may now find an expression for the population inversion density $N^* = N_2 - (g_2/g_1)N_1$ by substituting for N_2 in eqn (5.4), and subtracting $(g_2/g_1) \times$ eqn (5.4) from (5.3). We find,

$$N^* = \frac{R_2 \tau_2 \left[1 - (g_2/g_1) A_{21} \tau_1\right] - (g_2/g_1) R_1 \tau_1}{1 + \sigma_{21} \frac{I}{\hbar \omega_L} \left[\tau_2 + (g_2/g_1) \tau_1 - (g_2/g_1) A_{21} \tau_1 \tau_2\right]}. \quad (5.5)$$

By setting $I \to 0$ in eqn (5.5) it is easy to see that the numerator is equal to the population inversion density $N^*(0)$ that would be generated in the absence of the radiation at ω_L.[2] Hence, we may write this last result in the more convenient form,

[2] Of course, we assume that the pumping and decay rates of the levels are unchanged if the radiation is switched off.

$$N^*(I) = \frac{N^*(0)}{1 + I/I_s(\omega_L - \omega_0)}, \quad (5.6)$$

$$I_s(\omega_L - \omega_0) = \frac{\hbar \omega_L}{\tau_R} \frac{1}{\sigma_{21}(\omega_L - \omega_0)}, \quad (5.7)$$

$$\tau_R = \tau_2 + \frac{g_2}{g_1} \tau_1 \left[1 - A_{21} \tau_2\right], \quad (5.8)$$

where $N^*(I)$ is the population inversion density in the presence of a beam of total intensity I. The parameters $I_s(\omega_L - \omega_0)$ and τ_R are known as the **saturation intensity** and **the recovery time** respectively.

We see that the increased rate of stimulated emission caused by the intense beam of radiation reduces, or 'burns down,' the population inversion from the unsaturated value, $N^*(0)$, by a factor of $[1 + I/I_s]$. The saturation intensity I_s (units of W m^{-2}) is a measure of the intensity required to reduce the population inversion significantly (i.e. by a factor of two) from the value achieved (by the same pumping) in the absence of radiation at the transition frequency. In other words the saturation intensity divides 'low' intensities from 'high' intensities: if $I \ll I_s$ the population inversion will not be greatly affected by the presence of the radiation; if $I \gg I_s$ the population inversion will be reduced significantly.

We have reinserted the frequency dependence of the optical gain cross-section in eqn (5.7) in order to emphasize that the saturation intensity of a homogeneously broadened laser transition depends on the relative detuning from the centre frequency of the transition. This is to be expected since near the line centre the gain cross-section is large; the atoms interact with the radiation strongly, and the populations of the upper and lower laser levels will be significantly perturbed at lower intensities. In contrast, if the frequency of the radiation is tuned by many homogeneous linewidths away from the centre frequency, the optical gain cross-section is small; the atoms will interact with the radiation only weakly, and the intensity of the radiation will have to be very much larger before the levels are perturbed to a significant extent.

The recovery time τ_R

Before proceeding it is worth commenting on the physical significance of recovery time. To see this, we note that we can rewrite eqns (5.1) and (5.2) in the form:

5.1 Saturation in a steady-state amplifier

$$\frac{dN_2}{dt} = R_2 - \frac{N^* I}{\tau_R I_s} - \frac{N_2}{\tau_2} \tag{5.9}$$

$$\frac{dN_1}{dt} = R_1 + \frac{N^* I}{\tau_R I_s} - \frac{N_1}{\tau_1} + N_2 A_{21}. \tag{5.10}$$

This is a rather satisfying form of the rate equations! We can see that the rate of stimulated emission (either out of the upper level, or into the lower level) is N^*/τ_R multiplied by the ratio of the intensity to the saturation intensity. Further insight can be obtained by subtracting (g_2/g_1) times eqn (5.10) from eqn (5.9):

$$\frac{dN^*}{dt} = R^* - \left(1 + \frac{g_2}{g_1}\right)\frac{N^* I}{\tau_R I_s} + \text{spontaneous terms}, \tag{5.11}$$

where $R^* = R_2 - (g_2/g_1)R_1$ is the net pumping rate of the population inversion density. We have not bothered to write out all the various spontaneous decay terms, such as N_2/τ_2, since for laser levels interacting with radiation at, or significantly above, the saturation intensity they will be relatively small. In this case the time taken for the population inversion to respond to a perturbation to the system, such as a change in pumping rates, is of the order of the recovery time (divided by $[1 + (g_2/g_1)(I/I_s)]$). So, in the presence of radiation with an intensity equal to the saturation intensity, the rate of stimulated emission is approximately N^*/τ_R, and the time taken for stimulated emission to re-establish a steady state after a perturbation to the system is of the order τ_R.

Approximate expressions for the saturation intensity

A number of useful approximations for the saturation intensity may be found. For example, in so-called four-level lasers (see Section 7.1.5) the fluorescence lifetime of the lower laser level is much shorter than that of the upper level. In this case the recovery time may be approximated by,

$$\tau_R = \tau_2 + (g_2/g_1)\tau_1[1 - A_{21}\tau_2] \approx \tau_2. \tag{5.12}$$

We find precisely the same result in the case when the upper laser level decays pre-dominantly on the laser transition itself and collisional quenching of the upper level is insignificant, since then $A_{21}\tau_2 \approx 1$.

For both of these cases the saturation intensity takes the particularly simple form:

$$I_s = \frac{\hbar\omega_L}{\sigma_{21}\tau_2} \quad \text{Special case: } \tau_1 \ll \tau_2 \text{ or } A_{21}\tau_2 \approx 1, \tag{5.13}$$

where here, and from now on, we will drop the explicit frequency dependence of the optical gain cross-section if no ambiguity is introduced by so doing.

The saturated gain coefficient

Weak probe

Suppose that the population inversion is saturated, to some extent, by a beam of radiation with total intensity I in a narrow band of frequencies close to ω_L as shown schematically in Fig. 5.2. We can now investigate the gain that would be measured by a weak probe beam with angular frequencies close to

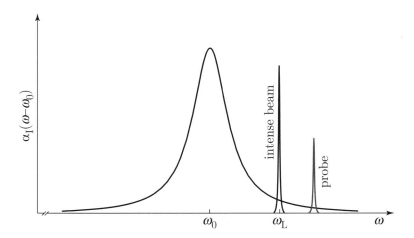

Fig. 5.2: Measurement of the saturated gain coefficient by a weak probe at ω in the presence of intense radiation at ω_L.

some general frequency ω. Since the medium is homogeneously broadened, the probe beam interacts with exactly the same atoms as does the saturating beam, and hence the population inversion experienced by the weak probe beam is also burnt down according to eqn (5.6). Thus, the gain coefficient measured by the probe beam is given by,

$$\alpha_I(\omega - \omega_0) = N^*(I)\sigma_{21}(\omega - \omega_0) = \frac{N^*(0)\sigma_{21}(\omega - \omega_0)}{1 + I/I_s(\omega_L - \omega_0)}. \quad (5.14)$$

In other words, the gain coefficient experienced by the probe beam is the **saturated gain coefficient**:[3]

$$\alpha_I(\omega - \omega_0) = \frac{\alpha_0(\omega - \omega_0)}{1 + I/I_s(\omega_L - \omega_0)},$$

Weak probe, saturating radiation at ω_L (5.15)

where $\alpha_0(\omega - \omega_0) = N^*(0)\sigma_{21}(\omega - \omega_0)$ is the small-signal gain coefficient that would be measured in the absence of the saturating radiation.[4] We see that the saturating radiation has the effect of reducing the measured gain coefficient by *the same factor*, $[1 + I/I_s(\omega_L - \omega_0)]$, as the population inversion density. This factor depends in an obvious way on the intensity of the intense radiation at ω_L; but it also depends on the detuning of ω_L from the centre frequency of the transition ω_0. Again: for a given intensity I the level of saturation decreases as the saturating radiation is detuned from the centre frequency of the transition.

For a saturating beam of fixed frequency and intensity, the denominator in eqn (5.15) is constant, and hence the frequency dependence of the gain coefficient $\alpha_0(\omega - \omega_0)$ as measured by the weak probe is exactly the same as measured in the absence of saturation; the gain is simply reduced *at all frequencies* by the same factor $[1 + I/I_s]$.

Intense probe

Suppose now that we have only one (intense) narrowband beam and that we use this to measure the frequency dependence of the gain coefficient. Now, the

---

[3] By 'saturated gain coefficient' we mean that, unlike the small-signal gain coefficient, the population inversion is reduced to some extent by an interaction with radiation. We do not mean to imply that the gain is necessarily heavily saturated, or saturated to a particular value.

[4] This follows immediately by letting $I \to 0$ in eqn (5.15).

probe beam takes the role of both the saturating beam and the probe beams in Fig. 5.2, so that the gain coefficient is given by eqn (5.15) with $\omega_L \to \omega$. Hence, the gain coefficient measured in this case is:

$$\alpha_I(\omega - \omega_0) = N^*(I)\sigma_{21}(\omega - \omega_0) = \frac{N^*(0)\sigma_{21}(\omega - \omega_0)}{1 + I/I_s(\omega - \omega_0)}$$

$$= \frac{\alpha_0(\omega - \omega_0)}{1 + I/I_s(\omega - \omega_0)}. \quad \textbf{Intense probe} \quad (5.16)$$

It should be clear that the frequency dependence of the gain measured with a non-weak probe beam will be severely distorted from that measured by a weak probe beam owing to the frequency dependence of both the optical gain cross-section and the saturation intensity. In general, the saturation of the population inversion will be most severe near the line centre, and be weak for large detuning from ω_0. Consequently, the measured gain profile will be reduced when the (intense) probe is tuned near to the centre frequency of the transition, but be relatively unaffected in the wings of the profile, leading to flattening and increased width of the measured gain lineshape. This process is known as **power broadening**. In Exercise 5.7 this process is considered in detail for the case when the homogeneous lineshape is Lorentzian.

In Fig. 5.3 we compare the frequency dependence of the gain coefficient for a homogeneously broadened laser transition as measured by weak and non-weak probe beams. In Fig. 5.3(b) is shown the gain coefficient measured by a weak probe beam in the presence of radiation at ω_L with an intensity equal to the saturation intensity at ω_L. By definition, under these conditions the population inversion will be reduced from the unsaturated value by a factor of 2; as a consequence for all frequencies ω the gain coefficient measured by a weak probe will also be reduced from the unsaturated value by a factor of 2. In contrast, Fig. 5.3(c) shows the gain coefficient recorded by an intense probe beam with a constant intensity $I_s(0)$. When this beam is tuned to the centre frequency, ω_0, its intensity will equal the saturation intensity and hence, by definition, the measured gain coefficient will be half of the small-signal

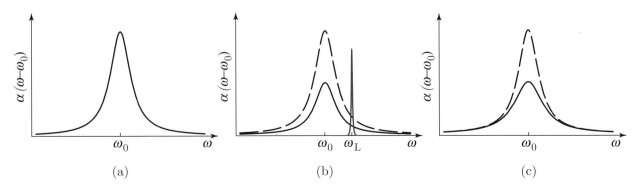

Fig. 5.3: The frequency dependence of the gain coefficient measured with (a) a weak probe beam; (b) a weak probe beam in the presence of intense, narrowband radiation at $\omega = \omega_L$ with a total intensity $I = I_s(\omega_L - \omega_0)$; (c) an intense probe beam with a constant intensity $I_s(0)$. In (b) and (c) the dashed curve shows for reference the small-signal gain coefficient of part (a). The lineshape of the transition has been assumed to be Lorentzian, but homogeneously broadened transitions with other lineshapes would show qualitatively similar behaviour.

value. As the intense probe beam is tuned away from the centre frequency the saturation intensity increases; consequently the probe beam will saturate the transition to a lesser extent, and the measured gain coefficient will be greater than half the small-signal value. In the limit of large detuning, the saturation intensity becomes very large, and the probe beam may be considered 'weak'— so that the measured gain will correspond to the small-signal value.

Beam growth in a steady-state, homogeneously broadened laser amplifier

We are now in a position to see how saturation affects the amplification of a beam of radiation as it propagates through a steady-state, homogeneously broadened amplifier. The intensity of narrow-bandwidth radiation will grow according to eqn (5.16):

$$\frac{1}{I}\frac{dI}{dz} = \alpha_I(\omega - \omega_0) = \frac{\alpha_0(\omega - \omega_0)}{1 + I/I_s(\omega - \omega_0)}, \tag{5.17}$$

which may be rearranged in the form,

$$\int_{I(0)}^{I(z)} \frac{1 + I/I_s}{I} dI = \int_0^z \alpha_0 dz, \tag{5.18}$$

where again we have dropped the frequency dependencies to avoid clutter. Equation (5.18) may be solved straightforwardly to give,

$$\ln\left[\frac{I(z)}{I(0)}\right] + \frac{I(z) - I(0)}{I_s} = \alpha_0 z. \tag{5.19}$$

Figure 5.4 shows the general solution to eqn (5.19). We may identify two extreme cases. The first occurs when the intensity of the beam is well below the saturation intensity, i.e. $I \ll I_s$, whereupon eqn (5.19) becomes,

$$I(z) \approx I(0) \exp(\alpha_0 z), \tag{5.20}$$

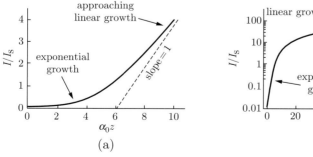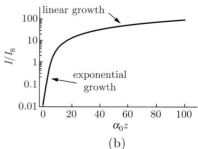

Fig. 5.4: The growth of a beam of narrowband radiation through a steady-state, homogeneously broadened amplifier shown on: (a) a linear scale; (b) a logarithmic intensity scale. The input beam is taken to have an intensity of $0.01 I_s$. The intensity grows exponentially with distance until $I \approx I_s$. At longer lengths, such that $I \gg I_s$, the intensity of the beam grows approximately linearly with distance. Notice that in (a) the growth is not quite yet linear even for the largest values of $\alpha_0 z$ plotted.

and the beam grows exponentially with distance, as we saw earlier. For this last result to be true the intensity must be much less than the saturation intensity *at all points in the amplifier*. However, since $I(z) > I(0)$, we can satisfy the condition provided that the *output* intensity is much smaller than the saturation intensity.

The second case, that of strong saturation, occurs when $I \gg I_s$ and eqn (5.19) becomes,

$$I(z) \approx I(0) + \alpha_0 I_s z = I(0) + \frac{N^*(0)\hbar\omega}{\tau_R} z. \qquad (5.21)$$

For this last expression to be valid the intensity must be much greater than the saturation intensity at all points in the amplifier and, following the logic above, this requires the *input* intensity to be much greater than the saturation intensity. We see that once the beam is amplified significantly above the saturation intensity it grows only linearly with propagation distance. The far right-hand side of eqn (5.21) illustrates once again the role of the recovery time, and emphasizes that under conditions of strong saturation the growth in intensity is independent of the optical gain cross-section. In fact, in this regime the probe beam grows in intensity at the maximum possible rate, limited only by the pumping and decay rates of the upper and lower laser levels. In this limit, atoms arriving in the upper laser level are stimulated down to the lower laser level before they can decay by other means.[5]

[5] See Exercise 5.8.

5.1.2 Inhomogeneous broadening†

As might be expected, the phenomenon of gain saturation is considerably more complicated when the laser transition is inhomogeneously broadened. We must now consider separately the extent to which the population inversion of each frequency class is saturated, and the contribution by each such class to the overall gain.

Suppose that in the absence of any saturating beam of radiation the population inversion is distributed over frequency classes according to the normalized lineshape $g_D(\omega_c - \omega_0)$, such that the population inversion for atoms with centre frequencies lying in the interval ω_c to $\omega_c + \delta\omega_c$ is,

$$\Delta N_2(\omega_c - \omega_0)\delta\omega_c - (g_2/g_1)\Delta N_1(\omega_c - \omega_0)\delta\omega_c = N^* g_D(\omega_c - \omega_0)\delta\omega_c, \qquad (5.22)$$

where $N^* = N_2 - (g_2/g_1)N_1$ is the *total* population inversion.

Let us now suppose that the gain medium is subjected to a beam of intense, narrow-bandwidth radiation with frequencies centered on ω_L. Each frequency class will be homogeneously broadened to some extent, indeed each class interacts with the saturating radiation at ω_L with a strength determined by the *homogeneous* gain cross-section $\sigma_{21}(\omega_L - \omega_c)$. Further, each class may be treated independently, and we may apply the results of Section 5.1.1 to find that the population inversion for atoms with centre frequencies in the interval ω_c to $\omega_c + \delta\omega_c$ is burnt down to,

$$\frac{N^*(0) g_D(\omega_c - \omega_0)\delta\omega_c}{1 + I/I_s(\omega_L - \omega_c)}.$$

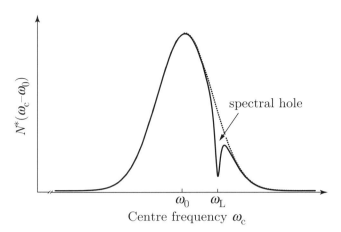

Fig. 5.5: The effect of a beam of intense, narrow-bandwidth radiation of total intensity I with frequencies close to ω_L on the distribution of the population inversion over centre frequencies. The dotted curve shows the unperturbed distribution for $I = 0$, and the solid curve the distribution for $I \approx I_s(0)$. For this plot $\Delta\omega_H/\Delta\omega_D = 0.1$.

Notice that the saturation intensity is given by eqn (5.7), and hence is *still* determined by the *homogeneous* gain cross-section. As we would expect, the saturation factor now depends on the detuning of the intense beam from the centre frequency of the particular class being considered, *not* the centre frequency ω_0 of the inhomogeneous lineshape. It should be clear that if the frequency of the saturating radiation is many homogeneous linewidths $\Delta\omega_H$ away from ω_c, the saturation intensity will be large and the denominator will remain close to unity. That is, the frequency class will interact with the radiation only weakly, and hence the population inversion will be burnt down only very slightly. In contrast, the frequency classes with $\omega_c \approx \omega_L$ will interact strongly with the radiation, and will be burnt down appreciably. As such, a **spectral hole** will be burnt into the population inversion, as illustrated in Fig. 5.5. Only those classes with central frequencies close to ω_L on the scale of $\Delta\omega_H$ will interact appreciably with the saturating radiation, and so the width of the spectral hole will be approximately $\Delta\omega_H$.

The saturated gain coefficient
Weak probe

We are now in a position to determine the gain coefficient experienced by a weak, narrowband probe beam of angular frequency ω. Those atoms with centre frequencies close to ω_c are detuned from the frequency of the probe beam by an amount $\omega - \omega_c$ and the intense beam at ω_L is detuned from the same atoms by an amount $\omega_L - \omega_c$. Hence, the contribution of that class of atoms to the gain is,

$$\frac{N^*(0)g_D(\omega_c - \omega_0)\delta\omega_c}{1 + I/I_s(\omega_L - \omega_c)}\sigma_{21}(\omega - \omega_c). \qquad (5.23)$$

The total gain experienced by the probe beam is given by integrating this contribution over the distribution of centre frequencies:

$$\alpha_I^D(\omega - \omega_0) = \int_0^\infty N^*(0)\frac{g_D(\omega_c - \omega_0)}{1 + I/I_s(\omega_L - \omega_c)}\sigma_{21}(\omega - \omega_c)d\omega_c. \qquad (5.24)$$

Notice that, once again, it is the homogeneous cross-section that appears in eqn (5.24).

For a truly inhomogeneously broadened laser transition the inhomogeneous linewidth is very much larger than the homogeneous linewidth, in which case the numerator in the integrand is strongly peaked at $\omega_c \approx \omega$, reflecting the sharply peaked nature of the homogeneous cross-section $\sigma_{21}(\omega - \omega_c)$. Turning our attention to the denominator, we see that for $\omega_c \approx \omega$ the denominator is close to unity for probe frequencies that are many homogeneous linewidths away from ω_L. Hence, well away from ω_L the gain coefficient is approximately the unsaturated value:

$$\alpha_I^D(\omega - \omega_0) \approx \int_0^\infty N^*(0) g_D(\omega_c - \omega_0) \sigma_{21}(\omega - \omega_c) d\omega_c \quad (5.25)$$

$$= \alpha_0^D(\omega - \omega_0) \quad (5.26)$$

$$= N^*(0) \frac{\hbar \omega}{c} B_{21} g(\omega - \omega_0), \quad (5.27)$$

where, just as in eqn (3.44), $g(\omega - \omega_0)$ is the convolution of the inhomogeneous and homogeneous lineshapes. Obviously, in the limit of extreme inhomogeneous broadening, $\Delta \omega_D \gg \Delta \omega_H$, the convoluted lineshape $g(\omega - \omega_0)$ is simply equal to the inhomogeneous lineshape $g_D(\omega - \omega_0)$.

For probe frequencies close to ω_L the denominator becomes large for $\omega_c \approx \omega$, leading to a strong reduction in the gain coefficient compared to that measured in the absence of the saturating radiation: a spectral hole is burnt in the frequency dependence of the gain coefficient that reflects the spectral hole in the population inversion. As explored in Exercise 5.9, in the limit of extreme inhomogeneous broadening ($\Delta \omega_D \gg \Delta \omega_H$), and if the homogeneous lineshape is Lorentzian, the spectral hole burnt in the gain profile is also Lorentzian, with a full width at half-maximum of,

$$\Delta \omega_{\text{hole}} = \Delta \omega_H \left[1 + \sqrt{1 + I/I_s(0)} \right]. \quad (5.28)$$

For this case we see that if the intensity of the saturating beam is small compared to the saturation intensity $\Delta \omega_{\text{hole}} \approx 2 \Delta \omega_H$. This result is consistent with our expectation that the spectral hole burnt in the population inversion is approximately $\Delta \omega_H$ wide, and the range of frequency classes with which the probe interacts is also approximately equal to $\Delta \omega_H$. Figure 5.6 shows how the spectral hole burnt in the gain profile of an inhomogeneously broadened laser transition varies as the intensity of the saturating radiation is increased.

Intense probe

We can find the gain coefficient experienced by a narrowband probe beam with an intensity that is not small compared to the saturation intensity by setting $\omega_L = \omega$ in eqn (5.24):

$$\alpha_I^D(\omega - \omega_0) = \int_0^\infty N^*(0) g_D(\omega_c - \omega_0) \frac{\sigma_{21}(\omega - \omega_c)}{1 + I/I_s(\omega - \omega_c)} d\omega_c. \quad (5.29)$$

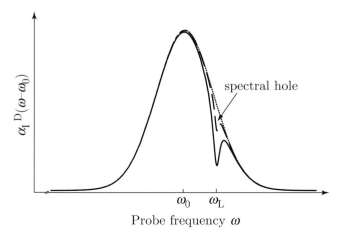

Fig. 5.6: Spectral hole-burning in the gain profile of an inhomogeneously broadened laser transition by a narrow-bandwidth beam of intensity I and with frequencies close to ω_L. The gain profile measured by a weak probe beam is shown for $I = 0$ (dotted), $I = I_s(0)$ (dashed), $I = 10 I_s(0)$ (solid). For this plot $\Delta\omega_H/\Delta\omega_D = 0.03$.

In the strongly inhomogeneous limit this may be written,

$$\alpha_I^D(\omega - \omega_0) = N^*(0) g_D(\omega - \omega_0) \int_0^\infty \frac{\sigma_{21}(\omega - \omega_c)}{1 + I/I_s(\omega - \omega_c)} d\omega_c, \quad (5.30)$$

since the inhomogeneous lineshape $g_D(\omega - \omega_0)$ is much broader than the remaining terms in the integral. We may further simplify eqn (5.30) by making the substitution $x = \omega - \omega_c$,

$$\alpha_I^D(\omega - \omega_0) \approx N^*(0) g_D(\omega - \omega_0) \int_{-\infty}^\infty \frac{\sigma_{21}(x)}{1 + I/I_s(x)} dx, \quad (5.31)$$

where we have set the lower limit to $-\infty$ since $\omega \gg \Delta\omega_H$ and the integrand is only significant for $x = \omega - \omega_c \approx 0$. The integral in eqn (5.31) is equal to a constant; in other words the gain coefficient measured by a strong probe has the same frequency dependence as that measured by a weak probe, and is simply equal to the small-signal value divided by a constant.

If the homogeneous lineshape is Lorentzian, it may be shown (see Exercise 5.10) that eqn (5.30) becomes,

$$\alpha_I^D(\omega - \omega_0) = \frac{\alpha_0^D(\omega - \omega_0)}{\sqrt{1 + I/I_s(0)}}. \quad (5.32)$$

Notice that this looks rather similar to eqn (5.15) that was obtained for homogeneous broadening. However, this similarity is very misleading since eqn (5.15) refers to the gain measured for a *homogeneously* broadened transition by a *weak* probe in the presence of a saturating beam at $\omega = \omega_L$; eqn (5.32) refers to the gain measured for an extreme *inhomogeneously* broadened transition by a *intense* probe in the absence of any other saturating radiation. The situations couldn't be more different!

To summarize: for homogeneously broadened gain media an intense, narrowband beam of radiation at $\omega = \omega_L$ causes the *entire* population inversion to be burnt down by a factor $[1 + I/I_s(\omega_L - \omega_0)]$ such that the gain profile

5.1 Saturation in a steady-state amplifier

Table 5.1 Summary of different cases of gain saturation.

	homogeneous	inhomogeneous
Weak probe, intense beam at ω_L	Population inversion reduced by const. factor; gain coeff. has same frequency dependence as small-signal case	Spectral hole appears at ω_L
Intense probe beam	Stronger reduction of gain from small-signal value near line centre than in wings	Population inversion reduced by const. factor; gain coeff. has same frequency dependence as small-signal case

measured by a weak probe beam has the same shape as that measured in the absence of the saturating beam, but with the gain being reduced by the factor $[1 + I/I_s(\omega_L - \omega_0)]$; for inhomogeneously broadened gain media a spectral hole is burnt into the population inversion, and the gain measured by a weak probe beam is only reduced for frequencies close to ω_L. For homogeneously broadened systems, the gain measured by an intense probe beam is reduced by a greater extent at frequencies close to the line centre, leading to a flattening and broadening of the measured gain profile; for strongly inhomogeneously broadened media the gain is everywhere reduced by the same factor, and the measured gain profile has the same shape as measured by a weak probe beam. These cases are also summarized in Table 5.1.

Beam growth in a steady-state, inhomogeneously broadened laser amplifier

We can use our last result to deduce how a narrowband beam is amplified in an amplifier that is strongly inhomogeneously broadened, allowing for saturation of the gain medium. If we assume that the homogeneous component of the line broadening has a Lorentzian lineshape, then from eqn (5.32) we have,

$$\frac{1}{I}\frac{dI}{dz} = \frac{\alpha_0^D}{\sqrt{1 + I/I_s(0)}}, \tag{5.33}$$

where, to avoid clutter, we have dropped the frequency dependence of the small-signal gain coefficient α_0^D. Hence,

$$\int_{I(0)}^{I(z)} \frac{\sqrt{1 + I/I_s(0)}}{I} dI = \alpha_0^D z. \tag{5.34}$$

This may be integrated to give a (somewhat complicated) transcendental equation for the intensity $I(z)$. The solution to eqn (5.34) is shown in Fig. 5.7. However, just as for the homogeneous case, we may identify two extreme cases that yield algebraic solutions for the intensity. The first occurs when the intensity of the beam is well below the saturation intensity, i.e. $I \ll I_s(0)$, in which case the intensity of the beam grows exponentially as:

$$I(z) \approx I(0) \exp\left(\alpha_0^D z\right). \tag{5.35}$$

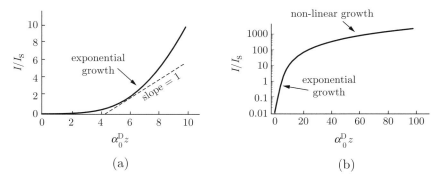

Fig. 5.7: The growth of a beam of narrowband radiation through a steady-state, inhomogeneously broadened amplifier. The input beam is taken to have an intensity of $0.01 I_s(0)$. Note that the intensity grows exponentially with distance up to $\alpha_0 z \approx 10$. At longer lengths, such that $I \gg I_s$, the intensity of the beam still grows non-linearly with distance.

Again, the equation above holds provided the output intensity is much smaller than the saturation intensity.

The second case, that of heavy saturation, occurs when $I \gg I_s(0)$ at all points in the amplifier and the solution to eqn (5.34) becomes,

$$\sqrt{I(z)} \approx \sqrt{I(0)} + \frac{1}{2}\sqrt{I_s(0)}\alpha_0^D z. \tag{5.36}$$

For this last expression to be valid the input intensity must be much greater than the saturation intensity. We see that once the beam is amplified significantly above the saturation intensity it grows *quadratically* with propagation distance. Note that our finding that the intensity of the beam grows *quadratically* follows from our assumption that the homogeneous lineshape is Lorentzian. If the homogeneous component has a different lineshape then the intensity of the beam will grow or not grow according to eqn (5.36). However, regardless of the form of the homogeneous lineshape, in the strongly saturated limit the intensity if the beam *will* grow non-linearly with propagation distance, in contrast to the purely homogeneously broadened case. For homogeneously broadened gain media the whole population inversion becomes saturated as the intensity of the beam grows, and the beam extracts power from the inverted atoms at the maximum possible rate. In contrast, for inhomogeneously broadened amplifiers a spectral hole is dug in the population inversion. As the intensity of the beam grows, both the width and depth of the spectral hole increase. Whilst those frequency classes very close to resonance with the beam will be strongly saturated, and only able to contribute a linear growth in intensity, frequency classes further detuned from the beam will not be fully saturated and hence will still be able to contribute additional gain, albeit small, by interacting with the beam via the wings of the homogeneous lineshape. It is the fact that the width of the spectral hole increases as the intensity of the beam grows that allows the intensity to grow faster than linear.

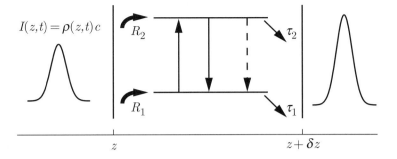

Fig. 5.8: Schematic diagram of the amplification of a pulse of laser radiation in a homogeneously broadened gain medium.

5.2 Saturation in a homogeneously broadened pulsed amplifier†

So far in this chapter our investigation of gain saturation has been restricted to laser transitions operating in the steady state. However, a very wide range of scientifically and commercially important lasers and laser amplifiers work not in the steady state but in a pulsed mode. In this section we consider how saturation affects the amplification of pulses of laser light, and compare the results to those found for the steady state. The underlying physics can be understood by considering the simplest case of a laser transition that is homogeneously broadened. Figure 5.8 shows schematically the amplification of a pulse of light as it passes through a slice of the amplifying medium lying between z and $z + \delta z$. The pump rates, level lifetimes, and optical gain cross-section of the laser transition are as described in Section 5.1.1.

Since the incident light is pulsed, it must have a finite frequency bandwidth. However, we will assume that the bandwidth is small compared to that of the laser transition, which is reasonable apart from the important exception of the amplification of short laser pulses that are close to being bandwidth limited.[6] As such, we consider the radiation to be in a narrow band around the laser frequency ω_L, with a total intensity $I(z, t)$ that now depends on position z and time t. The rate equations for the upper and lower laser levels are then just as in Section 5.1.1:

$$\frac{dN_2(z,t)}{dt} = R_2(t) - N^*(z,t)\sigma_{21}\frac{I(z,t)}{\hbar\omega_L} - \frac{N_2(z,t)}{\tau_2} \quad (5.37)$$

$$\frac{dN_1(z,t)}{dt} = R_1(t) + N^*(z,t)\sigma_{21}\frac{I(z,t)}{\hbar\omega_L} - \frac{N_1(z,t)}{\tau_1} \quad (5.38)$$
$$+ N_2(z,t)A_{21},$$

where we have emphasized that the pump rates may depend on time and that the level populations and the laser intensity will depend on both time and position.

We can determine the growth equation for the intensity of the pulse as follows. Let the energy density of the pulse be $\rho(z, t)$ such that $I(z, t) = \rho(z, t)c$. The rate of increase of the electromagnetic energy lying between z and $z + \delta z$

[6] As discussed in Section 17.1 the product of the bandwidth $\Delta\omega_{RMS}$ and duration Δt_{RMS} of a pulse cannot be smaller than π. Our analysis is therefore restricted to pulses with $\Delta t \gg 1/\Delta\omega$. The path to a more general analysis should be apparent from the results in this section.

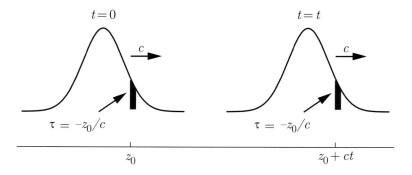

Fig. 5.9: Illustrating the travelling-wave coordinate system (z, τ). At $t = 0$ a certain part of the pulse is at $z = z_0$. Sometime later, $t = t$, that part of the pulse will have moved to $z = z_0 + ct$, but the value of $\tau = t - z/c$ will be unchanged.

may be written as:

$$\frac{\partial \rho(z, t)}{\partial t} A \delta z = [I(z, t) - I(z + \delta z, t)] A + N^* \sigma_{21} \frac{I}{\hbar \omega_{\rm L}} \times \hbar \omega_{\rm L} \times A \delta z, \tag{5.39}$$

in which the first term on the right-hand side gives the net rate at which the beam transports energy into the slice of medium by the beam, and the second term is the rate of increase of electromagnetic energy arising from stimulated emission. From this last result we find,[7]

$$\frac{\partial I(z, t)}{\partial z} + \frac{1}{c} \frac{\partial I(z, t)}{\partial t} = N^* \sigma_{21} \frac{I}{N^* \sigma_{21}}. \tag{5.40}$$

[7] Note that in the steady state this reduces to the growth equation we derived in Section 4.1

This last equation is difficult to solve since it depends on partial derivatives in z as well as t. This difficulty is removed by transforming to **travelling-wave coordinates** (z, τ), where τ is defined by $\tau = t - \frac{z}{c}$. In other words, we now solve the equations in a frame of reference that travels with the light pulse such that a given value of τ describes a certain 'slice' of the pulse, as illustrated in Fig. 5.9. The partial derivatives in the new coordinates are related to those in the old coordinates by:

$$\left(\frac{\partial}{\partial z}\right)_\tau = \frac{1}{c}\left(\frac{\partial}{\partial t}\right)_z + \left(\frac{\partial}{\partial z}\right)_t \tag{5.41}$$

$$\left(\frac{\partial}{\partial \tau}\right)_z = \left(\frac{\partial}{\partial t}\right)_z. \tag{5.42}$$

It is straightforward to show that in the travelling-wave coordinate system the growth equation for the intensity of the laser pulse becomes:

$$\left(\frac{\partial I}{\partial z}\right)_\tau = N^* \sigma_{21} I(z, \tau), \tag{5.43}$$

and the rate equations are unchanged apart from the substitution $t \to \tau$. Equations (5.37), (5.38), and (5.43) fully describe the problem, but in general require numerical solution. In order to gain further insight, we now make further simplifying assumptions that will allow us to reach an algebraic solution:

- We ignore pumping of the upper and lower laser levels during the passage of the laser pulse through each slice of material. As such we will consider

the case when a population inversion on the laser transition has been produced by the pumping before the arrival of the laser pulse, and that further pumping during the passage of the laser pulse is negligible. Hence, we set $R_1(t) = R_2(t) = 0$ during the times of interest.
- We assume that the rate of spontaneous transitions from the upper laser level is negligible compared to the rate of stimulated emission, and so set $N_2(t)/\tau_2 = 0$.
- We then consider two idealized cases:[8]
    - Rapid decay of the lower level: Here, the lower laser level is assumed to decay so quickly that even when there is rapid stimulated emission to the lower level $N_1(t) \approx 0$. In this case $N^*(t) = N_2(t)$.
    - Slow decay of the lower level: In this case the rate of spontaneous decay of the lower laser level is negligible so that stimulated emission causes the population in the lower laser level to build up, and the total population in the upper and lower levels is constant, i.e. $\partial(N_1(t) + N_2(t))/\partial t = 0$. This situation is known as **bottlenecking**.

[8]The approach employed in this section follows the adaptation by Siegman (Siegman, 1986) of an analysis originally performed by (Frantz and Nodvik (1963)).

With these assumptions the rate equations and beam growth equation reduce to,

$$\left(\frac{\partial N^*}{\partial \tau}\right)_z = -\frac{\beta \sigma_{21}}{\hbar \omega_L} N^*(z,\tau) I(z,\tau) \tag{5.44}$$

$$\left(\frac{\partial I}{\partial z}\right)_\tau = \sigma_{21} N^*(z,\tau) I(z,\tau), \tag{5.45}$$

where,

$$\beta = \begin{cases} 1 & \textbf{Rapid decay of lower level} \\ 1 + \frac{g_2}{g_1}. & \textbf{Bottlenecking of lower level} \end{cases} \tag{5.46}$$

For a fixed slice of the pulse (i.e. constant τ) we can integrate eqn (5.45) over length, to give,

$$I_{\text{out}}(\tau) = G(\tau) I_{\text{in}}(\tau), \tag{5.47}$$

where the single-pass gain is given by,

$$G(\tau) = \exp\left(\int_0^\ell N^* \sigma_{21} \mathrm{d}z\right). \tag{5.48}$$

Equation (5.44) may be solved by substituting eqn (5.45) into its right-hand side and integrating over z to give:[9]

[9]As explored in Exercise 5.12.

$$\Gamma_{\text{out}}(\tau) = \int_{-\infty}^{\tau} I_{\text{out}}(\tau) \mathrm{d}\tau = \Gamma_s \ln\left[\frac{G_0 - 1}{G(\tau) - 1}\right] \tag{5.49}$$

$$\Gamma_s = \frac{\hbar \omega_L}{\beta \sigma_{21}} \tag{5.50}$$

$$G(\tau) = \frac{G_0}{G_0 - [G_0 - 1] \exp\left[-\frac{\Gamma_{\text{in}}(\tau)}{\Gamma_s}\right]}, \tag{5.51}$$

Fig. 5.10: Frantz–Nodvik analysis of the amplification of laser pulses with a Gaussian temporal profile for various values of $G_0 \Gamma_{\rm in}(\infty)/\Gamma_{\rm s}$. For small values of $G_0 \Gamma_{\rm in}(\infty)/\Gamma_{\rm s}$ the pulse is amplified with little temporal distortion. However, for $G_0 \Gamma_{\rm in}(\infty)/\Gamma_{\rm s} \geq 1$ significant distortion occurs, especially towards the back of the laser pulse (larger values of τ) that sees a reduced gain following saturation by the front of the laser pulse. Note that for these plots $\Gamma_{\rm s}$ was varied with G_0 and $\Gamma_{\rm in}(\infty)$ fixed so that in the absence of saturation all the curves would lie on top of each other.

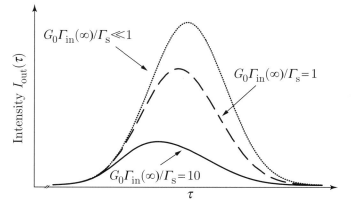

[10] This fact is a consequence of us considering the extreme cases of rapid decay of or strong bottlenecking in the lower laser level. In the more general case the amplification will depend on the pulse shape since for finite rates of spontaneous decay from the upper and lower laser levels the populations in these levels will depend, to some extent, on the shape of the incident pulse.

[11] See Exercise 5.11.

where G_0 is the initial single-pass gain, i.e. before the passage of the pulse through the medium. The amplification is described in terms of the **fluence** Γ (units of $\mathrm{J\,m^{-2}}$) of the pulses entering and leaving the amplifier, where the fluence is equal to the integrated intensity over time. Note that the gain $G(\tau)$ experienced by each slice of the pulse depends only on the fluence $\Gamma_{\rm in}(\tau)$ of the incident pulse, and not on its temporal shape.[10] The saturation parameter is now a **saturation fluence**, $\Gamma_{\rm s}$, which is related closely to the saturation intensity we derived earlier. The factor of $1/\beta$ appearing in the expression for the saturation fluence highlights the different behaviour of transitions in which the lower level decays rapidly and those that experience strong bottlenecking. If the lower level decays rapidly, each stimulated emission event reduces the total population inversion by unity; in contrast, for a bottlenecked lower level the decrease in population inversion for each stimulated emission event is equal to $1 + g_2/g_1 = \beta$, and as a consequence the transition saturates more easily.[11]

Equation (5.51) shows that the gain experienced by each slice of the pulse decreases with τ, that is from the front to the back of the pulse. This behaviour simply reflects the burning down of the population inversion by the earlier parts of the pulse. The extent to which the burning down of the population affects the gain is determined by the ratio of the input fluence to the saturation fluence. For an input fluence small compared to $\Gamma_{\rm s}$ the gain remains close to G_0 throughout the pulse, and hence the output pulse will have essentially the same temporal profile as the input pulse, but will be more intense. In contrast, for pulses with an input fluence that is comparable to, or greater, than $\Gamma_{\rm s}$ the gain experienced by the pulse will change during the pulse, leading to distortion. Figure 5.10 shows the input and output pulses for three example cases.

The total energy that can be extracted from the amplifier can be investigated by substituting eqn (5.51) into eqn (5.49) to give,

$$\Gamma_{\rm out}(\tau) = \Gamma_{\rm s} \ln\left(G_0 \left\{ \exp\left[\frac{\Gamma_{\rm in}(\tau)}{\Gamma_{\rm s}}\right] - 1 \right\} + 1 \right). \tag{5.52}$$

It is now clear that the greatest possible output fluence is achieved when $\Gamma_{\text{in}}(\tau)/\Gamma_s \to \infty$. In this limit the output fluence tends to a maximum given by,

$$\Gamma_{\text{out}}^{\max}(\tau) = \Gamma_{\text{in}}(\tau) + \Gamma_s \ln G_0. \tag{5.53}$$

The maximum fluence that can be extracted is then,

$$\Gamma_{\text{ext}}^{\max}(\infty) = \Gamma_{\text{out}}^{\max}(\infty) - \Gamma_{\text{in}}(\infty) = \Gamma_s \ln G_0 = \frac{N_0^*}{\beta} \times \hbar\omega_{\text{L}}, \tag{5.54}$$

where N_0^* is the initial population inversion. This should make sense! For a laser in which the lower level decays rapidly the maximum possible energy that can be extracted corresponds to extracting one laser photon for each inverted atom; for a laser in which the lower level exhibits severe bottlenecking the energy that can be extracted is reduced by a factor of β.[12]

[12] See Exercise 5.11.

5.3 Design of laser amplifiers

From the sections above it should be clear that saturation plays a key role in the amplification of light by stimulated emission (it is also of fundamental importance in the operation of laser oscillators, as we shall see later). The design of real amplifier systems can be involved. However, the ideas we have developed allow us to make some general observations.

For steady-state (pulsed) amplifiers it is desirable that the saturation intensity (fluence) of the laser transition is as high as possible. If we want the amplifier to operate in a linear regime—i.e. for a steady-state amplifier the output intensity is proportional to the input intensity, and for a pulsed amplifier the output pulse is not temporally distorted—the radiation leaving the amplifier must be below the saturation intensity (fluence). Consequently, for a high output we require a high saturation intensity (fluence).

However, it is also the case that the saturation intensity (fluence) should be large if we don't care about distortion and we want to extract as much power (energy) as possible from the amplifier. The reasoning is as follows: The small-signal gain over the length of the amplifier must usually be kept below a certain value to avoid losses by parasitic laser oscillation or strong amplification of spontaneous emission on the laser transition. For a fixed single-pass gain, reducing the optical gain cross-section increases the saturation intensity (fluence) and the population inversion density. Thus, the energy stored in the population inversion increases, allowing more power (energy) to be extracted, as confirmed by eqns (5.21) and (5.54).

Finally, we note that it is often desirable to operate steady-state amplifiers under conditions of heavy saturation because in this regime the intensity of the amplified beam is almost independent of the intensity of the input beam, and hence the relative fluctuations in the intensity of the amplified beam will be very much smaller than those of the input. In contrast, if the amplifier operates well below saturation the fluctuations in the intensity of the input beam will be strongly amplified. Similar considerations apply to pulsed amplifiers, but with the complication that operating under conditions of heavy saturation will lead to (potentially severe) distortion of the temporal profile of the laser pulse. Whether or not this is an issue will depend on the application.

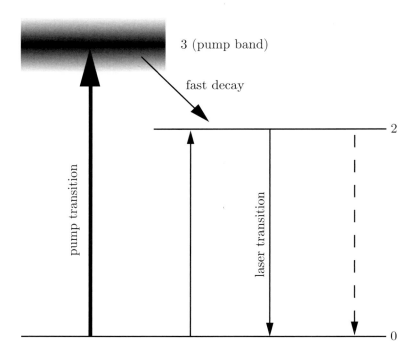

Fig. 5.11: Parameters used in Exercise 5.2 to calculate the saturation intensity of an ideal three-level laser system.

Exercises

(5.1) **General ideas**

(a) Derive a simplified expression for the saturation intensity by defining it to be the intensity that produces a rate of stimulated emission from the upper laser level equal to the rate of spontaneous decay (characterized by a fluorescence lifetime τ_2) from that level. Comment on the conditions under which this approximation expression agrees with eqn (5.7).

(b) Consider the expression for the recovery time given in eqn (5.8). Show that if the laser levels are able to sustain steady-state gain,

$$\tau_R > \frac{g_2}{g_1}\tau_1.$$

[Hint: See Section 2.2.2.]

(5.2) **Saturation in three-level laser systems**[13]

Here, we consider the modifications required to the theory presented in Section 5.1.1 when the laser transition is a three-level system (as discussed in Section 7.1.5). In an ideal three-level laser, pumping from the ground state, 1, transfers population to level 3. This level decays very rapidly to the upper laser level 2. If the pumping is sufficiently strong, lasing can occur from level 2 to the ground state; these processes are illustrated schematically in Fig. 5.11. Now, the derivation of the saturation intensity given in Section 5.1.1 assumed that the pump rates of the upper and lower laser levels were independent of the populations in those levels, which is clearly not true in this case. We will assume that the $3 \to 2$ transition is very rapid such that $N_3 \approx 0$. Further, we assume that the upper laser level decays pre-dominantly on the laser transition (i.e. $A_{21} \approx 1/\tau_2$) so that significant population exists only in the two laser levels. In this case we may

---

[13] As discussed in Section 7.1.5, extremely high pump powers are required to achieve steady-state laser oscillation in a three-level laser system and consequently continuous-wave lasing in such systems is a curiosity rather than being practically important.

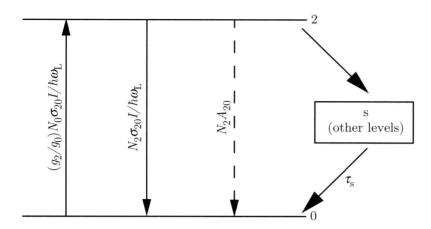

Fig. 5.12: Parameters used in Exercise 5.3 to calculate the saturation intensity for absorption.

write $N_1 + N_2 = N_T$, where N_T is the total density of atoms or ions of the laser species.

The rate of pumping of level 3, and hence of level 2, is proportional to the density of ions in the ground state and so we may write $R_2 = R_3 = N_1/\tau_p$, where τ_p is a parameter describing the rate of pumping.

(a) Show that within these assumptions the rate equations of this system may be written in the form,

$$\frac{dN_2}{dt} = \frac{N_1}{\tau_p} - N^*\sigma_{21}\frac{I}{\hbar\omega_L} - \frac{N_2}{\tau_2}$$

$$\frac{dN_1}{dt} = -\frac{dN_2}{dt},$$

where the remaining parameters are as defined in Section 5.1.1.

(b) Hence, show that in the steady state the population inversion density may be written in the form

$$N^*(I) = \frac{N^*(0)}{1 + I/I_s},$$

where,

$$N^*(0) = \frac{\tau_2/\tau_p - g_2/g_1}{\tau_2/\tau_p + 1} N_T,$$

and

$$I_s = \frac{\hbar\omega_L}{\sigma_{21}\tau_R},$$

and where the recovery time is given by

$$\frac{1}{\tau_R} = \frac{1}{\beta}\left(\frac{1}{\tau_p} + \frac{1}{\tau_2}\right),$$

in which $\beta = 1 + g_2/g_1$.

(c) Show that for the population inversion to be positive

$$\frac{1}{g_2}\frac{1}{\tau_p} > \frac{1}{g_1}\frac{1}{\tau_2},$$

i.e. the pump rate *per state* of the upper laser level must be greater than the rate per state at which the lower laser level is populated by fluorescence from the upper laser level.

(d) Hence, argue that for a good three-level laser system the recovery time, and therefore the saturation intensity, is determined primarily by the pump rate τ_p^{-1}.

(e) Comment briefly on how the pump rate $1/\tau_p$ will depend on the intensity of the pump radiation.

(5.3) **Saturation of absorption**

In this problem we consider another case where the theory of Section 5.1.1 must be modified: that of saturation of optical absorption. Consider an experiment to measure the absorption from the ground state 0 to an upper level 2, as illustrated schematically in Fig. 5.12. In an experiment of this type there will not normally be any pumping of the levels other than that which results from the probe beam. Absorption of radiation on the transition of interest, $2 \leftarrow 0$, transfers population to the upper level. This level has a fluorescence lifetime τ_2 against decay (either radiative or non-radiative) to all lower levels (including the ground state). Rather than worry about the distribution of population over these levels, we group all the levels between 2 and the ground state into a block of levels denoted by s. The population density of atoms in this group of levels is taken to be N_s, and we suppose that in the steady-state transitions from these levels returns population to the ground state at a rate N_s/τ_s. Population can only reside in levels 0, 2, or s and consequently we may write $N_0 + N_2 + N_s = N_T$, where N_T is the total density of atoms or ions of the laser species.

(a) Show that the rate equations of this system take the form:

$$\frac{dN_2}{dt} = -N^* \frac{\sigma_{20} I}{\hbar \omega_L} - \frac{N_2}{\tau_2}$$

$$\frac{dN_s}{dt} = N_2 \left(\frac{1}{\tau_2} - A_{20} \right) - \frac{N_s}{\tau_s}$$

$$\frac{dN_0}{dt} = +N^* \frac{\sigma_{20} I}{\hbar \omega_L} + N_2 A_{20} + \frac{N_s}{\tau_s},$$

where A_{20} is the Einstein A coefficient for spontaneous emission on the $2 \to 0$ transition, and $N^* = N_2 - (g_2/g_0) N_0$.

(b) Show that in the steady-state

$$\frac{N_s}{\tau_s} = N_2 \left(\frac{1}{\tau_2} - A_{20} \right).$$

(c) Hence, show that in the steady-state,

$$N_2 = -N^* \sigma_{20} \frac{I}{\hbar \omega_L} \tau_2$$

$$N_0 = N_T + N^* \sigma_{20} \frac{I}{\hbar \omega_L} \tau_2 \left(1 + \frac{\tau_s}{\tau_2} - A_{20} \tau_s \right).$$

(d) Hence, show that the 'population inversion' $N^* = N_2 - (g_2/g_0) N_0$ produced by the absorption may be written in the form,

$$N^*(I) = -\frac{(g_2/g_0) N_T}{1 + I/I_s},$$

where,

$$I_s = \frac{\hbar \omega_L}{\sigma_{20} \tau_R},$$

and the recovery time for the absorption is given by,

$$\tau_R = \tau_2 \left(1 + \frac{g_2}{g_0} \right) + \frac{g_2}{g_0} \tau_s (1 - A_{20} \tau_2).$$

Comment on the value of $N^*(I)$ in the limit $I \to 0$.

(e) Show that the absorption coefficient $\kappa_I = -N^*(I) \sigma_{20}$ may be written as,

$$\kappa_I = \frac{\kappa_0}{1 + I/I_s},$$

where κ_0 is the unsaturated absorption coefficient. Write down expressions for κ_0 in terms of the total population density N_T and: (i) the optical gain cross-section σ_{20}; (ii) the optical absorption cross-section σ_{02} (see Section 4.5).

(f) Comment briefly on the factors that determine the recovery time of the absorption.[14]

(5.4) **A saturated steady-state amplifier**
A saturated amplifier: A steady-state laser amplifier operates on the homogeneously broadened transition between two levels of equal degeneracy. Population is pumped exclusively into the upper level at a rate of 1.0×10^{18} s^{-1} cm^{-3}. The lifetimes of the upper and lower levels are 5 ns and 0.1 ns, respectively. A collimated beam of radiation enters the 2 m long amplifier with an initial intensity $I(0)$. The gain cross-section of the medium is 4×10^{-12} cm^2 at the 400 nm wavelength of the monochromatic beam. Calculate the intensity of the beam at the exit of the amplifier when:

(a) $I(0) = 0.1$ W cm^{-2};

(b) $I(0) = 500$ W cm^{-2};

(c) $I(0) = 50$ W cm^{-2}. [Hint: You will need to use a numerical method to find a solution.]

(5.5) **More steady-state amplifiers**
The two amplifiers considered in the problem both operate on homogeneously broadened transitions and under steady-state conditions.

(a) When a signal of intensity 3.0 kW m^{-2} is applied at the input of an amplifier of length 10 m, the output signal intensity is 36 kW m^{-2}. When the input signal is reduced to 1.0 kW m^{-2} the output is 20 kW m^{-2}. Calculate the value of the saturation intensity of the amplifier and the small-signal gain coefficient.

---

[14] The recovery time is a measure of how long the absorption takes to return to its unsaturated value after the (intense) probe radiation is switched off. As such it plays an important role in distinguishing between the 'slow' and 'fast' saturable absorbers that are used in different types of modelocking (see Section 8.3.4).

(b) Another laser amplifier gives an output of $5I_0$ when it receives input intensity I_0 tuned to ω_0, the centre frequency of the laser transition. When the input radiation is tuned to a nearby angular frequency ω_1, an input intensity of $3I_0$ is amplified to an output intensity of $7I_0$. What is the ratio of the gain cross-sections at the two frequencies?

(5.6) A useful result

First, we prove a result that will be useful for several of the problems below.

(a) Show that for a homogeneously broadened transition with a Lorentzian lineshape of full width at half-maximum of $\Delta\omega_H$,

$$\frac{\sigma_{21}(\omega - \omega_0)}{1 + I/I_s(\omega - \omega_0)}$$
$$= \frac{1}{\sqrt{1 + I/I_s(0)}} \frac{\hbar\omega}{c} B_{21} g_s(\omega - \omega_0),$$

where $g_s(\omega - \omega_0)$ is a normalized Lorentzian with a full width at half-maximum of

$$\Delta\omega_s = \Delta\omega_H \sqrt{1 + I/I_s(0)}.$$

(5.7) Frequency dependence of saturated gain coefficient for homogeneous broadening with a Lorentzian lineshape

(a) Use the result of Exercise 5.6 to show that for the special case of homogeneous broadening with a Lorentzian lineshape the gain coefficient measured by an intense, narrowband probe beam of constant intensity I is given by,

$$\alpha_I(\omega - \omega_0)$$
$$= \frac{N^*(0)}{\sqrt{1 + I/I_s(0)}} \frac{\hbar\omega}{c} B_{21} g_s(\omega - \omega_0). \quad (5.55)$$

(b) Comment on the form of this result.

(c) Confirm that this last result is consistent with the gain coefficient being given by,

$$\alpha_I(0) = \frac{\alpha_0(0)}{1 + I/I_s(0)},$$

when the probe is tuned to the line centre.

(d) Show that for large detuning from the line centre eqn (5.55) yields $\alpha_I(\omega - \omega_0) \approx \alpha_0(\omega - \omega_0)$.

(e) On the same graph, sketch $\alpha_I(\omega - \omega_0)$ for the two cases $I \ll I_s(0)$ and $I = 3I_s(0)$.

(5.8) Extreme saturation of a steady-state, homogeneously broadened amplifier

In this problem we show that under conditions of extreme saturation the amplified beam extracts energy from the inverted medium at the maximum possible rate.

(a) Use eqn (5.6) to show that in the steady state the net rate R_s of stimulated emission is given by,

$$R_s = N^*(I)\sigma_{21}\frac{I}{\hbar\omega_L} = \frac{\frac{I}{I_s}}{1 + \frac{I}{I_s}} \frac{N^*(0)}{\tau_R}.$$

(b) Sketch graphs of R_s and $N^*(I)$ as a function of I/I_s and comment on their form.

(c) Hence, show that the maximum possible rate per unit volume at which energy can be transferred to the amplified beam is

$$W_{\max} = \frac{N^*(0)}{\tau_R} \hbar\omega_L = \alpha_0 I_s(0).$$

What is the population inversion in this limit?

(d) Hence show that when the beam extracts energy from the gain medium at the maximum possible rate, the intensity of the beam will grow according to,

$$\frac{dI}{dz} = \alpha_0 I_s(0),$$

and compare this to eqn (5.21).

(5.9) †Spectral hole-burning in an inhomogeneously broadened laser amplifier

Here, we investigate the shape of the spectral hole that is burnt into the spectral profile of the gain of an inhomogeneously broadened laser amplifier. Each frequency class of the gain medium is assumed to be homogeneously broadened with a full width at half-maximum of $\Delta\omega_H$.

(a) Use eqn (5.24) and the result of Exercise 5.6 to show that the *change* $\Delta\alpha_I^D(\omega)$ in the gain coefficient measured by a weak probe beam when an intense, narrow-bandwidth beam at frequencies close to ω_L and of total intensity I is applied is given by,

$$\Delta\alpha_I^D(\omega - \omega_0) = -\frac{\pi}{2}\frac{\Delta\omega_H^2}{\Delta\omega_s}\frac{I}{I_s(0)}N^*(0)\frac{\hbar\omega}{c}B_{21}$$
$$\times \int_0^\infty g_D(\omega_c - \omega_0) g_L(\omega - \omega_c) g_s(\omega_L - \omega_c) d\omega_c,$$

where $g_L(\omega - \omega_0)$ and $g_s(\omega - \omega_0)$ are normalized Lorentzian functions with full width at half-maxima equal to $\Delta\omega_H$ and $\Delta\omega_s = \Delta\omega_H\sqrt{1 + I/I_s(0)}$, respectively.

(b) Show that, in the strongly inhomogeneous limit ($\Delta\omega_D \gg \Delta\omega_H$), this result may be written as

$$\Delta\alpha_I^D(\omega - \omega_0) = -\frac{\pi}{2}\frac{\Delta\omega_H^2}{\Delta\omega_s}\frac{I}{I_s(0)}\alpha_I^D(\omega - \omega_0)$$

$$\times \int_0^\infty g_L(\omega - \omega_c)g_s(\omega_L - \omega_c)d\omega_c,$$

where $\alpha_0^D(\omega - \omega_0)$ is the gain coefficient that would be measured in the absence of the saturating beam.

(c) Using the result that the convolution of two normalized Lorentzian functions centred at $\omega = 0$ and $\omega = \omega_0$ is equal to a normalized Lorentzian function centred at $\omega = \omega_0$ with a full width equal to the sum of the widths of the two functions, show that

$$\Delta\alpha_I^D(\omega - \omega_0) = -\alpha_0^D(\omega - \omega_0)\frac{\pi}{2}\Delta\omega_H$$

$$\times \frac{I/I_s(0)}{\sqrt{1 + I/I_s(0)}}$$

$$\times \frac{1}{\pi}\frac{\Delta\omega_{\text{hole}}/2}{(\omega - \omega_L)^2 + (\Delta\omega_{\text{hole}}/2)^2},$$

where

$$\Delta\omega_{\text{hole}} = \Delta\omega_H\left(1 + \sqrt{1 + I/I_s(0)}\right).$$

(d) Show that if the saturating beam has a low intensity ($I \ll I_s(0)$), $\Delta\omega_{\text{hole}} \approx 2\Delta\omega_H$, and that the depth of the hole is proportional to I.

(e) Discuss how the width and depth of the spectral hole varies as the intensity of the saturating beam is increased beyond this level.

(5.10) Use the result proved in Exercise 5.6 to show that if the homogeneous lineshape is Lorentzian, eqn (5.31) reduces to,

$$\alpha_I^D(\omega - \omega_0) = \frac{\alpha_0^D(\omega - \omega_0)}{\sqrt{1 + I/I_s(0)}}.$$

(5.11) Bottlenecking

(a) Suppose stimulated emission of Δn photons per unit volume occurs on the laser transition $2 \to 1$. Find the corresponding change in the population inversion density, ΔN^*, for the cases:
(i) rapid decay of the lower laser level;
(ii) strong bottlenecking of the lower laser level.

(b) Hence, show that

$$\Delta n = -\frac{\Delta N^*}{\beta},$$

where β is defined in eqn (5.46).

(5.12) †Frantz–Nodvik analysis of pulse amplification

(a) Derive eqns (5.41) and (5.42), and hence show that in travelling-wave coordinates the growth equation for the intensity of the pulse is given by eqn (5.43).

(b) Show that for a transition that exhibits strong bottlenecking, and if all spontaneous decay processes can be ignored,

$$\frac{\partial N^*(z,t)}{\partial t} = \beta\frac{\partial N_2(z,t)}{\partial t},$$

and hence derive eqn (5.44).

(c) Integrate eqn (5.45) over z to derive eqns (5.47) and (5.48).

(d) By substituting eqn (5.45) into eqn (5.44), and integrating over z, show that

$$\frac{d}{d\tau}\ln[G(\tau)] = \frac{1}{\Gamma_s}\left[I_{\text{out}}(\tau) - I_{\text{in}}(\tau)\right].$$

(e) By eliminating $I_{\text{in}}(\tau)$, show that the output fluence is given by eqn (5.49).

(f) Use a similar approach to show that the input fluence is related to the gain by,

$$\Gamma^{\text{in}}(\tau) = \Gamma_s \ln\left[\frac{1 - 1/G_0}{1 - 1/G(\tau)}\right],$$

and hence derive eqn (5.51), which gives the gain as a function of the input fluence of the laser pulse. This completes the missing steps in the Frantz–Nodvik analysis.

(g) Show that if the pulse is not to be distorted we require $[G_0 - 1]\Gamma_{\text{in}}(\infty) \ll \Gamma_s$.

(h) Use these results to sketch the output pulse shapes for a rectangular input pulse such that
(i) $[G_0 - 1]\Gamma_{\text{in}}(\infty) \ll \Gamma_s$
(ii) $[G_0 - 1]\Gamma_{\text{in}}(\infty) \approx \Gamma_s$.

The laser oscillator

6.1 Introduction

In the previous chapter we described how stimulated emission could be used to amplify a beam of radiation with frequencies lying within the linewidth of an inverted transition. *Laser amplifiers* of this type are used to increase the power of an input beam. That beam is itself usually[1] derived from a *laser oscillator* which—at its simplest—comprises a gain medium located within an optical cavity. In this chapter we describe the key features of laser oscillators.

6.2 Amplified spontaneous emission (ASE) lasers

To illustrate the basic features of a laser oscillator, we first discuss what happens in a gain medium that is *not* located within an optical cavity. Let us imagine trying to generate an intense beam of radiation by pumping a long cylindrical rod of material to produce a population inversion on some transition of interest, as illustrated in Fig. 6.1. Atoms in the upper laser level can emit radiation in all directions with a spectrum determined by the linewidth of the transition. Most of this light will be emitted at large angles to the axis of the cylinder and so will leave the rod after propagating only a short distance through the inverted medium. As such, this radiation will experience very little amplification, will be of low intensity, and will have a spectrum proportional to the lineshape of the transition.

Consider now radiation emitted from near the end of the rod at $z = 0$. A very small fraction of this—approximately equal to $\Omega/4\pi$, where Ω is the solid angle subtended by the other end of the cylinder—will remain within the cylinder as it propagates along its entire length. This

6.1 Introduction	83
6.2 Amplified spontaneous emission (ASE) lasers	83
6.3 Optical cavities	85
6.4 Beam quality[†]	103
6.5 The approach to laser oscillation	106
6.6 Laser oscillation above threshold	111
6.7 Output power	117
Further reading	123
Exercises	123

[1] Although see Chapter 18 and note 4 in this chapter.

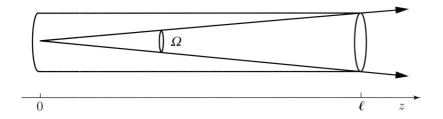

Fig. 6.1: A cylindrical rod of material in which a population inversion exists on some transition. The cylinder is taken to be of length ℓ, and the solid angle subtended at one end by the other end is Ω. Radiation travelling at small angles to the axis of the rod will be amplified to form a beam of amplified spontaneous emission within a solid angle of approximately Ω.

radiation will be amplified as it propagates, and is therefore known as **amplified spontaneous emission** (ASE). Of course, spontaneous radiation emitted *anywhere* within the rod and happening to propagate close to the axis will also be amplified, but the intensity of radiation emitted from one end of the cylinder will be dominated by the contribution of ASE from atoms near the other end that has therefore been amplified over a longer distance.

Provided there is sufficient gain, radiation emitted from the rod will therefore be highly anisotropic: radiation emitted from the sides of the rod will be pre-dominantly from spontaneous emission, and hence be rather weak; radiation emitted from the ends will be dominated by stimulated emission, and form a more or less well-defined conical beam with a solid angle of approximately Ω. A similar beam of ASE will leave the other end of the rod.[2]

This configuration can, provided the optical gain is sufficiently high, generate intense beams of radiation from each end of the rod. Ignoring saturation, radiation emitted spontaneously from a point z along the rod, travelling towards positive z and remaining within the rod, will be amplified by a factor of $\exp[\alpha(\omega - \omega_0)(\ell - z)]$. Since the small-signal gain coefficient, $\alpha(\omega - \omega_0)$ will be sharply peaked at a frequency ω_0, the spectrum of the radiation will exhibit gain narrowing as discussed in Sections 4.1.4 and 18.3.3.

The amplified spontaneous emission from a long, narrow region of optical gain produces a laser-like beam of radiation that can be intense, of relatively narrow divergence, and with a spectral width significantly narrower than the linewidth of the transition. By extending the length of the rod the output will increase in intensity and the divergence of the beam will decrease.[3] While this beam clearly exhibits **l**ight **a**mplification by the **s**timulated **e**mission of **r**adiation, it would not usually be regarded as laser radiation. Instead, the device is known as a **mirror-less laser** or an **ASE laser**.[4] True laser radiation is emitted by a laser *oscillator*, in which the gain medium is enclosed in an optical cavity. The use of an optical cavity:

1. Allows stimulated emission to build up by repeated reflections of the beam through the gain region. Hence, unlike the system discussed above, lasing can occur even if the single-pass gain is small.
2. The cavity restricts oscillation to one or more **longitudinal modes** that have frequencies determined by the properties of the optical cavity. The spectral width of the oscillating modes can be very much narrower than the linewidth of the transition.
3. The cavity also restricts oscillation to one or more **transverse modes** that determine the transverse spatial profile and divergence of the emitted beam. The transverse profile of the beam will usually be significantly smoother, and the divergence lower, than in the absence of the cavity.

---

[2] For transient pumping, this 'backward-going' beam of ASE can be greatly reduced in intensity by implementing **travelling-wave pumping**, as discussed in Section 18.3.1. In this configuration the pumping is pulsed, starts at (say) $z = 0$, and propagates along the rod towards positive z at the speed of light. As a consequence, radiation propagating towards negative z experiences greatly reduced gain, leading to intense output from one end of the gain medium only.

[3] The behaviour of the width of the spectrum as the length of the gain region is increased is a complex matter, and depends on whether the medium is homogeneously or inhomogeneously broadened. This point is discussed further in Chapter 18.

[4] ASE lasers are particularly important in the extreme ultraviolet and soft X-ray regions of the spectrum, as discussed in Chapter 18.

6.3 Optical cavities

6.3.1 General considerations

Closed cuboidal cavity

In order to develop some general ideas on laser cavities, we first return to the simple cuboidal cavity considered in Section 1.2. For a given angular frequency ω, the electic-field distribution of the allowed modes of such a cavity is given by:

$$\begin{aligned}
\mathbf{E}(\mathbf{r}, t) = {} & E_{0x}\mathbf{i} \cos\left(\frac{\pi}{L_x}lx\right) \sin\left(\frac{\pi}{L_y}my\right) \sin\left(\frac{\pi}{L_z}pz\right) \\
& + E_{0y}\mathbf{j} \sin\left(\frac{\pi}{L_x}lx\right) \cos\left(\frac{\pi}{L_y}my\right) \sin\left(\frac{\pi}{L_z}pz\right) \\
& + E_{0z}\mathbf{k} \sin\left(\frac{\pi}{L_x}lx\right) \sin\left(\frac{\pi}{L_y}my\right) \cos\left(\frac{\pi}{L_z}pz\right),
\end{aligned}$$

where the integers (l, m, p) define the mode. For a given frequency, the mode only can exist if the dispersion relation (1.10) of the cavity is obeyed:

$$k_x^2 + k_y^2 + k_z^2 = \left(\frac{\pi}{L_x}l\right)^2 + \left(\frac{\pi}{L_y}m\right)^2 + \left(\frac{\pi}{L_z}p\right)^2 = \frac{\omega_{lmp}^2}{c^2}. \tag{6.1}$$

This last result can be more conveniently written in terms of the wavelength of the mode λ_{lmp}, given by $\omega_{lmp} = 2\pi c/\lambda_{lmp}$:

$$\left(\frac{\lambda_{lmp}}{2L_x}l\right)^2 + \left(\frac{\lambda_{lmp}}{2L_y}m\right)^2 + \left(\frac{\lambda_{lmp}}{2L_x}p\right)^2 = 1, \tag{6.2}$$

from which we see that if the dimensions of the cavity are not large compared to the wavelength, only a handful of modes are allowed. For example, if $L_x < \lambda_{lmp}/2$ and $L_y, L_z = \lambda_{lmp}/\sqrt{2}$, only the $(0, 1, 1)$ mode of the cavity obeys eqn (6.2) and hence can oscillate.

Spontaneous and stimulated emission from atoms within the cavity must be into one of the cavity modes. Laser oscillation can only occur for those modes for which the energy losses of the cavity—arising, for example, from the finite reflectivity from the cavity walls—are compensated for by laser gain. This requirement imposes a **threshold condition**, as discussed in Section 6.5.2.

Applying these ideas to lasers operating in the visible range of the spectrum, we see immediately that in order to restrict the radiation field to a few modes, a closed cavity of this type would have to be only a few optical wavelengths across. Such cavities would have several disadvantages: the volume of laser material enclosed would be so small that it would be difficult to reach the threshold for laser oscillation; even if threshold were reached, the power that could be usefully extracted in the form of laser radiation would be tiny; they would be difficult to manufacture.

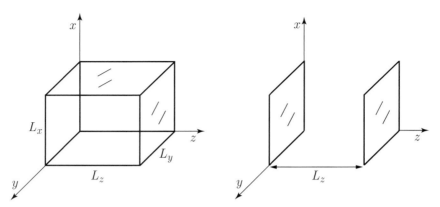

Fig. 6.2: Two types of optical cavity: (a) a closed cuboidal cavity; (b) an open cavity formed by removing all the walls of cavity (a) except from those at $z = 0$ and $z = L_z$.

A way forward was pointed out in a pioneering paper by Schawlow and Townes (1958). In that they considered an 'optical maser' operating in an *open* cavity of the type shown in Fig. 6.2(b). For the moment we defer a more detailed discussion of the modes of open cavities. Here, we outline Schawlow and Townes's arguments that structures of this type have modes that are of sufficiently low loss that oscillation can occur.

Open cuboidal cavity

We first note that the $\sin(k_x x)$ and $\cos(k_x x)$ terms in eqn (6.1) can be written in the form $[\exp(ik_x x) - \exp(-ik_x x)]/2i$ and $[\exp(ik_x x) + \exp(-ik_x x)]/2$, with analogous expressions for the y- and z- components. Consequently, the solution can be written as the sum of plane, travelling waves of the form $\exp(i[\mathbf{k} \cdot \mathbf{r} - \omega t])$, where the wave vector can take one of eight possibilities: $\mathbf{k} = \pm k_x \mathbf{i} \pm k_y \mathbf{j} \pm k_z \mathbf{k}$.

Let us now consider the cavities shown in Fig. 6.2: (a) which is long in the z-direction, but closed; (b) which is identical to (a), but with all the walls removed apart from those at $z = 0$ and $z = L_z$. The modes of the closed cavity are given by eqn (6.1). The modes of the open cavity will differ from those of the closed cavity in one important respect: its modes will be *lossy*, even if the walls are perfectly reflecting, since radiation can escape through its open sides. As such, for an open cavity any electromagnetic field will decay with time. The key questions for laser cavities are: *how* lossy are the modes and how quickly does the radiation field decay?

Schawlow and Townes argued that the open cavity would have *low-loss* modes that are similar to those of the closed cavity, but with the restriction that $k_x, k_y \ll k_z$ since such modes constitute plane waves travelling at small angles to the z-axis. The allowed frequencies of these low-loss modes are given by the dispersion relation of eqn (6.1) with the restriction that $k_x, k_y \ll k_z$. We find,[5]

[5] See Exercise 6.1.

$$\nu_{lmp} = \frac{c}{2L_c} p \left[1 + \frac{l^2 + m^2}{2p^2} \left(\frac{L_c}{a} \right)^2 + \ldots \right], \tag{6.3}$$

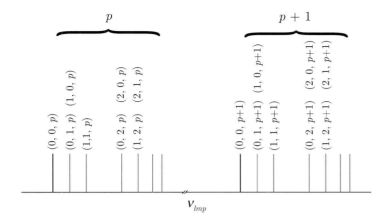

Fig. 6.3: Frequencies ν_{lmp} of the modes of the open cavity illustrated in Fig. 6.2(b), according to eqn (6.3). Note the break in scale between the set of transverse modes with mode number p and those with mode number $p+1$. Note also the degeneracy in frequency of those transverse modes with a common value of $l^2 + m^2$.

where $\nu_{lmp} = \omega_{lmp}/2\pi$ is the frequency of the mode, and we have taken $L_x = L_y = a$ and $L_z = L_c$; that is, the cavity is formed by a pair of parallel square reflectors of side a separated by a distance L_c.

We see that the frequencies of the modes depend on all three indices l, m, and p, as illustrated schematically in Fig. 6.3. However, it is clear that the mode index p plays a distinctly different role from that of the indices l and m, reflecting the fact that the cavity is much longer in the z-direction than in the x- and y-directions. For example, to zero order the frequency of the mode depends only on p:

$$\nu_{lmp} \approx \frac{c}{2L_c} p, \tag{6.4}$$

which can be rearranged in the form,

$$L_c = \left(\frac{\lambda_{lmp}}{2}\right) p. \tag{6.5}$$

In other words, in this approximation, the allowed wavelengths of the cavity are those for which an integer number of half-wavelengths fit into the length of the cavity. This result agrees with the picture that the low-loss modes of the open cavity comprise waves travelling at small angles to the z-axis: equal-amplitude waves travelling to positive and negative z will interfere to form a standing wave with a separation between nodes equal to $\lambda_{lmp}/2$. Equation (6.5) is also consistent with the boundary condition that the mode must have nodes at the two cavity mirrors.

As we shall see below, this difference in the roles of p and l, m goes further. It is found that the transverse spatial profile of the modes depends only on l and m. Consequently, modes with the same values of l and m are said to be in the same **transverse mode**. Similarly, the spatial distribution along the axis of the cavity depends only on p, so that modes with the same value of p are said to be in the same **longitudinal mode**. Of course, to describe a single mode we must specify all three indices, in which case there would be little point in artificially dividing the mode into separate longitudinal and transverse components. Very often, however, we are only concerned with either the transverse or longitudinal behaviour of the modes. For example, we may wish to restrict laser oscillation to the lowest transverse mode; that is we wish to ensure that only modes with $l = m = 0$ oscillate, but are not concerned which—or even how many—of the $(00p)$ modes do so. In this case we may control the transverse modes by using an aperture to modify the transverse profile of the laser radiation within the cavity. This has no effect on the longitudinal modes, and in general several of these, with different frequencies, can oscillate.

It is worth noting here that the frequency difference between two adjacent longitudinal modes of the plane–plane cavity of Fig. 6.2(b) is, from eqn (6.3):

$$\Delta \nu_{p,p-1} = \frac{c}{2L_\mathrm{c}}. \tag{6.6}$$

As shown in Exercise 6.1, for a given longitudinal mode, the difference in frequency between two transverse modes of the cavity of Fig. 6.2(b) is of order $(\lambda/a)^2$ of the resonant frequency. For example, for a laser operating in the visible part of the spectrum ($\lambda \approx 500$ nm), with a cavity 0.5 m long, the frequency separation of adjacent longitudinal modes is approximately 300 MHz, whilst the frequency separation of adjacent transverse modes is of order 100 Hz.

Modes in open cavities

The discussion above gives some insight into how an open cavity can have modes in which the radiation field comprises two components travelling in opposite directions along the axis of the cavity, and how the boundary conditions imposed by the cavity cause only certain frequencies to be allowed.

Before considering the detailed form of the cavity modes, we first provide a more general definition of a cavity mode. A mode of a cavity can be defined as a spatial distribution of the radiation field that is reproduced in one trip around the cavity. Consider the situation illustrated in Fig. 6.4 in which a wave is launched from some point P in an optical cavity. If the wave is in one of the cavity modes the amplitude *and phase* of the wave after propagating round the cavity once—i.e. reflecting once from each mirror and returning to P—must be unchanged. Hence, for a cavity mode, the total phase accumulated by the wave in propagating round the cavity once must satisfy:

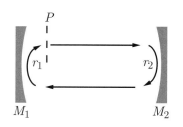

Fig. 6.4: General definition of a cavity mode.

$$\phi_\mathrm{rt} = 2\pi p \quad p = 0, 1, 2, 3, \ldots . \quad \textbf{Condition for cavity mode} \tag{6.7}$$

Here, ϕ_rt is the total phase accumulated by the wave in propagating once round the cavity; it will include contributions arising straightforwardly from propagation, plus any phase shifts imposed by the cavity mirrors and optical components (such as a gain medium) placed in the cavity.

We emphasize that if eqn (6.7) is satisfied the field components of the wave at each point in the cavity can oscillate with an amplitude that is independent of time. Of course, in a real cavity the amplitude of the launched wave will generally decrease steadily with each round trip owing to the finite coefficient of reflection of the cavity mirrors. In a laser oscillator this loss, plus any others, must be made up by optical gain in the laser medium. The point is that even if the cavity losses are countered by optical gain, if eqn (6.7) is not met it would not be possible to achieve a steady-state distribution of optical fields within the cavity since the phases of the waves at each point would change on each round trip.

Diffraction losses in open cavities

It should be clear that any 'beam-like' radiation field propagating along the axis of the cavity will diffract away from the axis. When this beam reaches the cavity mirrors, radiation lying outside the (finite) area of the mirror will be lost from the cavity, and only that part of the beam within the mirror will be returned along the axis. This loss, caused by the finite area of the cavity mirrors is known as **diffraction loss**.

It is useful to define two types of optical cavity. In the first, low-loss cavities, the cavity can support modes in which the radiation is confined close to the cavity axis and hence the amplitudes of the modes are small at the edges of the mirrors. As such, the diffraction losses are low, and the detailed transverse distribution of the modes is essentially independent of the lateral dimensions of the cavity mirrors. In contrast, the modes in high-loss cavities are not confined to be close to the axis, the diffraction losses are high, and the transverse distribution of the modes depends critically on the transverse dimensions of the cavity mirrors. These two types of cavity are discussed in more detail below.

6.3.2 Low-loss (or 'stable') optical cavities

Lowest-order mode

The diffraction losses of an open optical cavity can be made small by using concave mirrors so that reflection from the mirrors refocuses the radiation and so counteracts the diffraction occuring in the space between the mirrors.

Figure 6.5 shows schematically a generalized two-mirror cavity comprising concave mirrors with radii of curvature R_1 and R_2, separated by a distance L_c. The modes of such a cavity are found to be[6] a superposition of waves propagating to positive and negative z, each of the form of a **Gauss-Hermite beam**. We consider first the *lowest-order* of these solutions, which takes the form:

$$U = U_0 \exp(\mathrm{i}[kz - \omega t]) \frac{\exp(\mathrm{i}kr^2/2q)}{q}, \quad (6.8)$$

where the **complex radius** of the beam is given by,

$$q = (z - z_0) - \mathrm{i}z_R, \quad (6.9)$$

in which z_R is known as the **Rayleigh range** or **confocal parameter**.[7] As discussed in Exercise 6.2, this solution looks like that of a spherical wave in the paraxial approximation, but with a complex radius of curvature q. The

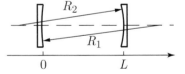

Fig. 6.5: A general low-loss optical cavity.

[6] See, for example, Brooker (2003).

[7] The name 'confocal parameter' arises from the fact that it is equal to half the length of that symmetric confocal cavity for which the beam is a lowest-order mode. See eqn (6.23). When called the confocal parameter, z_R is often given the symbol b.

form of the solution becomes somewhat clearer if we rewrite it in the form (see Exercise 6.2)

$$U = i\frac{U_0}{z_R} \exp(i[kz - \omega t])$$
$$\times \frac{w(z_0)}{w(z)} \exp\{-[r/w(z)]^2\}$$
$$\times \exp\{ikr^2/2R(z)\}$$
$$\times \exp\{-i\alpha(z)\}, \quad \textbf{Lowest-order Gaussian beam} \quad (6.10)$$

where,

$$w(z) = w(z_0)\sqrt{1 + \left(\frac{z - z_0}{z_R}\right)^2} \quad (6.11)$$

$$R(z) = z - z_0 + \frac{z_R^2}{z - z_0} \quad (6.12)$$

$$z_R = \frac{\pi w(z_0)^2}{\lambda} \quad (6.13)$$

$$\tan \alpha(z) = \frac{z - z_0}{z_R}. \quad \textbf{Gaussian beam parameters} \quad (6.14)$$

The first line in eqn (6.10) tells us that the wave propagates towards positive z (if k is positive). The second line multiplies the wave by a Gaussian function of the distance r from the propagation axis, the radial extent of this distribution being determined by $w(z)$. Note that the factor of $w(z_0)/w(z)$ ensures that the energy of the beam passing through a plane perpendicular to the z-axis is constant. It is this Gaussian shape that keeps the amplitude of the wave small at all points in the cavity—and in particular at the edges of the cavity mirrors, thereby reducing the diffraction losses. The third line in eqn (6.10) tells us that, for a given point along the z-axis, the wavefronts have a radius of curvature $R(z)$. The final factor is a phase factor that, as we shall see, plays a significant role in determining the resonant frequencies of the cavity. Figure 6.6 illustrates schematically the main features of the lowest-order Gauss–Hermite beam.

Two key parameters of the beam are the **spot size** (or **spot radius**) $w(z)$ and the **radius of curvature** $R(z)$. The spot size determines the transverse scale over which the amplitude of the beam is significant. Clearly the spot size is smallest at $z = z_0$, which represents a focus of the beam. This focus is known as the **beam waist**, and the spot size at this point, $w(0)$, is the **waist size**. Away from the focus the spot size—and hence the transverse scale of the beam—increases according to eqn (6.11). It is clear from this that the spot size increases by a factor $\sqrt{2}$ after the beam propagates a distance z_R in either direction away from the focus. The parameter z_R therefore gives the length

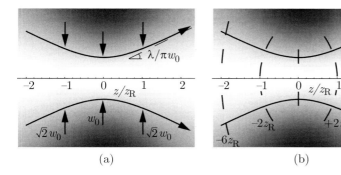

Fig. 6.6: The main features of a lowest-order Gauss–Hermite beam of focal spot size w_0 and Rayleigh range z_R: (a) the variation of the spot size of the beam; (b) the variation of the radius of curvature of the wavefronts. For both plots, white represents the highest intensity, black the lowest. The solid curves show the locus of points at a radial distance of one spot size from the propagation axis. Note that for $z \gg z_R$ these points diverge from the propagation axis at an angle of $\theta = \lambda/\pi w_0$.

scale over which the beam diffracts significantly. Note that for $z \gg z_R$ the spot size increases linearly with propagation distance z, such that the contours of constant amplitude (or intensity) are straight lines diverging from the axis at an angle of $w(z_0)/z_R = \lambda/\pi w(z_0)$. This limit corresponds to Fraunhofer diffraction, in which the transverse profile of the beam scales linearly with propagation distance.[8]

Higher-order modes†

Having established the form of the lowest-order transverse mode of a low-loss laser cavity, we now quote the general form of the Gauss–Hermite beams:

$$U = i\frac{U_0}{z_R} \exp(i[kz - \omega t])$$

$$\times \frac{w(z_0)}{w(z)} \exp\{-[r/w(z)]^2\}$$

$$\times H_l\left[\sqrt{2}\frac{x}{w(z)}\right] H_m\left[\sqrt{2}\frac{y}{w(z)}\right]$$

$$\times \exp[ikr^2/2R(z)]$$

$$\times \exp[-i(l + m + 1)\alpha]. \quad (6.15)$$

We see that the more general solution corresponds to the lowest-order solution discussed above, multiplied by: (i) the Hermite polynomials[9] $H_l(\sqrt{2}\frac{x}{w})$ and $H_m(\sqrt{2}\frac{y}{w})$; (ii) an *additional* phase factor of $\exp[-i(l + m)\alpha]$. The different transverse modes of the cavity, labelled by the integers l and m, therefore have different transverse structure in addition to the Gaussian shape discussed above. Each such mode is usually referred to as the TEM$_{lm}$ mode, where the TEM stands for 'transverse electric and magnetic field'.[10] Note that the argument of the Hermite polynomials is scaled by the spot size $w(z)$, and consequently for all the transverse modes the transverse beam profile scales with propagation distance z in the same way, i.e. according to eqn (6.11). Figures 6.7 and 6.8 show the transverse structure of the first few modes. Notice how the transverse extent of the beam increases as the order of the mode increases.

[8] This should make sense. Our knowledge of Fraunhofer diffraction leads us to expect that a wave restricted to a transverse size of $w(z_0)$ must diffract in the far field with a half angle of order $\lambda/w(z_0)$, the extra factor of π found for the Gaussian beam is a detail arising from the Gaussian profile.

[9] See Table 13.1 for a list of the first few Hermite polynomials. Comprehensive tables may be found in Gradshteyn et al, (2000).

[10] This name is unfortunate since the electromagnetic fields in the beam are not purely transverse, although the longitudinal components are small. See, for example, Brooker (2003).

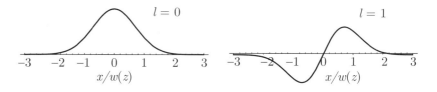

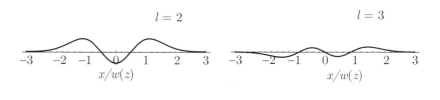

Fig. 6.7: Some Gauss–Hermite functions $\exp(-[x/w(z)]^2) H_l(\sqrt{2}x/w(z))$ for various modes l. The functions have been normalized to contain the same power, which emphasizes the fact that the transverse extent of the mode increases—and the peak height decreases—with l.

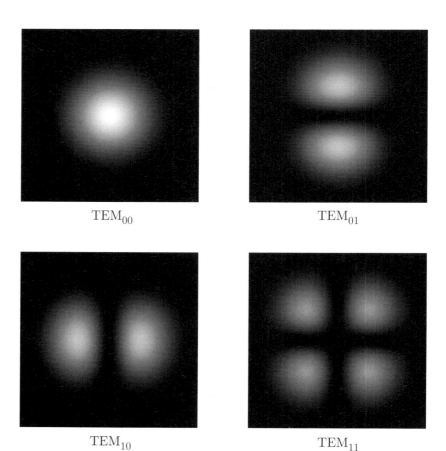

Fig. 6.8: Transverse intensity profiles of some Gauss–Hermite beams TEM_{lm}. All plots extend to ± 2 spot sizes in the x- and y-directions, and are normalized to the same total power.

Effect of cavity parameters on the modes

Although we have established the general form of the modes of the cavity, we have not yet related this to the physical parameters of the cavity, i.e. the radii of curvature of the mirrors, and their separation. There are a number of ways of fitting the general Gauss–Hermite functions to the particular cavity under consideration.[11] Here, we outline one of these.

Consider again the general low-loss cavity shown in Fig. 6.5. The condition for a Gauss–Hermite beam to be a mode of the cavity is simply that at each mirror the radius of curvature of the wavefronts is equal to that of the mirror, so that the beam is returned along the path by which it arrived. If this is achieved at both mirrors, the spatial distribution of radiation within the cavity will be stationary in time (apart from the oscillation at the frequency of the radiation): the radiation will be in one of the cavity modes. The condition for matching the radii of curvature of the wavefronts and mirrors is simply:

$$-R_1 = z_1 + \frac{z_R^2}{z_1} \tag{6.16}$$

$$R_2 = z_2 + \frac{z_R^2}{z_2} \tag{6.17}$$

$$L_c = z_2 - z_1. \tag{6.18}$$

[11] For example the use of 'ABCD' matrices can be used to find the size and location of the beam waist, and to establish whether or not the cavity is 'stable'. See, for example, Brooker (2003).

Note that the minus sign in the first equation arises from the fact that the radius of curvature is defined to be positive for concave mirrors irrespective of whether the centre of curvature lies to the right or left of the mirror. In contrast, for the beam the radius of curvature is positive if the centre of curvature lies to the left of the wavefront, negative if it lies to the right. Note also that we have defined the waist of the Gauss–Hermite beam to lie at $z = 0$.

We now have three equations for three unknowns: the positions of the mirrors z_1 and z_2; and the Rayleigh range or confocal parameter z_R, which suffices to determine the spot size of the beam at the waist once the wavelength λ is specified. Exercise 6.3 outlines a method for solving these three equations. We find that the Rayleigh range of the modes matched to the cavity is given by,

$$z_R^2 = \frac{g_1 g_2 (1 - g_1 g_2)}{[g_1(1 - g_2) + g_2(1 - g_1)]^2} L_c^2, \tag{6.19}$$

where the dimensionless quantities g_1 and g_2 are defined as:

$$g_1 = 1 - \frac{L_c}{R_1} \tag{6.20}$$

$$g_2 = 1 - \frac{L_c}{R_2}. \tag{6.21}$$

Now, if the modes are to be low-loss modes they must have a real waist so that the amplitude of each mode decreases exponentially with radial distance from the cavity axis. This in turn requires the Rayleigh range to be positive.

Hence, for the mode to be a low-loss mode the numerator in eqn (6.19) must be positive, which leads to

$$0 < g_1 g_2 < 1. \qquad \textbf{Condition for low-loss/stable cavity} \qquad (6.22)$$

This condition for low-loss modes to exist is plotted in Fig. 6.9, and is is often said to be the condition for the cavity to be **stable**; cavities that do not obey the condition are said to be 'unstable'. This convention is unfortunate in that it *does not* mean that stable (unstable) cavities are less (more) sensitive to perturbations or misalignment; the difference is entirely one of the relative loss of the cavity modes.

Figure 6.10 illustrates some important types of laser cavity; these are considered below.[12]

[12] Also see Exercise 6.4

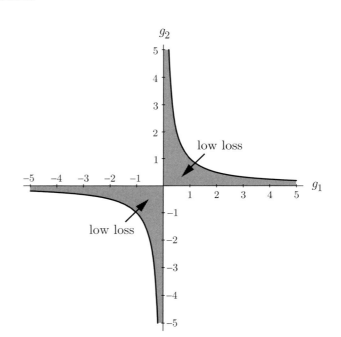

Fig. 6.9: Plot of the condition for an optical cavity to support a low-loss mode.

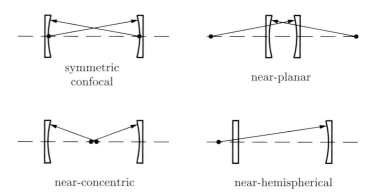

Fig. 6.10: Schematic diagram of some important types of laser cavity. The filled circles marks the centres of curvature of the mirrors.

Symmetric confocal ($R_1 = R_2 = L_c$)

A confocal cavity is any cavity for which the focal points of the two mirrors (given by $R/2$) coincide; for a symmetric confocal cavity the two foci are located at the centre of the cavity.

The spot sizes at the centre (w_0) and at the two end mirrors (w_1 and w_2) of the lowest-order Gaussian beam that is matched to the cavity are given by

$$w_0^2 = \frac{L_c \lambda}{2\pi} \quad \text{and} \quad w_1^2 = w_2^2 = \frac{L_c \lambda}{\pi}. \tag{6.23}$$

Note that the spot sizes are small throughout the cavity, of order $\sqrt{L_c \lambda}$, which is in the range 100 μm – 1 mm for typical cavity lengths and visible wavelengths. This is usually small compared to the diameter of the region of gain in the laser, making operation in a single transverse mode inefficient.

The symmetric confocal resonator is rarely used in practical laser cavities since the transverse modes are highly degenerate in frequency.[13] In contrast to the comb of mode frequencies illustrated in Fig. 6.3, for the symmetric confocal resonator the mode frequencies are $(c/4L_c)p'$, where p' is an integer. This corresponds to a set of modes at the usual longitudinal mode frequencies $(c/2L_c)p$, where p is an integer, interleaved by a second set of modes with frequencies of $(c/2L_c)(p+1/2)$. Any linear combination of degenerate modes will have the same frequency, but a different transverse spatial profile. In practice, this means that with a symmetric confocal cavity the laser can oscillate in several combinations of the degenerate modes,[14] and may switch between them essentially randomly, causing the output beam profile and power to fluctuate.

[13] For further discussion of this point see Brooker (2003).

[14] Indeed, for the intra-cavity beam to fill the gain region it will have to.

Near-planar ($R_1 \approx R_2 \approx \infty$)

The optical cavities used in many early lasers employed mirrors with radii of curvature that were large compared to the cavity length. Assuming $R_1 = R_2 = R$, the spot sizes of the lowest-order cavity mode are given by

$$w_0^2 \approx w_1^2 \approx w_2^2 = \frac{L_c \lambda}{\pi} \sqrt{\frac{R}{2L_c}}, \tag{6.24}$$

which is larger than the spot sizes in a symmetric confocal cavity by a factor of order $(R/L_c)^{1/4}$. For practical values of the mirror radii of curvature this increase in spot size is only modest. Further, the cavity is very sensitive to misalignment of the mirrors since their centres of curvature lie well outside the cavity.[15] For these reasons near-planar cavities are rarely used in low-gain, continuous-wave lasers—although they are often used in pulsed, high-power lasers for which the effective number of round trips is small, and hence cavity alignment is less critical.

[15] See Exercise 6.6.

Near-concentric ($L_c \approx R_1 + R_2$)

In a near-concentric cavity the cavity length is only slightly shorter than the sum of the two radii of curvature and hence the two centres of curvature almost coincide. In particular, in the symmetric case $R_1 \approx R_2 \approx R = L_c/2 + \Delta L_c$ the spot sizes of the lowest-order cavity mode are given by,

$$w_0^2 = \frac{L_c \lambda}{2\pi} \sqrt{\frac{2\Delta L_c}{L_c}} \quad \text{and} \quad w_1^2 = w_2^2 = \frac{L_c \lambda}{2\pi} \sqrt{\frac{L_c}{2\Delta L_c}}. \tag{6.25}$$

It is clear that in this case the spot size at the centre of the cavity is vanishingly small, but those at the two mirrors are larger than that in the symmetric confocal resonator by a factor of order $(L_c/\Delta L_c)^{1/4}$. This factor can be large for small ΔL_c, and can be controlled by adjusting the cavity length.

The disadvantage of this cavity is its high sensitivity to misalignment of the mirrors; since the centres of curvature of the two mirrors lie very close to each other, a small angular misalignment of one of the mirrors causes the axis of the cavity to rotate by a large amount.[16]

[16]See Exercise 6.6.

Near-hemispherical ($R_1 = \infty$, $R_2 = L_c + \Delta L_c$)

This is essentially half a near-concentric cavity, with a plane mirror placed just inside the centre of curvature of the other mirror. The spot sizes of the lowest-order mode are now given by,

$$w_0^2 = w_1^2 \approx \frac{L_c \lambda}{\pi} \sqrt{\frac{\Delta L_c}{L_c}} \quad \text{and} \quad w_2^2 = \frac{L_c \lambda}{\pi} \sqrt{\frac{L_c}{\Delta L_c}}. \tag{6.26}$$

Just as for the near-concentric cavity, the spot size on mirror 2 may be made large by adjusting the cavity length. This allows power to be extracted efficiently from a gain medium located near the curved mirror where the spot size is large.

The near-hemispherical resonator is much less sensitive to angular misalignment of the mirrors than the near-concentric resonator. Angular rotation of the curved mirror merely has the effect of translating the axis of the cavity by a small amount. In practice, therefore, such cavities are often aligned accurately by *translating* the curved mirror in directions perpendicular to the desired axis, as shown schematically in Fig. 6.11.

These desirable features of the hemispherical resonator make it one of the most common low-loss cavities used in practice.

Selection of the lowest-order transverse mode

It is often desirable to restrict laser oscillation to the lowest transverse mode of the cavity since this has a smooth and well-characterized beam profile.

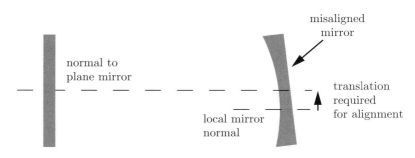

Fig. 6.11: Schematic diagram showing how small *angular* misalignment of the concave mirror in a near-hemispherical cavity may be corrected by *translation* of the mirror. Similar considerations show that in principle *any* cavity comprised of curved mirrors may be aligned to a given axis by translation of the mirrors in the plane normal to that axis.

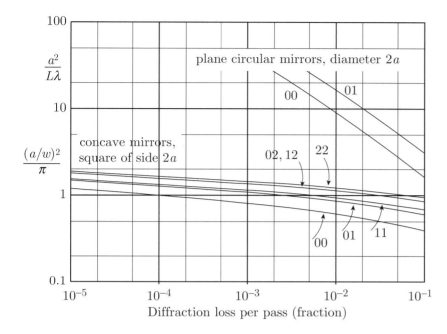

Fig. 6.12: The diffraction loss for: two configurations of an 'unstable' cavity formed from two plane circular mirrors of diameter $2a$ and separation L; and several cavity eigenfunctions of 'stable' cavity modes formed from concave square mirrors of side $2a$ and separation L. The plane-mirror loss is determined by $a^2/(L\lambda)$. The concave-mirror loss is determined by $(1/\pi)(a/w)^2$, where w is the spot size on the mirror; this is the same as $a^2/(L\lambda)$ for the special case of confocal mirrors. (After Boyd and Gordon (1961).)

For continuous-wave lasers it is usually straightforward to restrict oscillation to the lowest-order mode by employing an aperture to cause high losses—sufficient to prevent laser oscillation—for the higher transverse modes, whilst ensuring only small losses for the lowest-order mode.[17] Such an aperture can be extremely effective owing to the fact that the higher modes have very much larger transverse profiles than the lowest-order mode. Figure 6.12 provides some indication of this process. Consider, for example, the case of a cavity formed by square, concave mirrors. If a square aperture is placed at one of the mirrors with a size (i.e. $2a$) adjusted so that it introduces a loss of around 1% for the lowest-order mode, the losses suffered by higher-order modes will be more than 10 times greater.[18] This large difference in the losses is sufficient to ensure that the laser oscillates on the lowest-order mode only.

The most convenient position for a mode-selecting aperture of this type is near one of the cavity mirrors, where the size of all modes is largest. In practice, the aperture may not be one that is deliberately introduced at all; it may be a component inherent to the laser, such as the edge of a discharge tube; or it may be a 'soft' aperture such as the finite transverse extent of the gain region in an optically pumped laser. Rather than adjust the size of the aperture, it is usually more convenient to adjust the cavity length—thereby varying the spot sizes of all the transverse modes—until the losses on the higher modes are sufficiently high to prevent them reaching the threshold for achieving laser oscillation.

6.3.3 High-loss (or 'unstable') optical cavities[†]

The modes in high-loss cavities, that is those that do not obey eqn (6.22), are more complicated than their low-loss counterparts and do not have analytic

[17] Selection of a single transverse mode is not always as straightforward in pulsed lasers since the number of cavity round trips made during the time for which there is gain may be relatively low. In such cases the losses experienced by the different modes may not be sufficiently different to prevent oscillation on higher-order modes.

[18] In principle, it should not matter where the aperture is placed since, from eqn (6.15) the transverse extent of all the modes scale as $w(z)$, and hence their relative size is independent of position along the z-axis.

solutions. This does not mean that high-loss, or unstable, cavities are not useful. The main advantage they have is that the transverse extent of their modes is much larger than those of low-loss cavities; indeed it is comparable to that of the cavity mirrors. High-loss cavities are therefore useful in extracting power from laser media with large gain volumes, but they are obviously limited to systems in which the gain is large enough to overcome the high cavity losses.

In addition to their large, controllable mode size unstable cavities have several other advantages:

- good transverse mode discrimination;
- no requirement for a partially transmitting mirror for coupling out the laser radiation;
- well-collimated output beams; and relative ease of alignment.

Unstable optical cavities fall into two categories according to the way in which they fail eqn (6.22): **negative-branch** cavities, for which $g_1 g_2 < 0$; and **positive-branch** cavities, for which $g_1 g_2 > 1$.

The analysis of the modes of unstable cavities is complex, owing to the fact that the transverse profile of the modes depends on the shape of the mirrors that form the cavity. However, useful insight may be achieved by using an argument first proposed by A. E. Siegman.[19] Our analysis of stable cavities showed that the modes were comprised of right- and left-going Hermite–Gauss beams of the form of eqn (6.15). The phase fronts of such waves are spherical, with a radius of curvature given by eqn (6.12), and the amplitude of the wave decreases exponentially with radial distance from the axis of propagation. It is the exponential decrease in the amplitude that keeps the losses of the mode low. As a first approximation, then, we can assume that in an unstable cavity the circulating beams also have spherical wavefronts, but that their amplitudes do *not* decrease exponentially in the transverse direction but are, instead, constant.

Figure 6.13 shows an important class of unstable cavity, the unstable confocal cavity. For this simple case it is clear that the origins of the right- and left-going spherical waves are coincident at the common focus of the two mirrors. For the negative-branch cavity we see that the left-going spherical wave is reflected to form a right-going parallel beam. If this parallel beam has a diameter D, after a further round trip of the cavity it would form another parallel beam with a diameter MD, where $M = |R_1/R_2| = |f_1/f_2|$.[20] For the

[19] Unstable optical cavities were first analysed by Siegman in 1965. A very clear and thorough discussion of the analysis and properties of this important class of optical resonator can be found in Siegman's book Siegman (1986).

[20] See Exercise 6.8.

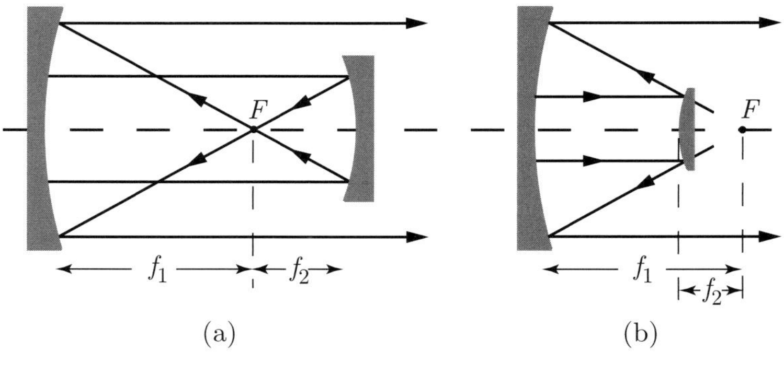

Fig. 6.13: Unstable confocal cavities: (a) negative branch; (b) positive branch. The rays shown depict a geometric optics approximation to the cavity modes in which the modes are assumed to take the form of spherical waves with an amplitude that is constant over the spherical wavefronts. For confocal resonators the foci of the two mirrors lie at a common point F, which is also the source of the spherical waves.

cavity shown, $R_1 > R_2$ and hence the diameter of the right-going, parallel beam increases with each round trip of the cavity. For this reason this spherical wave and parallel beam comprise what is known as the **magnifying wave** of the cavity. The right-going *spherical* wave (not shown in Fig. 6.13) generates a left-going parallel beam that decreases with each round trip; this is the **demagnifying wave**.[21]

Both the magnifying and demagnifying waves are important in the behaviour of unstable cavities. As discussed below, the *demagnifying* wave plays an important role in that the laser output originates from the spontaneous emission into this wave, whereas it is the parallel beam of the *magnifying* wave that forms the laser output. Indeed, it is found that the *magnifying* wave is a reasonable approximation to the true lowest-order mode of the cavity.

In practice, the magnifying wave is coupled out of the cavity by making the diameter of one of the mirrors sufficiently small that the outer part of the parallel beam goes past it, as shown in Fig. 6.13, and only a proportion of the wave is reflected back into the cavity.[22] When used in this way, the output beam from an unstable cavity is a ring, with an inner diameter approximately equal to the diameter of the right-hand mirror and an outer diameter of M times this. As such, the near-field[23] beam from an unstable resonator will be determined by Fresnel diffraction of the beam as it passes the output mirror and may exhibit an axial spot known as Poisson's spot. Although an output beam of this profile may be disadvantageous for some applications, it is important to remember that such beams can still be focused to a well-defined spot with a sharp axial maximum, albeit possibly with some outer diffraction rings caused by the hard edges of the mirrors.

Further considerations

We now consider briefly how the laser output beam is formed from a seed of spontaneous emission, and outline how the design of an unstable resonator may be optimized for a given laser system.[24]

As the population inversion is first established, photons will be emitted spontaneously in all directions into both the magnifying and demagnifying waves. As in any laser, spontaneously emitted photons form seed radiation that is amplified by stimulated emission to form an output beam. This process occurs throughout the duration of the gain, but that emission occurring soon after the gain is established is most important since this has the opportunity to experience greater amplification.

The laser output is associated with the magnifying wave, and hence it might be thought that the seed radiation is formed by spontaneous emission into this wave. In fact this is not the case, for the following reasons. After the gain is switched on, the intensity of the demagnifying wave increases much more rapidly than that of the magnifying wave because its diameter becomes smaller on each round trip, whereas the demagnifying wave has a growing diameter and suffers much higher losses. As a consequence, fewer round trips are required for the intensity of the demagnifying wave to grow to the point that stimulated emission dominates spontaneous emission. Hence, it is spontaneous radiation emitted into the demagnifying wave that is able to provide the seed for the output laser radiation.

[21] The positive-branch confocal cavity may be analysed in the same way. As drawn it is the left-going spherical wave that is part of the magnifying wave; the right-going spherical wave that forms the demagnifying wave. Note that for the negative-branch confocal unstable cavity the magnifying and demagnifying waves exhibit a real focus. This property is generally true of negative-branch unstable cavities, which makes them undesirable for use with high-power laser systems owing to problems of optically induced breakdown or damage near the focal points.

[22] Unstable cavities can therefore use mirrors with a reflectivity of 100%, which is an advantage in regions of the spectrum—such as the infra-red or ultraviolet—where it is difficult to make mirrors that are partially transmitting.

[23] In general, for a beam of wavelength λ and approximate diameter D one may define an approximate Rayleigh range $z_R \approx D^2/\lambda$. The **near field** extends to distances short compared to z_R, and is characterized by Fresnel diffraction of the beam; the **far field** corresponds to distances significantly beyond z_R, for which the shape of the beam is determined by Fraunhofer diffraction (see, for example Brooker (2003).)

[24] We use the approach first described by Zemskov et al. (1974) and Isaev et al. (1974).

It is clear that the demagnifying wave grows in intensity much faster than the magnifying wave, but how does this radiation leave the cavity? After all, if the diameter of the demagnifying wave decreases with each round trip it can never 'spill past' the output coupler. However, at some point the diameter of the demagnifying wave becomes small enough for diffraction to become significant. It may be shown that diffraction causes an initially demagnifying wave to resemble a magnifying wave; diffraction is also responsible for establishing oscillation on the lowest-order mode of the cavity, at which point the beam may be said to be diffraction-limited.[25] The number of round trips required to reach this diffraction limit may be estimated as[26]

$$m_d = 1 + \frac{\ln M_0}{\ln M}; \qquad M_0 = \frac{D_2^2}{|\lambda R_1|}, \qquad (6.27)$$

where λ is the wavelength of the laser radiation and D_2 is the diameter of the smaller mirror.

For a pulsed laser there is only time for a finite number of round trips, suggesting that m_d should be as small as possible and hence that M should be as large as possible.[27] However, there is another consideration: in the time taken for the initially demagnifying wave to reach the diffraction limit the (unwanted) initially magnifying wave will continue to be amplified, and the intensity of this wave must not reach the saturation intensity or it will burn down the population inversion density before the (wanted) initially demagnifying wave has the opportunity to do so. This sets an upper limit on the small-signal gain coefficient, α_0^{cr}; if the gain exceeds this, the laser output will be far from diffraction-limited and will comprise amplified spontaneous emission in both the magnifying and demagnifying beams. It may be shown that the small-signal gain coefficient must satisfy the condition:[28]

$$\alpha_0 \ell_g < \alpha_0^{cr} \ell_g = \left[\frac{\ln\left(\frac{4\pi}{\Omega}\right)}{\ln M_0} + 1 \right] \ln M, \qquad (6.28)$$

where Ω is the solid angle subtended by one end of the gain medium when viewed from its other end.[29] Equations (6.27) and (6.28) are helpful in designing an unstable cavity for a particular laser system.

Wave analysis of unstable cavities

It is clear that the transverse dimensions of the magnifying wave will increase until they become comparable to the size of the cavity mirrors. As such, unlike the case of a stable cavity, the edges of the mirrors play an important role in determining the precise nature of the cavity modes. This complication means that in general it is not possible to find analytical expressions for the modes of an unstable optical cavity, and numerical techniques must be employed.

Numerical studies of the modes of stable and unstable cavities were pioneered by Fox and Li (Fox and Li 1961). Similar techniques are still used today, and are generally known as 'Fox–Li' calculations.

The Fox–Li method is illustrated in Fig. 6.14 for a plane–plane cavity. We suppose that at some point the radiation field has some arbitrary amplitude

[25] See Exercise 6.8.
[26] See Isaev et al. (1974) and Exercise 6.8.

[27] It should make sense that the number of round trips required to reach the diffraction limit decreases as the magnification of the cavity increases.

[28] See Zemskov et al. (1974) and Exercise 6.9.

[29] That is, for a cylindrical gain medium of diameter D_g and length ℓ_g, $\Omega \approx D_g^2/4\pi\ell_g$. Note that various expressions for Ω are given in the literature. However, since it appears within a logarithm the deduced critical gain coefficient does not depend strongly on the value of Ω used.

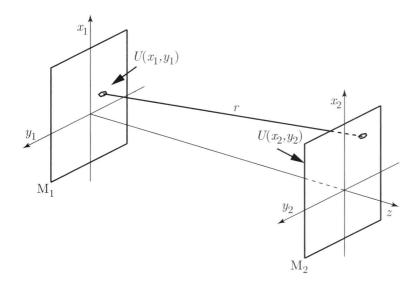

Fig. 6.14: Illustration of the Fox–Li method used to calculate the modes of a plane-parallel optical cavity.

distribution $U(x_1, y_1)$ on the left-hand mirror M_1. In principle, we can then calculate the field at a point (x_2, y_2) on the right-hand mirror M_2 with the aid of the Kirchhoff diffraction integral,[30]

$$U(x_2, y_2) = \frac{1}{i\lambda} \int_{M_1} U(x_1, y_1) \frac{1 + \cos\theta}{2} \frac{\exp(ikr)}{r} dx_1 dx_2, \qquad (6.29)$$

where r is the distance between (x_2, y_2) and (x_1, y_1), and θ the angle between **r** and the cavity axis.

[30] See, for example, Brooker (2003).

Any radiation lying outside the area of M_2 will be lost from the cavity, whilst that lying within the area of the mirror will be reflected back towards M_1. The distribution of the light returning to M_1 can be found with the aid of a second Kirchhoff diffraction integral. This process can be repeated until the transverse distribution of the radiation at the two mirrors is essentially unchanged with further round trips, only the amplitude changing by a fixed proportion on each round trip. The radiation is then in one of the cavity modes, the change in the square of the amplitude giving the diffraction loss of the mode. Of course, the Fox–Li calculation mimics the process that occurs when a laser is switched on and the radiation in the cavity builds up from spontaneous and stimulated emission.

The symmetry of the mode that is calculated depends on the symmetry of the initial distribution of radiation that is assumed. For example, if the cavity mirrors are square or rectangular and the initial distribution is symmetric with respect to reflection in either the (x, z) or (y, z) planes, the calculated mode will be the lowest-order mode with the same symmetry: the TEM_{00} mode. Conversely, if the initial distribution is antisymmetric with respect to reflections in (say) the (y, z) plane, the calculated mode will be the TEM_{10} mode. In this way it is possible to calculate the transverse profiles and diffraction losses of the lowest symmetric and antisymmetric modes of the cavity. It is possible to calculate higher-order modes using more advanced techniques, although these higher modes are not usually of as much interest.

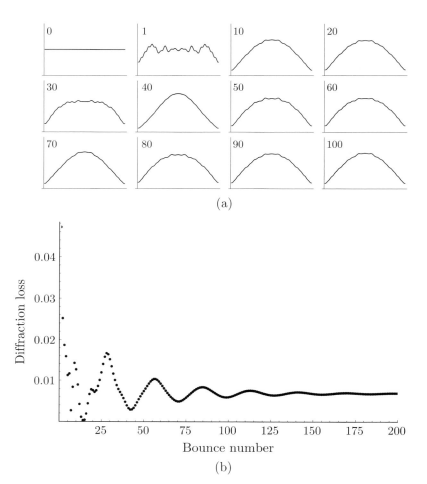

Fig. 6.15: Results of the Fox–Li calculation described in Exercise 6.10 for a cavity comprised of two plane, square mirrors of side $2a$ separated by a distance L_c. In (a) is shown the distribution of the amplitude of radiation along the x-axis of the mirror surface after various numbers of passes across the cavity. In (b) is shown the diffraction loss per pass as a function of the number of completed bounces between the mirrors for the case of a one-dimensional 'strip mirror' of height $2a$; the diffraction loss per pass of a square mirror of side $2a$ is approximately twice the plotted value. For these calculations the Fresnel number of the cavity, $N_F = (2a)^2/\lambda L_c$, was taken to be 25.

Figure 6.15(a) shows the calculated one-dimensional spatial variation of the amplitude of the radiation after various passes across the cavity for the case of two plane mirrors, as discussed in Exercise 6.10. The evolution of the mode profile from an initially uniform distribution is apparent. In Fig. 6.15(b) is plotted the calculated diffraction loss per pass as a function of the number of passes completed.

Finally, we should emphasize that Fox–Li calculations are also performed for *stable* optical cavities.[31] These calculations give the diffraction losses and transverse modes of real laser cavities with mirrors of finite size. Exercise 6.10 shows how to set up and run a Fox–Li calculation.

Variable-reflectivity mirrors

Finally in this section, we note that the properties of the beams generated by lasers with unstable cavities may be improved by the use of so-called **variable-reflectivity mirrors (VRM)**.[32] In this approach the reflectivity of the output mirror decreases in a smooth way with radius from a maximum R_0 (which will

[31] See, for example, Fox and Li (1963).

[32] See Zucker (1970).

not in general be 100%). Two commonly used profiles are Gaussian, in which the reflectivity varies as $R(r) = R_0 \exp(-2[r/w_m]^2)$; or 'super-Gaussian', for which $R(r) = R_0 \exp(-2[r/w_m]^n)$, where $n > 2$.[33] The use of variable-reflectivity mirrors tends to fill in the 'hole' produced by finite-size mirrors of 100% reflectivity, and reduces or removes structure on the output beam arising from diffraction at the edges of such mirrors. By carefully tuning the value of R_0 and the way in which the reflectivity decreases with radial distance from the cavity axis, it is possible to achieve output beams with smooth transverse profiles in both the near and far fields. Alternatively, beams with a 'flat-top' near-field profile can be generated. These are of interest in a number of applications, such as producing a uniform region of gain in solid-state lasers optically pumped by the beam, or frequency doubling the beam in a non-linear crystal.

[33] The use of mirrors with super-Gaussian reflectivities was first considered by De Silvestri et al. (1988).

6.4 Beam quality[†]

It should be apparent from the discussion above that in many circumstances the output beam of a laser system will not be a lowest-order Gaussian beam, i.e. of the form of eqn (6.10), but may instead be a superposition of several (or many) modes. Here, we briefly discuss how this affects the propagation of a beam and, the reverse problem, whether one can gain insight into the mode composition of a beam from simple measurements of the beam properties.

Let us first consider the form of a higher-order Hermite–Gaussian mode, with an amplitude given by eqn (6.15). For convenience we will restrict our discussion to one dimension. The most important feature of the higher-order modes is that their amplitude distribution contains a factor of the form $H_n[\sqrt{2}x/w(z)]$. This introduces n zeros in the amplitude profile and increases the width of the beam from that of the lowest-order mode. It is shown in Exercise 6.7 that an nth-order Hermite–Gaussian mode contains n zeros and that for reasonably large n the width of the beam is $W(z) \approx \sqrt{n} w(z)$. These features are illustrated in Fig. 6.16 for the case $n = 10$.

The presence of higher-order modes not only causes the beam to be wider, it also diffracts more quickly than would be expected from its width. This can be seen from the following argument. Suppose at its waist the beam has a width $W(0)$. Since the mode contains n zeros along the x-axis, the width of each 'lobe' will be of order $w_{\text{lobe}} \approx W(0)/n$. Our experience of Fraunhofer diffraction tells us that a beam with amplitude or phase structure with a size of w_{lobe} will diffract with a far-field angular divergence of approximately $\lambda/w_{\text{lobe}} = n\lambda/W(0)$.[34] This is a factor of n greater than we would expect for a uniform beam of width $W(0)$.[35]

[34] As an illustration, consider a diffraction grating comprised of N slits of width a and spacing d. The range of angles over which the Fraunhofer diffraction has significant intensity is determined by the 'diffraction envelope' that may be shown to be of the form $\text{sinc}^2[\pi(a/\lambda)\sin\theta]$, where $\text{sinc}(x) = \sin(x)/x$. This has an angular width of order λ/a, and hence we see that the extent of the diffraction pattern is determined by the width of each grating element (a), not the total width of the grating (Nd). We therefore can expect that the far-field divergence of a beam will be inversely proportional to the scale of the 'fine structure' in the amplitude or phase of the beam, not the total beam width.

[35] The far-field divergence of Hermite–Gaussian beams is also explored in Exercise 6.7.

6.4.1 The M^2 beam-propagation factor

In general, rapid variation in the transverse plane of the amplitude and/or phase of a beam will cause it to diffract more quickly than would be the

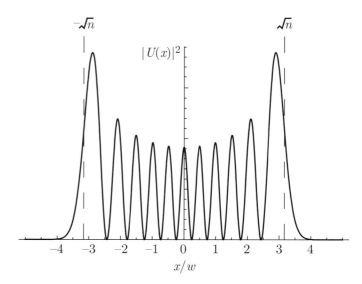

Fig. 6.16: The transverse intensity profile of the Gauss–Hermite mode $U(x) = \exp(-[x/w(z)]^2)H_n[\sqrt{2}x/w(z)]$ for the case $n = 10$.

case for a 'smooth' beam of similar dimensions. Here, we outline one method of characterizing such beams: the **'M-squared' beam propagation factor**.

The first problem to be addressed is one of finding a sufficiently flexible definition of the width of a beam that it can be applied unambiguously to a beam of any transverse intensity profile. One such measurement is the variance (or 'second moment') of the intensity profile. Along the x-axis, for example, this is given by

$$\sigma_x^2 = \frac{\int_{-\infty}^{\infty}(x - x_0)^2 I(x, y)\mathrm{d}x\mathrm{d}y}{\int_{-\infty}^{\infty} I(x, y)\mathrm{d}x\mathrm{d}y}, \tag{6.30}$$

where x_0 is the centre of gravity of the beam and $I(x, y)$ is the intensity of the beam in the x, y-plane.[36] The variance in the y-direction may be defined in an analogous way.

[36] The centre of gravity of the beam, x_0, is defined by $\int_{-\infty}^{\infty}(x - x_0)I(x, y)\mathrm{d}x\mathrm{d}y = 0$.

An important feature of σ_x is that it turns out that for *any* beam it varies with position z along the axis of propagation of the beam according to

$$\sigma_x^2(z) = \sigma_x^2(z_0) + \sigma_\theta^2(z)(z - z_0)^2, \tag{6.31}$$

where z_0 is the position of the focus of the beam, and $\sigma_\theta^2(z)$ is the variance of the angular spread of the beam.

It is trivial to show that for a lowest-order Gaussian beam the spot size $w_x(z) = 2\sigma_x(z)$. It is therefore convenient to define the width $W_x(z)$ of *any* beam in terms of its variance: $W_x(z) = 2\sigma_x(z)$, where we use a capital letter to indicate that the beam may not be Gaussian. Equation (6.31) then yields straightforwardly

$$W_x^2(z) = W_x^2(z_0) + 4\sigma_\theta^2(z)(z - z_0)^2.$$

We are free to rewrite this in the form

$$W_x^2(z) = W_x^2(z_0) + M_x^4 \left(\frac{\lambda}{\pi W_x(z_0)}\right)^2 (z - z_0)^2, \quad (6.32)$$

$$= W_x^2(z_0) + M_x^4 \left(\frac{z - z_0}{z_R}\right)^2, \quad (6.33)$$

where $z_R = \pi W_x(z_0)^2/\lambda$. This result is of the same form as for a lowest-order Gaussian beam (eqn (6.11)), but with an additional factor of M_x^4 multiplying the second term on the right-hand side.

Note that in order to measure the M^2 value of a beam, it is necessary to measure the intensity profile $I(x, y)$ in at least two points along the z-axis, to yield two values of $W_x(z)$; these may be substituted into eqn (6.32) to find M_x^2. Such measurements are often made at the beam waist, to yield $W_x(z_0)$, and far from the waist (i.e. $|z - z_0| \gg z_R$). It is easy to show from eqn (6.32) that

$$W_x(z) W_x(z_0) = M_x^2 \frac{\lambda}{\pi} |z - z_0| \qquad (|z - z_0| \gg z_R). \quad (6.34)$$

This result may be used in a number of ways. The far-field divergence may be defined as

$$\theta_x = \lim_{z \to \infty} \left(\frac{W_x(z)}{|z - z_0|}\right) = M_x^2 \frac{\lambda}{\pi W_x(z_0)}. \quad (6.35)$$

Alternatively, one can find the waist $W_x(0)$ formed by a lens of focal length f of a beam with a spot size in the plane of the lens equal to W_L. To do this we set $z_0 = 0$, $z = -f$ and $W_x(z) \to W_L$, and find

$$W_x(0) = M_x^2 \frac{\lambda f}{\pi W_L}. \quad (6.36)$$

The right-hand sides of both eqn (6.35) and eqn (6.36) are larger than those of a lowest-order Gaussian beam by a factor of M_x^2. For these reasons the beam is sometimes said to be 'N-times diffraction-limited,' where N is the value of M_x^2.

The embedded Gaussian

It is useful to realize that eqn (6.32) may be rewritten in the form

$$\left[\frac{W_x(z)}{M_x}\right]^2 = \left[\frac{W_x(z_0)}{M_x}\right]^2 + \left[\frac{\lambda}{\pi W_x(z_0)/M_x}\right]^2 (z - z_0)^2. \quad (6.37)$$

This is identical to the equation describing the variation of the spot-size of a lowest-order Gaussian beam with a spot-size that is everywhere equal to $w_x(z) = W_x(z)/M_x$. Thus, the variation of the size $W_x(z)$ of any beam can be calculated by considering the behaviour of an 'embedded Gaussian' with a waist $w_x(z) = W_x(z)/M_x$. Powerful methods[37] exist for calculating the propagation of lowest-order Gaussian beams through optical systems—including lenses and mirrors—so, provided the value of M_x^2 is known, tracking

[37] These include matrix methods. See, for example, Brooker (2003).

the propagation of a real beam via an embedded Gaussian can give useful information about the propagation of complex beams. Of course, it should be emphasized that this technique only calculates one parameter of the real beam, its variance; details of the intensity or phase profile require full modelling of the beam.

This description is in accord with our earlier observation that the width of an nth-order Hermite–Gaussian beam is approximately $W(z) \approx \sqrt{n} w(z)$, where $w(z)$ is the spot size of the lowest-order Gaussian beam. Thus, we expect that a description of the real beam in terms of a superposition of Hermite–Gaussian beams will contain modes with orders up to n, where $\sqrt{n} \approx M_x$.[38]

To summarize:

- For a lowest-order Gaussian beam $M_x^2 = 1$; in general $M_x^2 \geq 1$.
- The value of M_x^2 is a measure of how many times diffraction-limited the beam is.
- A beam with a focal spot N times bigger than expected, based on the size of the beam at the focusing optic, is said to be 'N-times diffraction-limited'. This factor is related to the M-squared value by $N \approx M_x^2$.[39]

Difficulties with the M^2 description

Defining the quality of a laser beam in terms of its value of M^2 has the advantage that this parameter is rigorously defined for any beam, and that some aspects of the propagation of the real beam may be deduced by calculating the propagation of the associated 'embedded Gaussian' beam.[40]

However, determining the value of M^2 is not always easy in practice. The principle difficulty is that the value of the variance, and hence $W_x(z)$, calculated from a measured intensity profile $I(x, y)$ depends strongly on the intensity in the wings of the beam owing to the factor $(x - x_0)^2$. The intensity here is usually much lower than near the beam axis, and consequently is sensitive to noise, variations in the background signal, and detector non-linearities. As a consequence, values of M^2 are often estimated by comparing the size of the focal spot with that expected given the size of the beam at the lens; or by comparing the Rayleigh range z_R with that expected for the measured focal spot size.

6.5 The approach to laser oscillation

6.5.1 The 'cold' cavity

We consider first injection of an external beam into a 'cold' laser cavity—that is, one that does not contain a gain medium—as illustrated in Fig. 6.17. To simplify matters, let us suppose that the injected beam corresponds to one of the modes of the cavity such that it excites within it a right- and left-going wave that might, for example, have the form of the low-loss modes given in eqn (6.15). The input beam, and the right- and left-going waves can then be written, respectively, as:

---

[38] That is, modes with $n \gtrsim M_x$ are expected to have a low amplitude. Note too that our treatment is one-dimensional: there will in general be approximately m such modes in the y-direction; for reasonably symmetric beams $m \approx n$.

[39] This does not contradict the fact that the size of the embedded Gaussian is everywhere $W(z)/M_x$. At the lens, the spot size of the embedded Gaussian will be W_L/M_x, which will form a waist equal to $w_0 = \lambda f / \pi (W_L/M_x)$. The waist size of the real beam will be $W_0 = M_x w_0 = M_x^2 \lambda f / \pi W_L$, in agreement with eqn (6.36).

[40] It should be noted that in some idealized cases the calculated value of M^2 is infinite, although the laser beam may have desirable qualities. For example, an ideal 'top hat' beam has an infinite second-order moment in the far field and hence M^2 is calculated to be infinite. This does not mean that the beam is of low quality; indeed laser beams with top-hat transverse profiles have been found to be very useful for laser machining and for pumping laser-amplifier stages.

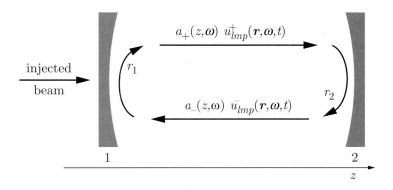

Fig. 6.17: Injection of a beam into a single mode of a laser cavity.

$$U_{\text{in}}(\mathbf{r}, \omega, t) = a_{\text{in}}(z, \omega) u_{lmp}^{+}(\mathbf{r}, \omega, t) \tag{6.38}$$

$$U_{+}(\mathbf{r}, \omega, t) = a_{+}(z, \omega) u_{lmp}^{+}(\mathbf{r}, \omega, t) \tag{6.39}$$

$$U_{-}(\mathbf{r}, \omega, t) = a_{-}(z, \omega) u_{lmp}^{-}(\mathbf{r}, \omega, t), \tag{6.40}$$

where $a_{\text{in}}(z, \omega)$ is the amplitude of the input beam just outside the cavity, and $a_{\pm}(z, \omega)$ are the amplitudes of the right- and left-going beams, which have a distribution described by $u_{lmp}^{\pm}(\mathbf{r}, \omega, t)$.[41]

Under steady-state conditions, the amplitude of the right-going wave near the surface of mirror M_1 must be given by

$$a_{+}(0, \omega) u_{lmp}^{+}(\mathbf{r}, \omega, t) = t_1 a_{\text{in}}(0, \omega) u_{lmp}^{+}(\mathbf{r}, \omega, t) \\ + r_1 \exp(i\phi_1) a_{-}(0, \omega) u_{lmp}^{-}(\mathbf{r}, \omega, t), \tag{6.41}$$

where t_1 is the amplitude transmission coefficient and $r_1 \exp(i\phi_1)$ is the amplitude reflection coefficient of M_1, ϕ_1 being the phase shift on reflection. Now, eqn (6.41) must be true for all points $\mathbf{r}$ on mirror 1, and hence it must be that on the surface of this mirror $u_{lmp}^{-}(\mathbf{r}, \omega, t) = u_{lmp}^{+}(\mathbf{r}, \omega, t)$. Hence eqn (6.41) becomes

$$a_{+}(0, \omega) = t_1 a_{\text{in}}(0, \omega) + r_1 \exp(i\phi_1) a_{-}(0, \omega). \tag{6.42}$$

[41] In anticipation of introducing a gain medium into the cavity, we have allowed the amplitudes of the right- and left-going beams to be functions of ω and z.

Now, in propagating to M_2 and back the right-going wave experiences: (i) a phase shift of $\delta_{\text{rt}} - \phi_1$, where δ_{rt} is the total round-trip phase change experienced by the wave—including the phase shifts ϕ_1 and ϕ_2 at the cavity mirrors; (ii) a change in amplitude due to the finite reflection coefficient of M_2. Hence, we can relate the amplitude of the left-going wave at M_1 to that of the right-going wave:

$$a_{-}(0, \omega) = a_{+}(0, \omega) \times r_2 \exp(i[\delta_{\text{rt}} - \phi_1]). \tag{6.43}$$

Armed with this, eqn (6.42) becomes,

$$a_{+}(0, \omega) = t_1 a_{\text{in}}(0, \omega) + a_{+}(0, \omega) r_1 r_2 \exp(i\delta_{\text{rt}}),$$

so

$$a_+(0, \omega) = \frac{t_1 a_{\text{in}}(0, \omega)}{1 - r_1 r_2 \exp(i\delta_{\text{rt}})}. \qquad (6.44)$$

The intensity $I_+(0, \omega)$ of the right-going wave at M_1 is proportional to $|a_+(0, \omega)|^2$:

$$I_+(0, \omega) \propto |a_+(0, \omega)|^2 = \left| \frac{1}{1 - r_1 r_2 \exp(i\delta_{\text{rt}})} t_1 a_{\text{in}}(0, \omega) \right|^2. \qquad (6.45)$$

As shown in Exercise 6.12 this last result may be simplified to:

$$\frac{I_+(0, \omega)}{I_{\text{in}}(0, \omega)} = \frac{T_1}{(1-R)^2} \frac{1}{1 + F \sin^2\left(\frac{\delta_{\text{rt}}}{2}\right)}, \qquad (6.46)$$

where

$$F = \frac{4R}{(1-R)^2}, \qquad (6.47)$$

and $R = r_1 r_2$ and $T_1 = |t_1|^2$. This last result is essentially the familiar result for the intensity *transmitted* through a Fabry–Perot etalon.[42]

The cavity finesse, Q, and lifetime τ_c

If the mirrors of the laser cavity have reasonably high reflectivities, the parameter F will be large compared to unity. Consequently, the intensity of the right-going wave will show strong peaks when $\delta_{\text{rt}} = 2\pi p$, i.e. when the total round-trip phase change is a multiple of 2π. This is *exactly* the condition for the frequency to equal the frequency of the cavity mode (l, m, p).[43]

The width of each peak is determined by the parameter F. For the peak at $\delta_{\text{rt}} = 2\pi p$, the points at which the intensity falls to half that at the peak are given by

$$F \sin^2\left(\frac{2\pi p \pm \frac{1}{2}\Delta\delta_{1/2}}{2}\right) = 1, \qquad (6.48)$$

where $\Delta\delta_{1/2}$, is the full width at half-maximum measured in terms of phase. Provided F is reasonably large, we find,[44]

$$\Delta\delta_{1/2} = \frac{4}{\sqrt{F}}. \qquad (6.49)$$

If within the cavity there is only a single, uniform medium of refractive index n the round-trip phase shift for light of wave vector k is given by

$$\delta_{\text{rt}} = 2kL_c + \epsilon, \qquad (6.50)$$

where ϵ is the sum of the phase shifts upon reflections from the cavity mirrors and any additional phase shifts arising from the properties of the cavity mode.[45] To find the width of the peak in terms of wave vector or frequency we note that[46]

$$\Delta\delta_{\text{rt}} = 2\Delta k L_c = \frac{2nL_c}{c}\Delta\omega. \qquad (6.51)$$

[42] If this is not familiar, consult a standard text on optics such as Brooker (2003). In the calculation of the transmission of a Fabry–Perot etalon we would need to evaluate the amplitude of the right-going wave at M_2 and multiply by the amplitude transmission coefficient t_2 of the right-hand mirror. This changes the first factor of eqn (6.46) to $T_1 T_2/(1-R)^2$.

[43] See eqn (6.7).

[44] See Exercise 6.12.

[45] For example, eqn (6.15) shows that in propagating between the planes $z = z_1$ and $z = z_2$, a Gaussian beam undergoes a phase shift of $-(l+m+1)[\alpha(z_2) - \alpha(z_1)]$ in addition to the phase shift of $k(z_2 - z_1)$ experienced by a plane wave.

[46] In general, ϵ will also be a function of k and ω, which will lead to corrections to eqn (6.51).

Setting $\Delta\delta_{\mathrm{rt}} = \Delta\delta_{1/2}$ in eqn (6.51) gives the angular frequency width of each peak:

$$\Delta\omega_{\mathrm{c}} = \frac{2c}{nL_{\mathrm{c}}\sqrt{F}} = \frac{c}{nL_{\mathrm{c}}}\frac{1-R}{\sqrt{R}}. \tag{6.52}$$

Similarly, in terms of phase, the spacing between adjacent peaks is simply $\Delta\delta_{\mathrm{rt}} = 2\pi$. Hence, from eqn (6.51) the spacing between adjacent peaks is given in terms of angular frequency by[47]

$$\Delta\omega_{p,p-1} = \frac{\pi c}{nL_{\mathrm{c}}}. \tag{6.53}$$

The ratio of the separation of the peaks to their width is known as the **finesse** $\mathcal{F}$ of the cavity:

$$\mathcal{F} = \frac{\Delta\omega_{p,p-1}}{\Delta\omega_{\mathrm{c}}} = \frac{\pi}{2}\sqrt{F} = \frac{\pi\sqrt{R}}{1-R}. \tag{6.54}$$

[47] For a Fabry–Perot etalon this quantity is known as the **free spectral range**. In frequency units this is given by $\Delta\nu_{\mathrm{FSR}} = (1/2\pi)\Delta\omega_{p,p-1} = c/2nL_{\mathrm{c}}$.

The intensity of the right-going wave is plotted as a function of the total round-trip phase in Fig. 6.18 for various values of the cavity finesse. It is seen that the intensity is peaked strongly about values of the round-trip phase equal to $2\pi p$, or, equivalently, at the modes of the cavity. We can also see that the peaks become sharper as the finesse of the cavity is increased. As such, the finesse is one measure of the 'quality' of the laser cavity.

Two other measures of the quality of the laser cavity are often considered. The **quality factor** Q of any resonant system may be defined as the ratio of the resonant frequencies to the width of the resonance. Hence, for our laser cavity

$$Q = \frac{\omega_p}{\Delta\omega_{\mathrm{c}}}. \tag{6.55}$$

As shown in Exercise 6.13, the Q of a laser cavity can be very high, of the order of 10^9.

A second parameter of interest, particularly for calculations of dynamic cavity effects such as Q-switching, is the **cavity lifetime** τ_{c}. Suppose that there

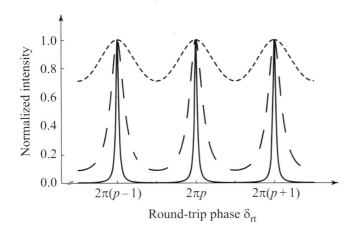

Fig. 6.18: Normalized intensity of the right-going wave as a function of the round-trip phase δ_{rt} for an optical cavity with a finesse equal to 1 (dotted line); 5 (dashed line); and 25 (solid line). The peaks occur when the round-trip phase is equal to an integer multiple of 2π, and correspond to a longitudinal mode of the cavity. The width of the peaks decreases as the finesse increases.

are n_ϕ photons per unit volume circulating within the cavity. After an interval equal to the time taken for one round trip, T_{rt}, each photon will have been reflected once from each mirror and hence—if no photons are added by any other process—the density of photons within the cavity will be $R_1 R_2 n_\phi$. We may therefore write

$$\frac{dn_\phi}{dt} \approx \frac{R_1 R_2 n_\phi - n_\phi}{T_{rt}} = -\frac{n_\phi}{\tau_c}, \qquad (6.56)$$

where

$$\tau_c = \frac{T_{rt}}{1 - R_1 R_2} \approx \frac{2L_c}{v_g} \frac{1}{1 - R_1 R_2}, \quad \textbf{Cavity lifetime} \qquad (6.57)$$

and the approximation holds for the simple case of a cavity of length L_c filled with a uniform medium in which light propagates with a group velocity v_g.

Comparison with eqn (6.52) shows that if $R_1 \approx R_2 \approx 1$ the cavity linewidth may be written in the form,[48]

[48] See Exercise 6.13.

$$\Delta \omega_c \approx 1/\tau_c. \quad \textbf{Cavity linewidth} \qquad (6.58)$$

6.5.2 The laser threshold condition

We now consider injection of an external beam into a laser cavity containing a medium that exhibits optical gain. Equation (6.42) still holds under these conditions, but eqn (6.43) must be modified to allow for amplification of the right- and left-going beams in the gain medium:

$$a_-(0, \omega) = a_+(0, \omega) \times r_2 \exp[i(\delta_{rt} - \phi_1)] \qquad (6.59)$$
$$\times \exp[\alpha_0(\omega)\ell_g] \exp[-\kappa(\omega)L_c], \qquad (6.60)$$

where $\alpha_0(\omega)$ is the small-signal gain coefficient, ℓ_g is the length of the gain medium, and $\exp[-\kappa(\omega)L_c]$ represents attenuation of the beams throughout the cavity by scattering or other losses.

It is important to note we use the *small-signal* gain coefficient α_0, rather than the saturated gain coefficient α_I, when considering the conditions required to reach the *threshold* for lasing. The reason is simply that close to threshold the intensity of the laser radiation within the laser cavity will be small compared to the saturation intensity. Note too that the amplitude of the beam increases by a factor of $\exp[\alpha_0(\omega)\ell_g]$, not $\exp[2\alpha_0(\omega)\ell_g]$, in one round trip through the gain medium since we are dealing here with amplitudes rather than intensities.

The derivation of the amplitude of the right-going wave at M_1 now proceeds exactly as before, and we find

$$a_+(0, \omega) = \frac{a_{in}(0, \omega)}{1 - r_1 r_2 \exp(i\delta_{rt}) \exp[\alpha_0(\omega)\ell_g] \exp[-\kappa(\omega)L_c]}. \qquad (6.61)$$

The introduction of gain into the laser cavity means that it is now possible for the amplitude of the right- (and left-) going beam to be finite even if no beam is injected into the cavity, i.e. $a_{in}(0, \omega) = 0$. In other words oscillation occurs. The **threshold condition** for oscillation is simply that the denominator in eqn (6.61) becomes zero, which occurs for some value of the small-signal gain $\alpha_0^{\text{th}}(\omega)$ given by:

$$r_1 r_2 \exp(i\delta_{\text{rt}}) \exp[\alpha_0^{\text{th}}(\omega)\ell_g] \exp[-\kappa(\omega)L_c] = 1. \tag{6.62}$$

Equation (6.62) is a complex equation, and hence may be divided into real and imaginary parts. The imaginary part of the right-hand side is zero; this condition, plus the fact that the left-hand side of eqn (6.62) must be positive, yields the condition:

$$\delta_{\text{rt}} = 2\pi p, \tag{6.63}$$

i.e. the frequency of the radiation must correspond to one of the cavity modes according to eqn (6.7). Equating the real parts of eqn (6.62) gives

$$r_1 r_2 \exp[\alpha_0^{\text{th}}(\omega)\ell_g] \exp[-\kappa(\omega)L_c] = 1, \tag{6.64}$$

or, taking the square-modulus of both sides

$$R_1 R_2 \exp[2\alpha_0^{\text{th}}(\omega)\ell_g] \exp[-2\kappa(\omega)L_c] = 1.$$
Threshold condition (6.65)

This last result has a straightforward interpretation: if we imagine launching a beam of unit intensity from within the cavity, after one round-trip it would have an intensity equal to the left-hand side of eqn (6.65); if this is less than unity the beam will decay in intensity with each round trip; if it is greater than unity, the beam will grow in intensity.

To summarize, laser oscillation can only occur if *for some cavity mode the unsaturated round-trip gain exceeds the round-trip loss*. The threshold condition determines—for a given laser cavity—a threshold value, α_0^{th}, for the gain coefficient. In turn, this defines threshold values for the population inversion and the rate of pumping of the upper laser level.

6.6 Laser oscillation above threshold

We now consider the behaviour of a steady-state laser oscillator as the pumping of the upper laser level is increased from below to above the threshold level. As the pumping is increased the population of the upper laser level and the rate of spontaneous emission on the laser transition will both steadily increase. As a consequence, the energy in those cavity modes with frequencies lying within the lineshape of the laser transition will increase. However, whilst the pumping remains below threshold the intensity of radiation within the cavity will be low, the number of photons per mode will typically be much less than unity; the output of the laser will be weak, will be dominated by spontaneous emission, and will have a spectrum corresponding to that of the transition lineshape modulated by the resonance response of the cavity. Under these conditions the 'laser' is nothing more than a classical incoherent light source located within an optical cavity.

Further increases in the pumping will bring the population inversion density and the small-signal gain coefficient to the point that the threshold condition (6.65) is satisfied for the cavity mode closest to the line centre. The laser is poised to oscillate. As clarified below, if the pumping is exactly at the threshold value the intensity of the laser radiation circulating within the laser cavity will be essentially zero. However, raising the pumping just above the threshold value enables laser oscillation to occur—increasing the energy density and number of photons in the oscillating mode by many orders of magnitude.[49] Increasing the pumping significantly above the threshold value will always increase the intensity of the radiation circulating within the cavity, and the radiation output by the laser, but the detailed behaviour depends on whether the laser transition is homogeneously or inhomogeneously broadened.

[49]The number of photons in *other* cavity modes will remain low.

6.6.1 Condition for steady-state laser oscillation

Before describing the detailed behaviour of lasers with homogeneously and inhomogeneously broadened gain medium, we first clarify the conditions that must be satisfied when a laser operates under steady-state conditions.

We emphasize that increasing the pumping—even above the threshold value—will *always* increase the unsaturated population inversion density $N^*(0)$ and the small-signal gain coefficient α_0. This must be so since these quantities depend only on the pump rates and the properties of the laser levels. However, once laser oscillation is established, the gain experienced by an oscillating mode is not the small-signal gain coefficient, but the *saturated* gain coefficient α_I. As a consequence, eqn (6.65) does not hold in a laser operating above threshold since the population inversion will be saturated, at least to some extent, by the intense laser radiation circulating within the cavity. What, then, *can* we say about a laser operating in the steady-state above threshold?

We noted that at the threshold pumping the round-trip gain experienced by one of the cavity modes must equal the round-trip loss. This energy balance must also be maintained in a laser operating above threshold. In other words:

> in a laser operating under steady-state conditions the round-tip gain experienced by an oscillating mode must equal the round-tip loss, and hence must remain clamped at the threshold value.[50]

[50]The refractive index of a medium depends, very slightly, on the distribution of population over the atomic levels. Thus, when a population inversion is generated the refractive index of the gain medium will undergo a small change. As a consequence, the frequencies of the modes of the cavity will change slightly. In general, it is found that the frequencies of the cavity modes are 'pulled' towards the centre frequency of the laser transition. The extent of this 'frequency pulling' is typically around 10^{-3} of the longitudinal mode spacing, and hence can usually be neglected. See Section 16.1.

In an operating laser that round-trip gain is not determined by the small-signal gain α_0, but by the *saturated* gain α_I.[51] Under steady-state conditions this round-trip gain experienced by an oscillating mode must balance the cavity losses, and hence must remain clamped at this level even if the pumping is increased beyond the threshold. The mechanism for this behaviour is simply that as the intensity of the oscillating mode increases, it burns the population inversion - and hence the saturated gain - down to the threshold value set by the cavity losses.

[51]In general, the gain experienced by an oscillating mode must be calculated by integrating equations describing the growth in beam intensity (i.e. equations of the form of eqn (5.17)) over one round trip of the cavity; in Sections 6.7.1 and 6.7.2 we show how this may be done in simple cases. Whether or not the system is amenable to calculation, the condition for steady-state operation fixes the total round-trip gain of an oscillating mode to a value determined by the cavity losses.

We therefore have the following picture: increasing the pumping always increases the unsaturated population inversion $N^*(0)$ and small-signal gain coefficient α_0. However, once the laser is above threshold, increasing the pumping also increases the intensity of the oscillating mode, burning down the population inversion to a level that ensures that the saturated gain coefficient α_I experienced by the oscillating mode is clamped at the threshold value.

These general observations apply to both homogeneously and inhomogeneously broadened transitions. However, the detailed behaviour depends on the class of broadening, as discussed below.

6.6.2 Homogeneously broadened systems

Figure 6.19 shows the behaviour of the saturated gain coefficient $\alpha_I(\omega - \omega_0)$ and the output power of the laser as the level of pumping is increased. In Fig. 6.19(a) the pumping is exactly at the threshold value so that the small-signal gain experienced by the longitudinal mode nearest the line centre is equal to the threshold value given by eqn (6.65). At this point the intensity of the radiation in this mode will be essentially zero since any significant intensity would reduce the saturated gain coefficient below the threshold value.

In Fig. 6.19(b) the pumping has been increased above the threshold value. Now, the intensity of the oscillating mode is high, giving significant laser output power at this frequency. However, despite the increased pumping, the saturated gain coefficient is unchanged from that shown in (a). This occurs because the gain experienced by the oscillating mode must remain clamped at

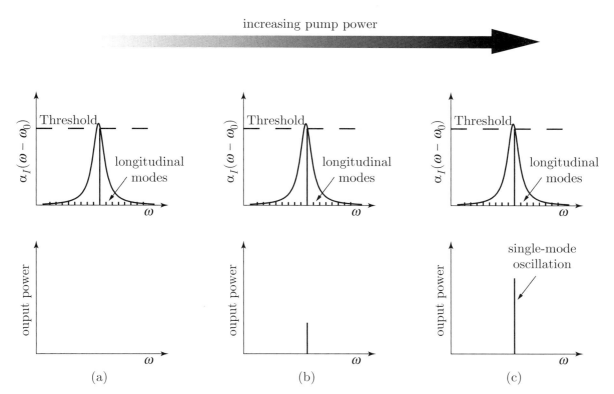

Fig. 6.19: Behaviour of the saturated gain coefficient $\alpha_I(\omega - \omega_0)$ and the output power of a steady-state *homogeneously* broadened laser as the pumping is increased: (a) pumping at the threshold value; (b) pumping above threshold; (c) pumping further above threshold.

the threshold value and—since the transition is homogeneously broadened—radiation on the oscillating mode interacts with all the atoms in the medium with the same strength. Thus, the population inversion must remain clamped at the threshold value.

Further increases in pumping, as shown in Fig. 6.19(c), will increase the intensity of the oscillating mode, but will not change the saturated gain coefficient. According to this picture, only the cavity mode with the largest gain can ever oscillate since the gain available to other cavity modes will always remain below the threshold for oscillation.

Spatial hole-burning

In practice, homogeneously broadened lasers *do* sometimes oscillate on more than one cavity mode. This behaviour arises from the fact that an oscillating mode forms a standing wave within the cavity, and consequently the spatial distribution of intensity is not uniform. Near the antinodes of the standing wave the population inversion will be burnt down to the threshold value; however, near the nodes the intensity will be low, and the population inversion will remain essentially unsaturated. The non-uniform burning down of the population inversion in a homogeneously broadened laser system is known as **spatial hole-burning**, by analogy with the spectral hole-burning discussed below. Spatial hole-burning allows other modes—with slightly different frequencies, and hence different positions of the nodes and antinodes—to feed off regions of unsaturated population inversion and reach the threshold for oscillation. In such circumstances more than one cavity mode can oscillate, leading to **multimode oscillation** at several frequencies within the linewidth of the transition. This process is illustrated schematically in Fig. 6.20.

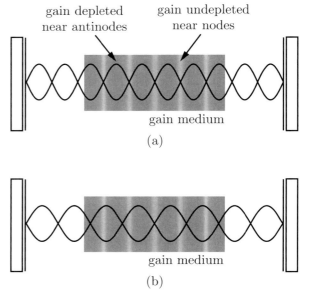

Fig. 6.20: Spatial hole-burning in a homogeneously broadened steady-state laser oscillator. In (a) the standing wave of an oscillating mode causes the gain to be depleted where the intensity is high. However, the gain is unaffected near the nodes of the standing wave since the intensity in these regions is low. This can allow other cavity modes—which will have a different frequency and hence different locations for the nodes and antinodes in their standing-wave pattern—to feed off the regions of unused gain, as shown in (b). In both plots, regions of heavily saturated gain are darker than those regions that are unsaturated or only lightly saturated.

6.6.3 Inhomogeneously broadened systems[†]

The behaviour of inhomogeneously broadened lasers is quite different. Once again, as the pumping is increased to the threshold value the mode with the largest gain will start to oscillate. If the pumping is increased further, the gain experienced by the oscillating mode must remain clamped at the threshold value. However, this mode only interacts with one particular class of atoms; the population inversion for *other* classes of atom will continue to increase, increasing the available gain at their centre frequencies. The saturated gain coefficient $\alpha_I^D(\omega - \omega_0)$ will therefore develop a **spectral hole** at the frequency of the oscillating mode as shown in Fig. 6.21(b).[52]

Further increases in pumping will therefore increase the gain at all other frequencies. Once the gain available to a cavity mode reaches the threshold value it can start to oscillate; further increases in the pumping increase the intensity of the oscillating mode but do not change the saturated gain at that frequency, which remains clamped at the threshold value, forming an additional spectral hole. Thus, as shown in Fig. 6.21(c), increases in pumping are accompanied by **multimode** oscillation on all cavity the modes that can reach threshold, with each oscillating mode drilling a spectral hole in the gain profile.

[52]As discussed in Section 5.1.2 the width of the spectral hole will be of the order of the homogeneous linewidth of the laser transition.

Spectral hole-burning in a Doppler-broadened medium

If the gain medium is inhomogeneously broadened by the Doppler effect, then our discussion above should be modified slightly.

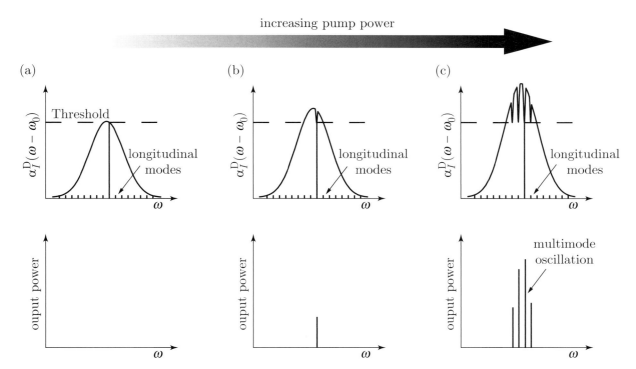

Fig. 6.21: Behaviour of the saturated gain coefficient $\alpha_I^D(\omega - \omega_0)$ and the output power of a steady-state *inhomogeneously* broadened laser as the pumping is increased: (a) pumping at the threshold value; (b) pumping above threshold; (c) pumping further above threshold.

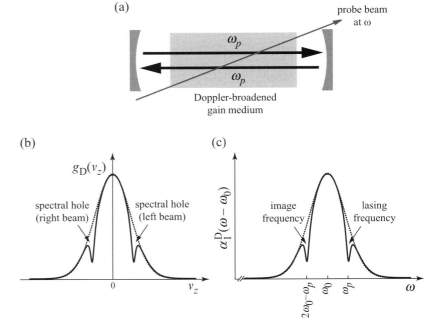

Fig. 6.22: Formation of a spectral hole at the image frequency for a Doppler-broadened gain medium. It is assumed that the frequency ω_p of the lasing mode is greater than the centre frequency of the transition. The right-going beam in (a) is resonant (that is, Doppler-shifted into resonance in the rest frame of the atom) with atoms with $v_z < 0$. This causes a hole to be burnt in the velocity distribution $g_D(v_z)$, as shown in (b). Similarly, the left-going beam burns a hole in the distribution for atoms with a velocity equal but opposite to those interacting with the right-going beam. Thus, *two* holes are burnt in the velocity distribution and, as shown in (c), this results in *two* spectral holes being burnt in the gain profile measured by a weak, right-going probe beam: one at the frequency of the lasing mode and one at the image frequency $2\omega_0 - \omega_p$.

Suppose that one mode is oscillating with a frequency ω_p, as shown in Fig. 6.22. The right-going beam will interact with those atoms with a velocity v_z given by

$$\omega_0 - \frac{v_z}{c}\omega_0 = \omega_p. \tag{6.66}$$

In contrast, the left-going beam will interact with those atoms for which,

$$\omega_0 + \frac{v_z'}{c}\omega_0 = \omega_p. \tag{6.67}$$

Hence, the oscillating mode will burn spectral holes in two velocity classes of atoms:

$$\frac{v_z}{c} = \frac{\omega_0 - \omega_p}{\omega_0} \tag{6.68}$$

$$\frac{v_z'}{c} = -\frac{\omega_0 - \omega_p}{\omega_0}. \tag{6.69}$$

If the gain profile were measured by a weak (probe) beam of tunable radiation travelling towards positive z, spectral holes would be detected at frequencies ω for which these two saturated classes are shifted into resonance, i.e. at frequencies:

$$\omega = \omega_0 - \frac{v_z}{c}\omega_0 = \omega_p \tag{6.70}$$

$$\omega = \omega_0 - \frac{v_z'}{c}\omega_0 = 2\omega_0 - \omega_p. \tag{6.71}$$

We see that spectral holes are burnt at the frequency of the oscillating mode ω_p, and at the **image frequency** $2\omega_0 - \omega_p$. Note that no laser oscillation occurs at the image frequency.

6.7 Output power

We now turn to the matter of the output power of the laser—a point that is clearly of great practical importance. In doing so, we restrict ourselves to the simplest cases only: steady-state lasers operating on homogeneously broadened laser transitions. Whilst this is clearly a severe restriction, useful conclusions may still be drawn for more general situations such as lasers operating in a pulsed mode.

6.7.1 Low-gain lasers

The parameters of the problem to be solved are illustrated schematically in Fig. 6.23.

We consider first the growth of the right- and left-going beams in the oscillating mode. The total intensity of the right-going beam will grow according to,

$$\frac{dI_+(z)}{dz} = \alpha_I I_+(z), \tag{6.72}$$

where α_I is the *saturated* gain coefficient at the frequency of the oscillating mode. A similar expression can be written for the intensity of the left-going wave.

If the laser is a low-gain system, the increase in the intensities of the beams as they propagate between the cavity mirrors is small, and consequently (assuming uniform pumping) the degree of saturation of the gain coefficient does not change appreciably with position z in the gain medium. Further, the change in intensity in one pass of the gain medium is small, so that we may write,

$$\frac{dI_+}{dz} \approx \frac{I_+(\ell_g) - I_+(0)}{\ell_g} = \alpha_I I_+. \tag{6.73}$$

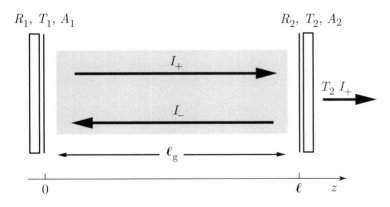

Fig. 6.23: Parameters relevant to the calculation of the output power of a low-gain continuous-wave laser.

If we launch a beam once round the cavity, the fractional increase in intensity—the **fractional round-trip gain**—would be,

$$\delta_{\text{gain}} = 2\frac{I_+(\ell_g) - I_+(0)}{I_+(0)} = 2\alpha_I \ell_g, \quad (6.74)$$

where the factor 2 accounts for the fact that the beam passes twice through the gain medium.

In an similar way, we can characterize the losses arising from absorption and the finite reflectivity of the cavity mirrors by considering the intensity of a beam after one round trip in a cavity with no gain:

$$\begin{aligned}\delta_{\text{total loss}} &= \frac{1 - R_1 R_2 \exp(-2\kappa_I L_c)}{1} = 1 - R_1 R_2(1 - 2\kappa_I L_c) \\ &\approx 1 - (1 - A_1 - T_1)(1 - A_2 - T_2)(1 - 2\kappa_I L_c) \\ &\approx (A_1 + A_2 + T_1 + 2\kappa_I L_c) + T_2, \end{aligned} \quad (6.75)$$

where we have used the fact that for each mirror $R + A + T = 1$, where A and T are the absorption and transmission coefficients. In the above we have distinguished between unwanted losses, which arise from absorption at the mirrors or in the medium between them, and the useful coupling of power through the right-hand mirror, T_2. It is useful, then, to define the **fractional round-trip loss** in terms of the unwanted loss only:

$$\delta_{\text{loss}} = A_1 + A_2 + T_1 + 2\kappa_I L_c. \quad (6.76)$$

Under steady-state conditions the round-trip gain must be balanced by the round-trip losses: $\delta_{\text{gain}} = \delta_{\text{loss}} + T_2$. Hence, from eqn (6.74), we may write:

$$2\alpha_I \ell_g = \delta_{\text{loss}} + T_2. \quad \textbf{Low-loss threshold condition} \quad (6.77)$$

The saturated gain coefficient is, from eqn (5.16), given by

$$\alpha_I = \frac{\alpha_0}{1 + I(z)/I_s}, \quad (6.78)$$

where $I(z) = I_+(z) + I_-(z)$ is the *total* intensity of the right- and left-going beams at a point z in the gain medium, and the saturation intensity is evaluated at the frequency of the oscillating mode. We have already established that the intensity of the right- and left-going beams will not vary strongly with position, and hence the two beams will have almost the same intensity. Thus, $I(z) \approx 2I_+(z) \approx 2I_+$ and the condition for steady-state oscillation then becomes

$$2\frac{\alpha_0}{1 + 2I_+/I_s}\ell_g = \delta_{\text{loss}} + T_2. \quad (6.79)$$

From this we find the intensity of the right-going beam to be

$$I_+ = \frac{1}{2}I_s\left(\frac{2\alpha_0 \ell_g}{T_2 + \delta_{\text{loss}}} - 1\right). \quad (6.80)$$

The power coupled out of the cavity is simply $T_2 I_+ A_{mode}$, where A_{mode} is the cross-sectional area of the beam. Hence, the output power of the laser is,

$$P = \frac{1}{2} I_s A_{mode} T_2 \left(\frac{2\alpha_0 \ell_g}{T_2 + \delta_{loss}} - 1 \right). \quad (6.81)$$

The general form of the variation of output power with T_2 (in this context usually referred to as the **output coupling**) is illustrated schematically in Fig. 6.24. The variation may be explained as follows:

- At large values of T_2 the losses of the cavity are larger than the maximum possible round-trip gain and consequently no laser oscillation occurs.[53]
- As T_2 is decreased, the threshold for laser oscillation is reached when $2\alpha_0 \ell_g = T_2^{th} + \delta_{loss}$.
- As T_2 is decreased below the threshold value the lower cavity losses mean that the required round-trip gain also decreases. This occurs through an increase in the intra-cavity intensity. Note that as T_2 is decreased below threshold the intra-cavity intensity *always* increases.
- Eventually, the decrease in the fraction of power coupled out through T_2 outweighs the increase in intra-cavity intensity and the output power decreases. Clearly the output power will be zero when $T_2 = 0$, although the intra-cavity intensity will then be at its greatest.

Finally, we note that the output power varies linearly with the small-signal gain coefficient α_0. The small-signal gain is proportional to the *unsaturated* population inversion, and this in turn will often vary approximately linearly with the rate of pumping of—or, equivalently, the pumping power supplied to—the upper level. As a consequence, the output power of a laser operating above threshold often increases linearly with the pumping (e.g. with the discharge current, the diode current, the flashlamp energy, etc.), as illustrated schematicaly in Fig. 6.25.[54]

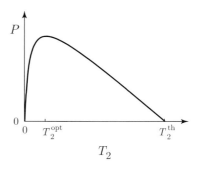

Fig. 6.24: Output power vs. the transmission T_2 of the output coupler for a low-gain laser system (eqn (6.81)).

[53] In this regime eqn (6.81) yields a negative output power which, of course, is unphysical.

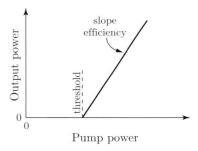

Fig. 6.25: Idealized plot showing the output power P of a laser oscillator as a function of pumping power P_{pump} of the upper laser level. Laser oscillation occurs for pump rates above a threshold value, and typically increases approximately linearly with the pump power. The gradient of the line, dP/dP_{pump}, is known as the **slope efficiency**.

[54] See, for example, eqn (9.39).

Optimum output coupling

The optimum output coupling T_2^{opt} is found straightforwardly by differentiating eqn (6.81):

$$T_2^{opt} = \sqrt{2\alpha_0 \ell_g \delta_{loss}} - \delta_{loss}, \quad (6.82)$$

which yields an optimum output power of

$$P_{opt} = (\alpha_0 \ell_g) I_s A_{mode} \left[1 - \sqrt{\frac{\delta_{loss}}{2\alpha_0 \ell_g}} \right]^2. \quad (6.83)$$

There are several points to note from this result. The maximum possible output is achieved when the losses are zero, whereupon the output intensity P_{opt}/A_{mode} is equal to $(\alpha_0 \ell_g) I_s$. Since we are considering low-gain systems, $\alpha_0 \ell_g$ cannot be large, and hence the largest output intensity must be of order the saturation intensity. It is also clear that the optimum output power is very sensitive to the losses. For example, if δ_{loss} is only 10% of the fractional round-trip gain, $2\alpha_0 \ell_g$, the maximum output power is reduced to less than 50% of that which could be achieved in the absence of unwanted losses. This sensitivity to cavity losses is typical of low-gain laser systems, and reflects the fact that the intensity of the radiation circulating round the laser cavity is very much higher than that of the output beam. As such, the absolute power scattered by a small intra-cavity loss can be a high proportion of the power coupled out of the cavity.

6.7.2 High-gain lasers: the Rigrod analysis[†]

For lasers in which the single-pass gain is not small, we cannot assume that the right- and left-going beams have near-constant intensity. Instead we must solve the equations describing the growth in intensity of the right- and left-going beams, subject to the boundary conditions imposed by the losses at the cavity mirrors.[55]

[55]This analysis was first described by Rigrod (1965).

For the right- and left-going beams we have,

$$\frac{dI_+(z)}{dz} = \alpha_I I_+(z) = \frac{\alpha_0 I_+(z)}{1 + [I_+(z) + I_-(z)]/I_s} \quad (6.84)$$

$$\frac{dI_-(z)}{dz} = -\alpha_I I_-(z) = -\frac{\alpha_0 I_-(z)}{1 + [I_+(z) + I_-(z)]/I_s}. \quad (6.85)$$

To reduce clutter, we define right- and left-going intensities normalized to the saturation intensity, $J_\pm(z) = I_\pm/I_s$, whereupon our equations become

$$\frac{dJ_+(z)}{dz} = \alpha_I J_+(z) = \frac{\alpha_0 J_+(z)}{1 + J_+(z) + J_-(z)} \quad (6.86)$$

$$\frac{dJ_-(z)}{dz} = -\alpha_I J_-(z) = -\frac{\alpha_0 J_-(z)}{1 + J_+(z) + J_-(z)}. \quad (6.87)$$

By adding eqns (6.86) and (6.87) we find

$$J_-(z)\frac{dJ_+(z)}{dz} + J_+(z)\frac{dJ_-(z)}{dz} = \frac{d}{dz}\left[J_-(z)J_+(z)\right] = 0. \quad (6.88)$$

Hence, we conclude that for all points in the gain medium

$$J_+(z)J_-(z) = C, \quad (6.89)$$

where C is a constant.

It is now straightforward to integrate eqns (6.86) and (6.87) over the length of the gain medium by using eqn (6.89) to eliminate the unwanted variable. We find

$$\ln\left[\frac{J_+(\ell_g)}{J_+(0)}\right] + [J_+(\ell_g) - J_+(0)] - C\left[\frac{1}{J_+(\ell_g)} - \frac{1}{J_+(0)}\right] = \alpha_0 \ell_g \quad (6.90)$$

$$\ln\left[\frac{J_-(\ell_g)}{J_-(0)}\right] + [J_-(\ell_g) - J_-(0)] - C\left[\frac{1}{J_-(\ell_g)} - \frac{1}{J_-(0)}\right] = -\alpha_0 \ell_g. \quad (6.91)$$

Subtracting these last two equations then yields

$$\ln\left[\frac{J_+(\ell_g)}{J_-(\ell_g)}\frac{J_-(0)}{J_+(0)}\right] + 2[J_+(\ell_g) - J_+(0) - J_-(\ell_g) + J_-(0)] = 2\alpha_0\ell_g, \tag{6.92}$$

where we have eliminated C using eqn (6.89).

In order to proceed any further we need the boundary conditions at the left- and right-hand mirrors:[56]

$$J_+(0) = R_1 J_-(0) \tag{6.93}$$

$$J_-(\ell_g) = R_2 J_+(\ell_g). \tag{6.94}$$

Substituting the boundary conditions into eqn (6.89) we then find,

$$C = J_+^2(\ell_g) R_2 = \frac{J_+^2(0)}{R_1}, \tag{6.95}$$

from which we find the relation between $J_+(0)$ and $J_+(\ell_g)$:

$$J_+(0) = \sqrt{R_1 R_2} J_+(\ell_g). \tag{6.96}$$

[56]Note that we *do not* assume that the gain medium fills the laser cavity; in any space between the end of the gain medium and the cavity mirrors, the beams propagate with constant intensity and return with an intensity reduced by the reflectivity of the mirror.

Substituting this last result, and the boundary conditions (6.93) and (6.94) into eqn (6.92) we find an expression for $J_+(\ell_g)$:

$$J_+(\ell_g) = \frac{\alpha_0\ell_g + \ln\sqrt{R_1 R_2}}{\left(1 - \sqrt{R_1 R_2}\right)\left(1 + \sqrt{R_2/R_1}\right)}. \tag{6.97}$$

The output power is given by $P = T_2 J_+(\ell_g) I_s A_{\text{mode}}$:

$$P = \frac{T_2 I_s A_{\text{mode}}}{\left(1 - \sqrt{R_1 R_2}\right)\left(1 + \sqrt{R_2/R_1}\right)} \left(\alpha_0\ell_g + \ln\sqrt{R_1 R_2}\right). \tag{6.98}$$

We may make several observations. For a finite output power the contents of the right-hand bracket must be positive, which requires $2\alpha_0^{\text{th}}\ell_g = -\ln R_1 R_2$. This is just the threshold condition (6.65) for the case when there are no intra-cavity losses, as assumed here.

If the reflectivity of the rear mirror is 100%, the output power is given by

$$P = \frac{T_2 I_s A_{\text{mode}}}{(1 - R_2)} \left[\alpha_0\ell_g + \ln\sqrt{R_2}\right]. \tag{6.99}$$

In this case the maximum output power clearly occurs when $R_2 \to 1$, assuming that the output coupler is also loss-less so that $T_2 = 1 - R_2$. The maximum possible output power from the laser is then

$$P_{\text{max}} = (\alpha_0\ell_g) I_s A_{\text{mode}}, \tag{6.100}$$

as we found in Section 6.7.1 for the low-gain case.

Figure 6.26 shows the calculated output powers for two values of $\alpha_0\ell_g$. It is seen that for the relatively large single-pass gains considered here, the output power shows a broad maximum with the output coupling T_2, and we conclude that for high-gain laser systems the output coupling is not critical. However, it is clear that even fairly small losses at the rear mirror cause the

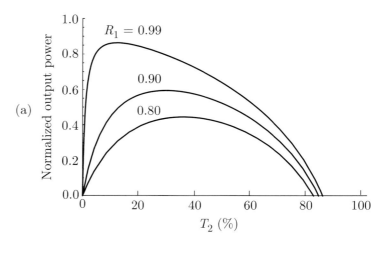

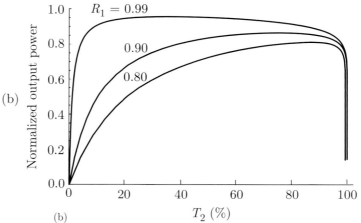

Fig. 6.26: Calculated output power using the Rigrod analysis for cavities with various reflectivity rear mirrors for small-signal, single-pass gains of: (a) $\alpha_0 \ell_g = 1$; (b) $\alpha_0 \ell_g = 10$. The output power has been normalized to the maximum possible output (eqn (6.100)).

output power to decrease significantly. By the same token, although it is not explicitly included in our analysis, non-saturable losses in the gain medium itself can cause substantial decreases in the output power of the laser—just as we found in the low-gain case.

It is possible to differentiate eqn (6.98) to derive an expression for the optimum output coupling. However, this procedure results in a transcendental equation for T_2 and is therefore not particularly useful. It is not difficult to plot Rigrod curves for a given laser system to estimate the optimum output coupling, but it is more common to determine the optimum output coupling experimentally by using a rear mirror with the highest reflectivity achievable at the laser wavelength and measuring the output power for several different output couplers.

Finally, we note that as the single-pass small-signal gain decreases, the power curves predicted by the Rigrod formula become identical with those predicted by eqn (6.81) for the case when the only cavity loss arises from the finite reflectivity of the mirrors.

6.7.3 Output power in other cases

The two analyses above allow us to draw some general conclusions. First, relatively small intra-cavity losses can have a significant effect on the output of the laser. Secondly, for large values of the small-signal, single-pass gain the output does not depend critically on the output coupling employed; in contrast, for low-gain systems the output is very sensitive to the choice of output coupling.

Further reading

A more detailed description of the properties of Gauss–Hermite beams is given in (Brooker, 2003). An extensive treatment of optical beams and resonators may be found in the textbook by Siegman (1986); the same author has also written a pair of excellent review papers on this subject (2000a, 2000b) and a very useful tutorial[57] on the M^2 parameter of a beam and how to measure it (Siegman, 1998).

An analysis of unstable cavities, and in particular the role of the magnifying and demagnifying waves in establishing laser oscillation, is given in a series of papers by Zemskov et al. (1974) and Isaev et al. (1974), (1975), (1977).

Exercises

(6.1) Schawlow–Townes treatment of open cavities

    (a) From eqn (6.1), show that the frequencies of modes with $k_x, k_y \ll k_z$ are given by eqn (6.3).

    (b) Find approximate values for the frequency of the pth longitudinal mode and for the frequency separation, $\Delta \nu_{\text{long}}$, of adjacent longitudinal modes. Show that the wavelength λ_p of this mode is related to the length of the cavity by,

$$L_c \approx \frac{\lambda_p}{2} p. \qquad (6.101)$$

    (c) Show that the frequency separation of adjacent transverse modes, $\Delta \nu_{\text{trans}}$, is of order $(\lambda_p/a)^2$ times the frequency of the pth longitudinal mode.

    (d) Hence, show that the ratio of the frequency separation of adjacent transverse modes to that of adjacent longitudinal modes is of order

$$\frac{\lambda_p}{a} \frac{L_c}{a}.$$

Comment on this result.

    (e) Estimate $\Delta \nu_{\text{long}}$ and $\Delta \nu_{\text{trans}}$ for a He-Ne laser operating at a wavelength of 633 nm in a 30-cm long cavity formed by mirrors of diameter 10 mm.

(6.2) Properties of Gauss–Hermite beams

    (a) We first consider a spherical wave of angular frequency ω originating at the point $(0, 0, z_0)$. The amplitude of the wave may be written in the form

$$U(x, y, z) = U_0' \frac{\exp\{i[kR(z) - \omega t]\}}{R(z)},$$

---

[57] A version of this paper is also available at www.stanford.edu/siegman/beam_quality_tutorial_osa.pdf.

where $R(z)^2 = (z - z_0)^2 + r^2$ is the radius of curvature of the wavefronts and $r^2 = x^2 + y^2$. Show that in the paraxial approximation (i.e. $r \ll |z - z_0|$) we may write

$$R(z) \approx z - z_0 + \frac{r^2}{2(z - z_0)} = q + \frac{r^2}{2q},$$

where we define $q = z - z_0$.

(b) Hence, show that within the paraxial approximation we may write the wave in the form

$$U(x, y, z) = U_0 \exp\{i[kz - \omega t]\}$$
$$\times \frac{\exp(ikr^2/2q)}{q},$$

where $U_0 = U_0' \exp(-ikz_0)$. Compare this to the form of the lowest-order Gauss–Hermite beam of eqn (6.8).

(c) Now consider a lowest-order Gauss–Hermite with a complex beam parameter $q = z - z_0 - iz_R$. By substituting for q, show that eqn (6.8) may be written in the form of eqns (6.10) to (6.14). Interpret the form of these equations.

(d) For the remainder of the question, take the waist of the beam to be located at $z = z_0 = 0$. Sketch the following on either side of the beam waist as a function of z/z_R:
   (i) the radius of curvature $R(z)$ of the wavefronts;
   (ii) the wavefronts of the beam;
   (iii) the spot size $w(z)$;
   (iv) $\tan\alpha(z)$ and $\alpha(z)$.

(e) What is the additional phase shift relative to a plane wave that the beam accumulates as it propagates between two planes located many Rayleigh ranges either side of the beam waist?[58] When might this phase shift be important?

(6.3) Matching a Gauss–Hermite beam to a cavity
In this problem we deduce whether it is possible to match a Gauss–Hermite beam to a given optical cavity and, if it is, deduce the properties of the modes in terms of those of the cavity.

(a) From eqns (6.16) and (6.17) show that

$$z_1^2 = -z_R^2 - R_1 z_1 \quad (6.102)$$
$$z_2^2 = -z_R^2 + R_2 z_2. \quad (6.103)$$

(b) Using the fact that $z_2^2 - z_1^2 \equiv (z_2 - z_1)(z_2 + z_1) = L_c(z_2 + z_1)$, show that the mirrors must be located at

$$z_1 = \frac{-g_2(1 - g_1)}{g_1(1 - g_2) + g_2(1 - g_1)} L_c \quad (6.104)$$

$$z_2 = \frac{g_1(1 - g_2)}{g_1(1 - g_2) + g_2(1 - g_1)} L_c, \quad (6.105)$$

where g_1 and g_2 are given by eqns (6.20) and (6.21).

(c) Hence, use eqn (6.104) or eqn (6.105) to show that the Rayleigh range z_R of the modes is given by eqn (6.19), and hence show that the spot size at the mode waist $w_0 \equiv w(0)$ is given by,

$$w_0^2 = \frac{\sqrt{g_1 g_2 (1 - g_1 g_2)}}{g_1(1 - g_2) + g_2(1 - g_1)} \frac{\lambda L_c}{\pi}. \quad (6.106)$$

(d) As discussed in the text, for the mode to be low loss the waist size and the Rayleigh range must both be real. Show that for this to be the case, eqn (6.22) must be satisfied.

(e) We can now calculate the spot sizes of the beam at the cavity mirrors. Use eqn (6.11), and the deduced values for z_1, z_2, and z_R to show that the spot sizes at the two mirrors $w_1 \equiv w(z_1)$ and $w_2 \equiv w(z_2)$ are given by:

$$w_1^2 = \sqrt{\frac{g_2}{g_1(1 - g_1 g_2)}} \frac{\lambda L_c}{\pi} \quad (6.107)$$

$$w_2^2 = \sqrt{\frac{g_1}{g_2(1 - g_1 g_2)}} \frac{\lambda L_c}{\pi}. \quad (6.108)$$

(f) Consider now a **symmetric cavity**, so that $R_1 = R_2 = R$ and $g_1 = g_2 = g$. Sketch the spot size as a function of g at the cavity mirrors, and comment on your results.

(6.4) Important cavity types
In this problem we consider several important types of low-loss cavity formed by mirrors of radii R_1 and R_2, separated by a cavity length L_c. We define the cavity parameters $g_1 = 1 - L_c/R_1$ and $g_2 = 1 - L_c/R_2$.

(a) Sketch a graph showing for which regions of the (g_1, g_2) plane cavities may be constructed that support at least one low-loss mode.

---

[58]This is an example of the **Guoy effect** in which a beam undergoes an extra phase shift in passing through a focus. For further information consult Siegman (1986)

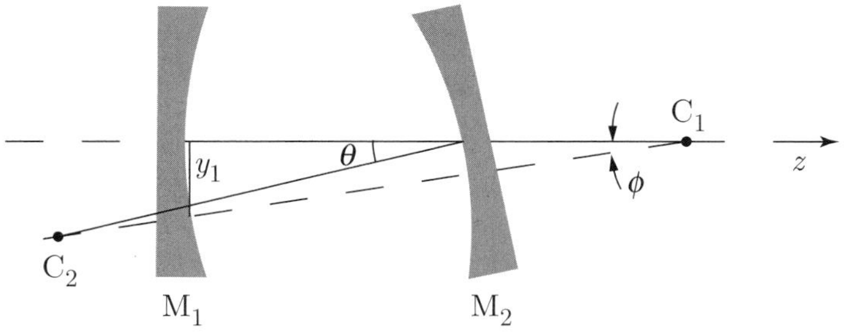

Fig. 6.27: Effect of angular misalignment of a laser cavity. The centres of curvatures of the two mirrors are labelled C_1 and C_2.

(b) Derive expressions for the spot size at the mode waist and at the two cavity mirrors for the following important cavities:
   (i) A symmetric confocal cavity with $R_1 = R_2 = L_c$.
   (ii) A near-planar cavity with $R_1 \approx R_2 \approx \infty$.
   (iii) A near-concentric cavity with $R_1 \approx R_2 = R = L_c/2 + \Delta L_c$.
   (iv) A near-hemispherical cavity with $R_1 = \infty$, $R_2 = L_c + \Delta L_c$.

   In each case find expressions for g_1 and g_2, and indicate on the sketch of part 6.4(a) where the cavity resides.

(c) Show that *asymmetric* confocal resonators do not support a low-loss mode, and indicate on the sketch of part 6.4(a) the locus of cavities of this type. Indicate which part of the locus represents the positive and negative branches.

(6.5) A diode-laser-pumped solid-state laser operating at 1030 nm is constructed using a thin disc of Yb:YAG laser crystal. A highly reflecting coating on one of the crystal faces acts as one of the mirrors to form a near-hemispherical cavity as shown in Fig. 6.28. The pump laser beam is focused to a waist of spot radius 400 μm inside the Yb:YAG crystal. For efficient operation it is desired to have a cavity mode diameter of 400 μm in the crystal.

(a) By considering the radius of curvature of the lowest-order Gaussian beam, show why the output coupling mirror must have a radius of curvature greater than 244 mm. (Neglect the thickness of the crystal.)

(b) Two concave output coupler mirrors M_1 and M_2 are available, with radii of curvature of 250 mm and 500 mm, respectively. For each output coupler mirror calculate the cavity lengths required to produce a stable cavity mode with a diameter of 400 μm in the laser crystal.

(c) As the pump power is increased well above threshold, thermal effects result in the crystal acting as a thin positive lens. Determine which of the cavities considered above is the more stable against this thermal lensing.

(6.6) **Cavity misalignment**

Here, we consider a laser cavity comprised of two mirrors M_1 and M_2, of radii of curvature R_1 and R_2, respectively, initially perfectly aligned so that their centres of curvature both lie along the z-axis.

(a) Suppose M_2 is rotated through an angle θ about the point at which the z-axis passes through M_2, as illustrated in Fig. 6.27. Explain why the axis of this 'misaligned' cavity must pass through the centres of curvature of both mirrors.

(b) Show that

$$\tan\phi = \frac{R_2 \sin\theta}{R_1 + R_2 \cos\theta - L_c} \quad (6.109)$$

$$y_1 = R_1 \sin\phi, \quad (6.110)$$

where ϕ is the angle the new axis makes with the z-axis, and y_1 is the distance from the z-axis of the new cavity axis at M_1.

For each of the following cases find expressions for ϕ and y_1 valid when θ is small, and comment on your results:

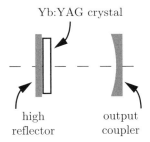

Fig. 6.28: Schematic diagram of the laser cavity discussed in Exercise 6.5.

(i) Symmetric cavity with large radii mirrors: $R_1 = R_2 = R \gg L_c$.
(ii) Symmetric confocal cavity: $R_1 = R_2 = 2L_c$.
(iii) Near-concentric cavity $R_1 + R_2 = L_c + \Delta L_c$, where ΔL_c is small.
(iv) Near-hemispherical cavity: $R_1 \to \infty$, $R_2 = L_c + \Delta L_c$, where ΔL_c is small.

(c) For the case of a cavity with $R_1 = R_2 = 10\,\text{m}$ and $L_c = 0.5\,\text{m}$, find the displacement y_1 for an angular misalignment of 1 mrad. Comment on this result.

(d) Explain how in principle any cavity comprised of curved mirrors may be aligned to a given axis by translation of each mirror in a plane normal to that axis.

(6.7) **Properties of high-order Hermite–Gaussian beams**
Here, we consider the properties of a Hermite–Gaussian beam of the form of eqn (6.15). For simplicity we will consider the problem in one dimension, that is a beam with an amplitude given by

$$U \propto H_n\left[\sqrt{2}\frac{x}{w(z)}\right]\exp\{-[x/w(z)]^2\}.$$

(a) Given that $H_0(x) = 1$, $H_1(x) = 2x$, and the recursion relation

$$H_{n+1}(x) = 2xH_n(x) - 2nH_{n-1}(x),$$

what is the highest power in x of the Hermite polynomial of order n?

(b) Sketch the variation of the beam intensity with x, and describe its general features. How many nodes does the intensity profile of the nth-order Hermite–Gaussian beam have along the x-axis?

(c) We now estimate the width of the beam along the x-axis, at a distance z from the beam waist, by estimating the position of the turning point nearest the edge of the beam. By arguing that near this turning point $H_n(x) \approx (2x)^n$, show that it occurs at $x_{\max}(z) \approx \sqrt{n/2}w(z)$. It turns out that a more accurate approximation is $x_{\max}(z) \approx \sqrt{n}w(z)$, which we will assume for the remainder of the question.

(d) How does the spot size $w(z)$ vary with distance z for large values of z? Hence, show that the far-field divergence is given by

$$\theta = \lim_{z \to \infty}\left(\frac{x_{\max}(z)}{z}\right) = \sqrt{n}\frac{\lambda}{\pi w(0)}.$$

(e) What is the approximate size, $W(0)$, of the beam at its waist (i.e. at $z = 0$)? Show that the far-field divergence of the beam is approximately n times that of a lowest-order Gaussian beam of waist $W(0)$.

(6.8) **Analysis of unstable cavities**†
Here, we show how diffraction is responsible for turning an initially demagnifying wave into a magnifying wave. For simplicity, we consider the negative-branch unstable cavity shown in Fig. 6.13. A demagnifying wave can be modelled as a point source located in the front focal plane of mirror 2, which generates a parallel beam travelling towards mirror 1. Figure 6.29 shows an unfolded version of the cavity, with the mirrors replaced by thin lenses of the same focal length. Note that propagation between the planes A and B corresponds to one trip round the cavity.

(a) Use geometric optics to show that in propagating between the planes A and B a parallel beam of diameter D is converted to a parallel beam of diameter D/M, where $M = f_1/f_2$.

(b) Assume that after the first round trip the parallel beam passes through plane A with a diameter D. Show that after m round trips the diameter of the beam passing plane A is D/M^{m-1}. Note that the maximum width of the beam decreases with each round trip—this is a demagnifying wave.

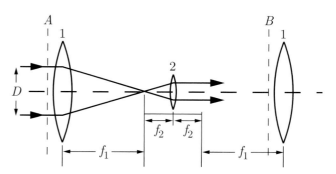

Fig. 6.29: The unfolded negative-branch unstable cavity considered in Exercise 6.8.

(c) We now consider the effects of diffraction. Argue that a beam of diameter D incident on lens 1 is focused not to a point but to a spot with a radius of order $f_1 \lambda / D$, where λ is the wavelength of the radiation.

(d) We can reason that the parallel beam striking lens 1 starts to looks more like radiation from a point source, and the 'point source' located in the front focal plane of lens 2 looks more like a parallel beam, when the beam incident on lens 1 and the spot to which it is focused have the same diameter. Show that this 'switch-over' occurs after m_s round trips, where

$$\frac{D}{M^{m_s - 1}} = \frac{2 f_1 \lambda M^{m_s - 1}}{D}. \qquad (6.111)$$

(e) Explain why it is appropriate to replace D with the diameter of M_2 in this formula, and hence show that

$$m_s = 1 + \frac{1}{2} \frac{\ln M_0}{\ln M}, \qquad (6.112)$$

where M_0 is given by eqn (6.27).

(f) After this point the diameters of the beams striking lenses 1 and 2 continue to decrease and increase, respectively. Can the wave still be said to be 'demagnifying'? Illustrate your answer with sketches of the wave propagation within the cavity before and after the transition.

(g) The wave will experience significant losses once the diameter of the beam striking lens 2 becomes comparable to the diameter of M_2. We can argue that at this point the waves propagating within the cavity will resemble the lowest-order cavity mode, and the beam will be diffraction-limited. Show that this will occur after approximately m_d round trips, where

$$\frac{2 f_1 \lambda M^{m_d - 1}}{D_2} = D_2, \qquad (6.113)$$

and hence derive eqn (6.27).

(h) Use these results to argue that the number of round trips required for the initially demagnifying wave to resemble the magnifying wave is approximately half the number of round trips required to establish a diffraction-limited beam.

Fig. 6.30: Part of an unfolded cavity showing a linear chain of regions of gain. For this part of the calculation the effect of the magnification of the unstable cavity is ignored.

(6.9) **Critical gain coefficient in an unstable cavity**†

We now derive the critical gain coefficient in an unstable cavity. To do this we again unfold the cavity and, initially, ignore the magnification arising from the cavity mirrors. The cavity and gain medium of length ℓ_g may then be replaced by a linear chain of regions of gain, as shown in Fig. 6.30.[59]

The laser radiation is derived from that spontaneous emission, which happens to be emitted into some critical solid angle Ω, which we will estimate below.

(a) Explain why the main contribution to the laser output will be formed by the amplification of spontaneous radiation emitted from the region within a length of order $1/\alpha_0$ from one end of the cylinder of gain.

(b) Show that the intensity of spontaneous radiation emitted from this region, and within a solid angle Ω, is approximately I_0 where

$$I_0 = \frac{N_2 \hbar \omega A_{21}}{\alpha_0} \frac{\Omega}{4\pi}, \qquad (6.114)$$

where N_2 is the population density of the upper laser level.

(c) Hence, show that the intensity of the amplified spontaneous emission a distance ℓ from one end of the gain medium is given by

$$I(\ell) \approx I_0 \exp(\alpha_0 \ell). \qquad (6.115)$$

(d) Use geometric optics to show that any ray striking mirror 2 at an angle θ_2^m will, after one further round trip, strike mirror 2 at an angle θ_2^{m+1}, where[60]

$$\frac{\theta_2^{m+1}}{\theta_2^m} = \frac{1}{M} \qquad (6.116)$$

---

[59]In this problem we derive an approximate expression for the intensity of amplified spontaneous emission from a long cylindrical gain medium (see Section 6.2). A more complete derivation of this intensity is considered in Exercise 18.1.
[60]You may want to consider the unfolded cavity illustrated in Fig. 6.29.

and $M = f_1/f_2$.

(e) Hence, argue that a suitable value for Ω is

$$\Omega \approx \frac{\pi D_g^2}{4 L_g^2}, \quad (6.117)$$

where D_g and L_g are the diameter and length of the gain medium within the cavity.

(f) By now taking into consideration the expansion of the magnifying beam on each round trip, show that after m round trips the intensity of the amplified spontaneous emission will be given by

$$I_m \approx I_0 \frac{1}{M^{2m}} \exp(\alpha_0 2 m \ell_g). \quad (6.118)$$

(g) Show that this reaches the saturation intensity when

$$2 m \ell_g \alpha_0 = \ln\left(\frac{N^*}{N_2} \frac{1}{A_{21} \tau_R} \frac{4\pi}{\Omega} M^{2m}\right) \quad (6.119)$$

$$\approx \ln\left(\frac{4\pi}{\Omega} M^{2m}\right). \quad (6.120)$$

(h) The critical small-signal gain coefficient α_0^{cr} is that for which the saturation intensity is reached after $m_d/2$ (see eqn (6.27)) round trips. Show that α_0^{cr} is given by eqn (6.28).

(6.10) **A Fox–Li calculation**†

Here, we show how to set up a Fox–Li calculation for the simple case of a cavity comprised of two rectangular plane mirrors. We use the geometry of Fig. 6.14.

(a) Show that the distance r from a point (x, y, z) on mirror 1 to the point (X, Y, Z) on mirror 2 may be written as

$$r \approx L_c - \frac{Xx}{L_c} + \frac{x^2}{2L_c} + \frac{X^2}{2L_c} - \frac{Yy}{L_c} + \frac{y^2}{2L_c} + \frac{Y^2}{2L_c},$$

where L_c is the separation of the mirrors.

(b) Hence, show that within this approximation for r the amplitude $U_2(X, Y, L_c)$ of the radiation on the surface of mirror 2 may be written in the form $U_2(X, Y, L_c) = -i \exp(ikL_c) f_2(X) h_2(Y)$, where,

$$f_2(X) = \sqrt{\frac{1}{\lambda L_c}} \exp(ikX^2/2L_c)$$

$$\times \int_{-a}^{a} f_1(x) \exp\left[ik\left(-Xx/L_c + x^2/2L_c\right)\right] dx,$$

and a similar expression may be written for $h_2(Y)$. As might be expected, the problem has been reduced to one dimension.

(c) It is useful to scale the problem as follows. By defining scaled variables $x' = x/(2a)$ and $X' = X/(2a)$ for mirror 1 and mirror 2, respectively, show that

$$f_2(X') = \sqrt{N_F} \exp(i\pi N_F X'^2)$$

$$\times \int_{-1/2}^{1/2} f_1(x') \exp\left[2\pi i N_F \left(-X'x' + x'^2/2\right)\right] dx',$$

$$(6.121)$$

where the *Fresnel number* is given by,

$$N_F = \frac{(2a)^2}{\lambda L_c}.$$

Note that this result depends only on the Fresnel number of the cavity. Consequently, the transverse mode profiles (in scaled units) and the diffraction losses must depend only on the Fresnel number of the cavity.

(d) We now consider how the integral of eqn (6.121) may be calculated numerically. Let us divide the x'- and X'-axes of the two mirrors into $2N_p - 1$ points separated by $\delta x'$. By considering the integrand, show that *provided*

$$\delta x' \ll \frac{1}{N_F},$$

eqn (6.121) may be represented by the sum,

$$f_2(j) = \sqrt{N_F} \exp[i\pi N_F (j - N_p)^2 \delta x'^2]$$

$$\times \delta x' \sum_{i=1}^{2N_p-1} f_1(i) \exp\left\{2\pi i N_F \delta x'^2\right.$$

$$\times \left[-(j - N_p)(i - N_p)\right.$$

$$\left.\left.+ (i - N_p)^2/2\right]\right\}, \quad (6.122)$$

where we have taken the centre of each mirror to be labelled by $i, j = N_p$.

(e) It is straightforward to evaluate the sum in eqn (6.122) using a programming language such as Mathematica, Maple, C, Fortran, etc. (especially if it can handle complex numbers). Write a programme to undertake the Fox–Li calculations outlined above using the following basic scheme:

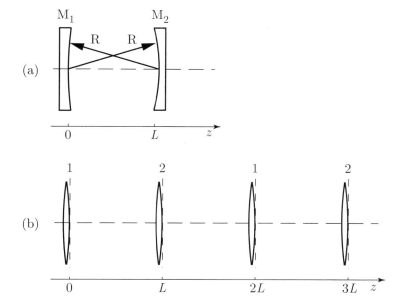

Fig. 6.31: Geometry of a Fox–Li calculation of the lowest-order mode of an optical cavity formed by two square, concave mirrors as discussed in Exercise 6.11. (a) shows the layout of the cavity; (b) shows the equivalent 'unfolded' cavity.

- Step 1. Determine a suitable value for $\delta x'$, and hence a value for N_p.
- Step 2. Define an array $f1[i]$ and fill it with an arbitrary amplitude. For example, let $f1[i] = 1$ for $i = 1$ to $2N_p - 1$.
- Step 3. Define an array $f2[j]$ and evaluate each member of the array using eqn (6.122). This represents the first pass of radiation from mirror 1 to mirror 2.
- Step 4. In order to calculate the amplitude of radiation after two passes of the cavity (i.e. after one complete round trip), let $f1[i] = f2[i]$ and repeat Step 3.[61]

You should find that any initial amplitude profile $f1[i]$ evolves to a stable profile after a few hundred bounces between the mirrors. As mentioned in the text, this process mimics what happens in a laser cavity when the intra-cavity radiation builds up from noise, although here we have neglected complications arising from amplification and gain saturation, etc.

(f) Use your programme to explore how the shape of the lowest-order mode varies with the Fresnel number of the cavity and discuss your findings.

(g) Extend your programme to calculate the diffraction loss per pass and plot this as a function of the number of completed passes. Compare your answers to those shown in Fig. 6.15(b), and explain the observed behaviour.

(h) Explore how the steady-state diffraction loss per pass varies with the Fresnel number. Remembering that the calculation performed here is one-dimensional, calculate also the diffraction loss per pass for a cavity formed by two square mirrors of side $2a$.

(i) Your programme has probably calculated the lowest-order *even*-symmetry mode. How can you calculate the transverse mode profile and diffraction losses for the lowest-order odd-symmetry mode?

(6.11) Fox–Li calculations for symmetric, low-loss cavities[†]
It is straightforward to extend the numerical routine developed in Exercise 6.10 to the case of a symmetric cavity formed by two square, concave mirrors of radius of curvature R, as shown schematically in Fig. 6.31(a). The easiest way to treat this system is to 'unfold' the cavity and replace each mirror with a thin lens of focal length $f = R/2$, as shown in Fig. 6.31(b). A single pass across the cavity is then equivalent to propagation between two planes each located immediately after a thin lens.

---

[61] It is useful to keep the energy of the radiation bouncing between the mirrors constant. To do this calculate a quantity proportional to the total radiation energy on the surfaces of the two mirrors, and multiply the array $f2[j]$ by the square root of this ratio before setting $f1[i] = f2[i]$.

(a) It is helpful first to deduce an expression for a spherical wave in the paraxial approximation. Show that the amplitude at a point (x, y, z) of a spherical wave originating from the point $(0, 0, z_0)$ may be written in the form,
$$u(x, y, z) = \frac{u_0}{|z - z_0|}$$
$$\times \exp[ik(z - z_0)] \exp\left(ik\frac{x^2 + y^2}{2|z - z_0|}\right),$$
provided that the paraxial approximation holds, i.e. $(x^2 + y^2) \ll \sqrt{|z - z_0|^3 \lambda}$.

(b) Hence, show that the wave may be written as
$$u(x, y, z) = \frac{u_0}{R_s} \exp[ik(z - z_0)]$$
$$\times \exp\left(ik\frac{x^2 + y^2}{2R_s}\right),$$
where R_s is the radius of curvature of the wave.

(c) Imagine a plane wave of amplitude $u_0 \exp(ikz)$ incident on a thin lens of focal length f. By considering the form of the wavefront, show that immediately after the lens the amplitude of the wave is given by
$$u(x, y, z) = u_0 \exp(ikz) \exp\left(ik\frac{x^2 + y^2}{-2f}\right),$$
and hence that in the paraxial approximation the effect of a thin lens of focal length f is to multiply the incident amplitude by the transfer function $\exp[-ik(x^2 + y^2)/2f]$.

(d) We are now in a position to allow for the mirrors in our Fox–Li calculation to have a finite radius of curvature. Show that if the mirrors have a radius of curvature R, eqn (6.122) becomes
$$f_2(j) = \sqrt{N_F} \exp\left[i\pi N_F\left(1 - \frac{2}{R'}\right)\right.$$
$$\left. \times (j - N_p)^2 \delta x'^2\right] \delta x'$$

$$\times \sum_{i=1}^{2N_p - 1} f_1(i) \exp\left\{2\pi i N_F \delta x'^2\right.$$
$$\times \left[-(j - N_p)(i - N_p)\right.$$
$$\left.\left. + (i - N_p)^2/2\right]\right\}, \quad (6.123)$$
where the normalized radius $R' = R/L_c$.

(e) It will also be useful to compare the lowest-order mode calculated by the Fox–Li approach with the lowest-order matched Gaussian beam calculated in Section 6.3.2. Use eqn (6.19) to show that the waist size of the Gaussian beam matched to the cavity is $w_0 = 2aw'_0$, where the normalized waist is given by
$$w'_0 = \sqrt{\frac{1}{2\pi N_F}} \left(\frac{1 + g}{1 - g}\right)^{1/4},$$
where $g = 1 - 1/R'$.

(f) Hence, find the spot size of the beam on the surface of the two mirrors. Adapt the Fox–Li programme developed in Exercise 6.10 to allow for finite radius of curvature of the mirrors and to compare the transverse profile of the calculated mode with the lowest-order Gaussian beam matched to the cavity. Consider a cavity with $N_F = 25$ and perform calculations for the following mirror curvatures:

  (i) $R = 20$;
  (ii) $R = -20$;
  (iii) $R = 0.6$;
  (iv) $R = 0.4$.

In each case comment on the the transverse profile of the lowest-order mode and the relative size of the diffraction loss.[62]

(6.12) **Fabry–Perot etalons**

(a) Show that the denominator in eqn (6.45) may be written as,
$$1 + (r_1 r_2)^2 - 2\cos\delta_{rt}.$$

(b) Hence, derive eqn (6.46) for the intensity of the transmitted beam.

---

[62]For some of these cavities you should find that the diffraction losses are much lower than for the plane-plane cavity explored in Exercise 6.10. In such cases it will take many more cavity bounces for the lowest-order mode to become dominant. The number of cavity bounces can be reduced to some extent by using an initial amplitude distribution on mirror 1 that is closer to the lowest-order mode; e.g. $f_1(x) = \cos^n x$, where n is an integer.

(c) Sketch the transmitted intensity predicted by eqn (6.46) as a function of the round-trip phase δ_{rt}, indicating clearly the values of δ_{rt} for which the transmission is a maximum.

(d) Show that the values of the round-trip phase for which the transmission is half the peak value are given by $\delta_{rt} = 2\pi p \pm \Delta\delta_{1/2}$, where

$$\sin^2\left(\frac{\Delta\delta_{1/2}}{2}\right) = \frac{1}{F}.$$

(e) Evaluate F for reasonable values of $R = r_1 r_2$, and hence deduce eqn (6.49).

(f) Complete the argument to show that the finesse of the cavity is given by eqn (6.54).

(6.13) **The finesse, Q and lifetime of an optical cavity**
In this problem we find approximate relationships between the finesse, Q, and lifetime of an optical cavity filled with a uniform medium of refractive index n. We neglect dispersion, so that we may write $v_g = c/n$, and assume that the reflectivities of the mirrors are high so that $R_1 \approx R_2 \approx 1$.

(a) Show that in this case the cavity lifetime (eqn (6.57)) may be written as

$$\tau_c \approx \frac{1}{2\pi} T_{rt} \mathcal{F}. \qquad (6.124)$$

From this result we see that the mean number of trips round the cavity taken by a photon is of the order of the finesse of the cavity.

(b) Show that

$$\tau_c \approx \frac{nL_c}{c} \frac{1}{1-R}, \qquad (6.125)$$

where $R = \sqrt{R_1 R_2}$ and hence show that

$$\Delta\omega_c \approx \frac{1}{\tau_c}, \qquad (6.126)$$

where $\Delta\omega_c$ is defined by eqn (6.52).

(c) By writing the Q of a laser cavity as

$$Q = \frac{\omega_p}{\Delta\omega_{p,p-1}} \frac{\Delta\omega_{p,p-1}}{\Delta\omega_c}, \qquad (6.127)$$

show that,

$$Q \approx \mathcal{F} p, \qquad (6.128)$$

where p is the order of the longitudinal mode.

(d) Estimate the longitudinal mode number p, and calculate the finesse, Q, and photon lifetime for the following cavities:
  (i) A cavity typical of a He-Ne laser operating at 633 nm: a 30-cm long cavity with mirror reflectivities of 100% and 90%.
  (ii) A cavity typical of a KrF laser operating at 248 nm: a 1-m long cavity with mirror reflectivities of 100% and 10%.
  (iii) A cavity typical of a GaAs semiconductor diode laser operating at 850 nm: a $250 - \mu m$ long cavity with mirror reflectivities of $R_1 = R_2 = 30\%$. Take the refractive index of GaAs to be 3.4.

(6.14) †**Low-loss limit of the Rigrod analysis**
(a) Show that when absorption within the laser medium may be ignored (i.e. $\kappa_I = 0$) and when the cavity is low-loss, the Rigrod formula for the output power (eqn (6.98)) is identical with the result for a low-loss cavity (eqn (6.81)). [Hint: Take the reflectivity of each mirror to be $R_i = 1 - A_i - T_i$ and consider the limit when A_i and T_i are small such that $\sqrt{R_i}$, etc. may be approximated by a binomial expansion.]

7

Solid-state lasers

7.1 General considerations 132
7.2 Nd^{3+}:YAG and other trivalent rare-earth systems 157
7.3 Ruby and other trivalent iron-group systems 169
Further reading 184
Exercises 184

A large number of technologically and scientifically important lasers operate on optically pumped transitions within ions doped as an impurity species into a solid host; indeed the first ever laser, the ruby laser, is one such example. The term **solid-state lasers** is usually reserved for systems of this type, lasers operating between levels of the electron band structure found in semiconductors being referred to as **semiconductor diode** or, simply, **diode** lasers.

Solid-state lasers have been realized in many ions, such as the trivalent rare earths (e.g. Nd^{3+}, Er^{3+}, Ho^{3+}), the divalent rare earths (e.g. Sm^{2+}, Dy^{2+}), and transition metals (e.g. Cr^{3+}, Ti^{2+}, Ni^{2+}); and in a wide variety of crystal hosts, including oxides, fluorides, and vanadates, as well as in various glasses. Laser oscillation has been achieved in hundreds of ion–host combinations with fundamental output wavelengths ranging from the mid-infra-red to the visible. Some solid-state lasers can output very narrow bandwidth radiation, whilst others are continuously tunable over very broad wavelength ranges, such that—especially when used in conjunction with non-linear techniques (see Chapter 15)—solid-state laser systems can cover an ever-increasing proportion of the electromagnetic spectrum between near-infra-red and ultraviolet wavelengths. The advantages they bring of compact and rugged construction has meant that they (and semiconductor diode lasers) are the preferred solution for a wide variety of applications, to the extent that lasers operating in other states of matter (gases, liquids, and plasma) are being replaced by solid-state or diode lasers unless: (i) they operate at very long or very short wavelengths, where solids are highly absorbing; or (ii) they output very high mean powers, which can be difficult to handle in solid media owing to problems associated with conducting away waste heat.

In the sections below we outline first some general considerations of laser oscillation in solid-state media, before illustrating these with some specific examples.

7.1 General considerations

7.1.1 Energy levels of ions doped in solid hosts[†]

Ions doped into a solid host differ from free ions in that they are subjected to the electric field—the **crystal field**—arising from the ions of the solid host. This field will, in general, shift and split the energy levels of the free ion by the Stark effect; sometimes to the extent that there is little correlation between the levels of the free ion and those found within a solid host.

Energy levels in the free ion

Before considering how the crystal field affects the energy levels of ions doped into a solid, it is useful to review the interactions responsible for determining energy levels in a free ion.

We start with the time-independent Scrödinger equation, which may be written in the form,

$$H\Psi = E\Psi, \qquad (7.1)$$

where H is the Hamiltonian and E the energy of the system. Ignoring for the moment relativistic effects, the Hamiltonian for an isolated atom or ion with N electrons may be written as,[1]

$$H = \sum_{i=1}^{N}\left\{\frac{-\hbar^2}{2m_e}\nabla_i^2 - \frac{Ze^2}{4\pi\epsilon_0 r_i} + \sum_{j>i}^{N}\frac{e^2}{4\pi\epsilon_0 r_{ij}}\right\}, \qquad (7.2)$$

where the charge of the nucleus is Ze, $\boldsymbol{r}_i$ is the vector from the nucleus to the ith electron, and $r_{ij} = |\boldsymbol{r}_i - \boldsymbol{r}_j|$. The first two terms in the Hamiltonian represent the total kinetic energy[2] of the N electrons and their total potential energy in the field of the nucleus. The third term arises from the mutual repulsion of the electrons, and it is this term that prevents a separation of the Schrödinger into N equivalent Schrödinger equations—one for each electron.

Equations (7.1) and (7.2) may be tackled once it is realized that much of the mutual repulsion contained in the third term is in the radial direction. This central component may be incorporated into an effective potential $V_{cf}(r_i)$ that depends only on the radial distance r_i and in which each electron moves independently. The Hamiltonian (7.2) may then be written as,

$$H = H_{cf} + H_{re}, \qquad (7.3)$$

where

$$H_{cf} = \sum_{i=1}^{N}\left\{\frac{-\hbar^2}{2m_e}\nabla_i^2 + V_{cf}(r_i)\right\}, \qquad (7.4)$$

and the **residual electrostatic interaction** is simply the difference between the electrostatic interactions contained in eqn (7.2) and the central potential:

$$H_{re} = \sum_{i=1}^{i=N}\left\{-\frac{Ze^2}{4\pi\epsilon_0 r_i} + \sum_{j>i}^{N}\frac{e^2}{4\pi\epsilon_0 r_{ij}} - V_{cf}(r_i)\right\}. \qquad (7.5)$$

We may now make the **central-field approximation** by assuming that the residual electrostatic interaction is small, and so may be treated as a perturbation. In this case, the Schrödinger equation becomes separable and the wave function, Ψ, is a product of one-electron wave functions, $\phi(\boldsymbol{r}_i)$, i.e. $\Psi(\boldsymbol{r}_1, \boldsymbol{r}_2, \boldsymbol{r}_3, \ldots, \boldsymbol{r}_N) = \phi(\boldsymbol{r}_1)\phi(\boldsymbol{r}_2)\phi(\boldsymbol{r}_3)\ldots\phi(\boldsymbol{r}_N)$, where the one-electron wave function is a solution of a one-electron Schrödinger equation:

$$\left\{\frac{-\hbar^2}{2m_e}\nabla_i^2 + V_{cf}(r_i)\right\}\phi(\boldsymbol{r}_i) = E_i\phi(\boldsymbol{r}_i), \qquad (7.6)$$

---

[1] Further details of the quantum mechanics of multielectron atoms and ions consult may be found in standard textbooks on atomic physics such as the book by Foot (2005).

[2] We have also ignored the kinetic energy of the nucleus that arises from its motion about the centre of mass. As discussed in standard texts, this may be taken into account by replacing the electron mass, m_e, with the reduced mass $\mu = m_e M/(M + m_e)$, where M is the mass of the nucleus.

and the energy of the atom or ion is given by the sum of the energies of each electron:

$$E = \sum_{i=1}^{N} E_i. \tag{7.7}$$

Since the potential appearing in eqn (7.6) is a central potential, the solutions are of the form,

$$\phi(\boldsymbol{r}_i) = R_{n_i}(r_i) Y_{\ell_i}^{m_i}(\theta, \phi), \tag{7.8}$$

where the form of the radial part of the wave function, $R_{n_i}(r_i)$ depends on the details of the central potential,[3] and $Y_{\ell_i}^{m_i}(\theta, \phi)$ are the well-known spherical harmonic functions. The quantum numbers of each electron are: n_i, the principal quantum number; and the orbital angular momentum quantum numbers, which give the square of the orbital angular momentum as $\ell_i(\ell_i + 1)\hbar^2$ and the z-component of the orbital angular momentum as $m_i \hbar$.

Hence, within the central-field approximation the electrons move independently within the spherically symmetric central potential, each with well-defined and constant values of n_i, ℓ_i, m_i. The energy of the atom depends on the values of n_i and ℓ_i for each electron, but not on the values of m_i since the Hamiltonian is unchanged by an arbitrary rotation of the axes. The energy levels of the atom are therefore described by specifying the values of $n_i \ell_i$ for each electron, a so-called **configuration**. By convention, the orbital angular quantum numbers $\ell = 0, 1, 2, 3, 4, 5, 6, \ldots$ are given the code letters s, p, d, f, g, h, i, ... For example, the ground-state configuration of neon has two electrons with $n = 0, \ell = 0$, two with $n = 1, \ell = 0$ and six with $n = 1, \ell = 1$, which may be written as $1s^2 2s^2 2p^6$. The various configurations of the atom, each of which has (in general) a different energy, comprise what is known as the **gross structure** (see Fig. 7.1).

In calculating the gross structure of the atom or ion we have neglected the residual electrostatic interaction—of energy ΔE_{re}—as well as a host of other interactions. The largest of these other terms is usually[4] the **spin-orbit** interaction, of energy $\Delta E_{\text{s-o}}$, which arises from the interaction between the magnetic dipole moment of an electron and the magnetic field it experiences in moving through the electric fields of the nucleus and other electrons. The Hamiltonian describing the spin-orbit interaction for a single electron may be written in the form,

$$H_{\text{s-o}} = \xi(r_i) \boldsymbol{\ell}_i \cdot \boldsymbol{s}_i, \tag{7.9}$$

where the parameter $\xi(r_i)$ is a function of the position of the electron, and depends on the central potential, and $\boldsymbol{\ell}_i$ and $\boldsymbol{s}_i$ are the orbital and spin angular momenta of the electron, respectively.

The neglected interactions may be treated as perturbations to the zeroth-order Hamiltonian (7.4), the perturbations being considered in the order of decreasing magnitude. For free ions the largest perturbation is often the residual electrostatic interaction. Remembering that this interaction arises from mutual repulsion between the electrons, we see that it cannot change the *total* orbital or spin angular momentum of the electrons. The formation of the total

---

[3] Of course, we have yet to determine the form of the central potential $V_{\text{cf}}(r_i)$! This may be achieved by using the self-consistent field method due to Hartree. The details are given in textbooks on atomic physics, but in essence the method is as follows: (i) assume a form for the central potential; (ii) calculate the one-electron wave functions for this potential from eqn (7.6); (iii) use these wave functions to calculate the charge distribution in the electron cloud surrounding the ion, and hence a new form for the central potential; (iv) iterate this procedure until convergence is achieved.

[4] This case with $\Delta E_{\text{re}} \gg \Delta E_{\text{s-o}}$ leads to LS, or 'Russell–Saunders', coupling. Other cases are possible of course. For example if $\Delta E_{\text{s-o}} \gg \Delta E_{\text{re}}$ then the orbital and spin angular momentum of each electron are coupled to form the total angular momentum of the electron, $\boldsymbol{j}_i = \boldsymbol{\ell}_i + \boldsymbol{s}_i$; the residual electrostatic interaction then couples the total angular momenta of the electrons in a scheme known as 'jj-coupling'. For further details consult, for example, Foot (2005).

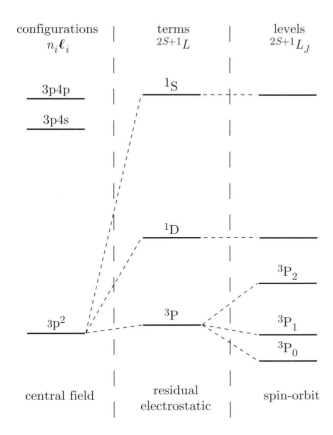

Fig. 7.1: Diagram showing schematically how the gross structure, terms, and fine structure are formed in the energy levels of a free atom or ion. The energy levels shown are schematic only, and are not drawn to scale. In moving from left to right the energy levels are shown on an increasingly finer detail. For illustration, examples of configurations, terms, and levels are given for the case of two valence electrons outside the closed subshells $1s^2 2s^2 2p^6 3s^2$.

orbital and spin angular momenta is shown in terms of the so-called vector model in Fig. 7.2. As a consequence, the residual electrostatic interaction shifts and splits each electron configuration into several **terms** described by the total orbital angular momentum $\mathbf{L} = \sum_{i=1}^{N} \boldsymbol{\ell}_i$ and the total spin angular momentum $\mathbf{S} = \sum_{i=1}^{N} \mathbf{s}_i$. The energy levels are then labelled by: the configuration plus the total orbital angular quantum number L, which gives the square of the total orbital angular momentum as $L(L+1)\hbar^2$; and the total spin angular quantum number S, which gives the square of the total spin angular momentum as $S(S+1)\hbar^2$. By convention, the notation employed is of the form $\gamma\ ^{2S+1}L$, where γ represents the configuration giving rise to the level. For example, the lowest three terms in the helium atom are described as:[5] $1s^2\ ^1S$, $1s2s\ ^3S$ and $1s2s\ ^1S$.

Finally, the spin-orbit interaction cannot change the *total* angular momentum (i.e. the orbital and spin angular momenta) of the electrons since it acts internally within the system of electrons. As a consequence the spin-orbit interaction shifts and splits the terms formed by the residual electrostatic interaction into a series of levels described by the total angular momentum $\mathbf{J} = \mathbf{L} + \mathbf{S}$. Formation of the total angular momentum $\mathbf{J}$ is illustrated by the vector model in Fig. 7.3. The resulting levels are then described by $\gamma\ ^{2S+1}L_J$,

[5] An equivalent code as for the configurations is used to give the total orbital angular momentum quantum number: $L = 0, 1, 2, 3, \ldots$ are written as S, P, D, F, ….

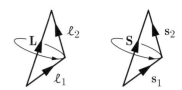

Fig. 7.2: The vector model of the coupling of orbital and spin angular momenta by the residual electrostatic interactions, shown for the simple case of two electrons. Within this model the individual electron momenta precess round the total orbital and spin momenta so that at all times $\mathbf{L} = \boldsymbol{\ell}_1 + \boldsymbol{\ell}_2$ and $\mathbf{S} = \mathbf{s}_1 + \mathbf{s}_2$. The rate of precession is proportional to the strength of the residual electrostatic interaction. For further details consult, for example, Foot (2005).

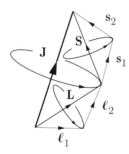

Fig. 7.3: The vector model of the coupling of the total orbital angular momentum **L** and the total spin angular momentum **S** to form the total angular momentum $\mathbf{J} = \mathbf{L} + \mathbf{S}$. If the spin-orbit interaction is weak the rate at which **L** and **S** precess about **J** is slow compared to the rate at which the individual electron orbital and spin momenta precess about **L** and **S**. For Further details consult, for example, Foot (2005).

[6] Named after the German physicist Johannes Stark (1874–1957).

where the square of the total angular momentum of the electrons is given by $J(J+1)\hbar^2$.

Figure 7.1 illustrates schematically the formation of the gross structure, terms, and fine structure in a free atom or ion.

Energy levels in an ion doped within a solid

In addition to the interactions described above, the electrons within an ion doped into a solid experience the electric field—the so-called **crystal field**—arising from the neighbouring ions. The interaction with the crystal field leads to an additional term in the Hamiltonian of the form,

$$H_\mathrm{c} = -eV_\mathrm{c}, \tag{7.10}$$

where V_c is an electrostatic potential describing the crystal field. In general, the effect of this interaction is to split and shift the energy levels from their positions in the absence of the crystal field, an effect known as the Stark effect.[6]

The crystal field has a symmetry reflecting that of the crystal lattice. Consequently, a proper treatment of the effect of the crystal field involves detailed consideration of this symmetry; indeed, as discussed below, in some circumstances these symmetry properties are used to label the energy levels arising from the interaction with the crystal field.

The nature of the influence of the crystal field on the energy-level structure depends critically on its strength relative to the other terms in the Hamiltonian. Three limiting cases may be identified:

- **Weak field:** $|H_\mathrm{c}| \ll |H_\mathrm{s-o}| \ll |H_\mathrm{re}|$. In this case the interaction of the crystal field is weak compared to the spin-orbit and residual electrostatic interactions (as well the interaction with the central field) so that the energy levels are changed only slightly from those in the free ion. The effect of the crystal field in this case is to cause a small splitting and shifting of the levels of the free ion to form a so-called **manifold** of closely spaced levels. Since the energy shifts are small, the labelling of the manifolds is simply that of the energy levels of the free ion. Further, in the weak-field case the energy-level structure of an ion will be almost independent of the crystal host in which it is embedded. A good example of the weak-field case are the energy levels formed by the $4f^n$ configurations in trivalent rare earth ions discussed in Section 7.2.
- **Intermediate field:** $|H_\mathrm{s-o}| \ll |H_\mathrm{c}| \ll |H_\mathrm{re}|$. In this situation the crystal-field interaction must be considered as a perturbation acting on the terms formed by the residual electrostatic interaction before the spin-orbit interaction is taken into account; the crystal field splits and shifts these terms. The spin-orbit interaction leads to further splitting of the energy levels. For intermediate crystal fields it is no longer possible to label the energy levels with the quantum numbers L, S, and J. The first-row transition-metal ions, such as Cr^{3+} frequently exhibit crystal fields of intermediate strength. However, it is more difficult to calculate the energy levels for intermediate crystal fields than for weak or strong fields. Frequently, the energy levels of ions experiencing intermediate crystal fields are labelled

with either a weak- or strong-field notation even though neither of these approximations is valid in this case.
- **Strong field**: $|H_{\text{s-o}}|, |H_{\text{re}}| \ll |H_{\text{c}}|$. Here, the spin-orbit or residual electrostatic interactions may be ignored (in the first approximation) and the crystal field acts on the single-electron orbitals of the central potential to give single-electron 'crystal-field orbitals' that reflect the symmetry of the crystal field. As discussed in Section 7.3.1, the energies of the orbitals depend on the spatial distribution of the orbital wave function with respect to the crystal lattice. The energy levels are labelled by the symmetry properties of the electron wave function using a notation derived from group theory. Strong crystal fields can arise in the second- and third-row transition-metal ions.

For each electron the relative strengths of the residual electrostatic, spin-orbit, and crystal field interactions depends very strongly on the radial distance r_i of the electron from its nucleus, and the distance d from the active ion to the neighbouring ions. For example, for neighbouring ions distributed with octahedral symmetry the energy of interaction is proportional to,[7]

$$\frac{\langle r_i^4 \rangle}{d^5},$$

where $\langle r_i^4 \rangle$ is the expectation value of r_i. In contrast, the energy shift arising from the spin-orbit interaction is proportional to,[8]

$$\left\langle \frac{1}{r_i^3} \right\rangle.$$

Hence, the ratio of the interaction with the crystal field to the spin-orbit interaction varies approximately as,

$$\frac{\Delta E_{\text{c}}}{\Delta E_{\text{s-o}}} \propto \frac{\langle r_i^7 \rangle}{d^5}. \tag{7.11}$$

As a consequence, very small changes in the relative size of the mean electron radius or the nearest-neighbour distance can change the relative strength of the interaction with the crystal field enormously.[9]

7.1.2 Radiative transitions†

For all the solid-state laser transitions discussed in this chapter, the laser transition occurs between two levels *within the same electron configuration*. The parity of the electron wave function does not change in such transitions, and hence they are forbidden by the selection rules of electric dipole radiation (see Section 2.3.2). Whilst this would indeed strictly be the case for transitions between the equivalent levels in a free ion, for ions doped into a solid these electric dipole transitions may occur for two reasons. The first is simply that the presence of the crystal field means that there is no longer inversion symmetry at the site of the active ion: in other words the electron wave function no longer has a definite parity and under the transformation $r_i \rightarrow -r_i$ the electron

[7] See Burns (1990).

[8] For further details consult, for example, Foot (2005).

[9] Frequently, it is said that the relative strength of an electron's interaction with the crystal field depends on the extent to which it is shielded by other electrons in the ion. For example, the 4f electrons in trivalent rare-earth ions are often said to be shielded by the 'outer' 5s and 5p electrons. Shielding effects of this type do occur, but calculation shows them to be approximately an order of magnitude smaller than the effects of changing mean electron radius and nearest-neighbour distance. For a discussion of this point see Burns (1990).

wave function can change magnitude (as well as possibly change sign). A *pure* electron configuration *must* have a well-defined parity since it is a solution of the Schrödinger equation in a spherically symmetric central field. However, the crystal-field interaction is not spherically symmetric and therefore introduces small admixtures of configurations of opposite parity to form a wave function that is no longer spherically symmetric, and no longer has a well-defined parity. These admixtures of configurations of opposite parity can allow transitions to occur between two levels that nominally have the same electron configuration.

A second mechanism by which electric dipole transitions may occur between two levels of the same configuration is a dynamically induced transition strength through vibrations of the crystal lattice that destroy the inversion symmetry at the site of the active ion. This second mechanism becomes important when the active ion is located at a crystal site exhibiting a high degree of symmetry. As might be expected, in such cases the radiative lifetime of the level depends on the temperature of the crystal.

Notwithstanding the above, the dipole-forbidden nature of transitions within a single configuration means that the transition rates are significantly slower than fully dipole-allowed transitions. The *radiative* lifetimes of the upper levels of visible transitions within ions doped into a solid are 3 to 6 orders of magnitude longer than fully dipole-allowed transitions in free atoms and ions.

Finally, we note that the strength of a radiative transition can depend on the orientation of the polarization of the radiation with respect to the axes of the crystal. This is true for both absorption and emission. In such cases the orientation of the crystal axes with respect to the axis of the laser cavity and the direction and polarization of the pump radiation can be important, and the laser output can be partially or totally polarized.

7.1.3 Non-radiative transitions[†]

Non-radiative transitions play a very important role in the operation of solid-state lasers: rapid non-radiative decay can provide an efficient route for feeding population from one or more excited levels into the upper laser level, and similarly can help to keep the population of the lower laser level low.

Phonon de-excitation

The energy levels of an active ion are coupled to the vibrations of the crystal lattice via the crystal-field interaction. Now, the lattice vibrations have a spectrum of normal modes; and the energy of the lattice vibrations is quantized, the unit of excitation being known as a **phonon**. For each kind of mode there is a maximum allowed frequency, and consequently for a particular crystal lattice there exists a maximum phonon frequency.[10]

Transitions between the energy levels of an ion may therefore occur not by emission of radiation, but by the emission or absorption of one or more lattice phonons. In general, the rate of this **phonon de-excitation** is found to decrease very rapidly with the number p of phonons involved in the transition. For example, for the case of weak electron-phonon coupling it may be shown

---

[10]There are two classes of these modes, known as acoustic and optical modes. A discussion of normal modes of crystals may be found in textbooks on solid-state physics, for example the books by Kittel and McEvan (2005), Ashcroft and Mermin (1976), or Singleton (2001).

that the rate of de-excitation on a transition of energy E_{21} by p phonons of energy $\hbar\omega_\text{eff}$ varies as[11]

$$W_\text{nr}^p(T) = C\left[\frac{\exp\left(\frac{\hbar\omega_\text{eff}}{k_\text{B}T}\right)}{\exp\left(\frac{\hbar\omega_\text{eff}}{k_\text{B}T}\right)-1}\right]^p \exp\left(-\beta\frac{E_{21}}{\hbar\omega_\text{eff}}\right), \qquad (7.12)$$

[11] See, for example, Powell (1998).

where the constants C and β depend on the properties of the host crystal, but not on the transition, and $\hbar\omega_\text{eff}$ is the energy of the effective phonon involved in the transition. In most cases this effective phonon is simply the highest-energy phonon available, since this reduces the number of phonons involved in the transition. In practice, the de-excitation rate is found to be small if the number of phonons required is greater than approximately 5. It is also worth noting that the rate of phonon de-excitation increases rapidly as the temperature T is increased, reflecting the greater thermal population of phonon modes at higher temperatures.

The rate of phonon de-excitation depends strongly on the strength of coupling between the electrons of the ions and the crystal field, and on the phonon spectrum of the lattice—and in particular on the maximum phonon energy. The de-excitation rates therefore vary widely for different combinations of impurity ions and hosts. For small energy differences the rate of non-radiative transitions can be extremely high: at room temperature the rate is of order 10^{11}–10^{12} s^{-1}.

The strong dependence of the rate of phonon de-excitation on the energy gap between the initial and final levels of the ion means that if a level is to be suitable as an upper laser level it should lie significantly above the nearest lower-lying level. This is necessary if the level is to have a long fluorescence lifetime, which allows a large population to build up in the level, and ensures that it decays pre dominantly by emission of radiation - which enables the population in the upper laser level to be used efficiently.

Concentration quenching and self-trapping

The fluorescence lifetime of an excited level in an active ion can also depend on the concentration of active ions and on the geometry of the sample.

As discussed in Section 4.5.3, self-trapping occurs on transitions to the ground state[12] for samples that are sufficiently large and dense. Under these conditions, a photon emitted by one ion can have a high probability of being reabsorbed by another ion before it escapes the crystal; to an observer it is as if the radiative decay of the first ion did not occur at all since no photons are observed to leave the crystal, and the number of ions in the initial level is unchanged. Consequently, the measured fluorescence lifetime of the upper level—which is found by measuring the rate of decay of fluorescence *from the whole crystal* on the transition of interest—will increase if self-trapping occurs. It is the fluorescence lifetime[13] of the level that is required in a rate equation description of a laser, since it is the fluorescence lifetime that gives the effective rate of decay of atoms or ions in the upper laser level.

Very frequently, even for levels that decay pre-dominantly to the ground state, the opposite effect to self-trapping is observed: an increase in the density of active ions is accompanied by a decrease in the fluorescence lifetime of

[12] Self-trapping can occur, in principle, on any transition. However, for self-trapping to become important the density of population in the lower level of the transition must be sufficiently high that the absorption depth is small compared to the dimensions of the sample. In practice, this nearly always means that self-trapping only occurs on transitions to the ground state, which has a high population. See Sections 4.5.2 and 4.5.3.

[13] This must be measured under the operating conditions of the laser.

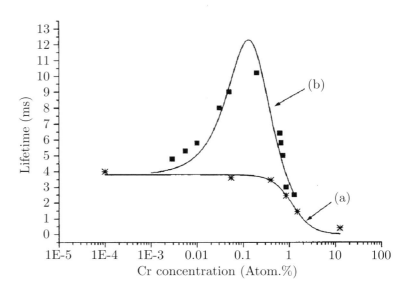

Fig. 7.4: Measured fluorescence lifetime of the upper level of the ruby (Cr:Al$_2$O$_3$) laser as a function of the concentration of Cr ions for: (a) a powdered sample; (b) a bulk crystal. For these measurements the crystal temperature was 77 K. Reprinted from *Journal of Luminescence*, Vol 94–95, F. Auzel et al., "The interplay of self-trapping and self-quenching for resonant transitions in solids; role of a cavity," pages 293–297, ©(2001), with permission from Elsevier.

the level. This effect is known as **concentration quenching**, and is frequently associated with a non-exponential decay of the population density in the upper level. Note that concentration quenching also differs from self-trapping in that it does not depend on the size or geometry of the sample.

Figure 7.4 shows the measured fluorescence lifetime of the upper level of the ruby (Cr:Al$_2$O$_3$) laser as a function of the concentration of Cr ions for powdered and bulk crystal samples. The upper laser level in ruby decays predominantly by radiative transitions to the ground state. We see that for the powdered sample the lifetime is almost constant until the atomic concentration reaches approximately 0.2%, whereupon the lifetime decreases rapidly as a result of concentration quenching. For the bulk sample, however, the fluorescence lifetime initially *increases* with the concentration of active ions as a result of self-trapping. Self-trapping increases the fluorescence lifetime by a factor of about 2.5 before concentration quenching causes the fluorescence lifetime to decrease at concentrations of the active ion above approximately 0.1%.

Concentration quenching is a manifestation of several types of energy transfer between ions that occur through dipole–dipole (or higher-order) interactions.[14] An active ion undergoing a transition to a lower level will develop an electric dipole moment p_1 that oscillates at the frequency of the transition. As is well known from the classical theory of electromagnetism,[15] at distances r from the dipole that are small compared to the wavelength of the transition, but large compared to the size of the dipole, the electric field is identical to that produced by a static dipole—except that it oscillates with a frequency equal to the frequency of oscillation of the dipole.[16] As such, the strength of the electric field E_1 induced by the dipole varies as r^{-3}. The energy of interaction of this electric field with a second ion, with an oscillating dipole moment p_2 is $H = -p_2 \cdot E_1$, and consequently this also varies as r^{-3}. This dipole–dipole interaction allows de-excitation of ion 1 to be accompanied by

[14] These interactions were first investigated for liquids by Förster (1949) and for solids by Dexter (1953).

[15] See, for example, Lorrain et al. (1988) or Jackson (1999).

[16] This is the so-called "static zone". At distances much greater than the wavelength—the "radiation zone"—the electromagnetic fields become transverse and decrease in amplitude as r^{-1}.

excitation of ion 2, *without the emission of radiation*, provided that the energy of the two transitions are sufficiently close. The rate of energy transfer is proportional to $|H|^2 \propto r^{-6}$, and hence is a very strong function of the ion spacing. Higher-order interactions of this type decrease even more quickly with ion separation, and consequently in general the rate of concentration quenching depends strongly on the densities of the donor (ion 1) and acceptor (ion 2) species.

As mentioned above, concentration quenching is often associated with non-exponential decay of the upper level. The reason for this is simply that within a crystal sample an active ion can exist in several different types of environment: those for which there is one or more neighbouring ions with which it can exchange energy, leading to rapid, non-radiative decay; those where there are no such ions, leading to a slower rate of decay. In such cases the intensity of the fluorescence observed from a bulk sample decays with a range of decay rates.

A wide variety of energy-transfer processes can occur, some of which are schematically illustrated in Fig. 7.5. In all cases, the ion undergoing de-excitation can be said to be quenched. In Fig. 7.5(a), ion A makes a transition from an excited level to the ground state, and ion B is excited from the ground state to an excited level. This process is particularly rapid if the two ions are of the same species since the energy levels will match perfectly. In this case process (a) is known as energy diffusion. Although there is no net change in the number of excited ions, and consequently there is no quenching of population, energy diffusion is an important mechanism by which energy is transferred from ions that would otherwise decay only slowly to ions for which the local environment is such that they are quenched rapidly.

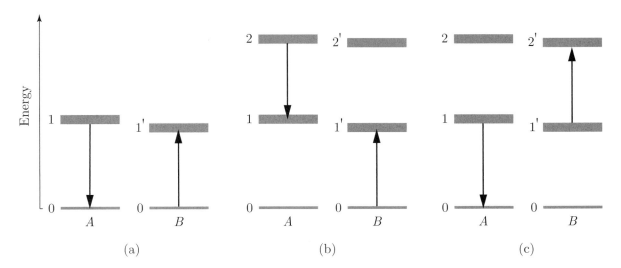

Fig. 7.5: Schematic diagram of some processes leading to non-radiative transfer of energy between ions A and B.

Figure 7.5(b) shows an energy transfer process in which ion A decays from an excited level 2 to an intermediate level 1, thereby causing ion B to be excited from the ground state to level $1'$. If the two ions are of the same species the ions are said to be self-quenched.

Figure 7.5(c) shows the reverse process to (b): ion A decays from an intermediate level to the ground state, causing ion B to be excited from an intermediate level to a higher-lying level. If the two ions are of the same species this process is known as **up-conversion**, and can proceed very rapidly if the energies of the $1 \rightarrow 0$ and $2 \leftarrow 1$ transitions are closely matched.

7.1.4 Line broadening[†]

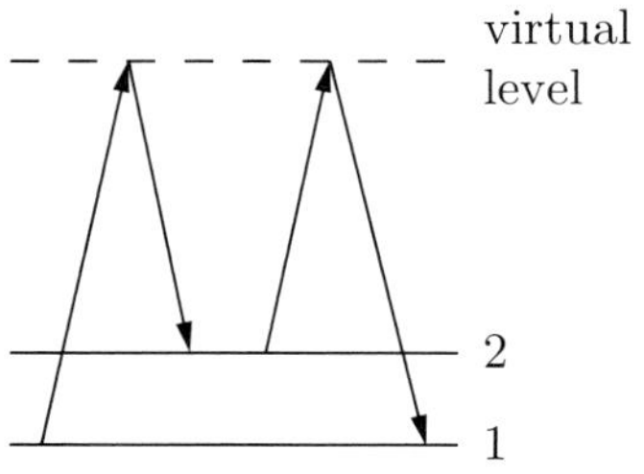

The natural linewidths of transitions between the levels of an impurity ion are typically very small, owing to the long radiative lifetimes of the levels. In addition, a temperature-dependent lifetime broadening arises from the increased rate of decay of the levels caused by non-radiative transitions. In practice, however, the measured linewidths of transitions are significantly broader than can be accounted for by the measured lifetimes of the upper and lower levels, and hence other processes must be responsible for the additional broadening.

A variety of mechanisms can cause this additional broadening. One of these, two-phonon Raman scattering is illustrated in Fig. 7.6.[17] In this process two phonons of different energy are successively absorbed and emitted so as to remove, and then return an ion from its energy level. Since the ion is returned to its initial level there is no change in the lifetime of the level, but a broadening occurs proportional to the rate of scattering. As might be expected, the rate of scattering, and hence its contribution to the linewidth, depends strongly on temperature.[18]

Phonons may also play a more direct role in determining the linewidth of a radiative transition. For some ion–host combinations radiative transitions can be accompanied by the emission or absorption of one or more phonons. These so-called vibronic transitions have very broad, temperature-dependent linewidths, as discussed in Section 7.3.1.

Fig. 7.6: Schematic diagram of the two-phonon Raman process leading to line broadening. In this process phonons of two different energies transfer ions from level 1 to level 2 via a virtual level, followed by the reverse process that returns the ion to level 1. The net result is no change in either the level occupied by the ion or the phonon spectrum but a broadening of the energy level by an amount proportional to the rate at which this process occurs.

[17]See Section 10.4.2 for a discussion of the Raman scattering of *photons*.

[18]It is also worth noting that continual absorption and emission of *virtual* phonons gives rise to a self-energy that causes a small temperature-dependent shift in the energy levels of the ion. At zero temperature this shift is analogous to the Lamb shift of levels found in atomic physics (Powell, 1998).

7.1.5 Three- and four-level systems

Figure 7.7 shows schematically the processes responsible in forming the population inversion in an optically pumped solid-state laser. The basic operation of this type of laser is as follows:

- The population inversion is achieved by optical pumping on broad **pump bands** $0 \rightarrow 3$. The broad nature of the pump bands reduces the difficulty of matching the frequency of the pump light to that of the pump transition, and also allows pumping by broad-band sources such as flashlamps. In practice, there may be more than one pump band involved in forming the population inversion.
- Level 3 decays rapidly by non-radiative processes to populate a range of lower levels. The proportion of decays from level 3 that populate the upper laser level, 2, is known as the **branching ratio** η_{branch}.

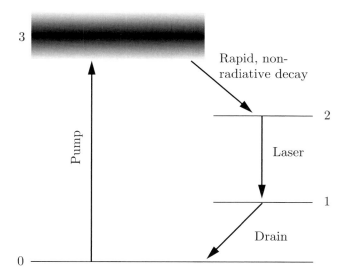

Fig. 7.7: Schematic diagram of the energy levels of an optically pumped solid-state laser system.

- If the pumping is sufficient to realize a population inversion, lasing occurs on the transition $2 \rightarrow 1$.
- If the lower laser level is the ground state, or if its energy above the ground state is small compared to $k_\mathrm{B} T$, the lower laser level will have a large thermal population. In this case there are essentially only 3 levels of importance (since levels 0 and 1 effectively form a single level), and the laser is classed as a **three-level laser**.
- In contrast, if the energy of the lower level above the ground is large compared to $k_\mathrm{B} T$ the laser is classed as a **four-level laser**. To avoid build-up of population in the lower laser level, known as **bottlenecking**, it is desirable for the lifetime of the lower level to be short.

Threshold pumping in three- and four-level systems

In general, the pump power or pump energy required to achieve a population inversion is several orders of magnitude higher for a three-level laser than for a four-level system.[19] The reason for this, as illustrated schematically in Fig. 7.8, is that in a three-level system a large fraction of the total population of active ions resides in the lower laser level prior to the onset of optical pumping. Essentially half[20] of this population must be excited to the upper laser level for the population inversion to become positive. Further pumping is required for the population inversion to be high enough for the round-trip gain to balance the cavity losses. In contrast, in a four-level laser system the population in the lower laser level prior to pumping is essentially zero and consequently it is only necessary to transfer sufficient ions into the upper laser level for the threshold population inversion density to be reached. In practice, the threshold population inversion density required to beat the cavity losses is orders of magnitude smaller than the total density of active ions, and consequently the pumping required to achieve lasing in a four-level laser is smaller than that required in a three-level laser by a similar factor.

[19] The classification of laser systems as three- or four-level is not restricted to solid-state lasers. However, it is convenient to introduce the concept here.

[20] This assumes that the degeneracies of the upper and lower laser levels are equal. Extension to the case when the degeneracies are different is straightforward, but the conclusions are not changed substantially.

Fig. 7.8: Illustrating why the pumping required to achieve lasing is much greater for a three-level laser system than for a four-level system. (a) In a three-level system a large number of ions must be *removed* from the lower laser level before the population inversion becomes positive; (b) in a four-level system any atoms excited to the upper laser level constitute a population inversion on the laser transition $2 \rightarrow 1$.

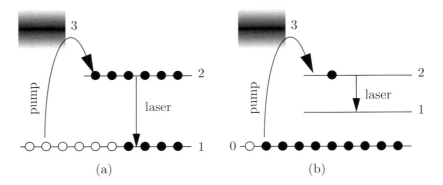

We can illustrate this difference by considering examples of each class of laser system. The threshold condition for laser oscillation, eqn (6.65), may be written as

$$2\alpha_0 \ell_g = 2\kappa(\omega) L_c - \ln(R_1 R_2)$$
$$\Rightarrow 2\sigma_{21} N_{\text{th}}^* \ell_g = 2\kappa(\omega) L_c - \ln(R_1 R_2), \tag{7.13}$$

where N_{th}^* is the threshold population inversion. Losses arising from the finite reflectivity of the mirrors and absorption and scattering losses therefore determine the minimum small-signal gain, and hence population inversion density, required for laser oscillation to occur.

In a three-level laser system the level populated by the optical pumping, level 3, decays rapidly to the upper laser level so that we have $N_3 \approx 0$. Within this approximation, the ions can only be in level 1 or level 2, and hence we may write,

$$N_2 + N_1 = N_T \tag{7.14}$$

$$\text{and} \qquad N_{\text{th}}^* = N_2^{\text{th}} - \frac{g_2}{g_1} N_1^{\text{th}}, \tag{7.15}$$

where N_T is the total density of active ions, and N_2^{th} and N_1^{th} are the population densities of the upper and lower laser levels when the threshold condition is reached. Rearranging eqn (7.15), we find straightforwardly that,

$$N_2^{\text{th}} = \frac{(g_2/g_1) N_T + N_{\text{th}}^*}{1 + (g_2/g_1)}. \tag{7.16}$$

As we shall see below, the threshold population inversion density, N_{th}^*, required to overcome the cavity losses is tiny compared to the *total* ion density N_T, and hence

$$N_2^{\text{th}} = \frac{N_T}{1 + (g_1/g_2)}. \qquad \textbf{Threshold condition: 3-level} \tag{7.17}$$

Note that if we take the simple case when $g_2 = g_1$, the threshold condition for lasing becomes $N_2^{\text{th}} = \frac{1}{2} N_T$, i.e. half the total number of active ions must be promoted to the upper laser level. Clearly even when $g_2 \neq g_1$ a large fraction of the total ion density must be excited to the upper laser level.

Having calculated the threshold upper-level population density, it is straightforward to estimate the energy required to do this. Taking the pump laser radiation to have a wavelength of λ_p, the energy required to raise each ion to the upper laser level is hc/λ_p. The pump energy required to reach threshold is therefore,

$$E_{\text{abs}}^{\text{th}} = \frac{N_T}{1 + (g_1/g_2)} V_{\text{gain}} \frac{hc}{\lambda_p}, \qquad (7.18)$$

where V_{gain} is the volume of the gain region.

Example: Consider the pump energy required to reach threshold in a ruby laser. We take the case of a ruby rod of 5 mm radius and length 20 mm, doped with a density of active (Cr^{3+}) ions equal to $N_T = 2 \times 10^{19}$ cm^{-3}. Assuming a mean pump wavelength of 500 nm, and taking $g_1 = g_2$ we find from eqn (7.18) that $E_{\text{abs}}^{\text{th}} \approx 6$ J.

The energy calculated from eqn (7.18) is that which must be absorbed by the laser rod. For flashlamp pumping the electrical energy supplied must be larger by a factor of about 70 as follows:

× **2** To achieve uniform pumping within the laser medium the doping and diameter must be such that only approximately 50% of the pump photons are absorbed. If the concentration of active ions is higher, or the rod diameter larger, population inversion is only achieved near the surface of the rod;

× **10** Only approximately 12% of the output of the flashlamp will lie in the pump bands;

× **2** Only about 50% of the pump light is geometrically coupled into the rod;

× **1.7** Only 60% of the electrical energy is converted into light.

In our numerical example, the threshold pump energy that must be supplied to the flashlamps is therefore of order 400 J. Note that, as discussed in Section 7.1.7, the electrical efficiency can be greatly enhanced by employing semiconductor diode lasers to provide the pump radiation—assuming that a diode laser is available with an output that is matched to the pump bands. When it is available, diode pumping reduces the electrical energy required to reach threshold by at least an order of magnitude, owing to the fact that all the output lies within the pump bands (saving a factor or 10 in the above analysis), and improved coupling into the laser rod (a further factor of up to 2).

Having established the energy required to reach threshold, it is straightforward to calculate the pump power required to achieve continuous operation of a three-level laser. The rate of decay of population density in the upper laser level is simply N_2/τ_2, and consequently the threshold pump power for c.w. operation is simply the threshold pump energy divided by the upper laser level lifetime. For ruby, $\tau_2 \approx 3$ ms and hence for the numerical example considered above we find a threshold electrical pump power for flashlamp pumping of order 100 kW! This is a large power, typical for a three-level laser, and consequently it is unusual for three-level lasers to be operated continuously.[21]

We may compare these results with the pumping required to reach the threshold for laser oscillation in a four-level laser system. The threshold population inversion is again given by eqn (7.13). However, for a four-level laser system we may assume that at threshold the lower laser level is essentially empty, and hence the threshold condition becomes,

$$N_2^{\text{th}} = N_{\text{th}}^*. \quad \textbf{Threshold condition: 4-level} \quad (7.19)$$

This population density is much smaller than the density of active ions, and consequently the required pump energy or power is greatly reduced.

[21] The ruby laser has been operated continuously (see Venkatesan and Muall (1977) and Nelson and Boyle (1962)) by reducing the diameter of the laser rod to a millimetre, or less, and cooling it to liquid-nitrogen temperature. However, these are technically difficult solutions. Very recently c.w. operation of a ruby laser has been demonstrated by optically pumping with semiconductor diode lasers.

> **Example:** As an example we take the case of a Nd:YAG laser rod of the same dimensions as the ruby rod considered above. If we assume the cavity losses on the right-hand side of eqn (7.13) to be equal to 10%, and taking the effective optical gain cross-section of the laser transition to be 3.7×10^{-19} cm^2 we find $N_{\text{th}}^* \approx 6.8 \times 10^{16}$ cm^{-3}. Note that this is more than two orders of magnitude smaller than the total density of active ions.
>
> Having estimated N_2^{th} calculation of the threshold pump energy proceeds as before. Taking $\lambda_p = 800$ nm we find $E_{\text{abs}} \approx 26$ mJ. The same conversion efficiencies that we used for the case of the ruby laser gives a threshold electrical energy for flashlamp pumping of order 1.8 J.
>
> Once again, the power required for continuous operation is given by dividing by the above energies by τ_2. The effective lifetime of the upper laser level in Nd:YAG is 230 μs, which for our numerical example yields an absorbed power of approximately 100 W and a total input electrical power of order 8 kW for flashlamp pumping. These parameters can be achieved quite easily and consequently this laser, as many other four-level laser systems, can be operated continuously.

7.1.6 Host materials

The host into which the active ion of a solid-state laser is doped plays an important role in the operation of the laser through its physical properties and its influence on the energy level structure of the ion. Desirable physical properties include: hardness, so that the material surface may be polished to an optical-quality finish; high thermal conductivity, so that waste heat may be conducted away rapidly; and ease of growth, so that the material may be grown to the required size, whilst maintaining high optical quality throughout the sample. It is also desirable that the crystal exhibits low thermal lensing[22] and low thermally induced birefringence.[23] At the microscopic level, the host material must be capable of accepting the active ion into its structure. Figure 7.9 shows a variety of doped materials that have been cut and polished into laser rods.

Solid-state hosts fall into three classes: crystals, glasses, and ceramics, as discussed below.

[22] Thermal lensing refers to a spatial variation of the refractive index of the laser gain medium arising from spatial variations in temperature and thermal stresses within the material. Frequently, the induced variation in refractive index – and, for a solid material, possible changes in the surface figure of the laser rod or disk – causes the gain medium to act as a thin lens of focal length f. For solid-state lasers f may be positive or negative, with a magnitude that can be as small as a few tens of centimetres. Further details may be found in Koechner and Bass (2003).

[23] The change in refractive index resulting from thermally induced strain within the crystal can be different for different polarizations. For example, in a laser rod the change in refractive index will, in general, be different for the radial and azimuthal polarization components.

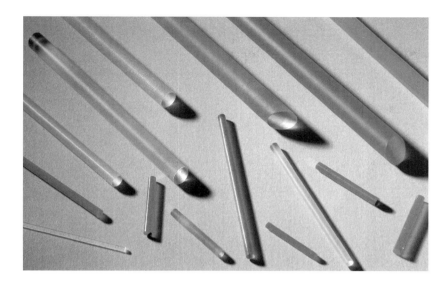

Fig. 7.9: A variety of laser materials, cut and polished into laser rods. (Courtesy Kentek Corporation.)

Crystalline hosts

Crystalline hosts have thermal conductivities that are typically an order of magnitude greater than in glass hosts, and can therefore sustain higher mean pumping powers before the thermal lensing of the laser rod becomes problematic. Laser transitions in crystal hosts are usually homogeneously broadened, which makes single-mode operation easier to achieve, and gives narrower linewidths for non-vibronic transitions (see Section 7.3.1). Disadvantages of crystal hosts are that they are often birefringent, making them polarization sensitive, and it can be difficult to grow large crystals of high optical quality.

Several hosts based on oxides have been employed:

- Sapphire (Al_2O_3) is very hard and has a high thermal conductivity; transition-metal ions may be substituted for the Al ion, but the rare-earths are too large to be incorporated at useful densities.
- A number of garnets have been used as host materials for rare-earth ions including yttrium aluminium garnet ($Y_3Al_5O_{12}$), gadolinium gallium garnet ($Gd_3Ga_5O_{12}$), and gadolinium scandium aluminium garnet ($Gd_3Sc_2Al_3O_{12}$)—usually abbreviated to YAG, GGG, and GSGG, respectively. The garnets are all stable, hard, have high thermal conductivities, and may be doped with the rare earth ions. YAG is by far the most commonly used since very high quality laser rods may be produced from it.
- Yttrium orthovanadate (YVO_4) has become an important host since the pump bands for Nd:YVO_4 overlap the output of semiconductor diode lasers.

Hosts based on fluorides are less commonly used since doping them with rare-earth ions requires charge compensation, which complicates crystal growth. The most important fluoride host is lithium yttrium fluoride ($YLiF_4$), usually abbreviated as YLF. This material is transparent down to 150 nm,

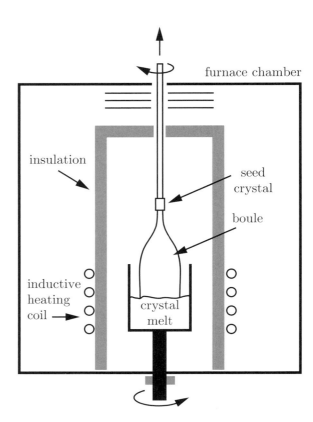

Fig. 7.10: Schematic diagram showing the Czochralski method for pulling a crystal boule from a melt.

and consequently the intense ultraviolet output of some flashlamps does not damage it. Thermal lensing and problems associated with induced birefringence are reduced compared to YAG, and these—together with other advantages for some types of laser—ensure that YLF is a common alternative host to YAG for lasers based on transitions in Nd^{3+} ions.

Laser crystals are often grown by the Czochralski technique illustrated in Fig. 7.10. In this method a seed crystal is rotated at typically (10–20 rev. min^{-1}) and slowly pulled from a melt of the constituent species. Since the temperature of the seed is lower than that of the molten material, the melt slowly crystallizes onto the seed. The single crystal—known as a 'boule'— pulled from the melt is usually quite large, perhaps 150 mm long with a mean diameter of 100 mm. Laser rods and disks are cut from the boule and the surfaces then polished to an optical-quality finish.

Glass hosts

Glass hosts have several advantages over crystals: glass can be cast into many forms—from optical fibres to metre-sized discs—with good homogeneity and high optical quality; glass is inexpensive; and it has low, or zero, birefringence. Further, to some extent, the physical properties of the glass and the properties of the laser transition can be controlled by adjusting the composition of the glass. The biggest drawback of glass is the relatively low thermal conductivity, which makes thermal management more difficult. Optical transitions within ions doped into a glass host are always inhomogeneously broadened, with linewidths several orders of magnitude larger than non-vibronic transitions in crystalline hosts. This offers advantages in providing broad pump bands, but means that the laser transition will also be broad. As a consequence, the optical gain cross-section will be low, making it more difficult to achieve laser oscillation in an oscillator owing to the lower round-trip gain (although this restriction does not apply to fibre lasers). The low gain cross-section has significant advantages, though, in laser *amplifiers*. As discussed in Section 5.3, in an amplifier an upper limit on the single-pass gain is set by the onset of parasitic laser oscillation caused by small Fresnel reflections from the ends or edges of the laser rod, or depletion of the gain by strong amplified spontaneous emission. The low gain cross-section obtained with glass laser materials allows a larger population inversion to be generated before this is a problem, and this allows more energy to be extracted by an injected laser pulse.

To summarize: the higher gain and good thermal conductivity of crystalline hosts means that they are often employed in laser oscillators and lower-power amplifiers; glass hosts are frequently used in high-energy pulse amplifiers (with low mean power) owing to the low optical gain cross-section and ease with which large rods and discs may be fabricated.

Ceramic hosts

Recently, there has been considerable interest in employing **ceramic** hosts. Ceramics are produced by forming a powder of nanometre-sized particles into the desired final shape and then heating to just below the melting point[24] to promote growth of crystals. The resulting ceramic is a mosaic of closely packed crystalline granules of $10-30\,\mu m$ size.

[24] This process is known as 'sintering'.

The level lifetimes and the absorption and emission spectra of active ions doped into a ceramic host are typically very similar to those of ions doped into a single crystal and the thermal conductivity is also comparable to that of a single crystal.[25]

[25] Note that ions sited close to the grain boundaries experience a different environment from those nearer the centre of the grain, and as such can have a different transition frequency. However, the proportion of such ions is of order 0.3% of the total number of doped ions, and hence this contribution to inhomogeneous broadening is negligible.

Ceramic laser materials are much easier and much cheaper to fabricate than single crystals, to the extent that they can be mass produced, and it is also possible to make laser rods and discs with large dimensions. Ceramics therefore combine the advantages of crystalline hosts with the ease of manufacture associated with glass materials.

7.1.7 Techniques for optical pumping

Sources

Solid-state lasers may be optically pumped by radiation from flashlamps or other lasers.

Flashlamps emit incoherent radiation from a moderate- or high-pressure discharge through a tube containing a noble gas, usually Xe. For continuous operation—when the device is more properly called an **arc lamp**—the discharge tube is usually filled to a pressure of several atmospheres, and

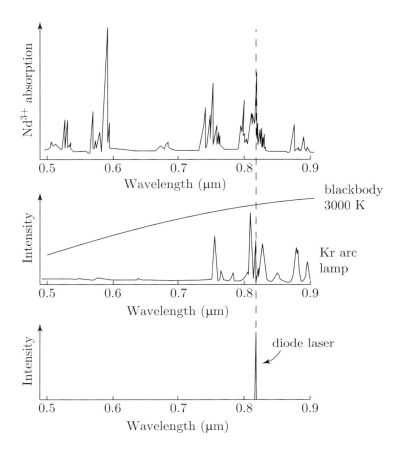

Fig. 7.11: Flashlamp and diode pumping solid-state lasers.

the discharge current is typically 20–50 A. The spectral output under such conditions comprises many closely spaced, pressure-broadened lines with wavelengths ranging from the ultraviolet to the infra-red, superimposed upon a relatively weak quasi-blackbody continuum. The wavelengths output by an arc lamp depend on the gas employed, and for this reason the gas employed is sometimes determined by the laser transition being pumped. For example, the spectral output of a Xe arc is poorly matched to the pump bands of Nd:YAG, and hence for this laser Kr arcs are employed, as shown in Fig. 7.11.

Pulsed flashlamps operate at several hundred millibars pressure, and employ much higher currents (typically hundreds of Amperes). The spectral output of pulsed flashlamps is dominated by a broad quasi-blackbody continuum arising from electron–ion recombination and bremsstrahlung radiation, although weak spectral features of the discharge gas may still be apparent. The duration of the emission is determined by the electrical circuit used to drive the discharge, and is typically between 100 µs and 10 ms. The lifetime of a continuously operating lamp is typically a few hundred hours; that of a pulsed flashlamp is 10^7 to 10^8 pulses.

There is considerable interest in employing semiconductor diode lasers to pump solid-state lasers since, as discussed in Chapter 9, they have very high

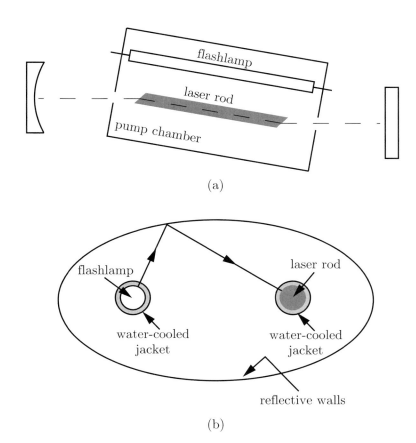

Fig. 7.12: Schematic diagram of a flashlamp-pumped solid-state laser. In (a) is shown the general arrangement of the pump chamber and the optical cavity. Note that the ends of the laser rod have been cut at Brewster's angle, which bends the cavity axis at the rod's ends. One design of pump chamber is shown in (b). In this arrangement the pump chamber has an elliptical cross-section and the flashlamp and laser rod are located near the foci of the ellipse.

wall-plug efficiencies. Several types of semiconductor diode lasers have been used to pump solid-state lasers, including AlGaAs, InGaAs, and GaInP. Linear arrays of diode lasers—known as 'diode bars'—and stacks of such bars may be used to give output powers as high as a few kilowatts. Continuously operating diode lasers have lifetimes of 10 000 h, or more, and pulsed laser diodes can provide 10^8 to 10^9 pulses before failing.

Flashlamps are inexpensive, simple to operate, and can convert electrical energy into optical radiation with an efficiency as high as 50%. However, the radiation they generate covers a wavelength range that is typically several hundred nanometres wide. Consequently, the spectral overlap with the pump bands of the laser is poor, and typically only of order 10% of the optical output can be absorbed on the crystal pump bands. Further, the output radiation is emitted into a solid angle of 4π steradians and hence it is not straightforward to collect all of the light emitted by the flashlamp and couple it into the laser rod. In practice, only approximately 50% of the light from a flashlamp can be coupled into the active medium.

In contrast, diode lasers are very much more expensive and considerably more complex to operate than flashlamps, although they have similar electrical efficiencies. The major advantage they offer is that, in principle, *all* of their spectral output can be matched to one of the pump bands of the crystal, as shown in Fig. 7.11. This typically increases the electrical efficiency of the solid-state laser by an order of magnitude compared to flashlamp pumping. In some circumstances the improved spectral overlap can offer a secondary advantage, i.e. reduced thermal loading of the gain medium. Flashlamp-pumping typically excites many manifolds above the upper laser level. The energy lost in the subsequent downwards cascade through the excited manifolds is deposited into the laser medium in the form of heat (phonons), resulting in thermal lensing and thermally induced stress. With diode laser pumping, however, it is often possible to pump to a single manifold lying only just above the upper laser level, reducing the thermal loading by as much as an order of magnitude. Finally, since their output is in the form of a beam, it is easier to couple all of the diode laser output into the laser rod.

It should be stressed, however, that achieving an overlap between the output of a diode laser and the available pump bands is not always straightforward. For example, in Nd:YAG the pump band at 808 nm has a width of order 2 nm. Whilst the output of a single diode laser might be as small as 1 nm, an *array* of such diodes can have a width that is several times this owing to variation in composition of the semiconductor material or temperature gradients within the array. As a consequence, it is necessary to have good control of the temperature and driving current of the diode array to ensure that good overlap with the pump band is maintained.

Several important solid-state lasers are routinely pumped by other types of laser, although since this usually complicates the laser system considerably it is only done when necessary. This might be the case, for example, if the pumped bands cannot be accessed by diode laser radiation and the lifetime of the upper laser level is too short for flashlamp radiation to be able to supply the threshold pump energy before population in the upper laser level decays. A good example is the Ti:sapphire laser, which is pumped by output from an argon-ion laser or from a frequency-doubled Nd:YAG laser.

Geometries for flashlamp-pumping

In the first ever laser, the ruby laser (see Fig. 1.2), optical pumping was provided by surrounding a laser rod with a coiled flashlamp, and surrounding the rod and flashlamp with a **pump chamber** with highly reflecting walls. Today, linear flashlamps are nearly always used and several different types of pump chamber have evolved.

Figure 7.12(a) shows the general layout of a flashlamp-pumped solid-state laser: a pump chamber contains one or more flashlamps and the laser rod; small holes in the pump chamber (usually sealed with windows to prevent dust or other material from entering the pump chamber and coating the rod) allow the laser radiation to propagate along the axis of the cavity. The flashlamps and laser rod are often cooled by flowing distilled water[26] in coaxial flow tubes, sealed at the ends of the flashlamp or laser rod.

Figure 7.12(b) shows in detail one geometry of pump chamber that is commonly employed: an elliptical chamber. In this approach the pump chamber has an elliptical cross-section and the laser rod and flashlamp are placed near to the foci of the ellipse. The properties of an ellipse are such that any ray leaving one focus would be reflected from the inner surface of the ellipse so as to pass

[26]The coolant may contain substances designed to reduce the build-up of films of material on the surfaces of the flashlamps and laser rod, which would reduce the optical pumping.

through the other focus. Consequently, this arrangement ensures that a very high proportion of the radiation leaving the flashlamp passes through the laser rod. Note that if the diameter of the flashlamp and rod were infinitely small, and the reflectivity of the inner surface of the ellipse was independent of angle, then the illumination of the laser rod would be symmetric about its axis. However, a rod of finite diameter will intercept rays propagating at a small angle to the major axis of the ellipse, increasing the illumination on the flashlamp side of the rod and decreasing the illumination on the far side; this causes the rod illumination, and hence the distribution of gain, to be asymmetric. The double elliptical pump chamber shown in Fig. 7.13(a) can alleviate this problem to some extent, as well as increasing the total optical power reaching the laser rod.

Figure. 7.13(b) shows an alternative arrangement, so-called close-coupling, in which the rod and flashlamp are mounted as close as possible to each

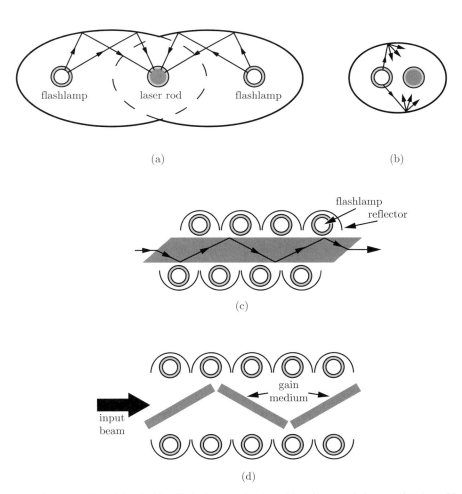

Fig. 7.13: Flashlamp pumping geometries: (a) a double-elliptical pump chamber; (b) a close-coupled pump chamber with diffusely reflecting walls; (c) a slab geometry; (d) a large-aperture plate amplifier.

other. Very often with this arrangement the walls of the pump chamber are chosen to be highly efficient diffuse (rather than specular) reflectors, such as $BaSO_4$ or white ceramic. In such cases very little pump radiation is absorbed in the pump chamber walls, and the pump radiation bounces round the pump-chamber until it is absorbed by the laser rod, is reabsorbed by the flashlamp, or escapes through the small holes at the end of the chamber. As a consequence the pumping can be very uniform. A potential problem, however, is radiant heating of the laser rod by the flashlamp; such heating can increase the laser threshold, and is also not uniform within the rod.

The operation of solid-state lasers at high powers is accompanied by an increasing deposition of heat into the host material, which must be removed in order to avoid problems associated with thermal lensing, thermally induced birefringence, or even damage. In such cases a rod geometry is not ideal since heat can only be removed from the rod surface, and the surface area to volume ratio is relatively low. A solution to this problem is the slab geometry illustrated in Fig. 7.13(c). With this arrangement slabs of laser material—typically with a rectangular cross-section with sides of order 10 mm and a length of order 100 mm—are irradiated from one side (or both) by flashlamp radiation. Heat may be removed from the crystal by flowing cooling fluid over its sides. The generated laser radiation takes a zig-zag path through the gain medium, being totally internally reflected at the inner surface of the slab. A path of this form serves to average out the effects of non-uniformities caused by thermal and stress gradients. Both flashlamp- and diode-pumped slab lasers—with outputs up to several kilowatts—have been developed.

Finally, Fig. 7.13(d) shows schematically a geometry that can be used in very large glass-based laser systems able to amplify pulses to very high energies.[27] In this arrangement, large plates of the gain medium are oriented at Brewster's angle so as to avoid Fresnel reflection losses for the amplified beam. This geometry allows beams with diameters up to about 1 m to be amplified.[28] Note that amplifiers of this type operate at very low pulse repetition rates (about one shot per hour), primarily owing to the time it takes to cool the disc and so remove thermal gradients in the gain medium prior to the next laser pulse. The large surface area of each plate helps with this, and allows efficient convective or forced-air cooling.

Geometries for diode pumping

Figure 7.14(a) shows the so-called **end-pumping** geometry for diode laser pumping a solid-state laser. With this approach radiation from the diode laser is focused into the gain medium by a lens system[29] that is coaxial with the laser rod. The pump beam must therefore pass through the rear mirror of the cavity of the solid-state laser, and hence the optical coating of this mirror must be designed to have a high transmission at the wavelength of the pump radiation, but be highly reflective at the wavelength of the solid-state laser. In the design shown in Fig. 7.14(a), an optional non-linear crystal is incorporated within the laser cavity for efficient intra-cavity frequency doubling (see Section 15.5).

---

[27] See page 488.

[28] Amplifiers of such large dimensions always use glass hosts since, even if it were possible, it would be forbiddingly expensive to grow crystals to this size. Glass plates, however, may be cast in almost any size.

[29] This lens system might include an anamorphic expansion system to make the divergence of the diode laser output more cylindrically symmetric, see Section 9.10.1.

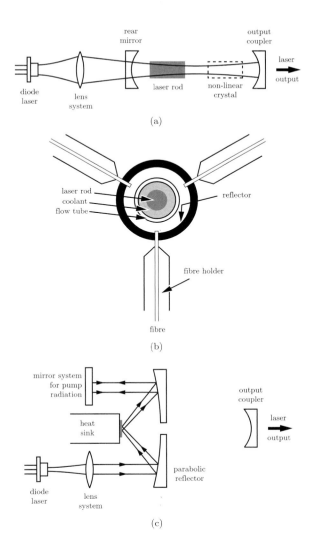

Fig. 7.14: Diode-pumping geometries: (a) End-pumping; (b) side-pumping with fibre-coupled diode lasers (after Golla et al. (1995)); (c) thin-disc pumping.

The advantage of the end-pumping geometry is that the pump energy is deposited in a volume that is symmetric about the cavity axis and that has a radius that is relatively small (100 µm to 1 mm). As a consequence it is relatively easy to restrict oscillation to the lowest-order mode of the laser cavity, and hence for the output beam to be of high quality.

Solid-state lasers pumped by diode laser radiation in the end-pumping geometry can provide mean output powers as high as 15 W. However, since all the pump radiation is delivered through one end of the laser rod in a beam of small diameter, it is not possible to achieve very high power operation with this geometry owing to damage to the laser rod. The pump power supplied to the laser rod may be increased by **side pumping**. One way of achieving this is to mount diode lasers in a cylindrically symmetric array around the laser rod.

However, this requires a large number of components (diode lasers, focusing optics, and cooling systems for the diode lasers) to be mounted close to the rod. A more convenient arrangement is shown in Fig. 7.14(b), in which optical fibres transmit the pump radiation from diode lasers located in a separate unit. Radiation from each fibre passes through the laser rod, unabsorbed pump radiation being reflected back onto the rod by a coaxial cylindrical reflector. Very uniform pumping of the laser rod can be achieved by adjusting the distance from the ends of the fibres to the rod appropriately. The centre–centre spacing of the fibres along the length of the rod is typically 2 mm. Side-pumped lasers of this type can generate output powers of order 100 W in the lowest-order cavity mode. Cooling of the rod is provided by flowing coolant along the length of the laser rod in a coaxial tube; the surfaces of the flow tube have an antireflection coating for the wavelength of the pump laser. We note that side pumping of slabs of laser material, as in Fig 7.13(c), has been achieved using essentially the same technique.[30]

[30] For an example, see Shine et al. (1995).

The rod geometry—whether end or side pumped—suffers from the fact that heat is removed in the radial direction, causing thermal gradients and hence thermal lensing in this direction. Figure 7.14(c) shows one solution to this problem: **the thin-disc laser**. In this approach a disc of the gain medium is mounted directly on a water-cooled heat sink. The face of the solid-state gain medium in contact with the heat sink is coated so as to be highly reflective at the wavelength of the pump laser and that of the generated output; the opposite face is coated so as to have low reflection at these wavelengths. The laser cavity is therefore formed between the rear face of the disc and an output coupler positioned on the axis of the disc.

Pump radiation from a diode array is focused onto the disc by a coaxial parabola; unabsorbed pump radiation is reflected from the rear of the disc, to provide a second pumping pass, and is collected and recollimated by the parabola. Additional pumping passes may be achieved by reflecting the unabsorbed pump radiation back onto the parabola: a plane mirror, for example, allows a total of four pump passes; more complicated arrangements can give as many as 16 passes of the pump radiation.[31]

[31] See, for example, Stewen et al. (2000).

[32] For a review of thin-disc lasers, see the article by Kemp et al. (2004).

The thin-disc geometry offers several advantages.[32] The surface area through which heat may be removed is large for the small volume of material. As a consequence, aggressive cooling can be provided whilst generating relatively low thermal gradients within the gain medium. Further, since the thermal gradients are in the axial direction the thermal lensing is very much reduced.[33]

[33] The axial temperature gradient can cause thermal lensing via bowing of the front face of the disc. However, this is much smaller than the lensing arising in laser rods from radial temperature gradients. Further, bowing of the disc can be reduced by clamping the disc to the heat sink with a plate made from a strong, transparent material such as sapphire (Liao, et al. 1999).

The thin disc typically has a diameter of about 10 mm and a thickness of just 250 μm. Very high output powers are possible with this geometry; for example, thin-disc Yb:YAG lasers have generated output powers as high as 2 kW with an optical efficiency of better than 50% (Stewen, et al., 2000). The major limitation of thin-disc lasers is that the output beam is usually of relatively poor beam quality, since the transverse dimensions of the pumped area (several millimetres) is large compared to the size of the lowest-order cavity mode. Typically, the output beams of thin-disc lasers diverge at 10–20 times diffraction limit, although this can be improved by better cavity design.[34]

[34] See, for example, Kemp et al. (2004).

7.2 Nd^{3+}:YAG and other trivalent rare-earth systems

A large number of solid-state lasers operate on $4f^n - 4f^n$ transitions in trivalent rare-earth ions, and in particular the lanthanide series—the actinides being radioactive, and hence difficult to work with.[35] Important trivalent rare-earth laser ions include: Ce^{3+}, Nd^{3+}, Ho^{3+}, and Er^{3+}.

7.2.1 Energy-level structure

Neutral atoms of these rare earths have a ground-state electronic configuration of the form [Xe] $4f^{n+1} 6s^2$, or [Xe] $4f^n 5d 6s^2$, where $n \geq 0$ and [Xe] represents the ground state configuration of xenon:

$$1s^2\, 2s^2 2p^6\, 3s^2 3p^6 3d^{10}\, 4s^2 4p^6 4d^{10}\, 5s^2 5p^6.$$

In forming the trivalent ion, the three loosest-bound electrons are lost:[36] the two 6s electrons and either one of the 4f electrons or the 5d electron, to give a ground-state configuration that is always of the form [Xe] $4f^n$. The absorption spectrum of the ions corresponds to transitions within the 4f configuration ($4f^n \leftarrow 4f^n$ transitions), or $4f^{n-1} 5d \leftarrow 4f^n$ transitions to the lowest-lying empty orbital, 5d.

Of key importance to the spectroscopy of the trivalent rare-earth ions in crystalline media is the fact that the wave function of the 4f orbital is more compact than those of the occupied 5s and 5p orbitals. As such, the mean radius of the 4f electrons are relatively small compared to the size of the ions, and hence to the nearest-neighbour distance. The interaction with the crystal field is therefore weak, producing a series of manifolds that are only slightly perturbed from those of the levels of the free ion. As a consequence, the energy-level structure of the 4f configurations in the trivalent rare earths is approximately independent of the crystal host, although the optical cross-sections of transitions may vary significantly.

The manifolds are labelled by the levels of the free ion from which they arise, i.e. $^{2S+1}L_J$. If the total spin quantum number S is an integer, the maximum number of levels within the manifold is given by $(2J + 1)$; if S is half-odd-integer the levels of the crystal field are all doubly degenerate,[37] and consequently the maximum number of levels within the manifold is reduced to $(2J + 1)/2$. It should be emphasized that these are the maximum number of non-degenerate levels; the actual number formed depends on the symmetry of the crystal field.

7.2.2 Transition linewidth

The natural linewidth of the $4f^n - 4f^n$ transitions is very narrow owing to the long radiative lifetimes of the levels. This linewidth is increased substantially by phonon collisions, as discussed in Section 7.1.4, to give a homogeneously

---

[35] The only actinide ion in which stimulated emission has been observed is U^{3+}, lasing being observed in 1960 Sorokin and Stevenson (1960) on a $5f^3 - 5f^3$ transition (analogous to the $4f^n - 4f^n$ transitions discussed in this section) in U:CaF$_2$ at a wavelength of approximately 2.5 μm. Interestingly, this was the *second* laser to be demonstrated.

[36] In the free ion the electrons are removed from the atom, in a solid they form bonds with the atoms of the host lattice.

[37] This is known as Kramers' degeneracy.

broadened transition with a linewidth that depends on temperature. At room temperature the linewidth is typically in the range 0.1–3 THz (10–100 cm^{-1}).

Inhomogeneous broadening can arise if the crystal field varies with position in the crystal. For example, the presence of crystal defects, strain, or impurity ions can cause variation in the crystal field and hence local shifts of the ion energy levels. For good-quality laser crystals these effects should be small, in which case the line broadening will be pre-dominantly homogeneous. In contrast, ions doped into glasses experience a very wide range of local environments. The energies, broadening, and even the number of levels varies significantly from site to site and consequently all transitions will be strongly inhomogeneously broadened. The inhomogeneous linewidth of the $4f^n - 4f^n$ transitions in ions doped into a glass host are of order 30 THz (1000 cm^{-1}).

7.2.3 Nd:YAG laser

Perhaps the most important example of a laser based on a $4f^n - 4f^n$ transition in a trivalent rare-earth ion is the Nd:YAG laser. Neodymium-based lasers are very widely used in science and industry. They are frequently used as pump lasers for dye lasers, or other solid-state lasers such as Ti:sapphire. In medicine they find applications in removing secondary cataracts or tissue removal. High-power Nd:YAG lasers can be used in laser drilling and welding.

Crystal properties

In Nd:YAG the Nd^{3+} ion replaces the Y^{3+} ion. Since the size of the Nd ion, which has a radius of 98 pm, is greater than that of the Y ion, of radius 90 pm, it is not possible to introduce Nd ions with an atomic concentration much above 1.5% without straining the crystal lattice unduly. It is worth noting that an atomic concentration[38] of Nd ions equal to 1% corresponds to a density of Nd ions of 1.386×10^{20} cm^{-3}.

Doped YAG crystals are grown using the Czochralski method illustrated in Fig. 7.10. For Nd:YAG the growth rate must be rather slow in order to avoid inhomogeneities within the finished rods, a typical pulling rate is 0.5 mm h^{-1}— i.e. it takes several weeks to grow one boule.

Energy levels

Figure 7.15 shows the energy levels of the Nd^{3+} ion and how these are split into manifolds by the crystal field of YAG. Broad pump bands are provided by several closely spaced manifolds with energies between approximately 12 000 and 33 000 cm^{-1} (300–800 nm). Particularly strong pumping occurs at 810 nm (12 300 cm^{-1}) and 750 nm (13 300 cm^{-1}), which can be accessed by diode lasers.

The manifolds excited by optical pumping on these bands are relatively closely spaced compared to the maximum phonon energy in YAG of 850 cm^{-1}. Consequently, excited ions cascade down through these manifolds by very rapid non-radiative transitions until the two levels (denoted R_1 and R_2) of

---

[38]Doping levels of crystals can be expressed in several different ways. The 'atomic concentration' (often abbreviated to 'at.%') is the proportion of available sites that are occupied by the active ion; the 'concentration by weight' (often abbreviated to 'wt%') is the proportion of the mass of the dopant species relative to that the *undoped* crystal. These differences are explored in Exercise 7.1.

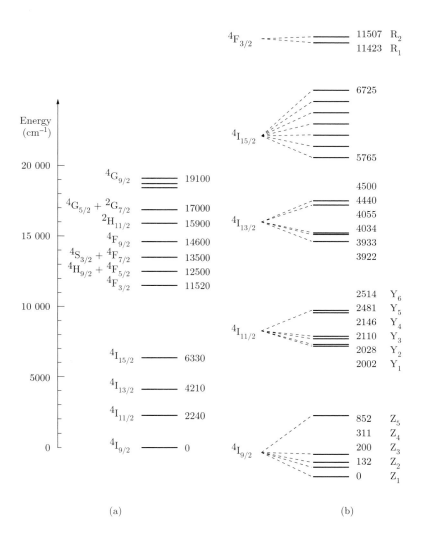

Fig. 7.15: Energy-level diagrams for the Nd^{3+} ion in YAG. (a) shows the lowest-lying levels of the free ion formed from the [Xe] $4f^3$ configuration; (b) illustrates how these levels are split by the crystal field in YAG to form manifolds of closely spaced levels. Note that in (b) the levels of each manifold are shown on an expanded energy scale, and the spacing between manifolds is not drawn to scale. The energies of the levels are given in cm^{-1}.

the $^4F_{3/2}$ manifold is reached. The levels of the $^4F_{3/2}$ manifold are metastable because the nearest lower-lying level—the top of the $^4I_{15/2}$ manifold—is separated by some $4698\,cm^{-1}$, corresponding to more than 5 phonons. The rate of non-radiative decay of the $^4F_{3/2}$ manifold is therefore slow, and it decays instead almost entirely by radiative transitions to levels of the ^4I manifolds, with a fluorescence lifetime of $230\,\mu s$. In contrast, the levels of the ^4I manifolds, are separated by less than $1500\,cm^{-1}$, and consequently decay non-radiatively with lifetimes of order $100\,ps$.

It can be seen, therefore, that the energy levels of Nd:YAG are very well suited to achieving laser oscillation: the levels of the $^4F_{3/2}$ manifold may be populated efficiently by rapid non-radiative transitions from levels accessed by broad pump bands, and they are metastable—ideal properties of an upper laser level. Suitable lower laser levels exist within the ^4I manifolds; these lie significantly above the ground state and decay very rapidly to it. The system is therefore an almost ideal example of a four-level laser system. Figure 7.16 shows a simplified energy level scheme for the strongest Nd:YAG laser transition; transitions to the levels of other ^{4}I manifolds are analogous, with the exception of those to the lowest-lying manifold, $I_{9/2}$. Lasers operating on transitions to this manifold behave quite differently from those operating on transitions to the higher-lying manifolds, since the levels of the $I_{9/2}$ manifold can have a significant thermal population. As such, these lasers behave more like three-level systems.

Lasing has been achieved from the $^4F_{3/2}$ manifold on more than 20 transitions to the $I_{13/2}$, $I_{11/2}$, and $I_{9/2}$ manifolds, with wavelengths near 1319 nm, 1064 nm, and 946 nm, respectively. Of these, the strongest—and certainly the most commonly used—transition is the $^4F_{3/2} \rightarrow {}^4I_{11/2}$ transition.[39]

[39]Indeed, in order to obtain lasing to the other manifolds it is necessary to employ frequency-selective cavities that suppress lasing to the $^4I_{11/2}$ manifold and enhance the feedback on the desired transition.

The R_1 and R_2 levels of the $^4F_{3/2}$ manifold are separated by 84 cm^{-1}. Rapid non-radiative transitions ensure that the populations of these two levels is maintained in thermal equilibrium and, importantly, the rate at which the population in these two levels is mixed is sufficiently fast that this remains

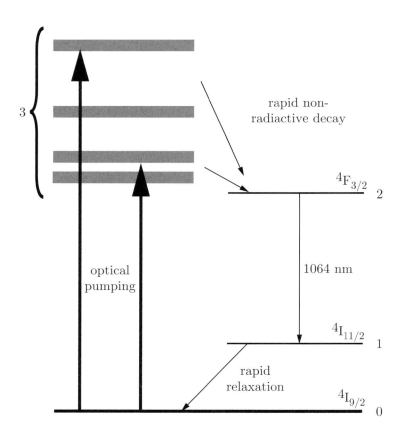

Fig. 7.16: Simplified energy level diagram of the $^4F_{3/2} \rightarrow {}^4I_{11/2}$ laser transition in Nd:YAG showing the four-level nature of the system.

the case even during laser oscillation. Hence, at room temperature the relative populations of these levels are approximately 60% and 40%, respectively. Of the possible $^4F_{3/2} \rightarrow\, ^4I_{11/2}$ transitions, the strongest is the ℓ_2 transition between the R_2 and Y_3 levels with a vacuum wavelength of 1064.15 nm. At room temperature there is sufficient population in the R_2 level for this transition to dominate the laser output. However, if the crystal is cooled to low temperatures almost all the population of the $^4F_{3/2}$ manifold resides in the R_1 level and lasing then occurs on the ℓ_1 transition from R_1 to the Y_2 level of $^4I_{11/2}$ at a vacuum wavelength of 1064.40 nm.

Broadening

In Nd:YAG the $^4F_{3/2} \rightarrow\, ^4I$ transitions are homogeneously broadened by phonon collisions, with a full width at half-maximum of approximately 190 GHz (6.5 cm^{-1}). Note that this is much larger than the natural broadening—dominated by the short lifetime of the lower laser level—which is of order 1 GHz.

Effective gain cross-section[†]

Calculation of the optical gain is complicated by the fact that both the ℓ_1 and ℓ_2 transitions in Nd:YAG can contribute to the gain. In such cases it is useful to use a single **effective cross-section** that takes into account the contribution to the gain from all the transitions involved. The effective cross-section can be defined in such a way that the gain coefficient is given by multiplying it by the *total* population of the manifold of levels, rather than the population of any single level.

We may develop an expression for the effective gain cross-section as follows. If we make the simplifying assumption that the population in the lower levels is essentially zero, the rate of stimulated emission caused by narrowband radiation of angular frequency ω and total intensity I is:

$$N_{R_1}\sigma_{R_1}(\omega - \omega_0^{R_1})\frac{I}{\hbar\omega} + N_{R_2}\sigma_{R_2}(\omega - \omega_0^{R_2})\frac{I}{\hbar\omega}, \quad (7.20)$$

where N_{R_i}, $\omega_0^{R_i}$, and $\sigma_{R_i}(\omega_L - \omega_0^{R_i})$ are the population density, centre frequency, and optical gain cross-section of the ℓ_i transition. In terms of the effective cross-section $\sigma_{\text{eff}}(\omega - \omega_0)$, this rate can be written as:

$$N_2 \sigma_{\text{eff}}(\omega - \omega_0)\frac{I}{\hbar\omega}, \quad (7.21)$$

where ω_0 is the centre frequency of the combination of transitions.

Comparison of (7.20) and (7.21) then yields the following expression for the effective gain cross-section:[40]

$$\sigma_{\text{eff}}(\omega - \omega_0) = f_{R_1}\sigma_{R_1}(\omega_L - \omega_0^{R_1}) + f_{R_2}\sigma_{R_2}(\omega_L - \omega_0^{R_2}), \quad (7.22)$$

where f_{R_i} is the fraction of the population in level R_i to the total population in the $^4F_{3/2}$ manifold.

Figure 7.17 shows the frequency dependence of the optical gain cross-sections for the ℓ_1 and ℓ_2 transitions and the effective cross-section derived

[40]It should be noted that other definitions of the effective gain cross-section are possible. For example, Davies (2000) defines the effective cross-section in terms of the population density in the R_2 level, i.e. the cross-section of (7.22) divided by f_{R_2}.

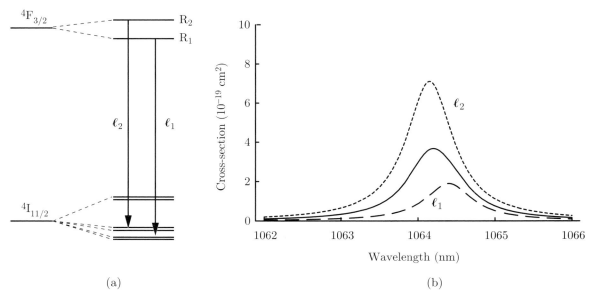

Fig. 7.17: Calculation of the effective optical gain cross-section in Nd:YAG. (a) shows the ℓ_1 and ℓ_2 transitions from the R_1 and R_2 levels of the $^4F_{3/2}$ manifold. (b) shows the contribution of the ℓ_1 transition (dashed) and ℓ_2 transitions (dotted) to the effective cross-section (solid) as defined by eqn (7.22).

from them. The peak optical gain cross-sections of the ℓ_1 and ℓ_2 transitions are 1.9×10^{-19} cm^2 and 7.1×10^{-19} cm^2, respectively. Hence, the dominant contribution comes from the ℓ_2 transition; if we neglect the contribution of the gain from the ℓ_1 transition the peak value of the effective gain cross-section is given by $\sigma_{\text{eff}}(0) \approx f_{R_2}\sigma_{R_2}(0) = 2.8 \times 10^{-19}$ cm^2. In Exercise 7.5 it is shown that the contribution of the ℓ_1 transition increases the effective cross-section to 3.7×10^{-19} cm^2.

Finally in this section, we note that the relative contributions of the two transitions depends on the relative populations in the R_1 and R_2 levels. Further, the broadening of the transitions will increase owing to an increased rate of phonon de-excitation as the temperature is raised. As a consequence, the value of the effective cross-section, and the frequency at which the gain is a maximum, depends on temperature. It is found that for temperatures within 60 °C of room temperature, the peak laser wavelength increases[41] with temperature at a rate (Zagumennyi, et al. 2004) of 5×10^{-3} nm K^{-1}.

Practical implementation

Nd:YAG lasers are one of the most common types of laser in use, and can be operated in a wide variety of different configurations: they can run continuously or in a pulsed mode; they can be pumped by flashlamps or laser diodes; and they can provide mean output powers from a few milliwatts up to several kilowatts. Very often, Nd:YAG lasers are frequency doubled, tripled, or quadrupled to generate radiation of wavelength 532, 355, or 266 nm, respectively.[42]

---

[41] Note that if the increased relative population of R_2 were the dominant effect, the centre wavelength of the composite transition would *decrease* with temperature, which is not what is observed.

[42] See Chapter 15.

Flashlamp-pumped systems employ pump chambers with single or multiple flashlamps, as in Fig. 7.13(a). In such systems the laser rod is typically of about 5 mm diameter and 30–150 mm long, and the optical cavity is of order 0.5 m in length. With pulsed pumping the laser is often Q-switched at a pulse repetition of 10–50 Hz to yield output pulses of 5–10 ns duration and energies of a few tens to several hundred millijoules. The addition of one or more Nd:YAG amplifier stages can increase the output pulse energy to several Joules.

Flashlamp-pumped systems can generate pulsed output with a much higher pulse repetition frequency by employing continuous pumping and employing acousto-optic modulators to Q-switch the cavity at a repetition rate of 10–20 kHz. The mean output power of this type of system can be as high as 15 W.

Nd:YAG lasers may be pumped by GaAs diode lasers operating at 808 nm, which closely matches the peak absorption corresponding to excitation of levels of the $^2H_{9/2}$ and $^4F_{5/2}$ manifolds. The excited levels lie only 900 cm^{-1} above the upper laser levels, and consequently much less energy is deposited in the crystal in the form of heat (phonons) than is the case for flashlamp pumping which excites all manifolds up to some 10 000 cm^{-1} above the upper laser levels.

Diode-pumped Nd:YAG lasers can generate continuous-wave output with powers of: 10 W or more with end pumping; above 100 W with side pumping; and of order 1 kW with slab pumping. Q-switching and modelocking is also possible. The slope efficiencies of diode-pumped Nd:YAG can be as high as 50–60% for end-pumped configurations, and 25–40% for side pumping. These figures are much higher than for flashlamp-pumped lasers, for which the slope efficiency is typically 3%.

7.2.4 Other crystalline hosts

Neodymium ions have exhibited laser oscillation in a wide variety of hosts.[43] Here, we briefly mention a few hosts of note. The properties of some Nd lasers are summarized in Table 7.1.

[43] A comprehensive catalogue of solid-state laser transitions and their properties may be found in the book by Weber (1982).

YLF

Lithium yttrium fluoride, LiYF$_4$, (usually written as YLF) is a common alternative host for Nd ions. The crystal exhibits lower thermal lensing and

Table 7.1 Important parameters of Nd:YAG, Nd:YLF, Nd:GdVO$_4$, and Nd:glass (Hoya LHG-5 phosphate) lasers.

	Nd:YAG	Nd:YLF	Nd:GdVO$_4$	Nd:glass
λ (nm)	1064	1053	1063	1054
τ_2 (μs)	230	450	90	290
τ_1 (ps)	100s	100s	100s	100s
$\Delta\nu$ (GHz)	160	380	180	5000
σ_{21} (10^{-20} cm^{-2})	37	18.7	76	4.1
κ (W m^{-1} K^{-1})	13.0	6.0	12.3	1.19

thermally induced birefringence than YAG, and consequently it is often used when beams of very high optical quality are required.

The dominant transitions between the Stark levels of the $^4F_{3/2}$ and $^4I_{11/2}$ manifolds are different for the two possible polarizations.[44] As a consequence, lasing with σ-polarized radiation occurs at 1053 nm, whilst lasing with π-polarized radiation occurs at 1047 nm. Hence by inserting an intra-cavity polarizer it is possible to restrict laser oscillation to one of these transitions. The linearly polarized output is helpful for subsequent frequency doubling and tripling or for electro-optic modelocking or Q-switching.

The 1053-nm transition is well matched to the peak gain in phosphate Nd:glass lasers; the linewidth is also some three times that of the same transition in YAG, allowing shorter pulses to be generated. As a consequence, Nd:YLF is often used in the modelocked oscillators and preamplifiers at the front end of the large Nd:glass laser systems used in fusion research.[45]

[44] Radiation with the electric field vector parallel to some axis—in this case the optical axis of the crystal—is described as π-polarized; that with the electric field vector perpendicular to the axis is said to be σ-polarized.

[45] See Section 17.3.3.

Vanadates

As discussed in Section 7.1.7, for diode pumping it is desirable to reduce the tolerance required on the wavelength of the diode laser by increasing the linewidth of the absorption transition, and to this end the vanadate crystals YVO_4 and $GdVO_4$ are frequently used as hosts for Nd ions. For example, the 808.4-nm pump band in Nd:$GdVO_4$ is 80% broader than the same pump band in Nd:YAG, and the absorption coefficient is some 7 times stronger. Neodymium-doped vanadate crystals are frequently employed in low- and medium-powered diode-pumped laser systems generating mean output powers of tens of watts.

7.2.5 Nd:glass laser

A wide variety of oxide-, fluoride-, and sulphide-based glasses have been developed as hosts for Nd ions. The wavelength of the $^4F_{3/2} \rightarrow {}^4I_{11/2}$ transition varies from approximately 1054 to 1062 nm depending on the glass host. The lifetime of the upper laser level is similar to that in YAG.

In glass hosts the laser transition is strongly inhomogeneously broadened, the linewidth increasing by a factor of approximately 50 to typically 6.5 THz ($\Delta\lambda \approx 25$ nm) compared to that found in Nd:YAG. Largely as a result of the greater linewidth, the gain cross-section of Nd.Glass is an order of magnitude smaller than that in Nd:YAG.

The small optical gain cross-section of Nd:glass allows a large population-inversion density to be generated without the onset of amplified spontaneous emission (see Section 5.3). The material is therefore frequently employed to amplify laser pulses to large energies, as discussed on page 488. The low thermal conductivity of Nd:Glass restricts the pulse repetition rate of amplifiers based on this material to no more than a few pulses per second; and for very high energy amplifiers, the maximum pulse repetition rate is much slower. Finally, we note that the large linewidth supports amplification of pulses as short as 100 fs.

7.2.6 Erbium lasers

Laser action has been achieved in erbium ions doped into a variety of garnet and fluoride crystalline hosts, as well as in several types of glass. There are two transitions of interest: the $^4I_{13/2} \rightarrow\, ^4I_{15/2}$ transition lases in ions doped into both crystalline and glass hosts, with wavelengths of approximately 1.64 μm and 1.54 μm, respectively; the $^4I_{11/2} \rightarrow\, ^4I_{13/2}$ transition lases only in crystalline hosts, and has a wavelength of 2.9 μm. Radiation at 2.9 μm is absorbed very strongly by water, and consequently erbium lasers operating on this transition have found applications in medicine. The 1.5-μm transition matches the third transparency window of optical fibres, and consequently is important for optical communications, particulary in the form of the **erbium-doped fibre amplifier (EDFA)** discussed in Section 10.2. Further, this wavelength falls into the so-called 'eye-safe' window,[46] and consequently erbium lasers may also be used in telemetry[47] and laser-ranging applications.

[46] Lasers operating between 1.45 and 1.70 μm are sometimes known as 'eye-safe' since radiation in this region is strongly absorbed by the cornea of the eye, and so cannot reach the retina. The term, however, is something of a misnomer since a sufficiently high-power laser operating in the 'eye-safe' region could still cause damage to the cornea or lens of the eye.

[47] That is, the process of obtaining measurements in one place and relaying them to a distant point.

Erbium energy levels

The electron configuration of Er^{3+} ions is $[Xe]\,4f^{11}$, that is three 4f electrons short of a full 4f subshell. This configuration may be viewed as comprising a full 4f subshell plus three 4f holes, and consequently the energy level structure is analogous to that of Nd^{3+} ions, which have a ground configuration of $[Xe]\,4f^3$, except that the energy ordering of the manifolds is reversed.

The energy levels of Er^{3+} ions in YAG and glass hosts are shown in Fig. 7.18 and described in more detail below.

Er:YAG

Figure 7.18(a) shows the low-lying manifolds of the Er^{3+} ion doped into a YAG host. In this host the $^4I_{11/2} \rightarrow\, ^4I_{13/2}$ and $^4I_{13/2} \rightarrow\, ^4I_{15/2}$ laser transitions have wavelengths of 2.94 μm and 1.64 μm, respectively. In a crystalline host such as YAG, these transitions are homogeneously broadened by phonon collisions to give a homogeneous width of order 300 GHz.

The lower level of the 2.94-μm transition lies more than 6000 cm^{-1} above the ground state. This offers the benefit of a low thermal population at room temperature, and so this laser transition may be described as a four-level system. However, the large energy gap also causes the rate of multiphonon decay from the levels of this manifold to be relatively small, and as a consequence the fluorescence lifetime is long, approximately 3 ms. In contrast, the upper level of this transition lies only 3450 cm^{-1} above the lower laser level; its rate of multiphonon decay is therefore somewhat higher, and consequently has a shorter fluorescence lifetime of order 0.1 ms. However, despite this highly unfavourable lifetime ratio—and the fact that the upper laser level decays rapidly by non-radiative transitions to the lower laser level—this laser transition may be operated continuously (particularly when diode pumped) in highly doped crystals. The reason for this is that at high doping levels the strong $Er^{3+} - Er^{3+}$ interaction allows the following up-conversion:

$$Er^{3+}\left(^4I_{13/2}\right) + Er^{3+}\left(^4I_{13/2}\right) \rightarrow Er^{3+}\left(^4I_{9/2}\right) + Er^{3+}\left(^4I_{15/2}\right). \quad (7.23)$$

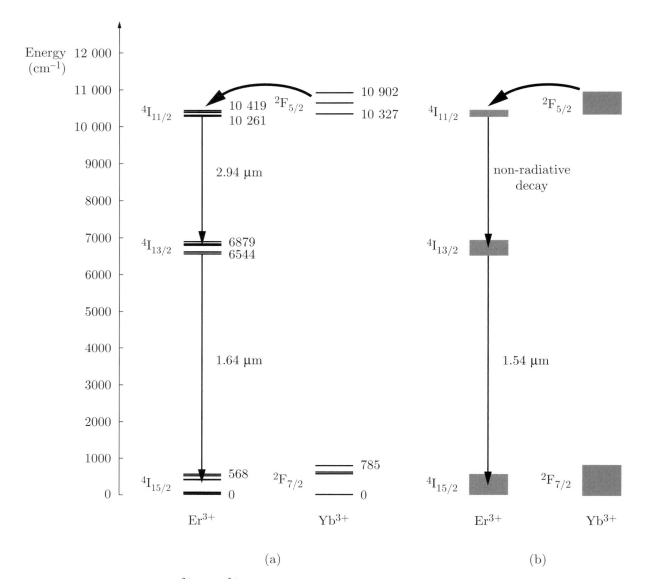

Fig. 7.18: Energy-level structure of: Er^{3+} and Yb^{3+} ions in: (a) a YAG host; (b) in a glass host. For the YAG host the energies of the highest and lowest levels in each manifold are given in cm^{-1}.

This interaction transfers ions out of the lower laser level to the ground manifold (from where they may be repumped) or to the $^4I_{9/2}$ manifold at approximately $15\,000\,cm^{-1}$, which decays rapidly to the upper laser level. In practice, sufficiently rapid up-conversion requires doping levels of 10–50%.

The operation of the 1.64 μm laser transition in Er:YAG is analogous to the 1.54-μm laser transition in Er:glass discussed below.

The 2.94-μm and 1.64-μm laser transition in Er:YAG may be either flashlamp or diode pumped. As for Nd:YAG, flashlamp pumping excites several higher-lying manifolds that decay rapidly to the $^4I_{11/2}$ manifold by non-radiative transitions. Diode pumping may be achieved with InGaAs/GaAs strained quantum-well lasers operating at 970 nm that transfer ions in the ground manifold to the upper levels of the $^4I_{11/2}$ manifold; for the 2.94-μm transition this is known as **in-band pumping** since the pumping occurs directly to the upper laser manifold.

Er:Glass

Figure 7.18(b) shows the energy levels of Er^{3+} ions in a glass host, and the important $^4I_{13/2} \rightarrow {}^4I_{15/2}$ transition at 1.54 μm. The upper and lower manifolds of levels are split by the crystal field into 7 and 8 degenerate levels, respectively. In a glass host the 56 possible transitions between these levels are broadened homogeneously by lifetime and phonon broadening, and inhomogeneously by spatial variations in the local environment. The broadening characteristics of the laser transition are therefore complex, and to some extent the relative importance of homogeneous and inhomogeneous broadening depends on the properties of the glass host employed. However, the linewidth of the laser transition is approximately an order of magnitude greater than in a crystalline environment and hence the transition may be considered to be pre-dominantly inhomogeneously broadened.[48]

Optical pumping at 980 nm (10 200 cm^{-1}) to the $^4I_{11/2}$ manifold can be achieved with InGaAs/GaAs diode lasers; alternatively, in-band pumping to the higher levels of the $^4I_{13/2}$ manifold can be realized using InGaAsP lasers operating at 1480 nm (6800 cm^{-1}). For either pump band, excitation is followed by rapid non-radiative decay[49] to the lowest level of the $^4I_{13/2}$ manifold, which acts as the upper laser level. As for the case of Er:YAG, since the $^4I_{13/2}$ level lies well above the next lower level—those of the ground manifold— the rate of multiphonon decay of this level is low and the level decays predominantly radiatively with a fluorescence lifetime of approximately 8 ms. Lasing occurs from the bottom of the $^4I_{13/2}$ manifold to the levels of the ground-state manifold, $^4I_{15/2}$.[50]

In order to increase the pumping efficiency, it is common to codope the glass with Yb^{3+} ions. Optical pumping of Yb^{3+} ions causes a large population density to be formed in the $^4F_{5/2}$ manifold, either by direct pumping at approximately 980 nm or by rapid non-radiative cascade from higher manifolds. Since the $^4F_{5/2}$ manifold of Yb^{3+} lies some 10 000 cm^{-1} above the ground state, the rate of multiphonon decay of this level is low and it decays predominantly radiatively with a fluorescence lifetime of approximately 1 ms. The $^4F_{5/2}$ manifold of Yb^{3+} therefore acts as an efficient reservoir of excitation. Further, the energy of this manifold is closely matched to the $^4I_{11/2}$ manifold of Er^{3+} and as a consequence rapid transfer of energy can occur between excited Yb^{3+} ions and Er^{3+} ions in the reaction:[51]

$$Yb^{3+}\left({}^4F_{5/2}\right) + Er^{3+}\left({}^4I_{15/2}\right) \rightarrow Yb^{3+}\left({}^4F_{7/2}\right) + Er^{3+}\left({}^4I_{11/2}\right). \quad (7.24)$$

The levels of the $^4I_{11/2}$ manifold of the Er^{3+} ion rapidly decay non-radiatively to the levels of the $^4I_{13/2}$ manifold. Note that it is important this decay of the $^4I_{11/2}$ manifold is rapid in order to avoid backtransfer of excitation in the reverse reaction[52] to eqn (7.24).

[48] The importance of inhomogeneous broadening in Er:glass is confirmed by the observation of spectral hole-burning in the gain profile. See, for example, Bigot et al. (2004) or Cordina (2004).

[49] The rates of non-radiative decay to the $^4I_{13/2}$ manifold from the $^4I_{11/2}$ manifold, and above, are between 10^5 s^{-1} and 10^7 s^{-1}.

[50] For the $^4I_{13/2} \rightarrow {}^4I_{15/2}$ transition the up-conversion mechanism described by eqn (7.23) depletes the upper laser level, and consequently the concentration of Er^{3+} ions must be kept low, generally to below 1 × 10^{20} cm^{-3}.

[51] More correctly, energy migrates through the host by rapid transfer between neighbouring Yb^{3+} ions until it reaches a Yb^{3+} ion near a ground-state Er^{3+} ion. Energy transfer then leads to excitation of the Er^{3+} ion.

[52] In this respect phosphate glass hosts confer some advantage since their higher maximum phonon energy (1325 cm^{-1}) than that of silica glasses (1190 cm^{-1}) increases the rate of multiphonon decay of the $^4I_{11/2}$ manifold of Er^{3+} (Laporta, et al. 1999).

The properties of some Er lasers are listed in Table 7.2. By far the most important application of Er:glass lasers is the erbium-doped fibre amplifier discussed in Section 10.2, but they have been operated in a variety of other configurations. For example, Fig. 7.19 shows an Er:glass microlaser pumped by radiation from an InGaAs diode laser. In this design the active medium is a thin disc of Yb:Er:glass approximately 2 mm thick. The concentration of Er^{3+} ions is in the range $0.1 - 2 \times 10^{19}$ cm^{-1}, and that of the Yb^{3+} ions is typically 1×10^{21} cm^{-1}. The rear face of the disc is coated to be highly reflective between 1530 and 1560 nm, but to be transparent to the pump radiation at 980 nm. The laser cavity is formed between the reflective rear surface of the Yb:Er:glass disc and a concave output coupler of 5–10 mm radius of curvature, coated to have a reflectivity of approximately 98% for the wavelengths of the $^4I_{13/2} \rightarrow {}^4I_{15/2}$ transition. These rather compact devices operate continuously with output powers of several tens of milliwatts. Single-frequency operation may be obtained by introducing into the cavity a thin (100−300 μm) uncoated plate that acts as a Fabry–Perot etalon. The output wavelength of the laser may be tuned in the range 1530−1565 nm, with a linewidth of less than 50 kHz, by adjusting the angle of the etalon.

Er:glass lasers have also been operated in pulsed mode.[53] Q-switched operation is able to generate pulses of nanosecond duration, which is of interest in eye-safe laser ranging. Alternatively, modelocked microlasers can generate pulses of approximately 20 ps duration with a pulse repetition rate as high as several GHz.

[53] A detailed discussion of the operation of both continuously operating and pulsed Er:glass lasers may be found in the review article by Laporta et al. (1999).

Table 7.2 Important parameters of Er:YAG and Er:glass lasers.

		Er:YAG		Er:glass
		$^4I_{11/2} \rightarrow {}^4I_{13/2}$	$^4I_{13/2} \rightarrow {}^4I_{15/2}$	$^4I_{13/2} \rightarrow {}^4I_{15/2}$
λ	(nm)	2940	1646	1540
τ_2	(μs)	100	7700	8000
τ_1	(μs)	7700	∞	∞
$\Delta\nu$	(GHz)	300	300	3500
σ_{21}	(10^{-20} cm^{-2})	2.6	0.5	0.7
κ	(W m^{-1} K^{-1})	13	13	1

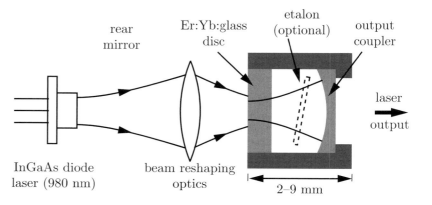

Fig. 7.19: Schematic diagram of a Yb:Er:glass microlaser. (After Laporta et al. (1999).)

7.2.7 Praseodymium ions

Space does not permit a comprehensive review of the many trivalent rare earths in which laser oscillation has been observed. Here, we briefly mention just one other: praseodymium, and in particular Pr^{3+}: YLF. This system is interesting since it is the only really efficient visible solid-state laser.

The trivalent Pr ion has a ground-state configuration of $[Xe]4f^2$, and a lowest-lying manifold of $^3H_4$. The absorption bands of Pr^{3+}: YLF are relatively narrow, making flashlamp pumping inefficient. Early work employed various argon-ion laser lines to pump the $^3P_2 \leftarrow ^3H_4$ band at approximately $22\,600\,\text{cm}^{-1}$. However, more recently the output of GaN semiconductor diode lasers operating at 444 nm have been successfully employed; these pump lasers are much more efficient, and more compact, than argon-ion lasers.[54]

Optical pumping of the levels of the $^3P_2$, or higher-lying, manifolds is followed by rapid non-radiative decay to the lowest levels of the $^3P_1$ and $^3P_0$ manifolds. These levels are thermally coupled, with a room-temperature fluorescence lifetime of 36 μs. Lasing can occur from either of these levels on more than 15 transitions to levels in the $^1G_4$, $^3F_{4,3,2}$, and $^3H_{6,5}$ manifolds, with output wavelengths ranging from 520 nm to 720 nm. Several of these transitions show vibronic structure[55] and hence are tunable. For example, the $^3P_0 \rightarrow ^3H_6$ transition at 613 nm has a linewidth of approximately 1 nm; Kerr-lens modelocking[56] of this transition has been demonstrated, yielding pulses as short as 400 fs.[57]

[54] Cornacchia et al. (2008).

[55] See Section 7.3.1.

[56] See page 218.
[57] Sutherland et al. (1996).

7.3 Ruby and other trivalent iron-group systems

7.3.1 Energy-level structure†

The ground-state configuration of neutral atoms in the iron group of the transition metals is of the form $[Ar]\,3d^{n+1}\,4s^2$ or $[Ar]\,3d^{n+2}\,4s$, and hence that of the trivalent ions is $[Ar]\,3d^n$. The spectroscopy of these ions is very different from that of the $4f^n$ configuration in the trivalent rare earths, owing to the fact that the 3d orbitals are larger than the 4f orbitals in the trivalent rare earths, and the ionic size (and hence d) is smaller for the iron group than the rare earths. As a consequence, the interaction with the crystal field is more than an order of magnitude bigger for the 3d electrons in the iron group than the 4f electrons in the rare earths, to the extent that the interaction with the crystal field is stronger than the spin-orbit interaction; the LS coupling scheme no longer applies.

It is highly instructive to consider in outline the form of the wave functions and energy levels in the strong crystal field limit. In this case the effect of the crystal field must be considered before the residual electrostatic and spin-orbit interactions, and consequently we determine the effect of the crystal field acting on the single-electron orbitals formed in the central potential of the active ion.

The crystal field prevents the electrons moving freely within the spatial distributions they would have in a free ion, and hence ℓ is no longer a good

[58] We note that other, generally lower, symmetries of the crystal field are often treated by considering the difference between the true symmetry and octahedral symmetry as a perturbation that acts on the energy levels found for the case of octahedral symmetry.

[59] A subscript 'g' (standing for 'gerade', the German for 'even') may also be included in the label to indicate that the orbitals have even parity, i.e. e_g and t_{2g}. Orbitals with odd parity can be labelled with a subscript 'u', standing for 'ungerade'.

quantum number. Instead, the modified electron states are labelled by its symmetry properties, and by the electron spin, which is unchanged by the interaction with the crystal field.

As an example, let us consider the case of a d-electron moving in a crystal field with octahedral symmetry.[58] Octahedral symmetry of the crystal field arises when the active ion is located at the centre of a cube, and identical neighbouring ions are located on each face of the cube, as illustrated in Fig. 7.20. In the absence of the crystal field the d-electron energy levels would have five-fold degeneracy, corresponding to $m = -2, -1, 0, 1, 2$. A crystal field with octahedral symmetry partially lifts this degeneracy to form a triply degenerate crystal field orbital labelled t_2, and a higher-lying doubly degenerate orbital e.[59] The spatial distributions of the t_2 and e orbitals are very different, as illustrated schematically in Fig. 7.20. The e wave functions have regions of high probability density centred on the axes connecting the active ion with the neighbouring ions, whereas for the t_2 orbitals these regions are centred between the axes. Consequently, electrons in the e orbitals are on average closer to the neighbouring ions than those in the t_2 orbitals and therefore their interaction with the neighbouring ions—that is, with the crystal field—causes their energy to be increased from the unperturbed energy by a greater amount than electrons in the t_2 orbitals.

We may now consider the configurations formed in the crystal field as the number of d-electrons is increased. The ground-state configuration of a single d-electron is simply t_2. Adding a second and third d-electron to the system

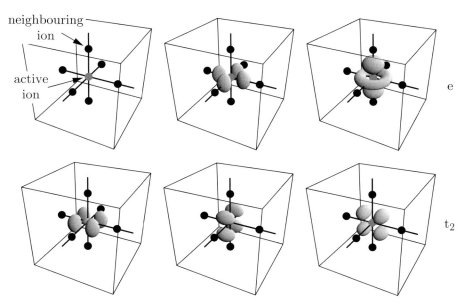

Fig. 7.20: Crystal-field orbitals of a d-electron in a field with octahedral symmetry. The top row shows the relative positions of the active ion and the neighbouring ions, and a surface of constant probability density for the two degenerate wave functions corresponding to the e_g orbitals. The bottom row shows equivalent plots for the three degenerate wave functions corresponding to the t_2 orbital. Note that the wave functions of the e orbitals have lobes pointing towards the neighbouring ions, and hence have higher energy than the t_2 orbitals for which the lobes in the wave function point between the neighbouring ions.

gives rise to ground-state configurations comprised of two or three t_2 electrons, respectively, since the t_2 orbital is three-fold degenerate. It is found that, after including the effects of the residual electrostatic and spin-orbit interactions, the total energy is minimized if the number of aligned spins is maximized,[60] the spins of the electrons will all point in the same direction.[61] A fourth d-electron added to the system may also occupy the t_2 orbital, but its spin must point in the opposite direction to satisfy the Pauli exclusion principle.[62] Considerations of this kind lead straightforwardly to the ground-state configurations illustrated schematically in Fig. 7.21.

The possible configurations of excited levels may be identified in a similar way. For example, for the case of three d-electrons two different excited configurations may be formed: one of the spins of the t_2 electrons may be reversed to give a configuration lying above the ground-state owing to the larger spin-orbit and residual electrostatic interactions; alternatively, one of the electrons can be promoted to the e orbital. The two types of excited configurations give rise to two types of transition, as illustrated in Fig. 7.22. In a **spin-flip** transition all the electrons remain in the same orbital, but the spin of one of the electrons is reversed, as illustrated in Fig. 7.22(a); in a **configuration transition** one electron makes a transition from an e orbital to a t_2 orbital.[63]

As we shall explore further below, these two types of transition have very different characteristics. In a spin-flip transition the spatial distribution of the total electronic wave function is unchanged. As a consequence, the strength of the interaction of the active ion with the neighbouring ion is unchanged and, as we will see, this results in narrowband transitions in which there is no change in the vibrational energy of the crystal lattice. Transitions of this type are known as **zero-phonon** transitions.[64]

In contrast, in a configuration transition the spatial distribution of the total electron wave function changes significantly. Since the interaction of the active ion with the neighbouring ions is very different for the upper and lower levels, the neighbouring ions move during the transition. Configuration transitions are therefore associated with changes in the vibrational motion of the lattice, and are characterized by broad-band, temperature-dependent lineshapes—the broad bandwidth arising from the emission or absorption of varying numbers

[60] This is an example of 'Hund's rules'.

[61] Hence the ground-state configurations of t_2^2 will have $S = 1$ (a triplet), and that of t_2^3 will have $S = 3/2$ (a quartet).

[62] We note that if the crystal field were weak, the ground-state configuration would be $t_2^3 e$, with all spins parallel since this minimizes the spin-orbit interaction; in other words the increase in energy associated with promoting an electron to the e orbital would be smaller than the increase in energy caused by reversing the spin of the electron.

[63] The spin of the electron may, or may not, also change.

[64] The $4f^n$–$4f^n$ transitions considered above are zero-phonon transitions since the interaction with the crystal field is weak for both the upper and lower levels.

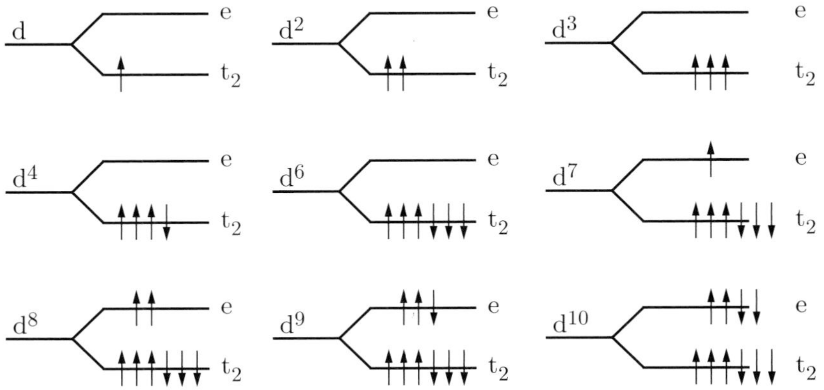

Fig. 7.21: Schematic diagram showing the ground-state configurations arising from splitting of a d^n configuration a strong crystal field with octahedral symmetry. The arrows indicate electron spin.

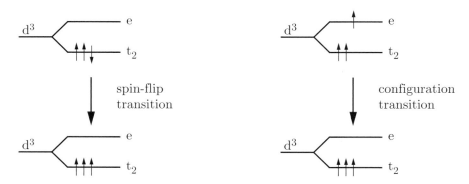

Fig. 7.22: Spin-flip and configuration transitions of a d-electron in a crystal field with octahedral symmetry.

and energies of lattice phonons. Transitions of this type are therefore known as **phonon-assisted** or **vibronic transitions** and are the basis of an important class of widely tunable solid-state laser.

Configuration coordinate diagrams

Further insight into the properties of energy levels in the strong-coupling limit is provided by the **configuration coordinate diagram**. In this model the active ion and the surrounding ions are treated as a single system. To simplify matters this multidimensional problem[65] is reduced to one dimension, and the distance between the active ion and the surrounding ions is parameterized by a 'configuration coordinate' Q. We can think of Q as representing some sort of average value of the distance between active and neighbouring ions.[66]

Within the configuration coordinate picture the Schrödinger equation for the active ion may be written in the form,

$$[T_n + T_e + V(Q, r)]\Psi(Q, r) = E\Psi(Q, r), \qquad (7.25)$$

where: $r \equiv r_1, r_2, \ldots r_N$ represents the coordinates of the electrons; T_n is the kinetic energy operator of the nuclei of the active and surrounding ions; T_e is the sum of the kinetic energy operators of the electrons in the active ion;[67] and $V(Q, r)$ is the electrostatic potential energy, which contains terms arising from the attraction of the electrons in the active and neighboring ions to the nuclei of the active and neighbouring ions, mutual repulsion between the electrons, and mutual repulsion between the nuclei.

Just as for the case of diatomic molecules discussed in Section 13.2, we may make the **Born–Oppenheimer approximation** and write the total wave function as a product of an electronic wave function $\psi_e(Q, r)$ and a nuclear (or 'vibrational') wave function $\psi_n(Q)$. The Schrödinger equation for the electronic wave function may then, in principle, be solved for fixed Q to yield an energy eigenvalue $U(Q)$ that is found to be the effective potential in which the nuclei move. For values of Q near its equilibrium value, Q_0, this potential will be approximately that of a harmonic oscillator,[68] and hence the active ion will exhibit a ladder of vibrational energy levels as shown in Fig. 7.23.[69]

---

[65] If there are N neighbouring ions there are in principle N different values of the spacing between the active ion and the nearest neighbours. Whether all N values are different depends on the nature of the lattice vibrations, as discussed in Note 69.

[66] Although see Note 69.

[67] We will ignore the electrons in the neighbouring ions since these will not be involved in any of the transitions of interest.

[68] This follows from expanding $U(Q)$ about the equilibrium position: $U(Q) = U(Q_0) + \mathrm{d}U/\mathrm{d}Q|_{Q_0} (Q - Q_0) + (1/2) \, \mathrm{d}^2U/\mathrm{d}Q^2\big|_{Q_0} (Q - Q_0)^2 + \ldots$. However, $\mathrm{d}U/\mathrm{d}Q|_{Q_0} = 0$ since by definition $U(Q)$ is a minimum at $Q = Q_0$.

[69] We see now that Q need not be thought of as the average value of the distance between active and neighbouring ions. Rather, when $Q = Q_0$ all the neighbouring ions are in their equilibrium positions; when $Q = Q_0 + \delta Q$ all the neighbouring ions have increased their separation from the active ion by a distance δQ. The configuration coordinate model are therefore effectively considers the lattice vibrations to be such that all the neighbouring ions change their separation from the active ion by the same amount, and consequently can be represented by a single value of Q.

If the spatial distribution of the electronic wave functions is similar for the upper and lower electronic levels then the interaction of the ion with the neighbouring ions will be similar for the two levels, and hence the upper and lower potential wells will have similar values of Q_0 and so lie above one another on a configuration coordinate diagram, as shown in Fig. 7.23(a). In contrast, if the spatial distributions of the wave functions for the upper and lower levels are very different, the interaction of the electrons of the active ion will be different, and consequently Q_0 will be different for the two levels. This situation is shown in Fig. 7.23(b).

We may now use the configuration coordinate picture to deduce the form of the absorption and emission spectra of active ions for the two types of transition, as illustrated in Fig. 7.23. To do this we note that, in analogy with the case of diatomic molecules, the strength of an optical transition is proportional to the square of the overlap integral of the vibrational wave functions of the upper and lower level.[70] Suppose that the lower electronic level corresponds to the electronic ground state, and let us consider optical excitation to the upper electronic level. The vibrational wave functions for the lowest vibrational level generally resemble a Gaussian curve, whilst those of high-lying levels are strongly peaked near the classical turning points.[71] Absorption from the lower electronic level will be dominated by transitions from the lowest vibrational level, since this will have by far the largest thermal population. The strongest absorption will therefore be from here to the lowest vibrational level of the upper electronic level, as shown in Fig. 7.23(a), since the overlap integral for

[70] See Section 13.3.

[71] For a given vibrational energy level the classical turning points are the values of Q for which the potential curve has an energy equal to the total vibrational energy. In classical mechanics the kinetic energy of a particle is zero at such points, and hence they represent the extrema of its oscillatory motion. Classically, the probability of finding a particle is greatest at the turning points, since there it is stationary. The correspondence principle therefore tells us to expect the wave functions of high-lying (i.e. with large energy) vibrational levels to be strongly peaked at these points.

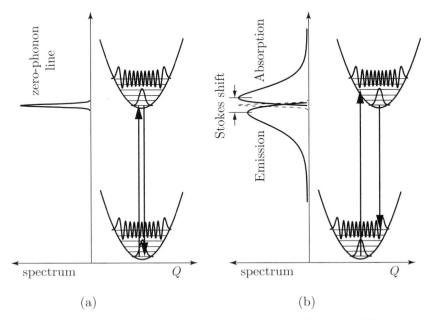

Fig. 7.23: Transitions between electronic levels: (a) when the equilibrium configuration coordinate is essentially the same for the upper and lower electronic levels; (b) when the equilibrium configuration coordinates are different for the two levels, causing the potential curves to be displaced. Vibrational wave functions are illustrated schematically for the lowest vibrational level and an excited vibrational level. In (b) the position of the zero-phonon line is shown by the dotted curve.

these two vibrational wave functions will be large. The overlap integral will be comparatively small for transitions to high-lying vibrational levels since the vibrational wave functions of such levels show rapid oscillations and are only large near the classical turning points, where the amplitude of the wave function of the lowest vibrational level of the ground state is small. In any case, excitation to excited vibrational levels of the upper electronic level will be followed by extremely rapid de-excitation to the lowest vibrational level by phonon de-excitation, as discussed in Section 7.1.3. The emission spectrum will therefore be dominated by emission from the lowest vibrational level of the upper electronic level. Just as for the absorption spectrum, and for the same reason, the emission will be dominated by transitions to the lowest vibrational level of the electronic ground state. In other words, when the equilibrium coordinates of the upper and lower electronic levels are similar—or, equivalently, when the spatial distribution of the wave functions of the upper and lower levels are similar, such as in spin-flip transitions—the absorption and emission spectra are dominated by a single, zero-phonon transition as we would expect from our earlier considerations.

In contrast, consider the case shown in Fig. 7.23(b) in which the spatial distribution of the electronic wave functions are different for the upper and lower levels, and consequently the equilibrium configuration coordinates are different. Now, absorption from the lowest vibrational level of the ground electronic state is distributed over many vibrational levels of the upper electronic level, leading to absorption extending to higher frequencies than that of the zero-phonon line. Absorption at frequencies above that of the zero-phonon line corresponds to electronic excitation accompanied by excitation of lattice vibrations. Following absorption, ions in excited vibrational levels of the upper electronic level will undergo rapid phonon de-excitation to the lowest-lying vibrational level. Just as for the absorption, emission will then occur to a wide range of vibrational levels of the ground electronic state, producing a broad emission spectrum shifted to frequencies below that of the zero-phonon line.[72] The absorption and emission spectra are widely separated, the separation in frequency between the peaks of the two spectra being known as the Stokes shift. Hence, as expected from our earlier discussion, transitions between electronic levels with different spatial distributions (and hence different values of Q_0), such as configuration transitions, are broad-band, vibronic transitions. Note that the width of the absorption and emission bands, and the magnitude of the Stoke's shift, both increase with the magnitude of the difference in the equilibrium configuration coordinates of the two levels.

[72]Note the similarities with the absorption and emission spectra observed in the dye lasers discussed in Section 14.3.

7.3.2 The ruby laser

The ruby laser is of enormous significance since it was the first optical maser. It is much less commonly used today than it once was, but it still finds applications in holography (where the relatively short wavelength for a solid-state laser is better suited to the response of photographic emulsions), pulsed interferometry, and some medical applications. It is also the best known example of a three-level laser.

Crystal properties

Ruby is formed by doping Cr^{3+} ions into a sapphire (Al_2O_3) host with a doping level[73] of approximately 0.05 wt%, corresponding to an active ion density of 1.58×10^{19} cm^{-3}. The chromium ions, of radius 62 pm, substitute for the aluminium ions, of radius 53 pm, with little distortion of the host lattice. Ruby may be grown by the Czochralski method, and has excellent mechanical, thermal, and chemical properties. Ruby laser rods typically have a diameter of between 3 and 25 mm and lengths up to 200 mm.

[73] In ruby gemstones the doping level of chromium ions is much higher, of order 1%, which gives them a rich red colour. Ruby laser rods typically appear pink.

Energy levels

Figure 7.24 shows the energy levels of Cr^{3+} ions doped in Al_2O_3 together with a simplified configuration coordinate diagram. In ruby the crystal-field is strong, and hence, as discussed in Section 7.3.1 the crystal field configuration of the ground state is t_2^3. This configuration forms crystal-field terms labelled[74] as $^4A_2$ (the ground-state), ^2E, $^2T_1$, and $^2T_2$ where the superscripts are equal to $2S + 1$. The lowest-lying excited configuration is $t_2^2 e$, which forms the terms $^4T_2$, $^4T_1$ and $^2A_1$. Of particular importance is the fact that since the crystal field is strong the $^4T_2$ and $^4T_1$ terms (which arise from the $t_2^2 e$ configuration) lie above the ^2E term (which arises from the t_2^3 configuration).[75]

The wave functions of the levels of the $t_2^2 e$ configuration have a very different spatial distribution from those of the t_2^3 configuration, and consequently,

[74] Just as for levels in a free atom or ion, single-electron orbitals are labelled with lower-case letters, crystal field terms are labelled with upper-case letters.

[75] The $^4T_2$ and $^4T_1$ levels may also be denoted in an alternative notation as $^4F_2$ and $^4F_1$, respectively.

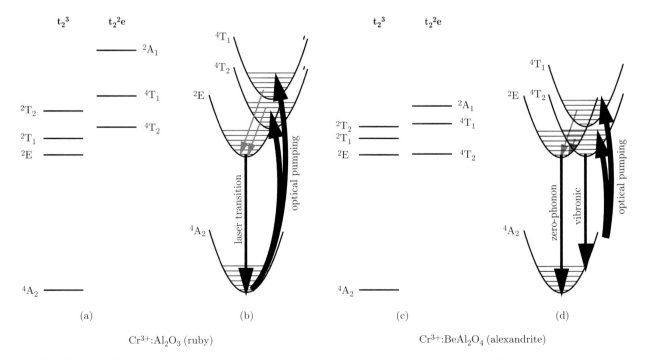

Fig. 7.24: The crystal-field configurations and energy levels (a and c) and simplified configuration coordinate diagrams (b and d) for ruby and alexandrite. Note that the diagrams are not to scale, and for clarity not all the energy levels shown in (a) and (c) are shown in the corresponding configuration coordinate diagram.

as shown in Fig. 7.24 the equilibrium configuration co-ordinate of the $^4T_2$ and $^4T_1$ levels is quite different from that of the $^4A_2$ and ^2E levels.

The absorption spectrum of ruby is dominated by the $^4T_2 \leftarrow {}^4A_2$ and $^4T_1 \leftarrow {}^4A_2$ transitions since these are fully spin-allowed. Since these are configuration transitions, as discussed in Section 7.3.1, they form broad, vibronic pump bands some 50 nm wide and centred at approximately 554 nm and 404 nm, respectively.[76] These may be pumped conveniently by flashlamp radiation. Absorption to the doublet levels is much weaker since they are spin-forbidden.

Absorption on the $^4T_{2,1} \leftarrow {}^4A_2$ pump bands is followed by rapid non-radiative relaxation to the lowest-lying excited level: the ^2E level. This level is metastable since it can only decay on the spin-forbidden $^2E \rightarrow {}^4A_2$ transition; in fact it has one of the longest fluorescence lifetimes of any level in the solid state, approximately 3 ms. Since the $^2E \rightarrow {}^4A_2$ transition involves no change in the crystal-field orbitals, it is a narrowband, zero-phonon transition.

The ^2E level is actually split[77] into two levels separated by only 29 cm^{-1}. This splitting results in two, closely spaced laser lines denoted R_1 and R_2 with wavelengths of 694.3 nm and 692.8 nm, respectively. Of these two transitions, the R_1 line at 694.3 nm usually dominates owing to the fact that it arises from the lower of the ^2E levels, and consequently has a higher thermal population.

Population is rapidly transferred between these two levels by phonon collisions, so that the levels may be considered to be in thermal equilibrium, with a slightly larger population in the lower of the two ^2E levels.

Laser parameters

The R_1 and R_2 transitions are homogeneously broadened by phonon interactions to give a temperature-dependent linewidth, which at room temperature is approximately equal to 450 GHz. The optical gain cross-section of the R_1 transition in ruby is approximately 2.5×10^{-20} cm^{-2}.

The energy levels of ruby therefore have many desirable features for a laser transition: broad pump bands allow for efficient absorption of broad-band pump radiation; almost all the ions excited on these transitions are transferred by rapid non-radiative transitions to the upper laser level; the upper laser level has a long fluorescence lifetime, and therefore acts as a storage level for excitation. The main difficulty, of course, is that the laser transition occurs to the ground state and consequently the laser is a three-level laser system, as illustrated schematically in Fig. 7.25.

In order to achieve a population inversion the optical pumping must be sufficiently strong to deplete the population of the ground state by approximately a half, as discussed in Section 7.1.5.

Practical implementation

Since ruby is a three-level laser, it is almost invariably operated in a pulsed mode. Optical pumping is usually provided by a medium-pressure xenon flashlamp.

The long lifetime of the upper laser level means that it is straightforward to Q-switch the laser, and indeed this is necessary if strong laser spiking is to be

---

[76] It is these strong absorption bands in the green and violet parts of the spectrum that are responsible for the striking colour of ruby gemstones.

[77] The splitting arises from a combination of the spin-orbit interaction and lowering of the symmetry of the crystal field from octagonal symmetry. The $^4A_2$ ground state is also split by the same interactions, but its splitting is even smaller. The split levels of the ^2E level are denoted $^2E_{3/2}$ and $^2E_{1/2}$, the latter being the lower. In the alternative notation mentioned in Note 75 these levels are denoted $\bar{E}$ and $2\bar{A}$.

7.3 *Ruby and other trivalent iron-group systems*   177

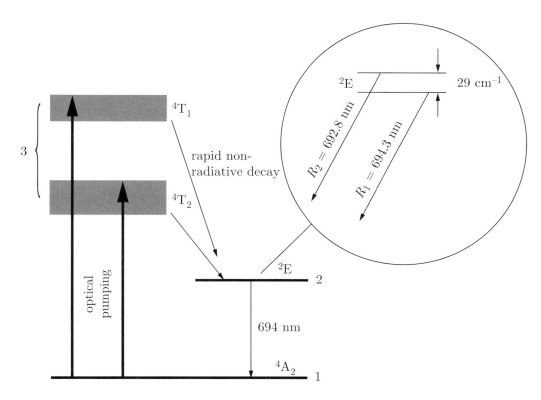

Fig. 7.25: Simplified energy-level scheme for the ruby laser. Inset: Splitting of the upper laser level.

avoided. Q-switched ruby lasers typically generate pulses with an energy of several Joules, and a duration between 10 and 50 ns. Ruby lasers may also be actively or passively modelocked to give output pulses of 5 to 10 ps duration.

7.3.3 Alexandrite laser

Alexandrite, or chromium-doped chrysoberyl (Cr^{3+}:$BeAl_2O_4$), was the first commercially important tunable solid-state laser. Alexandrite lasers have proved useful in applications which take advantage of their wide tunability, such as detection of pollutant species in the atmosphere and in several medical applications.

Crystal properties

The aluminium ions of the host material occupy two types of sites within the lattice: one in which the crystal field has a mirror symmetry, and one for which the field exhibits inversion symmetry. Typical doping levels are 0.04–0.1 at.%, the chromium ions substituting for aluminium ions, with a significant majority (78%) going into the sites with mirror symmetry.

The host is optically biaxial, leading to a strong polarization dependence on its optical properties that dominates thermal or stress-induced birefringence. The host has a very high thermal conductivity, a high optical damage threshold, is chemically stable, and has good mechanical strength and hardness. Laser rods of high optical quality may be produced with diameters of about 5 mm diameter and lengths up to 100 mm.

Energy levels

Figures 7.24(c) and (d) show the energy levels of alexandrite. In this material the crystal field is of intermediate strength and consequently the $^4T_1$ and $^4T_2$ levels are closer to the ^2E level than is the case for Cr^{3+} ions in Al_2O_3 (i.e. ruby). For example, in alexandrite the $^4T_2$ level lies only 800 cm^{-1} above the ^2E level, compared to approximately 1500 cm^{-1} in ruby.

As in the case of ruby, broad pump bands are provided by the vibronic $^4T_2 \leftarrow {}^4A_2$ and $^4T_1 \leftarrow {}^4A_2$ transitions lying at 550–650 nm and 350–450 nm, respectively. These may be pumped by xenon or mercury flashlamps, or by argon and krypton ion lasers.

Absorption on the $^4T_2 \leftarrow {}^4A_2$ pump bands is followed by rapid non-radiative relaxation to the lowest vibrational level of the $^4T_2$ level and to the ^2E level. Since the $^4T_2$ level lies only 800 cm^{-1} above the ^2E level, population is transferred between these levels by rapid non-radiative transitions to form a thermal distribution over the two levels. The radiative lifetime of the ^2E level is similar to that of the same level in ruby: 2.3 ms; that of the $^4T_2$ level is much shorter, 7.1 μs, since it may decay to the ground state via a spin-allowed transition. Since the population in these levels is shared, the effective lifetime of ions in either level depends on the relative population in each level, and is therefore strongly temperature dependent.[78] At room temperature the effective lifetime of either level is approximately 300 μs.

[78] Calculation of the effective lifetime, and its dependence on temperature, is explored in Exercise 7.4

Laser parameters

Lasing can occur in alexandrite on the analogous transition to the ruby laser transition: the zero-phonon $^2E \rightarrow {}^4A_2$ transition at 680.4 nm, which operates as a three-level scheme. However, of more importance is the $^4T_2 \rightarrow {}^4A_2$ transition since this is a vibronic transition and hence is widely tunable between approximately 700 nm and 818 nm.

The vibronic laser transition operates as a four-level laser: Pumping on the broad-band $^4T_{2,1} \leftarrow {}^4A_2$ transitions is followed by rapid non-radiative relaxation to the bottom of the $^4T_2$ level and to the ^2E level; the ^2E level acts as a population, storage level, increasing the effective lifetime of the upper laser level considerably; lasing can then occur to excited vibrational levels of the $^4A_2$ ground electronic level, followed by rapid non-radiative decay to the lowest vibrational level of the ground electronic state. A simplified energy level diagram of the operation of the vibronic laser transition in alexandrite is shown in Fig. 7.26.

Since the populations of the $^4T_2$ and ^2E levels are closely coupled, it is appropriate to use an effective optical gain cross-section together with the total population in the two levels, as discussed on page 161 for the case of the Nd:YAG laser. The effective cross-section for the vibronic transition in alexandrite is approximately 1×10^{-20} cm^2, although this is strongly dependent on temperature.

7.3 Ruby and other trivalent iron-group systems

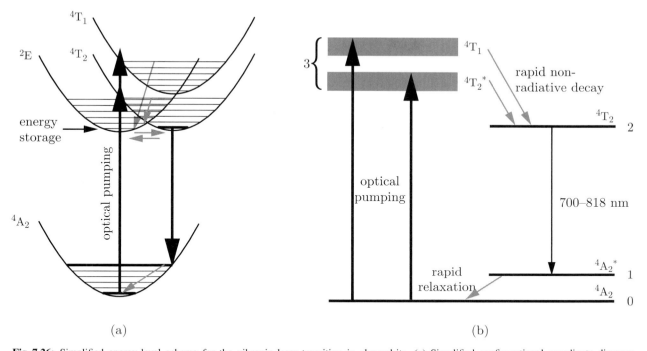

Fig. 7.26: Simplified energy-level scheme for the vibronic laser transition in alexandrite. (a) Simplified configurational coordinate diagram illustrating how the lowest vibrational level of the ^2E level acts as a storage level and is in thermal equilibrium with the lowest vibrational level of $^4T_2$; (b) Energy-level scheme showing the four-level nature of the laser transition. Note in (b) an asterisk indicates that the level is an excited vibrational level.

Table 7.3 Important parameters of the ruby, alexandrite, and Ti:sapphire laser transitions.

		Ruby	Alexandrite		Ti:sapphire
λ	(nm)	694.3	680.4	700–820	660–1180
τ_2	(μs)	3000	1500	260	3.8
τ_1		∞	∞	≈ 100 ps	≈ 100 ps
σ_{21}	(10^{-20} cm^2)	2.5	30	0.7^a	28
$\Delta\nu$	(GHz)	330	330	26 000	100 000
Doping level	(%)	0.05	0.1	0.1	0.1
Pump bands	(nm)	375–425, 525–575	380–630		380–620

The gain cross-section of Alexandrite increases to 2×10^{-20} cm^2 as the temperature is increased from 300 K to 475 K. See, Walling et al. (1980).

The main parameters of the laser transitions in ruby and alexandrite are given in Table 7.3

Practical implementation

Alexandrite lasers operate in a similar configuration to Nd:YAG lasers. The main difference is that the laser rod may be heated to between 50 and 70 °C in order to increase the relative population in the $^4T_2$ level.[79]

Alexandrite lasers are usually flashlamp pumped. Although they may in principle be operated continuously, the relatively low gain cross-section means

[79] Heating to higher temperatures would increase the population of the $^4T_2$ level still further. However, raising the temperature of the laser rod also increases the thermal population of the lower laser level, and hence the optimum temperature is below that which would maximize the thermal population of the upper laser level.

that the threshold population inversion density is high, and so pulsed operation is more usual. Pulse repetition rates between 10 and 100 Hz are typical, with output pulse energies of up to about 1 J.

Alexandrite lasers naturally produce output pulses with a duration close to the effective lifetime of the upper laser level, i.e. of 200–300 μs. Q-switching reduces the duration of the output pulse to typically 50 ns.

7.3.4 Cr:LiSAF and Cr:LiCAF

Two other lasers operating in Cr^{3+} ions should be mentioned: Cr^{3+}:$LiSrAlF_6$ (usually abbreviated to Cr:LiSAF) and Cr^{3+}:$LiCaAlF_6$ (Cr:LiCAF).

Both of these host materials have a colquirite structure, which is uniaxial and hence the optical absorption and emission spectra are strongly polarization dependent. For both hosts the Cr^{3+} ions substitute for the Al ions, and doping levels as high as 15% are possible without the lifetime of the upper laser level being reduced by concentration quenching.

Broad pump bands are provided by the $^4T_2 \leftarrow {}^4A_2$ and $^4T_1 \leftarrow {}^4A_2$ transitions; for Cr:LiSAF, for example, these bands lie at approximately 550–700 nm and 400–500 nm, respectively. The crystal field in LiSAF and LiCAF is weaker than in alexandrite to the extent that the ^2E level lies above the lowest vibrational level of the $^4T_2$ electronic level. As a consequence, after pumping the ions undergo rapid non-radiative transitions to the lowest vibrational level of the $^4T_2$ electronic level, and little population is stored in the long-lived ^2E level. The lifetime of the upper laser level in Cr:LiSAF and Cr:LiCAF is therefore somewhat shorter (67 μs and 170 μs, respectively) than in alexandrite, and is much less dependent on temperature.

Lasing occurs on the vibronic $^4T_2 \rightarrow {}^4A_2$ transition, which offers wide tunability: Cr:LiSAF may be tuned from 780 to 1010 nm; Cr:LiCAF from[80] 720 to 840 nm.

[80]The smaller tuning range in Cr:LiCAF is a result of excited-state absorption in this material.

The larger optical gain cross-section, upper-level lifetime, and tuning range for Cr:LiSAF means that it is generally preferred to Cr:LiCAF. These lasers may be operated as c.w. devices, giving outputs of about 1 W, or as larger flashlamp-pumped pulsed lasers delivering pulse energies up to 10 J. The large bandwidth of the laser transition enables short pulses to be generated via Kerr-lens modelocking,[81] and in this context Cr:LiSAF is of particular interest since it may be pumped by GaInP/AlGaInP quantum-well laser diodes operating at 670 nm, enabling an all-solid-state femtosecond oscillator[82] capable of generating pulses as short as 10 fs.

[81]See page 218.

[82]See, for example, Uemura and Torizuka (2003).

7.3.5 Ti:sapphire

Crystal properties

In titanium-doped sapphire, Ti:Al_2O_3, titanium ions (ionic radius 74.5 pm) substitute for aluminium ions (ionic radius 53 pm) with a typical concentration of 0.1 wt%. High-quality crystals may be grown by the Czochralski technique.

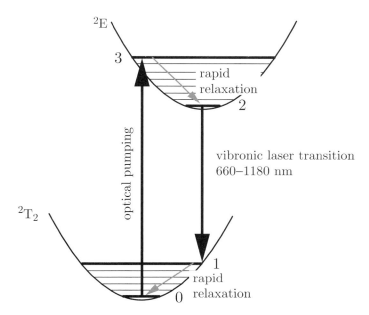

Fig. 7.27: Simplified energy level structure and laser transitions of the Ti^{3+} ion in Al_2O_3 (Ti:sapphire).

Energy levels

The Ti^{3+} ion has only a single d-electron, and consequently its energy-level structure is rather simple: the ground crystal-field configuration is t_2, which forms a $^2T_2$ term; the only other possible crystal-field orbital for a d-electron is e, which gives rise to a ^2E term, as shown in Fig. 7.27.

Radiative transitions between these two electronic levels obviously involve a change in the crystal-field configuration, and are therefore broad, vibronic transitions. It is worth noting that the sparse energy-level structure of Ti:sapphire means that losses arising from excited-state absorption are negligible.

Figure 7.28 shows the absorption and emission spectra of Ti:sapphire. The very broad bands, and their wide separation is a consequence of the large difference in equilibrium configuration coordinates for the upper and lower levels. Absorption from the ground electronic level occurs on the strong vibronic $^2E \leftarrow {}^2T_2$ transitions to the excited vibrational levels of the ^2E level, with wavelengths between 630 nm and 400 nm. Ions in excited vibrational levels decay rapidly by non-radiative transitions to the lowest vibrational level of the ^2E level, which forms the upper level of the laser transition. At room temperature, ions in this level decay pre-dominantly radiatively to the vibrational levels of the $^2T_2$ level, with a fluorescence lifetime of approximately 3.2 μs, the relatively short lifetime reflecting the fact that the transition to the ground state is spin-allowed.

Laser parameters

The Ti:sapphire laser is a four-level laser: lasing may occur from the lowest vibrational level of the ^2E level to excited vibrational levels of the $^2T_2$ level, at

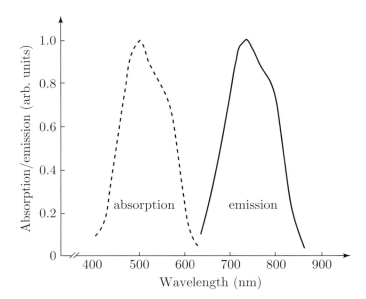

Fig. 7.28: Absorption and emission bands of the Ti:sapphire laser.

wavelengths tunable between 660 nm and 1180 nm. Indeed Ti:sapphire has the largest bandwidth of any laser.

Although the bandwidth of the laser transition is very large, the large radiative decay rate of the laser transition ensures that the optical gain cross-section remains relatively high for a solid-state laser: approximately 3×10^{-19} cm^2, which is comparable to that in Nd:YAG.

Practical implementation

A major difference between Ti:sapphire and all the lasers discussed so far in this chapter is the short fluorescence lifetime of the upper laser level. Flashlamp pumping is therefore not usually very efficient owing to the much longer duration of the flashlamp pulse. Instead, laser pumping is usually used, the most commonly employed pump lasers being frequency-doubled Nd lasers (YAG, YLF or vanadate).

Figure 7.29 shows the construction of two common types of Ti:sapphire laser oscillator. In both designs the Ti:sapphire laser rod is 2–10 mm long, with end faces cut at Brewster's angle so as to eliminate reflection losses for one polarization. The z-folded cavity compensates for astigmatism. This arises because it is necessary to focus the pump radiation to a small cross-sectional area in order to achieve sufficient population inversion. The design of the optical cavity design must therefore be such that the transverse modes are also brought to a focus at the Ti:sapphire crystal. However, the Brewster-cut faces of the crystal introduce astigmatism to the intra-cavity beam—i.e. the divergence of the beam becomes different in the horizontal and vertical planes. In the cavity shown in Fig. 7.29 this is corrected by the two concave mirrors used away from normal incidence, which introduce astigmatism of the opposite sign to that caused by refraction at the laser rod.

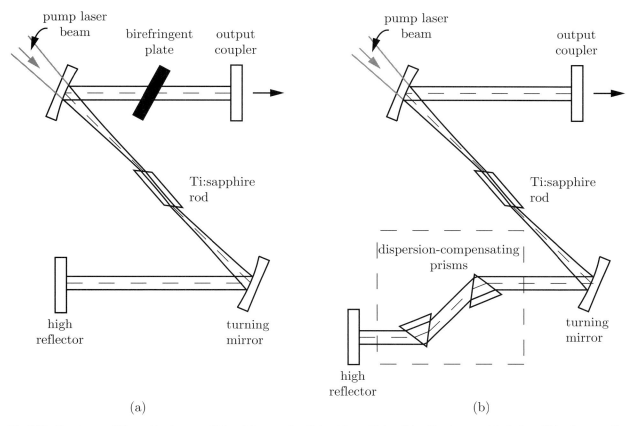

Fig. 7.29: Two types of Ti:sapphire laser oscillator: (a) a wavelength-tunable oscillator; (b) a Kerr-lens modelocked oscillator incorporating intra-cavity dispersion control (see page 218 and Section 17.3.1).

Continuously tunable c.w. Ti:sapphire lasers may be tuned between approximately[83] 700 nm and 1000 nm, with output powers of 1–2 W. Wavelength tuning may be accomplished by a variety of techniques. In the design illustrated in Fig. 7.29(a), tuning is achieved by adjusting a birefringent filter.[84] In order to reduce losses, the filter can be mounted at Brewster's angle, and tuning achieved by rotating the filter about an axis parallel to its normal. Other techniques, such as prisms or gratings, may be employed for selecting the output wavelength. For the oscillator shown in Fig. 7.29(a) the linewidth of the laser output is typically about 40 GHz; this may be reduced to better than

[83] The upper and lower limits to the tuning curve may be extended to approximately 1100 nm and 675 nm, respectively, by employing cavity mirrors optimized for these wavelengths. However, it is not possible continuously to tune from 675 nm to 1100 nm without changing the cavity mirrors.

[84] At its simplest a birefringent filter comprises a plate of birefringent material with its optical axis in the plane of the plate. The normal to the plate is usually oriented at Brewster's angle with respect to the incident light. The retardance introduced by the plate depends on the wavelength of light and on the angle between the optic axis and the direction of propagation within the plate. Those wavelengths for which the plate acts as a full-wave plate are transmitted with no change of polarization state and zero reflection losses. For other wavelengths the polarization state is changed and the transmission losses suffered at the rear surface of the plate, and at other polarization-sensitive elements (such as the laser rod), are finite. The filter may be tuned by rotating the plate about an axis parallel to its normal. In practice, several such plates may be employed, with different free spectral ranges (as for the etalons discussed in Section 16.2.2), so as to select a single wavelength. For further details consult Hodgkinson and Vukusic (1978).

1 GHz by employing an intra-cavity etalon. Further reduction of the linewidth to around 10 MHz is possible by employing a ring cavity, and by locking the operating frequency to an external cavity (as discussed in Section 16.4) the linewidth can be reduced to as little as 50 kHz.

Ti:sapphire lasers are widely used to generate very short optical pulses by the Kerr modelocking technique discussed in Section 8.3.4. The cavity design shown[85] in Fig. 7.29(b) is typical of modelocked Ti:sapphire lasers, and incorporates a pair of prisms for dispersion control as discussed in Section 17.2. The train of modelocked laser pulses from such a system typically has a pulse repetition rate of 80 MHz, a mean power of approximately 0.5 W (i.e. $\approx$ 5 nJ per pulse), with pulse durations of 35–50 fs. Modelocked oscillators of this type can also be wavelength tuned, although the tuning range is more restricted: 780–820 nm is common.

[85] In designing suitable cavities it is desirable to maximize the ratio of the round-trip losses under c.w. operation to those when Kerr-lens modelocking occurs. See, for example, Cerrullo et al. (1994).

Further reading

A more detailed discussion of the physics of solid-state laser materials, pumping geometries, and techniques for implementing modelocking and Q-switching may be found in the book by Koechner and Bass (2003).

A very comprehensive discussion of the theory underlying the physics of solid-state laser materials may be found in the book by Powell (1998). A recent review of developments in solid-state laser materials may be found in an article by Kaminskii (2003). Further information on ceramic laser hosts may be found in the article by Wisdom (2004).

Exercises

(7.1) Calculation of doping concentration
In this simple problem we explore the different ways that the doping of Nd ions in a YAG crystal may be expressed.

(a) Given that the density of pure YAG is 4.56 g cm^{-3}, calculate the density of unit cells in the crystal.

(b) Hence, calculate the density of sites available for Nd ions (this is *not* the same as the answer above!).

(c) Hence, show that a doping of 1 at.% corresponds to a density of Nd ions equal to 1.39×10^{20} cm^{-3}.

(d) Show that the concentration of Nd by weight will be smaller than the atomic concentration by a factor of 0.729.

(e) Nd:YAG crystals are grown from a molten mixture of pure YAG and Nd_2O_3 (rather than pure Nd). Show that the proportion by weight of Nd_2O_3 required is 0.850% for a doping level of 1 at.%.

[Take the molar masses of the constituents of YAG to be $M_Y = 88.91$ g mol^{-1}, $M_{Al} = 26.98$ g mol^{-1}, $M_O = 16.00$ g mol^{-1}, $M_{Nd} = 144.2$ g mol^{-1}]

(7.2) Threshold pump power in Nd:YAG
A Nd:YAG laser rod, 50 mm long and of diameter 4 mm, is placed in a two-mirror optical cavity with mirror reflectivities of 100% and 90%. The laser rod is optically pumped by the output from a semiconductor laser operating at 809 nm.

In this question take the effective optical gain cross-section to be 3.7×10^{-19} cm^2, and the fluorescence lifetime of the upper laser level to be 230 µs.

(a) Calculate the threshold population inversion density.

(b) Estimate the power of the pump laser required to reach the threshold condition for continuous-wave oscillation on the Nd:YAG laser transition at 1064 nm.

(c) Discuss how the power and spectrum of the laser output behaves as the pump power of the laser rod is increased above the threshold value.

(d) How would this behaviour differ if the host for the Nd^{3+} ions were glass rather than a crystalline material?

(7.3) **Threshold pump power in ruby**

(a) Calculate the population inversion density required for a ruby laser to reach the threshold for oscillation. Assume the dimensions of the laser rod and optical cavity are as in Exercise 7.2.

In this question take the effective optical gain cross-section to be 2.5×10^{-20} cm^2, and the fluorescence lifetime of the upper laser level to be 3 ms.

(b) A typical value for the density of Cr^{3+} ions in a ruby laser rod is 1×10^{19} cm^{-3}. How does the threshold population-inversion density compare with this? Hence, find the threshold density of ions that must be pumped to the upper laser level for laser oscillation to occur.

(c) Assuming that the pump radiation has a mean wavelength of 500 nm, calculate the pump energy required for the threshold to be reached.

(d) Calculate the pump power that would be required for continuous-wave oscillation, and compare this with the value calculated in Exercise 7.2.

(e) Comment briefly on the factors that determine how much *electrical* power must be supplied to achieve laser oscillation.

(7.4) **Effective fluorescence lifetime**

Here, we calculate the effective fluorescence lifetime of two closely spaced levels that decay at different rates. In Exercise 7.5 this method is extended to the general case of many levels.

As an example, we will consider the case of the closely spaced ^{2}E and ^{4}T$_2$ levels in alexandrite, as discussed in Section 7.3.3. Figure 7.30 illustrates the collisional and radiative processes of interest. We suppose that the two levels are excited at rates R_T and R_E, and that phonon collisions transfer population from ^{4}T$_2$ to ^{2}E at a rate R_{TE} and from ^{2}E to ^{4}T$_2$ at a rate R_{ET}. In deriving an effective lifetime, we will assume that transfer of population between the two levels is sufficiently rapid that their relative population is maintained in thermal equilibrium with a temperature T. We will also take the degeneracies of the two levels to be equal.

(a) Show that the fraction, f_T, of the total population that resides in the ^{4}T$_2$ level is given by

$$f_T = \frac{\exp(-\Delta E/k_B T)}{1 + \exp(-\Delta E/k_B T)}$$

where ΔE is the energy by which the ^{4}T$_2$ level lies above the ^{2}E level.

(b) Deduce an equivalent expression for the fraction, f_E, of the total population that resides the ^{2}E level.

(c) Evaluate these fractions for temperatures of 300 K and 77 K.

(d) For both the ^{4}T$_2$ and ^{2}E levels, write down the rate equations describing the processes illustrated in Fig. 7.30. By considering the total rate of decay of population from these levels, show that the total population decays with an effective fluorescence lifetime τ_{eff} where,

$$\frac{1}{\tau_{\text{eff}}} = \frac{f_E}{\tau_E} + \frac{f_T}{\tau_T}.$$

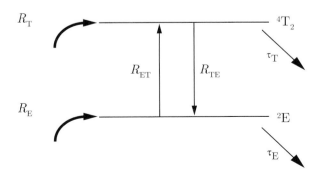

Fig. 7.30: Schematic diagram of the collisional and radiative processes relevant to the calculation of the effective fluorescence lifetime considered in Exercise 7.4.

(e) Show that provided the relative populations in the two levels are maintained in thermal equilibrium, the population in *either* level also decays with an effective fluorescence lifetime equal to τ_{eff}.

(f) Given that $\tau_T = 7.1\,\mu s$ and $\tau_E = 2.3\,\text{ms}$, find the effective fluorescence lifetime of these levels for temperatures of 300 K and 77 K, and comment on your results.

(g) Show that provided $\Delta E \ll k_B T$, the effective lifetime of the levels is given by the approximate expression,

$$\frac{1}{\tau_{\text{eff}}} \approx \frac{1}{2}\left(\frac{1}{\tau_E} + \frac{1}{\tau_T}\right),$$

(7.5) **Effective fluorescence lifetime and effective gain cross-section**

We now extend the case considered in Exercise 7.4 to the case of an arbitrary number of levels with differing degeneracies, and consider also how to derive an effective gain cross-section that is appropriate when the population of the lower laser level is negligible.

We suppose that the upper manifold has a total populations density N_2 distributed over levels of energy $E_2^1, E_2^2, E_2^3, \ldots$, with population densities $n_2^1, n_2^2, n_2^3, \ldots$.

(a) Show that if the population, of the levels of the manifold are maintained in thermal equilibrium at a temperature T, then the population density in the jth level of the upper manifold of levels may be written as $n_2^i = f_2^i N_2$ where,

$$f_2^j = \frac{g_j \exp\left(-\frac{E_2^j}{k_B T}\right)}{\sum_m g_m \exp\left(-\frac{E_2^m}{k_B T}\right)},$$

and g_m is the degeneracy of level m.

(b) Suppose that each of the levels of the upper manifold has a fluorescence lifetime τ_2^j. Show that the total rate of spontaneous decay from the upper manifold may be written as,

$$\frac{N_2}{\tau_2^{\text{eff}}},$$

where,

$$\frac{1}{\tau_2^{\text{eff}}} = \sum_j \frac{f_2^j}{\tau_2^j}.$$

(You may ignore spontaneous emission *within* the manifold of levels.)

(c) Show that *provided the population in the lower manifold of levels may be ignored* the total rate of stimulated emission from the upper manifold of levels arising from narrowband radiation of angular frequency ω may be written in the form,

$$N_2 \sigma_{\text{eff}}(\omega - \omega_0) \frac{I}{\hbar \omega},$$

where the effective gain cross-section is defined as,

$$\sigma_{\text{eff}}(\omega - \omega_0) = \sum_j f_2^j \sum_i \sigma_{ji}(\omega - \omega_{ji}),$$

where $\sigma_{ji}(\omega - \omega_{ji})$ and ω_{ji} are the optical gain cross-section and centre frequency of transitions between level j in the upper manifold and level i in the lower manifold.

(7.6) Here, we use the results of the previous problem to calculate the effective gain cross-section of the Nd:YAG laser. Data for the ℓ_1 and ℓ_2 transitions in Nd:YAG are given in Table 7.4. Note that the degeneracy of all the sublevels involved is the same ($g = 2$), and so the degeneracy factors may be ignored.

Table 7.4 Parameters of the ℓ_1 and ℓ_2 laser transitions in Nd:YAG. (Data from Zagumennyi (2004).)

			ℓ_1	ℓ_2
Centre vacuum wavelength		(nm)	1 064.40	1 064.15
Linewidth	$\Delta \tilde{\nu}$	(cm^{-1})	6.5	6.5
Peak cross-section	$\sigma_p^{\ell_i}$	(10^{-19} cm^2)	1.9	7.1

(a) Calculate the proportions $f_2^{R_1}$ and $f_2^{R_2}$ of the total population of the $^4F_{3/2}$ manifold residing in the R_1 and R_2 levels at a temperature of 300 K.

(b) Assuming that the ℓ_1 and ℓ_2 transitions have a Lorentzian lineshape of full width at half-maximum $\Delta \tilde{\nu}$, show that the contribution of the ℓ_1 transition to the gain available at the centre wavelength of the ℓ_2 transition is relatively small. Hence, argue that the effective cross-section is approximately given by $\sigma_{\text{eff}} \approx f_2^{R_2} \sigma_p^{\ell_2}$, where $f_2^{R_2}$.

(c) †A more accurate value of the effective cross-section may be obtained by plotting eqn (7.22) as a function of wavelength, and finding the peak effective cross-section numerically. Use a numerical software package to do this and compare your answer with Fig. 7.17.

(7.7) Calculation of quantum yield and fluorescence lifetime

(a) An excited level has lifetimes against radiative and non-radiative decay of τ_r and τ_{nr} respectively. Show that the fluorescence lifetime of the level is given by,

$$\frac{1}{\tau_2} = \frac{1}{\tau_r} + \frac{1}{\tau_{nr}}. \qquad (7.26)$$

(b) Show that the **quantum yield**—that is the proportion of ions excited to the level that decay by emitting radiation is given by,

$$\eta = \frac{\tau_{nr}}{\tau_{nr} + \tau_r}. \qquad (7.27)$$

(c) For the $^4F_{3/2}$ level of Nd:YAG $\tau_r = 250\,\mu s$ and $\tau_{nr} = 20\,ms$. Calculate the quantum yield and fluorescence lifetime of the level.

(d) Explain why these values are favourable for the upper level of a laser transition.

8

Dynamic cavity effects

8.1 Laser spiking and relaxation oscillations	188
8.2 Q-switching	193
8.3 Modelocking	203
8.4 Other forms of pulsed output	221
Further reading	222
Exercises	222

In this chapter we consider several effects that can arise from the interplay between stimulated emission in the laser gain medium and the feedback provided by the optical cavity. This interplay can, under certain conditions, cause unwanted fluctuations in the output power and mode structure of the laser. On the other hand, it can also be used to good effect by deliberately changing the optical feedback dynamically. This forms the basis of two techniques used very widely in lasers: **Q-switching** and **modelocking**, which are discussed in detail below.

8.1 Laser spiking and relaxation oscillations

We first discuss two types of transient behaviour that are frequently observed in lasers:

1. **Laser spiking:** This describes the large-amplitude, irregularly spaced, pulses of laser radiation that are sometimes output by a laser when it is first turned on or if it suffers a large disturbance to its steady-state running conditions.
2. **Relaxation oscillations:** These are more regular oscillations in the output power of the laser, the amplitude of which typically decays exponentially. Relaxation oscillations can occur in the tail of a series of laser spikes, or can follow a small perturbation to the laser system, such as the disturbance of a cavity mirror or a change in the pumping power.

As might be expected the two phenomena are related: they both occur when the population inversion and the intra-cavity laser intensity respond to changes in the operating conditions on significantly different time scales.

Consider first the case of spiking in a laser that is pumped continuously. Immediately after the pumping is switched on, the population in the upper laser level will grow from zero to its steady-state value in a time of the order of the upper level lifetime τ_2. At this point the density of photons in the cavity will be very low since the photons arise from spontaneous emission only. At some point the population inversion density will reach the threshold for laser oscillation, N_{th}^*, but laser oscillation will not commence since the density of photons in the relevant cavity mode will be low. As pumping continues, the population inversion grows beyond the threshold value, and the density of

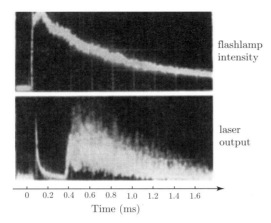

Fig. 8.1: Laser spiking in the first ruby laser. The top trace shows the flashlamp pulse, the lower the output from the ruby laser as recorded by a photodiode. Note that the sharp feature at $t = 0$ is caused by electrical noise from the trigger circuit. Reprinted figure with permission from T. H. Maiman, R. H. Hoskins, I. J. D'Haenens, C. K. Asawa and V. Evtuhov, *Physical Review* **123** 1151 (1961). ©(1961) by the American Physical Society. http://link.aps.org/doi/10.1103/PhysRev.123.1151

photons in the laser mode will build up by stimulated emission. Eventually, the density of photons in the laser mode will be sufficiently high to burn the population inversion down towards N_{th}^*, increasing, as it does so, the rate of stimulated emission and the photon density. Inevitably, the population inversion will be driven well below the threshold value, leading to reduction in the photon density and the cessation of lasing. The dramatic growth and rapid decline of the intra-cavity photon density correspond to a spike in the output power of the laser. As pumping continues, the population inversion will start to grow again and the cycle can repeat. Laser spiking of this type will not usually continue indefinitely, even in continuously pumped lasers, owing to the fact that the population inversion and photon densities do not decay to zero after the spike. As a consequence, after each spike these quantities tend to end up closer to their steady-state values, in which case the spiking will gradually decay. Figure. 8.1 shows an example of laser spiking.

It should be noted that laser spiking is often associated with rapid jumps in the longitudinal and/or transverse mode of the laser. This occurs for essentially the same reasons as spatial hole-burning: immediately after a laser spike, significant population inversion may remain in regions near the nodes of the standing-wave pattern of the cavity mode, or in regions where the transverse profile of the mode of the previous pulse was low in intensity. This gain can be exploited by other modes that, as pumping continues, may therefore reach threshold sooner than the mode (or modes) of the previous spike.

As suggested by the discussion above, laser spiking tends to occur when the time scale on which the population inversion responds to the pumping—typically of order τ_2—is slow compared to that in which the photon density changes in the cavity—which is of the order of the cavity lifetime τ_c. Hence, we expect that laser spiking will *not* occur if

$$\tau_2 \ll \tau_c. \quad \textbf{Condition for no laser spiking} \quad (8.1)$$

8.1.1 Rate-equation analysis

We can analyse the situation quantitatively by considering the rate equations for the upper and lower laser levels. These may be written simply as,

$$\frac{dN_2}{dt} = R_2(t) - N^*(t)\sigma_{21}\frac{I(t)}{\hbar\omega_L} - \frac{N_2(t)}{\tau_2} \tag{8.2}$$

$$\frac{dN_1}{dt} = R_1(t) + N^*(t)\sigma_{21}\frac{I(t)}{\hbar\omega_L} - \frac{N_1(t)}{\tau_1} + N_2(t)A_{21}, \tag{8.3}$$

where R_i is the pump rate of level i, and ω_L and $I(t)$ are the angular frequency and total intensity of the oscillating laser mode.[1]

[1] This is clearly a one-dimensional treatment in that we ignore transverse variations of the oscillating mode(s). We also ignore longitudinal variations of the intensity arising from amplification in the gain medium or from the standing-wave pattern of the oscillating mode. The quantity $I(t)$ is therefore a spatially averaged intensity.

Note that in the absence of stimulated emission ($I(t) = 0$) the population of the upper laser level grows in a time of order τ_2.

The rate per unit volume at which photons are stimulated into the oscillating mode is simply $N^*\sigma_{21}I/\hbar\omega_L$. Hence, if we suppose that the fraction of the volume of the oscillating mode lying within the gain medium is equal to f_c, the rate equation for the density of photons in the cavity is simply,

$$\frac{dn_\phi}{dt} = f_c N^*(t)\sigma_{21}\frac{I(t)}{\hbar\omega_L} - \frac{n_\phi(t)}{\tau_c}, \tag{8.4}$$

where the last term is the rate of loss of photons per unit volume from the cavity owing to coupling through the output mirror (plus any other losses).[2]

[2] See page 488 and eqn (6.57).

This equation may be simplified by noting that $I(t) = n_\phi(t) c \hbar\omega_L$. Then,

$$\frac{dn_\phi}{dt} = \left(\frac{N^*}{N^*_{th}} - 1\right)\frac{n_\phi}{\tau_c}, \tag{8.5}$$

where,

$$N^*_{th} = \frac{1}{f_c \sigma_{21} c \tau_c} \quad \textbf{Threshold inversion density} \tag{8.6}$$

and we have dropped the time dependence of $N^*(t)$ and $n_\phi(t)$ to avoid clutter. We see that N^*_{th} is the threshold population inversion density since we must have $N^* > N^*_{th}$ for the photon density to grow (by stimulated emission). Equation (8.5) also shows that the time scale in which the photon density changes is τ_c—to within a factor that depends on the extent to which the population inversion exceeds N^*_{th}.

8.1.2 Analysis of relaxation oscillations

Relaxation oscillations, being somewhat gentler than laser spiking, are amenable to algebraic analysis. We consider the simple example of an ideal four-level laser system in which the population of the lower laser level is essentially zero at all times. In this case, eqn (8.2) is also the rate equation for the population inversion density:

$$\frac{dN^*}{dt} = \frac{dN_2}{dt} = R_2 - N^*\sigma_{21}\frac{I}{\hbar\omega_L} - \frac{N_2}{\tau_2} \quad (8.7)$$

$$= R_2 - \frac{1}{f_c}\frac{N^*}{N_{th}^*}\frac{n_\phi}{\tau_c} - \frac{N^*}{\tau_2}. \quad (8.8)$$

Now, under steady-state conditions $N^*(t) = N_{th}^*$, and substituting this into eqn (8.8) then yields an expression for the steady-state photon density,

$$n_\phi^{ss} = \left(R_2 - \frac{N_{th}^*}{\tau_2}\right)f_c\tau_c = (r-1)\frac{N_{th}^* f_c \tau_c}{\tau_2}, \quad (8.9)$$

where the parameter $r = R_2\tau_2/N_{th}^*$ is the ratio of the population inversion that would be produced in the absence of any stimulated emission to the threshold population inversion for lasing, N_{th}^*. This **overpumping ratio** will also appear in our discussion of Q-switching.

We now suppose that having established steady-state conditions, the laser is subject to a small perturbation that shifts the population inversion and photon densities to:

$$N^*(t) = N_{th}^* + \Delta N^*(t) \quad (8.10)$$

$$n_\phi(t) = n_\phi^{ss} + \Delta n_\phi(t). \quad (8.11)$$

We may then *linearize* the differential equations for $N^*(t)$ and $n_\phi(t)$ by substituting eqns (8.10) and (8.11) into (8.5) and (8.8), and ignoring the small product terms in $\Delta N^*(t)\Delta n_\phi(t)$. We find:

$$\frac{d\Delta N^*}{dt} = -r\frac{\Delta N^*}{\tau_2} - \frac{1}{f_c}\frac{\Delta n_\phi}{\tau_c} \quad (8.12)$$

$$\frac{d\Delta n_\phi}{dt} = (r-1)f_c\frac{\Delta N^*}{\tau_2}. \quad (8.13)$$

We now assume solutions of the form,

$$\Delta N^*(t) = a\exp(mt) \quad (8.14)$$

$$\Delta n_\phi(t) = b\exp(mt). \quad (8.15)$$

Substitution into eqns (8.12) and (8.13) then gives,

$$\left(m + \frac{r}{\tau_2}\right)a + \frac{1}{f_c\tau_c}b = 0 \quad (8.16)$$

$$(r-1)\frac{f_c}{\tau_2}a - mb = 0. \quad (8.17)$$

These two simultaneous equations can be solved by the usual methods, yielding

$$m = -\frac{r}{2\tau_2} \pm \sqrt{\left(\frac{r}{2\tau_2}\right)^2 - \frac{r-1}{\tau_2\tau_c}}. \quad (8.18)$$

Clearly, relaxation oscillations will *not* occur if m is real, i.e. if

$$\tau_2 < \frac{r^2}{r-1}\frac{\tau_c}{4}. \qquad \textbf{Condition for no relaxation oscillations} \qquad (8.19)$$

In other words, if the upper-level lifetime is sufficiently short, relaxation oscillations do not occur and perturbations to $N^*(t)$ and $n(t)$ decay back to their steady-state solutions exponentially. We can see that this condition is consistent with the approximate condition for laser spiking not to occur given by eqn (8.1).

In contrast, if the above condition is not met, relaxation oscillations will occur with an angular frequency

$$\omega_{\text{ro}} = \sqrt{\frac{r-1}{\tau_2 \tau_c} - \left(\frac{r}{2\tau_2}\right)^2} \approx \sqrt{\frac{r-1}{\tau_2 \tau_c}}, \qquad (8.20)$$

where the approximation holds if $\tau_2 \gg \tau_c$. Note that typically $r - 1 \approx 1$, in which case ω_{ro} is close to the geometric mean of the upper level and cavity decay rates.

The condition (8.19) explains why relaxation oscillations (and also laser spiking) are unusual in gas lasers. Such systems typically operate on allowed transitions with upper-level lifetimes shorter than a few hundred nanoseconds,[3] which is comparable to, or shorter, than a typical cavity lifetime. In contrast, the upper-level lifetime for many solid-state laser transitions is of the order of microseconds or even milliseconds. In such cases laser spiking and relaxation oscillations are to be expected.

8.1.3 Numerical analysis of laser spiking

The rather more violent nature of laser spiking means that it is not amenable to algebraic analysis. However, it is quite straightforward to solve the problem numerically.

As in our analysis of relaxation oscillations above, we consider the case of an ideal four-level laser in which the lower laser level is essentially empty at all times. The population inversion and photon densities are then given by eqns (8.5) and (8.8). Figure 8.2 shows numerical solutions to these equations for a system with a filling factor f_c of unity pumped with an overpumping ratio $r = 4$. In each case, the pumping would, in the absence of stimulated emission, produce a population-inversion density that was a factor of r greater than the threshold value. However, once the population inversion rises above N^*_{th} the rate of stimulated emission exceeds the rate of loss of photons from the cavity, increasing the density of photons within the cavity, and causing in turn the population inversion to burn down towards the threshold value. In all the cases shown the population inversion and photon densities tend to their steady-state values N^*_{th} and n^{ss}_ϕ, respectively, at long times after the pumping is switched on at $t = 0$. Substantial laser spiking is observed when $\tau_2/\tau_c \gg 1$, consistent with our discussion above, and the number and amplitude of the spikes in the laser

---

[3] A notable exception is the CO_2 laser that under its operating conditions has an upper-level fluorescence lifetime of several hundred microseconds. As such, the CO_2 laser *can* exhibit relaxation oscillations.

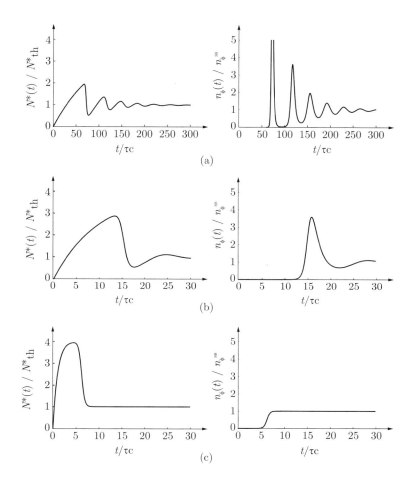

Fig. 8.2: Numerical analysis of laser spiking in an ideal four-level laser with a filling factor f_c of unity pumped with an over-pumping ratio $r = 4$ and: (a) $\tau_2/\tau_c = 100$; (b) $\tau_2/\tau_c = 10$; (c) $\tau_2/\tau_c = 1$. In each case, the plots show the temporal evolution of the population-inversion density and the photon density normalized to the steady-state value, n_ϕ^{ss}.

output is seen to decrease as the ratio τ_2/τ_c is reduced. In particular, for the case $\tau_2/\tau_c = 1$ there is essentially no spike in the photon density. Instead, the population-inversion density photon density reaches rN_{th} before the smoothly rising photon density burns the population inversion down to the steady-state value.

8.2 Q-switching

Q-switching is a technique that generates pulses of laser light with a peak power very much greater than the mean power that can be produced in steady-state operation. The high peak power—ranging from MW to GW—is particularly useful in applications that require a high peak intensity, such as laser cutting or drilling or pumping other laser systems. The output of a Q-switched laser takes the form of a train of *controlled* pulses and hence eliminates problems in lasers that might otherwise show uncontrolled spiking.

194 Dynamic cavity effects

In a Q-switched laser the cavity is deliberately spoiled—imagine, for the moment, that one of the cavity mirrors is blocked —whilst the pumping builds up the population inversion. Since the losses of the modified cavity are so high, the population inversion can build up to much greater levels than the threshold inversion for lasing in the normal cavity without any lasing occurring. Once the population inversion has grown to a large value the mirror is unblocked, thereby switching the 'Q' of the cavity from a low value to a high value. High rates of stimulated emission then cause the density of photons in the cavity to grow very rapidly, burning the population inversion density well below the threshold value. The output from the laser takes the form of a giant pulse of radiation—essentially a single laser spike of the type discussed above.

Q-switching is most effective when the lifetime of the upper laser level is long, since then a large population inversion may be built up slowly without a significant fraction of atoms in the upper laser level decaying by spontaneous emission. As such, Q-switching is very commonly employed in solid-state lasers, which typically have upper-level lifetimes of hundreds of microseconds; less so in gas lasers operating at visible, or shorter, wavelengths since for these systems the lifetime of the upper laser level is usually in the nanosecond range.

Note that the pumping in Q-switched lasers may be pulsed or continuous. With pulsed pumping the ratio of the peak population inversion to the threshold value can be larger than can be achieved with continuous pumping, leading to a higher peak output power.

8.2.1 Techniques for Q-switching

Rotating mirror

One of the earliest methods used to achieve Q-switching was to use a high-speed (20 000 to 60 000 r.p.m.) motor to rotate one of the cavity mirrors about an axis parallel to the mirror surface.[4] The cavity would then be completed briefly whenever the mirror was rotated into alignment. An improvement on this approach is to use a roof prism, rather than a plane mirror, as shown schematically in Fig. 8.3. A roof prism has the useful property that any ray incident in a plane perpendicular to the axis of the roof is reflected in the antiparallel direction to the incident ray. Hence, alignment in one plane (the vertical plane for the system illustrated in Fig. 8.3) is assured, which is useful given the inevitable vibration in a mechanical system. Alignment with the axis of the cavity is then achieved once per rotation.

The disadvantages of this approach are rather obvious: the switching is rather slow and the timing of the output pulse has a relatively high uncertainty. In order to estimate the switching time, we suppose that the mirror has to be aligned to an angular tolerance about equal to the divergence of the oscillating cavity mode, which is typically of the order of 1 mrad. Hence, for a motor running at 24 000 r.p.m. the switching time is about 400 ns; this is slow compared to other techniques. The disadvantages are such that this technique is now used only very rarely.

[4] In fact, the first demonstration of Q-switching employed a Kerr cell, as discussed below. See McClung and Hellwarth (1962).

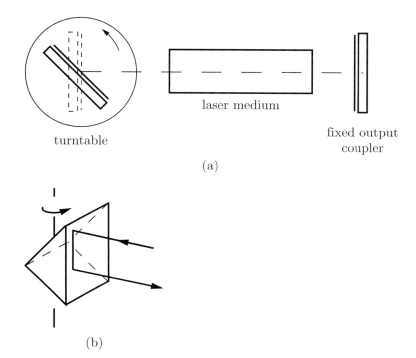

Fig. 8.3: Q-switching with a rotating mirror: (a) schematic diagram of the layout of the optical cavity; (b) use of a roof prism as the rotating element. A roof prism has the advantage of automatically ensuring alignment of the reflected beam in one plane (in this case the vertical plane).

Electro-optic switching

A more modern approach is the use of an electro-optic switch, as illustrated schematically in Fig. 8.4. The refractive index of an electro-optic material changes when it is subjected to an external electric field. In the **Pockels effect**,[5] observed in non-centro-symmetric materials, the induced refractive index varies linearly with the applied electric field. Suitable materials are KD*P or lithium niobate for visible to near-infra-red wavelengths, or cadmium telluride for the mid-infra-red. In the **Kerr effect**[6] the change in refractive index is proportional to the square of the electric field.[7] Larger electric fields are usually required with the Kerr effect since it arises from a higher-order process than the Pockels effect (via the third-order susceptibility rather than the second-order one). Since the electric field must be switched quickly for Q-switching to be effective, it is more usual to use the Pockels effect if a non-linear material with sufficient transmission at the laser wavelength is available.

We can treat electro-optic devices of this type as wave plates with a birefringence that depends on the strength of the applied electric field. In the arrangement shown in Fig. 8.4 a Pockels cell is oriented with its axis at 45° to the axis of a polarizer. The cavity is held in a low-Q state by applying a certain voltage (the 'quarter-wave' voltage) to the Pockels cell, which then behaves as a quarter-wave plate. Consequently, vertically polarized light transmitted by the polarizer becomes circularly polarized after its transmission through the Pockels cell. Upon reflection from the end mirror of the cavity the handedness of the circular polarization changes, and hence on passing again through the Pockels cell it is converted to horizontally polarized light and so is blocked

[5] Named after the German physicist F. C. A. Pockels (1865–1913).

[6] Named after the Scottish physicist John Kerr (1824–1907).

[7] Note that *all* materials exhibit the Kerr effect to some extent.

196 *Dynamic cavity effects*

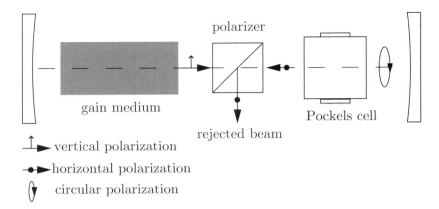

Fig. 8.4: Electro-optic Q-switching.: The system is shown with the Pockels cell voltage set so as to reject the beam reflected from the right-hand mirror.

or reflected out of the cavity by the polarizer. Hence, with the quarter-wave voltage applied to the Pockels cell the end mirror of the cavity is effectively blocked.

The mirror can be unblocked by switching the voltage across the Pockels cell to zero, which reduces the birefringence of the Pockels cell to zero.[8] Vertically polarized light can now pass through the polarizer and Pockels cell with low loss, and the Q-switched pulse can build up. A Kerr cell would be employed in essentially the same way.

Electro-optic switching has the advantages of fast switching times (of the order of 10 ns) and a high hold-off ratio, which allows the population inversion to be built up to many times the threshold value for the high-Q cavity. The disadvantages are that the technique requires several optics to be placed in the cavity (each with some insertion loss and subject to possible damage), and the devices and associated electronics are relatively expensive. Nevertheless, the technique is very widely used.

[8]It would also be possible to devise an arrangement in which the low-loss condition was reached by switching *on* the Pockels cell. However, it is usually technically easier rapidly to switch off high voltages—a technique known as 'crowbaring'—than to switch them on.

Acousto-optic switching

If an acoustic wave is launched through an acousto-optic crystal[9] the refractive index is changed slightly at the peaks and troughs as a result of the local expansion or contraction of the crystal lattice. The resulting periodic variation in the refractive index diffracts incident radiation. We can think of this as Bragg reflection from the high- and low-index planes established in the crystal.

Figure 8.5 shows an acousto-optic switch employed in a Q-switched laser. With the radio-frequency (RF) voltage applied to the crystal an acoustic wave propagates along the length of the crystal and is damped at the opposite end, the damping often being achieved by cutting the end of the crystal to form a wedge. Some fraction of the laser radiation incident on the crystal will be Bragg reflected away from the cavity axis and hence lost from the cavity mode. With the RF voltage applied, then, the acousto-optic crystal increases the losses of the laser cavity. Turning off the RF voltage removes this additional loss, and sets the cavity to a high-Q condition.

[9]A wide variety of acousto-optic materials are available, including: fused silica, tellurium dioxide and lithium niobate for visible wavelengths; gallium arsenide for use in the infra-red; and silicon for radiation in the 10-μm region.

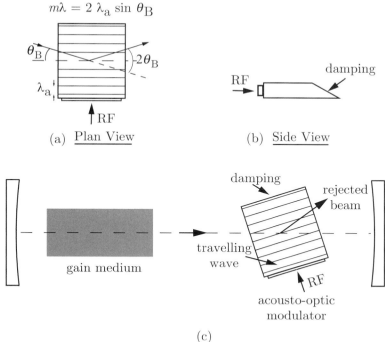

Fig. 8.5: Acousto-optic Q-switching. (a) Plan view of the acousto-optic crystal showing the Bragg condition for reflection of a beam of radiation by incident at an angle θ_B on an acoustic wave of wavelength λ_a. In contrast to the use of acousto-optic crystals in modelocking, as discussed in Section 8.3.3, it is important to prevent the formation of standing waves by reflection of the acoustic wave at the end of the crystal. One way of doing this is shown in the side view drawn in (b): cutting the end of the crystal so as to form a wedge. Q-switching with an acousto-optic modulator is shown schematically in (c). Note that the Bragg angle θ_B has been greatly exaggerated; a value closer to $1°$ would be more typical. Note also that the insertion loss of the acousto-optic crystal can be eliminated (for one polarization) by rotating the crystal out of the plane of the paper through Brewster's angle.

The advantages of acousto-optic switching are that it is relatively cheap, and insertion losses may be made low by orienting the crystal at Brewster's angle. However the hold-off ratio is low, and hence the technique is usually employed in continuously pumped Q-switched lasers for which the overpumping ratio is relatively modest. The switching times achieved are also relatively slow, the switching time corresponding to the time taken for the acoustic wave to propagate out of the region of the cavity mode. Taking the speed of sound in the crystal to be 5 km s^{-1}, we can estimate the switching time to be 200 ns for a 1-mm diameter beam. In order to decrease the switching time, the cavity mode is often focused through the acousto-optic crystal.

Saturable absorbers

A very simple, and quite widely used, method for achieving Q-switching is to insert a saturable absorber into the laser cavity. The absorber can take the form of a thin film on a substrate material, or a liquid contained in a cell. During the pump pulse the absorber will cause the laser mode to experience a high loss and lasing is prevented. However, once the population inversion reaches a high value the absorber will be subjected to an ever-increasing flux of spontaneous and weak stimulated emission. As population is transferred to the upper level of the absorbing species the absorption starts to saturate, increasing the feedback from the cavity mirror and so increasing the rate of stimulated emission. This positive feedback leads to rapid saturation (or 'bleaching') of the absorber until, at full saturation, it represents essentially zero loss. The cavity is now in a high-Q state, and a giant pulse develops.

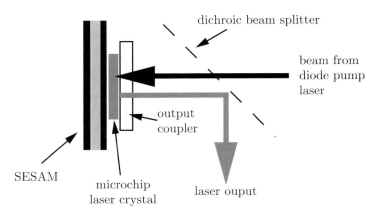

Fig. 8.6: Schematic diagram of a Q-switched microchip laser employing a semiconductor saturable absorber mirror (SESAM).

It is clear that for this technique to work the absorber must have a strong absorption transition at the laser wavelength. It is also the case that the lifetime of the upper level must be reasonably long so that the bleached state is maintained for the duration of the Q-switched pulse. On the other hand, the upper-level lifetime should be short compared to the desired time between Q-switched pulses.

Many types of saturable absorbers have been used, such as solutions of dyes, and doped crystals such as Cr^{4+}:YAG. The semiconductor saturable absorber mirrors (SESAMs) discussed in Section 8.3.4 are increasingly employed for passive Q-switching since their absorption properties may be tailored to the particular laser system. For example, SESAMs are frequently used in so-called microchip lasers of the type shown schematically in Fig. 8.6. Lasers of this type comprise a very short length of active medium, typically less than 1 mm, mounted directly on a SESAM. The very short cavity length means that the cavity round-trip time and the cavity lifetime are only a few picoseconds, and consequently Q-switched pulses may be generated with pulse durations down to a few tens of picoseconds. The repetition rate of the Q-switched pulses is determined largely by the power of the pump laser; by varying this it is possible to adjust the pulse repetition rate from the kilohertz to the megahertz region.

Saturable absorbers are a simple (there is no external drive circuit) and cheap way of achieving Q-switching. Further, the relatively slow turn on corresponds to many cavity round trips, which allows good mode discrimination. As a consequence, it is quite straightforward to achieve Q-switched lasing on a single transverse mode of the cavity. The disadvantages are relatively large timing jitter and shot-to-shot fluctuations in the output energy, and that the absorber can degrade over time.

8.2.2 Rate-equation analysis of Q-switching

A rate-equation analysis of Q-switching is relatively straightforward, providing we make a few simplifying assumptions: we neglect pumping during the development of the Q-switched pulse, and ignore spontaneous transitions. With these assumptions, the rate equations (8.2) and (8.3) become:

$$\frac{dN_2}{dt} = -N^*(t)\sigma_{21}\frac{I}{\hbar\omega_L} \quad (8.21)$$

$$\frac{dN_1}{dt} = +N^*(t)\sigma_{21}\frac{I}{\hbar\omega_L}. \quad (8.22)$$

The equations above correspond to the case of **severe bottlenecking** in which the rate of spontaneous decay of the lower laser level is negligible. As such, stimulated emission causes the population to build up in the lower laser level. The rate equation for the evolution of the population inversion during the Q-switched pulse can be found simply by subtraction:

$$\frac{dN^*}{dt} = \frac{dN_2}{dt} - \frac{g_2}{g_1}\frac{dN_1}{dt} \quad (8.23)$$

$$= -\beta N^*(t)\sigma_{21}\frac{I(t)}{\hbar\omega_L}, \quad (8.24)$$

where $\beta = 1 + g_2/g_1$.

A second case is also of interest, however: that of an ideal four-level laser. In these systems the lower level decays so quickly that throughout the Q-switched pulse $N_1(t) \approx 0$. In this case $N^*(t) \approx N_2(t)$ and hence the rate equation for the population inversion is identical to that for $N_2(t)$.

It is convenient to combine these two cases into one set of equations for the population inversion and the photon density:

$$\frac{dN^*}{dt} = -\frac{\beta}{f_c}\frac{N^*}{N_{th}^*}\frac{n_\phi}{\tau_c} \quad (8.25)$$

$$\frac{dn_\phi}{dt} = \left(\frac{N^*}{N_{th}^*} - 1\right)\frac{n_\phi}{\tau_c}, \quad (8.26)$$

where again we have used $I(t) = n_\phi(t) c \hbar \omega_L$ and eqn (8.6), and where[10]

$$\beta = \begin{cases} 1 & \text{ideal 4-level} \\ 1 + \frac{g_2}{g_1} & \text{severe bottlenecking.} \end{cases} \quad (8.27)$$

[10] See Exercise 5.11.

The general form of the solution to these equations is shown in Fig. 8.7. Note that the population inversion density is exactly equal to the threshold value, N_{th}^*, when the photon density reaches its peak—this must be the case since the peak in the photon density occurs when $dn_\phi/dt = 0$, which is the same condition that defines the threshold population inversion. Since stimulated emission will continue to occur whilst the photon density falls, the population inversion can be burnt down to a level well below the threshold value N_{th}^*.

If we integrate eqn (8.26) over the complete duration of the pulse we find,

$$\int_{-\infty}^{\infty} dn_\phi = \int_{-\infty}^{\infty}\frac{N^*}{N_{th}^*}\frac{n_\phi}{\tau_c}dt - \int_{-\infty}^{\infty}\frac{n_\phi}{\tau_c}dt = 0, \quad (8.28)$$

where we have used the fact that the first integral is zero since the initial and final densities of photons in the cavity are both zero.

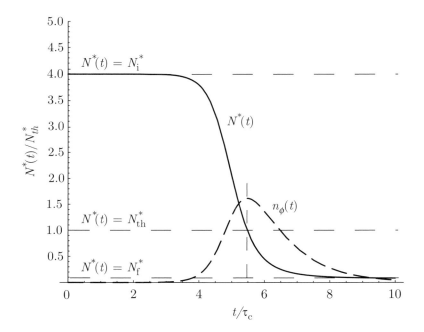

Fig. 8.7: Temporal evolution of the population inversion density $N^*(t)$ and photon density $n_\phi(t)$ during a Q-switched laser pulse for the case $N_i^* = 4N_{th}^*$. The Q-switch is opened at $t = 0$. Note that it takes several cavity lifetimes for the Q-switched pulse to develop, and that the peak in the photon density occurs when the population inversion density passes through the threshold value N_{th}^*.

Now, the rate at which photons are coupled out of the laser is equal to nV_c/τ_c, where V_c is the volume taken up by the oscillating mode in the cavity. Hence, the power output from the laser is,

$$P = \hbar\omega_L V_c \frac{n_\phi}{\tau_c}. \tag{8.29}$$

Applying this result to eqn (8.28), the energy of the output pulse is given by

$$\begin{aligned} E &= \hbar\omega_L V_c \int_{-\infty}^{\infty} \frac{n_\phi}{\tau_c} dt \\ &= \hbar\omega_L V_c \int_{-\infty}^{\infty} \frac{N^*}{N_{th}^*} \frac{n_\phi}{\tau_c} dt \\ &= -\hbar\omega_L V_c \frac{f_c}{\beta} \int_{N_i^*}^{N_f^*} dN^* \\ &= \frac{\hbar\omega_L}{\beta} V_g (N_i^* - N_f^*), \end{aligned} \tag{8.30}$$

where we have used eqn (8.25), N_i^* and N_f^* are the initial and final population inversion densities, and $V_g = f_c V_c$ is the volume of the oscillating mode in the gain medium. We therefore can write,

$$E = \eta N_i^* V_g \hbar\omega_L \quad \textbf{Q-switch pulse energy} \tag{8.31}$$

$$\eta = \frac{N_i^* - N_f^*}{\beta N_i^*}, \quad \textbf{Energy utilization factor} \tag{8.32}$$

where η is the **energy utilization factor**.

The interpretation of this result is straightforward: the initial number of inverted atoms is equal to $N_i^* V_g$; for an ideal four-level laser transition the number of photons emitted in the pulse is simply $(N_i^* - N_f^*) V_g$; whereas in the case of severe bottlenecking this number is reduced by a factor of β since for these systems each stimulated emission event reduces the population inversion by β.[11]

[11] See Exercise 5.11.

We may gain further information by dividing eqn (8.26) by (8.25) to give,

$$\frac{dn_\phi}{dN^*} = \frac{dn_\phi/dt}{dN^*/dt} = \frac{f_c}{\beta}\left(\frac{N_{th}^*}{N^*} - 1\right)$$

$$\Rightarrow \int_{-\infty}^{t} dn_\phi = \frac{f_c}{\beta} \int_{-\infty}^{t} \left(\frac{N_{th}^*}{N^*} - 1\right) dN^*, \quad (8.33)$$

which gives a useful relation between the photon and population inversion densities at all times during the Q-switched pulse:

$$n_\phi(t) = \frac{f_c}{\beta}\left[N_{th}^* \ln\left(\frac{N^*(t)}{N_i^*}\right) + N_i^* - N^*(t)\right]. \quad (8.34)$$

After the pulse, the photon density is zero, and hence setting $t \to \infty$ in the above yields,

$$N_i^* - N_f^* = -N_{th}^* \ln\left(\frac{N_f^*}{N_i^*}\right)$$

$$\Rightarrow \frac{N_i^* - N_f^*}{\beta N_i^*} = -\frac{N_{th}^*}{\beta N_i^*} \ln\left(\frac{N_f^*}{N_i^*}\right). \quad (8.35)$$

As in Section 8.1.2, we define the **overpumping ratio** as the ratio of the initial population inversion density to the threshold value:

$$r = \frac{N_i^*}{N_{th}^*}. \quad (8.36)$$

Equation (8.35) may then be rearranged to give a relation between the overpumping ratio and the energy utilization factor:

$$r = -\frac{\ln(1-\beta\eta)}{\beta\eta}. \quad (8.37)$$

[12] In Section 8.1.2 we stated that $r - 1 \approx 1$. The result here confirms that this is reasonable.

This tells us, for example, that to extract 80% of the energy stored in the initial population inversion in an ideal four-level laser requires pumping the initial population inversion to twice the threshold value.[12] Figure 8.8 plots $\beta\eta$ as a function of the overpumping ratio. We see that for $r \gtrsim 5$ virtually all the available energy is extracted from the population inversion.

Peak output power and pulse duration

The peak density of photons in the cavity occurs when $dn_\phi/dt = 0$, or, $N^* = N_{th}^*$. Equation (8.34) then gives the peak photon density as,

$$n_\phi^p = \frac{f_c}{\beta} N_{th}^* (r - 1 - \ln r). \quad (8.38)$$

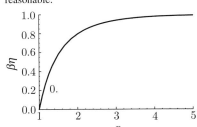

Fig. 8.8: The product of the energy utilization factor η and β plotted as a function of the overpumping ratio r.

From eqn (8.29) the peak output power is then,

$$P_p = \frac{n_\phi^p V_c \hbar \omega_L}{\tau_c} = \frac{N_{th} \hbar \omega_L V_g}{\beta \tau_c}(r - 1 - \ln r).$$

Q-switched pulse: Peak power (8.39)

We may then *estimate* the duration $\Delta \tau_{Qs}$ of the Q-switched pulse by dividing the pulse energy E by the peak power, to give:

$$\Delta \tau_{Qs} = \beta \eta \frac{r}{r - 1 - \ln r} \tau_c.$$ **Q-switched pulse: Duration** (8.40)

Pulse build-up time

It takes a finite time for the Q-switched pulse to develop after the Q-switch is opened, and it is useful to develop an understanding of how this time varies with the parameters of the laser. It is not possible to determine an exact analytical expression for this build-up time, but an approximate expression may be found as follows.

Immediately after the Q-switch is opened the rate of increase in the photon density is, from eqn (8.26),

$$\frac{dn_\phi}{dt} = \left(\frac{N^*}{N_{th}^*} - 1\right)\frac{n_\phi}{\tau_c}$$

$$= (r-1)\frac{n}{\tau_c}, \quad (8.41)$$

since initially $N^* = rN_{th}^*$. The initial phase of the growth in the photon density in the cavity is therefore described by,

$$n_\phi(t) = n_\phi(0) \exp\left[\frac{(r-1)t}{\tau_c}\right], \quad (8.42)$$

where $n_\phi(0)$ is the initial density of photons.

Of course, the growth in the photon density will be slower than described above once the population inversion density is burnt down significantly from its initial value. However, if we ignore this complication, the build-up time τ_b from the opening of the Q-switch to the peak of the Q-switched pulse is given by

$$n_\phi^p = n\phi(0) \exp\left[\frac{(r-1)\tau_b}{\tau_c}\right]$$

$$\Rightarrow \tau_b = \frac{\tau_c}{r-1} \ln\left[\frac{n_\phi^p}{n\phi(0)}\right]. \quad (8.43)$$

The initial density of photons in the cavity, $n_\phi(0)$, is not well known. A reasonable guess is that initially there is approximately 1 photon in the lasing mode within the volume of the laser mode, i.e. $n_\phi(0) \approx 1/V_c$ where V_c is the volume of the lasing mode in the cavity. With this assumption, and eqn (8.38), we may write

$$\tau_b \approx \frac{\tau_c}{r-1} \ln\left[\frac{V_g N_{th}^*}{\beta}(r - 1 - \ln r)\right],$$

Q-switched pulse: Build-up time (8.44)

where $V_g = f_c V_c$ is the volume of the lasing mode within the gain medium.

The logarithm on the right-hand side of eqn (8.44) is relatively insensitive to the values of laser parameters. For example, taking f_c to lie in the range 0.01 to 1, the optical gain cross-section to be between 10^{-20} and 10^{-16} cm^2, and τ_c to range from 10 to 1000 ns, the argument of the logarithm varies between approximately 10^{10} and 10^{19} for a gain volume of 0.1 cm^3 and an overpumping ratio of about 4. Hence, we expect,

$$\tau_b \approx \frac{30 \pm 10}{r-1}\tau_c. \quad (8.45)$$

In other words, the build-up time will be approximately an order of magnitude longer than the cavity lifetime and will decrease as the overpumping ratio is increased.

8.2.3 Comparison with numerical simulations

The analysis presented in this section allows us to draw several conclusions about the behaviour of the Q-switched pulse. As the overpumping ratio is increased, the fraction of the initial energy stored in the population inversion that is extracted by the optical pulse will increase, and consequently N_f^* will decrease. The rate of stimulated emission at the moment that the Q-switch is opened will increase as the overpumping ratio is increased, and consequently the build-up time and duration of the Q-switched pulse will both decrease. Quantitatively, in the limit $r \to \infty$, $\beta\eta \to 1$ and hence $\Delta\tau_{Qs} \to \tau_c$. Note that the temporal profile of the Q-switched pulse is not, in general, symmetric. For example, for large values of the overpumping ratio the Q-switched pulse rises very rapidly owing to the sudden onset of a high rate of stimulated—the population inversion is extracted almost instantaneously. However, the rate at which the these photons can leave the cavity is limited by the cavity lifetime τ_c, and hence in this limit the trailing edge of the Q-switched pulse will follow an exponential decay of lifetime τ_c.

These conclusions are in good agreement with the results of numerical solutions to eqns (8.25) and (8.26) shown in Fig. 8.9.[13]

8.3 Modelocking

Modelocking is an important technique that can be applied to a wide variety of laser systems. Its value lies in the ability to generate pulses of laser radiation that have a short duration—typically picoseconds (1 ps $\equiv 10^{-12}$ s) to femtoseconds (1 fs $\equiv 10^{-15}$ s)—and a peak power many orders of magnitude greater than the mean power that can be extracted from the laser medium.

[13] We should point out that the reason that the build-up times given in Fig. 8.9 are in such close agreement with the simulations is that the initial photon density used in the simulations was exactly $1/V_c$. As such, any differences between the results of the simulation and eqn (8.44) reflect only the assumptions underlying this result: that the population inversion is approximately constant during the build-up of the Q-switched pulse. Comparison of eqn (8.44) with experimental results cannot be expected to be as good since the initial density of photons in the cavity is not well known.

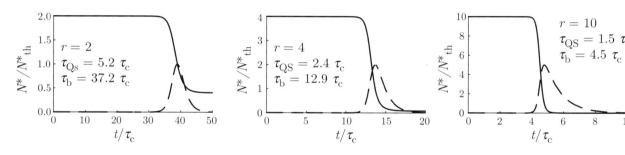

Fig. 8.9: Calculated evolution of the population-inversion density N^* (solid lines) and photon density (in arbitrary units–dashed lines) after opening of a Q-switch at $t = 0$ for several values of the overpumping ratio r. For each plot the duration of the Q-switched pulse $\Delta\tau_{Qs}$ and the build-up time calculated from eqns (8.40) and (8.44) are also given.

Pulses from modelocked lasers have found applications in a wide variety of scientific and technical applications[14] including: imaging techniques, such as two-photon laser-scanning fluorescence imaging, ballistic imaging, and optical coherence tomography; ultrafast chemistry, including coherent control of chemical reactions; pump-probe measurements of solid-state materials; and short-pulse laser machining and processing of materials.

[14] A brief review of the applications of short laser pulses is provided by Reid (2004).

8.3.1 General ideas

We recall from eqn (6.6) that the longitudinal modes of a laser cavity of length L are spaced in angular frequency by an amount,[15]

$$\Delta\omega_{p,p-1} = 2\pi \frac{c}{2L} = \frac{2\pi}{T_c}, \tag{8.46}$$

where T_c is the time taken for light to travel once round the cavity. In order to avoid clutter, it will be convenient to let $\Delta\omega \equiv \Delta\omega_{p,p-1}$ for the remainder of this chapter.

We may therefore write the frequencies of the cavity modes in the form,

$$\omega_p = \omega_{ce} + p\Delta\omega \quad p = 0, 1, 2, 3, \ldots, \tag{8.47}$$

where ω_{ce} is known as the **off-set frequency**.[16]

The right- and left-going waves of a cavity mode may be written in the simplified form,

$$E(z, t) = a_p \exp\left[i(k_p z \mp \omega_p t + \phi_p)\right], \tag{8.48}$$

where a_p is the amplitude of the wave, ϕ_p is the phase of the wave, the upper (lower) sign refers to waves propagating towards positive (negative) z, and we have ignored the details of the transverse profile of the waves.

Suppose that the laser oscillates simultaneously on N such longitudinal modes $p = q_0, q_0 + 1, q_0 + 2, \ldots$. The amplitude of the laser radiation emerging from the output coupler would have the form,

$$E(z, t) = \sum_{p=q_0}^{q_0+N-1} a_p \exp\left[i(k_p z - \omega_p t + \phi_p)\right]. \tag{8.49}$$

[15] Here, we have assumed that the laser radiation propagates everywhere with a constant speed c that, given that there must be some active gain medium within the cavity cannot strictly be the case. The effects of dispersion within the cavity are discussed in more detail in Sections 16.5 and 17.3.1.

[16] It might be thought that the frequencies of the longitudinal modes would be given by $\omega_p = p\Delta\omega$, i.e. that $\omega_{ce} = 0$. In general, this is not the case owing to dispersion in the laser cavity, as discussed further in Section 16.5.

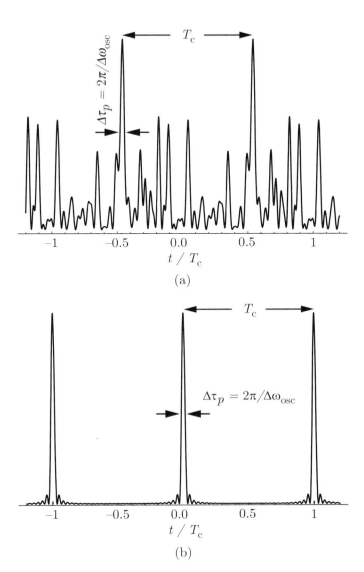

Fig. 8.10: Calculated intensity as a function of time emerging from a laser with $N = 61$ modes oscillating simultaneously with equal amplitudes and (a) random (but fixed) phases; (b) identical phases. Note that for both plots the pattern repeats every cavity round-trip time T_c, and that the sharpest features have approximately the same width. It should also be emphasized that the vertical scales of the two plots are quite different: the peak intensity of the modelocked pulses are approximately N times greater than the mean intensity in (a).

Figure 8.10(a) shows a calculation of the output *intensity* of a laser oscillating simultaneously on 61 modes with equal amplitudes, but with random—although constant—relative phases ϕ_p. The output has several key features,

1. Within an interval of T_c the output contains sharp fluctuations in intensity that are distributed randomly.
2. The sharpest of these peaks has a duration of order $2\pi/\Delta\omega_{\text{osc}}$, where $\omega_{\text{osc}} = (N-1)\Delta\omega$ is the total bandwidth occupied by the oscillating modes.
3. The pattern is periodic with a period equal to the cavity round-trip time T_c.

In contrast, Fig. 8.10(b) shows the calculated intensity when the modes all have the same phase ($\phi_p = 0$). Now, the output consists of a train of short pulses separated by the cavity round-trip time. This situation is known as **modelocking**.

We have considered the operation of a modelocked oscillator from the point of view of the laser output. However, it should be clear that if the output consists of a train of short pulses separated by the cavity round-trip time T_c, then there must be a single short pulse circulating *within* the cavity: every time this pulse reaches the output coupler a fraction of it is transmitted to supply one of the pulses in the output pulse train.

8.3.2 Simple treatment of modelocking

To gain further insight, let us calculate the output for the case where all the modes have the same amplitude, E_0. For convenience, we will assume that all the modes have the same phase, which we will take to be $\phi_p = 0$.[17] Further, for convenience, we also consider the wave at $z = 0$. Equation (8.49) then becomes

$$E(0, t) = E_0 \sum_{p=q_0}^{q_0+N-1} \exp\left[-i(\omega_{ce} + p\Delta\omega)t\right]. \tag{8.50}$$

[17] It can be shown quite simply that for the more general case when $\phi_p = \phi_0 + p\Delta\phi$, the peaks of all the modelocked pulses shift in time by $\Delta\phi/\Delta\omega$, but that otherwise the train is unchanged. See Exercise 8.3.

The sum in the expression above is a geometric progression, which may be evaluated in the usual way to give

$$E(0, t) = E_0 \exp(-i\omega_{ce}t) \exp(-iq_0\Delta\omega t) \frac{1 - \exp(-iN\Delta\omega t)}{1 - \exp(-i\Delta\omega t)}. \tag{8.51}$$

Tidying yields,

$$E(0, t) = E_0 \exp(-i\omega_{ce}t) \exp(-i\overline{p}\Delta\omega t) \frac{\sin\left(\frac{N}{2}\Delta\omega t\right)}{\sin\left(\frac{1}{2}\Delta\omega t\right)} \tag{8.52}$$

$$= E_0 \exp(-i\overline{\omega}t) \frac{\sin\left(\frac{N}{2}\Delta\omega t\right)}{\sin\left(\frac{1}{2}\Delta\omega t\right)}, \tag{8.53}$$

where $\overline{p} = q_0 + (N-1)/2$ and $\overline{\omega} = \omega_{ce} + \overline{p}\Delta\omega$ are, respectively, the mean mode number and mean frequency of the oscillating modes. We see that the output takes the form of a (travelling) wave of mean frequency $\overline{\omega}$ multiplied by an envelope function $\frac{\sin\left(\frac{N}{2}\Delta\omega t\right)}{\sin\left(\frac{1}{2}\Delta\omega t\right)}$.

Taking the intensity to be given by $|E(0, t)|^2$ (i.e. ignoring constants of proportionality):

$$I(t) = E_0^2 \frac{\sin^2\left(\frac{N}{2}\Delta\omega t\right)}{\sin^2\left(\frac{1}{2}\Delta\omega t\right)}. \tag{8.54}$$

This is the function plotted in Fig. 8.10(b).

The temporal behaviour calculated above corresponds to a train of intense modelocked pulses. Note that the peak intensity of each pulse is equal to $N^2 E_0^2$, and hence the peak intensity is a factor of N greater than the mean intensity of N modes oscillating with random phases.[18]

Pulses occur at times given by $\Delta\omega t_m = 2\pi m$, where m is an integer, and hence the separation in time of pulses is simply $2\pi/\Delta\omega = T_c$, i.e. the cavity round-trip time. The width, $\Delta\tau_p$, of the pulses may be taken to be the interval from the peak of a modelocked pulse to the first zero. It is straightforward to show that this is equal to:

$$\Delta\tau_p = \frac{2\pi}{N\Delta\omega} = \frac{T_c}{N} \approx \frac{2\pi}{\Delta\omega_{\mathrm{osc}}}, \tag{8.55}$$

where $\Delta\omega_{\mathrm{osc}} = (N-1)\Delta\omega$ is the (angular) frequency bandwidth covered by the oscillating modes.

In general, the pulses in the train are not identical. From eqn (8.51) we see that the phase of each pulse in the train is shifted from that of the previous pulse by an amount equal to[19]

$$\phi_{\mathrm{slip}} = \omega_{\mathrm{ce}} T_c = 2\pi\left(\frac{\omega_{\mathrm{ce}}}{\Delta\omega}\right). \tag{8.56}$$

This phase slip is important when the pulses are very short since then the phase of the carrier wave relative to the pulse envelope has a significant effect on the peak electric field within the pulse. It is also important in the theory of optical frequency combs discussed in Section 16.5. Figure 8.11 illustrates the effect of a non-zero phase slip between pulses.

It should be clear from the above that if we were able to lock the phase of N longitudinal modes of a laser we could generate a stream of pulses of high peak intensity and short duration. In practice, the number of modes that can be locked ranges from several hundred to several thousand, leading to a huge increase in peak power. The duration of the pulses is determined by the frequency width over which lasing can occur, and hence depends on the linewidth of the laser transition; transitions with broader linewidths can (in principle) generate shorter pulses. In practice, pulses with durations from a few picoseconds to a few femtoseconds can be generated.[20]

[18] Note that the *mean* intensity is the same whether or not the N modes are modelocked or not. Modelocking is an interference phenomenon, and interference neither destroys or creates energy!

[19] See Exercise 8.3.

[20] A very active area of current research is the development of techniques related to modelocking to generate pulses in the attosecond range (1 as $\equiv 10^{-18}$ s).

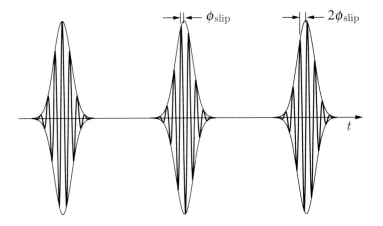

Fig. 8.11: Slip in phase between the carrier wave and the pulse envelope that occurs in a train of modelocked pulses when $\omega_{\mathrm{ce}} \neq 0$.

In the general discussion above we did not consider in any detail the properties of the laser gain medium. As we might expect, the modelocking behaviour of a laser depends on whether the laser transition is homogeneously or inhomogeneously broadened. We have already seen that an inhomogeneously broadened laser transition can oscillate simultaneously on all longitudinal modes for which the round-trip gain exceeds the round-trip loss. The discussion above applies quite well to such lasers, and in such cases the modelocking technique only needs to lock the phases of modes that are already oscillating simultaneously. As a first approximation, the duration of modelocked pulses output from a laser operating on an inhomogeneously broadened transition will be $\Delta \tau_p \approx 2\pi/\Delta\omega_D$, where $\Delta\omega_D$ is the inhomogeneous linewidth of the transition.

Homogeneously broadened laser transitions, however, usually only oscillate on a single longitudinal mode (unless spatial holeburning occurs). The modelocking must therefore play a stronger role, and encourage laser oscillation on many modes. As such, the dependence of the pulse duration on the 'strength' of the modelocking will be different from the case of inhomogeneous broadening, and the duration of the modelocked pulses will generally be longer than $2\pi/\Delta\omega_H$, where $\Delta\omega_H$ is the homogeneous linewidth.

Modelocking techniques can be divided into two classes: **active** and **passive**, as described in the following sections.

8.3.3 Active modelocking techniques

Active modelocking techniques fall into two classes: **amplitude modulation (AM)** and **frequency modulation (FM)**. We first note that we usually require a *single* modelocked pulse to circulate within the laser cavity. For a linear laser cavity this is achieved by locating the active component near one of the end mirrors, and driving the component at the longitudinal mode spacing $\Delta\omega$. However, it is possible to generate M such modelocked pulses within the laser cavity by locating the modulator suitably and driving it at $M\Delta\omega$. **Harmonic modelocking** of this type is occasionally used to increase the pulse repetition frequency. In principle, it can also be used to generate shorter pulses in homogeneously broadened lasers.[21]

[21] See eqn (8.62).

Before describing the various techniques employed to achieve active modelocking, we note that all these methods require that a modulator of some kind is driven with a frequency that is in close resonance with the frequency separation between adjacent longitudinal modes, $\Delta\omega$. The presence of long-term thermal drift, however, means that the frequency of the modulator is liable to drift out of resonance with the laser cavity. A nice solution to this problem is so-called **regenerative modelocking**. With this method a small fraction of the output of the laser is detected by a fast photodiode. Since the output of a photodiode is proportional to the *intensity* of the incident radiation, the output of the photodiode will contain beats with frequencies equal to the differences between the frequencies of the oscillating laser modes. Consequently, by filtering the photodiode output it is possible to obtain a signal at *exactly* the frequency difference between adjacent longitudinal modes. This signal may be amplified and used to drive the modulator directly.

AM modelocking

Once modelocking is achieved, the intra-cavity radiation takes the form of a short pulse that circulates round the laser cavity. An obvious way, therefore, to achieve modelocking is to introduce some sort of 'shutter' into the cavity that lets through short pulses, but that blocks—or increases the loss of—longer pulses. Radiation that happens to reach the shutter when it is open will pass through with low loss, will experience net round-trip gain, and hence will grow. In contrast, the radiation arriving at the shutter outside this interval will experience a round-trip loss that is greater than the round-trip gain, and so will be absorbed.

After many round-trips a pulse will form that propagates round the cavity in synchronization with the shutter, as illustrated schematically in Fig. 8.12. The duration of the modelocked pulse so formed is determined by an equilibrium between two competing effects. The shutter tends to decrease the duration of the pulse (and hence increase the bandwidth) since the losses of the leading and trailing edges of the pulse are greater than those experienced by the centre of the pulse. However, the gain medium can only amplify a finite-frequency bandwidth, which limits the minimum possible duration of the modelocked pulse.

The above description is from the point of view of the *time domain*. It is also possible to consider the problem from the perspective of the *frequency domain*. To this end we suppose that the amplitude transmission of the shutter varies harmonically as[22]

$$T(t) = T_0 + \frac{\delta}{2} \cos \Delta \omega t.$$

[22]Here, T represents the *amplitude* transmission coefficient of the shutter. In optics it would be more usual to use the symbol t to denote this quantity, with T reserved for the intensity transmission. Here, of course, t denotes time.

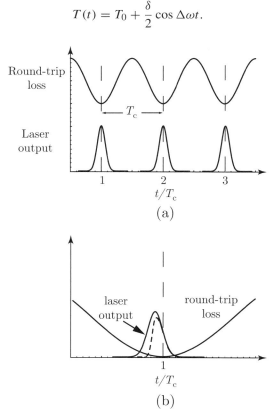

Fig. 8.12: Schematic diagram of AM modelocking. In (a) the pulses arrive perfectly synchronized with the shutter and hence experiences minimum loss. The pulse shown in (b) arrives slightly too early, and hence the leading edge is reduced in intensity. The effect is to move the peak of the pulse to later times, closer to synchronization with the shutter, and to decrease the duration of the pulse. The dashed pulse illustrates the form of the pulse after it has passed through the shutter.

Suppose that a wave of angular frequency ω_p is incident on the shutter, such that we may write its amplitude as the real part of,

$$E(t) = E_0 \exp(-i\omega_p t). \tag{8.57}$$

The amplitude of the transmitted wave will be:

$$E(t) = E_0 T_0 \exp(-i\omega_p t)$$
$$+ \frac{E_0 \delta}{4} \exp(-i[\omega_p - \Delta\omega]t) + \frac{E_0 \delta}{4} \exp(-i[\omega_p + \Delta\omega]t),$$

which follows from writing $\cos \Delta\omega t$ in exponential form.

We see that the amplitude modulation adds **frequency sidebands** at $\omega_p \pm \Delta\omega$. Since the modulation is driven at the frequency difference between longitudinal modes, the frequencies of the sidebands of a longitudinal mode p lie exactly at the frequencies of the adjacent longitudinal modes $p \pm 1$.

This frequency-domain picture provides insight into how it is possible to achieve modelocking in lasers operating on homogeneously broadened transitions. For each longitudinal mode, the amplitude modulation couples energy into the adjacent modes; and it does so with a definite phase. As such, we can consider the sidebands as 'injection seeding' the adjacent longitudinal modes. The system is clearly complex, since each of many modes is coupled to its adjacent modes, but it is clear that the amplitude modulation causes energy to flow from the modes closest to the line centre out to the wings, and this coupling can influence the phases of adjacent modes. The amplitude modulation therefore tends to force the laser out of single-mode oscillation, and broaden the frequency spectrum. A steady state is reached when this broadening is balanced by the frequency narrowing arising from the finite frequency response of the gain medium.

We note that for a homogeneously broadened laser transition the round-trip gain of the modes near the line centre is actually greater than the round-trip loss caused by the cavity alone; in the wings of the modelocked pulse the round-trip gain is less than the round-trip loss of the cavity, as shown schematically in Fig. 8.13. In both cases the difference in energy is made up by energy transfer to or from adjacent modes so that in the steady state the *net* gain and losses experienced by each mode are balanced.

Finally, we note that for both homogeneously and inhomogeneously laser transitions the system adopts the lowestloss configuration, which corresponds to simultaneous oscillation on a large number of longitudinal modes, phase locked so as to form a short pulse circulating round the cavity.

Experimental implementation

Amplitude modulation may be achieved by using a Pockels cell in much the same configuration as is used to generate Q-switched pulses. The Pockels cell is driven with a sinusoidal voltage of frequency $\Delta\omega$. Since the amplitude modulation does not need to be 100%, the driving voltage need only be some fraction of the quarter-wave voltage.

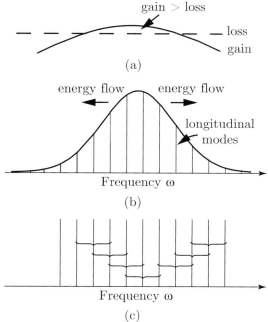

Fig. 8.13: Energy flow and coupling of modes in an AM modelocked laser operating on a homogeneously broadened transition. The frequency dependence of the optical gain and loss *in the absence of the amplitude modulation* is shown in (a). As illustrated in (b), the amplitude modulation causes energy to flow from those modes for which the round-trip gain exceeds the round-trip loss to modes in the wings of the gain profile for which the gain is lower than the loss. This energy transfer allows steady-state laser oscillation to be maintained simultaneously on many modes. The amplitude modulation causes each mode to develop sidebands at frequencies equal to those of the adjacent longitudinal modes, and hence couples adjacent modes as illustrated schematically in (c).

It is also possible to use an acousto-optic modulator to implement AM modelocking, using a similar configuration to that used for Q-switching, as shown in Fig. 8.14. However, for modelocking the faces of the acousto-optic block are cut parallel, and the length of the block is cut to be an integer number of half (acoustic) -wavelengths long. A standing wave can then be driven in the acousto-optic block, rather than the travelling acoustic wave used to Q-switch. The amplitude of the standing wave passes through zero twice per period, corresponding to windows of minimum loss. Hence, if the block is driven with an angular frequency ω_a, AM modelocking will be achieved if $2\omega_a = \Delta\omega$.

FM modelocking

Frequency-modulation modelocking is achieved by modulating the refractive index of an intra-cavity component at the mode-spacing frequency $\Delta\omega$, as illustrated schematically in Fig. 8.15. As we will see, this affects the frequency spectrum of an incident wave and hence the active device is known as a **frequency modulator**.

We consider first the effect on a wave of angular frequency ω of a modulator of length L_{mod} and time-dependent refractive index $n(t)$. The incident wave will emerge with an amplitude equal to the real part of,

$$E(L_{\mathrm{mod}}, t) = E_0 \exp\left(i\omega\left[n(t)\frac{L_{\mathrm{mod}}}{c} - t\right]\right). \tag{8.58}$$

Let us now suppose that the refractive index is modulated harmonically, such that,

$$n(t) = n_0 + \frac{\delta}{2}\sin\Omega t.$$

Fig. 8.14: Amplitude modulation mode-locking with an acousto-optic modulator. In (a) the Bragg condition for reflection of a beam of radiation being incident at an angle θ_B on an acoustic wave of wavelength λ_a. In contrast to the use of acousto-optic crystals in Q-switching, as discussed in Section 8.2.1, in order to achieve modelocking the cavity losses must be modulated at $\Delta\omega$. This is achieved by cutting the acousto-optic crystal so that its two ends are parallel and separated by an integer of half-wavelengths when the modulator is driven at $\Delta\omega/2$. This configuration allows a standing wave to be set up with an amplitude that passes through zero with a frequency $\Delta\omega$. A typical layout of the optical cavity is shown in (c). Note that the Bragg angle θ_B has been greatly exaggerated. Note also that the insertion loss of the acousto-optic crystal can be eliminated (for one polarization) by rotating the crystal out of the plane of the paper through Brewster's angle.

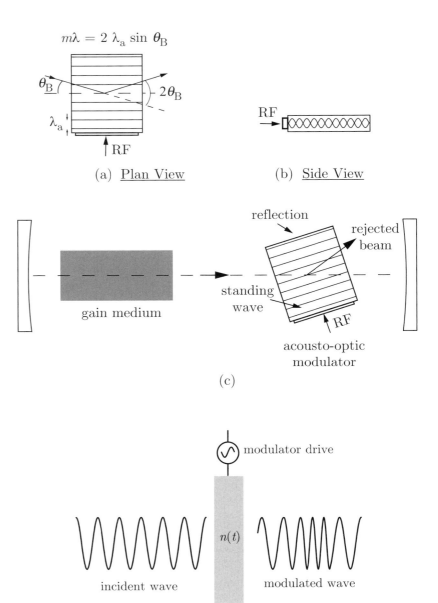

Fig. 8.15: Schematic diagram of a frequency modulator used in FM modelocking.

[23] Consult any text on mathematical methods, for example Riley et al. (2002).

The transmitted wave will be

$$E(L_{\text{mod}}, t) = E_0 \exp(-i\omega t) \exp(i\phi_0) \exp(i\beta \sin \Omega t), \quad (8.59)$$

where $\phi_0 = \omega n_0 L_{\text{mod}}/c$ is the (constant) phase shift experienced by light of angular frequency ω, and $\beta = \omega \delta L_{\text{mod}}/2c$ is the amplitude of the modulation in phase introduced by the modulator.

Now, we may use the identity,[23]

$$\exp(i\beta \sin \Omega t) \equiv \sum_{m=-\infty}^{\infty} J_m(\beta) \exp(im\Omega t), \tag{8.60}$$

where $J_m(x)$ is the Bessel function of the first kind of order m.

Thus, we may write the amplitude of the transmitted wave as,

$$E(L_{\mathrm{mod}}, t) = E_0 \exp(-i\omega t) \exp(i\phi_0) \sum_{m=-\infty}^{\infty} J_m(\beta) \exp(im\Omega t), \tag{8.61}$$

and we see that the modulator introduces a series of sidebands at frequencies of $\omega \pm m\Omega$, where m is an integer.

Applying these ideas to an intra-cavity modulator driven at the mode-spacing frequency $\Delta\omega$, we see that for each mode of frequency ω_p, the modulator generates frequency sidebands at frequencies of $\omega_p \pm m\Delta\omega$, thereby coupling the longitudinal modes in a similar way to amplitude modulation. The width of the pulse generated in this way is determined by the balance between the bandwidth broadening caused by the modulator and the frequency narrowing produced by amplification in the gain medium.

As shown schematically in Fig 8.16, the modelocked pulses are found to be located at the maxima and minima of $n(t)$, when the modulator produces no shift in frequency. This may be understood as follows. The effect of the phase modulation introduced by the frequency modulator is equivalent to modulating the cavity length by driving one of the cavity mirrors sinusoidally. In general, a pulse reflected from the moving cavity mirror will be Doppler shifted in frequency and, since the mirror is driven with a frequency of $\Delta\omega$, when it returns one round trip later it will be further shifted in frequency by the same amount. Eventually, the pulse will be shifted outside the bandwidth of the amplifying transition and will no longer be amplified. However, this progressive frequency shift is minimized for radiation that strikes the mirror at one of the extrema of its motion or, equivalently, if it passes through the modulator near the maxima or minima in $n(t)$.

Note that there are *two* stable positions for the modelocked pulse. An undesirable consequence of this is that the inevitable disturbances present in a real

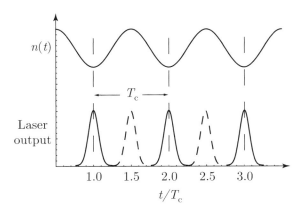

Fig. 8.16: Location of the modelocked pulses with FM modelocking. The train of modelocked pulses can coincide with either the maxima or minima of the refractive index of the frequency modulator.

laser can cause the modelocking to flip between these two states, or even that two modelocked pulses circulate within the cavity.

Experimental implementation

Frequency modulation is usually achieved by orienting a Pockels cell with its optical axis parallel to the polarization of the intra-cavity laser radiation. Applying a voltage to the Pockels cell will then change the refractive index experienced by the radiation, but will not alter its polarization state.

Synchronous pumping

Active modelocking may also be achieved by synchronous pumping in which the gain is modulated at the longitudinal mode spacing frequency $\Delta \omega$. With this technique modelocking arises by processes analogous to the AM modelocking discussed above. It should be clear that in order for the gain of the laser transition to respond to the modulation, the population inversion must decay in a time that is short compared to the cavity round-trip time.

For example, dye lasers can be modelocked by optically pumping the dye with pulses from a laser that is itself modelocked.[24] Synchronous pumping can also be achieved in semiconductor lasers by modulating the driving current.

[24] Obviously, care must be taken to ensure that the repetition frequency of the modelocked pump pulses corresponds to the longitudinal mode spacing of the dye laser: in other words the cavity round-trip times must be the same for the two lasers.

Pulse duration of actively modelocked, homogeneously broadened lasers

Calculation of the duration of the pulses produced by an actively modelocked laser is a complex matter.[25] Here, we note that it may be shown that the duration of the pulses generated by AM modelocking a c.w. laser operating on a homogeneously broadened laser transition is given by,

$$\Delta \tau_p \approx 0.45 \left(\frac{2\alpha_I \ell_g}{\delta} \right)^{1/4} \frac{2\pi}{\sqrt{\Delta \omega_{\text{mod}} \Delta \omega_H}}, \qquad (8.62)$$

[25] A description of the methods used to estimate the duration of the output pulses may be found in Siegman (Siegman, 1986, Chapter 27).

where α_I is the (saturated) gain coefficient, ℓ_g the length of the gain medium, $\Delta \omega_{\text{mod}}$ is the driving frequency of the modulator (usually set equal to $\Delta \omega$, but it may be a harmonic of this), and $\Delta \omega_H$ is the homogeneous linewidth. Note that the pulse duration decreases with increasing modulation, δ, but only rather slowly. In fact, for most systems the first bracketed term will be close to unity, and hence

$$\Delta \tau_p \approx 0.45 \frac{2\pi}{\sqrt{\Delta \omega_{\text{mod}} \Delta \omega_H}}. \qquad (8.63)$$

A similar result can be obtained for FM modelocking. Note that since $\Delta \omega_H \gg \Delta \omega$ the pulses output by an actively modelocked laser operating on a homogeneously broadened laser will be significantly longer than obtained using an inhomogeneously broadened transition of the same linewidth.

8.3.4 Passive modelocking techniques

Modelocking can also be achieved by inserting into the laser cavity an element that is not actively modulated, but that instead has transmission properties that depend on the intensity of the incident radiation. Modelocking achieved in this way is said to be passive, and has the advantage that it avoids the need to drive high-frequency modulators and is therefore simpler to implement practically. Passive modelocking techniques are perhaps of more practical importance than active techniques. They are also able to generate the shortest possible laser pulses.

It is more difficult to calculate the pulse duration of the modelocked pulse than in active modelocking, and treatments tend to be specific to the particular case under consideration. In any case, the lower limit of the pulse duration will always be given by approximately by $2\pi/\Delta\omega_{\mathrm{H,D}}$, where $\Delta\omega_{\mathrm{H,D}}$ is the linewidth of the laser transition.

Saturable absorbers

An absorption transition exhibits saturation of the absorption in an analogous way to the saturation of the gain presented in Section 5.1.[26] As for the gain, the key parameters describing the saturation of an absorption transition are the saturation intensity, I_s, and the saturation fluence, Γ_s:

$$I_s = \frac{\hbar\omega}{\sigma_{20}\tau_R}, \tag{8.64}$$

$$\Gamma_s = \frac{\hbar\omega}{\sigma_{20}}, \tag{8.65}$$

[26] See Exercise 5.3.

where τ_R is the recovery time of the transition, and σ_{20} is the optical cross-section of the $2 \leftarrow 0$ absorption transition.[27] Of key importance here is the fact that after an intense, saturating beam of incident radiation is switched off, the time taken for the absorption to return to its high, unsaturated value is of the order of the recovery time.

[27] Note that we have defined this cross-section in the *same* way as the optical gain cross-section: $\sigma_{20} = \frac{\hbar\omega}{c}B_{20}g(\omega-\omega_0)$.

The saturation intensity is relevant to steady-state conditions, or situations in which the duration of an incident pulse is much longer than the recovery time. In this case, I_s divides into two regions: low intensities for which the incident radiation experiences high, unsaturated absorption; and high intensities, in which the absorption is strongly reduced by saturation of the transition.

The situation is somewhat different in pulsed conditions. If the duration of an incident laser pulses is short compared to the recovery time, the absorption experienced at any time t during the pulse depends only on the intensity integrated over time to that point, i.e. on $\int_{-\infty}^{t} I(t')\mathrm{d}t'$. The reason for this is simply that atoms transferred to the upper level of the absorption transition will not decay (to any significant degree) during the incident laser pulse, and hence the density of atoms in the upper level will depend only on the *number* of incident photons per unit area, not on the details of the pulse shape. As such, the absorption will start to saturate after a certain number of incident photons per unit area, or, equivalently after a certain incident fluence (energy per unit area). In this case, then, the behaviour is determined by the *fluence* of the incident pulse relative to the saturation fluence.

Fast saturable absorbers

In the context of modelocking, a **fast saturable absorber** is a medium in which the recovery time is very much shorter than the cavity round-trip time, and in

[28] See Fig. 8.49(b).

fact is short compared to the typical spacing of the 'noise' spikes observed in the non-modelocked, multimode laser.[28] Typically, then, the recovery time of a fast saturable absorber will be a few tens of picoseconds, or less.

If a fast saturable absorber is placed within a laser cavity, near one of the laser mirrors, the round-trip loss will be lower for intense noise spikes than for low-intensity spikes. Consequently, the most intense spike will tend to grow at the expense of the others, and will start to saturate the available gain. The result is a single, intense pulse propagating round the laser cavity: modelocking. We can think of the saturable absorber as acting as a shutter that is activated by the pulse itself, as illustrated in Fig. 8.17.

Increasingly, so-called **semiconductor saturable absorber mirrors (SESAMs)** are employed as fast saturable absorbers. A SESAM comprises a semiconductor material sandwiched between two mirrors, as shown schematically in Fig. 8.18. Semiconductors, particularly composite semiconductors such as multiple quantum-well materials,[29] are attractive for making saturable absorbers since their absorption properties may be controlled. For example, for the strained InGaAs-GaAs superlattice illustrated in Fig. 8.18 the wavelength of the absorption edge may, in principle, be varied between 900 nm and 2 μm by adjusting the proportion of In in the InGaAs layers. Further, it is also possible to vary the relaxation time τ_R of the absorption transition through control of the conditions under which the semiconductor is grown. However, most semicon-

[29] See Chapter 9.

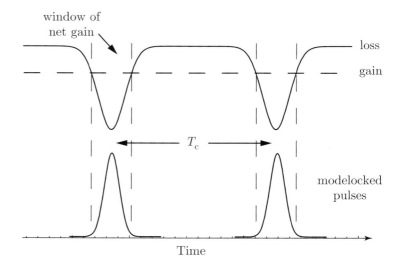

Fig. 8.17: Passive modelocking with a fast saturable absorber. The temporal variation of the round-trip loss in the cavity is shown relative to the round-trip gain. Saturation of a fast absorber causes the loss to drop below the gain when the intense modelocked pulse passes through the absorber.

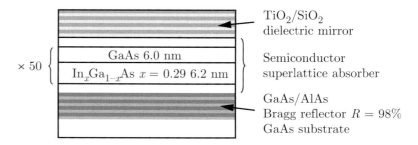

Fig. 8.18: Schematic diagram of an antiresonant Fabry–Perot SESAM for use with wavelengths close to 1 μm. (After Brovelli and Keller (1995).)

ductor materials cannot be used directly as a saturable absorber owing to their low saturation intensity and low damage threshold. In a SESAM the semiconductor is sandwiched between two mirrors spaced so that the round-trip phase is equal to $(2p + 1)\pi/2$, where p is an integer, to form a so-called antiresonant Fabry–Perot. With this arrangement the intensity within the Fabry–Perot—and hence the intensity to which the semiconductor is exposed—is only a small fraction, f, of the incident intensity. Furthermore, providing the reflectivities of the two mirrors is reasonably high, the proportion of incident radiation reflected from an antiresonant Fabry–Perot is large and almost independent of wavelength. For example, for mirror reflectivities of 100% and 98% $f \approx 0.01$ and hence the intensity within the semiconductor is some two orders of magnitude lower than the incident intensity. In effect, the saturation intensity and saturation fluence of the saturable absorber are increased by a factor of $1/f$. The effective saturation intensity and fluence may be controlled by varying the parameters of the Fabry–Perot resonator and this—together with control of the absorption properties of the semiconductor—allows the saturation properties of the SESAM to be tailored to the laser system of interest.

Slow saturable absorbers

Under some circumstances it is also possible to achieve modelocking using a **slow saturable absorber**, for which the recovery time is comparable to—although it must still be less than—the cavity round-trip time.

Figure 8.19 illustrates schematically how modelocking is achieved with a slow absorber. For simplicity we imagine that the saturable absorber and the gain medium lie in the same plane. The leading edge of the modelocked pulse will experience a net round-trip loss, owing to the fact that the (unsaturated) absorption is higher than the available gain, and hence will be eroded. As the pulse passes through the absorber, the absorption will gradually decrease as population is transferred to the upper level of the absorption transition. At some point the round-trip loss introduced by the saturable absorber will fall below the round-trip gain, and the pulse will experience amplification. In order for the generated pulse to be short, however, it is essential that shortly after this moment the growing pulse saturates the round-trip gain below the

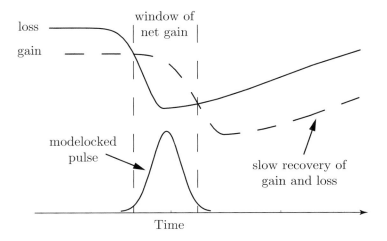

Fig. 8.19: Schematic diagram of modelocking with a slow absorber. The interplay of saturation of the gain and loss ensures that net gain only exists in a short window. The saturated gain and loss must recover within a cavity round-trip time T_c.

(now saturated) round-trip loss so that the trailing edge of the pulse is also attenuated. Further, both the gain and the absorption must recover to their initial values in less than a cavity round-trip time if the system is to settle down to a steady state in which a single, modelocked pulse circulates within the cavity.

It should be appreciated, therefore, that modelocking with a slow saturable absorber requires a delicate balance between the energy of the modelocked pulse, and the saturation fluences and recovery times of both the absorber and the gain medium. The requirement that the gain can be returned to its initial value by the pumping mechanism in less than one round-trip time means that the technique is limited to lasers with a relatively short upper-level lifetime, such as dye lasers or semiconductor lasers.

A particular implementation of this technique, known as colliding-pulse modelocking (CPM), was the first that was able to generate pulses shorter than 100 fs. In this approach the gain medium and saturable absorber are thin (approximately 10 μm) sheets of organic dye[30] solution produced by a slit jet.[31] In a CPM system a ring cavity is employed, which allows for the possibility of two counter-propagating pulses arriving at the absorbing dye jet simultaneously. If this occurs the absorption experienced by each pulse will be lower owing to the increased saturation of the dye, and as such the CPM geometry provides deeper modulation of the transmission of the circulating laser pulses, thereby assisting the generation of shorter pulses.

Kerr-lens modelocking

Kerr-lens modelocking (KLM) is one of the most important types of modelocking used today, and is able to generate the shortest optical pulses obtainable directly from a laser oscillator.

All materials exhibit the Kerr effect, in which the refractive index depends on the square of any applied electric field. If an intense optical pulse is incident on a medium, the **optical Kerr effect** will cause the refractive index to be modified by the electric field of the pulse itself. Since the square of the electric field is proportional to the intensity, I, of the pulse, we may write the refractive index of a material as

$$n(I) = n_0 + n_2 I. \tag{8.66}$$

It is found that n_2 is positive for nearly all optical materials, with values typically of order 10^{-16} cm^2 W^{-1}.

The optical Kerr effect is responsible for a variety of phenomena, such as self-focusing and self-phase modulation.[32] The former lies at the heart of Kerr-lens modelocking. Figure 8.20 shows the propagation of a beam with a Gaussian transverse profile through an optical Kerr medium. The intensity of the beam will be greater on the propagation axis than in the transverse wings, and hence the refractive index of the medium in the axial region will be greater than in the wings. The optical Kerr medium therefore behaves as a thin lens that focuses the beam slightly as it propagates through it.

Figure 8.21 shows schematically how the optical Kerr effect can cause modelocking. The optical Kerr medium is usually the (solid) laser gain medium itself. The laser cavity is designed so that in the absence of any self-focusing

---

[30] See Section 14.1.

[31] For example, the gain medium might be a jet of the dye rhodamine-6G dissolved in ethylene glycol, and the absorber the dye DOdcI (3,3 diethyloxadicarbocyanine iodide) dissolved in the same solvent.

[32] See Section 17.1.5.

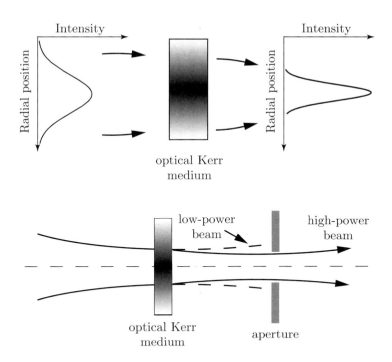

Fig. 8.20: Schematic diagram of self-focusing arising from the optical Kerr effect. Darker regions in the optical Kerr medium indicate regions of higher refractive index $n(I)$ induced by the intensity of the beam passing through it.

Fig. 8.21: Schematic diagram of Kerr-lens modelocking.

the transverse dimensions of the lowest-order cavity mode are comparable to, or larger than, the size of an aperture placed within the cavity. As such, low-intensity radiation experiences a relatively high round-trip loss. Higher-intensity radiation, however, will be self-focused by the Kerr effect and so experience lower round-trip losses.[33] Modelocking then arises by essentially the same processes as with a fast saturable absorber: the optical Kerr medium acts as a fast, self-activated shutter.

In order for KLM to be effective, the laser cavity must be operated close to being 'unstable' in order that the weak Kerr lens causes a large change in the transverse extent of the mode. The critical aperture in a Kerr-lens modelocked laser may be the 'soft' aperture formed by the finite transverse extent of the pump laser, or a 'hard' aperture deliberately introduced into the laser cavity.

Examples of modelocked lasers

Table 8.1 summarizes the parameters of some modelocked lasers. Several trends may be noted. In c.w. lasers, active modelocking is usually implemented with acousto-optic modulators owing to their lower insertion loss compared to a Pockels cell; passive modelocking may be achieved with slow saturable absorbers for lasers with reasonably fast upper-level lifetimes, or by Kerr-lens modelocking.

Pulsed modelocking

In the discussion of modelocking above we have tacitly assumed that the *pumping* of the modelocked laser is continuous. Pulsed pumping of modelocked lasers is also possible, but this brings additional considerations.

[33] It is worth noting that when modelocking occurs the peak intensity of the pulse within the cavity can be six orders of magnitude greater than the intensity when the laser is operated c.w.

Table 8.1 Parameters of some modelocked lasers.

Laser	Pumping	$\Delta\tau_p$	Method
He–Ne	c.w.	1 ns	acousto-optic
CO_2	c.w.	10–20 ns	acousto-optic
Ar^+	c.w.	150 ps	acousto-optic
Nd:YAG	c.w.	10 ps	acousto-optic
He–Ne	c.w.	300 ps	slow sat. abs.
GaAs	c.w.	5 ps	slow sat. abs.
Dye	c.w.	25 fs	slow sat. abs.
Ti:sapphire	c.w.	10 fs	KLM
Nd:YAG	pulsed	40 ps	electro-optic
CO_2	pulsed	1 ns	fast sat. abs
Ruby	pulsed	10 ps	fast sat. abs
Nd:YAG	pulsed	40 ps	fast sat. abs
Nd:glass	pulsed	5 ps	fast sat. abs
Dye	pulsed	1 ps	fast sat. abs

Continuous pumping of a modelocked laser is inefficient if the lifetime of the upper laser level is short compared to the cavity round-trip time, since then most atoms pumped to the upper laser level will decay spontaneously before they have the opportunity to be stimulated to emit by the circulating modelocked pulse. A solution is to **synchronously pump** the modelocking: in this technique pump pulses are delivered at exactly the repetition rate of the modelocked pulses, so that gain is generated just as the modelocked pulse arrives at the gain medium. Very often no further steps are required to achieve modelocking since the modulation of the gain promotes the formation of a modelocked pulse in an analogous way to the operation of a saturable absorber. It should be clear that in synchronously pumped modelocked lasers the repetition rate of the pump pulses must be closely matched to the round-trip time of the modelocked laser; this requires that the cavity length of the modelocked laser is actively controlled, or that the frequencies of the pump pulses are synchronized to the train of modelocked pulses.

If the lifetime of the upper laser level is long compared to the cavity round-trip time, strong pulsed pumping can lead to laser spiking.[34] In such cases it is usual also to Q-switch the modelocked laser; this results in a burst of modelocked pulses, the duration of the burst being approximately equal to that of the pulse produced by Q-switching alone. With active modelocking, the formation of the modelocked pulse typically occurs over hundreds or thousands of round trips, and this may be longer than the build-up time of the Q-switched pulse. A solution to this is to allow **pre-lasing**, by partially opening the Q-switch whilst the modelocker is running; this allows weak lasing and the formation of a circulating modelocked pulse of low energy. This pulse is rapidly amplified when the Q-switch is opened fully, generating an output burst of modelocked pulses.

The higher gain found in pulsed lasers means that it is possible to employ an electro-optic switch to achieve active modelocking, in either AM or FM mode. Passive modelocking of pulsed lasers may be achieved with fast saturable absorbers.

[34] See Section 8.1.

8.4 Other forms of pulsed output

For completeness, we mention briefly some other ways in which the output of a laser may be pulsed.

Gain switching

If the pumping of the population inversion is very rapid, the inversion can build up to values well above the threshold value before stimulated emission has time to burn the population down. The laser system then finds itself in a state equivalent to the moment at which the cavity is returned to a high-Q state in a Q-switched laser. The result is the same: a high-power output pulse, and depletion of the population inversion well below the threshold value.

Gain switching often occurs in solid-state lasers optically pumped by short laser pulses, since the lifetime of the upper laser level is then long, whilst the short pump pulse generates an almost instantaneous population inversion.[35]

Gain switching is perhaps most commonly used in semiconductor lasers, which may be modulated by modulating the current. Since the cavity length of a semiconductor laser is only a few hundred micrometres the cavity lifetime is typically only a few picoseconds and consequently very high modulation frequencies—up to the GHz range—are possible.

[35] In this case, the short duration of the pumping means that the population inversion is established in a time much *shorter* than the lifetime of the upper laser level, τ_2. Of course, it would take a time of order τ_2 to establish a *steady-state* population inversion, but the steady state is never reached in this case.

Cavity dumping

In this approach, laser oscillation is allowed to build up in a laser cavity with no output coupling. The low losses of such a cavity ensure that steady-state lasing is achieved with the population inversion burnt down to very low values by the large intra-cavity intensity. The very high circulating power can then be switched out of the cavity using an electro-optic switch in a similar arrangement to that shown in Fig. 8.4, or with an acousto-optic modulator, as illustrated in Fig. 8.22, to yield an output pulse with a duration approximately equal to the time taken for one cavity round trip—typically in the nanosecond range.

Cavity dumping can be driven repeatedly. With a proper choice of the repetition rate the mean output power can approach the mean output obtained when the laser is operated continuously with the optimum output coupling; what cavity dumping brings, however, is a much higher peak power. It is especially useful for lasers that have a short upper-level lifetime since it is difficult to achieve efficient Q-switching in this case. For example, cavity dumping is often used to generate short (approximately 10 ns) laser pulses in continuously pumped argon-ion lasers.

Cavity dumping can also be used to switch out single, or trains, of mode-locked pulses. This has the advantage that the repetition rate of the output is determined by the repetition rate of the cavity dumper rather than the (much higher) frequency of the output of the modelocked pulses, which is equal to the reciprocal of the cavity round-trip time. For example, typical dumper frequencies might be 100 kHz–1 MHz, slow compared to the usual modelocking frequency of 10s or 100s of MHz.

Finally, we note that it is useful for the cavity dumping to be less than 100% efficient; then the weak remaining intra-cavity radiation can seed the

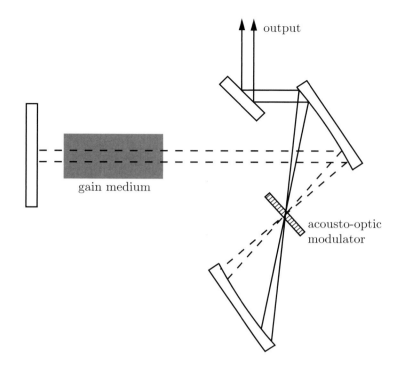

Fig. 8.22: Cavity dumping with an acousto-optic modulator. With the modulator turned off, the circulating laser radiation takes the dotted path shown. When the modulator is switched on, Bragg reflection from the acoustic waves deflects the radiation to the path, indicated by the solid lines, thereby switching the radiation out of the optical cavity. The two concave mirrors in the cavity ensure that the beam is brought to a focus at the modulator, which decreases the switching time.

next pulse, which helps to ensure that the subsequent output is in a single transverse mode.

Further reading

A more advanced treatment of modelocking may be found in the book by Siegman (1986). A review of the application of SESAMs to generating Q-switched and modelocked pulses from solid-state lasers is provided by Keller et al. (1996).

For further information on techniques for generating short laser pulses, see Chapter 17 and the reviews by Ostendorf (2004) and Reid (2004).

Exercises

(8.1) Q-switching

(a) What is meant by the term *Q-switching*? Include in your discussion sketches of the gain, loss and photon flux as functions of time for a Q-switched laser, as well as a description of the elements needed for such a laser, and a sketch of the cavity layout

(b) Write down the rate equations for the population-inversion density and photon density that are satisfied immediately after the Q-switch is opened and hence show that the energy that is extracted in the Q-switched pulse is given by eqns (8.31) and (8.32). Interpret this last result for the cases of: (i) an ideal four-level laser; (ii) severe bottlenecking.

(c) Show that the photon density at the peak of the Q-switched pulse is given by eqn (8.38) and hence derive approximate expressions for the pulse build-up time and the duration of the Q-switched laser pulse.

(d) The cavity of a Q-switched Nd:YAG laser is formed by two mirrors with reflectivities $R_1 = 0.98$ and $R_2 = 0.80$, separated by 0.3 m. The laser rod is 50 mm long, and when lasing the diameter of the laser beam is 4 mm throughout the gain medium. The wavelength and optical gain cross-section of the laser transition are 1064 nm and 10^{-19} cm^2, respectively, and the degeneracies of the upper and lower laser levels may be taken to be equal.
Calculate the cavity lifetime, τ_c and the threshold population inversion for this laser.

(e) Calculate the overpumping ratio required to extract 90% of the maximum possible energy that can be extracted from the population inversion, assuming that the lower laser level is strongly bottlenecked on the time scale of the Q-switched pulse.

(f) Hence, calculate the build-up time and the energy, duration, and peak power of the Q-switched pulse.

(g) Discuss how the build-up time and the duration of the Q-switched pulse vary with the overpumping ratio r. How does the energy of the Q-switched pulse vary with r in this limit?

(8.2) Relaxation oscillations
For the following laser system, discuss whether relaxation oscillations are likely to occur, and, if so, estimate their frequency:

(a) A Nd:YAG laser. Take the parameters of the laser cavity to be as in Exercise 8.1 and the lifetime of the upper laser level to be 230 μs.

(b) A monolithic, non-planar Yb:YAG ring laser. In this device a ring cavity is formed by internal reflections within a block of YAG of refractive index 1.82. The perimeter of the ring is approximately $L_p = 15$ mm;
for simplicity the cavity may be treated as a simple linear cavity of length $L_p/2$ with mirror reflectivities of 100% and 98%. The lifetime of the upper laser level is 1.16 ms.[36]

(c) An argon-ion laser. Take the cavity to be formed by mirrors with reflectivities of 99% and 95%, separated by 1 m. Assume that the lifetime of the upper laser level is 10 ns.

(d) A semiconductor diode laser. In such systems the cavity is formed by the Fresnel reflections[37] from the cleaved end faces of the semiconductor crystal, and has a length typically equal to 250 μm. Assume that the lifetime of the upper laser level is 1 ns, and that the refractive index of the semiconductor is equal to 3.6.

(8.3) Modelocking: more general treatment
We consider a more general case in which the frequencies and phases of the cavity modes are given by[38]

$$\omega_p = \omega_{ce} + p\Delta\omega$$
$$\phi_p = \phi_0 + p\Delta\phi \qquad p = 0, 1, 2, 3, \ldots.$$

(a) Show that if N modes oscillate with $p = q_0, q_0 + 1, q_0 + 2, \ldots$ then at $z = 0$ the amplitude of the positive-going wave may be written

$$E(0, t) = \exp\left[-i\left(\omega_{ce}t - \phi_0\right)\right]$$
$$\times \sum_{p=q_0}^{q_0+N-1} a_p \exp\left[-ip\left(\Delta\omega t - \Delta\phi\right)\right].$$

(b) By introducing a new time variable $t' = t - \Delta\phi/\Delta\omega$, show that the amplitude of the wave may be written as

$$E(0, t) = \exp\left\{-i\left[\omega_{ce}\left(t' + \frac{\Delta\phi}{\Delta\omega}\right) - \phi_0\right]\right\}$$
$$\times \sum_{p=q_0}^{q_0+N-1} a_p \exp\left[-ip\Delta\omega t'\right]. \qquad (8.67)$$

---

[36]These parameters are similar to the laser used to yield the data of Fig. 16.7. Consult this figure and the associated text.
[37]See, for example, Brooker (2003).
[38]See also Exercise 17.2.

(c) Show that this last result may be rewritten as

$$E(0, t) = \exp(-i\omega_{ce}t') \exp\left\{i\left[\phi_0 - \frac{\Delta\phi}{2\pi}\omega_{ce}T_c\right]\right\}$$

$$\times \sum_{p=q_0}^{q_0+N-1} a_p \exp\left[-i2\pi p\left(\frac{t'}{T_c}\right)\right],$$

where $\Delta\omega = 2\pi/T_c$.

(d) Use this result to show that the separation between adjacent modelocked pulses is T_c and that for successive pulses the phase of the carrier wave relative to the pulse envelope shifts by

$$\phi_{\text{slip}} = \omega_{ce}T_c = 2\pi\left(\frac{\omega_{ce}}{\Delta\omega}\right).$$

[If you wish, you may assume that all the modes have the same amplitude a_p.]

(e) Summarize the effects on the train of modelocked pulses of finite values of ω_{ce}, ϕ_0, and $\Delta\phi$.

(8.4) **Modelocking with a Gaussian spectral distribution**
Here, we extend the analysis of the previous problem to the more realistic case in which the locked modes have a Gaussian spectrum, that is the amplitudes of the oscillating modes are given by,

$$a_p = \exp\left\{-\left[\frac{(p-P_0)\Delta\omega}{\Delta\omega'}\right]^2\right\},$$

where P_0 is the number of the centre mode. To avoid clutter, we will assume also that the modes all oscillate with the same phase ϕ_0. i.e. that $\Delta\phi = 0$.

(a) In terms of $\Delta\omega'$, find the full width at half-maximum, $\Delta\omega_{\text{FWHM}}$, of the *power* spectrum of the oscillating modes.

(b) Using eqn (8.67), show that at the point $z = 0$ the amplitude of the positive-going wave is given by,

$$E(0, t) = \exp\{-i[\omega_{ce}t - \phi_0]\}$$

$$\times \sum_{p=q_0}^{q_0+N-1} \exp\left(-\left[\frac{(p-P_0)\Delta\omega}{\Delta\omega'}\right]^2\right) \exp[-ip\Delta\omega t].$$

(8.68)

(c) By defining $m = p - P_0$, show that the amplitude of the wave may be written as

$$E(0, t) = f(t) \exp\{-i[(P_0\Delta\omega + \omega_{ce})t - \phi_0]\},$$

(8.69)

where

$$f(t) = \sum_m \exp\left[-\left(\frac{m\Delta\omega}{\Delta\omega'}\right)^2\right] \exp[-im\Delta\omega t].$$

Identify the envelope and carrier frequency of the wave.

(d) By approximating this sum to an integral, show that the modelocked pulse has a Gaussian envelope in time, and find the full width at half-maximum, Δt_{FWHM}, of the *intensity* profile of the modelocked pulse.

[You may require the identity:

$$\int_{-\infty}^{\infty} \exp(-a^2\omega^2) \exp(-i\omega t) d\omega$$

$$= \sqrt{(\pi/a^2)} \exp(-t^2/4a^2) \quad \text{if } \Re(a^2) > 0.\Big]$$

(e) Hence show that the **time-bandwidth product** of the modelocked pulse obeys,

$$\Delta\omega_{\text{FWHM}} \Delta t_{\text{FWHM}} = 4\ln 2.$$

(f) What bandwidth (in real frequency units, i.e. Hz) would be needed to generate a pulse of 10 as duration? Estimate the longest mean wavelength of such a pulse. [1 as = 10^{-18} s]

(8.5) **FM modelocking in limit of small modulation amplitude**

After transmission through a frequency modulator, the amplitude of a single longitudinal mode may be written in the form (eqn (8.58)),

$$E(L_{\text{mod}}, t) = E_0 \exp\left\{i\omega_p\left[n(t)\frac{L_{\text{mod}}}{c} - t\right]\right\}.$$

Here, we consider the form of the transmitted amplitude in the limit where the frequency modulation is small. We assume that the refractive index of the frequency modulator may be written as,

$$n(t) = n_0 + \frac{\Delta n}{2} \cos \Delta\omega t.$$

(a) Show that the amplitude of the transmitted wave may be written as,

$$E(L_{\text{mod}}, t) = E_0 \exp(-i\omega_p t) \exp(i\phi_0)$$

$$\times \exp(i\Delta\phi \cos \Delta\omega t),$$

and find expressions for ϕ_0 and $\Delta\phi$.

(b) Show that in the limit where the amplitude of the phase modulation is small the amplitude of the transmitted wave becomes,

$$E(L_{\text{mod}}, t) = E_0 \exp[-i(\omega_p t + \phi_0)]$$
$$\times (1 + i\Delta\phi \cos \Delta\omega t).$$

(c) Hence, show that the transmitted wave may be written in the form,

$$E(L_{\text{mod}}, t) = E_0 \exp(i\phi_0)$$
$$\times \left[\exp(-i\omega_p t) + i\frac{\Delta\phi}{2} \exp[-i(\omega_p - \Delta\omega)t] \right.$$
$$\left. + i\frac{\Delta\phi}{2} \exp[-i(\omega_p + \Delta\omega)t] \right],$$

and comment on the form of this result.

(d) Discuss what happens to the frequency spectrum of the transmitted pulse as the amplitude of the refractive-index modulation is increased.

(8.6) **Kerr-lens modelocking**

Here, we estimate the peak intensity and electric field in an intra-cavity modelocked pulse and its effect on the refractive index of the gain medium. Consider a modelocked Ti:sapphire laser producing pulses of 30 fs duration with a pulse repetition rate of 100 MHz. Suppose that the mean output power of the laser is 0.5 W.

(a) Ignoring other losses, calculate the energy of the pulses circulating within the cavity if the transmission of the output coupler is 5%. What is the peak power P of the pulses?

(b) Suppose that within the Ti:sapphire crystal the beam diameter is focused to a spot size $w = 50\,\mu\text{m}$. Find the peak pulse intensity within the crystal.

(c) Given that the non-linear refractive index of Ti:sapphire is $n_2 = 3 \times 10^{-16}\,\text{cm}^2\,\text{W}^{-1}$ (see Section 17.1.5 and Table 17.2), calculate the difference in refractive index experienced by the peak of the laser pulse on axis with that experienced in the temporal or transverse wings of the pulse. Calculate the additional phase lag resulting from the non-linear refractive index for a crystal of thickness $L_g = 4$ mm.

(8.7) †**Focal length of a Kerr lens**

Consider a Gaussian beam focused to a waist w_0 at the centre of a thin medium with a non-linear refractive index n_2. We will suppose that the thickness, L, of the medium is small compared to the Rayleigh range of the focus so that within the medium the transverse intensity profile of the beam may be taken as $I(r) = I_0 \exp[-2(r/w_0)^2]$, where r is the radial distance from the beam axis.[39]

(a) Show that near the beam axis the non-linear refractive index causes an additional phase shift

$$\phi_{\text{NL}}(r) \approx \frac{2\pi}{\lambda_0} n_2 L I_0 \left[1 - 2\left(\frac{r}{w_0}\right)^2 \right], \quad (8.70)$$

where λ_0 is the vacuum wavelength.

(b) Use the approach of Exercise 6.2 to find the radius of curvature of the wavefronts of the beam, and hence show that the focal length of the induced Kerr lens is given by,

$$f = \frac{\pi w^4}{8 n_2 P L}, \quad (8.71)$$

where P is the power of the beam.

(c) Calculate f for the parameters of Exercise 8.6.

---

[39]The case of thick non-linear media, and corrections to the formula for the focal length derived in this problem, are considered by Sheik-Bahae et al. (1991).

9 Semiconductor lasers

9.1 Basic features of a typical semiconductor diode laser 226
9.2 Review of semiconductor physics 228
9.3 Radiative transitions in semiconductors 235
9.4 Gain at a p-i-n junction 236
9.5 Gain in diode lasers 238
9.6 Carrier and photon confinement: the double heterostructure 241
9.7 Laser materials 243
9.8 Quantum-well lasers[†] 244
9.9 Laser threshold 247
9.10 Diode laser beam properties 250
9.11 Diode laser output power[†] 257
9.12 VCSEL lasers[†] 259
9.13 Strained-layer lasers 261
9.14 Quantum cascade lasers[†] 262
Further reading 264
Exercises 264

[1] See Basov et al. (1959) and Dumke (1962).
[2] See Hall et al. (1962).
[3] Excellent accounts of the early days of the semiconductor laser may be found in articles by Du Puis (2004) and Hecht (2007).

[4] See Section 9.9.

There can be little doubt that semiconductor diode lasers (usually referred to simply as diode lasers) constitute the most important class of lasers—both in terms of the sheer numbers of such devices produced per year and the value of that production. This is still true despite the fact that many diode lasers produced for consumer items have a price tag of only a few pennies. Diode lasers are truly ubiquitous; they are found in all CD and DVD players and recorders, in telecommunications systems, barcode readers, and a whole range of 'smart' devices where some form of optical sensing is employed. Moreover, because they can convert electrical energy into visible or near-visible radiation with high efficiency (typically $\gtrsim 50\%$) they are increasingly important in providing the pump radiation for many types of c.w. and pulsed solid-state lasers that originally relied on flashlamp excitation.

9.1 Basic features of a typical semiconductor diode laser

Although the idea of laser action in semiconductor materials operating on transitions between electrons in the conduction band recombining with holes in the valence band goes back to the earliest days of the laser[1] it took the best part of a decade from the first demonstration[2] of laser action in a semiconductor until reliable devices were finally developed.[3]

Early devices could only be operated in a pulsed mode at cryogenic temperatures, and had very short operating lifetimes before the propagation of dislocations led to device failure. Modern semiconductor lasers operate at room temperature with lifetimes in excess of 100 000 h. Because they are low-voltage devices they tend to produce less electrical noise and represent a lower electrical hazard than flashlamps, but the converse is that they are much more susceptible to electrical damage by static than their gas-discharge-based counterparts.

As shown in Fig. 9.1, the gain region of a typical diode laser is formed on a semiconductor chip at the junction between a p-type region and an n-type region some 100 µm wide and ~ 1 µm high. The length of the gain region L_z, may be only some 100–300 µm. This, combined with the fact that the laser cavity is usually formed by the Fresnel reflection[4] at the cleaved facets at each end of the semiconductor crystal, means that extremely high

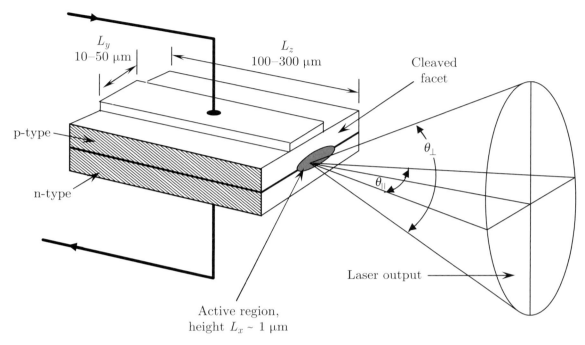

Fig. 9.1: Schematic diagram of an early edge-emitting diode laser. The device shown is a simple homostructure p-n junction; modern devices employ heterostructure architectures (see Section 9.6). Excitation current is applied by electrodes on the top and bottom surfaces. Note that the direction of current flow (the x-direction) is perpendicular to the axis of the gain length (the z-direction). Laser output diverges at full cone angles $\theta_\perp$ in the vertical (x, z) plane and $\theta_\parallel$ in the horizontal (y, z) plane. In this geometry $\theta_\perp > \theta_\parallel$.

gain coefficients (10 to 100 cm^{-1}) are required to reach the threshold for lasing. The characteristic feature of this **edge-emitting** geometry is that the direction of excitation current flow (the x-direction) is perpendicular to that of the laser output beam. The radiation profile inside the lasing cavity is laterally confined to some extent by the waveguide structure formed by the refractive-index differences between the laser gain region (higher refractive index) and the surrounding crystal (lower refractive index). The refractive-index difference is achieved by growing alloys with different compositions for the various regions.[5] Because of its letterbox shape at the end mirrors, the radiation mode leaving the laser is not a simple Gaussian mode, but has a large angular divergence in the direction parallel to L_x, the height of the lasing stripe, and a somewhat smaller divergence in the direction parallel to L_y, its width.

[5] See Section 9.7.

Although individual diode laser stripes produce powers of only a few hundred milliwatts they can be fabricated in linear arrays, and the arrays arranged in vertical stacks to form devices with total output powers of hundreds of Watts.

The wavelength at which a particular diode laser operates is, as we shall see, determined by the energy gap E_g between the valence and conduction bands of the semiconductor material of the active region, such that the laser photon energy $\hbar\omega \gtrsim E_g$.

9.2 Review of semiconductor physics

9.2.1 Band structure

The theory underlying the operation of diode lasers can only be explained in terms of the electronic band structure of semiconductors. A detailed account of band theory is beyond the scope of the present text.[6] We will attempt only to give a highly simplified outline of some of the principles of band theory relevant to diode laser operation.[7]

According to the tight-binding model of crystals[8] the overlapping of the outermost orbitals of atoms as their interatomic separations are progressively reduced from large values (where their energy levels are those of free atoms) to small values (as in the crystal) causes the initially discrete energy levels to merge and broaden into energy bands. At absolute zero all the bands are either empty or fully occupied: the highest-lying occupied band is known as the **valence band**, and the band above this in energy—which at zero temperature will be empty—is called the **conduction band**, and the next lowest is the valence band. Semiconductors are materials characterized by having an energy gap separating the highest part of the valence band from the lowest part of the conduction band of about $0.5 - 3.5$ eV.

The different bands are labelled by the subscript j. Electrons in each of these bands are subject to a periodic potential due to the regular spacings of the ions, and their wave functions—known as **Bloch wave functions**—may be written as

$$\psi_{j,\mathbf{k}}(\mathbf{r}) = \exp(i\mathbf{k} \cdot \mathbf{r})\, u_{j,\mathbf{k}}(\mathbf{r}), \tag{9.1}$$

where $\mathbf{k}$ is the wave vector of the wave function and the function $u_{j,\mathbf{k}}(\mathbf{r})$ reflects the underlying periodicity of the lattice and is therefore unchanged by a translation $\mathbf{T}$ so that:

$$u_{j,\mathbf{k}}(\mathbf{r} + \mathbf{T}) = u_{j,\mathbf{k}}(\mathbf{r}), \tag{9.2}$$

where

$$\mathbf{T} = n_1 \mathbf{a_1} + n_2 \mathbf{a_2} + n_3 \mathbf{a_3}, \tag{9.3}$$

and $\mathbf{a_1}, \mathbf{a_2}, \mathbf{a_3}$ are primitive lattice translation vectors.

The energy associated with $\psi_{j,\mathbf{k}}(\mathbf{r})$ *within* each band j is a continuous function of $\mathbf{k}$, but the map of $E_j(\mathbf{k})$ for any particular semiconductor material is a very complicated one. The example shown in Fig. 9.2 is part of the diagram of $E_j(k)$ for the compound GaAs, for two particular directions of $\mathbf{k}$. The case of GaAs is typical of the sort of semiconductor most frequently used for optoelectronic devices because it is an example of a **direct-gap** semiconductor, i.e. one in which the minimum in the conduction-band energy occurs at the same value of $\mathbf{k}$ as the maximum in the valence-band energy.[9] This means that photons can be absorbed or emitted without the simultaneous exchange of energy from the lattice vibrations in the form of phonons. As shown in Fig. 9.2 there are three branches of the valence band that come close to providing the minimum energy gap in the neighborhood of the Γ-point (the origin 0,0,0) in k-space.

---

[6] The reader will find suitable introductions to this topic in texts such as those by Singleton (2001), Fox (2001), or Rosenberg (1989).

[7] For a more detailed account of the physics of diode laser operation the reader is referred to texts by Coldren and Corzine (1995), Khan et al. (2001), or Dutta (2004).

[8] See, for example, the book by Singleton (2001, Chapter 4).

[9] Optical transitions in semiconductors for which the conduction-band-minimum does not occur at the same k value as the valence-band maximum in the Brillouin zone are called indirect transitions and involve the assistance of a phonon to balance energy and momentum overall. Such indirect transitions generally occur with much lower transition probability than transitions in direct-gap semiconductors (see, for example, Fig. 3.8 of Fox (2001)) and are not of interest for the semiconductor laser devices described in the present chapter. Further, the matrix element for an optical transition (see Section 9.5) can be represented as $\int \phi_2^* \mathbf{r} \cdot \mathbf{E}_0 \exp(i\mathbf{k}_{\text{opt}} \cdot \mathbf{r})\phi_1 d\tau$. Writing the electron wave functions in the Bloch form, this becomes $\int u_2^* \exp[i(\mathbf{k}_{\text{opt}} - \mathbf{k}_2 + \mathbf{k}_1) \cdot \mathbf{r}]\mathbf{r} \cdot \mathbf{E}_0 u_1 d\tau$. The integrand will oscillate rapidly, and hence sum to zero, unless $\mathbf{k}_{\text{opt}} - \mathbf{k}_2 + \mathbf{k}_1 \approx 0$; given that $\mathbf{k}_{\text{opt}}$ is very small, this condition becomes $\mathbf{k}_2 \approx \mathbf{k}_1$. Thus, all the transitions of interest in the remainder of this chapter will be represented by a vertical arrow in E versus k diagrams.

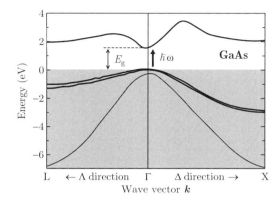

Fig. 9.2: A highly simplified version of part of the diagram of energy versus k for GaAs near the band gap. By convention, the zone centre (0,0,0) in k-space is labelled as the Γ-point. For GaAs (which has a zinc-blende lattice structure) the X- and L-points are at $k = \frac{2\pi}{a}(1, 0, 0)$ and $\frac{\pi}{a}(1, 1, 1)$, respectively. The right-hand half of the diagram shows the variation of energy with k in the Δ-direction (connecting the Γ point to the X-point in k-space) while the left-hand half shows the variation along the Λ-direction (connecting the Γ-point to the L-point). The valence bands, which at absolute zero are full of electrons, are indicated by the shaded part of the diagram below the Fermi level. Energies are expressed in eV with respect to the Fermi level.

If we were dealing with *free* electrons, familiar quantum mechanics would predict that their energy $E_{\text{free}}(k)$ is a simple parabolic function of k:

$$E_{\text{free}}(k) = \frac{p^2}{2m_e} = \frac{\hbar^2 k^2}{2m_e}, \quad (9.4)$$

where m_e is the electron mass and $\boldsymbol{p} = \hbar\boldsymbol{k}$ is the momentum of the free electron.

However, in the case of electrons moving in a periodic potential the dependence of $E_j(k)$ on k is not a simple parabolic one. Indeed there are values of $\boldsymbol{k}$ for which no wave functions exist, corresponding to the gaps in energy between the various bands. The reason for this is that we *cannot*, in general, regard $\hbar\boldsymbol{k}$ as 'the momentum of an electron', rather we should regard it as the momentum of the electron/crystal system as a whole and think of k and j as quantum numbers that describe a particular state of the composite electron/crystal system.

For GaAs (and many other direct-gap-semiconductors) the dependence of $E_c(k)$ in the conduction band can be approximated, as shown in Fig. 9.3, to a parabolic function of k:

$$E_c(k) = E_{c0} + \frac{\hbar^2 k^2}{2m_c^*}, \quad (9.5)$$

where we denote the conduction band by $j = c$, E_{c0} is the energy at the bottom of the conduction band and m_c^* is the effective mass[10] of an electron in the conduction band. In the valence band (denoted by $j = v$), there are three branches of the $E_v(k)$ versus k curve in the neighbourhood of the minimum energy gap. All three of these curves are approximately parabolic. However, in this case the dependencies are inverted, so that

$$E_v(k) = E_{v0} - \frac{\hbar^2 k^2}{2|m_v^*|}, \quad (9.6)$$

with each branch having its own characteristic value of the effective electron mass m_v^*.

[10] In quantum mechanics the velocity of a particle is given by the group velocity $v_g = \frac{1}{\hbar}\frac{dE}{dk}$, and the acceleration by $\frac{dv_g}{dt} = \frac{1}{\hbar}\frac{d}{dt}(\frac{dE}{dk}) = \frac{1}{\hbar}(\frac{dk}{dt})\frac{d}{dk}(\frac{dE}{dk}) = \frac{1}{\hbar}(\frac{dk}{dt})(\frac{d^2E}{dk^2})$. Thus, if an electron is subject to an external force F (in addition to the internal crystal fields that give rise to the E vs. k dependence) the rate of work done on the electron will be given by $\frac{dE}{dt} = Fv_g$. Hence, we may write Newton's second law, $F = m^*\left(\frac{dv_g}{dt}\right)$, in the form $\frac{1}{v_g}\left(\frac{dE}{dt}\right) = m^*\left(\frac{dv_g}{dt}\right)$ from which we find $\frac{1}{v_g}\left(\frac{dE}{dk}\right)\left(\frac{dk}{dt}\right) = m^*\frac{1}{\hbar}\left(\frac{dk}{dt}\right)\left(\frac{d^2E}{dk^2}\right)$. Hence, the quantity that plays the part of mass in the normal statement of Newton's second law is $m^* = \hbar^2\left(\frac{d^2E}{dk^2}\right)^{-1}$. This quantity can be taken as the definition of m^*, the effective mass of the electron when under the influence of both the external field and the various internal fields of the crystal. It is the curvature of the E vs. k curve at the relevant point in k-space that determines the effective mass—the sharper the curve the lower the effective mass. Note that for *electrons* with energies near the top of the valence band, m_v^* is therefore an intrinsically negative quantity.

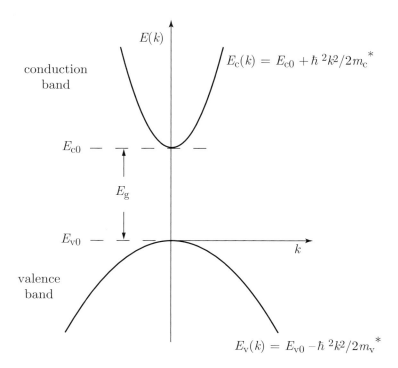

Fig. 9.3: Schematic diagram of the energies of the conduction and valence bands of a direct-gap semiconductor in the neighbourhood of the band gap. Note that, unlike some authors, we define increasing energy to be in the same direction for the conduction and valence bands.

Now, as indicated in Fig. 9.2, in the case of GaAs, the states within the valence band would normally be completely filled with electrons. A band that is completely full cannot contribute to the conduction of current since for every electron that has a given value of k_z there is another one with $-k_z$. If we now consider the situation where the valence band is full except for a vacancy for an electron in state l then, because of the shape of the $E_v(k)$ curve given by eqn (9.6), all of the remaining electrons in the band experience an acceleration in the direction of an applied electric field (instead of in the opposite direction, as would be the case for free electrons). The vacancy in the occupied states at state l also moves in the *same* direction as the rest of the electrons. It is therefore convenient to think of the carriers in this case as holes that have a positive charge equal in magnitude to the electron charge and a positive mass equal in magnitude to the effective mass of the electrons in the band.

The uppermost of the three bands comprising the valence band in Fig. 9.2 is called the **heavy-hole band** and the effective mass of holes in this band is denoted by m_{hh}^*. The next band, just slightly lower in energy is called the **light-hole band** and holes in this band have effective mass m_{lh}^*. The lowest of the three bands is the **split-off band**, holes in this band having an effective mass m_{so}^* For simplicity, in the discussion that follows we will represent the three branches of the valence band as a single parabolic curve, with carriers of an appropriately averaged effective mass m_v^*.

9.2.2 Density of states and the Fermi energy ($T = 0K$)

Now, for the moment we will assume we are dealing with a sample at a temperature of absolute zero. Suppose that we could, somehow, add an extra density of electrons n_e to our idealized semiconductor. As indicated in Fig. 9.4 the valence band is already full, (electrons are fermions and hence obey the Pauli exclusion principle) so the extra electrons will start to fill up the available states in the conduction band from the bottom upwards.

Again, because of the Pauli exclusion principle the electrons fill the available states up to a maximum energy E_F, which depends on n_e. To calculate the relationship between the Fermi energy E_F and the excess electron density n_e we have to know how many available states there are between the energy at the bottom of the conduction band E_{c0} and the highest filled state at E_F. This, in turn, depends on knowing the density of states for the conduction band, $g_c(k)$, which is defined as the number of states per unit volume with a value of k lying between k and $k + dk$.

The calculation of $g_c(k)$ is analogous to the calculation of the density of blackbody radiation modes.[11] Since electrons have a spin of one-half, two electrons can occupy a state with a given spatial wave function, and hence a factor of two arises in the calculation of the density of states in the same was as the factor of two arising from the two orthogonal polarizations of the radiation field. When described in terms of k, the density of electron states is therefore the same as the density of radiation modes

[11] See eqn. (1.25).

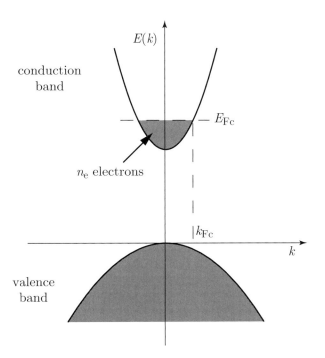

Fig. 9.4: When an extra density n_e is added to a sample of GaAs maintained at absolute zero temperature the conduction band is filled up to a maximum energy E_{Fc}.

$$g_c(k)\,\mathrm{d}k = \frac{k^2}{\pi^2}\,\mathrm{d}k. \tag{9.7}$$

The density of states may be rewritten in terms of energy using eqn (9.5):

$$g_c(E)\,\mathrm{d}E = g_c(k)\frac{\mathrm{d}k}{\mathrm{d}E_c}\,\mathrm{d}E_c = \frac{1}{2\pi^2}\left(\frac{2m_c}{\hbar^2}\right)^{\frac{3}{2}}(E_c - E_{c0})^{\frac{1}{2}}\,\mathrm{d}E. \tag{9.8}$$

The Fermi energy ($T = 0\,\mathrm{K}$)

We may calculate the Fermi energy by using the fact that the n_e electrons per unit volume will occupy all the states with k values up to some maximum $k = k_F$, where $E_F = \hbar^2 k_F^2 / 2m_c^*$. This condition may be written as,

$$n_e = \int_0^{k_F} g_c(k)\,\mathrm{d}k = \int_0^{k_F} \frac{k^2}{\pi^2}\,\mathrm{d}k = \frac{k_F^3}{3\pi^2}. \quad (T = 0\,\mathrm{K}) \tag{9.9}$$

Hence, we can write Fermi k-vector in terms of n_e as

$$k_F = (3\pi^2 n_e)^{\frac{1}{3}}. \quad (T = 0\,\mathrm{K}) \tag{9.10}$$

Having found the Fermi k-vector, it is straightforward to find the Fermi energy from eqn (9.5):

$$E_{Fc} = E_{c0} + \frac{\hbar^2}{2m_c^*}\left(3\pi^2 n_e\right)^{\frac{2}{3}}. \quad (T = 0\,\mathrm{K}) \tag{9.11}$$

9.2.3 The Fermi–Dirac distribution ($T \neq 0\,\mathrm{K}$)

We now turn to the question of how the electron distribution is altered if we raise the temperature above 0 K. Because electrons are fermions they distribute themselves across the lowest available energy states according to the Fermi–Dirac distribution function $f(E)$ that gives the probability that an electron occupies a state with energy between E and $E + \mathrm{d}E$ as

$$f(E) = \frac{1}{\exp\left(\frac{E - E_F}{k_B T}\right) + 1}. \tag{9.12}$$

The Fermi–Dirac function is shown in Fig. 9.5(a). As discussed in Section 9.2.2, for the case of $T = 0\,\mathrm{K}$, the function $f(E) = 1$ for $E < E_F$, and $f(E) = 0$ for $E > E_F$, i.e. all states up to an energy of E_F are filled and those with energies greater than E_F are empty. At finite temperatures the function $f(E)$ falls below 1 in the region near $E = E_F$, reaching $f(E_F) = \frac{1}{2}$ at the Fermi energy.[12] For energies greater than E_F the value of $f(E)$ tends to zero in a backwards S-shaped curve that is approximately symmetric about $E = E_F$.

Figure 9.5(b) shows the density of states $g_c(E)$, i.e. eqn (9.8). The distribution of population over states of energy E is given by multiplying the density of states, $g_c(E)$, by the probability of occupation, $f(E)$, as shown in Fig. 9.5(c).

[12]This is therefore a more general way of describing the significance of the Fermi energy at any temperature—it is the energy of those states for which the probability of occupation is equal to $\frac{1}{2}$.

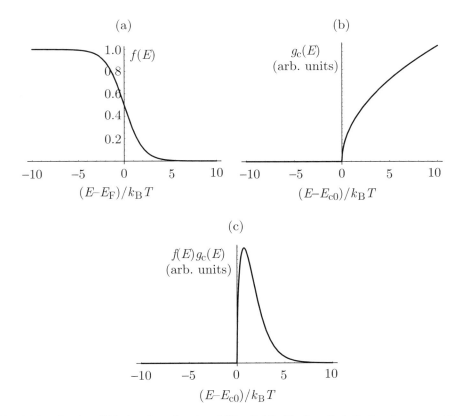

Fig. 9.5: (a) The Fermi–Dirac function $f(E)$ plotted as a function of $(E - E_F)/k_B T$, where E_F is the Fermi energy and T the temperature, for $T > 0$ K; (b) the density of states $g(E)$ (arbitrary units) plotted as a function of $(E - E_{c0})/k_B T$, where E_{c0} is the energy at the bottom of the band; (c) the population distribution $dn_e = g(E)f(E)dE$ as a function of $(E - E_{c0})/k_B T$; for convenience we have taken the particular case when the density of electrons introduced into the conduction band is such that the Fermi level is at the bottom of the band (i.e. $E_F = E_{c0}$).

9.2.4 Doped semiconductors

Lightly-doped semiconductors

In an ultrapure[13] sample of a semiconductor compound such as GaAs the number of electrons excited across the energy gap up into the conduction band at room temperature is very small,[14] and the corresponding intrinsic electrical conductivity would therefore be very small indeed. As indicated in Fig. 9.6(a) for such an ultrapure semiconductor the Fermi energy would lie approximately[15] half-way up the energy gap between the top of the valence band and the bottom of the conduction band, corresponding to the energy for which the probability of occupation is $\frac{1}{2}$.

Since the number of electrons excited thermally across the gap is so insignificant, for virtually all practical applications, and especially for optoelectronic devices, higher concentrations of electrons in the conduction band are required. This is achieved by adding carefully controlled concentrations of particular doping species to the bulk semiconductor compound. Now, GaAs is a III-V compound, so-called because Ga is in group III and As is in group V of the

[13] By 'ultrapure' we mean that fewer than 1 in 10^{12} atoms are impurity atoms Rosenberg (1989).

[14] The thermally excited population density of electrons is only $10^6 - 10^7$ cm^{-3} at 300 K in the case of GaAs, for which $E_g = 1.43$ eV. This compares with the value of $n_e \approx 10^{22}$ cm^{-3} in the conduction band of a good conductor such as copper.

[15] The precise position of the Fermi energy depends on the ratio m_c^*/m_v^* and the temperature T.

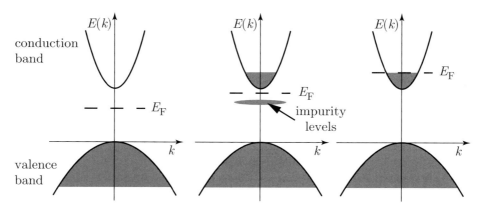

Fig. 9.6: Occupancy of conduction and valence bands in a semiconductor. In (a) the idealized case of an ultrapure semiconductor is shown; in this case the Fermi energy lies in the middle of the energy gap. The effect of introducing a *small* density of an electron donor impurity species is shown in (b). Now, the presence of impurity levels just below the bottom of the conduction band causes the Fermi energy to lie between the impurity levels and the bottom of the conduction band. The effect of introducing a *large* density of electron donor species results in the lower levels of the conduction band being filled with electrons, as shown in (c). The Fermi energy now lies well within the conduction band.

[16] Si belongs to group IV and substitutes for atoms originally on Ga sites. In the case of n-type GaAs, it is Si that is used most commonly.

[17] Te belongs to group VI and substitutes for atoms originally on As sites.

periodic table. Crystals of GaAs have the face-centred cubic structure of zinc-blende (ZnS) type with exactly equal numbers of Ga atoms and As atoms per unit volume. By adding *small* proportions of Si atoms[16] or Te atoms[17] we introduce a new energy level arising from the impurity species that lies just below the conduction band of the host material, as indicated schematically in Fig. 9.6(b). Atoms of these particular added species (Si or Te) have one more electron in their outer shells than the atoms they replace. They can therefore act as donor species by providing a source of electrons that go into the conduction band because the gap between the impurity level and the conduction band of the host material is small and the impurity atoms are easily ionized at room temperature. This type of material, with an excess of electrons in the conduction band is called n-type, and at *light* doping levels provides a number density of electrons in the conduction band that, although much larger than in the case of the intrinsic undoped material, nevertheless depends strongly on temperature and tends to zero at low temperatures. In this case, the Fermi energy lies between the impurity level and the bottom of the conduction band.

Degenerate semiconductors

By contrast, if the same n-type material is *heavily* doped the impurity level merges into the conduction band, and the Fermi level moves up into the conduction band to sit close to the top of the occupied states in the conduction band, as indicated in Fig. 9.6(c). In an analogous way, by doping GaAs with electron acceptor atoms[18] such as Be, Zn or C, p-type materials with excess holes in the valence band can be produced and at heavy doping levels the Fermi level of such materials moves down into the valence band.

[18] Be belongs to group IIA, Zn belongs to group IIB, and C belongs to group IVB.

Such heavily doped materials are examples of **degenerate semiconductors**. Since they have partially filled bands (like metals) they remain electrically conducting even at very low temperatures. Such heavily doped materials (both n-type and p-type) are important components in the construction of semiconductor laser devices, and throughout the remainder of this chapter we will assume that any doped materials are degenerate semiconductors.

9.3 Radiative transitions in semiconductors

There are real differences between the theory describing radiative transitions of free atoms (and molecules) on the one hand and those of semiconductors on the other. The differences arise because in the case of free atoms (and molecules) we are dealing with transitions from an initial energy level that is *occupied* by an electron to a final state of the same atom that is initially *empty* because the electron in a particular atom initially has to be in one state or the other. The probability of spontaneous emission for an excited atom can be described by an Einstein A coefficient that is simply a constant for the transition in question. It does not depend on the population of the lower level in other atoms of the same species elsewhere in the assembly.[19]

In the case of semiconductors (or metals) it is not enough that a population exists in a particular state in the conduction band for spontaneous emission to occur; there must also be a vacancy (i.e. a hole) at the same value of k in the valence band. For example, if we take the case of the degenerately doped n-type material illustrated in Fig. 9.6(c) there is plenty of population in the excited states in the conduction band, but the sample does not continuously emit spontaneously since the valence band is full and if an electron tried to make a transition to a state in the valence band it would find nowhere to accommodate it.

Owing to this fundamental difference it is not possible to describe the population dynamics in semiconductors with the same Einstein treatment we applied to the case of free atoms and molecules, and we must start to examine the conditions for radiative transitions from first principles. This would take us beyond the scope of the present text, but there are some conclusions we can draw without the need for a full treatment. For example, we expect the net rate $R_{c \to v}$ of stimulated transitions from a level in the conduction band to one in the valence band to be proportional to the product of three factors:

- $\varrho(\omega)$, the radiation density in the appropriate frequency range;
- $f_c(\boldsymbol{k})$, the probability that a state (at a particular value of the k-vector) in the conduction band *is* occupied;
- $[1 - f_v(\boldsymbol{k})]$, the probability that a level in the valence band (at the *same* value of the k-vector) is *not* occupied.

Hence,

$$R_{c \to v} = B'_{cv} f_c(\boldsymbol{k})[1 - f_v(\boldsymbol{k})]g_r(E)\varrho(\omega)\mathrm{d}\omega, \quad (9.13)$$

where $\varrho(\omega)\mathrm{d}\omega$ is the energy density of radiation in the range ω to $\omega + \mathrm{d}\omega$, and B'_{cv} is a coefficient describing the stimulated emission process. Analogous considerations lead to a similar expression for $R_{v \to c}$, the rate of absorption of photons:

$$R_{v \to c} = B'_{vc} f_v(\boldsymbol{k})[1 - f_c(\boldsymbol{k})]g_r(E)\varrho(\omega)\mathrm{d}\omega. \quad (9.14)$$

[19] While effects such as radiation trapping (see Section 4.5.3) do indeed affect the overall rate of decay of population for the assembly of atoms as a whole, and do depend on the lower-level population of atoms throughout the assembly, they do not determine the rate of spontaneous emission of individual atoms.

[20] For a more rigorous derivation of the transition probabilities we should take account of the fact that Fermi's golden rule assumes that the electron initially occupies a single state, and makes a transition to one of a large number of final sates. In a semiconductor both final *and* initial states of the electron belong to a dense manifold of states, and hence we need to calculate the density of transition pairs of initial and final states (with the same transition energy between them) per unit transition energy range. For a discussion of this point the reader is referred to Chapter 4 and Appendix 6 of the book by Coldren and Corzine (1995).

[21] Or from Section A.3.

[22] This is not the only possibility. We could use doped n-type material for the active region, cladding it with an n-type material of higher energy gap on one side and a p-type material on the other. Alternatively, we could use a p-type material as the active layer material cladding it with a p-type material of higher energy gap on one side and an n-type material on the other. Such layers of n- and p-type materials of high energy gap are denoted as N and P layers, respectively.

[23] This constitutes a population inversion corresponding to that required in atomic or molecular laser systems.

The factor $g_r(E)$ appearing in eqns (9.13) and (9.14) is the so-called **joint density of states**.[20] The role played by N_2, the population density of the upper level, in the definition B_{21} for atoms (see Section 2.1) is played in eqn (9.13) by the 'density of stimulated emission transitions per unit energy range'—the product of $f_c(k)[1 - f_v(k)]$ with the density of states $g_r(E)$. Here, it suffices to use an appropriately averaged value of $g_r(E)$ defined such that $\frac{1}{g_r} = \frac{1}{g_c} + \frac{1}{g_v}$, as discussed in Section 9.5.

Although we have not space here to give a detailed proof it should not appear unreasonable that the constants of proportionality B'_{cv} and B'_{vc} appearing in eqns (9.13) and (9.14) should be equal since we are dealing with processes that in the analogous case of atomic systems are those of stimulated emission and absorption. The equality follows from the same arguments used in Section 2.1.1 to find the relationships between the Einstein B coefficients for atoms.[21]

9.4 Gain at a p-i-n junction

In a bulk sample of semiconductor (even a heavily doped one) in thermal equilibrium at a finite temperature it is clear from Fig. 9.7 that there is no possibility that the rate of stimulated emission could exceed the absorption rate since although there may be a large population of electrons in the conduction band there are no corresponding vacancies in the valence band. For this to happen we need (i) to consider a junction region between n-type and p-type degenerate semiconductors and (ii) to induce a non-equilibrium situation by injecting electrons and holes from an outside source. For simplicity [22] we will consider a scheme in which a narrow layer of intrinsic semiconductor is sandwiched between degenerately doped sections of p-type and n-type material as indicated in Fig. 9.7(a) to form a p-i-n junction device (i.e. p-type/intrinsic/n-type material layers).

The way that the electron populations are distributed in the three regions is shown in Fig. 9.7(b) for the case of no applied bias voltage. The Fermi levels of the two degenerately doped regions align themselves to be equal so that there is no net transport of electrons across the gap—i.e. no net current flow. As is evident from Fig. 9.7(b), in none of the three regions is there an electron population in the conduction band *and* a population of holes in the valence band. However, if a sufficiently large forward voltage bias is applied across the junction, then the whole band structure in the p-type region moves downward as shown in Fig. 9.7(c) and the Fermi levels (strictly, the quasi-Fermi levels since this is a non-equilibrium situation) of the conduction and valence bands now bend across the intrinsic region to connect up with the unperturbed values in the bulk regions some distance away from the junction.

Within the intrinsic layer there are now places where a significant electron population exists in the conduction band at the same point as a significant population of holes in the valence band,[23] and the transitions corresponding to the recombination of electrons and holes in this region can be stimulated to provide optical gain at the wavelength of the recombination emission.

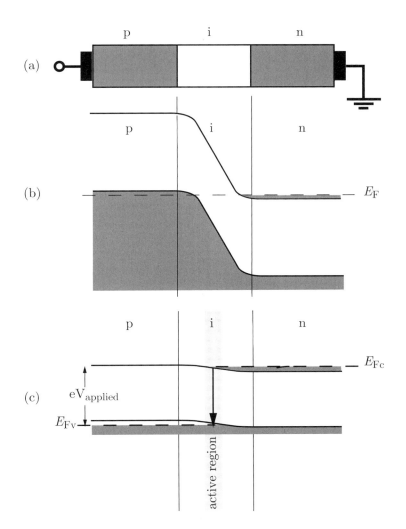

Fig. 9.7: (a) p-i-n diode structure. (b) Occupancy of the valence band of p-type material and conduction band of n-type material of a p-i-n diode with no applied bias. Note that there is a single value of the Fermi energy throughout the structure. (c) When a positive bias is applied the Fermi energy of the conduction band of the p-type material is lowered relative to that of the n-type material by the applied voltage. In the active region between the two there are electrons in the conduction band together with holes in the valence band of the intrinsic material of this region.

Figure 9.8 illustrates the non-equilibrium situation that exists in the intrinsic medium when forward bias is applied. If we consider emission from an upper level at an energy of E_2 to a lower level of energy E_1, then the photon energy of the transition is given by[24]

$$\hbar\omega = E_2 - E_1. \tag{9.15}$$

Since E_2 must represent a state in the populated part of the conduction band, while E_1 must likewise represent a state in the valence band not populated by electrons then clearly

$$\hbar\omega > E_g. \tag{9.16}$$

The condition for gain to exist can be derived from eqns (9.15) and (9.16) as follows. We require:

$$\frac{R_{c \to v}}{R_{v \to c}} = \frac{f_c(E_2)[1 - f_v(E_1)]}{f_v(E_1)[1 - f_c(E_2)]} > 1, \tag{9.17}$$

[24] We remind the reader that in the convention we have adopted the direction of increasing energy is the same for both the conduction and valance bands. See Fig. 9.3.

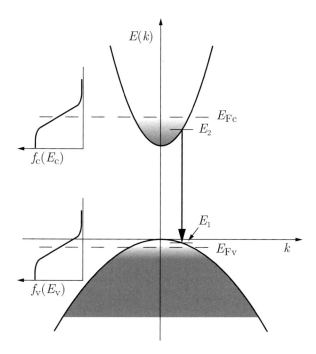

Fig. 9.8: Occupancy of the conduction band and valence band in the intrinsic region of a p-i-n diode under an applied forward bias voltage.

which implies,

$$f_c(E_2) > f_v(E_1). \tag{9.18}$$

Now,

$$f_c(E_2) = \frac{1}{\exp\left(\frac{E_2 - E_{F_c}}{k_B T}\right) + 1}, \tag{9.19}$$

and similarly for $f_v(E_1)$. Hence, the inequalities (9.16) and (9.18) yield

$$\boxed{E_{F_c} - E_{F_v} > \hbar\omega > E_g.} \quad \textbf{Bernard–Duraffourg condition} \tag{9.20}$$

This condition[25] was first derived by Bernard and Duraffourg in 1961.[26]

[25] Some authors prefer to use an energy scale referred to the top of the valence band as origin with downward as positive in the valence band and from the bottom of the conduction band upward as positive in the conduction band. The Bernard and Duraffourg condition with these definitions becomes

$$E_{F_c} + E_{F_v} > \hbar\omega - E_g.$$

[26] Bernard and Duraffourg (1961).

[27] Readers who wish to explore this subject further will find an excellent discussion in Coldren and Corzine (1995), in particular Chapter 4 and Appendices 6–10.

9.5 Gain in diode lasers

As we have noted in the preceding section, a rigorous and detailed derivation of the expression for small-signal gain of semiconductor diode lasers is beyond the scope of the present text.[27] We present here only the final results and the analogies that can be drawn to the corresponding results for atoms and molecules.

The expression for the small-signal gain coefficient $\alpha_0(\omega)$ is:

$$\alpha_0(\omega) = \frac{\lambda_0^2 \hbar}{4n^2 \tau_{sp}^{21}} g_r(\hbar\omega) \left[f_c(E_2) - f_v(E_1)\right], \tag{9.21}$$

in which the effective population difference is represented by the product of $g_r(\hbar\omega)$ the joint density of states with $[f_c(E_2) - f_v(E_1)]$ the difference of Fermi functions for the conduction and valence bands. The joint density of states is given by

$$g_r(\hbar\omega) = \frac{1}{2\pi^2}\left(\frac{2m_r^*}{\hbar^2}\right)^{3/2}(\hbar\omega - E_g)^{1/2}, \qquad (9.22)$$

which is similar to the expression of eqn (9.8) for the density of states in the conduction band, except that m_c^* (the effective mass of electrons in the conduction band) is here replaced by m_r^*—a reduced mass that includes contributions from both m_c^* and m_v^*, (the effective mass of holes in the valence band) combined as follows:

$$\frac{1}{m_r^*} \equiv \frac{1}{m_c^*} + \frac{1}{m_v^*}. \qquad (9.23)$$

The factor, n^2, (where n represents the refractive index of the medium at the laser wavelength)[28] that appears in the denominator on the right-hand side of eqn (9.21) arises from the fact that hitherto in this chapter our discussion has been in terms of the wave number k in the medium, which is related to the vacuum wavelength λ_0 by $k = 2\pi/\lambda = 2\pi n/\lambda_0$.

In eqn (9.21) all of the physics concerned with the matrix element for the transition is swept up into the factor $(\tau_{sp}^{21})^{-1}$. This is the **spontaneous rate constant**–or **two-level lifetime** value[29] for transitions from one state within the conduction band at E_2 to one in the valence band at E_1.

Given the near impossibility of calculating accurate values of τ_{sp}^{21} from theory, in practice the gain is measured as a function of excitation current and the constant K in the expression:

$$\alpha_0(\omega) = K\left(\hbar\omega - E_g\right)^{1/2}[f_c(E_2) - f_v(E_1)] \qquad (9.24)$$

is derived by curve fitting.

The behaviour of the gain as a function of applied bias voltage (and hence excitation current) can be understood from eqn (9.24). With no pumping—no injected carriers—the value of $f_c(E_2)$ is close to zero, while $f_v(E_1) \approx 1/2$ since the lower level must be somewhere in the very narrow range of unoccupied levels between E_{Fv} and the top of the valence band. The term in square brackets in eqn (9.24) is therefore negative and the semiconductor is absorbing at a photon energy of $\hbar\omega$. As the pump rate is increased, E_{Fc} moves upward into the conduction band, while E_{Fv} decreases and moves downward within the valence band. When $E_{Fc} - E_{Fv} = E_g$ the medium just becomes transparent at $\hbar\omega = E_g$.

As shown in Fig. 9.9 transparency occurs at an injected carrier density of approximately 1.5×10^{18} cm^{-3} in the case of GaAs. With further increases in the excitation current E_{Fc} moves higher and E_{Fv} lower, so that gain is available over a range of photon energies determined by eqn. (9.20). For example, according to Fig. 9.9, at a carrier density of 2.5×10^{18} cm^{-3} gain is available for photon energies ranging from 1.424 eV (the value of E_g in GaAs at room temperature[30]) up to about 1.480 eV. This corresponds to wavelengths from

[28] The value of refractive index in the spectral region of the laser transition is itself affected by the number of injected carriers and thus depends on whether the medium is absorbing, transparent or exhibits gain at and around the laser wavelength. However, the effect is typically at the 10% level, the refractive index of GaAs varying from its average value of 3.42 over the spectral region of interest to reach a maximum of 3.6 for a carrier density of 5×10^{16} cm^{-3}. The effect of this carrier-dependent refractive index manifests itself in the dynamics of the 'laser chirp'—a time-varying output frequency in pulsed lasers (see Section 17.1.3).

[29] Coldren and Corzine (1995) are careful to distinguish the constant $(\tau_{sp}^{21})^{-1}$ from the 'band-to-band' spontaneous lifetime and the inverse of the spontaneous linewidth for the overall spontaneous emission, since in a semiconductor there will be many pairs of levels within the conduction and valence bands that can contribute to spontaneous emission and gain at a given photon frequency.

[30] The value of E_g in GaAs at $T = 0$ K is 1.52 eV. The fact that the band gap depends on temperature is a consequence of the anharmonic nature of the potential versus displacement curve in which the ions of the lattice move—the same feature responsible for the thermal expansion of the material. By curve fitting experimental data the temperature dependence of E_g in GaAs is found to be $E_g(T) = E_g(0) - (5.41 \times 10^{-4}T^2/(T + 204)$, where E_g is in electron volts and T is in Kelvin. See Bart van Zeghbroek (1996).

240 *Semiconductor lasers*

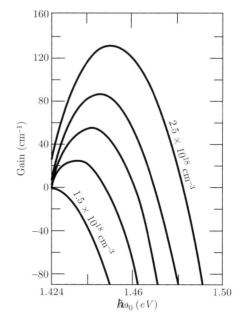

Fig. 9.9: The calculated small-signal gain coefficient (cm^{-1}) versus photon energy in GaAs at various values of injected carrier density at room temperature. The injected carrier density in the active region is directly proportional to the current density—see eqn (9.25). For injected carrier densities smaller than 1.5×10^{18} cm^{-3} the sample shows loss at all wavelengths. (After Yariv (1989).)

871 nm to 838 nm, although in practice losses within the laser cavity reduce the range of wavelengths over which the device can lase.

The expressions for the gain coefficient given in eqns (9.21) and (9.24) involve f_c and f_v, rather than the injected carrier density. To understand how f_c and f_v are related to the carrier density n_e and thus to I, the current flowing through the junction, we recall that in an intrinsic semiconductor material (such as that forming the active region of our simple p-i-n diode) the density of electrons is equal to that of holes, i.e. $n_e = n_p$. However, the contribution to the total current made by the motion of electrons is considerably greater than that due to holes because the electrons have a much greater mobility[31] than holes. In steady state, and in the absence of stimulated emission, the rate of injection of free electrons in the active region will equal their rate of loss and we can write

$$\frac{n_e}{\tau_{rec}} = \eta_{inj} \frac{I}{|e| V_{act}}, \qquad (9.25)$$

in which V_{act} denotes the volume of the active region. The injection efficiency η_{inj} accounts for the fact that a small fraction of the current is carried by holes and, more importantly, some of the current may flow through shunt paths around the active region or give rise to carriers that recombine outside the active region. The recombination lifetime τ_{rec} characterizes the total rate of decay[32] of a small excess density of electrons injected into the active region by all processes other than stimulated emission.

Having a value for n_e from eqn (9.25) in terms of I, we can now relate this to f_c, and finally to the value of E_{F_c} since from eqns (9.8) and (9.12) n_e is an implicit function of E_{F_c} via,

[31] The mobility of electrons (drift velocity per unit applied electric field) in GaAs is 0.8 m^2 V^{-1} s^{-1}, while that of holes is only 0.03 m^2 V^{-1} s^{-1}, i.e. some 27 times smaller. Thus, 96.4% of the total current is carried by electrons and only 3.6% by holes.

[32] The process of electron–hole recombination in semiconductors is mediated by 'recombination centres' or 'traps'. We are here concerned with the behaviour of a small excess electron density Δn_e in a region where the ambient electron density is n_e. It is usual to define a quantity $\tau_{rec}^{-1} = R\Delta n_e$, so that $d\Delta n_e/dt = -\Delta n_e/\tau_{rec}$. The recombination rate R is constant only over a small range of steady-state ambient electron density. Moreover, it should also be remembered that included in the value of τ_{rec} are the effects of many other loss processes for carriers (such as non-radiative decay, loss via leakage of carriers from the active volume, recombination at interfaces and even Auger processes) besides useful recombination leading to emission—see, for example, Coldren and Corzine (1995). A detailed discussion of recombination processes in semiconductors will be found in Shockley and Read (1952).

$$n_{\mathrm{e}} = \int_{E_{\mathrm{c0}}}^{\infty} g_{\mathrm{c}}(E) f_{\mathrm{c}}(E) \, \mathrm{d}E$$

$$= \frac{1}{2\pi^2} \cdot \left(\frac{2m_{\mathrm{c}}}{\hbar^2}\right)^{3/2} \int_{E_{\mathrm{c0}}}^{\infty} (E - E_{\mathrm{c0}})^{1/2} \left[\exp\left(\frac{E - E_{\mathrm{Fc}}}{k_{\mathrm{B}}T}\right) + 1\right]^{-1} \mathrm{d}E.$$

(9.26)

In a similar way we can find f_{v} and the value of $E_{\mathrm{F_v}}$ using the fact that in the intrinsic active region $n_{\mathrm{e}} = n_{\mathrm{p}}$.

9.6 Carrier and photon confinement: the double heterostructure

Side-emitting diode lasers are so-called because their gain axis is perpendicular to the direction of current flow. Early devices of this type (such as that illustrated in Fig. 9.1) were based on a simple p-n junction made by joining n- and p-doped slabs of the same semiconductor alloy. Such devices are called **homostructures** and could only be made to work at cryogenic temperatures. The lack of any way to confine the current to the gain region (and the ease with which carriers could diffuse away from their point of generation to recombine elsewhere) made the gain very low. Modern diode lasers are constructed with spatially confined gain regions surrounded by alloys whose properties (band gaps and refractive indices) are different from these of the undoped (intrinsic) material of the gain region. Typically, the gain region is sandwiched between cladding layers of p-type and n-type material above and below. Devices constructed with such combinations of different semiconductor alloys are known as **heterostructures**. An example of a double-heterostructure diode laser architecture is shown in Fig. 9.10. In this case—the so-called **buried mesa** design—the gain stripe is sandwiched between cladding layers of materials with wider band gaps not only above and below the gain stripe (as in earlier double heterostructure devices) but also on each side of it.

Since the gain of the device depends on the carrier generation rate, and hence on the excitation current density, it is advantageous to force the current to flow only across the gain stripe. The arrangement of the insulating oxide layer shown on Fig. 9.10 prevents the lateral spread of current flow outside the active region.

As well as providing current confinement, this type of double-heterostructure device also provides photon confinement and carrier confinement. As indicated in Fig. 9.11(c) the higher refractive index of the material of the gain stripe relative to that of the cladding layers above and below it forms a waveguide, concentrating the optical field distribution close to the axis of the gain region as shown in Fig. 9.11(d). However, because the field of the laser mode extends appreciably into the unexcited regions on either side, where there is no population inversion, it is important that the material in these regions does not absorb at the operating wavelength of the laser. As indicated in Fig. 9.11(b), this is ensured by choosing cladding materials with band gaps larger than the laser photon energy.

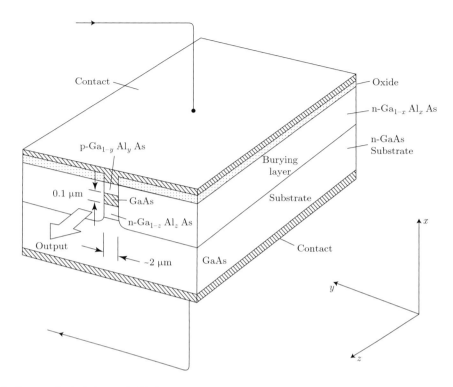

Fig. 9.10: Schematic diagram of a buried mesa double-heterostructure GaAs diode laser. (After Yariv (1989).)

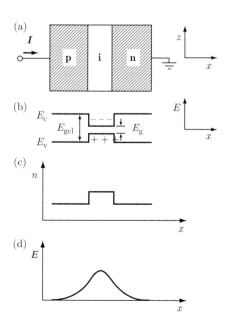

Fig. 9.11: Aspects of the design of double-heterostructure diode lasers: (a) schematic diagram of the material structure; (b) energy diagram of the conduction and valence bands vs. distance in the x-direction, i.e. the layer growth direction. The energy gap in the active region is E_g, while that of the cladding layers is denoted by E_{gcl}; (c) refractive-index profile, showing the refractive index in the cladding layers as lower than that of the active region; (d) the electric-field profile for a mode travelling along the active region in the z-direction. Note that in this figure the axes have been rotated with respect to those of Fig. 9.10 in which the x-direction (the direction of current flow) is vertical.

9.7 Laser materials

The selection of materials from which to make the cladding layers is a crucial one. As argued in Section 9.6 we need materials whose band gap is wider than that of the intrisic material but they must also be capable of being deposited via epitaxial growth techniques[33] on the active region to form crystals free from defects without grain boundaries or regions of dislocation. This means that the alloy used for cladding must have a lattice constant very closely matched to that of the intrinsic material.

As shown in Fig. 9.12, GaAs and AlAs have very similar lattice constants. It is therefore possible to form the ternary alloy $Ga_{1-x}Al_xAs$, with any atomic proportion x of aluminium between 0 and 1, with only small changes in the lattice constant and hence without unduly straining the crystal lattice. This is, however, a happy circumstance unique to the GaAs/AlAs system. If we started with a different intrinsic semiconductor we would in general have to resort to using quaternary (four-component) compounds in order to have the freedom to increase the band-gap energy at a constant value of the lattice spacing. The precise formulation of the semiconductor compound combinations used in the structure of the diode laser is thus a complex compromise between several competing considerations. Table 9.1 lists some of the combinations that have been used.

Our discussion hitherto has concerned semiconductor lasers with band gaps of order $0.5 - 2\,\text{eV}$ that emit in the red and near-infra-red regions (remember that $\hbar\omega_L \approx E_g$). A breakthrough that led to the development of lasers with output in the violet and near ultraviolet end of the visible spectrum around 400 nm ($E_g \approx 3.1\,\text{eV}$, well beyond the top of the range shown in Fig. 9.12) was made by Shujii Nakamura of Nichia Chemical Industries in 1995. It had been known for some years that the ternary compound InGaN had a direct bandgap varying from 2 to 3.5 eV, but no suitable means of doping GaN to produce p-type material had been found. Nakamura found a new technique to activate the holes in p-type GaN by annealing the layers in nitrogen

[33] There are various techniques of epitaxy (the lattice-matched growth of one semiconductor material over another). The most commonly used are liquid-phase epitaxy (LPE), vapour-phase epitaxy (VPE) and molecular (beam) epitaxy (MBE). In LPE, the new material is grown by cooling a saturated solution that is in contact with the substrate. In VPE (sometimes called chemical vapour deposition—CVD) the epitaxial layer is grown by reacting elements or compound in the gas phase at the surface of a heated substrate. A widely used variant of this technique is metal organic chemical vapour deposition (MOCVD) in which volatile organometallic compounds are the feedstock material. In MBE the epitaxial layer is grown by reactions of atoms of the required constituents in the form of atomic or molecular beams impinging on the heated crystalline substrate in a high-vacuum apparatus.

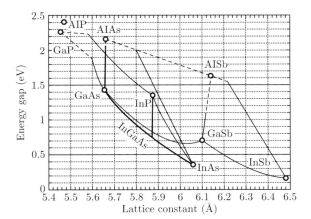

Fig. 9.12: Band gap versus lattice constant for several families of III-V semiconductors. These are the materials of choice for lasers that emit in the $0.7 - 1.6\,\mu\text{m}$ range that includes the optical fibre communication bands at 0.85, 1.31 and 1.55 μm as well as the Nd:YAG pump band at 0.81 μm and the lasers used in CD players at 0.78 μm. The solid lines on the diagram represent ternary compounds (e.g. the line from GaAs to InAs represents InGaAs with increasing proportions of In to Ga) for which the energy gap is direct. The dashed lines represent regions of indirect gap. Note 1 Ångstrom = 0.1 nm. (After Coldren and Corzine (1995).)

Table 9.1 Diode laser materials and wavelengths of operation.

Active region	Cladding	Wavelengths (nm)	Substrate
GaAs	(AlGa)As	800–900	GaAs
GaInP	(AlGa)InP	630–650	GaAs
InGaAs	GaAs	900–1000	GaAs
InGaAs	InGaAsP	1550	InP
InGaAsP	InGaAsP	1300–1550	InP
InAsP	InGaAsP	1060–1400	InP
InGaAs	InP	1550	InP
InGaN	AlGaN, GaN	370–460	Al_2O_3, GaN

at 700°C, thus opening a path to the development of semiconductor lasers operating in the spectral region from about 370 to 460 nm and white-light LEDs.

In addition to the difficulty of producing p-type GaN material, Nakamura also faced the problem of finding a suitable substrate on which to grow the devices. Bulk GaN can be used, but is not currently available in the form of large (8 inch or more) wafers needed for mass production. Silicon carbide has a lattice constant very close to that of GaN at 0.3 nm, but is very expensive. Sapphire is widely used as a substrate for GaN devices despite the fact that it has a 15% lattice mismatch since, with appropriate buffering layers, material can be grown on it with sufficiently low densities of defects to make long-lived devices. The field of GaN device development is one of extremely lively research interest at the moment, not only for filling in parts of the spectrum where high-power diode lasers are presently lacking but also for the enormously important goal of developing highly efficient white-light LEDs for domestic lighting.

9.8 Quantum-well lasers[†]

Radical improvements in the design and capabilities of edge-emitting diode lasers came about with the introduction of techniques of 'band-gap engineering' in the 1980s, when it became possible to grow epitaxial layers of thickness[34] less than 10 nm. Prior to this the thickness of the active layer in simple double-heterostructure edge-emitting lasers was typically 100–500 nm.

The change in material properties at the boundaries of the active layer with the cladding material of the double heterostructure not only constrains the electric field of the mode of the laser radiation as indicated in Fig. 9.11, but also constrains the wave functions of the carriers in the active region to be those appropriate to a potential energy well rather than the Bloch functions of the bulk material, since in the x–direction the de Broglie wavelength of the carriers is now comparable to the well width. The energy levels of the carriers in this region can be obtained by separating the Hamiltonian into one

[34] The thickness referred to here is that in the x-direction in Figs. 9.10 and 9.11, i.e. the height of the active stripe.

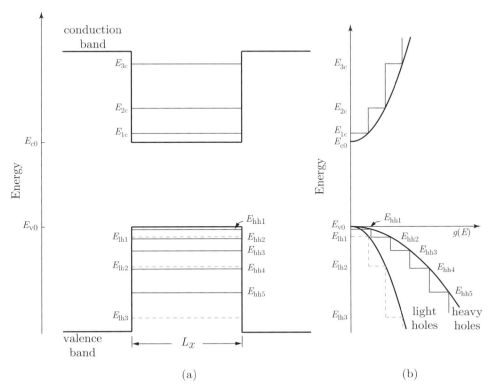

Fig. 9.13: Schematic diagram of the energy levels and density of states in a quantum-well structure. (a) The energies of the electrons in the potential well in the conduction band, and of holes within the potential hump in the valence band, are determined by the well thickness L_x. To that extent they are externally engineered, rather than being properties of the materials. (b) Density of states for a quantum-well heterostructure. The curves originating from the conduction-band edge E_{c0} and valence-band edge E_{v0} correspond to the density of states in a bulk sample. The step-like densities of states are characteristic of the two-dimensional nature of the quantum-well structure; note that *within a given band* the step in $g(E)$ is of a fixed size, as given by eqn (9.31). Note that for the 3D case (parabolic curve) $g(E)$ has dimensions m^{-3}, while that for the 2D case (stepped variation) has dimensions m^{-2}. In order to make the comparison meaningful, we need to consider a quantum well of effective length $L_{x,\text{eff}}$ so that for the 2D case the quantity plotted on the x-axis is $g(E) = (m^*/\pi\hbar^2)/L_{x,\text{eff}}$.

component along the x-direction, and those in the y- and z-directions that give rise to the usual Bloch functions. The energy eigenvalues are now given by

$$E(n, k_y, k_z) = E_{c0} + E_n + \frac{\hbar^2}{2m^*}(k_y^2 + k_z^2), \quad (9.27)$$

where E_{c0} is the energy at the bottom of the well in the conduction band, E_n is the nth confined-particle energy level for carrier motion in the x-direction within the well, m^* is the effective mass of the carriers and k_y, k_z are the y- and z-components of the usual Bloch wave vectors $\boldsymbol{k}$ in eqn (9.1).

The values of E_n are shown schematically in Fig. 9.13(a). The electron energies within the potential well in the conduction band (the so-called subband energies) are denoted by $E_{1c}, E_{2c}, E_{3c}, \ldots$, etc. Since, in the valence band, the 'well' is actually a potential 'hump', the subband energy levels E_n in this region number downwards from the top of the valence band, and because they depend on the mass of the hole (see Section 9.2.1) the energies E_{1hh}, E_{2hh}, E_{3hh} for the heavy holes will be slightly different from those for the corresponding light holes E_{1lh}, E_{2lh}, E_{3lh}.

For a potential well with infinitely high walls, quantum mechanics yields the magnitude of E_n as

$$|E_n| = \frac{\pi^2 \hbar^2}{2m^* L_x^2} n^2 \qquad n = 1, 2, 3, \ldots. \qquad (9.28)$$

Since energies of levels in the potential hump are referred to the energy of the top of the band E_{v0}, the energy of the nth level in the potential hump is $E_{v0} - E_n$ using the same convention we have used earlier (e.g. as in deriving eqn (9.6)). Thus, the energy of a photon emitted in the transition from the lowest conduction band level E_{1c} to E_{1hh} (the highest valence band level) would be

$$\hbar\omega = E_{c0} + E_{1c} - (E_{v0} - E_{1hh}) = E_g + \frac{\pi^2 \hbar^2}{2L_x^2}\left(\frac{1}{m_c^*} + \frac{1}{m_v^*}\right) \qquad (9.29)$$

$$= E_g + \frac{\pi^2 \hbar^2}{2m_r^* L_x^2}, \qquad (9.30)$$

where m_c^* is the effective mass of an electron in the conduction band, m_v^* is the effective mass of a hole in the valence band, and m_r^* is the reduced mass given by eqn (9.23).

Although in reality the walls are not infinite, it is clear from this reasoning that by introducing a quantum-well structure, the energy of the emitted photons can be chosen to some extent by a judicious selection of the well width L_x. In realistic cases the Schrödinger equation appropriate to a well of finite depth must be solved to obtain more accurate values for the subband energies. For the case of a well of width $L_x = 10$ nm in GaAs clad with $Al_{0.3}Ga_{0.7}As$, the energies calculated by the infinite square-well model are about twice as large as those for the finite well, as summarized in Table 9.2.

The discrete energy levels of the quantum well modify the density of states to a stepwise variation characteristic of the 2D nature of the carrier motion instead of the parabolic form[35] of the bulk material where the carriers can move in a 3D space, as indicated in Fig. 9.13. For the bulk material the density of states (per unit volume) depends on $(E - E_{c0})^{1/2}$, as given by eqn (9.8); in the case of the quantum well each energy level in the conduction band has a constant (independent of energy) density of states (per unit area) given by[36]

$$g(E) dE = \frac{m_e^*}{\pi \hbar^2} dE. \qquad (9.31)$$

A similar relationship applies to the density of states in the well levels of the valence band with the appropriate hole masses substituting for the electron mass. The introduction of the quantum well modifies the dependence of density of states on energy for the bulk material to the stepwise variation shown in Fig. 9.13(b). The introduction of a quantum-well structure allows the wave-

Table 9.2 Bound states of a 10-nm $GaAs/Al_{0.3}Ga_{0.7}As$ quantum-well calculated using the finite and infinite well models. The states are labelled as in Fig. 9.13. The energies are in meV. (Data from Fox (2001).)

State	Finite well	Infinite well
E_{1e}	32	57
E_{2e}	120	227
E_{3e}	247	510
E_{hh1}	7	11
E_{hh2}	30	44
E_{hh3}	66	100
E_{hh4}	112	177
E_{lh1}	21	40
E_{lh2}	78	160

[35] The density of states $g(E)$ for a bulk semiconductor varies as $(E - E_{c0})^{1/2}$ according to eqn (9.8). Thus, when plotted with $g(E)$ as the x-axis and E as the y-axis as in Fig. 9.13, the curve is parabolic.

[36] See Exercise 9.9.

length of laser operation to be chosen (at least to a limited extent) and usually[37] shifted to shorter[38] wavelengths than that of the bulk material.

Because of the selection rule that only transitions from the nth level of the conduction band to the nth level of the valence band are allowed,[39] the recombination transitions in a quantum-well heterostructure proceed from a block of electrons, all in principle at a nearly[40] fixed energy (say at E_1 in eqn (9.27)) with a similar block of holes also at a nearly fixed energy (say at E_{hh1} above E_{v0}). Figure 9.13(b) illustrates another very important and advantageous feature of quantum-well heterostructures over bulk materials. In the bulk sample the recombining carriers are distributed in energy over parabolically varying densities of states, which are small at the edges of both conduction and valence bands. Thus, in principle the electrons and holes cannot all be located at fixed energies and so cannot then recombine, giving rise to a spectrally narrow wavelength band, unlike the case for quantum-well emission. Further, if carriers are injected a sample at energies well above the bottom of the conduction band, they can be thermalized (scattered downwards in energy) by multiple phonon collisions ending up near the bottom of the band.[41] The rate of this process is proportional to the final density of states that, near the bottom of the band, is low in the case of bulk samples and acts to slow down the thermalization process. There is no such limitation in the case of the quasi-2D quantum-well system having a constant density of states, so that the basic electron–phonon scattering interaction is enhanced, and all the recombination radiation can occur at essentially the same energy.

The net result of these effects is to confer on the quantum-well heterostructure laser several advantages including lower threshold current, higher efficiency, and the ability of the laser to be modulated over an increased range of frequencies.[42]

With the degree of control in fabrication that modern epitaxial growth techniques permit, the idea of fabricating several quantum wells in the same device has become a practical reality. Semiconductor lasers in which several quantum wells of thickness L_x (say $\approx$ 10 nm of GaAs separated by layers AlGaAs of thickness T_x (say $\approx$ 1000 nm) are laid down in sequence on top of one another are called **multi-quantum-well (MQW) devices**. The concept can be carried even further, to the point where the value of the thickness of the separating layer T_x is reduced to a size comparable to that of the active layers L_x, in which case it is usual to refer to the structure as a 'superlattice'.

The discussion of this and previous sections has focused on the example of GaAs/AlGaAs lasers, but clearly the lessons drawn are applicable to semiconductor lasers fabricated in many other combinations of materials.

[37] It might seem obvious from Fig. 9.13(b) that the introduction of a quantum-well structure must always shift the operating wavelength to shorter values compared to the bulk material. However, the fact that the carrier thermalization by phonon collisions in the quantum well case tends to be much more efficient, means that an electron can be transferred before recombination to states below E_{1e} and can in fact lead to laser operation at photon energies $\hbar\omega < E_g$. See Holonyak et al. (1980).

[38] For quantum wells fabricated in $In_{0.53}Ga_{0.47}As$ the laser wavelength for $L_x = 100$ nm is 1.70 μm. For $L_x = 10$ nm the wavelength shifts to 1.55 μm, while for $L_x = 8$ nm the wavelength is 1.50 μm.

[39] See Exercise 9.8.

[40] Although the exact transition energy depends (via eqn (9.27)) on the values of k_y and k_z for the electron in the conduction band before recombination, this has only a very small effect on the transition energy because of the tendency of the electron energies to thermalize to the bottom of the band where the $k_{y,z}$ values are small. The conservation of momentum and the negligible k vector of the emitted photon ensures that the same is true for the state of the hole in the valence band involved in the recombination process so that there is little room for variation in the transition energy to occur.

[41] Holonyak et al. (1980).

[42] Since the volume of the active region in quantum-well lasers is considerably smaller than that of conventional edge-emitting lasers, the injected carrier density is higher for a given current that accounts for its lower threshold current. Similarly, its small size minimizes both capacitance and transit time effects, helping to improve its high-frequency modulation response.

9.9 Laser threshold

From the same considerations as that used to derive eqn (6.65) we might expect to be able to write the threshold condition for a semiconductor laser in a simple cavity of length L bounded by two identical end mirrors each of reflectivity R as

$$R^2 \exp\left[2(\alpha_0^{\text{th}}(\omega-\omega_0)-\kappa)L\right]=1, \qquad (9.32)$$

in which $\alpha_0^{\text{th}}(\omega-\omega_0)$ is the small-signal gain coefficient at threshold and κ is the corresponding distributed loss coefficient. In the case where mirrors of unequal reflectivities R_1 and R_1 are used, the effective value of R is $\sqrt{R_1 R_2}$. If, as is frequently the case for semiconductor diode lasers, the cavity is formed by two uncoated cleaved surfaces then R is simply the Fresnel reflection coefficient:

$$R = \frac{|n_a - 1|^2}{|n_a + 1|^2}, \qquad (9.33)$$

in which n_a is the refractive index of the active medium. In the case of GaAs $n_a \approx 3.4$ so $R \approx 0.3$. Clearly, eqn (9.32) can be written in the form

$$\alpha_0^{\text{th}}(\omega-\omega_0) = \kappa + \frac{1}{L}\ln\left(\frac{1}{R}\right). \qquad (9.34)$$

However, eqn (9.34) is an oversimplification for the case of semiconductor lasers because an appreciable fraction of the energy circulating in the cavity is in the form of the electric field of the evanescent wave in the region immediately outside the active gain stripe. A more realistic version of eqn (9.34) is therefore:

$$\left[\alpha_0^{\text{th}}(\omega-\omega_0)-\kappa_{\text{fc}}\right]\Gamma_a = \kappa_n\Gamma_n + \kappa_p\Gamma_p + \kappa_s + \frac{1}{L}\ln\left(\frac{1}{R}\right). \qquad (9.35)$$

In this equation the individual sources of loss[43] making up the overall distributed loss coefficient κ in eqns (9.32) and (9.34) are identified: κ_{fc} represents the coefficient of loss within the active stripe due to free carriers and unpumped material, κ_n and κ_p represent loss coefficients for the cladding n- and p-type materials, respectively, and κ_s is the overall loss caused by surface and interfacial roughness.

The other important parameters in eqn (9.35) are the so-called 'confinement factors' $\Gamma_{a,n,p}$ that represent the fraction of the total energy of the transverse mode of the radiation field in the active and n- and p-type cladding regions, respectively. It is clear from Fig. 9.11(d) that a considerable fraction of the field energy lies within the cladding region on either side. If, as in Fig. 9.11, the active stripe extends from $x = -L_x/2$ to $x = +L_x/2$ and $E(x)$ is the electric field of the mode[44] then Γ_a is defined by

$$\Gamma_a = \frac{\int_{-L_x/2}^{L_x/2}|E(x)|^2\,dx}{\int_{-\infty}^{\infty}|E(x)|^2\,dx}, \qquad (9.36)$$

with analogous definitions for Γ_n and Γ_p.

Once the threshold excitation current I_{th} has been reached, if every electron injected were to recombine usefully and give rise to a laser photon we might expect the laser output power P_L to be given by

$$P_L = \frac{I - I_{\text{th}}}{e}\hbar\omega, \qquad (9.37)$$

[43] It should be borne in mind that all four of the coefficients κ_{fc}, κ_n, κ_p, and κ_s are, in general, frequency dependent, and also depend on the concentration of carriers present but for simplicity of notation we shall consider them to have the values appropriate to threshold conditions.

[44] Since the (normalized) transverse shape of the radiation mode will not vary strongly with position along the length of the active stripe, the quantities Γ_a, Γ_n, and Γ_p will not depend significantly on the value of z, and it is sufficient for calculation to use their average values.

since the rate at which (useful) electrons are injected into the active volume is simply the excess current above the threshold value divided by $|e|$, the magnitude[45] of the charge on the electron, and the laser power is simply that rate multiplied by the energy per photon. However, we have to take account of the fact that there are ways in which injected carriers can be lost other than by making useful recombination events that give rise to laser photons.

As discussed in Section 9.5 the efficiency with which electrons are injected into the active volume is expressed via the parameter η_{inj}. However, we must now take into account the fact that of the electrons injected into the active volume only a fraction η_{rad} of them disappear as a result of useful recombination events, the others being lost by non-radiative recombination or other events that do not contribute usefully. In addition, there are ways in which the laser photons themselves can be lost other than by contributing to the output of the laser. If we represent the useful transmission loss of the cavity T_{loss} as an equivalent distributed loss coefficient $T_{\text{loss}} = \ln(1/R^2)$ and, with $\kappa = \kappa_s + \kappa_n \Gamma_n + \kappa_p \Gamma_p + \kappa_{\text{fc}} \Gamma_a$, the total round trip loss for a cavity will be $T_{\text{loss}} + 2\kappa L$. An optical efficiency factor η_{opt} can thus be defined as

$$\eta_{\text{opt}} = \frac{\text{Useful transmission loss}}{\text{Total loss}}$$

$$= \frac{T_{\text{loss}}}{T_{\text{loss}} + 2\kappa L} = \frac{\ln\left(\frac{1}{R^2}\right)}{\ln\left(\frac{1}{R^2}\right) + 2\kappa L}$$

$$= \frac{\ln R}{\ln R - \kappa L}. \tag{9.38}$$

[45] Near the top of the valence band, the effective mass of a hole is an intrinsically positive quantity so that eqn (9.6) applies equally well to holes, although in this case the modulus sign is redundant.

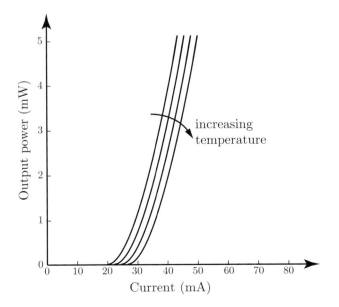

Fig. 9.14: Typical variation of the output power of a semiconductor diode laser as a function of the current.

A more realistic expression for laser output power than eqn (9.37) is therefore:

$$P_L = \eta_{\text{inj}} \eta_{\text{rad}} \eta_{\text{opt}} \frac{I - I_{\text{th}}}{e} \hbar \omega. \qquad (9.39)$$

In practice, the value of η_{inj} is found by fitting a curve of the form of eqn (9.39) to experimental data for laser output power versus excitation current. The typical form of the output power of a diode laser as a function of current is shown in Fig. 9.14.

9.10 Diode laser beam properties

9.10.1 Beam shape

For edge-emitting diode lasers with the single gain stripe geometry shown in Fig. 9.1 the output beam is elliptical in cross-section with a divergence in the x-direction typically some 2–3 times that in the y-direction (e.g. $\theta_\parallel \sim 6°, \theta_\perp \sim 20°$).

Various schemes have been devised to render the beam more nearly circular after it leaves the cavity. One such scheme, shown in Fig. 9.15, uses prisms to expand the beam in one dimension. Other approaches include the use of anamorphic lenses (which produce different magnifications in the x- and y-directions); or mirrors, which cut the beam into several sections vertically and then restack it horizontally.

9.10.2 Transverse modes of edge-emitting lasers

Whilst the height L_x of the active stripe is typically only 1–2 times the wavelength, its width L_y may be several tens or even hundreds of wavelengths across, so that the field distribution of the transverse mode of such a device may show a complicated structure in the y-direction that may vary with excitation current.

As shown in Fig. 9.16 the transverse mode pattern of a single stripe laser of width $L_y = 10$ μm has a single lobe, but the near-field pattern of wider stripes shows one or more reversals of field direction across the y-dimension,

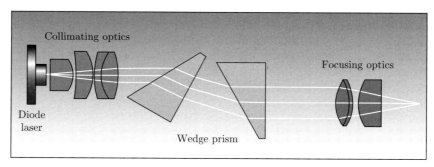

Fig. 9.15: Diode laser beam shaper. The prisms expand the width of the beam in the plane of the diagram, but not the height of the beam perpendicular to that plane.

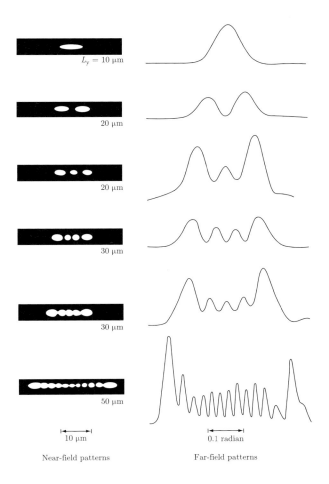

Fig. 9.16: Transverse modes of single-stripe edge-emitting diode lasers of various widths L_y. Reproduced with permission from Yonezu et al. (1973).

the number of such reversals increasing with excitation current. Thus, while semiconductor lasers show very high values of electrical to optical conversion efficiencies, the ability to bring their beams to a small focus is limited especially at high output power levels. This has led to the development of technologies such as diode-pumped solid-state lasers (DPSSLs) or cladding-pumped fibre lasers in which the diode laser plays the part of providing the primary conversion from electrical energy to radiation,[46] while one or other of these optical-to-optical conversion schemes is employed to improve the quality of the final output beam, even to the point of reducing it to a single transverse mode.

[46] See Sections 7.1.7, 10.5 and 10.6.

9.10.3 Longitudinal modes of diode lasers

For edge-emitting lasers operating in simple Fabry–Perot cavities the transitions can access a wide range of levels near the bottom of the conduction band and the top of the valence band that provide gain above cavity loss. Although, as indicated in Fig. 9.9, the range of wavelengths over which oscillation is possible increases with injected carrier density, the effect on the wavelength spread of oscillating axial modes is just the opposite.

Figure 9.17 shows the spectrum of an edge-emitting diode laser as a function of increasing excitation currrent. It is seen that near threshold the laser mode spectrum has the wide bandwidth characteristic of an LED, but as the excitation current is increased a dominant axial mode establishes itself. This competes more and more effectively for the available inversion as the current increases so that the bandwidth spanned by the oscillating modes decreases as the current increases. From the form of the mode pattern at low currents, it is evident that although the broadening of the linewidth is to a large extent inhomogeneous, the homogeneous component is still large enough to make the few modes near the line centre completely dominant at high excitation current.

The wavelength spacing between adjacent modes is given approximately by[47]

$$\Delta\lambda \approx \frac{\lambda_0^2}{2n_a L_c}, \qquad (9.40)$$

where n_a is the refractive index of the active medium and λ_0 is the wavelength measured in vacuum.

[47] The reason why this relationship is only approximate is that we have not taken into account the effect of the variation of n_a with frequency over the range of interest. A better approximation would be to replace n_a in this expression with $n_{a,\text{eff}} = \frac{d(k_0 n_a)}{dk_0} = n_a(\lambda_0) - \lambda_0 \left.\frac{dn_a}{d\lambda}\right|_{\lambda_0}$. See Exercise 9.6.

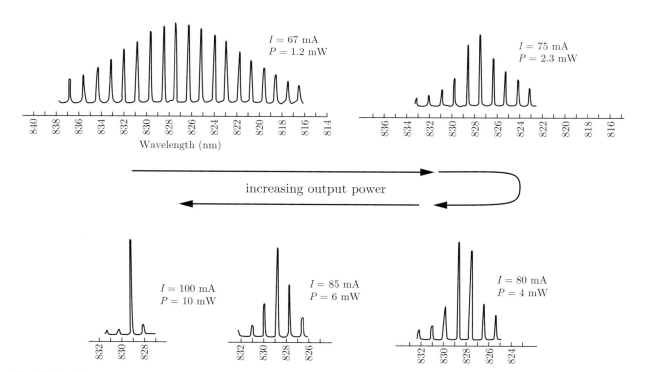

Fig. 9.17: Axial-mode spectrum of an edge-emitting diode laser with a simple Fabry–Perot cavity as a function of increasing excitation current. Note that the scale of the output power has been reduced by a factor of 100 between the bottom and top traces in this figure. (After Kressel et al. (1980).)

For a cavity of length L_c of 100 μm and a refractive index of 3.4 for GaAs at a wavelength of 830 nm, the intermode spacing is $\Delta\lambda \approx 1$ nm–easily resolvable with a simple grating monochromator.[48]

9.10.4 Single longitudinal mode diode lasers

As evidenced in Fig. 9.17 the output of typical semiconductor lasers can approach single longitudinal mode operation at sufficiently high currents. However, the achievement of truly single longitudinal operation at a predetermined frequency requires inclusion of a frequency-selective element within the cavity. The most widely used methods make use of the frequency-selective properties of Bragg gratings in one form or another. As illustrated in Fig. 9.18 one approach is to coat the crystal facet at the output end of the active stripe of a simple laser diode with an antireflection coating[49] and to couple the emerging radiation into a short length of single-mode optical fibre (known as a pig-tail) containing a section of **fibre Bragg grating (FBG)**. The FBG is written into the fibre[50] so that its reflectivity curve is peaked at the desired output wavelength, but so that it still has some 10–50% transmission at this wavelength. Since the cavity provides feedback at only the frequency determined by the external FBG, it oscillates on a single longitudinal mode since even the nearest-neighbour modes lie well outside the narrow band at which the FBG has high reflectivity.

The Bragg condition[51] for a strong reflection peak of order m for diffraction of X-rays of wavelength λ from a 3D array of scattering atoms arranged in parallel sets of planes spaced apart by separation d is usually written in the form $m\lambda = 2n_X d \sin\theta$, where θ is the angle between the direction of propagation

[48] This compares to the case of lasers with cavity lengths of order 1 m, whose axial modes are separated by wavelength intervals of order 0.34 pm at 830 nm, (i.e. frequency intervals of 150 MHz) and can only be resolved using high-finesse interferometers or via the radio-frequency beats they produce in the output noise spectrum of photodiodes or photomultipliers.

[49] See, for example, Section 6.7 of the book by Brooker (2003).

[50] The FBG is manufactured by exposing a short length of the fibre to a transverse pattern of two-beam interference fringes generated by an ultraviolet laser that permanently writes a grating in the form of a set of parallel slices of high-index material in the UV photosensitive core. The writing mechanism involves the UV breaking chemical bonds in the Ge-doped silica fibre core that compacts the core material locally.

[51] See, for example, Chapter 2 of the book by Rosenberg (1989).

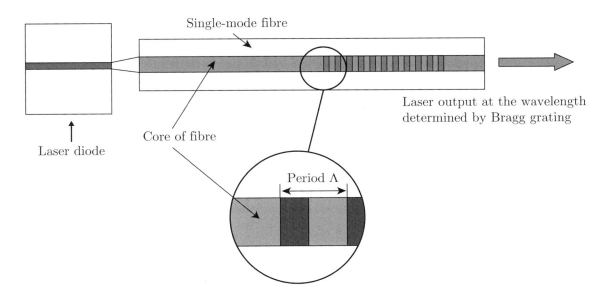

Fig. 9.18: Schematic diagram of a single-mode diode laser stabilized by an external fibre Bragg grating.

of the incident X-rays and the plane containing the scattering atoms, and n_X is the refractive index at the X-ray wavelength.

In the case of the FBG, we have a set of planes repeating with spatial period Λ at which n_B (the refractive index of the fibre core) undergoes a small change from the ambient value. Each such discontinuity gives rise to a small reflection for the beam of light within the laser cavity that travels along the axis of the fibre at $\theta = 90°$ to the planes. The Bragg law tells us that the cumulative effect of the array of weakly reflecting planes repeating at period Λ will amount to a *strong* first-order[52] reflection peak, provided

$$\lambda_0 = 2 n_B \Lambda, \qquad (9.41)$$

where λ_0 is the wavelength of the radiation in vacuum.[53]

Two alternative methods of constraining the diode laser to oscillate in a single axial mode are the **distributed feedback (DFB)** and the **distributed Bragg reflector (DBR)** approaches shown in Fig. 9.19. In the DFB type, a Bragg grating in the form of a surface corrugation with the appropriate period Λ in the z-direction is etched into the waveguide comprising the active region within the laser. The corrugations are in the form of washboard stripes across the width of the gain stripe in the y-direction and are etched into the crystal layers during the photolithographic fabrication process. In the DBR approach, similar corrugations forming Bragg gratings at each end are etched into a continuation of the waveguide outside the central electrically excited zone. These two Bragg reflectors in the waveguide strip at each end of the gain region form frequency-selective mirrors for the laser cavity, and force it to oscillate in the axial mode whose wavelength satisfies the Bragg condition of eqn (9.41).

9.10.5 Diode laser linewidth

The frequency of oscillation of a diode laser depends critically on both the excitation current (via the dependence of gain on injected carrier density—see Fig. 9.9) and operating temperature (via the effect on the Fermi–Dirac factors f_c and f_v.[54] To achieve narrow linewidth output the laser must therefore be operated with very stable current and temperature conditions.

---

[52] Gratings can be designed to operate at higher order numbers but since the UV interferometer setup for writing gratings can conveniently generate gratings with Λ appropriate to first order at the typical diode laser wavelengths there is no advantage in using higher orders–in fact gratings are designed to operate at the lowest order practicable, to prevent the occurrence of pass-bands at unwanted wavelengths within the diode's range of potential oscillation.

[53] It is worth emphasizing the connection between the operation of a fibre Bragg grating, or a Bragg mirror, with the 'quarter-wave stack' often used in high-reflectivity mirrors. In a Bragg reflector the phase difference between light reflected from the first surface of the first layer, and the last surface of the second layer, is an integer times 2π. A quarter-wave stack is a special case of the Bragg reflector, and comprises alternate layers of material, each with a thickness equal to one-quarter of that of the wavelength of the light in that medium. Thus, the phase shift experienced by the light in a single pass through each layer is $\pi/2$, and hence each pair of quarter-wave layers satisfies the Bragg condition. Finally, we note that for a pair of layers $i = 1, 2$ of thickness t_i and refractive index n_i, the Bragg condition amounts to $(2\pi/\lambda_0)(n_1 t_1 + n_2 t_2) = 2\pi p$, where p is an integer. Equation (9.41) follows for $p = 1$ and when $n_1 \approx n_2$.

[54] See eqns (9.12) and (9.24).

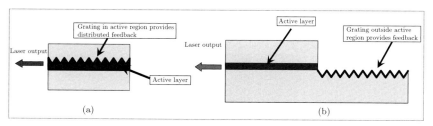

Fig. 9.19: Single longitudinal mode diode lasers. (a) Distributed feedback (DFB) type employing an internal grating and (b) Distributed Bragg reflector (DBR) type, in which frequency selective feedback is provided by a Bragg grating etched in the cladding layer external to the active region.

As in other types of laser, the linewidth of a single-mode diode laser is ultimately determined by the internal noise arising from spontaneous emission into the lasing mode, assuming that competition from all other modes is adequately suppressed. We would therefore expect that the residual linewidth $\Delta\nu$ to be given by the Schawlow–Townes formula.[55] However, in the case of diode lasers there is an extra factor[56] of typically 5–50 multiplying the Schawlow–Townes value of $\Delta\nu_{ST}$. Typical values of $\Delta\nu$ measured for diode lasers operating in the $1.3-1.55\,\mu\text{m}$ wavelength range at output powers of a few mW are of order several hundred kHz to a few MHz. Such lasers are simple monolithic devices with no elaborate external stabilization. In laser designs with elaborate care taken to reduce sources of broadening, values as low as 3.6 kHz have been achieved Okai et al. (1994).

According to the Schawlow–Townes formula, eqn (16.26), one might expect that the limiting value of $\Delta\nu$ could approach zero as the power tends to very large values. However, in practice it is found that the value of $\Delta\nu$ tends to a power-independent linewidth, which in typical cases[57] is 600 to 900 kHz. The origins of this linewidth have been attributed to power-independent carrier–density fluctuations, fast thermal fluctuations of electronic-state occupancy, flicker noise induced by carrier-mobility fluctuations, and other factors.

[55] See eqn (16.26).

[56] This factor, introduced by Henry (1982), takes account of the contribution to fluctuations of the refractive index of the active medium arising from fluctuations of the carrier density in the active medium.

[57] An extensive discussion of the contributions to semiconductor laser linewidth broadening will be found in the book by Ohtsu (1992).

9.10.6 Tunable diode laser cavities[†]

Diode lasers with a wavelength-selective element (such as a Bragg grating) included in the cavity can be tuned by varying the peak transmission or reflection wavelength of the selective element, but without proper care in the design this can lead to sudden jumps in the output wavelength even when the selective element's response is altered in a smooth fashion. This is because the cavity may suddenly change the longitudinal mode on which it oscillates from mode number p to a neighbouring mode $p \pm 1$ if this has lower loss. The wavelength selected by the Bragg grating element for maximum feedback is λ_0, the condition for there to be an integer number p of half-wavelengths along the length of the cavity is

$$p\lambda_0 = 2n_{\text{eff}}L_c, \qquad (9.42)$$

where $n_{\text{eff}}L_c$ is the overall effective optical length of all parts of the optical path between the mirrors—including all the separate $n_i L_i$ contributions from the gain region and the end regions, etc. If some mechanism can be provided so that the value of $n_{\text{eff}}L_c$ varies linearly with λ_0 then the wavelength may be tuned continuously without the mode number p changing, i.e. without mode hopping.

A monolithic three-section DBR laser that satisfies this difficult requirement is shown schematically in Fig. 9.20. Three individually regulated currents are injected at separate electrodes on the top of the device into the three regions that are grown simultaneously on the same substrate. The active region on the right-hand side is excited by injection of current I_{ex}. The active

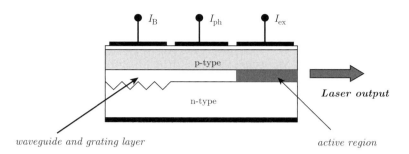

Fig. 9.20: Schematic diagram of a monolithic three-section tunable DBR diode laser.

region shares the waveguide with the Bragg reflector of period Λ etched into the cladding layer that forms the cavity mirror at the left-hand end of the waveguide. Because the local refractive index depends on the injected carrier density, injection of current I_B into this region controls the refractive index $n_B(I_B)$ and allows the reflection peak of the DBR section to be tuned to the required value of λ_0 as described by eqn (9.41). If the active region is of length L_a and has refractive index $n_a(I_{ex})$, the grating region has length L_B and refractive index $n_B(I_B)$ and the middle section (the phase adjusting section) has length L_{ph} and refractive index $n_{ph}(I_{ph})$, then eqn (9.42) can be written as

$$p\lambda_0 = 2n_{eff}L_c = 2n_B(I_B)L_B + n_{ph}(I_{ph})L_{ph} + n_a(I_{ex})L_a. \qquad (9.43)$$

Having set I_{ex} to give the output power required from the laser, and I_B to select the required wavelength, the overall effective optical length of the cavity is adjusted[58] to satisfy eqn (9.43) with a constant value of p as the wavelength is tuned by adjusting I_{ph} to give the required value of n_{ph}. The values of I_{ph} are pre-programmed into the current-control circuitry to ensure that this is done automatically. Devices of this type can be tuned continuously near 1550 nm over a range of 5 nm—or over a range of 8 nm if mode hopping can be accepted.

[58] It might seem possible to maintain a constant optical path without the phase-adjusting section. However, without the phase-adjusting section, altering I_B to tune the wavelength to the desired value, and adjusting I_{ex} to compensate the optical path change would lead to unacceptable output power variations across the tuning range.

If very wide tuning ranges are needed the cavity shown in Fig. 9.21 provides one possible solution. It employs a macroscopic diffraction grating in a first-order Littrow-type configuration. In order to utilize the maximum resolving power available from the grating some form of beam-expansion technique with an intra-cavity telescope or prisms is required. With careful choice of the position of the rotation axis of the grating the overall effective cavity length can be maintained constant as the grating is rotated to tune λ_0, which is given by

$$\lambda_0 = 2d\sin\phi, \qquad (9.44)$$

where d is the period of the diffraction grating and ϕ is the angle of rotation of the grating from the cavity axis. Very large tuning ranges are possible with such a system; for example Bagley et al. (1990) report a tuning range of 240 nm in the 1.3–1.55 μm region.

An alternative, which does not require intra-cavity beam-expanding optics, is the Littman-type cavity described Section 14.5 in connection with dye lasers.

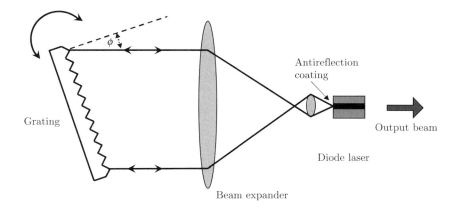

Fig. 9.21: Schematic diagram of an external cavity tuned diode laser.

For example, commercial instruments based on this design are available to tune diode lasers to the Cs or Rb resonance transitions at 852/894 and 780/795 nm, respectively.

9.11 Diode laser output power[†]

Although diode lasers can be extremely efficient (50–70%) at converting electrical energy to laser output, the residual heat energy inevitably increases the operating temperature of the device with increasing excitation current. Increasing the temperature of the device leads to an increase of the non-radiative recombination processes and a reduced carrier confinement in the active region; this causes the output power to roll over at large excitation currents.

Another effect that limits the power capability is the deterioration of the beam quality resulting from a phenomenon known as filamentation arising from the interaction between the optical field and the electronic carriers in the active region. The spatial hole-burning[59] produced by the optical field locally increases the refractive index transverse to the beam propagation direction and leads to self-focusing of the optical wave. In addition, damage of the end faces can be a problem at high powers, but with the use of sophisticated coatings this effect can be ameliorated or even eliminated. In practical terms this means that narrow (4–8 μm wide) single-stripe lasers producing single transverse mode output beams (see Fig. 9.16) are limited to output powers of a few hundred milliwatts.

Higher output powers can be reached without incurring facet damage by increasing the stripe width in the y-direction, thus reducing the optical power density by spreading the beam over a larger area at the end facet. However, without specific measures to prevent it, this can be done only at the expense of allowing the output beam to break up into higher transverse modes (see Fig. 9.16). For some applications of diode lasers—such as broad that area illumination or pumping solid-state lasers—rather poor beam quality may

[59] See Section 6.6.2.

[60] One approach employs a graded-index separate confinement heterostructure (GRINSCH) above and below the active region. Another approach is to use a step change in the refractive index at the boundaries of the vertical dimension of the optical waveguide to confine the electric field.

[61] See Unger (2004).

be entirely acceptable. However, there are more sophisticated approaches[60] that allow the beam area to be increased by using waveguides that are larger in the vertical (x-direction) than the 1–2 μm typical of edge-emitting lasers without sacrificing beam quality. In such a height-extended waveguide both lowest- and second-order transverse modes can propagate but the second-order mode (unlike the lowest-order mode) has an electric-field profile that extends appreciably into the cladding layers where there is no gain (and indeed some loss) so it does not compete effectively with the lowest-order mode.

An alternative, but more complex, way of achieving good mode quality in a high-power diode laser employs a master-oscillator power-amplifier (MOPA) configuration in which a conventional narrow-stripe oscillator section injects power into a tapered semiconductor amplifier. This allows the output beam to have a high power, but with the low-order transverse mode associated with the small aperture of the oscillator. Note that both oscillator and amplifier can be integrated on a single monolithic chip.[61]

A typical diode laser operating at 810 nm with a single gain stripe of 100 μm width and 1.0 μm height produces multi-transverse mode output. Rising from a threshold current of 0.25 A the output increases linearly with increasing current reaching a maximum of 1 W at a current of 1.3 A. Monolithic integration of laser diodes into one-dimensional linear arrays can be achieved during the growth process as indicated in Fig. 9.22. In order to confine the excitation current to the width of the gain stripes, the regions between the stripes are subjected to proton implantation during the growth process in order to raise the resistivity in these regions and thus to channel the current distribution to where it is needed. If the distance between the emitting stripes is large enough there is no optical coupling between adjacent stripes, and the emission of each laser stripe is independent of its neighbours. Linear arrays 1 cm wide with output powers of 50 W and operating currents of 70 A are commercially available for pumping Nd:YAG lasers at 808 nm.

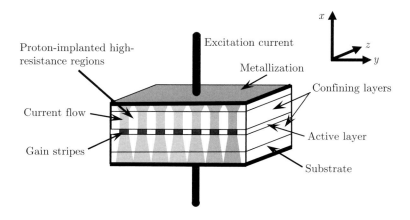

Fig. 9.22: A multistripe diode laser array.

When narrow gain stripes producing lowest-order transverse mode output are arranged in close proximity to one another, the optical field generated by one stripe couples to that of other elements, leading to a phase-locked array. Optical coupling can occur either via the evanescent field in the regions outside the waveguide walls or 'leaky' radiation modes.[62] By designing the array so that stripe coupling occurs from one active stripe to the next via a leaky mode, the neighbouring stripes tend to oscillate in phase with one another and the array exhibits a centrally peaked single-mode pattern in the far field. For evanescent-wave coupling the individual stripes tend to oscillate out-of-phase with their nearest neighbours, which produces a double-lobed beam in the far field.

If even larger output powers are required, linear arrays of diode lasers can be packaged into vertical stacks (i.e. in the x-direction) with appropriate arrangements for water cooling the layers. Diode stacks with output power in the multi-kW range have been demonstrated and are available commercially. One manufacturer (Osram) specifies a 12-bar stack producing over 1 kW. The threshold current is 20 A and 1 kW output is reached at a current of 92 A. Even higher output powers can be obtained by increasing the number of bars in the stack–for example up to 60 in a single device.

[62] Evanescent-wave arrays have propagation constants between the values corresponding to the low- and high-index material. Leaky modes have values of the propagation constant below the value corresponding to the low- index material. The nature of the coupling depends strongly on the geometry of the waveguides and the effective refractive-index step from the waveguide regions to the regions that mediate the coupling between the waveguides.

9.12 VCSEL lasers†

In all of the edge-emitting diode laser geometries discussed in this chapter so far, the direction of epitaxial growth (the x-direction) is perpendicular to the gain axis of the laser (the z-direction). In 1988 Iga introduced a radically different geometry for diode lasers in which the laser beam axis is along the same direction as that of epitaxial growth and the excitation current as indicated in Fig. 9.23. In common with other earlier types of 'surface-emitting' laser geometries[63] this allows the laser beam to emerge from the surface of the laser chip in a direction perpendicular to the plane of the substrate, but in this case the entire laser cavity including the mirror sections is grown layer by layer in the vertical direction. Lasers of this type are known as **vertical cavity surface-emitting lasers (VCSELs)**.

In the case of VCSELs based on InGaAs and operating in the 800–900 nm region, the mirror stacks are grown as alternating quarter-wave layers of GaAlAs ($n \approx 3.5$) as the high refractive index material and AlAs ($n \approx 2.9$) as the low-index material. The structure repeats with period Λ and forms a DBR[64] with peak reflectance at the wavelength given by eqn (9.41). Since the gain length of typical edge-emitting laser is 100–300 μm, but that of a VCSEL cannot be made very long—the thickness of a few quantum wells[65]—the round-trip gain will only be a small fraction of that of a typical edge-emitting laser. VCSEL mirrors must therefore have much higher reflectivity than the 30% or so typical of facet-cleaved reflectance values of edge-emitting types. To achieve the reflectance of 98% required for the output coupling mirror, the number of periods required is 12–15, while 20 or more periods may be

[63] Previous designs of edge-emitting diode lasers, also called 'surface-emitting' lasers, incorporated a grating or prism in the monolithic device structure in order to turn the emitted beam through a right angle so that it emerged perpendicular to the plane of the substrate.

[64] In this case, there is little or no distinction between a distributed Bragg reflector (DBR) and a conventional high reflectance quarter-wave stack. In cases where the step change in refractive index is very small (such as in the FBG of Fig. 9.18) and several tens or hundreds of periods are needed to attain high reflectance, the structure is clearly a DBR. Where materials can be chosen to have a large difference in refractive index between the high- and low-index layers, only a small number of layers are needed to obtain a high overall reflectance as in a conventional quarter-wave stack. In the case of VCSELs the choice of materials is somewhat restricted and it is usual to refer to the structure as a DBR despite the fact that it is close to being a conventional quarter-wave stack.

[65] See Section 9.8.

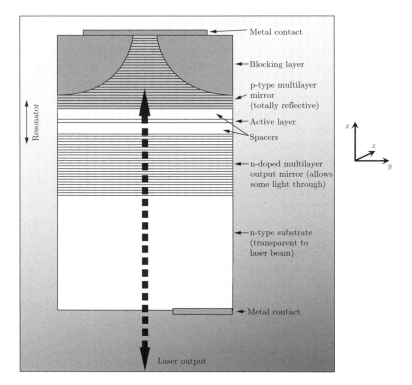

Fig. 9.23: A vertical cavity surface-emitting laser (VCSEL). In the design shown here, the upper electrode is opaque and the output laser beam emerges via the transparent substrate. In other designs the high reflectance mirror stack is on the bottom, and the output beam emerges from the top via a transparent top electrode.

required for the high reflectance stack. One great advantage of the VCSEL concept is that despite the fact that the growth procedure is relatively intricate, it is capable of a high degree of process control so that the percentage yield of good-quality devices is high. This contrasts with the situation for edge-emitting lasers where the facet-cleaving step can only be carried out at a late stage in the manufacturing process. Since the quality of the device is determined by the quality of the cleave, this can lead to an undesirably high incidence of rejects at this stage in the process. Further, because the mirror reflectivities can be chosen by the designer, the VCSEL can be tailored to have maximum output power or lowest current threshold. Finally, the device can be tested while it is still on the wafer.

The spacer layers above and below the active layer are each of order one half-wavelength thick so that despite the finite thickness of the DBR layers the effective cavity length of a typical VCSEL is only some 1–2 μm. The spacing of longitudinal modes of a laser operating with an output wavelength $\lambda_0 \approx 830$ nm is given by eqn (9.40) as $\Delta\lambda_0 \approx 100$ nm. Since the DBR high reflectance band extends only some ±50 nm either side of its peak, the VCSEL oscillates on a single axial mode—and does not mode hop as its wavelength is tuned[66]—without any additional components.

The circular symmetry about the laser axis means that the output beam of a VCSEL is well suited to coupling into optical fibres. The typical output divergence of 10° can be reduced to 2° by lenses integrated with the laser during

[66]Wavelength tuning is achieved by varying the current or temperature.

manufacture. Low-power VCSELs generally produce beams with lowest-order transverse modes, but higher-power devices tend to produce multitransverse mode output. Instantaneous output powers of a few milliwatts are typical of VCSELs of 20 μm diameter operated in a quasi-c.w. manner, while 50 mW can be obtained from devices of 100 μm diameter, albeit in multitransverse mode.

Because the excitation current is routed through the DBR stacks there is a tendency of the VCSEL to overheat at high input power levels—especially in c.w. operation. This not only leads to the wavelength wandering away from its design value but also causes the output power to saturate and turnover. However, one unique advantage that VCSELs possess is the ability to be manufactured in the form of 2D addressable arrays, containing many hundreds of separate lasers, each individually tuned, on a single wafer.

9.13 Strained-layer lasers

Up until the late 1980s, conventional wisdom held that reliable laser devices could only be made from lattice-matched materials grown epitaxially. The wavelengths available from lattice-matched conventional AlGaAs-GaAs double heterostructures ranged from 0.65 to 0.88 μm. Likewise, the range of wavelengths available from lattice-matched conventional double heterostructure and multi-quantum-well heterostructure InGaAsP-InP lasers is from 1.10 to 1.6 μm. These wavelength ranges are sufficient for many important applications such as the use of modulated InGaAsP-InP lasers at $\lambda \approx 1.55$ μm for low-loss optical fibre networks (see Section 10.3) and the use of high-power AlGaAs-GaAs laser arrays at $\lambda \approx 0.82$ μm for pumping Nd:YAG solid-state lasers. However, radiation in the wavelength gap from 0.88 to 1.10 μm is not readily available from any lattice-matched heterostructure system with III-V materials. The all-important wavelength of $\lambda \approx 0.98$ μm needed for pumping Er-doped fibre amplifiers (EDFAs) for telecommunications unfortunately lies within this gap and so, by the early 1980s, researchers had begun to consider using non-lattice-matched structures to extend the wavelength range that could be covered.

The unit cell of $In_xGa_{1-x}As$ can be as much as 3.6% larger (for $x = 0.5$) than that of GaAs, in contrast to $Al_xGa_{1-x}As$ that has a unit cell never more than about 0.13% larger than the GaAs unit cell. Assuming a thick GaAs host, the $In_xGa_{1-x}As$ cell is shortened in both directions parallel to the interface (biaxial compression) and elongated in the direction normal to the interface (uniaxial tension). If this force exceeds the tension in a dislocation line, migration of a threading dislocation results in a single misfit dislocation.[67] For sufficiently thin layers, however, the strain can be contained without giving rise to the propagation of a dislocation.

[67] See Matthews and Blakeslee (1974).

For an $In_xGa_{1-x}As$ strained-layer quantum well bounded by GaAs, the critical thickness for compositions up to $x \approx 0.3$ is greater than 10.0 nm, allowing the laser designer a sufficiently wide range of quantum-well thickness to achieve useful changes in operating wavelength.

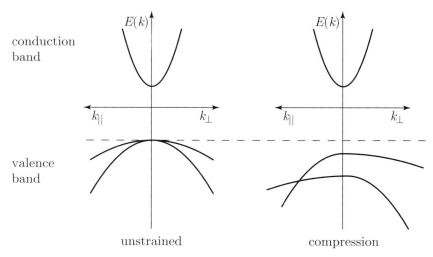

Fig. 9.24: Schematic diagram of the conduction and valence band E vs. k diagrams for unstrained and compressively strained semiconductor layers. (After Coleman (2000).)

In such a strained-layer lattice-mismatched system, biaxial strain affects more than just the positions of the band gaps at $k = 0$. As shown in Fig. 9.24, there are changes in both the heavy-hole and light-hole valence bands as a function of k-vector, normalized to $2\pi/a_0$, where a_0 is the GaAs lattice constant. The salient feature of Fig. 9.24 is that the upper valence band (hh) has the heavy-hole effective mass in the direction perpendicular ($\perp$) to the plane of the interfaces but is no longer symmetric and has a much lighter effective mass in the direction parallel ($\parallel$) to the growth interfaces. A typical laser structure grown over an n-type GaAs substrate comprises a multi-quantum-well (MQW) structure in which four $In_xGa_{1-x}As$ quantum wells are separated by GaAs barrier layers and sandwiched between n- and p-type $Al_{0.3}Ga_{0.7}As$ cladding layers. The emission wavelength depends upon the indium fraction x. As x increases the emission wavelength increases, but for x larger than the critical value (≈ 0.3) the strain is too high to yield long lived, defect-free material. For $x \approx 0.20$ the emission wavelength is near 0.98 μm, as required for EDFA pumping. Threshold current densities as low as 47 A cm^{-2} have been reported for $In_{0.2}Ga_{0.8}As$/GaAs strained MQW lasers, and single-mode output powers of 400 mW have been demonstrated.[68]

[68] See Dutta (2004).

9.14 Quantum cascade lasers[†]

Lasers based on transitions across the energy gap from conduction to valence bands (as described in the preceding sections of this chapter) are available to provide output at specific wavelengths in the spectral region from 380 nm to 1.8 μm. In 1994, Faist et al. introduced a new type of semiconductor laser that operates on transitions from one sublevel to another sublevel *within the conduction band* rather than by recombination of an electron from the conduction band with a hole in the valence band. These lasers are known as quantum cascade lasers (QCL), and have been demonstrated to provide output at selected wavelengths in two regions: 3.5–24 µm in the near-infra-red, and 66–87 µm in the far-infra-red. This ability to design the laser so as to provide radiation at a chosen wavelength in the spectral region where gas molecules show absorption from one vibrational level to another is of prime interest for the manufacture of sensors employed for the spectroscopic detection of specific gas species.

The introduction of these lasers carries one stage further the idea of band-gap engineering, introduced in the previous section in the form of superlattices, to the extent that the laser wavelength is no longer determined by the natural energy gap of a specific material, but to the difference between energy levels of quantum wells formed by thin layers of two different semiconductor compounds.

Quantum cascade lasers comprise many tens of active regions each with three or four adjacent quantum wells made up of alternating layers of high- and low-gap semiconductors each a few atoms thick. The wells are separated by very thin barriers of high-gap material. The low-gap material could be GaAs, while the high-gap material either side of it could be $Al_xGa_{1-x}As$. The operation of the quantum cascade laser can be understood from the highly simplified diagram of Fig. 9.25. Electrons are injected into the conduction-band superlattice where, as shown in Fig. 9.25(b), they drift from left to right under the influence of the applied electric field. The electrons injected into subband 3 in the left-hand well can undergo a radiative transition that returns them to subband 1 with emission of a laser photon in the terahertz region. From subband 1 they tunnel into the neighbouring wells to the right eventually arriving at the next active region in subband 3 because of the downward slope of the conduction band under the influence of the electric field. In the next active region to the right the whole process is repeated.

The main difficulty in the realization of an intersubband laser like this is the achievement of a population inversion that does not arise automatically once electrons are injected as in conventional semiconductor lasers operating

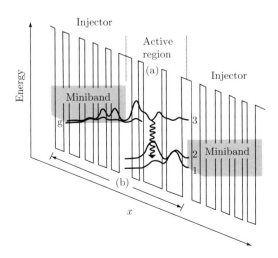

Fig. 9.25: Variation of potential within the conduction band of a quantum cascade laser: (a) one quantum well and its subbands; (b) one period of the multiwell structure including four quantum wells separated by very narrow barrier regions. In a typical device there may be as many as 30 or more such periods. The overall slope of the potential is due to the applied electric field of order 30–50 kV cm^{-1}. The wavy arrow in the active region indicates the laser transition, in the terahertz region of the mid-infra-red. (After Dutta (2004).)

on transitions across interband energy gaps. In the QCL the non-radiative lifetimes against scattering losses from the upper subbands as well as their lifetimes against decay to level 1 are critical. The competition with non-radiative processes means that the threshold current densities required for lasing in QCL are somewhat higher than in conventional MQW devices but the fact that the electron involved in the optical transition is still available after a laser photon has been emitted means that it is possible to cascade many (e.g. 10–30) such stages in sequence. One electron passing through the structure can thus give rise to the emission of many laser photons—a process that has been called 'electron recycling'. This not only improves the efficiency but also boosts the laser output capability to several hundred milliwatts, depending on the temperature, mode of operation (pulsed or c.w.) and the material of laser construction.

Further reading

For further information on solid-state physics, see Rosenberg (1989) and Chapters 4–7 of Singleton (2001).

Further information on laser diodes and LEDs may be found in Fox (2001), Khan et al. (2001), and Coldren and Corzine (1995). Detailed information on specific diode laser systems is available in Chapters B2.1 to B2.8 of Webb and Jones (1964).

Exercises

(9.1) (a) Derive eqn (9.8) from eqn (9.7).

(9.2) This problem is adapted from Q5.9 of Fox ((2001)).

(a) Show that in the case of an undoped (i.e. an intrinsic) semiconductor with no extra carriers injected and $E_g \gg k_B T$, use the 'classical' approximation $(\exp[E - E_F]/k_B T \gg 1)$ in eqn (9.26) to show that the number density of electrons thermally excited into the conduction band is given by

$$n_e = \exp\left(\frac{E_F - E_{c0}}{k_B T}\right)$$

$$\times \frac{1}{2\pi^2}\left(\frac{2m_c^* k_B T}{\hbar^2}\right)^{3/2} \int_0^\infty x^{1/2} \exp(-x) dx.$$

(9.45)

(b) Given that

$$\int_0^\infty \sqrt{x} \exp(-x) dx = \frac{\sqrt{\pi}}{2},$$

estimate the thermal population of electrons in the case of undoped GaAs at $T = 300$ K using the approximation that E_F lies exactly midway between the top of the valence band and the bottom of the conduction band. (For GaAs $E_g = 1.424$ eV at $T = 300$ K, $m_c^* = 0.067\, m_e$.)

(c) A more sophisticated treatment, which does not assume the location of E_F, gives

$$n_e = 2\left(\frac{2\pi k_B T}{h^2}\right)^{3/2} (m_c^* m_v^*)^{3/4} \exp\left(-\frac{E_g}{2k_B T}\right).$$

(9.46)

Use this result to calculate a more accurate value for n_e given that for GaAs $m_v^* = 0.50\, m_e$.

(9.3) This problem is adapted from Q5.11 of Fox (2001).

(a) Use the 'classsical' result of eqn (9.45) to calculate E_F for intrinsic GaAs that has a density of electrons in the conduction band (created, for example, by absorption of laser radiation) of $n_e = 10^{20}\,\text{m}^{-3}$. Repeat the calculation for $n_e = 10^{24}\,\text{m}^{-3}$. In each case comment on whether your answer is consistent with the assumptions made.

(9.4) (a) Derive eqn (9.25).

(b) A GaAs laser of the buried mesa design shown in Fig. 9.10 has an active volume of dimensions: length 500 µm, width 3.0 µm (normal to the direction of current flow), and height 0.10 µm. Assuming $\eta_{\text{inj}} = 80\%$ and $\tau_{\text{rec}} = 4.0\,\text{ns}$, calculate:

  (i) I_{trans}, the current needed to create an electron density of $1.55 \times 10^{18}\,\text{cm}^{-3}$ (sufficient to bring the GaAs to transparancy);

  (ii) $I_{2.5}$, the current corresponding to an electron density of $2.50 \times 10^{18}\,\text{cm}^{-3}$ (at which, according to the data of Fig. 9.9, the small-signal gain coefficient is $130\,\text{cm}^{-1}$ at a photon energy of 1.45 eV).

(9.5) This problem is adapted from Q5.15 of Fox (2001).

(a) GaN and InN are direct-gap semiconductors with band gaps of 3.4 eV and 1.9 eV, respectively. A light-emitting diode made from the alloy $\text{Ga}_x\text{InN}_{1-x}$ is found to emit at 500 nm. Estimate the composition of the alloy on the assumption that E_g varies linearly with x.

(9.6) This problem is adapted from Q5.16 of Fox (2001).

(a) Show that the frequency separation between adjacent longitudinal mode frequencies of a GaAs laser (whose gain medium of refractive index n completely fills the cavity of length L_c) is given by

$$\Delta\nu = \left(\frac{c}{2L_c}\right)\left[n + \nu\frac{dn}{d\nu}\right]^{-1}. \quad (9.47)$$

(b) Derive the equivalent expression for $\Delta\lambda_0$, the mode spacing in terms of the vacuum wavelength λ_0 and $dn/d\lambda_0$.

(c) Estimate a value of $\Delta\lambda_0$ for a laser of this type with $L_c = 100\,\mu\text{m}$ operating at 850 nm and 103 K. Assume that under these conditions $n = 3.59$ and $\lambda_0(dn/d\lambda_0) \approx -2.00$.

(d) Calculate the reflectivity of the air/GaAs interface.

(e) One end of the diode laser is now coated so that it has a reflectivity of 90%; the other end remains uncoated. Assuming that other losses are negligible, calculate the threshold gain coefficient α_0^{th}.

(9.7) This problem is adapted from Q5.17 of Fox (2001).

(a) A laser diode emits at 820 nm when operated at an injection current of 120 mA. Calculate the maximum possible power that can be emitted by the device.

(b) Calculate the power conversion efficiency if the actual power output is 60 mW and the operating voltage is 1.9 V.

(c) The threshold current of the laser is 45 mA. What is the slope efficiency and the effective quantum efficiency $\eta_Q = \eta_{\text{inj}}\eta_{\text{rad}}\eta_{\text{opt}}$? [69]

(9.8) Selection rules in quantum wells

Light polarized in the (y, z) plane is incident on a semiconductor material containing quantum wells in the conduction and valence bands of width d in the x-direction. The electric dipole matrix element for transitions between the nth energy level of an infinite quantum well in the valence band and the n'th level of an infinite quantum well in the conduction band can be written as the product of M_{cv} (the valence–conduction-band dipole moment) and the overlap integral

$$M_{nn'} = \int_{-\infty}^{\infty} \varphi^*_{en'}(x)\varphi_{hn}(x)\,dx.$$

(a) Show that:

$$M_{nn'} = \frac{2}{d}\int_{-d/2}^{d/2} \sin\left(k_n x + \frac{n\pi}{2}\right)\sin\left(k_{n'}x + \frac{n'\pi}{2}\right)dx. \quad (9.48)$$

(b) Hence, derive the selection rule for transitions between such energy levels.[70]

(9.9) (a) Consider a gas of spin $\frac{1}{2}$ particles of mass m^* moving in a two-dimensional y, z layer. Apply

---

[69] See eqn (9.39).
[70] Strictly, the selection rule you have derived applies only in the case of infinite quantum wells. However, it is a good approximation to finite quantum wells in most cases of practical interest.

the Born–von Karman boundary conditions (i.e. $\exp(iky) = \exp[ik(y+L)]$, etc. where L is a macroscopic length) to show that the density of states in the $k_{y,z}$ space is $1/(2\pi)^2$.

(b) By considering the annular area enclosed by two concentric circles in k-space differing in radius by dk show that the number of states with k vectors between k and $k+$dk is given by $g(k)$d$k = (k/2\pi)$dk.

(c) Hence, show that if the energy dispersion is given by $E(k) = \hbar^2 k^2/2m^*$, the density of states (per unit area) is given by,

$$g(E)\,dE = \frac{m^*}{\pi \hbar^2}\,dE.$$

Fibre lasers

10

10.1 Optical fibres

10.1.1 The importance of optical-fibre technology

One of the applications foreseen by the pioneers of laser technology was that of optical communications. The enormous bandwidth available at optical frequencies beckoned like a vast new continent waiting to be opened up just at the time when the radio-frequency spectrum seemed to be reaching its limits of capacity for accommodating new channels. However, it was quickly realized that point-to-point transmission of information carried by laser beams along open paths at low levels in the atmosphere has several limitations. Fog, snow, and even rain severely limited the quality of any transmission path longer than a few kilometres. The idea of enclosing the light path in straight runs of buried pipe, although pursued to the point of feasibility trials at Bell Labs in the 1960s, also proved impractical because of the near impossibility of purchasing the necessary straight-line rights-of-way along new highways built in open country let alone along existing city streets.

Popular lecture demonstrations in the nineteenth century showed that light beams could be guided along curved paths by water jets and glass rods. However, early experiments using glass fibres to guide the propagation of laser beams over useful distances encountered problems because of the high optical loss in the types of glass available in the early 1960s. At that time, the clearest glass was characterized by a loss[1] of 1 dB per metre, so that after transmission through a 20-metre length of fibre only one per cent of the original intensity emerged. Eventually, after a great deal of development work at the laboratories of the Corning company and elsewhere, ultrapure fused silica (SiO_2) was identified as the most promising material. Based on that development, present-day silica optical fibres typically have a loss in the range of 0.15–0.2 dB/km at wavelengths around 1550 nm where they show minimum loss.[2] It is therefore possible to transmit optical signals at this wavelength through approximately 170 km of fibre before an attenuation of 30 dB is reached and some form of amplifier is needed to boost the signal back to its launched level.

Today, optical-fibre technology plays a leading role in all forms of telecommunication networks. Apart from the last few metres of connection to the home, virtually all long-distance telephone, video and data transmissions are carried over optical fibres. There are plans to eliminate even the last copper-wire connection by fibre-to-the-home links in the foreseeable future. The continents are linked by undersea cables comprising bundles of many separate optical fibres, each capable of transmitting many data streams simultaneously

10.1 Optical fibres	267
10.2 Wavelength bands for fibre-optic telecommunications	280
10.3 Erbium-doped fibre amplifiers	282
10.4 Fibre Raman amplifiers	285
10.5 High-power fibre lasers	289
10.6 High-power pulsed fibre lasers	293
10.7 Applications of high-power fibre lasers	295
Further reading	296
Exercises	296

[1] Because the intensity of a signal varies with distance z along a transmission line as $I_{\text{out}} = I_{\text{in}} \exp(-\gamma z)$ the losses of transmission lines are most conveniently characterized in decibels, where the loss in dB $= -10 \log_{10}\left(\frac{\text{Power out}}{\text{Power in}}\right)$. See Exercise 10.1

[2] See, for example, Midwinter and Guo (1992), Stewart (2004) and Section 10.2.

by interleaving the digital signals in time (TDM – time division multiplexing), and by transmitting signals at several closely spaced carrier wavelengths along the same fibre[3] (WDM – wavelength division multiplexing). The development of the internet, with its world-wide coverage, has been made possible by the availability of such broadband optical communication links spanning the continents.

[3] The devices for tuning the multiple source lasers and separating out the various channels at the receiving end by wavelength employ the technology of fibre Bragg gratings (FBGs) discussed in Sections 9.10.4 and 10.1.5, respectively.

Lasers play a vital role in the implementation of this technology in several ways. First, the outputs of semiconductor diode lasers operating at wavelengths in the region of 1550 nm are modulated to provide the digitally encoded signals representing the data stream. Secondly, if the distance between repeater stations (where the optical signal can be detected and turned back into electronic form) exceeds 150 km ($\approx$ 100 miles), an in-line laser amplifier is required so as to boost the signal back up to the power level it had at launch. Thirdly, the excitation energy for these in-line optical amplifiers is itself supplied by semiconductor pump lasers operating at shorter wavelengths. A major part of the present chapter will be devoted to the theory and operation of the erbium-doped fibre amplifiers that form the basis of the in-line laser amplifiers.

Another way in which the technologies of optical fibres and lasers have come together most significantly is that of high-power fibre laser oscillators. In the first few years of the present century such lasers have seen a remarkable growth in output power capability, to the extent that they rival and may soon outstrip CO_2 lasers and Nd:YAG lasers that have traditionally dominated the market for applications in materials processing—for example the cutting and welding of thick steel sheet. The theory and operating characteristics of these powerful fibre-laser oscillators will therefore constitute the second major topic to be addressed in the present chapter.

10.1.2 Optical-fibre properties: Ray optics

The reason why optical fibres are able to guide light beams with minimal loss is that the mechanism confining the beam to the core of the fibre is that of total internal reflection that occurs, as its name suggests, without any loss. A ray striking the boundary between two dielectric media is totally internally reflected provided its angle of incidence exceeds the critical angle. When this is not the case, and a ray travelling in a medium of refractive index n_1 is incident at an angle of θ_1 (less than the critical angle) to the normal at the boundary with a medium of refractive index n_2, part of the incident intensity is reflected at angle θ_1, and part is transmitted as a ray at θ_2 into the second medium as shown in Fig. 10.1. The relationship between θ_1 and θ_2 is given by Snell's Law:[4]

[4] Named after one of its discoverers, the Dutch mathematician Willebrord Snellius (1580–1626).

$$n_1 \sin \theta_1 = n_2 \sin \theta_2. \quad (10.1)$$

If $n_1 > n_2$, then for angles of incidence θ_1 greater than the critical angle θ_c there are no real values for θ_2, and *all* of the intensity in the incident ray is reflected back into the medium of incidence. The critical angle is given by

$$\sin \theta_c = \frac{n_2}{n_1}. \quad (10.2)$$

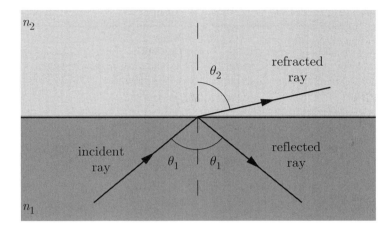

Fig. 10.1: Refraction at the plane boundary between a medium of refractive index n_1 and a second medium of refractive index n_2. For the case illustrated $n_1 > n_2$.

It might be thought that a thin glass fibre in air with no external cladding would suffice to act as a light guide (since $n_1 \approx 1.5$ for glass and $n_2 \approx 1.0$ for air), but there are two good reasons why this is impractical for all but a very small range of specialized applications. First, such a fibre would be very fragile and, when subjected to the slightest strain, would fracture very easily via cracks initiated from the multitude of unavoidable imperfections on its surface. Secondly, if any object were to come into contact with it anywhere along its surface, the condition for total internal reflection would be invalidated at the point of contact, and light would leak from that point. Moreover, if two bare fibres were to come into contact then there would be the possibility of "cross-talk" as the light from the optical signal in one fibre would leak into the other. For these reasons optical fibres are fabricated as shown in Fig. 10.2 with a small core of high refractive index material surrounded by a much larger cladding layer of lower refractive index. Typical values for fibres used in communications (at 1550 nm) are as summarized in Table 10.1.

As well as the 'step-index' profile of refractive index shown in Fig. 10.2, it is also possible to make fibres in which the refractive index gradually reduces

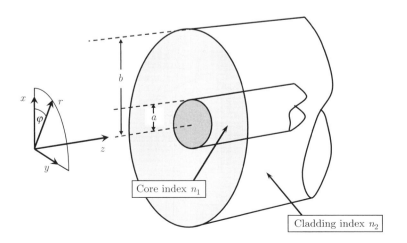

Fig. 10.2: Step-index optical fibre. Note that a protective external plastic coating (not shown because it plays no part in guiding the optical fields) is applied to the fibre at the time of manufacture, to prevent the initiation of cracks from microscopic imperfections on the outer surface of the cladding layer.

Table 10.1 Typical parameters for single- and multimode step-index optical fibres.

			Comments
Cladding index:	n_2	1.4500	pure fused silica
Core index:	n_1	1.4645	fused silica lightly doped with GeO_2
Core diameter:	$2a$	≈ 5 μm	single-mode fibre
		≈ 50 μm	multimode fibre
Cladding diameter:	$2b$	≈ 125 μm	

[5] See Exercise 10.6.

in an axially symmetric fashion from a maximum value at the axis towards lower values in the outer regions. Such fibres are known as 'graded-index' or as 'grin' fibres.[5]

As illustrated in Fig. 10.3, the maximum angle θ_0 (to the normal at the entrance face) from the medium of incidence with refractive index n_0, for light that can be confined by total internal reflection at the core–cladding boundary, determines the **numerical aperture** (N A) defined as:

$$NA = n_0 \sin \theta_0. \tag{10.3}$$

It is left as an exercise for the reader to show that if the medium of incidence is air ($n_0 \cong 1.000$)

$$NA = \sqrt{(n_1^2 - n_2^2)}. \tag{10.4}$$

For the case of a silica-based fibre, i.e. with $n_1 = 1.4645$ and $n_2 = 1.4500$, this yields NA = 0.2055, and thus $\theta_0 \approx 12°$.

The combination of this relatively small angle of acceptance and the small diameter of the fibre core (of order 10 μm) explains why laser sources are ideally suited to optical-fibre technology for the efficient transport of optical power.

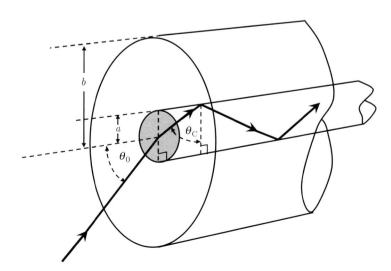

Fig. 10.3: Numerical aperture of a step-index fibre. In the meridional plane, the angle of incidence θ_0 of a ray that reaches the boundary between the core and the cladding at the critical angle θ_c determines the numerical aperture (NA) of the fibre.

10.1.3 Optical-fibre properties: Wave optics

Depending on the relative magnitudes of core radius a and the free-space wave vector $k_0 = 2\pi/\lambda_0$, only certain transverse distributions (modes) of the electric- and magnetic-field components can be transmitted by a step-index fibre without change in their form. In principle, the transverse modes may be found by solving Maxwell's equations, and matching the boundary conditions at the core cladding and cladding air boundaries. However, these derivations are complex, and beyond the scope of the present text.[6]

In practice, it is sufficient to consider the diameter of the cladding layer to be infinite; within this approximation the modes can be described as: transverse electric (TE), for which $E_z = 0$; transverse magnetic (TM), for which $H_z = 0$; and hybrid (HE and EH) for which both E_z and H_z are non-zero. In this notation the lowest-order mode is designated HE_{11}. However, description of the modes is rather complicated mathematically and is not easy to apply to practical cases.

A more amenable approach is that introduced by Gloge (1971) for the situation that applies to most fibres used in practice, namely that

$$\Delta = \frac{n_1 - n_2}{n_2} \ll 1. \tag{10.5}$$

Gloge was able to show that expressions can be derived for modes that propagate with very small field components along the z-direction. Such modes are designated as *linearly polarized* (LP) modes since the fields in the fibre are effectively linearly polarized in two orthogonal directions essentially perpendicular to z.[7] With reference to the coordinate system defined in Fig. 10.2, the amplitudes of the transverse electric fields in the core and cladding have the form:[8]

$$E_{\text{core}} = E_l \frac{J_l(Ur/a)}{J_l(U)} \cos(l\phi) \qquad (0 < r < a) \tag{10.6}$$

$$E_{\text{clad}} = E_l \frac{K_l(Wr/a)}{K_l(W)} \cos(l\phi) \qquad (a < r < \infty), \tag{10.7}$$

in which $l = 0, 1, 2\ldots$, J_l is a Bessel function of the first kind of order l, K_l is the modified Bessel function of order l, E_l is the transverse electric field at the core cladding interface. The parameters U and W are defined by

$$U = a\sqrt{n_1^2 k_0^2 - k_z^2}, \tag{10.8}$$

$$W = a\sqrt{k_z^2 - n_2^2 k_0^2}. \tag{10.9}$$

An important parameter that determines which modes are allowed to propagate is the so-called normalized frequency V defined as:

$$V = k_0 a \sqrt{n_1^2 - n_2^2} = k_0 a \text{NA}. \tag{10.10}$$

From eqns (10.8), (10.9), and (10.10) it is clear that

$$V^2 = U^2 + W^2. \tag{10.11}$$

[6] The interested reader is referred to specialized texts such as Midwinter and Guo (1992) or Stewart (2004).

[7] This result is not too surprising in view of the fact that, given $(n_1 - n_2)/n_2 \ll 1$, in the ray picture the ray must propagate at a very small angle to the axis of the fibre if it is to be totally internally reflected.

[8] There are also solutions in which $\cos(l\phi)$ on the right-hand side of eqns (10.6) and (10.7) is replaced by $\sin(l\phi)$. There are thus two identical patterns of electric field around any diameter, but rotated from one another by an angle π/sl for $l > 0$. In the case of $l = 0$ there is perfect rotational symmetry, and there can be only one set of patterns.

Once all of the other quantities involved (n_1, n_2, a and k_0) have been specified there still remains one undetermined quantity k_z, the wave number of the propagating wave. To find the value of k_z it is necessary to solve the following equation numerically:

$$U\frac{J_{l-1}(U)}{J_l(U)} = -W\frac{K_{l-1}(W)}{K_l(W)}, \qquad (10.12)$$

which follows, after somewhat lengthy manipulation, from the boundary conditions that the tangential components of E and H must be continuous at the core cladding interface.[9]

[9] See Gloge (1971).

The boundary conditions at $r = a$ determine how the solutions $J_l(Ur/a)$, which hold for $r < a$ (see Fig. 10.4(a)) match onto the solutions $K_l(Wr/a)$ for $r > a$ (see Fig. 10.4(b)). Clearly, for a given value of l there are several ways in which this matching can be done, depending on how many times the $J_l(Ur/a)$ function for small r/a crosses the zero line before $r = a$.

The number of times that the function $J_l(Ur/a)$ crosses the zero line for $r < a$ is denoted by the parameter m so that the modes can be designated LP$_{lm}$. The cut-off[10] is the value of V below which the mode just ceases to be guided and occurs when k_z, the wave number of the propagating mode in the core, is equal to the wave number of the signal in the cladding $n_2 k_0$. This means that for cut-off, $W = 0$ or equivalently that $V = U$. From eqn (10.12) the general cut-off condition can therefore be written as $J_{l-1}(V) = 0$. The cut-off conditions for the various modes are therefore:

[10] The word 'cut-off' is somewhat ambiguous in this context since it applies to the critical value of V above which propagation in the mode in question is allowed, so it might more logically be called the 'cut-on' condition. The word 'cut-off' is however such a widespread nomenclature in the literature of this subject that we have adopted it here.

Mode family		Comment	Cut-off cond.
LP$_{0m}$ modes:	$J_{-1}(V) = -J_1(V) = 0$		
LP$_{01}$	$m = 1$	first zero of J_1	$V = 0$
LP$_{02}$	$m = 2$	second zero of J_1	$V = 3.832$
etc.			
LP$_{1m}$ modes:	$J_0(V) = 0$		
LP$_{11}$	$m = 1$	first zero of J_0	$V = 2.405$
LP$_{12}$	$m = 2$	second zero of J_0	$V = 5.52$
etc.			

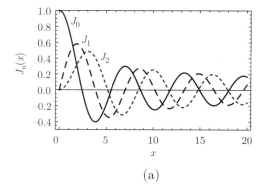

(a)

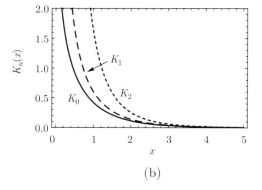

(b)

Fig. 10.4: The Bessel functions (a) $J_n(x)$ and (b) $K_n(x)$.

10.1 *Optical fibres* 273

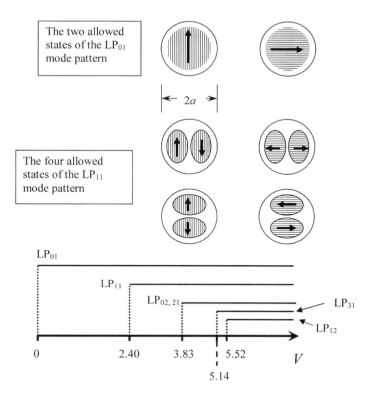

Fig. 10.5: Field distributions in the fibre core region for the allowed polarization states of the LP_{01} and the LP_{11} modes. In all cases the fields have peak amplitudes at the centre of the shaded regions, grading to lower values at the edges. The lower part of the diagram shows the ranges of V for which the lowest-order LP modes are allowed. Hatch lines indicate polarizations. Note that because of the two forms of solution for $l > 0$ (with $\cos(l\phi)$ and $\sin(l\phi)$ dependencies, respectively), and the two orthogonally polarized solutions for all values of l, there are two forms of field distribution for the LP_{0m} modes but four in the case of all other LP_{lm} modes.

From this, the condition for single-mode operation, i.e. the condition that only the LP_{01} mode should be guided is $0 < V < 2.405$.

Figures 10.5 and Fig. 10.6 show the field and intensity distributions of various LP_{lm} modes. The general form of a high-order LP_{lm} mode consists

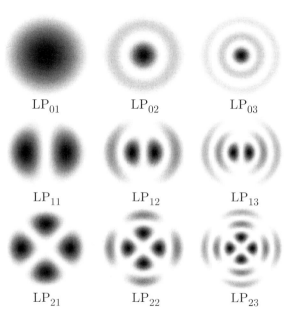

Fig. 10.6: Intensity variation of various higher-order LP modes.

of a symmetric pattern of bright spots having m spots along any radius connecting to a ring of $2l$ spots near the circumference (see, for example, Fig. 10.6). Because the dispersion characteristics of the various modes are different, a pulse transmitted over any appreciable distance along a fibre supporting many transverse modes will become distorted as the Fourier components making up the pulse will be transmitted with different speeds. It is therefore very important that an optical fibre used for long-distance optical communications should support only the lowest-order mode: the LP_{01} mode.

For a fibre with given values of n_1, n_2 and a, the value of the normalized frequency V determines which modes can be supported. The allowed LP modes for different ranges of V are stated in Table 10.2 and also indicated in the lower part of Fig. 10.5.

10.1.4 Dispersion in optical fibres

Intermodal dispersion

Any signal transmitted down a length of multimode fibre will be subject to distortion because different fibre modes have different transit times. This can be seen clearly from the ray-optics analogy of step-index fibres.[11] Rays enter the fibre at different angles to the axis within the cone defined by the numerical aperture of the fibre. Rays that enter with a shallow angle travel closer to the axis and take a more direct path than those entering at higher angles, which reflect many more times at the core cladding boundary. This **intermodal dispersion** limits the useable bandwidth of multimode fibres. For example, a 50-μm core diameter step-index fibre is limited to a (bandwidth × distance) product of about 20 MHz km, i.e. a 1-km long fibre has a useable bandwidth of 20 MHz. Intermodal dispersion is considerable smaller in comparable multimode graded-index fibres.[12] For example the (bandwidth × distance) product is typically 1 GHz km for a graded-index fibre of 50 μm core diameter.

The obvious way of eliminating intermodal dispersion is to use single-mode fibres. However, even when this is done there still remain other sources of dispersion, as discussed below.

[11] See Exercise 10.3.

[12] See Exercise 10.6.

Table 10.2 Allowed LP modes.

$0 < V < 2.40$	LP_{01}
$2.40 < V < 3.83$	LP_{01}, LP_{11}
$3.83 < V < 5.14$	$LP_{01}, LP_{11}, LP_{02}, LP_{21}$
$5.14 < V < 5.52$	$LP_{01}, LP_{11}, LP_{02}, LP_{21}, LP_{31}$

Table 10.3 Values of x for which $J_l(x)$ has its mth zero (after $x = 0$).

Values of $l \Rightarrow$	0	1	2	3
$m = 1$	2.40	3.83	5.14	6.40
$m = 2$	5.52	7.01	8.40	9.70
$m = 3$	8.65	10.20	11.60	13.00

Material dispersion

Even in a single-mode fibre, the various Fourier components of the signal will be transitted with different transit times because the refactive indices n_1 of the core material and n_2 of the cladding material are both functions of frequency. In calculating this contribution to the time spread it usually suffices to approximate the dispersion of both core and cladding to that of the bulk material (usually silica).

The group velocity in a material of refractive index $n(\omega)$ is given by

$$v_g = \frac{d\omega}{dk} = c[\frac{d}{d\omega}(\omega n(\omega))]^{-1} = \frac{c}{n(\omega) + \omega \frac{dn}{d\omega}} = \frac{c}{n_g}, \qquad (10.13)$$

where the bulk medium group velocity index $n_g = n(\omega) + \omega \frac{dn}{d\omega} = n - \lambda_0 dn/d\lambda_0$. The transit time T for a pulse transmitted through a medium of length L is therefore $T = n_g(L/c)$, and the spread of arrival times due to the spread of wavelengths $\Delta \lambda_0$ present in the light source is

$$\Delta T = \frac{L}{c} \frac{dn_g}{d\lambda_0} \Delta \lambda_0 = -\frac{L}{c} \lambda_0 \frac{d^2 n}{d\lambda_0^2} \Delta \lambda_0. \qquad (10.14)$$

We can therefore define a **material dispersion coefficient** D_m such that $\Delta T = D_m L \Delta \lambda_0$, where

$$D_m = -\frac{\lambda_0}{c} \frac{d^2 n}{d\lambda_0^2}. \qquad (10.15)$$

The quantity D_m is conventionally expressed in units of $\text{ps nm}^{-1} \text{km}^{-1}$.

Waveguide dispersion

In the case of a guided wave propagating in a fibre, a given mode will be described by terms of the form $\exp[i(k_z z - \omega t)]$ in which k_z is the propagation constant introduced in eqns (10.8) and (10.9). It is thus possible to define a quantity $v_g = \frac{d\omega}{dk_z}$ for that mode. From this we can define a dispersion parameter D_{total} such that,[13]

$$D_{total} = -\frac{\lambda_0}{c} \frac{d^2 n_e}{d\lambda_0^2}, \qquad (10.16)$$

where n_e, the effective index of the propagation mode, is defined in terms of the wave number of the propagating mode k_z as in eqns (10.8) and (10.9) such that

$$k_z = n_e k_0. \qquad (10.17)$$

The significance of D_{total} is that it includes the effects of *both* material and waveguide dispersion such that $|D_{total}|$ gives the asymptotic pulse broadening per unit bandwidth $\Delta \lambda$ per unit distance travelled. Including now the effects of both material and waveguide dispersion, the spread in arrival times becomes $\Delta T = D_{total} L \Delta \lambda_0$.[14] The total dispersion D_{total} can be represented as the sum of contributions from the bulk material D_m and the contribution of the waveguide (fibre) D_{wg}, as shown in Fig. 10.7.[15] By judicious choice of the fibre

[13] See, for example, Stewart (2004).

[14] The group-velocity dispersion defined here for an optical fibre is analogous to that defined in Section 17.1.3 for a dispersive medium, except that in a fibre the effects of waveguide dispersion are also included.

[15] An explicit value of D_{wg} can be derived from properties of the fibre neglecting the medium dispersion, but it is simpler to work with expression (10.16) since this automatically includes both.

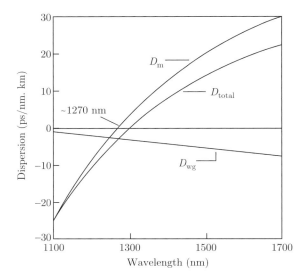

Fig. 10.7: Material, waveguide and total dispersion for a conventional single-mode fibre. (After Stewart (2004).)

parameters it is possible to balance the effects of waveguide dispersion against those of material dispersion to achieve near-zero values of D over a range of wavelengths around the design wavelength. In the example shown in Fig. 10.7 the D_{total} curve crosses the zero dispersion line at a wavelength of 1270 nm, so for wavelengths around this value the fibre is virtually dispersionless, and will transmit signals with minimum distortion.

10.1.5 Fabrication of optical fibres

The process of manufacturing optical fibres proceeds in two separate steps. Firstly, a large-diameter 'preform' is constructed in the form of a solid cylindrical rod of the cladding glass with a slim pencil of higher-index core glass running along its axis. In a second step, the preform is heated in a platinum crucible to its softening temperature and 'pulled' under tension (rather like toffee or chewing-gum) through a small hole in the bottom of the crucible. Such fibres typically have cladding and core diameters of 120–150 μm and 8 μm, respectively, and have a continuous length of several kilometres.

As indicated in Fig. 10.8 a layer of resin is applied to the newly formed fibre before it is finally wound on a take-up spool. The plastic layer greatly enhances the strength of the fibre by preventing the initiation of cracks that could otherwise propagate from small imperfections or microfissures on the glass surface.

Several techniques have been used to make preforms, but all are based on the use of ultrapure fused silica as the basic material. The addition of small quantities of the oxides TiO_2, GeO_2, P_2O_5, or Al_2O_3 increases the refractive index, while B_2O_3 and F lower it. Preforms can therefore be made with a pure silica cladding on a germanium-doped silica core, or alternatively with a pure silica core surrounded by a boron-doped silica cladding.

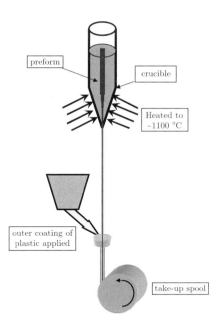

Fig. 10.8: Manufacture of optical fibres (schematic). The preform is heated to its softening temperature in a crucible near the top of a fibre-pulling tower. A viscous stream of the material is pulled under tension from a hole in the bottom of the crucible. The fibre cools progressively as it moves down the tower before being wound on a take-up spool at the foot of the tower.

An example of a doped-core technique for manufacturing preforms is the **inside vapour deposition** method, in which a tube of ultrapure fused silica (e.g. 8 mm ID and 10 mm OD and 40 cm long) is slowly rotated in a horizontal glass lathe and heated by the flame of a hydrogen burner to a temperature of 1900 K.[16] A slow flow of the gases silicon tetrachloride ($SiCl_4$) and germanium tetrachloride ($GeCl_4$) together with oxygen is maintained inside the tube. At this temperature the reactions:

$$SiCl_4 + O_2 \rightarrow SiO_2 + 2Cl_2 \qquad (10.18)$$

$$GeCl_4 + O_2 \rightarrow GeO_2 + 2Cl_2 \qquad (10.19)$$

[16] This is in contrast to an earlier technique (called simply 'chemical vapour deposition') in which a lower temperature was used, and the reactions could occur only on the tube walls. For this reason the current technique is also called *modified* chemical vapour deposition (MCVD).

can occur efficiently in the gas phase throughout the volume of the tube. The oxide particles agglomerate in chains and deposit on the tube's inner wall as 'soot'. Multiple passes of the gas torch along the tube sinter the soot particles into a clear glass layer bonded to the tube. Finally, the tube is evacuated and the temperature raised to approximately 2250 K, which causes the tube to collapse inward to create the preform as a solid cylindrical rod with the higher refractive index core material along its axis.

10.1.6 Fibre-optic components

In order to be able to manipulate light beams confined to optical fibres with the same flexibility as is available in normal 'open-beam' optical systems, fibre-optical components have been developed to perform the functions of beam splitters, bandpass filters, etc.

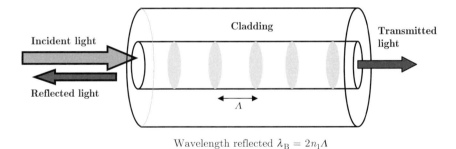

Wavelength reflected $\lambda_B = 2n_1\Lambda$

Fig. 10.9: Fibre Bragg grating, schematic.

Light sources

For most common applications in fibre optics, semiconductor lasers are the light sources of choice. The single-mode laser illustrated in Fig. 9.18 is representative of this class of laser. In this example the laser has a short length of fibre (the pig-tail) bonded directly onto the semiconductor laser chip. The external fibre-optic system is connected to the laser by clamping the end section of the pig-tail fibre and the external fibre accurately in alignment in a special jig, and fusing them together with the help of a radio-frequency discharge plasma.

Bandpass and bandstop filters

As indicated in Fig. 10.9, a fibre Bragg grating (FBG) comprises a length of fibre in which the core refractive index is modulated periodically along the fibre axis.[17] FBGs can be written directly into the Ge-doped core of the step-index fibre by exposing (side-on) a length of fibre to the interference pattern formed at the intersection of two beams from the same moderately powerful UV laser operating in the spectral region 248–255 nm. In the high-intensity UV stripes the Ge–O bonds are broken, and the local refractive index of the core is permanently modified.

If the FBG is written along only a short length of fibre (say only a few tens of μm) so that it comprises only a relatively small number of periods, then the transmission peak will be relatively broad and the peak reflectivity rather low. However, if the FBG is inscribed over a length of several mm or even cm then, as illustrated in Fig. 10.10, the stopband peak can be very narrow.

[17] Fibre Bragg gratings are analogous to the distributed Bragg reflectors discussed in Section 9.10.4.

Directional couplers

The component that in fibre-optic systems plays the role of a beam splitter in conventional optics is the directional coupler. This device allows a fraction of the optical power conducted along one fibre to give rise to a signal at the same frequency in one or more other fibres. The design of such a directional coupler is shown schematically in Fig. 10.11.

The device is formed by stripping off the outer protective plastic coating over a short length of two fibres that are brought into close contact with one

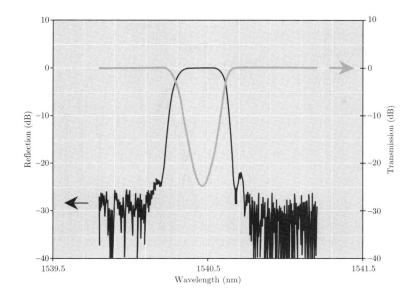

Fig. 10.10: Transmission and reflection properties of a fibre Bragg grating written into a Ge-doped single-mode fibre by a frequency-doubled copper-vapour laser at 255 nm. The stopband is 50 GHz wide at 1540.4 nm. (Courtesy Oxford Lasers Ltd.).

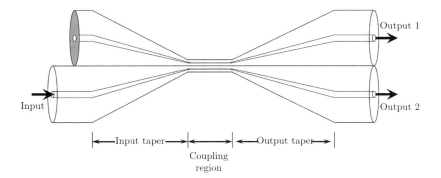

Fig. 10.11: Schematic diagram of a fibre directional coupler.

another over a region to which heat is applied to soften the silica. At the same time, the fibres are subjected to a carefully controlled tension causing them to stretch and taper down. Ultimately, the fibres fuse together over the central part of the heated region. In the workstation described by Cronin et al. (2005), the heat is applied by a CO_2 laser via a diffractive optical element that ensures an even distribution of the 10.6-μm laser radiation, strongly absorbed by silica.

The evanescent wave in the narrowed cladding of the input fibre is responsible for the leakage of radiation fields into the narrowed cladding of the other fibre and the growth of a propagating radiation mode in the core of the second fibre.

Since the power coupled to the two output ports is monitored while the process of fibre stretching is under way, any particular output coupling ratio

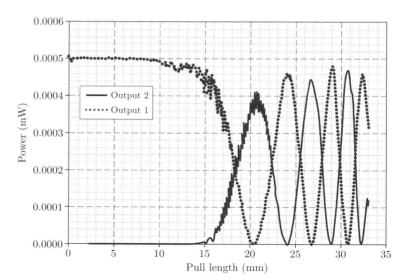

Fig. 10.12: The available power at the two fibre outputs of the device shown in Fig. 10.11 as a function of the elongation of the fibres in the heated region. It should be noted that in Fig. 10.11 the fibres are twisted together in the contact region before heat is applied, so that the fibre on the top on the right-hand side is the continuation of the fibre on bottom on the left-hand side. Hence Output 1 is, as expected, 100% at zero pulling force, since it is the continuation of the input fibre. The crossing is necessary to maintain good contact between the fibres in the heated region. The authors are grateful to Prof. T. Glynn of NUI Galway for pointing this out. Figure reproduced from Cronin et al. (2005) courtesy of SPIE.

in the finished product can be achieved by stopping the pulling process when the desired ratio is reached. To calculate the characteristic length period over which the two fibres must be in contact in order for the travelling-wave mode in the input fibre to excite a similar mode in the output fibre requires the solution of Maxwell's equation for the five-layer structure (cladding-core 1-cladding-core 2-cladding) with appropriate field-matching conditions applied at each of the boundaries.[18] If, as in Fig. 10.11, the two fibres are identical, then $P_2(z)$, the power induced in fibre 2 at a distance z from the start of the overlap region is related to $P_1(z)$, the power in fibre 1, by:

$$P_2(z) = P_1(0) \sin^2 \kappa z \qquad (10.20)$$

$$P_1(z) = P_1(0) \cos^2 \kappa z, \qquad (10.21)$$

[18]See, for example, Stewart (2004).

where κ is a coupling coefficient determined by the overlap of the fields in the two fibres, which can in principle be calculated numerically, but is usually determined empirically as described above. The oscillation of $P_1(z)$ and $P_2(z)$ is evident in Fig. 10.12.

Couplers providing any desired ratio of splitting between 50:50 and 15:1 are available commercially, with an excess loss less than 1 dB. The fused region of the coupler can be packaged in a housing of less than 3 mm diameter and 25 mm length.

10.2 Wavelength bands for fibre-optic telecommunications

First-generation optical-fibre networks, operating over relatively short distances, tended to use wavelengths in the range of 850 nm, where GaAs-based laser sources were available. However, as shown in Fig. 10.13 the absorption

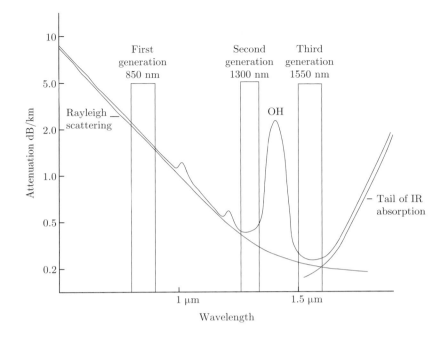

Fig. 10.13: Losses in silica fibres for the 3 main wavelength regions used in optical-fibre networks. (After Stewart (2004)).

of silica fibre is relatively high over this wavelength range ($\approx 2\,\text{dB}\,\text{km}^{-1}$) due to Rayleigh scattering within the fibre.

Second-generation optical networks employed wavelengths around 1300 nm where Rayleigh scattering losses are considerably lower—although not as low as ultimately proved attainable in the 1550 nm range, once the effect of the OH absorption peak at 1380 nm had been largely eliminated by careful selection and preparation of materials. Not only is the attenuation in modern silica fibre at its lowest around 1550 nm (the effect of absorption in the infrared bands of silica becomes increasingly important towards even longer wavelengths), but the availability at this wavelength of erbium-doped fibre amplifier (EDFA) technology to boost signals ensured that this would become the wavelength range of choice for long-haul optical telecommunications systems. The invention of the EDFA in 1987 was announced in publications by David Payne and coworkers at the University of Southampton (Mears et al. 1987) and independently by Emmanuel Desurvire et al. (1987) then at Bell Labs.

As indicated in Fig. 10.14, in addition to the optically pumped fibre-laser amplifiers based on Er^{3+} and Th^{3+} for the 1550-nm band, analogous devices based on Pr^{3+} or Nd^{3+} have been developed for the 1300-nm band.

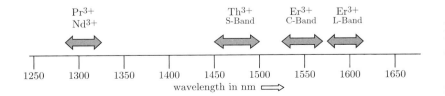

Fig. 10.14: Optically pumped fibre amplifiers and their regions of operation. The main telecommunications band centred on 1550 nm is subdivided into 'short', 'central' and 'long' wavelength bands indicated by the prefixes S-, C-, and L-, respectively.

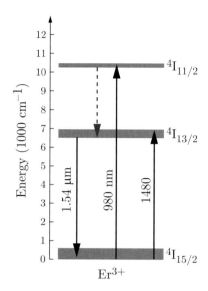

Fig. 10.15: Schematic diagram showing the energy levels of Er^{3+} ions relevant to the erbium-doped fibre amplifier and the two possible pump bands at 980 and 1480 nm. The dotted arrow indicates a non-radiative transition.

In the next two sections we discuss in detail two important types of fibre laser used in telecommunications: the erbium-doped fibre amplifier, and fibre Raman amplifiers.

10.3 Erbium-doped fibre amplifiers

10.3.1 Energy levels and pumping schemes

The energy levels of Er^{3+} ions doped into crystalline and glass hosts are described in detail in Section 7.2.6. For convenience, the relevant energy levels in Er:Glass are also shown schematically in Fig. 10.15.

Although the EDFA can be pumped at visible wavelengths, it is clear that to obtain the best overall laser efficiency the pump wavelength should be close to that of the laser transition. As discussed in Section 7.2.6, optical pumping at 980 nm excites ground-state ions to the $^4I_{11/2}$ level from which they decay non-radiatively to the $^4I_{13/2}$ upper laser level. Alternatively, so-called in-band pumping at 1480 nm promotes ions to the upper levels of the $^4I_{13/2}$ manifold, which is followed by rapid non-radiative relaxation to the bottom of the $^4I_{13/2}$ manifold.

10.3.2 Gain spectra

Figure 10.16(a) shows the wavelength dependence of the absorption cross-section $\sigma_a(\lambda)$ of erbium ions in a glass fibre, in the absence of any pumping. Also shown is the gain cross-section $\sigma_g(\lambda)$ that would apply if the lower laser level were empty. Since these two curves overlap, the behaviour of the Er.glass laser is complex, and in particular the bandwidth over which gain can be achieved depends on the pumping rate of the upper laser level.

These effects are evident in Fig. 10.16(b), which shows the net gain coefficient as a function of wavelength for different pump intensities. It is seen that at long wavelengths positive gain is achieved at low levels of pumping, which is characteristic of four-level laser behaviour. In contrast, at shorter wavelengths gain only occurs for high pump intensities, which is associated with three-level lasers. This behaviour arises from the fact that the lower manifold of states extends over approximately 850 cm^{-1}, compared to $k_B T \approx 210$ cm^{-1}. Lasing at long wavelengths occurs to the top of the $^4I_{15/2}$ manifold that has only a low thermal population and that undergoes rapid relaxation. However, lasing at shorter wavelengths occurs to lower-lying Stark levels that have a large thermal population; this population must be significantly reduced by the pumping before a population inversion is achieved.

This description may be made quantitative by defining a net cross-section $\sigma_{\text{net}}(\lambda)$ by

$$\sigma_{\text{net}}(\lambda) = \frac{N_2}{N}\sigma_g(\lambda) - \frac{N_1}{N}\sigma_a(\lambda), \qquad (10.22)$$

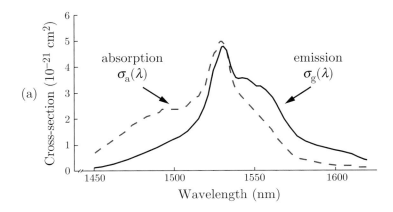

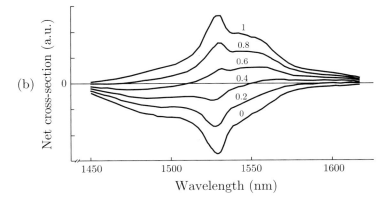

Fig. 10.16: Net gain in an EDFA. (a) the wavelength dependence of the gain and absorption cross-sections of Er^{3+} ions doped in a glass host. (b) the net gain cross-section $\sigma_{\text{net}}(\lambda)$ for various values of the inversion ratio N_2/N, as determined by eqn (10.22). (After Cordina (2004).)

where N_2 and N_1 are the population densities of the upper and lower laser levels, respectively, and N is the total density of Er ions. The net gain *coefficient* is then equal to $N\sigma_{\text{net}}(\lambda)$. Increased pumping increases N_2 and decreases N_1, thereby extending the region of gain to shorter wavelengths, as observed in Fig. 10.16(b).

In practice, inversion ratios up to a maximum $N_2/N \approx 0.8$ may be achieved. From Fig. 10.16 we see that an inversion ratio of this magnitude produces sufficient net gain in the C-band region from 1528 nm to 1560 nm in fibre amplifiers only a few metres long. Although it is clear that gain is also available at wavelengths outside this region, particularly at the long-wavelength end, this is not usable since for telecommunications applications the gain vs. wavelength curve must be relatively flat over the entire band. In practice, this is achieved by inserting gain-flattening filters after the amplifier section of fibre to bring the highest gain down to a value consistent with that in the pedestal region 1540 nm to 1560 nm. Hence, in order to exploit the gain available in the L-band region from 1565 nm to 1610 nm it is necessary to construct much longer (up to 100 m) fibre amplifiers than is the case for the C-band.[19]

[19] Excited-state absorption eventually sets a limit to the amplifier length that can be used.

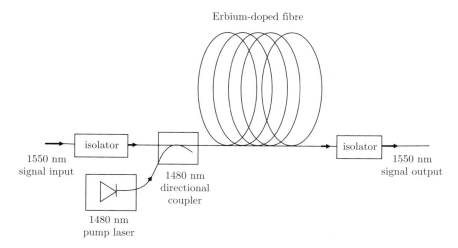

Fig. 10.17: Schematic diagram of an erbium-doped fibre amplifier (EDFA) operating at 1550 nm.

The small-signal gain G (expressed in dB) of a given length of fibre laser described by a *four*-level scheme in which the lower state is effectively empty, can be related to the pump power absorbed, P_p, in the same length of fibre by

$$G[\text{dB}] \approx 4.3 \times \frac{\phi(\text{NA})^2 P_p}{2\pi(2.4)^2 \Delta\nu h\nu_P}, \qquad (10.23)$$

where $h\nu_P$ is the energy of the pump photons, and ϕ is the fraction of the net pumping rate that reaches the upper laser level.[20] Here, $\Delta\nu$ is the FWHM width of the laser gain cross-section; the gain cross-section over the frequency range of interest is approximated by half its peak value; and the mode area for both pump and laser radiation is assumed equal to the core area πa^2, where the core area is related to the fibre's numerical aperture (NA) by eqn (10.23), with V chosen to have maximum value ($V = 2.4$) consistent with single-mode operation.

[20] See Exercise 10.5.

Equation (10.23) cannot be applied directly to the case of the Er-doped amplifier since it is not a simple four-level scheme, as discussed above. An analytical treatment of the three-level case is given by Desurvire et al.,[21] but it is found that for pump radiation intensities of only a few mW, the three-level case reverts to that described by eqn (10.23) for a four-level system.[22] The reason why the pump power needed to achieve high intensities is quite modest is that the cross-sectional area of the core is very small ($10^{-11} - 10^{-10}$ m^2). We note that EDFAs typically have a single-pass small-signal gain of order 10 dB per mW of pump power.

[21] See Desurvire et al. (1987).

[22] See, for example, Hanna (2004).

10.3.3 EDFA design and layout

The key elements of an erbium-doped fibre amplifier are shown in Fig. 10.17. Radiation from a semiconductor pump laser operating at 1480 nm is coupled into the core of the erbium-doped fibre via a directional coupler[23] that allows

[23] Because directional couplers are wavelength-sensitive devices, they are sometimes referred to as 'wavelength division multiplexers' or WDMs.

transmission at 1480 nm but not at 1550 nm. The coil of erbium-doped fibre (comprising typically 1–10 m of fibre for C-band amplifiers, but up to 100 m for L-band devices) may be pumped at the output end as well as the input end. Optical isolators[24] are used for a number of reasons: the isolator at the input end excludes backward-travelling noise leaving the amplifier; and that at the output end prevents backward-travelling light in the amplifier band from entering the amplifier and competing with the signal. In practice, commercial amplifiers employ several other components such as gain-flattening filters to equalize the gain across the available bandwidth and diodes to monitor the power at various points along the amplifier.

[24] Optical isolators are devices that have a low loss for travelling waves propagating in the forward direction but show high loss for waves propagating in the reverse direction. See Sections 14.6.2, Fig. 14.12, and 16.2.3.

10.3.4 Fabrication of erbium-doped fibre amplifiers

One technique used to make erbium-doped fibre amplifiers (and other analogous rare-earth doped devices) is simply an extension of the MCVD technique described in Section 10.1.5. A volatile compound of erbium is introduced along with the germanium tetrachloride, silicon tetrachloride and oxygen stream, and causes trivalent erbium to be deposited as a component of the soot. The technique allows for very accurate control of the dopant concentration and profile since it is possible to build up the core layers with a sequence of depositions of controlled composition.

However, since rare-earth compounds are relatively expensive, an alternative and more economical approach starts by carrying out the process of laying down the core material in the form of soot on the inside wall of the tube via the first stage of the MCVD process described in Section 10.1.5. At that point the tube is filled with an aqueous solution containing the rare-earth salt. Gentle heating of the tube after the liquid has been decanted leaves the rare-earth ions deposited in the voids of the soot. The temperature is then raised so that the porous coating inside the tube is compacted and sintered to the tube wall, and finally the temperature is raised once again so that the tube collapses to a solid-rod preform.

10.4 Fibre Raman amplifiers

10.4.1 Introduction

In addition to the optically pumped fibre laser amplifiers employing trivalent rare-earth-doped fibres discussed in Section 10.3, amplifiers using the stimulated Raman effect have recently begun to assume equal importance for long-haul telecommunications networks. Theoretical and practical aspects of these devices are discussed in the next two subsections.

10.4.2 Raman scattering

The process of spontaneous Raman scattering[25] is illustrated in Fig. 10.18. A photon of angular frequency ω_p is incident upon a ground-state molecule

[25] Spontaneous Raman scattering was discovered by C. V. Raman in 1928. A detailed discussion of this phenomenon will be found in the texts by Hayes and Loudon (1978) or D. A. Long (1977).

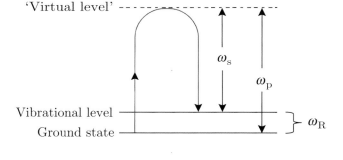

Fig. 10.18: Raman scattering. In a single interaction a ground-state molecule absorbs a photon of frequency ω_p and simultaneously emits a photon of frequency ω_s leaving the molecule in an excited state of energy $\hbar\omega_R$. The off-resonance transition proceeds via a so-called 'virtual' level that corresponds to the far wing of a real level of parity opposite to that of the ground state.

and in a single interaction both excites it to a virtual state of energy $\hbar\omega_p$ and de-excites it with the emission of a photon of $\hbar\omega_s$, leaving the molecule in a vibrationally excited state $\hbar\omega_R$. As shown here, the emitted photon at the signal frequency ω_s has lower energy than the incident pump photon of frequency ω_p, so the process produces a positive[26] Stokes shift. Provided the photons incident on the molecule are pre-dominantly at the pump frequency, while those at the signal frequency escape in all directions, the rate per unit volume of material at which photons are spontaneously Raman scattered into solid angle $d\Omega$ is given by

$$\text{rate of scattering per unit vol.} = N\phi_p \left(\frac{d\sigma^R}{d\Omega}\right) d\Omega, \tag{10.24}$$

where N is the number of scattering molecules per unit volume and ϕ_p is the incident flux of photons at the pump frequency. Equation (10.24) may be considered as the definition of $\frac{d\sigma^R}{d\Omega}$, the differential Raman scattering cross-section per molecule.

[26] To the extent that the energy of the excited vibrational state $\hbar\omega_R$ is sufficiently high that its thermal population at ordinary temperatures is negligible, there is little possibility for anti-Stokes scattering to occur. In the case of anti-Stokes scattering the molecule would have to start in the state of energy $\hbar\omega_R$ and proceed via a virtual state of energy ($\hbar\omega_p + \hbar\omega_R$) back to the ground state with emission of a photon at the anti-Stokes frequency ($\omega_p + \omega_R$).

10.4.3 Fibre Raman amplifiers

In the case where there is a collimated beam of strong intensity I_p at the pump frequency ω_p *and* also a strong beam of intensity I_s at the (Stokes) signal frequency ω_s (where $\omega_s = \omega_p - \omega_R$) then the interaction can proceed as a stimulated process. Thus, in an optical fibre, where the modal cross-section of the pump and signal beams are assumed to be equal, the evolution of intensity of the two beams is governed by:

$$\frac{dI_s}{dz} = g_R I_p I_s - \alpha_s I_s, \tag{10.25}$$

and

$$\frac{dI_p}{dz} = -\frac{\omega_p}{\omega_s} g_R I_p I_s - \alpha_p I_p, \tag{10.26}$$

where the absorption coefficients of the fibre at the signal and pump wavelengths are α_s and α_p, respectively. All of the details of the Raman interaction are swept up into the effective Raman gain coefficient g_R, given by

$$g_R = \frac{2\lambda_p \lambda_s^2 N}{\hbar c} n_s^2 \Delta\omega_R \left(\frac{d\sigma^R}{d\Omega}\right), \quad (10.27)$$

in which $\Delta\omega_R$ is the FWHM of the Raman spectral line and n_s is the refractive index of the fibre at the signal frequency. The differential Raman cross-section $\frac{d\sigma^R}{d\Omega}$ can be expressed in terms of the fundamental atomic properties of the medium,[27] but in practice values of g_R (and hence $\frac{d\sigma^R}{d\Omega}$) are found by experiment. Further, since the bulk material of optical fibres used for optical materials is silica (SiO$_2$), it is usual to express the Raman cross-sections normalized to that of SiO$_2$, as in Table 10.4.[28]

In the steady-state regime (where the pump pulse duration is much longer than the Raman mode dephasing time), and for the case where the depletion and absorption of the pump signal are both negligible, eqn (10.25) can be integrated to yield the (small-signal) growth of the intensity at the signal frequency with the distance travelled z

$$I_s(z) = I_s(0) \exp\left(g_R I_p z - \alpha_s z\right). \quad (10.28)$$

In reality, the situation is more complicated since the effects of pump depletion and conversion into higher Stokes components, as well as power in both forward and backward travelling waves, must be considered. However, eqn (10.28) underlines the importance of minimizing the fibre loss.

Fibre Raman amplifiers offer the possibility of extending the wavelength range over which amplification can be obtained. The Raman gain profile for silica has a bandwidth of about 7 THz FWHM and a shift of 13.2 THz, so that by selecting a pump wavelength of 1450 nm Raman amplification at the centre of the C band at 1550 nm becomes possible. Thus, the transmission fibre itself can be used as an amplifying medium simply by coupling directly to the core of the fibre an intense source of pump radiation at the appropriate wavelength. However, as indicated in Fig. 10.19, by using multiple dopants, or multiple pump wavelengths with a single dopant it is possible to provide Raman gain over an extended wavelength range. This provides a way to increase the information-carrying capacity of a single fibre by exploiting the technique of dense wavelength division multiplexing (DWDM) in which signals at many closely separated carrier frequencies are propagated over the same fibre link.

[27] In SI units, $\left(\frac{d\sigma^R}{d\Omega}\right) = \frac{1}{(4\pi)^2} \frac{(\hbar\omega_s c)^4}{2m_r \omega_R} \left(\frac{\partial\alpha}{\partial q}\right)^2$, where m_r is the reduced mass, α is the polarizability and q is the representative bond length between the molecules forming the bond whose vibrational frequency is ω_R. See, for example, Basiev and Powell (2004).

[28] Data from Headley (2004).

Table 10.4 Relative peak Raman cross-sections of the most commonly used glass materials and the frequency shift $\nu_R = \omega_R/2\pi$ at which the peak occurs.

Dopant	Rel. peak cross-section	Frequency shift ν_R (THz)
SiO$_2$	1	13.2
GeO$_2$	7.4	12.6
B$_2$O$_3$	4.6	24.2
P$_2$O$_5$	4.9	19.2
P$_2$O$_5$	3.0	41.7

10.4.4 Long-haul optical transmission systems

It is clearly much easier and cheaper to service the pump lasers if they are all situated in easily accessible shore-based stations than if they lie on the ocean floor. For this reason repeaterless links are favoured for short stretches of undersea fibre. As indicated in Fig. 10.20, by using a combination of EDFAs

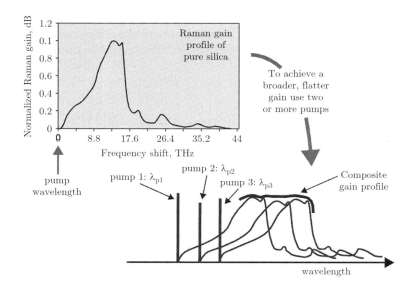

Fig. 10.19: Normalized Raman gain profile for pure silica and the composite gain profile that can be achieved by pumping at multiple pump wavelengths. (After Urquhart (2004)).

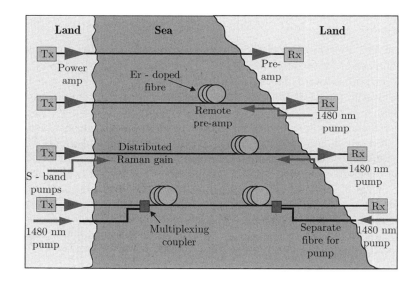

Fig. 10.20: Some of the possible amplifier configurations available for repeaterless optical-fibre links. Gain can be provided by amplifiers at the shore-based terminal stations, remote sections of erbium-doped fibre, distributed Raman gain in the transmission fibre, or by various combinations of these. Tx indicates transmitter and Rx receiver. The approaches illustrated schematically here can span up to 200–400 km. For longer undersea links repeaters must be used, which implies that the pump lasers and amplifiers must be located in submerged pods, as shown in Fig. 10.21. (After Urquhart (2004).).

and Raman fibre amplifiers it is possible to span distances of up to 200–400 km using undersea fibre links with all of the powerful pump lasers and electronics located in shore-based stations at either end or both.

Although this solution is adequate for island-hopping links it is clearly not relevant to the enormous distances involved in transatlantic (6 000 km) or transpacific (11 000 km) spans. In these cases there is no alternative but to provide submerged repeater units at intervals along the length of the multifibre cable and integral to its construction. Electrical power for the pump lasers is supplied via conventional electrical conductors included in the cable bundle. As shown in Fig. 10.21, with all of its protective steel and plastic outer sheathing, the

Fig. 10.21: Undersea fibre-optic cable repeaters in the hold of the cable laying ship 'Intrepid' in 2008 prior to being laid on the sea bed as part of the Southeast Alaska communications network. Photograph by Hall Anderson, courtesy of the Ketchikan Daily News, Alaska.

cable can exceed 15 cm in diameter with sections housing repeaters having diameters of 30–40 cm.

10.5 High-power fibre lasers

10.5.1 The revolution in fibre-laser performance

As we have seen, optically pumped fibre-laser amplifiers have been in widespread use since the early 1990s for boosting weak signals to extend the range of optical communications networks over long distances. However, advances in the early years of the twenty-first century have made it possible to realize fibre lasers with powers in the range of kilowatts to tens of kilowatts. Such devices have found many applications in engineering, including cutting and welding sheet steel.

The problem that had to be overcome was that the power necessary to excite the active ions in rare-earth-doped fibre amplifiers can, in practical systems, only be obtained from *multielement* arrays of semiconductor lasers whose output is rather divergent and incapable of being focused to dimensions compatible with the 5–12 μm diameter of the fibre core. It is thus not possible to couple efficiently the high power from such an array directly into a tiny fibre core using an *end-on* pumping geometry.

The way to solve this problem was demonstrated by Eli Snitzer and colleagues at Polaroid Corporation who invented the cladding-pumped fibre laser in 1988.[29] The key concept is to surround the active core with *two* layers of cladding. The inner, undoped, cladding layer has a refractive index intermediate between that of the core and the outer cladding. The intermediate cladding layer can be more than 100 μm across, and easily accepts many hundreds of watts of moderately divergent pump radiation from a semiconductor laser array. All of this light, confined to the intermediate layer by total internal

[29]Snitzer et al. (1980a).

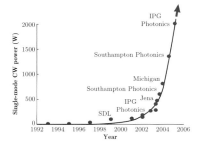

Fig. 10.22: Growth of single-mode c.w. output power capability of high-power fibre lasers. (Data courtesy of Prof. Almantas Galvanauskas, University of Michigan.)

[30] See, for example, Graydon (2006).

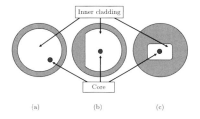

Fig. 10.24: Some possible cladding-pumped fibre laser geometries: (a) off-centre core, (b) D-shaped inner cladding, (c) rectangular inner cladding. The outer cladding is indicated by the dark grey shading, and the final protective plastic coating is omitted.

reflection, must eventually intercept the core somewhere along its length and thus contribute to pumping the fibre laser. By adding reflectors at each end of a length of the fibre, the amplifier may be turned into a very powerful oscillator.

Since 1992 commercial devices of this type have shown a rapid increase in power output capability, as illustrated in Fig. 10.22. Reported[30] c.w. output powers reached 36 kW in 2006.

10.5.2 Cladding-pumped fibre-laser design

As indicated in Fig. 10.23, the pump radiation is confined to the guide bounded by the outer cladding layer and progresses down the length of the guide until it is completely absorbed by the active core. However, the radiation profile excited in the inner cladding guide must always comprises a superposition of high-order LP_{lm} modes (see Section 10.1.3) many of which have zero intensity along the cylindrical guide axis.

While the LP_{0m} modes exhibit a maximum of intensity along the axis, as shown in Fig. 10.6, the higher-order modes have zero intensity along the axis. Clearly, if the active-mode core were situated exactly along the axis of a guide with perfect cylindrical symmetry, then a good fraction of the pump radiation would never be intercepted by the core. In order to avoid this problem several different arrangements of fibre core, inner and outer cladding geometry have been employed.

Although, as indicated in Fig. 10.24, the earliest scheme involved simply siting the core of the fibre off-centre with respect to the inner and outer cladding layers, it has proven more practical for quality control in the mass production of these devices to employ a centrally located core but to ensure scrambling of high-order modes in the inner cladding layer by forming its boundary with the outer cladding in some non-circular shape, two examples of which are illustrated in Fig. 10.24. To further promote a thorough degree of mode mixing, the fibre itself is usually wound on a kidney-shaped former rather than as a straightforward circular coil. In terms of ray optics then,

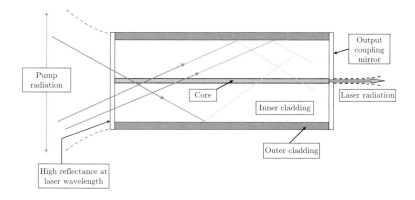

Fig. 10.23: Schematic diagram of a cladding-pumped fibre laser. Only the fibre core, inner cladding and outer cladding glass layers are shown. The outer protective plastic jacket is omitted. Note that in practice the fibre may be up to several metres in length.

one can consider that rays of pump radiation simply bounce around in the space bounded by the outer cladding, intercepting the core at intervals. Thus, sooner or later along the length of the fibre, they are completely absorbed by the core.

The cladding-pumped fibre laser efficiently converts laser radiation from one wavelength to another, slightly longer, wavelength. If that was all that it did then it would not be very important. The main reason it is of interest is that it can efficiently convert semiconductor laser radiation of very poor beam quality into an output beam of high quality, often in a single transverse mode. Moreover, the output is delivered via a flexible coupling to the workpiece, and hence this type of laser possesses most of the desirable qualities for high-power c.w. laser machining systems.

10.5.3 Materials and mechanisms of cladding-pumped fibre-laser systems

Several species of trivalent rare-earth ions such as Nd^{3+}, Er^{3+} and Tm^{3+} can form the basis of high-power cladding-pumped fibre-lasers; however, by far the most important system is that based on trivalent ytterbium in germano-alumino-silicate glass.

Figure 10.25(a) shows the energy-level scheme of Yb^{3+} in a germano-alumino-silicate glass host. The Yb^{3+} system is particularly suitable for fibre lasers since its pump band at 915 nm lies in a region where powerful semiconductor lasers are available. Further, there are no energy levels higher than those shown in Fig. 10.25(a) until the UV region is reached, so there is no possibility of absorption to excited states causing internal loss. While cooperative transfer of excitation[31] can occur, its only effect is to transfer excitation from one of the Stark sublevels within the upper-level manifold to one slightly lower in energy within the same manifold. The observed fluorescence lifetime of 700 ns is essentially radiative with negligible non-radiative contributions.

The cross-sections and decay times for the various levels identified in the simplified level scheme of Fig. 10.25(b) are listed in Table 10.5.

10.5.4 High-power fibre lasers: Linewidth considerations

It is evident from the variation of gain cross-section with wavelength shown in Fig. 10.25(c) that gain on the 1080-nm transition of Yb^{3+} is spread over a bandwidth of approximately 50 nm. The main effect responsible for the gross broadening is the random nature of the Stark effect on the sublevels of Yb^{3+} ions in their various sites within the glass matrix. From this we might be led to conclude that the broadening is inhomogeneous in nature, suggesting that it would be difficult to extract power from the entire inversion density in the form of narrowband 'single-longitudinal mode' output. While it is true that high-power fibre lasers not provided with extra line-narrowing components tend to produce radiation spread over bandwidths of order 5–10 nm FWHM, it

[31] In rare-earth ion systems (such as Yb^{3+} in glass) at high levels of doping, one excited ion can pass on its excitation to a neighbour and that in turn becomes excited and can transfer its excitation to another neighbouring ground-state ion and so on. See Section 7.2.6, Fig. 7.18, and Hanna (2004).

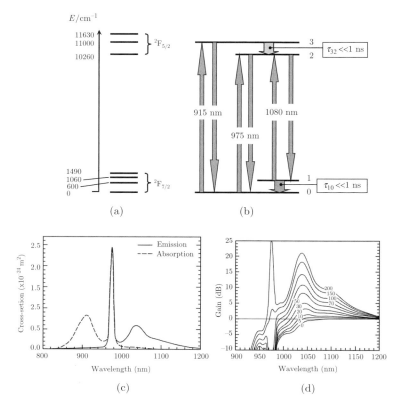

Fig. 10.25: (a) The energy levels of the Yb^{3+} ion in a germano-alumino-silicate glass host showing the three Stark sublevels of the $^2F_{5/2}$ upper-level manifold and the 4 Stark sublevels of the $^2F_{7/2}$ ground-state manifold. (b) A simplified 4-level scheme that can be used for rate-equation modelling. (c) The emission and absorption cross-sections (after Pask et al. (1995)). (d) Calculated gain/loss versus wavelength for a 1-m long fibre of core diameter 3.75 μm with 550 ppm doping of Yb^{3+}. The curves are labelled by the launched pump power in mW. (After Pask et al. (1995).)

Table 10.5 Cross-sections and decay times for Yb^{3+} in germano-alumino-silicate glass. (Data from Paschotta et al. (1997) and Kelson and Hardy (1998).)

Wavelength (nm)	Cross-section (10^{-20} cm^2)		Decay times (μs)
915	$\sigma_{03} = 0.6$	$\sigma_{30} = 0.025$	$\tau_{32} \ll 0.001$
975	$\sigma_{02} = 2.7$	$\sigma_{20} = 2.7$	$\tau_{20} = 770$
1080	$\sigma_{21} = 0.25$	$\sigma_{12} = 0.014$	$\tau_{10} \ll 0.001$

is possible to reduce the oscillating bandwidth of lower-power 1550 nm Er^{3+} lasers down to values of order 30 kHz by including FBG filters along the length of the fibre to provide distributed feedback.[32]

Experimentally, it is found that the broadening of the Yb^{3+} laser transition is sufficiently homogenous over the 50-nm range where gain is possible, that power can be extracted efficiently from such amplifiers even with narrowband input.

Erbium-doped fibre lasers of up to 100 mW output power are commercially available[33] that use gas lasers stabilized against well-defined absolute frequency references to control the fibre laser output and generate output beams

[32] See, for example, Voo et al (2004).

[33] DiCOS Technologies OFS-3100 Datasheet, www.dicostech.com.

of $< 2\,\text{kHz}$ linewidth with $\pm 80\,\text{MHz}$ absolute frequency stability at various wavelengths in the 1550-nm region.[34]

[34] See Section 16.4.1.

10.6 High-power pulsed fibre lasers

10.6.1 Large mode area (LMA) fibres

All of the fibre lasers discussed in this chapter so far have been c.w. devices. They can achieve single transverse mode operation (or at least M^2 numbers near unity) in step-index fibres of small core size and relatively large numerical aperture, at power levels of hundreds of watts. However, new factors have to be taken into account when considering the design of pulsed devices of high peak power. First, if the power density at the fibre exit face exceeds the optical damage threshold of silica, permanent damage at the silica air interface may result—i.e. quite simply the tip of the fibre may be blown off. Even before this limit is encountered, there is another problem that has to be overcome before Q-switched operation of fibre lasers becomes a practical proposition. Due to the high gains encountered in conventional single-mode fibre lasers, the rapid build-up of amplified spontaneous emission, and ultimately the onset of mirrorless lasing,[35] limit the energy that may be stored in the population inversion of a fibre amplifier to a few tens of microjoules. This is very much smaller than the several millijoules typically stored in the gain media of crystalline lasers in Q-switched operation.[36]

[35] See Section 6.2.

[36] See Section 8.2.

One way to decrease the effect of ASE would be to increase the cross-sectional area of the core of the fibre, but as discussed in Section 10.1.3 this would have a deleterious effect on the optical quality of the output beam since it would allow a large number of high-order transverse modes to lase in a standard fibre. Taverner et al.[37] showed it was possible to increase the mode area while still constraining operation to a single transverse mode by lowering the NA of the fibre, as may be seen from eqn (10.10).

[37] Taverner et al. (1997).

An increase of area by a factor of ten over conventional designs for a purely single-mode output has been demonstrated; however, this approach is limited by the increased losses incurred when the fibre is bent. A more robust approach is to tailor carefully the radial distribution of the active ions within the core during the formation of the preform so that gain is restricted to an annular region that, although not optimum for the LP_{01} mode, is even more deleterious for higher-order modes (see Fig. 10.6).

In the case shown in Fig. 10.26, the inner cladding of the LMA double-clad fibre was chosen to be rectangular in cross-section (as in Fig. 10.24(c)) to avoid helical ray paths in the cladding and thus ensure high efficiency of coupling of pump radiation to the Yb-doped core. A protective outer coating of silicone rubber was applied at the fibre pulling stage, and has the elliptical cross-section indicated in the inset to Fig. 10.26, and created a NA of 0.4 for the inner cladding. The NA of the fibre core was 0.075. The large core mode area of $\approx 1500\,\mu\text{m}^2$ allowed a stored energy of the order of 0.5 mJ to be achieved, compared to the $\approx 25\,\mu\text{J}$ typical of standard fibres of $\approx 9\,\mu\text{m}$ core diameter.

Fig. 10.26: Index profile of the inner cladding of an LMA fibre. As shown in the inset the cladding was machined to a rectangular cross-section at the preform stage. In the finally drawn fibre, the cladding measures 350 μm by 175 μm—as indicated by dashed lines. The outer ring (of radius 45 μm) of raised index increases the mode area and reduces the bend loss of the lowest-order mode. During manufacture ytterbium is incorporated at the position of the inner ring of radius 22 μm. (After J. A. Alvarez-Chavez et al. (2000).)

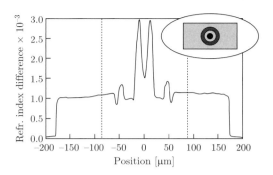

10.6.2 Q-switched fibre lasers

A scheme for achieving Q-switching in high-power fibre lasers, devised by researchers at the Optics Research Centre at the University of Southampton, is illustrated in Fig. 10.27. The gain element comprises a 36-m length of the Yb-doped LMA fibre described in Section 10.6.1. The laser is pumped by a 915-nm diode laser bar whose output is rendered into an approximately circular cross-section by use of a beam shaper (see Section 9.10.1), and the system is Q-switched by an acousto-optic modulator (see page 196). With 35 W of pump power, output pulses with energies of 1.6 mJ at a repetition rate of 1 kHz and 1 mJ at 5 kHz were obtained. Pulse duration decreased with increasing pulse energy, ranging from several microseconds down to 100 ns at the highest pulse energies.

10.6.3 Oscillator–amplifier pulsed fibre lasers

Although Q-switching does enable high-power fibre lasers to operate in the short-pulse regime required for many industrial applications, it only does so at the expense of sacrificing some of the fibre laser's most attractive features—in particular its simplicity and ruggedness. A more practical approach to the design of short-pulsed fibre lasers intended for industrial applications is to use a low-power semiconductor laser as a short-pulse oscillator, following it by a fibre laser to amplify its output to the high peak powers needed for materials processing.

As oscillators, semiconductor lasers operating at 1.06 μm are available, but semiconductor laser oscillators operating at 1.09 μm are also used for the

Fig. 10.27: Q-switched fibre laser employing Yb-doped LMA fibre. Q-switching is performed by the acousto-optic modulator AOM, which deflects the intra-cavity beam to meet the end mirror at normal incidence when an acoustic pulse is applied (see Section 8.2.1). (After J. A. Alvarez-Chavez et al. (2000).)

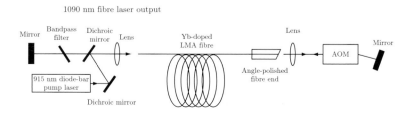

highest power ranges. In both cases the excitation current to the semiconductor laser is modulated to achieve the required output pulse format. As can be seen from Figs 10.25(c) and (d), gain is available at both these wavelengths in Yb-doped cladding pumped fibre-laser amplifiers. At the time of writing commercial versions[38] of these oscillator–amplifier combinations are available that can deliver up to 0.5 mJ per pulse either in the form of individual pulses or as continuous trains with repetition rates up to 200 kHz. Pulse lengths are adjustable between 10 and 200 ns. At even higher average power levels, versions[39] are available that offer pulse repetition rates up to 50 kHz and pulse lengths adjustable from 10 µs to fully c.w., with average powers up to 200 W. The beam at 1090 nm has quality close to the diffraction limit ($M^2 < 1.1$).

[38] Woods and Flinn (2006).

[39] See data sheets from, for example, Southampton Photonics Inc.

10.7 Applications of high-power fibre lasers

Fibre lasers are not only highly efficient, they are also extremely robust and tolerant to a degree of vibration in their environments, which would play havoc with the cavity alignment of conventional lasers. They are also very compact, which is often a prime consideration in the congested space of the factory shop floor. Moreover, it is also relatively straightforward to couple their output to the workpiece via a flexible fibre-optic system. All of these qualities make them ideally suited to many applications in industry and in laser surgery.

Devices in the 1–10 W average power range are already established in applications such as marking and etching where their ability to ablate glass, ceramics and steel makes them attractive for use in labelling and bar-coding, as well as in the traditional printing and graphic image processing industry in which metal surfaces must be textured to hold and release printing inks in a precisely controlled manner. Lasers at the lower end of this range have also been used as replacements for conventional solid-state Nd:YAG systems for aesthetic medical procedures such as wrinkle treatment. In the electronics industry fibre lasers up to 200 W are used to cut solder masks for circuit boards where precision and small kerf width of the cut are the important features. In the same industry sector they are also carving out a niche in welding, bending, aligning, stress relieving, soldering and heat-treating applications.

However, it is the automotive and aerospace industry where fibre lasers in the kilowatt average power class are having a major impact and are providing intense competition for traditional CO_2 industrial systems for welding and cutting metals. Not only does their shorter wavelength (1.09 µm compared to 10.6 µm) lead to better absorption in metals, but the very much smaller spot size of the focused beam allows much greater precision in cutting and welding as well as in the extent of the surrounding heat-affected zone. Such factors, in addition to their compactness, long-term reliability, capital cost and the cost of maintenance, will determine whether these systems will replace CO_2 as the major player in these markets but at the time of writing there seems every indication that they will.

Further reading

An excellent account of the application of fibre optics to telecommunications may be found in the book by Hecht (2003). Further information on fibre lasers may be found in Midwinter and Guo (1992) and in the following articles in *The Handbook of Laser Technology and Applications*: Stewart (2004); Basiev and Powell (2004); Hanna (2004); Tunnermann (2004); Headley (2004); Cordina (2004); and Urquhart (2004).

Exercises

(10.1) **Attenuation in optical fibres**

  (a) As mentioned in Section 10.1.1, before the development of modern optical fibres, typical values of attenuation in glass were of order 1 dB m^{-1}. Calculate the fractional power transmission for fibres of length:
    (i) 1 m;
    (ii) 1 km.

  (b) What is the longest such a fibre could be if at least one photon per second were to be transmitted when 1 mW of 1.3 μm wavelength light were launched into it?

(10.2) **Numerical aperture**

  (a) Show that for a step-index fibre in air

  $$\mathrm{NA} = \sqrt{(n_1^2 - n_2^2)},$$

  where n_1 is the refractive index of the core and n_2 is that of the cladding.

(10.3) **Intermodal dispersion**

  (a) A *meridional* ray is one that propagates in a plane containing the axis of the fibre (e.g. the (x, z)- or (y, z) planes for a fibre lying along the z-axis). By considering how axial and meridional rays propagate in a step-index fibre, explain what is meant by intermodal dispersion and, using the notation of Exercise 10.2, show that the difference in time Δt for the extreme meridional and axial rays to travel a distance L along the fibre is given approximately by

  $$\Delta t = \frac{L\mathrm{NA}^2}{(2n_1 c)},$$

  provided $(n_1 - n_2) \ll n_1$.

  (b) Explain briefly the importance of intermodal dispersion in optical fibres used for communication systems and discuss qualitatively how fibres may be designed to alleviate this problem.

(10.4) **Single-mode fibres**

  A step-index fibre is made with a core of refractive index $n_1 = 1.4645$ and cladding of refractive index $n_2 = 1.450$ over the wavelength range of interest. The core diameter is 5.60 μm.

  (a) State with reasons whether the fibre can support a single-mode only for radiation whose air wavelength is:
    (i) 1.55 μm;
    (ii) 1.30 μm.

(10.5) (a) Derive eqn (10.23).
  [Hint: $(I_{\mathrm{out}}/I_{\mathrm{in}}) = \exp(N_2 \sigma z)$. Thus, $G = 10(N_2 \sigma z) \log_{10}(e) = 4.34 N_2 \sigma z$], where G is expressed in decibels.

(10.6) A graded-index fibre lies along the z-axis, and has a refractive index given by $n(r) = n_0\left(1 - \beta^2 r^2/2\right)$, where r is the radial distance from the axis.

  (a) Show that a ray launched parallel to the fibre's axis from $r = r_0$, $z = 0$ propagates according to

  $$r(z) = r_0 \cos \beta z.$$

(b) Show that the time taken by such a ray to pass through a length L of fibre is independent of r_0 up to terms of order $\beta^2 r_0^2$, provided $L \gg \beta r_0^2$.

(c) Discuss the significance of this result in the context of modal dispersion.

[You may assume that the path of the ray has radius of curvature R, where $1/R = (1/n)\,dn/dr$ and that for a curve in a single plane given by $y = f(x)$, the length along the curve from $x = a$ to $x = b$ (in a region where the curve is everywhere above the x-axis) is given by

$$s = \int_a^b \sqrt{1 + \left(\frac{dy}{dx}\right)^2}\,dx,$$

and that $1/R = \dfrac{d^2 y}{dx^2}\left(1 + \left(\dfrac{dy}{dx}\right)^2\right)^{-3/2}$.

(10.7) **Raman fibre amplifier**

A laser operating at 1450 nm is used to pump a Raman fibre amplifier that is required to optical gain centred at 1640 nm. Using the data of Table 10.2, identify a suitable dopant material for the glass of the fibre.

11

Atomic gas lasers

11.1 Discharge physics interlude 298
11.2 The helium-neon laser 314
11.3 The argon-ion laser 321
Further reading 329
Exercises 329

Atomic gas lasers such as the helium-neon laser and the argon-ion laser played a very important part in the early development of laser technology and applications—indeed, despite the fact that their replacement by all-solid-state devices has been widely predicted for many years, they continue to find applications and are still manufactured in large numbers. With a few notable exceptions[1] (e.g. semiconductor lasers—see Chapter 9), nearly all types of lasers depend upon gas discharges for the primary conversion of electrical energy into radiation. For example, until the advent of high-power semiconductor laser arrays, most crystalline solid-state lasers were pumped by radiation from a flashtube—a gas-discharge device. In the case of most gas lasers the active medium of the laser is of course itself a gas discharge.

In order to provide some insight into the mechanisms of these devices[2] it is therefore worthwhile to digress a little at this stage to present a very brief review of the relevant topics in discharge physics.

[1] Perhaps the most important exception is the class of chemical lasers of which the COIL (chemical oxygen iodine laser) is a prime example. These extremely high power c.w. lasers involve the mixing of two reactive chemical species to produce a population inversion without any other source of energy and are the basis of proposed airborne laser weapons.

[2] This discussion is also highly relevant to understanding the processes involved in sodium lamps and mercury fluorescent lamps that account for a major portion of the output of the lighting industry worldwide; see for example Coaton and Marsden (1997)

11.1 Discharge physics interlude

11.1.1 Low-pressure and high-pressure discharges

In a gas containing molecules of a single species maintained at a uniform temperature T_{gas} the number density N of molecules is related to the pressure p by:[3]

$$p = Nk_B T_{\text{gas}}. \tag{11.1}$$

[3] In SI units, p is in Pascals (N m^{-2}), N is in molecules m^{-3}, T_{gas} is in K and k_B is Boltzmann's constant. However, among discharge physicists a number of non-SI units are in common use: 1 Torr = 1 mm Hg, so that 760 Torr = 1 atmosphere = 1.013 bar where 1 bar = 1.000×10^5 Pa. Thus, 1 Torr = 1.33 mbar = 133 Pa. Another useful parameter is Loschmidt's number, the number density of molecules at STP: $N_L = 2.69 \times 10^{25}$ m^{-3}.

The properties of discharges at low gas pressure are quite different from those of discharges run at high pressure. At gas pressures above 0.1 Torr and below 30–100 Torr it is possible to run stable, steady-state discharges in tubes of diameter 100 mm or less between electrodes at each end of the tube. As the pressure is raised above typically 300–500 Torr the breakdown voltage needed to start and sustain such longitudinal discharges becomes very high indeed, and the discharge column quickly becomes contorted and unstable. As we shall see in Chapter 12, under such high-pressure conditions it is more useful to excite the gas by a pair of *transverse* electrodes whose separation is much smaller than their length. The tendency of such high-pressure transverse discharges to collapse from a distributed glow into a spatially non-uniform array of arcs is mitigated by launching the discharge in as uniform a way as possible, and by ensuring that the energy is fed from the external circuit into the discharge in a time short compared to this collapse. It is therefore no coincidence that high-pressure transverse discharge excitation is employed for large-volume *pulsed* gas lasers, while low-pressure longitudinal discharges are used for c.w. gas lasers.

The basic physics that gives rise to this difference arises from the nature of the dominant ion loss mechanism. In low-pressure discharges, ions are lost as a result of their diffusive motion radially outwards until they recombine with electrons on the tube wall. Loss by ion–electron recombination in the volume is negligible. However, at the opposite extreme of high pressure, volume processes[4] pre-dominate and processes at the walls play only a minor role.

[4] Processes such as the recombination of an ion with an electron in the presence of a third body, or attachment of electrons to molecules, followed by recombination of positive and negative ions, act to remove ions from the discharge volume.

11.1.2 Low-pressure glow discharge

When a DC discharge is run at low current density (say, less than $1\,\mathrm{A\,cm^{-2}}$) in a tube containing gas at low pressure with cold electrodes at either end, several regions may be distinguished that characterize the structure of the discharge. As indicated in Fig. 11.1, they are:

The cathode and negative glow regions

Immediately in front of the cold cathode (negative electrode) surface there may be a thin dark space (the Aston dark space) then a thin layer of cathode glow and then another dark space (the cathode dark space), and then a more extensive negative glow. All of these parts of the discharge occur typically within a few mm of the cathode surface and, except for the negative glow, are not shown in Fig. 11.1. They exist because of the steep potential gradient that develops in front of a cathode. This potential gradient is created because of the need to accelerate positive ions to energies high enough that when they bang into the cathode they cause the required[5] current to flow via the ejection of electrons out of the cold cathode. For this secondary-emission process the electrons in the cathode have to overcome the difference in potential between the Fermi energy inside the metal and that of the outside world.

[5] The current is set jointly by the power supply voltage, the ballast resistance, and the current–voltage characteristic of the discharge—see Section 11.1.4.

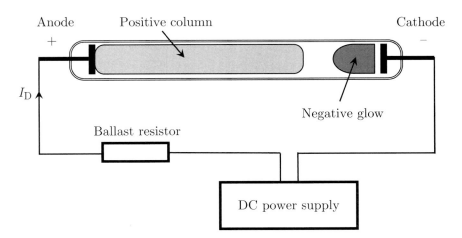

Fig. 11.1: DC discharge in low-pressure gas, indicating the positive column and negative glow regions.

In the negative glow region the newly liberated electrons are accelerated away from the cathode and may reach kinetic energies of up to approximately 300 eV. In effect, the negative-glow region acts as an electron gun and the glow is excited by the 'beam' of electrons losing their energy by successive inelastic collisions with gas atoms until their energy is reduced to the few electron volts typical of the positive column. The negative glow[6] acts as if it were stuck to the cathode and moves with it if the cathode is moved about. The negative-glow region has been used to excite gas lasers—indeed the mechanisms of so-called 'hollow-cathode discharge' (HCD) lasers depend on the presence of these 'beam' electrons.[7] However, apart from a few such specialized applications, the negative regions have not been widely applied to laser excitation, and we shall not discuss them further.

The positive column

Proceeding towards the anode (positive electrode) there is one more dark space (the Faraday dark space) before the start of the luminous positive column that fills almost all of the remaining length of the discharge. The positive column is a region in which a more or less uniform, but comparatively low, axial potential gradient exists. It may be lengthened almost indefinitely by elongating the discharge tube, and is the type of discharge region familiar in neon advertising signs or in mercury fluorescent lamps. The positive column is a region of true plasma, i.e. it is virtually charge neutral, with equal densities of electrons and positive ions.[8] Apart from the negative glow and other cathode regions, the positive column in a low-pressure discharge fills the remaining volume of the tube, independent of the position and orientation of the electrodes.

The positive column of a low-pressure (less than 30–100 Torr) glow discharge in a cylindrical tube (0.5–10 mm diameter) is commonly used for the active medium of low- and medium-power gas lasers and will be the subject of the remainder of this chapter. The ionized fraction is usually quite small, typically 1 part in 10^4 or 10^5, although in argon-ion lasers the ionized fraction can approach 1%. Despite the fact that in the positive column the number densities of electrons and positive ions are equal, virtually all the current is carried by the electrons whose mobility[9] is very much higher than that of the positive ions.

11.1.3 Temperatures

To understand how excitation and ionization occur in the plasma of the positive column, it is important to realize that in this region several different 'temperatures' can coexist—although only in some approximate sense since a true temperature can only be defined for a system in thermal equilibrium. Thus, although all of the temperatures would have the same value in a truly thermal equilibrium situation, we can speak of separate temperatures for each of these systems that are in good thermal contact internally, but in poorer contact with each other.

A *kinetic* temperature T_{gas} is one that can be applied to indicate the mean energy of the velocity distribution of the motion of electrons, ions or atoms. Strictly, the velocity distribution can only be described by a single parameter

---

[6] The really widespread use of the negative glow is for 'mains on' neon-filled indicator lamps.

[7] See, for example, Gill and Webb (1977).

[8] It is really the number of positive and negative charges per unit volume that are equal. In discharges containing electron-attaching species such as O_2, negative ions may be present, and in this case charge neutrality demands that the sum of the electron and negative ion densities equal to that of the positive ions. Likewise in discharges not containing attaching species, but with a significant number of doubly charged positive ion species, it is the sum of the singly charged species density plus twice the density of doubly charged ions that equals the electron density.

[9] The mobility is defined as the drift velocity per unit applied electric field. Electron drift velocities are not always exactly linearly proportional to the applied electric field (see, for example, von Engel (1965)), so that the concept of electron mobility is one of rather limited validity. However, it remains true that the drift velocity of electrons increases with increasing E/N (electric field to gas number density ratio) and that electron drift velocities are always very much larger than those of positive-ion species.

(temperature) when it is truly a Maxwellian distribution, and this is unlikely to be so for the whole distribution in practical cases. For example, the distribution of kinetic energy among the free electrons in a plasma may be close to Maxwellian only up to a certain critical energy above which a serious depletion of electrons in the high-energy tail may be apparent.

A *distribution-over-states* temperature (T_{exn}) arises because the neutral atoms and positive ions have states of internal excitation energy. The distribution of atoms over their various excited states is one to which we may be able to apply (in a very restricted sense, since it usually applies to only a few states at a time) some kind of 'distribution-over-states temperature' T_{exn} defined from the populations in terms of the Boltzmann factor:

$$\frac{N_j}{N_i} = \frac{g_j}{g_i} \exp\left(-\frac{E_j - E_i}{k_B T_{\text{exn}}}\right). \tag{11.2}$$

In a low-pressure, low-current discharge such as that in the He-Ne laser, we might have, for example:[10]

Kinetic temperature of atoms: $T_{\text{gas}} = 350$ K

Kinetic temperature of ions: $T_{\text{ion}} = 400$ K

Kinetic temperature of electrons: $T_e = 90\,000$ K

Distribution over states: $T_{\text{exn}} = 20\,000$ K

These differences in kinetic temperature among different types of particles in collisional contact with one another in the same discharge can exist because of the different processes contributing to the energy balance in each species. The atoms do not directly experience the axial electric field E of the discharge but the ions and electrons both do, and since they bear equal charges both species acquire equal momentum gains in equal times. Energy is transferred to the gas atoms from electrons and ions when they make elastic and inelastic collisions.[11]

As indicated in Fig. 11.2, *elastic* collisions are ones in which kinetic energy is conserved overall *as kinetic energy*. Of course, kinetic energy can be *transferred* from one partner to the other. When hard-sphere projectile particles of mass m collide elastically with target particles of mass M a fraction Δ of the original kinetic energy is transferred where

$$\Delta_{\text{elastic}} = \frac{\text{Kinetic energy lost by incident particle}}{\text{Initial kinetic energy of incident particle}} \tag{11.3}$$

The average value of Δ_{elastic} is given by[12]

$$\langle \Delta_{\text{elastic}} \rangle = \frac{2mM}{(m+M)^2}, \tag{11.4}$$

where the average is over all impact parameters out to the maximum 'hard-sphere' radius.

Since the mass of an ion is similar to that of an atom of the same species[13] eqn (11.4) tells us that for ions colliding elastically with atoms:

$$\langle \Delta_{\text{elastic}} \rangle_{\text{ion-atom}} \approx \frac{1}{2}, \tag{11.5}$$

[10]The distribution-over-states temperature is typical for a pair of neighbouring states in neutral neon in low-current, low-pressure discharges such as that of the He-Ne laser.

[11]Effectively all of the ionization and excitation is brought about by the inelastic collisions of electrons rather than those of ions. The number of ions that can reach energies high enough to cause excitation or ionization is negligible—see Exercise 11.1.

[12]For a derivation of eqn (11.3) see, for example, Chapter 3 of Lieberman and Lichtenberg (2005).

[13]In the laser examples we shall consider here (Ar$^+$ ions moving in Ar gas in the case of the argon-ion laser or, He$^+$ ions moving in He gas in the case of the He-Ne laser) the difference between atom mass and ion mass is only that of one electron. Even for Ne$^+$ colliding with He or He$^+$ colliding with Ne the value of Δ_{elastic} reduces from 0.5 to 0.28. On the other hand, low-energy electrons can only 'bounce' elastically off atoms without losing much of their energy—think of a ping-pong ball colliding with a ten-pin bowling ball!

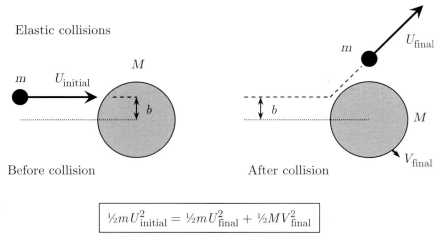

Fig. 11.2: An elastic collision of an electron of mass m_e with a neutral gas atom of mass M. The impact parameter b is the off-centre distance by which the electron's initial trajectory, if projected, would miss the centre of the atom.

but for electrons colliding elastically with atoms:

$$\langle \Delta_{\text{elastic}} \rangle_{\text{electron-atom}} \lesssim 10^{-3}. \tag{11.6}$$

Thus, the ions, as soon as they start moving at all, are able to get rid of large fractions of their energy by elastic collisions with gas atoms (which are by far the most abundant kind of targets around). The ions are therefore in good 'thermal contact' with the gas, the tube walls and the world at large. In contrast, the electrons can share their energy among their own kind, but not effectively with the atoms or ions, at least not until the electrons have enough energy to start making *inelastic* (excitation or ionization) collisions.

All of the excitation and ionization in the gas discharge thus arises from the impacts that electrons at the very high energy tail of the electron energy distribution make with gas atoms, since it is only these electrons that have energies of the magnitude required. To the extent that electrons exchange energy amongst themselves by making rapid collisions (and since they interact via a repulsive Coulomb field, their effective cross-sections are very large ($\approx 10^{-14}$ cm^2), they will tend to have a Maxwellian distribution of velocities. However, because of the drain on the high-energy end of the distribution imposed by the inelastic collisions with gas atoms, quite serious departures from a true Maxwellian electron-energy distribution can occur in practical cases—especially for energies above the first excitation potential of the majority gas species.

Since each of the individual species (electrons, ions and neutral atoms) within the discharge is in better collisional contact with others of its own kind than it is with members of a different species, it is easy to understand why each species can have its own kinetic temperature. Only the electrons and ions gain energy directly from the electric field of the discharge, however—for the reasons outlined above—the ion temperature is dragged down towards the gas-atom temperature much more strongly than the electron temperature. The gas-atom temperature is raised slightly above the wall temperature by virtue of the heat input the atoms receive from collisions with ions and, to a smaller extent, electrons.

11.1.4 The steady-state positive column

The operating parameters of the steady-state discharge are governed by a few quantities that we are free to choose externally—parameters such as type of gas, N the number density of gas, d the discharge tube internal diameter, and I_D the discharge current.

Note that for a low-pressure gas discharge it is effectively the discharge current we specify—rather than the voltage drop across the discharge. This is because the voltage versus current characteristic curve of a glow discharge is typically rather flat, usually with a negative slope. For a given voltage of the external power supply, the ballast resistor R_{ballast} in series with the discharge limits the discharge current I_D to a value consistent with the resistor load line and the I–V discharge characteristic. A more efficient way to stop the potential runaway growth of current is via the use of a current-limited power supply that delivers a fixed current regardless of the impedance of the external circuit.

Figure 11.3 shows the measured relationship between the overall tube voltage and discharge current for a small argon-ion laser, but its form is typical of many low-pressure glow discharges. The tube, of 2 mm diameter and length 300 mm, has a heated cathode and is filled with 0.5 Torr argon. The overall tube voltage decreases as the current is increased. With a series ballast resistor of 70 Ω, and a power supply of open circuit voltage 600 V, I_D is constrained to be 5.8 A.

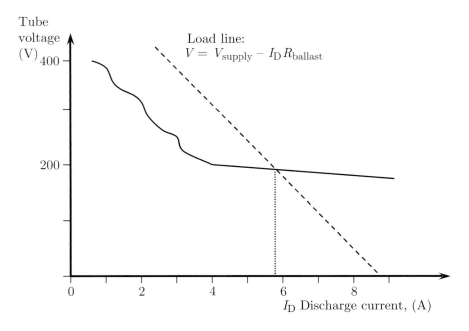

Fig. 11.3: Schematic current–voltage characteristic of a low-pressure discharge.

Once the external parameters have been set, the *internal* parameters of the discharge adopt values determined by the need for the discharge to satisfy certain constraints. The internal parameters are:

1. the electron density n_e;
2. the density of positive ions N_+ (equal to n_e in the plasma of the positive column);[14]
3. the axial electric field E;
4. the electron temperature T_e.

[14] Although see Note 8.

The four principal constraints for the parameters of the positive column are:

Constraint I. Discharge current:

$$I_D = \text{value set by external circuit.}$$

Now, the current density is approximately given by[15]

$$j_D \approx \frac{4I_D}{\pi d^2} = n_e e v_{\text{drift}}. \quad (11.7)$$

[15] Because the current density is not uniform across the diameter of the tube, the expression for current density averaged across the tube diameter is written as an approximate equality in eqn. (11.7). An exact expression would have a constant (of order unity) multiplying the RHS whose value depends on the profile of electron density with radius. For a discussion of this and related points see, for example, Webb et al. (1975) or Tonks et al. (1929).

The value of v_{drift}, the electron drift velocity in the positive column, is determined by electron–atom scattering and is thus a function of E/N since the acceleration between collisions depends on E, and the frequency of collisions is inversely proportional to N. Equation (11.7) effectively sets n_e given I_D and E/N.

Constraint II. Charge neutrality:

$$N_+ = n_e.$$

In the first few tens of nanoseconds following the initiation of the discharge, the electrons tend to move radially outwards to the wall leaving the slow-moving ions behind. However, the excess of positive charge in the volume and negative charge on the wall sets up a radial field that retards the further outflow of electrons (and slightly enhances that of the positive ions) so that the outward flux of ions and electrons quickly equalizes. The steady-state charge density imbalance in the volume is thus restricted by this self-regulating mechanism to negligible values.

Constraint III. Ion generation/loss rate:

$$\text{Ion generation rate} = \text{ion loss rate} = N_+/\tau_+.$$

[16] The loss rate of ions can be characterized by an effective lifetime τ_+, the average time taken for an ion to move from its point of generation to the wall of the tube under the influence of the radial component of electric field. The value of τ_+ is determined by whether the ions make many collisions with gas atoms before they reach the wall, or whether the gas density is low enough that the ions move freely to the wall without suffering collisions. See references cited in note 15. Typical values of τ_+ tend to be in the range of 1–100 μs.

To be viable as a steady-state discharge, ions lost by diffusion to the walls (with a decay rate characterized[16] by an effective lifetime τ_+) must be replaced at an equal rate by generation of new ions in the volume of the tube. This requirement effectively sets the value of T_e, since the ionization rate is determined by the number of electrons in the tail of the distribution above the ionization potential of the gas, and hence depends on the number and mean energy of the electrons in the electron energy distribution.

Constraint IV. Power balance:

$$\text{Power dissipated per unit length} = E I_D,$$

since the right-hand side is equal to the power input per unit length of tube. At low pressures the dominant dissipation is in the form of ionization. This energy appears in the form of a flux of ions falling on the walls, which then yield up an energy equal to the ionization energy when they recombine with free electrons on the wall. The outward radial flux of ions depends on the T_e value that determines the radial potential gradient. This constraint therefore imposes a relationship between the values of E and T_e.

In satisfying this system of constraints, the discharge establishes a set of n_e, E and T_e values that are consistent with each of these competing requirements. Figure 11.4 indicates schematically how the positive column of a glow discharge 'decides' upon values of n_e, E and T_e that are consistent with the external parameters N, d, and I_D.

The discharge current I_D (set externally) and the electron drift velocity value determine what electron density is required, and hence (via charge neutrality) what the ion density must be. The ion loss rate and this value of N_+ in turn decide what value T_e must have in order that electron–ion pairs should be created at a rate sufficient to replace the loss. The power balance relates the electric field to the electron temperature and E in turn decides the electron drift velocity via the E/N dependence of drift velocity. The system of constraints therefore forms a closed loop and although in practical cases the values of the relevant cross-sections may not be known with sufficient accuracy to enable a good computational model to be made, the diagram does provide some indication of the balancing act performed by the interacting processes within the discharge.

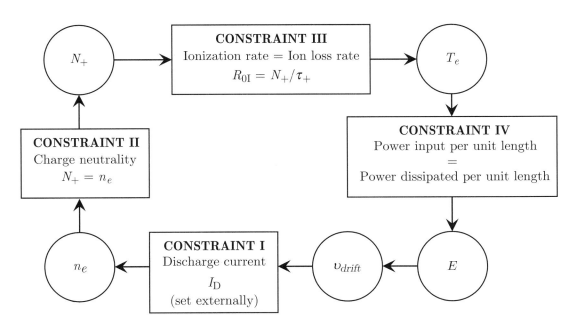

Fig. 11.4: Steady-state parameters of glow discharge positive column. Constraints are indicated by rectangular boxes; the internal parameters n_e, N_+, T_e, E, and v_{drift} are indicated by circles.

11.1.5 Ionization rates

We now need to examine in more detail the relationship between the electron-energy distribution and the rate of ionization by electron collisions needed to satisfy Constraint III in the previous section.

As indicated in Fig. 11.5, an electron with kinetic energy greater than the ionization energy ϵ_I colliding with an atom of species X can ionize X to give two electrons and an ion of species X$^+$ exit from the collision region.

For simplicity, we will consider first the rate of ion generation by a monoenergetic beam of electrons of energy ϵ_B.

The cross-section $Q_{0\text{I}}(\epsilon)$ for ionization of ground-state atoms by collisions of electrons of energy ϵ has the typical variation[17] shown in Fig. 11.6, rising from the threshold at ϵ_I (the ionization potential), to peak at an energy between approximately $2\epsilon_\text{I}$ and $4\epsilon_\text{I}$, and thereafter decreasing slowly. Typical values of $Q_{0\text{I}}^{\max}$ at the peak are of order $10^{-16} - 10^{-15}$ cm^2.

[17] See, for example, Allen (1963).

For electrons of velocity $v_\text{B} = \sqrt{2\epsilon_\text{B}/m_\text{e}}$ the frequency ν_coll of ionizing collisions of *one* electron with N_0 targets per unit volume is given by

$$\nu_\text{coll} = N_0 Q_{0\text{I}}(\epsilon_\text{B}) v_\text{B} = N_0 Q_{0\text{I}}(\epsilon_\text{B}) \sqrt{\frac{2\epsilon_\text{B}}{m_\text{e}}}, \qquad (11.8)$$

and so $R_{0\text{I}}^\text{beam}$, the *total* rate of ionizing collisions per second per m^3 of gas is given by

$$R_{0\text{I}}^\text{beam} = n_\text{e} N_0 Q_{0\text{I}}(\epsilon_\text{B}) v_\text{B}, \qquad \textbf{Monoenergetic electrons}, \qquad (11.9)$$

where n_e is the number density of electrons in the monoenergetic beam, which has a flux $\Phi = n_\text{e} v_\text{B}$ electrons m^{-2} s^{-1}.

Now we turn our attention to the practical case in which the electrons can no longer be assumed to have one velocity and energy. The fact that in the positive column of a glow discharge the electrons have a distribution in velocity must be taken into account by averaging the velocity dependent quantities in eqn (11.9) appropriately over all velocities, i.e.

$$R_{0\text{I}} = n_\text{e} N_0 \langle Q_{0\text{I}}(\epsilon) v \rangle. \qquad \textbf{Non-monoenergetic electrons} \qquad (11.10)$$

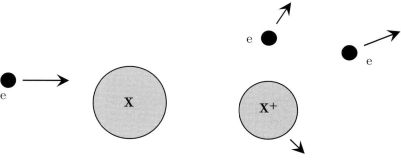

Fig. 11.5: Schematic representation of an ionizing inelastic collision of an electron with a gas atom of species X. The electron's initial kinetic energy is ϵ. When $\epsilon > \epsilon_\text{I}$, the ionization energy of species X, the collision can result in the ionization of the atom with two electrons emerging from the collision.

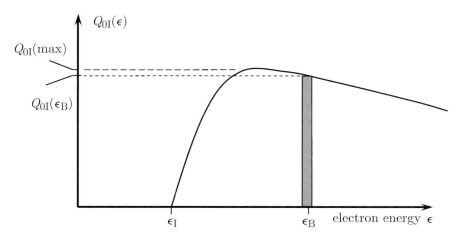

Fig. 11.6: Schematic representation of the typical variation of $Q_{0I}(\epsilon)$ the ionization cross-section of ground-state atoms by electrons of energy ϵ. For electrons at the beam energy ϵ_B, the cross-section is $Q_{0I}(\epsilon_B)$.

This average can be written explicitly in terms of the electron-energy distribution function $f(\epsilon)$ as follows:[18]

$$R_{0I} = n_e N_0 \int_{\epsilon_I}^{\infty} \sqrt{\frac{2\epsilon}{m_e}} Q_{0I}(\epsilon) f(\epsilon) d\epsilon = n_e N_0 s_{0I}, \quad (11.11)$$

[18] By definition, $f(\epsilon)d\epsilon$ is the proportion of electrons whose energies lie in the range ϵ to $\epsilon + d\epsilon$. Thus, we must have $\int_0^{\infty} f(\epsilon) d\epsilon = 1$.

where

$$s_{0I} = \langle Q_{0I}(\epsilon) v \rangle = \int_{\epsilon_I}^{\infty} \sqrt{\frac{2\epsilon}{m_e}} Q_{0I}(\epsilon) f(\epsilon) d\epsilon. \quad (11.12)$$

As indicated in Fig. 11.7, in the particular case of a Maxwellian distribution

$$f(\epsilon) = 2\sqrt{\frac{\epsilon}{\pi}} (k_B T_e)^{-\frac{3}{2}} \exp\left(-\frac{\epsilon}{k_B T_e}\right), \quad \textbf{Maxwellian distribution}$$

(11.13)

in which case values of $s_{0I}(T_e)$ can be evaluated once and for all if $Q_{0I}(\epsilon)$ is known.

11.1.6 Excitation rates

In a similar way we can write the rate of excitation of an excited level j of an atom from the ground state 0 by electron collisions in terms of the appropriate excitation cross-section $Q_{0j}(\epsilon)$ as:

$$R_{0j} = n_e N_0 s_{0j} \quad (11.14)$$
$$= n_e N_0 \langle Q_{0j} v \rangle \quad (11.15)$$
$$= n_e N_0 \int_{\epsilon_j}^{\infty} \sqrt{\frac{2\epsilon}{m_e}} Q_{0j}(\epsilon) f(\epsilon) d\epsilon. \quad (11.16)$$

308 *Atomic gas lasers*

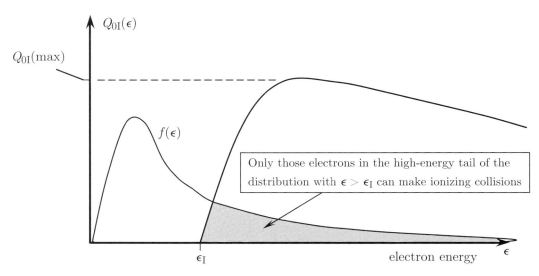

Fig. 11.7: Rate of ionization by collisions of electrons with a distribution of energies $f(\epsilon)$. Only the electrons in the high-energy tail of the distribution are effective in making ionizing collisions.

[19] The behaviour of the collision cross-sections in *ions* can be quite different. See Note 29 on page 517.

[20] If $Q_{0j}(\epsilon)$ is represented by a simple functional dependence on ϵ such as a step function, a linear rise and exponential fall or a inverse dependence on ϵ for energies above the threshold ϵ_j, the integration over energy of eqn (11.8) can be performed analytically to yield the rate of excitation in the form:

$$R_{0j} = n_e N_0 s_{0j}$$

$$= n_e N_0 2\sqrt{\frac{2\epsilon_j}{\pi m_e}}$$

$$\times Q_{max} \phi\left(\frac{\epsilon_j}{k_B T_e}\right) \exp\left(-\frac{\epsilon_j}{k_B T_e}\right),$$

in which the only factor depending on the details of the functional dependence of $Q_{0j}(\epsilon)$ upon ϵ is the slowly varying preexponential function $\phi\left(\frac{\epsilon_j}{k_B T_e}\right)$. The really sensitive dependence of R_{0j} upon T_e arises from the exponential factor that is common to all three types of functional dependence—see Exercise 11.2 and Green and Webb (1975).

[21] See, for example, Burgess and Tully (1978).

There is no simple approximation for the dependence of the excitation cross-section of $Q_{0j}(\epsilon)$ on ϵ. The magnitude and form of $Q_{0j}(\epsilon)$ will depend on whether the corresponding optical transition is allowed or forbidden and, if forbidden, on the way in which it is forbidden.

As shown schematically in Fig. 11.8 the value of $Q_{0j}(\epsilon)$ for allowed transitions in neutral atoms[19] typically rises from threshold to peak at a value of order $10^{-15} - 10^{-16}$ cm^2 at an impact energy about twice threshold or somewhat higher.[20] At very high energies the cross-section asymptotes to a more or less constant value representative of the finite oscillator strength of the corresponding optical transition. It might at first seem strange that the optical selection rules and oscillator strengths should play any part in determining the probability of excitation in a collision. However, this is no coincidence. As seen from the frame of the atom, the projectile electron produces a pulse of electric field as it whizzes by, starting out at low values while the electron is a long way away, rising to a peak at the distance of closest approach to the atom, and then dying away to zero as the electron recedes to infinity. If the electron has a sufficiently high energy, the pulse of electric field experienced by the atom will contain Fourier components in the same frequency range as the radiative transition, and according to the Bethe and Born approximations, the optical selection rules will govern the matrix element for the excitation cross-section.[21] The fact that the asymptotic behaviour of the excitation cross-section at high energies reflects the optical oscillator strength is the basis of one method of measuring oscillator strengths.

For excitation by electron impact on a transition that is optically forbidden only because it involves a change of spin quantum number S, the value of $Q_{0j}(\epsilon)$ may rise sharply for energies just above the threshold energy ϵ_j. This happens because the projectile electron may simply change places with the

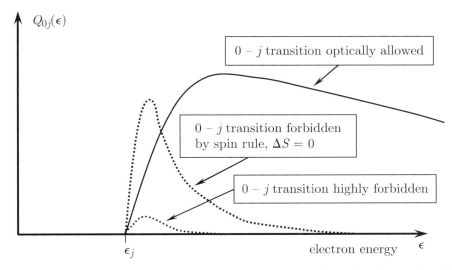

Fig. 11.8: Schematic representation of the energy dependence of the cross-section of a neutral atom from its ground state 0 to an excited state j when the corresponding optical transition $0 \to j$ is: (a) fully allowed; (b) forbidden only to the extent that it involves a change of total spin quantum number S; or (c) highly forbidden by the optical selection rules.

optical electron in the target atom, permitting a change of S by unity.[22] At high electron energies the cross-section falls to zero, again reflecting the value of the corresponding optical transition's oscillator strength.

If the reason why the optical transition is forbidden is that it involves some very low probability process,[23] then the cross-section will be much smaller at all energies than those for allowed transitions. However, for impact energies within several eV or tens of eV of threshold, the optical selection rules are no longer a good guide to the excitation cross-section—except in the very general sense that highly forbidden processes will tend to correspond to small cross-sections.

Of course, cross-sections for processes starting from levels j other than the ground state leading to ionization $Q_{jI}(\epsilon)$, or excitation of still higher levels k, $Q_{jk}(\epsilon)$, will usually have lower-energy thresholds ($\epsilon_I - \epsilon_j$) than the those starting on the ground state. Also, since they involve target atoms where the optical electron is already excited into an orbit of greater mean radius than in the ground state, values of Q_{jI} will usually be considerably higher than Q_{0I} values for ground-state ionization. The rate of ionization from an excited state j can be written, by analogy with eqn (11.16) as:

[22] For example, singlet to triplet.

[23] For example, the promotion of an inner-shell electron, or a transition in which two electrons would have to change their quantum numbers.

$$R_{jI} = n_e N_j \langle Q_{jI} v \rangle = n_e N_j s_{jI}$$
$$= n_e N_j \int_{\epsilon_I - \epsilon_j}^{\infty} \sqrt{\frac{2\epsilon}{m_e}} Q_{jI}(\epsilon) f(\epsilon) d\epsilon.$$

General excitation/ionization rate (11.17)

11.1.7 Second-kind or superelastic collisions

Up to now we have discussed inelastic collision of the first kind: those in which kinetic energy of the projectile electron is partially converted into internal energy (or ionization) of the target atom as a result of the collision. An example would be the excitation of an atom of species X from state 0 to state j by the collision of a fast electron, i.e. one whose kinetic energy exceeds the threshold energy for excitation ϵ_j. Such an excitation reaction is represented by

$$X_0 + e_{\text{fast}} \to X_j + e_{\text{slow}}. \tag{11.18}$$

The thermodynamic inverse of the reaction 11.18 is the time-reversed process in which a slow electron collides with an excited atom, de-excites it to the ground state,[24] and picks up the energy of excitation as kinetic energy. This process may be written as

$$X_j + e_{\text{slow}} \to X_0 + e_{\text{fast}}. \tag{11.19}$$

The type of collision shown schematically in Fig. 11.9 is an example of a *collision of the second kind*, nowadays more commonly called a *superelastic collision*.

Because the superelastic process is simply a time-reversed version of the inelastic excitation process, there is a relationship—the **Klein–Rosseland relationship**[25]—between the cross-section $Q_{0j}(\epsilon + \epsilon_j)$ for the up process at energy $\epsilon + \epsilon_j$ with the cross-section for corresponding down process $Q_{j0}(\epsilon)$ at energy ϵ:

$$Q_{j0}(\epsilon) = \frac{g_0}{g_j}\frac{\epsilon + \epsilon_j}{\epsilon}Q_{0j}(\epsilon + \epsilon_j), \quad \textbf{Klein–Rosseland relationship} \tag{11.20}$$

[24] In eqn (11.18) we have chosen to use de-excitation back to the ground state 0 to provide an example of a superelastic collision process, but clearly the process is entirely general and the final state could belong to any energy level lower in energy than the initial excited state.

[25] See Exercise 11.3.

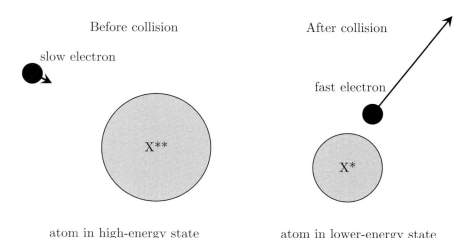

Fig. 11.9: Schematic representation of a superelastic collision (collision of the second kind) in which a slow electron de-excites an atom from a high-energy state to a state of lower energy. The electron takes up the difference in internal energy as an increase in its kinetic energy.

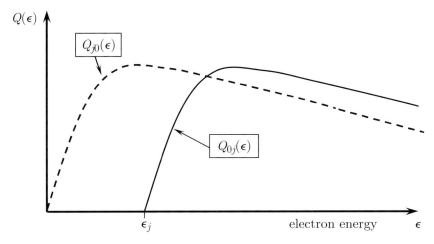

Fig. 11.10: Relationship between $Q_{j0}(\epsilon)$, the cross-section for de-excitation from excited level j to the ground level 0 by electrons of energy ϵ and the corresponding excitation cross-section $Q_{0j}(\epsilon)$ for the case $g_j = g_0$.

in which g_j and g_0 represent the statistical weights (degeneracies) of the excited level j and the ground level 0.

The variation with energy of the cross-sections for excitation and de-excitation is shown schematically in Fig. 11.10. It is evident that there is no energy threshold for the de-excitation process.

By analogy with eqn (11.16), the rate of de-excitation from an excited level j back to the ground state can be written as:

$$R_{j0} = n_e N_j s_{j0} = n_e N_j \langle Q_{j0} v \rangle$$

$$= n_e N_j \int_0^\infty \sqrt{\frac{2\epsilon}{m_e}} Q_{j0}(\epsilon) f(\epsilon) d\epsilon. \quad \textbf{Superelastic rate} \quad (11.21)$$

It is left as an exercise for the reader[26] to show that in the special case of a Maxwellian electron-energy distribution[27]

$$\frac{s_{j0}}{s_{0j}} = \frac{g_0}{g_j} \exp\left(+\frac{\epsilon_j}{k_B T_e}\right). \quad \textbf{Maxwellian only} \quad (11.22)$$

[26] See Exercise 11.2.

[27] The positive exponential factor in eqn (11.22) exactly cancels the effect of the negative exponential factor arising from the Maxwellian integral of note 20, so that the only dependence of R_{j0} upon T_e arises from the relatively slowly varying pre-exponential function $\phi\left(\frac{\epsilon_j}{k_B T_e}\right)$.

11.1.8 Excited-state populations in low-pressure discharges

The rate at which level j receives population by electron collisional excitation from the ground state is given by R_{0j}, which is a rapidly increasing function of T_e according to eqn (11.16). As illustrated in Fig. 11.11(a), if we neglect the rate of removal of atoms in level j by collisional ionization, the total rate of depopulation of level j is made up of R_{j0}, the rate of superelastic collisions given by eqn (11.21) together with the rate of loss process that do not depend on the electron density n_e. These comprise the rate of spontaneous emission $N_j A_j$ (where A_j is the *sum* of the Einstein coefficients for emission on all

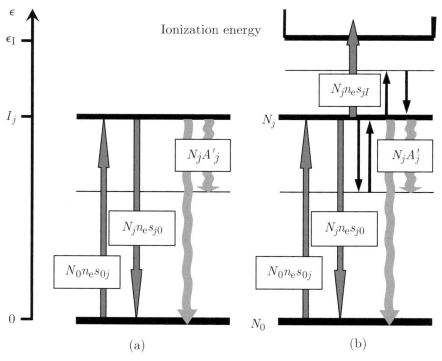

Fig. 11.11: Schematic representation of the processes affecting the populations of excited energy levels in low-pressure gas discharges in the absence of stimulated emission. Collisional process are represented by straight arrows, radiative and diffusive rates by wavy arrows. (a) Collisional and radiative processes populating and depopulating the level j ignoring collisional ionization. (b) Effect of including collisional ionization of level j. Note that since recombination of ions and electrons in low-pressure discharges occurs exclusively on the walls of the discharge tube, there is no reverse rate populating level j by recombination in the volume of the discharge.

transitions out of level j to all levels lower in energy) and the rate of destruction of atoms in level j by collisions with the tube wall. By solving the diffusion equation this rate can be expressed in terms of an average diffusion lifetime τ_D, so that rate of destruction of excited atoms by this process is N_j/τ_D. The total rate of destruction of atoms in level j by processes that do not depend on n_e can be represented by an effective Einstein coefficient A'_j such that

$$A'_j = A_j + \frac{1}{\tau_D}. \tag{11.23}$$

Under this set of assumptions, equating the rates of populating and depopulating rates in *steady state* leads to

$$N_0 n_e s_{0j} = N_j (n_e s_{j0} + A'_j), \tag{11.24}$$

so that,

$$\frac{N_j}{N_0} = \frac{n_e s_{0j}}{n_e s_{j0} + A'_j}. \tag{11.25}$$

In the idealized case of a Maxwellian electron energy distribution, in the limit of high electron densities, $n_e s_{j0} \gg A'_j$ and N_j tends towards a saturated value of,

$$N_j(\text{sat}) = \lim_{n_e \to \infty} \left(N_0 \frac{n_e s_{0j}}{n_e s_{j0} + A'_j} \right) \tag{11.26}$$

$$= N_0 \frac{s_{0j}}{s_{j0}} = N_0 \frac{g_j}{g_0} \exp\left(-\frac{\epsilon_j}{k_B T_e}\right), \tag{11.27}$$

in which the relationship eqn (11.22) has been used.

This is a particularly pleasing result because it shows that if the agency that dominates the populating and depopulating rates is collisional contact with the plasma electron-energy distribution (which is a 'heat bath' characterized by a *kinetic* temperature T_e) then the distribution-over-states temperature of level j is pulled into thermal equilibrium at a temperature T_e with respect to the ground-state population N_0. Such a situation, in which the populations achieve their full Boltzmann values and the same temperature prevails throughout all of the kinetic energy and population distributions, is known as **local thermodynamic equilibrium (LTE)**.[28]

Now, the electron drift velocity v_{drift} depends on E/N, the ratio of electric field to gas number density, since the electrons accelerate in the electric field between making collisions at a rate proportional to gas density. However, E/N, as is evident from Fig. 11.3, does not vary greatly with discharge current over most of the range of practical interest. We can therefore assume that in eqn (11.7) the value of v_{drift} is almost constant to first order and that n_e is directly proportional to the discharge current I_D. Thus, if we were to plot the measured population in a particular excited level against discharge current we might expect that if the population tends towards a saturated value in the high-current limit that this would signal that N_j has attained its full LTE value with respect to the ground-state population N_0. Such an assumption would, however, be wrong by several orders of magnitude in the case of populations of excited levels in low-pressure discharges.

The reason for the discrepancy can be traced to the neglect of a very important process in eqn (11.27). As indicated in Fig. 11.11(b), the ionization of excited atoms by electron collision constitutes a major destruction process for excited atoms; this proceeds at a rate R_{jI} (eqn (11.17)) that can exceed R_{j0}, the superelastic collision rate returning atoms to the ground state, by several orders of magnitude—as discussed in Section 11.1.6. With the inclusion of this term, eqn (11.27) becomes

$$\frac{N_j}{N_0} = \frac{n_e s_{0j}}{n_e s_{j0} + n_e s_{jI} + A'_j}. \tag{11.28}$$

In the limit $n_e \to \infty$,

$$N_j(\text{sat}) = \lim_{n_e \to \infty} N_0 \frac{n_e s_{0j}}{n_e s_{j0} + n_e s_{jI} + A'_j} \tag{11.29}$$

$$= N_0 \frac{s_{0j}}{(s_{j0} + s_{jI})}. \tag{11.30}$$

[28] The thermal equilibrium discussed here is 'local' in the sense that whilst the distribution over states is in equilibrium with the electron-energy distribution at the electron temperature, the kinetic temperatures of the atoms and ions are quite different from that of the electrons, and neither the atoms nor the electrons are in thermal equilibrium with the radiation emitted by the plasma. The system is far removed from true thermal equilibrium.

314 *Atomic gas lasers*

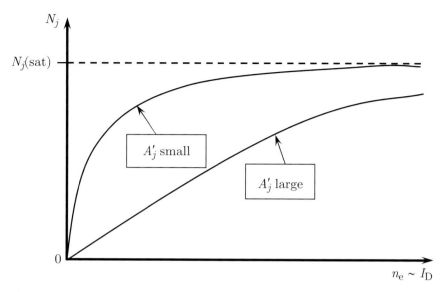

Fig. 11.12: Saturation of population of excited level j with increasing electron density (or discharge current). For low-pressure discharges in rare gases $N_j(\text{sat})$ is typically some 100 times smaller than the full Boltzmann value in LTE with the ground state. For metastable levels (A'_j small) the population saturates at relatively low n_e values. For levels that can decay rapidly by spontaneous emission (A'_j large), much higher values of n_e are required before N_j approaches its saturated value. For simplicity, any possible dependence of T_e upon I_D has been neglected.

The variation of N_j with n_e predicted by eqn (11.30) is shown in Fig. 11.12 for the case of A'_j large (i.e. for a level that can decay rapidly by emission of radiation) and A'_j small (i.e. for a metastable level). Both tend towards saturation a high n_e, but the approach to saturation is slower for large A'_j. However, even at saturation the population is a long way below the LTE value.

It is the absence of volume recombination process repopulating level j from above that accounts for this gross departure from LTE. Ionization of excited-state populations by electron collisions in the volume of the discharge represents a one-way street. The processes occurring at the tube wall play a dominant role in determining the behaviour of the low-pressure discharge in rare gases and, as we shall see in the next section, ultimately set a limit to the c.w. power that can be extracted from a He-Ne laser.

11.2 The helium-neon laser

11.2.1 Introduction

The helium-neon laser was the first gas laser to be developed and the first truly continuous-wave (c.w.) laser of any kind. It is a textbook example of light amplification by stimulated emission of radiation from free atoms, in which population inversion occurs by selective excitation of the upper level through resonant energy transfer.

The initial demonstration in 1960 of laser oscillation on the infra-red transition of Ne at 1.152 μm by Javan, Bennett and Herriott[29] of Bell Telephone Laboratories, was followed shortly thereafter by the demonstration of laser

[29] Javan et al. (1961).

oscillation on the 632.8 nm red transition by White and Rigden,[30] also at Bell Labs. Since then, the helium-neon (He-Ne) laser has been the workhorse for many scientific, industrial, medical, and commercial laser applications. The properties of typical He-Ne lasers are listed in Tables 11.1 and 11.2.

The oscillating fields in the He-Ne laser match very closely the classical description of light modes in a resonant optical cavity, and its output beam when operating in its lowest transverse mode[31] approximates very closely that of a coherent Gaussian beam with an M^2 value[32] typically smaller than 1.02. This is largely a consequence of the optical homogeneity of gas-phase light-amplifying media at low pressure and low power dissipation.

[30] White and Rigden (1962).

[31] See Section 6.3.2.
[32] See Section 6.4.

Table 11.1 Important parameters of the two most important transitions of the He-Ne laser.

			632.8 nm (red)	543.3 nm (green)
Mode of operation			\multicolumn{2}{c}{c.w.}	
Gain medium			\multicolumn{2}{c}{Air cooled, low-pressure discharge}	
Discharge tube bore diameter		(mm)	\multicolumn{2}{c}{1–2}	
DC discharge current		(mA)	\multicolumn{2}{c}{10–50}	
Active length		(mm)	\multicolumn{2}{c}{120–300}	
He-Ne ratio			\multicolumn{2}{c}{6:1 (approx.)}	
Pressure × tube diameter	pd	(Torr mm)	\multicolumn{2}{c}{3.6}	
Gas temperature	T_{gas}	(K)	\multicolumn{2}{c}{350–400}	
Transition probability[a]	A_{21}	(s^{-1})	3.4×10^6	2.8×10^5
Doppler-broadened linewidth	$\Delta\nu_D$	(GHz)	1.5	1.7
Effective upper-level lifetime[a]	τ_2	(ns)	110	110
Effective lower-level lifetime[a]	τ_1	(ns)	19	27
Gain cross-section	σ_{21}^D	(cm^2)	3.4×10^{-13}	1.8×10^{-14}
Homogeneous cross-section[b]	σ_{21}	(cm^2)	1.2×10^{-12}	7.0×10^{-14}
Typical round-trip gain		(%)	2–4	0.2–0.5

[a] Values of A_{21}, τ_2, and τ_1 are from Inatsugu and Holmes (1973).
[b] The homogeneous cross-sections σ_{21} have been calculated assuming $\Delta\nu_H \approx 300$ MHz, as is typical of pressure broadening under laser conditions. See Smith (1966), although note that Smith's values of $\Delta\nu_H$ are half-width at half-maximum values, not the FWHM.

Table 11.2 Wavelengths and typical output powers of the most important transitions in the He-Ne laser.

Wavelength		Transition (Paschen notation)	Output power[a] (mW)
3.391 μm	near-infra-red	$3s_2 \to 3p_4$	0.3–5
1.523 μm	near-infra-red	$2s_2 \to 2p_1$	0.3–1
1.152 μm	near-infra-red	$2s_2 \to 2p_4$	0.3–10
632.8 nm	red	$3s_2 \to 2p_4$	0.3–40
611.8 nm	orange	$3s_2 \to 2p_6$	0.3–7
593.9 nm	yellow	$3s_2 \to 2p_8$	0.3–4
543.3 nm	green	$3s_2 \to 2p_{10}$	0.1–3

[a] Since nearly all of these transitions compete with one another for the available population inversion, the output powers listed are those obtained with cavities whose mirror reflectivities are optimized to allow oscillation on the wanted transition only.

[33]See, for example, White and Tsufura (2004) and Willet (1974).

In this chapter we shall deal primarily with the 632.8-nm red laser since this is by far the most common type. However, the considerations for He-Ne lasers operating on other wavelengths listed in Table 11.2 are broadly similar. About a dozen visible transitions, and perhaps 15 times as many infra-red transitions, of the neon atom have been reported as showing either laser oscillation or optical gain. We have listed in Table 11.2 only the standard wavelengths for which He-Ne lasers are (or have been) available commercially.[33]

11.2.2 Energy levels, transitions and excitation mechanisms

In the He-Ne laser all of the laser transitions are between excited levels of *neon*—the helium is only there to provide selective excitation. The energy levels of He are a good example of LS coupling (also known as Russell–Saunders coupling).[34] The ground configuration $1s^2$ has two electrons in the 1s orbital, and forms only a singlet level $1\,^1S_0$ since the Pauli exclusion principle requires their spins to be antiparallel. The first excited configuration is 1s2s, i.e. one of the electrons remains in the 1s orbital while the other is excited to the 2s orbital. Because the two electrons differ in their principal quantum numbers the Pauli principle allows both the spin-parallel triplet level, $2\,^3S_1$, as well as the spin-antiparallel level, $2\,^1S_0$, to arise from this (and higher) configurations. The non-central residual electrostatic interaction between the two electrons is responsible for the fact that the $2\,^3S_1$ lies slightly lower in energy than the $2\,^1S_0$ level, as shown in Fig. 11.13. Electric dipole transitions from states belonging to either of these levels to the ground state are highly forbidden since electric dipole radiation requires a change of parity, and both $1s^2$ and 1s2s have even parity.[35] Because atoms in $2\,^3S_1$ and $2\,^1S_0$ cannot decay by emission of radiation they are said to be metastable. In practice, the lifetimes of these levels are determined by collisional processes.

[34]The absence of any singlet–triplet transitions (i.e. violations of the $\Delta S = 0$ rule) in the spectrum of He indicates that S is indeed a good quantum number in this case. For a discussion of coupling schemes in atoms see, for example, Woodgate (1980) or Foot (2005).

[35]The parity of the wave functions of all the levels belonging to a given configuration is even or odd according to whether the algebraic sum of the ℓ-values of all the electrons (outside closed shells) is even or odd, respectively. The parity selection rule is discussed in Section 2.3.2 and in Exercise 2.4.3.

The coupling scheme in Ne (and all other rare gases apart from He) is a complicated intermediate $(j - \ell)$ form. The ground state—which arises from the configuration $2p^6$—can be described as $^1S_0$ in LS notation. However, none of the other levels can be properly described by the LS coupling notation, and hence it is still customary to use the arbitrary labelling system introduced by Paschen. As shown in Fig. 11.13, in Paschen notation the Ne ground state is denoted $1s_1$, and levels belonging to the first manifold of excited states (arising from the configuration $2p^5 3s$) are labelled $1s_2 - 1s_5$.

In Paschen notation the common upper laser level of the 3.391-μm and 632.8-nm transitions in Ne is $3s_2$ (arising from the configuration $2p^5 5s$). The upper level of the 1.15-μm transition is $2s_2$ (arising from the configuration $2p^5 4s$). Both upper levels have quite short radiative lifetimes (10–20 ns) since both can decay to the ground state via fast electric dipole transitions in the vacuum ultraviolet. However, at neon pressures above 0.05 Torr the probability of resonance radiation from these transitions being reabsorbed by ground-state atoms is so high that the decay paths in the transitions from $3s_2$ and $2s_2$ to the

11.2 *The helium-neon laser* 317

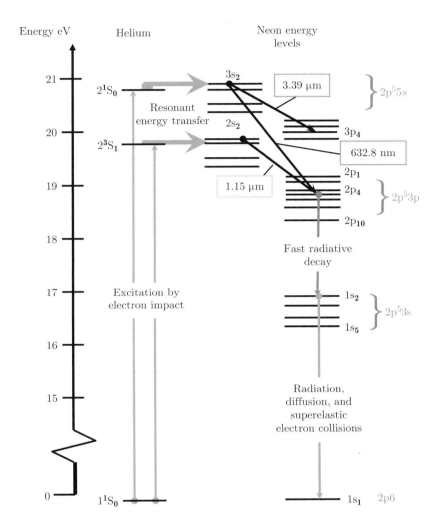

Fig. 11.13: Partial energy-level diagrams for He and Ne drawn on the same energy scale. The electron configurations giving rise to the various levels of Ne are noted in light grey on the right-hand side. Note that neither atom has excited levels below 16 eV. The excitation mechanism of the 1.15 μm He-Ne laser involves resonant transfer of energy from metastable He level $2\,^3S_1$ that lies very close in energy to the Ne level labelled $2s_2$, while that of the 633-nm He-Ne laser involves a similar transfer from the metastable He level $2\,^1S_0$ (which lies very close in energy to the Ne level labelled $3s_2$).

ground state may be considered as fully blocked.[36] Thus, the effective lifetimes of $3s_2$ and $2s_2$ are determined entirely by their decay in slower visible and infra-red transitions to lower levels well above the ground state ($\tau_2 \approx 136$ ns for $2s_2$, and $\tau_2 \approx 110$ ns for $3s_2$). The two lower laser levels $3p_4$ and $2p_4$ can decay by reasonably fast UV transitions to levels in the $1s_2 - 1s_5$ manifold with lifetimes of order $\tau_1 \approx 20$ ns.

In terms of eqn (2.15), which must be satisfied in order for a steady-state inversion to exist, it is clearly helpful that the three laser transitions at 623.8 nm, 1.152 μm and 3.391 μm all exhibit a very favourable ratio ($\tau_2/\tau_1 \approx 5$) of fluorescent lifetimes under the prevailing discharge conditions.

The other desirable factor, selective excitation of the upper laser levels, is ensured by the presence of the helium. The two levels of He that are important are the two metastable levels 2^3S_1 and 2^1S_0 and by two sheer (and unrelated)

[36] The increase in the fluorescence lifetime of an atomic level due to radiation trapping is discussed in Section 4.5.3. The related phenomenon of self-absorption broadening (also called opacity broadening) is discussed in Sections 4.5.2.

accidents of nature the energy of $2\,^3S_1$ lies within 313 cm^{-1} of the 2s$_2$ level of Ne (a coincidence of 1 part in 500) and that of 2^1S_1 of He coincides with the 3s$_2$ level of Ne to within 386 cm^{-1} (better than 1 part in 400). When a discharge is run in pure He at low pressure (1–3 Torr, say) electron collisions with ground-state He will excite some atoms directly (and by cascade from higher levels) to both metastable levels, from which they can only escape by collision with the wall or by making an ionizing collision with an electron.

With sufficient discharge current a reservoir of excited He atoms is created,[37] reaching steady-state values of order $10^{11} - 10^{12}$ atoms cm^{-3} in these two long-lived levels. If now a small quantity of Ne is added to the discharge (not enough to change its character as a discharge governed by the properties of He) *selective* excitation of Ne to the 2s$_2$ and 3s$_2$ levels can occur because of the energy resonance in the reactions between He metastable atoms and ground-state Ne atoms:

[37] See Section 11.1.8.

$$\text{He}\,(2\,^3S_1) + \text{Ne}_{\text{g.s.}} \rightarrow \text{He}_{\text{g.s.}} + \text{Ne}\,(2s_2) + 0.04\text{ eV} \qquad (11.31)$$

$$\text{He}\,(2^1S_0) + \text{Ne}_{\text{g.s.}} \rightarrow \text{He}_{\text{g.s.}} + \text{Ne}\,(3s_2) - 0.05\text{ eV}. \qquad (11.32)$$

Since atom–atom interactions occur via short-range forces, transfer of excitation will be appreciable only for those states of Ne close enough to exact resonance that many atoms in the Maxwellian distribution at thermal energy ($k_B T_{\text{gas}} \approx 0.03$ eV at 400 K) can make up the difference. Thus, because of the accidental energy resonances, the transfer of excitation[38] is selective in exciting pre-dominantly the 3s$_2$ and 2s$_2$ levels of neon, thus creating inversions of population with respect to the 3p$_4$ and 2p$_4$ lower levels.

[38] The reactions (11.31) and (11.32) can also go backwards, but since neon atoms in 3s$_2$ and 2s$_2$ either radiate spontaneously or are stimulated out of these levels fairly promptly, the backward reaction rate is negligible compared to the forward rate.

11.2.3 Laser construction and operating parameters

The construction of a typical He-Ne laser designed for operation at about 1 mW on the 632.8-nm transition is indicated in Fig. 11.14. Typical operating parameters are listed in Table 11.1.

Early He-Ne lasers were constructed with cavity mirrors external to the gas envelope. Since the gain per pass is only 1–3% per round trip, a Fresnel reflection loss of 4% (intensity) per interface for glass windows if used at normal incidence would represent a 28% round-trip loss and would prevent threshold being reached. One way of avoiding this problem is to close the gas envelope with windows tilted at Brewster's angle ($= \arctan n_{\text{glass}}$) so that for one plane of polarization (TM) the transmission is 100%. The output of such a laser is completely plane polarized.

Most modern He-Ne lasers are constructed as indicated in Fig. 11.14 with laser cavity mirrors sealed directly onto the ends of the discharge tube, permanently aligning them during the tube-fabrication process. The output of such a laser is unpolarized, unless an internal Brewster-angle window is fitted, as indicated in Fig. 11.14. The cavity is usually formed by curved mirrors used in a stable, near-confocal arrangement. One mirror would be a high

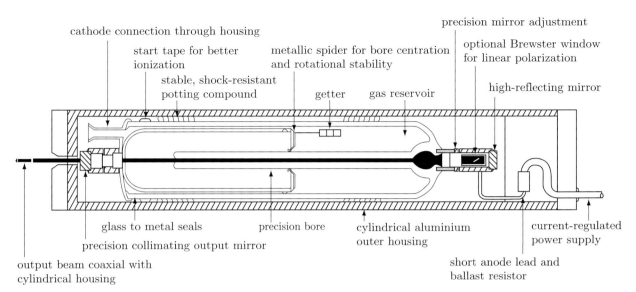

Fig. 11.14: Cross-sectional view of a He-Ne laser head (courtesy CVI Melles Griot). The positive column discharge runs from the anode ring (near the high reflector) along the length of the capillary tube to the negative glow covering the inside of the large bore hollow cathode structure at the left-hand end. A metal cap (on which the output mirror element is mounted) is in contact with the cold-cathode structure, and closes off the left-hand end of the glass gas envelope via a glass-to-metal seal. The capillary discharge tube is supported by a metallic 'spider' that ensures that it is concentric with the gas envelope tube. A 'getter' is included inside the gas volume to clean up any residual traces of molecular gases outgassed. The discharge is ignited by applying a high-voltage pulse to the 'start tape'. The entire discharge tube is sealed with shock-resistant potting compound inside, and accurately concentric with, an aluminium alloy tube housing. (See White and Tsufura (2004).)

reflector coated for maximum reflectivity at the laser wavelength, while the other transmits 0.5 to 1%.

Scrupulous cleanliness is essential during tube manufacture and filling since only a few parts per thousand of contaminant gas will strongly affect the properties of the positive column—in particular T_e. Tube failure is usually due either to the accumulation of outgassed or leaked contaminants, burial of neon (as Ne ions) in the cathode or loss of He by diffusion through the glass envelope of the tube. With careful attention to cleanliness of materials, tube life is typically greater than 10 000 h, and many systems have performed 50 000 h or more without a tube replacement.[39]

[39] See, for example, White and Tsufura (2004).

11.2.4 Output-power limitations of the He-Ne laser

For the reasons explained in Section 11.1.8, at high current densities, the populations of both the $2\,^3S_1$ and $2\,^1S_0$ levels of He tend towards saturated values (i.e. independent of discharge current). It is therefore not surprising that the excitation rates of the Ne $3s_2$ and $2s_2$ upper laser levels also tend to saturate with increasing discharge current, since they are almost exclusively excited by resonant transfer of excitation in reactions (11.31) and (11.32), respectively.

Fig. 11.15: Populations of various energy levels in a He-Ne laser discharge as a function of the discharge current. Open circles are populations of the He $2\,^1S_0$ metastable level (right-hand scale) measured via absorption of radiation on a transition from the $2\,^1S_0$ level in He. The relative populations of the $3s_2$ upper laser level and the $2p_4$ and $3p_4$ lower laser levels in Ne are monitored via the relative intensity of spontaneous emission on the 632.8, 609.6 and 359.3 nm transitions of Ne (left-hand scale). Reprinted figure with permission from A. D. White and E. I. Gordon, Appl. Phys. Lett., 3, pp 197–199, (1963), http://link.aip.org/link/?APPLAB/3/197/1, Copyright (1963) by the American Institute of Physics.

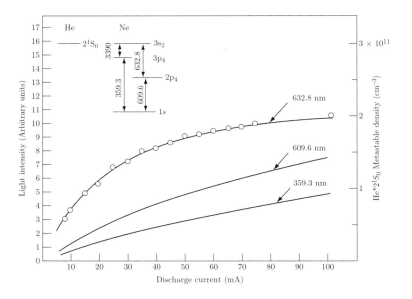

An elegant experiment showing this effect was performed by White and Gordon in 1963, their results being shown in Fig. 11.15. When the data points for the measured He 2^1S_0 population as a function of discharge current are superimposed on the curves showing the relative intensity of emission from the Ne $3s_2$ upper laser level the fit is excellent, showing that indeed the He-Ne atom–atom energy transfer in reaction (11.32) is the dominant excitation process for Ne $3s_2$.

The population of Ne $2p_4$ (the lower level of the 632.8-nm laser transition) however, shows a variation with discharge current that is much more nearly linear than that of the upper level, indicating that the contribution made by direct excitation of the Ne $2p_4$ levels from the Ne $1s_1$ ground state by plasma electron collisions is comparable to the radiative cascade feeding in the laser transition itself. The laser operates best when the inversion (the gap between the two curves for 632.8 and 609.6 nm emission) is largest, and for the particular laser tube used in Fig. 11.15 this occurs at a discharge current of 50 mA. Higher discharge currents tend to decrease the gap between the curves, to the extent that lasing is extinguished completely at high discharge currents.

Unfortunately, little can be done to overcome this problem associated with saturation of the pump species. The saturated He $2\,^1S_0$ metastable population could be increased by reducing the tube diameter since this increases T_e and thus the excitation rate of He $2\,^1S_0$. However, reducing the tube diameter reduces the laser volume and hence the extractable power. Increasing the Ne concentration in the discharge in an attempt to use more fully the available saturated He $2\,^1S_0$ population does not pay dividends either; increasing the Ne density lowers T_e since the discharge switches over to making Ne$^+$ ions directly, ignoring the He and adjusting its value of T_e to suit the more-easily ionized Ne.

11.2.5 Applications of He-Ne lasers

He-Ne lasers are small (relative to other gas lasers), convection cooled, and can be operated from a normal household electrical supply. Lower-power lasers can be operated easily from batteries, making them ideal sources for field equipment including theodolites, laser levels, and other construction alignment equipment. He-Ne lasers are relatively low in cost—typically under £100 each if bought individually.

The main drawback of the He-Ne laser in some applications is the output power. He-Ne lasers are low-power devices, with output ranging from 0.1 mW or less to the largest (for devices of practical size) of around 40 mW. This limitation makes them unsuitable for most materials processing applications or for pumping other laser sources. In the late 1980s, more than 750 000 He-Ne lasers were sold each year for applications as diverse as holography, barcode scanning, biological cell counting, metrology, construction alignment, printing and reprographics, medical and industrial pointing, and light shows. The overall numbers have dropped considerably, but even today, many years after the He-Ne was predicted to become obsolete as a result of advances in semiconductor diode laser technology, approximately 200 000 are sold annually–more than all other lasers combined (excluding, of course, diode lasers).[40]

[40] White and Tsufura (2004).

11.3 The argon-ion laser

11.3.1 Introduction

The c.w. argon-ion laser (and its close relatives, the Kr^+ and Xe^+ systems) also operates in the positive column of a low-pressure glow discharge. Unlike the He-Ne laser, the rare-gas ion lasers do not rely on a selective excitation process to excite the upper laser levels, but in view of the extremely favourable lifetime ratios that exist between upper and lower laser levels a selective excitation mechanism is not required.

Early in 1964, Bridges and Convert et al. reported pulsed laser oscillation on ten transitions of singly ionized argon in the blue-green wavelength region 450–530 nm. This was followed shortly after by reports from Bennett et al. of quasi-continuous operation and from Gordon et al. of truly continuous laser oscillation on these transitions.[41]

For many years the argon-ion laser was the only powerful source of c.w. laser radiation in the ultraviolet, visible and near-infra-red. Continuous-wave powers of several tens of watts in the visible (450–530 nm) and of several watts in the ultraviolet (229–363 nm) are available from commercial devices providing both high beam quality and spectral refinement. Experimental argon ion lasers have been demonstrated that generate hundreds of watts of continuous-wave power in the blue-green spectral region.

Despite their very poor electrical to optical conversion efficiency, rare-gas ion lasers are still in service despite the rapid development in recent years of diode-laser-pumped solid-state lasers. The wide range of output wavelengths available from ion lasers is a capability not yet possessed by solid-state continuous-wave sources.

[41] See Bridges (1964), Convert et al. (1964), Bennett et al. (1964), and Gordon et al. (1964).

Table 11.3 Parameters of two of the most important transitions of the argon-ion laser. Data from Dunn and Ross (1976) and references therein.

			488.0 nm	514.5 nm
Mode of operation			c.w.	
Gain medium			Water cooled, high current, low-pressure discharge	
Discharge tube bore diameter[a]		(mm)	1–2.5	
DC discharge current		(A)	up to 65	
Active length		(m)	0.5–1	
Gas fill[b] at room temperature		(Torr of pure Ar)	0.1–1	
Optimum pressure × tube diameter	pd	(Torr mm)	0.5 (for 488.0-nm line)	
Gas temperature	T_{gas}	(K)	2000–3000	
Ion temperature	T_{ion}	(K)	3000–4000	
Transition probability	A_{21}	(s^{-1})	8.5×10^7	9.2×10^6
Doppler-broadened linewidth[c]	$\Delta \nu_D$	(GHz)	4.1	4.0
Homogeneous linewidth[c]	$\Delta \nu_H$	(MHz)	800	800
Effective upper-level lifetime	τ_2	(ns)	9	6
Effective lower-level lifetime	τ_1	(ns)	0.4	0.4
Gain cross-section	$\sigma_{21}^D(0)$	(cm^2)	1.9×10^{-12}	0.23×10^{-12}
Homogeneous cross-section	$\sigma_{21}(0)$	(cm^2)	6.4×10^{-12}	7.7×10^{-13}
Typical round-trip gain			5–25%	5–25%

[a] In the case of discharge-tube designs in which the plasma confining structure comprises a series of metal discs, as in Fig. 11.9 the diameter of interest is that of the central hole in each metal disc rather than that of the outer gas envelope.
[b] The gas density in the active region when the laser is in operation is reduced by (almost) the ratio T_{room}/T_{gas} from the value corresponding to the filling pressure at room temperature.
[c] Both the homogeneous and inhomogeneous linewidths increase with the discharge current density. In particular, the homogeneous width includes contributions from Stark broadening in addition to natural broadening. See Webb et al. (1968).

11.3.2 Energy levels, transitions and excitation mechanisms

The neutral argon atom has a ground-state electron configuration comprising $3s^23p^6$ outside completely filled 1s and 2s shells. The ground state can be represented in LS notation as $^1S_0$ in LS notation or $1s_1$ in Paschen notation. The closed-shell structure ensures that the first ionization energy is high: 15.76 eV. Part of the term diagram of singly ionized argon (Ar$^+$, or in spectroscopic notation Ar II) is shown in Fig. 11.16.

The ground configuration of Ar$^+$ is $3s^23p^5$ and therefore has *odd* parity. The lower laser levels, at about 16 eV above the Ar$^+$ ground state, arise from the configurations $3s^23p^44s$ and $3s^23p^43d$ and are therefore of *even* parity. The next-highest configuration, $3s^23p^44p$, is of even parity, and gives rise to the upper laser levels at about 18 eV above the Ar$^+$ ground state.

The reasons for the favourable lifetime ratio can be understood from the term diagram of Ar$^+$ and the parity selection rule for electric dipole radiation. Since the upper laser levels in Ar$^+$ belong to configurations of *odd* parity, as does the Ar$^+$ ground state, their *only* allowed optical decay paths are in transitions to the lower laser levels, all of which are of *even* parity. In all cases the transitions

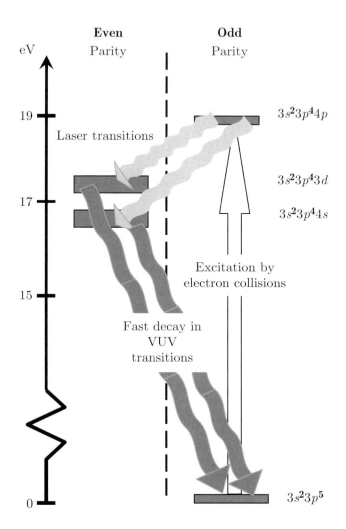

Fig. 11.16: Electron configurations giving rise to some of the energy levels involved in laser action in singly ionized argon. The energy scale is referred to the ground state of Ar$^+$, which is 15.76 eV above that of ground-state Ar.

occur by the emission of *visible* photons (2.0–2.5 eV transition energy). In contrast, the lower laser levels are connected by fast transitions in the vacuum UV spectral region (photon energies 16–17 eV). Since, for transitions of a given oscillator strength, the Einstein A coefficient is proportional to the square of the transition frequency[42] it is not surprising that the lifetime of a typical upper laser level is of order 7–10 ns, while that of a typical lower level is only 0.4 ns (see Fig. 11.3). The laser levels therefore have a very favourable lifetime ratio ($\tau_2/\tau_1 \approx 20$) and from eqn (2.15), we see that provided the lower laser levels are not excited at rates more than twenty times faster than those of the upper level, inversion of population is bound to occur.[43]

As shown in Fig. 11.17, c.w. inversion may be obtained simultaneously on about ten transitions in the blue and blue-green regions of the spectrum. Output powers available from the transitions most commonly produced by commercial argon-ion lasers are listed in Table 11.5, although it should be emphasized that this is not a complete listing of all possible argon-ion wavelengths.

[42] See Section 18.2.1.

[43] Note that in this case the excitation of the upper laser levels from the ground state occurs via an optically forbidden transition. For *neutral* atoms we would, from Section 11.1.6, expect the collisional cross-section for such transitions to be lower than that for excitation of the lower laser levels. However, the cross-sections for collisional excitation of levels in *ions* behave very differently from in neutral atoms, as explained in Note 29 on page 517.

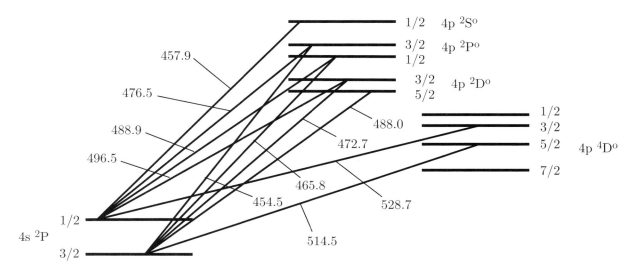

Fig. 11.17: Some of the strongest c.w. transitions of the Ar$^+$ laser, from levels of the $3s^23p^44p$ configuration (odd parity denoted by the superscript O) to levels of the $3s^23p^44s$ (even parity) configuration. Note that both doublet and quartet terms can arise from the $3s^23p^44p$ configuration, and that despite the fact that the 528.7- and 514.4-nm transitions both involve a change of spin they are, nevertheless, among the strongest c.w. laser transitions.

[44] From the point of view of establishing the route of excitation to the ion laser transitions the exact contributions of the various stepwise processes by which ground-state Ar$^+$ ions are formed are not important This is because the ground-state Ar$^+$ ion is a stable species, and the constraint of charge neutrality in the positive column ensures that, whatever the mechanisms responsible for its formation, its steady-state population is equal to the electron density. This remains true for the range of current densities of interest for laser operation on transitions of the singly ionized species. For a detailed discussion of the excitation mechanisms of the Ar$^+$ laser see (Dunn and Guttierrez 2004) and (Dunn and Ross 1976)

[45] See, for example, Allen (1963).

[46] These levels are metastable because they have quartet multiplicity and their decay to the doublet ground state is forbidden by the spin rule. See Dunn et al. (1976).

As indicated in Fig. 11.18 excitation of the upper laser levels of Ar$^+$ occurs in at least two distinct steps. Despite the high loss rate of ions from the volume of the discharge, the Ar$^+$ ground state, density is maintained at values of order 10^{13} cm^{-3} by a very rapid rate of collisional ionization of ground-and excited-state Ar atoms.[44] Only the electrons in the high-energy tail of the distribution (which has typically a mean energy in the range 3–5 eV) have enough energy to participate in this process. Excitation of the upper (and lower) laser levels of Ar$^+$ also involves collisions of high-energy electrons—but this time with ground-state Ar$^+$ ions. The cross-sections for electron collisional excitation of ions tend to be one or two orders of magnitude higher than those for excitation of neutral atoms (because of the long-range Coulomb attraction between ion and electron). They also show a quite different dependence on energy above threshold from that shown in Fig. 11.8 for neutral atoms, reaching their maximum values at energies near threshold.[45] Of the different excitation routes that were initially proposed, the three that are consistent with experimental evidence are those shown in Fig. 11.18, in which the second step involves excitation directly from the Ar$^+$ ground state to the upper laser levels, or feeding by radiative cascade from higher-lying levels that are themselves excited by electron collisions on ground state Ar$^+$, or by stepwise excitation via metastable levels of even parity belonging to the $3s^23p^43d$ and $3s^23p^44s$ configurations[46] lying just below the lower laser levels.

Thus, whether the excitation route to the upper laser levels is (a) direct from the ion ground state or (b) via radiative cascade from higher levels, the expected variation of excitation rate of excited level j is of the form

$$R_j = n_e N_0^+ s_{0j}^+ (T_e) \approx n_e^2 s_{0j}^+ (T_e), \qquad (11.33)$$

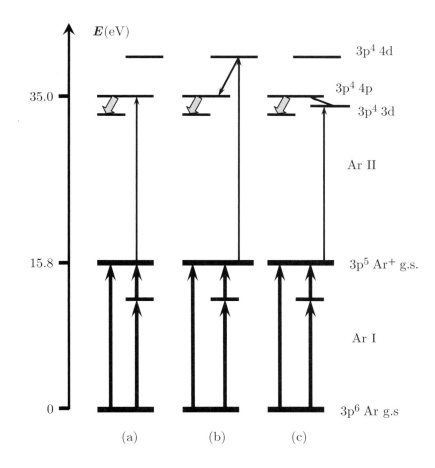

Fig. 11.18: Excitation routes to the upper laser levels of the argon ion laser. Ground-state Ar$^+$ ions are formed by collisions of electrons with ground-state Ar atoms and Ar atoms in excited levels. From there the ions can be excited to the upper laser levels of the $3p^4 4p$ configuration by electron collisions either (a) directly, or (b) to levels of the $3p^4 4d$ configuration that decay radiatively to the upper laser levels, or (c) to metastable levels of the $3p^4 3d$ configuration from which a further electron collision step excites them to the upper laser levels. It is possible that all three routes contribute to the overall excitation rate.

since the requirement of charge neutrality, and the fact that the overwhelming bulk of the ions are in the ground state, demands that $n_e \approx N_0^+$. Even if part of R_j involves excitation via an intermediate metastable ion level, the expected form of dependence of R_j upon n_e is still quadratic rather than cubic. This is because, at high current densities, the populations of the metastable Ar$^+$ levels saturate with respect to the *ion* ground-state population for the same reason that metastable levels of the neutral atom tend towards saturation with respect to the neutral ground-state population The population of ion metastable levels is therefore an approximately constant fraction of N_0^+ the ion ground-state population.

11.3.3 Laser construction and operating parameters

The construction of early argon-ion lasers resembled that of helium-neon lasers except that they were made of a refractory material (silica lined with graphite or ceramic), were provided with an oxide-coated thermionic cathode to handle the large currents, and a water jacket for cooling. In addition, a gas-return path connecting the anode and cathode bulbs was provided to prevent the gas from

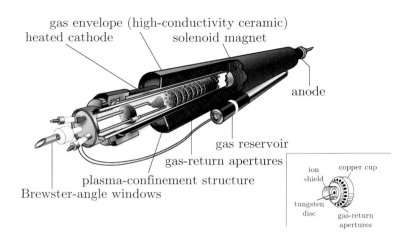

Fig. 11.19: Cutaway drawing of an argon-ion laser tube (courtesy Coherent Inc.). The inset diagram at the bottom right-hand side shows details of the plasma-confinement structure. It comprises a series of individual copper cups cooled by thermal contact with the high-conductivity ceramic gas envelope, each bearing a tungsten disc with a central aperture that defines the diameter of the discharge.

[47] Curiously, the gas in the tube is pumped towards the *anode* rather than towards the *cathode*. One might have expected the gas to be pushed mainly by the Ar^+ ions that tend move towards the cathode under the longitudinal electric field of the discharge. The explanation of this phenomenon was provided by A. N. Chester (1968) who showed that the electrons on average manage to transfer more longitudinal momentum per unit time (but not per collision) to the neutral atoms than do the ions because the ion motion under the large radial electric field is almost normal to the tube axis.

[48] Because the coating of the cathode is made of such chemically reactive material, if the tube is ever opened and exposed to air, the emission properties of the cathode will almost certainly be destroyed because of reactions with the water-vapour content of the atmosphere.

[49] Crystalline quartz is less prone to F-centre formation than fused silica, and suffers far less from the compaction damage caused by breakage of chemical bonds.

accumulating at one end of the tube due to the very fast pumping action of the discharge.[47]

The discharge-confinement structure of a typical commercial Ar^+ laser tube (see Fig. 11.19) might have a diameter of 3 mm. The discharge is run at a discharge current of 20–50 A at a voltage of about 400 V. The input power of over 10 kW is deposited in a plasma whose volume is of order 3 cm³. If the plasma tube itself were the confining structure, as in early laser designs, there is a severe problem in finding refractory pure materials that can withstand the high ion bombardment and high operating temperature of the discharge. Materials that have been employed as discharge tubes or linings for discharge tubes include pyrolytic graphite and beryllia (BeO).

As shown in Fig. 11.19 a modern argon-ion laser tube has a gas envelope made from ceramic of high thermal conductivity. The plasma-confinement structure comprises a set of accurately aligned apertures in tungsten discs each of which is supported by a copper cup. The copper cups are electrically isolated from one another but are in good thermal contact with the inner wall of the gas envelope tube. Near the periphery of each cup is a ring of holes that ensures gas-pressure equalization between neighbouring sections. A cold cathode could not supply even a small fraction of the current required for laser operation, and so a thermionic emitter cathode coated with a low work function material[48] is needed. The Brewster-angle windows at each end of the tube are made from crystalline quartz that is much more resistant to damage[49] by the deep UV emitted by the plasma than the (chemically similar) fused silica used in early lasers. The initial (cold) filling pressure of spectrally pure argon gas is about 0.5–0.7 Torr. In operation, the pressure of gas gradually decreases over a period of months as argon ions are driven into the walls of the tube structures. A reservoir filled with argon and fitted with an electrically operated valve is connected to the tube so that the gas lost can be replaced from time to time.

The operating parameters of typical large-frame and small-frame argon-ion lasers are given in Table 11.4.

Argon-ion lasers are available with fundamental wavelengths that range from 275.4 to 1090 nm. Lasers generating deep-UV wavelengths, obtained by

Table 11.4 Operating parameters of small- and large-frame argon-ion lasers.

Laser parameter		Small frame	Large frame
Maximum tube current	(A)	60	65
Nominal tube voltage	(V)	235	535
Active tube length	(m)	0.41	1.11
Bore diameter	(mm)	3.15	3.25
Current density	(A cm^{-2})	779	785
Input power	(kW)	13	33
Output power (457.9–514.5 nm)	(W)	10	30
Output power (333.3–363.8 nm)	(W)	2	10
Efficiency (all visible lines)	(%)	0.07	0.08

frequency doubling the fundamental wavelengths, to cover the 229.0–264.3 nm range are also commercially available.[50]

[50] See Dunn and Guttierrez (2004).

11.3.4 Argon-ion laser: Power limitations

If the current density in the laser discharge is increased to values exceeding 1 kA cm^{-2}, trapping of the resonance radiation in the 16-eV transitions from the lower laser levels to the ion ground state may become appreciable. However, this does not become a dominant trend until the c.w. laser power on the Ar$^+$ laser transitions exceeds 50–100 W. Ultimately, if the current is increased sufficiently, strong tapping sets in and both upward and downward transfer among the various Ar$^+$ levels becomes dominated by electron-collision processes, and the population inversion in Ar$^+$ will be destroyed. Because of the importance of wall processes, the populations of excited states of the neutral and singly ionized species are still a long way from LTE, despite being fully saturated with respect to their own ground-state densities. At about the same current density, states of doubly ionized argon Ar^{++} become increasingly populated and laser action (in the near-ultraviolet) becomes possible on transitions between excited levels of Ar^{++} for reasons similar to those giving rise to the inversion in Ar$^+$. Indeed, the laser output on the half dozen or so lines between 363.8 and 275.4 nm available at very high values of excitation current (near the limit for operation) in commercial large-frame argon-ion lasers (see Table 11.5) originate from transitions of Ar^{++}.[51]

[51] The spectroscopic designation of Ar^{++} is Ar III.

The argon-ion laser is not efficient—indeed it cannot be efficient. This is not simply because many levels of Ar$^+$ are excited by electrons just to get the wanted transitions inverted, it has to do with the requirement to have a high value of T_e and n_e in order to optimize the excitation rate. This requires that the ion loss rate should be high since Constraint III requires the T_e value to be high in order to create new ions at a rate sufficient to make up for the rate of loss of ions. Thus, over 99% of the power input to the argon-ion laser tube goes into making ions that promptly smash into the wall and give up energy directly to the plasma-confinement structure, and hence ultimately to the cooling water.

It is found that an axial magnetic field (of order 0.1 T) can improve laser efficiency by a factor of 2–5. However, the processes by which the magnetic field causes the enhancement are certainly more complicated than a simple

Table 11.5 Typical output powers of commercial argon-ion lasers (courtesy of Coherent Inc.).

Wavelength (nm)	Designation	Individual line output (W)	
		Main frame (25.0 W multiline)	Small air-cooled (1.00 W multiline)
528.7	$4p\,^4D^o_{3/2} \to 4s\,^2P_{1/2}$	1.8	0
514.5	$4p\,^4D^o_{5/2} \to 4s\,^2P_{3/2}$	10	0.350
501.7	$(^1D)\,4p\,^2F^o_{5/2} \to 3d\,^2D_{3/2}$	1.8	0
496.5	$4p\,^2D^o_{3/2} \to 4s\,^2P_{1/2}$	3.0	0
488.0	$4p\,^2D^o_{5/2} \to 4s\,^2P_{3/2}$	8.0	0.200
476.5	$4p\,^2P^o_{3/2} \to 4s\,^2P_{1/2}$	3.0	0
472.7	$4p\,^2D^o_{3/2} \to 4s\,^2P_{3/2}$	1.3	0
465.8	$4p\,^2P^o_{1/2} \to 4s^2P_{3/2}$	0.8	0
457.9	$4p\,^2S^o_{1/2} \to 4s\,^2P_{1/2}$	1.5	0.80[a]
454.5	$4p\,^2P^o_{3/2} \to 4s\,^2P_{3/2}$	0.8	0
333.6–363.8 combined	ArIII	7.0	(UV multiline 100 mW)[a]
363.8	ArIII	1.7	0
351.1	ArIII	1.8	0
351.1–385.8 combined	ArIII	4.4	0
300.3–335.8 combined	ArIII	3.0	0
275.4–305.5 combined	ArIII	1.6	0
334.5	ArIII	0.5	0
302.4	ArIII	0.38	0
275.4	ArIII	0.18	0

[a] Available with an appropriate UV mirror cavity.

pinch effect compressing the radial profile of the plasma, and are yet to be explained in detail. Unpublished data by one of the authors (CEW) shows that although at low magnetic fields the radial profiles of Ar^+ emission lines do indeed narrow, as the field is further increased they reach a minimum width and then start to widen again. At magnetic-field values typical of laser operation the ion profiles are wider than those in zero magnetic field. This suggests that the effect of magnetic field on laser operation may have to do with creating anisotropy in the electron-energy distribution so that the electron energies along the lines of magnetic field are enhanced at the expense of those in the transverse direction.

11.3.5 Krypton-ion lasers

Kr^+-ion lasers are similar in most respects to Ar^+ lasers, but typically provide less output power from devices of comparable size and, particularly in early devices, tended to have a somewhat shorter working life than their Ar^+ counterparts.[52] The strongest Kr^+ laser transition is at 647.1 nm. Since the Ar^+ laser system lacks a strong output in the red region, the availability of this transition (as well as two with wavelengths in the near-UV at 337.5 and 356.4 nm) are the main reason why Kr^+ lasers are sometimes appropriate for applications that Ar^+ lasers cannot address. If a wide range of output wavelengths from a single device is required, then it is possible to

---

[52] A possible reason for this is the greater rate of damage that the plasma confinement and cathode structure may sustain as a result of bombardment with Kr^+ ions (mass 84 AMU) compared to that of Ar^+ ions (of mass 40 AMU).

fill the laser with a mixture of argon and krypton and obtain simultaneous output on both on both Ar$^+$ and Kr$^+$ transitions, albeit at somewhat reduced efficiency.

11.3.6 Applications of ion lasers

While solid-state lasers have replaced ion lasers in many traditional applications, such as pumping Ti:sapphire lasers and dye lasers, there is not yet a viable alternative to ion lasers for pumping dyes in the UV. Similarly, ion lasers are used in scientific and commercial applications, such as laser spectroscopy in the UV, deep-UV, and red wavelengths, and violet wavelengths, where powerful all-solid-state c.w. lasers systems are not readily available.

The ability of the ion laser to provide c.w. output at multiple wavelengths simultaneously is important for its use in two- or three-colour laser Doppler velocimetry (LDV) in which the velocity components of fluid flow are measured via the scattering of light from particles entrained in the flow.[53]

[53]Tropea (2004).

Industrial applications for ion lasers include: the recording of CD and DVD master discs (Milster 2004); the manufacture of fibre Bragg gratings Jackson and Webb (2004); as well as the automated inspection and counting of biological cells via the process of flow cytometry (Wachsmann-Hogiu *et al.* 2004). Ion lasers are used to make holograms (Herzig 2004) and for the inspection of optical components using interferometric techniques. The entertainment industry uses ion-laser systems for light shows. Many of these applications require the use of argon, krypton, and argon/krypton mixed-gas lasers, as well as frequency-doubled argon-ion systems.

Further reading

Comprehensive discussions of the mechanisms responsible for population inversions in gas lasers may be found in the text by Willett (1974) and the article by Webb (1975).

More detailed information on He-Ne and argon-ion lasers may be found in the articles by White and Tsufura (2004) and Dunn and Guitarrez (2004).

Exercises

(11.1) A He$^+$ ion travelling with velocity u and mass M makes a head-on collision with a stationary He atom. Show that the minimum value of the kinetic energy $\frac{1}{2}Mu^2$ for the incident ion to excite the atom to its first excited level of energy E_1 is $2E_1$. [Since the value of E_1 for He is about 20 eV this shows that effectively none of the atomic excitation in a He discharge is provided by ion collisions, since the number of ions in a Maxwellian distribution (of mean energy not very far above $k_B T_{gas}$) that have kinetic energies that exceed 40 eV is completely negligible.

(11.2) The rate of excitation of level j from the ground state 0 by electrons in a discharge that have a Maxwellian distribution of energies characterized by a temperature T_e is:

$$R_{0j} = n_e N_0 s_{0j} = n_e N_0 \int_{\epsilon_j}^{\infty} \left(\frac{2\epsilon}{m_e}\right)^{\frac{1}{2}} Q_{0j}(\epsilon) f(\epsilon) d\epsilon,$$

where, from eqn (11.13),

$$f(\epsilon) = 2(k_B T_e)^{-3/2} \left(\frac{\epsilon}{\pi}\right)^{1/2} \exp\left(-\frac{\epsilon}{k_B T_e}\right).$$

(a) Show that, in the particular case where $Q_{0j}(\epsilon)$ is in the form of a step function such that:

$$Q_{0j}(\epsilon) = \begin{cases} 0 & \text{for } \epsilon < \epsilon_j \\ Q_0 & \text{for } \epsilon \geq \epsilon_j. \end{cases}$$

R_{0j} is given in practical units by:

$$R_{0j} = 6.7 \times 10^7 n_e N_0 Q_0 (k_B T_e)^{1/2}$$
$$\times \left(1 + \frac{\epsilon_j}{k_B T_e}\right) \exp\left(-\frac{\epsilon_j}{k_B T_e}\right) \text{ cm}^{-3} \text{ s}^{-1},$$

where n_e and N_0 are in cm^{-3}, $k_B T_e$ is in eV, and Q_{0j} is in cm^2.

(11.3) The rate of de-excitation by superelastic collisions is given by:

$$R_{j0} = n_e N_j s_{j0} = n_e N_j \int_0^{\infty} \sqrt{\frac{2\epsilon}{m_e}} Q_{j0}(\epsilon) f(\epsilon) d\epsilon.$$

(a) By integrating the above expression directly using the Klein–Rosseland relationship between $Q_{j0}(\epsilon)$ and $Q_{0j}(\epsilon + \epsilon_j)$ (eqn (11.20)), show that when $f(\epsilon)$ is Maxwellian:

$$\frac{s_{j0}}{s_{0j}} = \frac{g_0}{g_j} \exp\left(+\frac{\epsilon_j}{k_B T_e}\right).$$

[Comment: If the only agencies populating and depopulating the excited state j are electron collisions then equating R_{0j} with R_{j0} leads to the result that N_j is in LTE with N_0 at the electron temperature, as we would expect from the principle of detailed-balance considerations. This is in fact the justification for the Klein–Rosseland relationship, which relates the up and down cross-sections, and can be applied under all conditions—not just at thermal equilibrium. In this sense the argument here is analogous to the argument used to relate the Einstein A and B coefficients in Section 2.1.1.]

(11.4) A discharge contains a single atomic species that has an excited energy level j that is 10 eV above the ground state. Level j has a degeneracy equal to that of the ground state and a step-function excitation cross-section (as defined in Exercise 11.2) with $Q_0 = 10^{-17}$ cm^2 for electron impact excitation from the ground state 0. Atoms in level j have a lifetime of 1 μs against spontaneous radiative decay back to the ground state. The number density of ground-state atoms $N_0 = 6.6 \times 10^{16}$ cm^{-3}, the value of $(k_B T_e)$ is 2.0 eV, and the electron density $n_e = 10^{14}$ cm^{-3}.

Calculate:

(a) the value N_j would take in full Boltzmann equilibrium at the electron temperature;

(b) the value N_j would take if there were no quenching by superelastic collisions and no radiation trapping;

(c) the value N_j would have if radiative decay and quenching by superelastic collisions are effective, but wall collisions, ionizing collisions and transfers to and from other levels can be ignored;

(d) the value N_j would take in the limit of large electron density if atoms in level j can be ionized by electron collisions with a step function cross-section $Q_{jI} = 3 \times 10^{-15}$ cm^2 and a threshold energy $\epsilon_{jI} = 4$ eV. Ignore volume recombination.

(11.5) A He-Ne laser has an active discharge length of 30 cm and operates with internal mirrors having reflectivities of 100% and 98.0% at 632.8 nm but low reflectivities at all other wavelengths.

(a) Using the data of Table 11.1, calculate the population and excitation rate of the 3s$_2$ level of Ne necessary to achieve threshold for laser action at 632.8 nm assuming that the population of the 2p$_4$ lower level is negligible.

(b) The two mirrors of the laser are now replaced by a pair, one of which has 100% reflectivity at 543.3 nm, and the other has 99.5% reflectivity at 543.3 nm. By what factor would the excitation rate of the 3s$_2$ level have to be increased to bring the laser to threshold at this new wavelength? Again you may assume that the population of the 2p$_{10}$ lower level is negligible.

(11.6) A large-frame argon-ion laser has a round-trip gain of 20% on the 488.0 nm transition when operated in a cavity of length 1.2 m comprising one high-reflectivity mirror and an output coupler with a reflectivity of 95%. Assuming that there are no frequency-selective elements within the cavity and that the transition is predominantly Doppler broadened with a FWHM of $\Delta \nu_D = 4.0$ GHz, calculate the maximum number of longitudinal modes on which the laser might be capable of oscillating. Suggest a reason why not all of these modes may be observed simultaneously in the output of the laser.

[You may assume a homogeneous linewidth due to Stark and natural broadening of $\Delta \nu_H \approx 800$ MHz.]

12 Infra-red molecular gas lasers

12.1 Efficiency considerations 332
12.2 Partial population inversion between vibrational energy levels of molecules 335
12.3 Physics of the CO_2 laser 338
12.4 CO_2 laser parameters 343
12.5 Low-pressure c.w. CO_2 lasers 344
12.6 High-pressure pulsed CO_2 lasers 346
12.7 Other types of CO_2 laser 349
12.8 Applications of CO_2 lasers 351
Further reading 352
Exercises 352

12.1 Efficiency considerations

12.1.1 Energy levels of atoms and molecules

The fact that the first excited energy levels of *atoms* are typically a few eV above the ground state implies that their strongest absorption transitions are in the visible or ultraviolet regions of the spectrum. In most cases, it is the difference in energy due to the difference in electron configuration that is responsible for the separation (gross structure) of the energy levels.[1]

In the same way, strongly allowed transitions in which ground-state *molecules* absorb photons of UV or visible radiation correspond to changes in the electronic configuration of the molecule. However, molecules are capable of storing internal energy in two additional forms: in the form of vibrations, in which the internuclear separation exhibits oscillations about its mean value; and secondly in rotation of the molecule about axes for which the molecule has an appreciable moment of inertia.

Spectra corresponding to changes in the state of vibration and/or rotation of molecules (within a given electronic state) lie in characteristically different regions of the spectrum.[2] Pure rotation spectra, i.e. absorption transitions between adjacent rotational levels of the same vibrational level, are very weak[3] and lie in the far-infra-red region of the spectrum.[4] Spectra corresponding to changes in the vibrational state lie in the near- or mid-infra-red regions.[5] If the vibrational motion of the nuclei were that of a harmonic oscillator we would expect the corresponding energy levels to be a ladder with equally spaced rungs, since the energy levels of a one-dimensional harmonic oscillator of angular frequency ω are given by $(v + \frac{1}{2})\hbar\omega$, where the vibrational quantum number v is an integer from 0 upwards. Since the selection rule for transitions

---

[1] There are exceptions such as the oxygen atom, in which the ground configuration $(1s^2 2s^2 2p^4)$ gives rise to terms $^1D_2$ and $^1S_0$ that lie well above the ground state $^3P_2$. However, transitions between these levels and the ground state are forbidden to electric dipole radiation since all terms arising from a given configuration have the same parity.

[2] We will postpone a discussion of rotational and vibrational structure of electronic transitions of molecules until Chapter 13 since in the present chapter we are concerned only with molecular transitions within the electronic ground state. For a full description of the features and theory of molecular spectra the reader is referred to standard works on the subject such as Herzberg (1989), Houghton and Smith (1966) or Bransden and Joachain (2003).

[3] Only molecules (such as HCl) with a permanent electric dipole moment can show pure-rotation spectra. For **homonuclear** diatomic molecules the two nuclei are the same atomic species (e.g. $^{16}O - ^{16}O$ or $^{16}O - ^{18}O$) have no such transitions. Emission spectra are weak because both the transition matrix elements and the frequencies (see eqn (2.24)) are much smaller than those of electronic transitions.

[4] The far-infra-red region spans the wavelength range from say 50 μm to 1000 μm, corresponding to the wave number range 200 to 1 cm^{-1} or the terahertz frequency range from 6×10^{12} down to 3×10^{11} Hz.

[5] The near-infra-red spans the wavelength range from say 0.8 μm to 5 μm, corresponding to the wave number range 12500 to 2000 cm^{-1}, or in frequency units, 3.8×10^{14} to 6×10^{13} Hz. The mid-infra-red spans the gap from 5 μm to 50 μm in wavelength, corresponding to the wave number range 2000 to 200 cm^{-1}, or in frequency terms from 6×10^{13} down to 6×10^{12} Hz. All of these divisions are somewhat arbitrary and different authors may choose different boundaries.

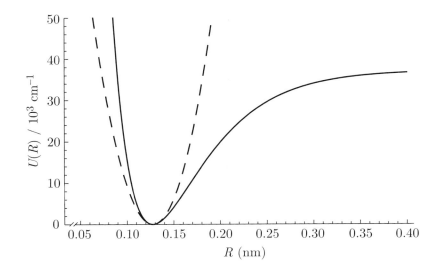

Fig. 12.1: The potential energy $U(R)$ of the ground electronic state of the HCl molecule as a function of internuclear distance R. The dashed curve is the simple harmonic approximation to the potential. The energy scales of several of the figures in this chapter are stated in cm^{-1}, a non-SI unit, but one used by most molecular spectroscopists. The conversion factor is $1\,\text{eV} = 8\,066\,\text{cm}^{-1}$.

between vibrational levels of the same electronic state is $\Delta v = \pm 1$, we might expect that the spectrum would consist of a single line.

However, as shown in Fig. 12.1, the effective potential well in which the nuclei move usually deviates substantially from a parabolic one so that neighbouring rungs on the ladder of vibrational energy levels become more closely spaced as we go up the ladder. The result is that transitions between neighbouring vibrational levels higher in the ladder correspond to frequencies progressively further into the infra-red, as shown in Fig. 12.2.

In addition to vibrational energy, a diatomic molecule can possess internal energy in the form of rotational energy E_J given by:

$$E_J = BJ(J+1), \qquad (12.1)$$

in which the J is the rotational quantum number ($J = 0, 1, 2\ldots$), and the rotational constant B is given by

$$B = \frac{\hbar^2}{2I}, \qquad (12.2)$$

where I is the moment of inertia. Each of the vibrational transitions within a given electronic state[6] has fine structure (a vibration–rotation band) because a change in rotational quantum number J can occur along with the change in v. For a diatomic molecule the vibration–rotation spectrum in the near-infra-red consists of two branches stretching either side of the centre of each vibrational transition.[7]

12.1.2 Quantum ratio

Gas lasers operating on transitions between vibrational levels of the electronic ground state of a molecule have much greater potential for high efficiency than do infra-red lasers operating on transitions between excited levels of atoms.

[6] In the case of the visible or ultraviolet spectra that arise when the molecule makes a transition from one electronic state to another, the vibration–rotation structure also shows up as fine structure leading to formation of a spectral band.

[7] For all stable diatomic molecules, there are just the two branches: the P-branch and the R-branch. The only exception among molecular species that remain in the gas phase at normal temperatures is nitric oxide (NO), which has an electronic ground state having (unusually for a diatomic molecule) a component of electronic angular momentum along the internuclear axis. This molecule has in addition to the P- and R-branches, a Q-branch corresponding to $\Delta J = 0$. The occurrence of Q-branches is quite common in the spectra of polyatomic molecules.

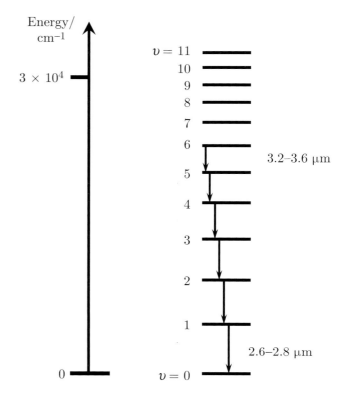

Fig. 12.2: Anharmonicity causes the separation of adjacent energy levels to decrease with increasing v. For example, the $v = 1$ to $v = 0$ transition in HF gives rise to a band in the region of 2.6–2.8 μm, but the $v = 6$ to $v = 5$ transition gives rise to a band in the region 3.2–3.6 μm.

This is because the **quantum ratio**, defined as the ratio of the output laser photon energy divided by the energy of the upper laser level, can be as high as 50% for molecular systems, but is seldom higher than 5% for an infra-red atomic laser. The quantum ratio sets an upper limit to the overall efficiency of the laser system because, even if all of the input energy could be channelled into exciting the upper laser level from the ground state, only the laser photon energy is recovered as useful output on each excitation event.

Atomic lasers (with only a few special exceptions) operate on electric dipole allowed transitions between excited levels that belong to configurations differing from that of the ground state. Usually, both upper and lower levels lie high above the ground state, as indicated on the left-hand side of Fig. 12.3. The quantum ratio of such systems is therefore limited to values of typically 10% or smaller. In contrast, the vibrational levels of the ground electronic state of a molecule offer a set of energy levels much lower in overall energy. As indicated on the right-hand side of Fig. 12.3, much higher quantum ratios are therefore possible for lasers operating on transitions between vibrational energy levels of a molecule within its ground electronic state. The overall efficiency of practical CO_2 and CO laser systems can be 40% or even higher. No other types of gas laser have efficiencies approaching this value, with the possible exception of the rare-gas-halide excimer laser systems.

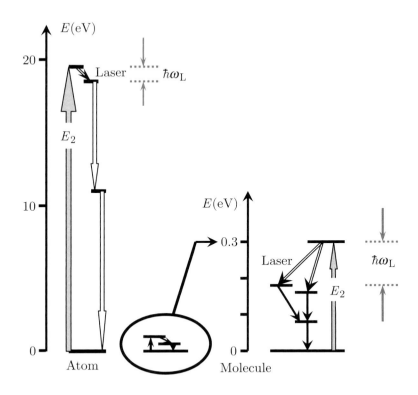

Fig. 12.3: Infra-red atomic gas lasers operate on transitions with rather low quantum ratios. In the case illustrated, a laser photon of energy 0.5 eV is small compared to the energy of approximately 19 eV invested to excite the upper laser level. In contrast, for gas lasers operating on transitions between vibrational levels of the electronic ground state of molecules, the energy recovered as a laser photon can be an appreciable fraction of the energy required to excite the upper laser level.

12.2 Partial population inversion between vibrational energy levels of molecules

In the case of a laser operating on a vibrational–rotational transition between two vibrational levels of a diatomic molecule, the selection rule for electric dipole radiation is $\Delta v = \pm 1$. The effect of rotation shows up as a dependence of the transition frequency of individual emission lines upon the rotational quantum number J, creating a vibration–rotation band. The selection rules dictate that there will usually[8] be only two branches to the band—the P- and R-branches, which are formed by lines connecting rotational levels J' in the upper[9] vibrational level to levels J'' in the lower vibrational level, such that *in emission*

$$\Delta J = +1 \quad \text{P-branch} \tag{12.3}$$

$$\Delta J = -1 \quad \text{R-branch.} \tag{12.4}$$

By convention, the numbering of the individual vibrational–rotational transitions is denoted by the rotational quantum number of the *lower* level of the transition (see Fig. 12.4).

[8] See note 7 on page 333.

[9] By convention, quantities associated with the upper level are denoted with a single prime; those with the lower level are denoted with a double prime.

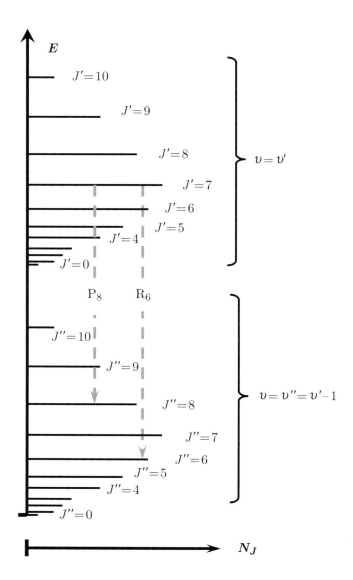

Fig. 12.4: The distribution of population over the rotational levels of two vibrational levels of a typical diatomic molecule. Transitions between the two vibrational levels give rise to the P- and R-branches of the band. The P_8 transition in emission originates on the $J' = 7$ rotational level of the upper vibrational level and terminates on the $J'' = 8$ rotational level of the lower vibrational level. The R_6 transition in emission originates on the $J' = 7$ rotational level of the upper vibrational level and terminates on the $J'' = 6$ rotational level of the lower vibrational level.

Owing to the rapidity of rotation-changing collisions, the distribution of population over rotational states $N_{J'}$ (within a given upper vibrational level) will be given by a Boltzmann distribution:

$$N_{J'} \propto g_{J'} \exp\left(-\frac{E_{J'}}{k_B T_r}\right) = (2J'+1) \exp\left(-\frac{BJ'(J'+1)}{k_B T_r}\right), \quad (12.5)$$

where B is the rotational constant[10] for the molecule and T_r is the rotational temperature. Note that $g_{J'}$ the degeneracy of the rotational level with quantum number J' is given by $(2J'+1)$. The effect of the selection rules and degeneracy factors gives rise to the interesting possibility that gain can occur on transitions between rotational levels of adjacent vibrational levels when the

[10] Many authors prefer to cite energy values in units of cm^{-1}. With B expressed in cm^{-1}, eqn (12.2) becomes $B = \frac{1}{100}\frac{\hbar}{4\pi c I}$ with all quantities on the RHS in SI units.

overall populations of the vibrational levels themselves are not inverted. For gain to exist we require

$$\frac{N_{J'}}{g_{J'}} > \frac{N_{J''}}{g_{J''}}. \qquad (12.6)$$

If the rotational temperature T_r is the same for all vibrational levels (as is usually the case),[11] and the overall populations of two neighbouring vibrational levels are equal, then the inequality (12.5) is satisfied for all P-branch transitions so that all the lines of the P-branch will show gain. Under the same conditions all of the R-branch transitions will show loss.[12]

It is evident from Fig. 12.5 that gain is possible on P-branch transitions for high J' numbers even when the overall population of the upper vibrational level is slightly less of that of the lower vibrational level. A population inversion of this type, i.e. within a restricted range of transitions despite there being no inversion between the overall populations of the vibrational levels involved, is known as a **partial inversion** of population. Laser oscillation is possible in

[11] Molecules in a gas of finite density with kinetic (translational) temperature T_{gas} make frequent glancing collisions. Such collisions can change a molecule's rotational state even though the energy exchanged may be $\ll k_B T_{gas}$. As a result, quasi-equilibrium can be established among the population of rotational states very rapidly.

[12] See Exercise 12.4.

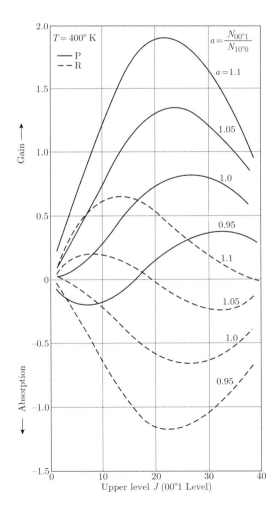

Fig. 12.5: Relative gain or loss on P-branch transitions (full curves) and R-branch transitions (dashed curves) as a function of the rotational quantum number J' of the upper level. The numbers beside each curve give the ratio of the overall population of the upper vibrational level to that of the lower vibrational level. The curves are appropriate to CO_2 at $T_r = 400$ K. Reprinted figure with permission from, C. K. N. Patel, Phys. Rev. Lett., 12, pp 588–590, (1964), http://prola.aps.org/abstract/PRL/v12/i21/p588_1, Copyright (1964) by the American Physical Society.

the usual way for those transitions exhibiting a partial population inversion. In contrast to the case for P-branch transitions, for R-branch transitions a substantial excess of overall population in the upper vibrational level may be necessary to achieve gain.

The value of $J'_{\max}$ for the rotational level having the maximum population is given by

$$J'_{\max} + \frac{1}{2} = \sqrt{\frac{k_B T_r}{2B}}. \qquad (12.7)$$

In the particular case[13] of CO_2 with $T_r = 400\,\text{K}$, eqn (12.7) gives $J'_{\max} = 19$, and the strongest inversion is expected on the P_{20} line. Partial inversions occur for several vibration–rotation transitions of CO_2 with J' values centred on $J' = 19$.

Because of the rapidity of energy transfer by collisions among the rotational levels, laser oscillation, once established on one particular line of the vibration–rotation band, can feed off population transferred from all the other rotational levels of the upper vibrational level—since the tendency towards establishing a new thermal-equilibrium population distribution will try to fill up the depleted level and may prevent any other transition in the vibration–rotation band from reaching threshold. To obtain laser oscillation on a weaker transition it may be necessary to introduce a tuning element (e.g. a diffraction grating) inside the laser cavity to suppress the oscillation on competing strong transitions.

[13] In the case of CO_2, the upper and lower laser levels have no component of angular momentum along the internuclear axis (see Section 12.3.1), and these transitions therefore lack Q-branches. The considerations of partial inversion on the P- and R-branches of these transitions therefore apply in the case of CO_2, just as if it were a diatomic molecule.

12.3 Physics of the CO_2 laser

12.3.1 Levels and lifetimes

As shown in Fig. 12.6, the CO_2 molecule—which has a linear structure—is capable of vibration in three different modes. In the symmetric-stretch mode (which has angular frequency ω_1), the atoms move along the internuclear axis with the bond lengths connecting the central O atom to the outer C atoms increasing and decreasing in phase with one another. In the bending mode the outer C atoms move out of line with the central O atom, either in the plane of the diagram or perpendicular to that plane. This mode (which occurs with angular frequency ω_2) is therefore doubly degenerate; the combination of the two vibrations in orthogonal planes can, with appropriate relative phase, give rise to a component of angular momentum along the internuclear axis. The third mode of vibration is the asymmetric stretch mode (angular frequency ω_3); it is one in which the atoms move along the internuclear axis with the bond lengths connecting the central O atom to the outer C atoms increasing and decreasing exactly in antiphase with one another.

The three numbers used to specify the vibrational state of the CO_2 molecule ($v_{1,2,3} = n_1 : n_2^\ell : n_3$) give the numbers of quanta of vibration in each of these three modes, respectively. The superscript ℓ on n_2 gives, in units of $\hbar$, the angular momentum about the internuclear axis arising from the orthogonal vibrations. Each of the vibrational modes gives rise to a ladder of excited vibrational levels, the rungs of which—at least for the first few

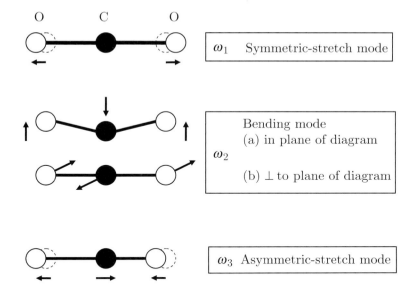

Fig. 12.6: The three modes of vibration of the CO_2 molecule.

members—are equally spaced in energy (by intervals of $\hbar\omega_{1,2,3}$) as one would expect for the energy levels of a harmonic oscillator.

As shown in Fig. 12.7, the strong CO_2 laser transitions (at about 10.6 μm) belong to the 10.4-μm vibration–rotation band comprising transitions from rotational levels of the (00^01) vibrational upper level to those of the (10^00) lower level. From the same upper level, lasing can also occur on the 9.4-μm band (at about 9.5 μm) to the (02^00) lower level. Direct excitation to both upper and lower laser levels (a nuisance effect in the case of the lower level) can occur as a result of collisions of fast electrons in the discharge:

$$CO_2\,(00^00) + e_{\text{fast}} \rightarrow CO_2\,(00^01) + e_{\text{slow}} \quad (12.8)$$

$$CO_2\,(00^00) + e_{\text{fast}} \rightarrow CO_2\,[(02^00) \text{ or } (10^00)] + e_{\text{slow}}. \quad (12.9)$$

The quantum ratio is about 41% for the 10.4-μm band and about 52% for the 9.4-μm band. The radiative lifetimes of all levels shown in the diagram are so long that, under conditions of any practical interest, the effective lifetimes are determined entirely by collisions with gas molecules. The effective decay rate of the upper level is set by collisions of the (00^01) CO_2 molecules with ground-state CO_2 molecules and is proportional to the density of ground-state CO_2. The rate is found to be:

$$\tau_2^{-1} = \begin{cases} 385\,\text{s}^{-1}\,\text{Torr}^{-1} \text{ at } 300\,\text{K} \\ 1300\,\text{s}^{-1}\,\text{Torr}^{-1} \text{ at } 500\,\text{K} \end{cases} \quad (12.10)$$

for collisional conversion from (00^01) to (00^00).

Now, both the lower laser levels, (10^00) of the 10.4-μm band and (02^00) of the 9.4-μm band, are depopulated by collisions with ground state CO_2 molecules at a rate faster than the rates of eqn (12.10) by an order of magnitude. This is because of the presence of the (01^10) level that, as might be expected,

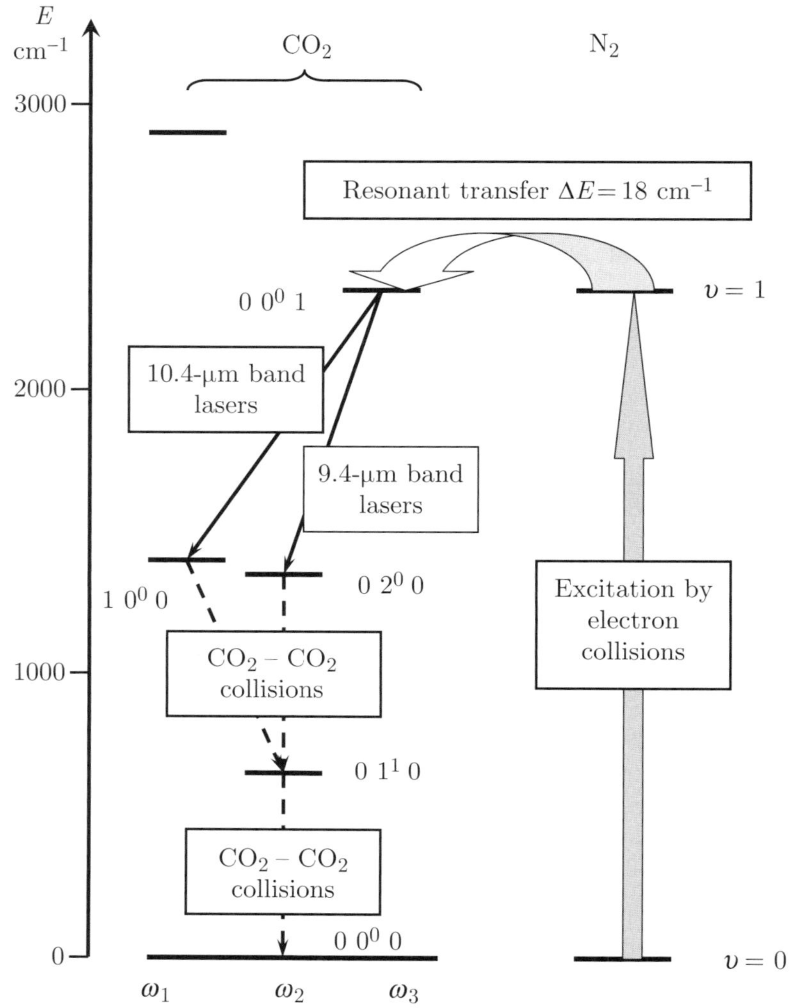

Fig. 12.7: Vibrational energy levels of the CO_2 molecule and the N_2 molecule on the same energy scale. The laser transitions of CO_2 are also shown.

[14] Collisions between two bodies in which no third body is ejected after the collision tend to occur with only very small cross-section if a large energy defect ΔE has to be taken up as kinetic energy after the collision. To eject the collision partners with much larger velocities than they had initially requires them to arrange themselves in a very special way during the collision so that the repulsion between their electron clouds can give them the required acceleration while conserving momentum overall. The requirement for 'energy resonance' is very much less stringent in the case where a third body—such as an electron—is involved.

lies half-way in energy between the (00^00) ground state and the (02^00) lower laser level, and is also fortuitously close to being half-way in energy between the (00^00) ground state and the (10^00) lower laser level. Thus, when CO_2 molecules in either of the two lower laser levels collide with one in the ground state, two molecules in the intermediate (01^10) level can be formed with very little conversion of vibrational to translational energy.[14] Collision processes that involve only a small amount of energy to be converted to or from kinetic energy are much more frequent than those in which a large amount of energy is converted. Therefore, a process such as

$$CO_2\,[(02^00)\text{ or }(10^00)] + CO_2\,(00^00) \rightarrow 2CO_2\,(01^10) + \Delta E \quad (12.11)$$

is efficient and frequent because ΔE is small. This process is illustrated schematically in Fig. 12.8.

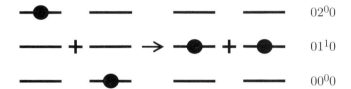

Fig. 12.8: Illustrating the collisional process of eqn (12.11) responsible for rapidly depopulating the lower laser levels of CO_2.

Thus, we have the makings of a good laser system: a high quantum ratio, and a favourable lifetime ratio between upper and lower levels. Unfortunately, however, the intermediate level (01^10) does not have a fast collisional decay path in pure CO_2, and in consequence decay from the lower level itself is bottlenecked, caused by build-up of population in the (01^10) level and back-transfer in the inverse reaction to eqn (12.11); this has serious consequences for laser power output unless measures are taken to overcome the problem (see Section 12.3.3).

12.3.2 The effect of adding N_2

The very first observations of c.w. laser action at 10.6 μm in CO_2 were made on simple positive-column discharges in pure CO_2 gas, in which the only excitation route for the upper level was via direct electron collisional excitation of the type represented by eqn (12.9). It was quickly established[15] that adding N_2 to the CO_2 gave rise to a tremendous increase in output power—from 1 mW to 10 W in early experiments.

The effect of adding the N_2 is almost entirely concerned with increasing the efficiency of excitation of the upper laser level, by increasing the fraction of the power input that is channelled into this wanted process. Because the spacing of the vibrational ladder of the diatomic molecule N_2 is almost exactly equal to that of the ($00^0 n_3$) modes of CO_2, the transfer of energy from N_2 occurs via the resonant process illustrated in Fig. 12.7:

$$N_2 (\upsilon = 1) + CO_2 (00^00) \rightarrow N_2 (\upsilon = 0) + CO_2 (00^01) - 18\,\text{cm}^{-1}. \quad (12.12)$$

Note that since the energy defect for reaction (12.12) is so small compared to thermal energies ($k_B T_{\text{gas}} \approx 200\,\text{cm}^{-1}$ for $T_{\text{gas}} \approx 300\,\text{K}$), the transfer is efficient and selective. Thus, any electron collisions that excite $N_2 (\upsilon = 1)$ from the ground-state $N_2 (\upsilon = 0)$ can potentially end up contributing to the laser excitation. Further, the ($\upsilon = 1$) vibrational level of N_2 is very long lived because it cannot decay by electric dipole radiation (the N_2 molecule lacks a permanent electric dipole moment). However, N_2 molecules in higher vibrational states $N_2 (\upsilon = n)$ can feed population into the ($\upsilon = 1$) level via near-resonant collisions with ground-state N_2 via:

$$N_2 (\upsilon = n) + N_2 (\upsilon = 0) \rightarrow N_2 (\upsilon = 1) + N_2 (\upsilon = n-1) - \Delta E. \quad (12.13)$$

For levels up to $N_2 (\upsilon = 4)$ the type of process described by eqn (12.12) is effective[16] since $\Delta E \lesssim k_B T_{\text{gas}}$. Moreover, these low-lying vibrational levels have large cross-sections[17] for excitation by discharge electrons of energy less than 1.5 eV. The overall result is that, whatever the vibrational state of N_2

[15] See (Patel, 1964).

[16] If the potential well of N_2 in the electronic ground state were exactly parabolic, then the vibrational levels of such a perfectly harmonic oscillator would be equally spaced in energy, and the energy defect ΔE in eqn (12.13) would be zero for all υ. However, because of the anharmonicity of the potential wells of real molecules (see, for example, Fig. 12.1) the spacing of neighbouring vibrational levels decreases with increasing energy, and for $\upsilon > 1$ the energy defect can only be made up by contributions from converted translational or rotational energy. This becomes rapidly more difficult as $\Delta E \gg k_B T_{\text{gas}}$.

[17] see (DeMaria, 1973).

initially excited by electron collisions in the laser discharge, the energy can end up by a series of selective and near-resonant processes making a useful contribution to the wanted excitation of the (00^01) level of CO_2. Further, when the upper-level population is drained by stimulated emission during laser operation the collisional processes tending to restore thermal equilibrium among the various N_2 and CO_2 levels act to make good the loss, and hence feed the excitation in the required direction.

12.3.3 Effect of adding He

In the early development of the CO_2 laser, it was quickly discovered that addition of He to the gas mixture of the laser discharge also had a beneficial effect (raising the output power to approximately 100 W c.w.). Addition of He has two distinct effects:

- The added gas allows the electron temperature to be reduced to values closer to the range appropriate for optimum efficiency for exciting the wanted levels of CO_2 and N_2.
- It helps to remove a bottleneck in the depopulation pathway back to the ground state from the (10^00) and (02^00) lower laser levels by reducing the rotational temperatures of the vibrational levels involved.

To understand how the addition of He helps to unblock this bottleneck in the relaxation pathway from the lower laser levels, we first need to consider the deleterious effect of allowing an increased rotational temperature $T_r (01^10)$ of the (01^10) level. As indicated schematically in Fig. 12.9, when T_r is high, the highest rotational levels of (01^10) become significantly populated. Because these levels are close in energy to the lower rotational levels of the (02^00) and (10^00) levels, molecules in high rotational levels of (01^10) can resonantly transfer population back to the lower laser levels in a variety of reactions of the form

$$CO_2 \, (01^10) \, J''_{\text{high}} + CO_2 \, (00^00) \rightarrow CO_2 \, (10^00) \, J'_{\text{low}} + CO_2 \, (00^00) + \Delta E, \quad (12.14)$$

all of which can lead to build up of population in the lower laser levels.

As more power is extracted from the laser transitions, more population is dumped into the intermediate (01^10) level; with no way to dispose of the energy[18] the rotational temperature of this level would rise until reactions such as (12.14) began to diminish the inversion and to limit the laser power. This problem is particularly acute at positions in the laser volume well away from the cooling influence of the tube wall. Addition of helium[19] drastically reduces the rotational temperature of the (01^10) level, by conducting 'heat' (rotational energy) from the excited CO_2 to the wall via rotational-to-translational energy converting collisions between CO_2 and He. Thus, if the discharge-tube wall is cooled the presence of helium allows this low temperature to be effectively

---

[18] Direct quenching of CO_2 molecules in the (01^10) level by collision with CO_2 ground-state molecules is inefficient because reactions such as: $CO_2 \, (01^10) + CO_2 \, (00^00) \rightarrow CO_2 \, (00^00) + 667 \, \text{cm}^{-1}$ involve a considerable conversion of vibrational to translational energy.

[19] Because the atomic mass of He is small, helium gas has a very high thermal conductivity. See Exercise 12.1.

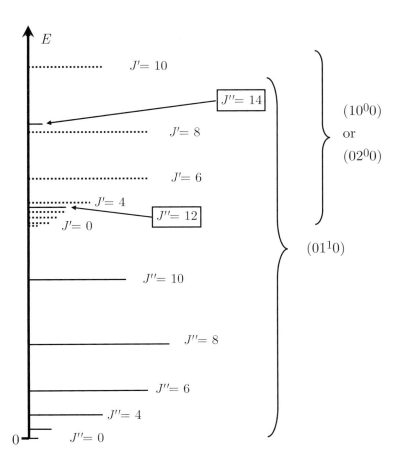

Fig. 12.9: Schematic illustration of the deleterious effect of increasing temperature upon the distribution of population among the rotational levels of two adjacent vibrational levels. Unlike the situation depicted in Fig. 12.4 for a lower temperature, there is now significant population in the $J'' = 12$ and $J'' = 14$ levels of the lower vibrational level that are within the range of energies of the lower rotational levels of the upper vibrational level. Note that in $C^{16}O_2$ the ^{16}O atoms have zero nuclear spin, and are therefore bosons. In order not to violate the Pauli principle, exchange of the ^{16}O bosons must leave the totally symmetric $^1\Sigma_g^+$ ground state unchanged. The overall effect for this electronic state is that all rotational levels with odd values of J'' are missing (see, for example, Herzberg (1990)). Similarly, for the antisymmetric stretch vibrational level (the upper laser level) that has Σ_u^+ symmetry, the opposite is true, i.e. all even numbered J'' levels are missing.

communicated to the rotational distributions of CO_2 throughout the volume of the laser tube. Direct quenching of (01^10) by He may also occur via:

$$CO_2\,(01^10) + He_{slow} \rightarrow CO_2\,(00^00) + He_{fast} + (667-800)\,\text{cm}^{-1}, \tag{12.15}$$

but this route for removal of (01^10) is not very efficient since it involves a very large conversion of internal energy to kinetic energy.

Another and even more effective solution to the problem of heat removal from the lower-level manifold is to provide for forced circulation of the gas mixture in a fast-flow, closed-circulation path that includes a heat exchanger. Even in such circulating-gas lasers, He is still a very beneficial additive to the gas mixture since it promotes better 'thermal contact' between the heat-exchanger surfaces and the rotational populations of the CO_2 molecule.

12.4 CO_2 laser parameters

The parameters of both low-pressure and high-pressure CO_2 laser are summarized in Table 12.1 below.

344 *Infra-red molecular gas lasers*

Table 12.1 Parameters of typical commercial CO_2 lasers.

	Low-pressure discharge	Atmospheric-pressure discharge
Commonest operating wavelength*	10.6 μm	10.6 μm
Mode of operation	c.w. or long pulse	Short pulse only
Output power/energy	50–5,000 W	1–30 J
Discharge cross-sectional area	1–20 mm diameter	25 mm × 25 mm
Discharge current	10–50 mA DC	multi-kiloamp pulse
Active length (cm)	12–30	30–100
Gas mixture $CO_2 : N_2 : He$	3:5:15	12:1:4
Total gas pressure (Torr)	30	760
Gas temperature T_{gas} (K)	280	300
Transition probability A_{21} s^{-1}	0.21	0.21
Doppler-broadened linewidth $\Delta \nu_D$ (MHz)	55	55
Homogeneous linewidth $\Delta \nu_H$ (GHz)	0.15	3.9
Effective upper-level lifetime τ_2^{eff} (μs)	650	4.8
Effective lower-level lifetime τ_1^{eff} (μs)	65	0.48
Gain cross-section (homogeneous) (10^{-18} cm^{-2})	39	1.5

*Other output wavelengths: 60–70 transitions in the 10.4-μm band, 40–50 in the 9.4-μm band.

12.5 Low-pressure c.w. CO_2 lasers

The CO_2 laser was invented by C.K.N. Patel at Bell Labs in 1964 Patel (1964). The first device used pure CO_2 as the laser medium, but a huge improvement in output power (from a few mW to several watts) was made when Patel used RF excitation of a stream of N_2 gas that was mixed with the CO_2 downstream in a discharge-free region of the tube forming the active region of the laser. This separation of discharge zone and active region proved to be unnecessary and early commercial versions of c.w. CO_2 lasers, like that shown in Fig. 12.10, employed a longitudinal DC discharge as the active medium. The He-N_2-CO_2 gas mixture was pumped out slowly at one end of the tube with new gas admitted at the other in order to remove unwanted molecular species

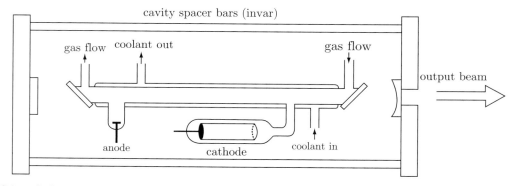

Fig. 12.10: Schematic diagram of a 30–50-W c.w. diffusion-cooled CO_2 laser. The discharge is operated in a slowly–flowing gas stream comprising 3 Torr CO_2, 5 Torr N_2, and 15 Torr of He. The discharge operates continuously with a running voltage of 8 kV, and drives a current of 25 mA through a discharge tube of 12 mm bore and 60 cm active length. The Brewster-angle windows are made from an infra-red-transmitting material such as ZnSe.

formed in the discharge. However, modern versions have largely solved the impurity build-up problem and can be operated as sealed-off devices. Heat removal is by radial diffusion to the walls that are cooled by water (or a refrigerated antifreeze mixture) circulated through the cooling jacket. The maximum c.w. power that can be extracted from such systems is about 200–300 W per metre of discharge.

The early approach to scaling lasers of this type to c.w. output powers of 500–1000 W was by stacking several individual diffusion-cooled discharge tubes end-to-end in a single optical cavity that often employed a 'folded' beam path to avoid an unwieldy overall device length. Nevertheless, such devices were huge and relatively inefficient.

In attempts to increase the performance of CO_2 lasers to the multi-kilowatt range the problem of cooling the gas has been solved by using forced convective cooling of the gas mixture. This involves using fans or blowers to circulate the gas mixture around a closed loop that includes one or more heat-exchanger units—for example, liquid-cooled honeycomb structures similar to car radiators.

In the fast axial-flow type of CO_2 laser design illustrated in Fig. 12.11(a) the DC discharge current flows in a cylindrical glass tube of a few centimetres diameter in the same direction as the gas flow. While the construction of this type of laser represents a relatively simple evolution of the diffusion-cooled design shown in Fig. 12.10, it does have the disadvantage of requiring rather special types of blower to achieve the required mass-transfer rates at the low gas pressures consistent with discharge uniformity.[20] Another factor that limits the operating pressure is that the voltage drop across the discharge increases to high values (many tens of kV) as the pressure is increased,[21] bringing with it the problem of corona losses and ozone formation at exposed terminals.

[20] As we will see in Section 13.8.3 there is a tendency for discharges to collapse from a uniform glow to a constricted arcs as the pressure is raised above a few tens of Torr, a problem that is particularly severe in molecular gases.

[21] The electric fields required to sustain a given current density in molecular gases tend to be considerably higher than those in atomic gases. In molecular gases the majority of electrons, including those in the lowest energy range, can suffer appreciable energy losses by making inelastic collisions that excite molecules to rotational or vibrational levels – processes that become possible when electrons have even a few meV of kinetic energy. This energy loss is in addition to that necessarily suffered by high-energy electrons in making ions that are lost to the walls by diffusion. In contrast, for atomic gases, inelastic losses are appreciable only for the small minority of electrons with several eV of energy.

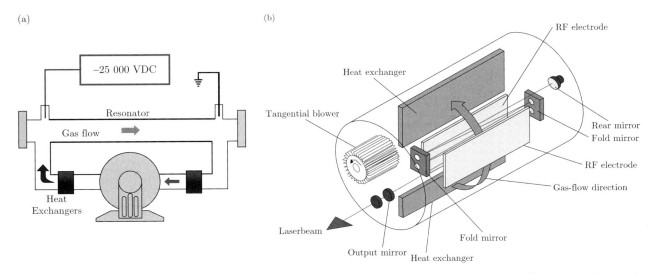

Fig. 12.11: (a) Longitudinally and (b) transversely excited CO_2 lasers with forced convective cooling. Heat is extracted from the gas mixture as it is circulated by electrically powered blowers around a closed loop that includes heat-exchanger units cooled by chilled antifreeze liquid. Figures kindly supplied by D. R. Hall. (Hall (2004)).

A way of reducing the voltage drop associated with a discharge of given volume and pressure is to use transverse electrodes as indicated in Fig. 12.11(b) so that the current flows across the width of the discharge rather than along its length. Typically, the electrodes are only 5–10 cm apart so that the same electric field corresponds to only 5–10% of the voltage drop per metre of longitudinal discharge. Another advantage of this arrangement is that the design of the fan needed to produce the required mass flow of gas mixture in the relatively open duct of the transverse gas-flow loop at pressures of 10–30 Torr (compatible with good discharge uniformity) is much simpler. In many lasers of this type, the excitation is applied as a DC current between transverse electrodes, but to prevent any tendency of the discharge to break up into a number of high current density constricted channels (the early stages of arc formation) one or both discharge electrodes are segmented into small sections, each section having its own ballast resistor. If the current in one segment should start to rise above the normal value for a uniform discharge, the extra current through the ballast resistor reduces the voltage applied to that segment and hence reduces the local current back towards the steady value.[22]

[22] Because of the influence of volumetric energy-loss processes for electrons arising from the inelastic collision processes at low energies mentioned in note 21, the V–I characteristic of the discharge at low currents in these molecular-gas mixtures tends to be much more nearly ohmic than is the case for atomic gases.

As an alternative to exciting the gas with a DC discharge, radio-frequency excitation (with RF frequencies in the 13 and 27 MHz ranges) has been successfully employed in transversely excited lasers of the type depicted in Fig. 12.11(b). Fully c.w. laser powers of up to 20 kW are obtainable from single laser units of this type, which are sometimes known as COFFEE (continuously operated fast flow electrically excited) lasers.

The physics and technology of radio-frequency-excited CO_2 lasers have been the subject of intensive studies by D. R. Hall and colleagues at Heriot Watt University over recent years. As a result, new types of compact high-power CO_2 laser have been developed exploiting the high thermal conductivity of the metal transverse electrodes themselves as an efficient way of diffusion-cooling the laser medium without the need for forced circulation of the gas. Continuous-wave powers up to 5 kW can be extracted from commercial lasers of this type measuring some $0.6\,m \times 1.7\,m \times 2.0\,m$ including power supply and gas handling.[23]

[23] For a detailed review of this work the reader is directed to the article by Hall (2004).

12.6 High-pressure pulsed CO_2 lasers

The design of low-pressure c.w. CO_2 lasers is constrained by two main considerations: (i) the need to extract heat efficiently from the laser medium; and (ii) the need to preserve discharge uniformity. Other things being equal, it is desirable to operate the laser at the highest possible pressure. This is because, according to eqn (6.81), the power that can be extracted from such a laser operating at a given round-trip gain and cavity loss is proportional to the saturation intensity that in the case of the CO_2 laser increases as the *square* of the gas density N_{gas} because:

- The linewidth $\Delta\omega_L$ is dominated by pressure broadening at CO_2 pressures above 5 Torr and thereafter increases linearly with density, so that from eqn (4.14) the gain cross-section at line centre $\sigma_{21}(0)$ is inversely proportional to CO_2 density over the same range.

- The effective lifetime τ_2 of the upper level is inversely proportional to pressure as indicated by eqn (12.10).

Now, from eqns (5.7) and (5.8) at line centre,

$$I_s = \frac{\hbar\omega}{\sigma_{21}(0)\tau_R}, \quad (12.16)$$

where,

$$\tau_R = \tau_2 + \frac{g_2}{g_1}\tau_1(1 - A_{21}\tau_2). \quad (12.17)$$

For the CO_2 laser $\tau_2 \gg \tau_1$, so that to a very good approximation $\tau_R \approx \tau_2$. Thus, other things being equal, we expect $I_s \propto N_{gas}^2$.

However, when the pressure of the CO_2 laser gas mixture is increased above 30–100 Torr the voltages required to strike and sustain a c.w. longitudinal discharge of 10–100 cm length rise to several megavolts. In addition, at these pressures the discharge shows a tendency to self-constrict so that the current is conducted in a narrow, snake-like channel that shifts unpredictably within the tube. To overcome the problem of running high-pressure CO_2 laser discharges (at least for pulse durations of fractions of a millisecond) Beaulieu (1970) developed the concept of the TEA (transversely excited atmospheric) laser. By operating with an electrode configuration in which the direction of the electric field is across, rather than along, the axis of the laser cavity (i.e. with electrode spacing of order 1–5 cm rather than 50–100 cm), the voltages required to achieve E/N values large enough to cause breakdown and to sustain a discharge are brought back to practical values.

In Beaulieu's original laser one of the electrodes comprised a conducting bar, while the opposite one consisted of a linear array of metal pins with each pin connected to a common conductor via its own series ballasting resistor providing the same ballasting action described in Section 12.5. Although Beaulieu's laser pointed the way to high-power pulsed lasers, it had the disadvantage that the array of narrow filamentary discharges did not fill the mode volume of the laser cavity very well.

A modern TEA laser is shown schematically in Fig. 12.12. The transverse metal electrodes, which are not ballasted, are shaped with a special cross-section profile[24] that ensures that the electric field is as uniform as possible over the centre region and the field in the edge regions is nowhere higher than in the central uniform-field region.

As indicated in Fig. 12.12, a basic version[25] of the discharge circuit employs a storage capacitor C that is charged to a high voltage (a few kV) via the combination of the series charging resistor R and inductor L. When the gas-filled switch S (a triggered spark gap or thyratron) is triggered, a high-voltage pulse appears on the top electrode, the charge of which cannot leak easily to ground in a time scale of less than milliseconds because of the inductor L. Prior to applying voltage across the electrode gap, the volume between the electrodes is irradiated with deep-UV radiation (around 115 nm seems to be the most effective band) from a series of high-current sparks between metal segments supported on an insulating board. This 'flashboard' (or 'sparkboard') seeds the main volume of the discharge with a fairly uniform concentration of

[24] Several designs for uniform-field electrode shapes are discussed in the literature. Among the classic analytical formulae for such designs are those by Rogowski and Bruce. Modern programs for generating such electrodes via numerically controlled machines milling machines are also available (see, for example, the article by Ernst (1984)).

[25] In practical systems several improvements to the basic circuit are often implemented. These include the use of saturable inductors to provide 'magnetic assist' to protect the thyratron from high current or the use of magnetic pulse-compression circuits using solid-state devices as the main switching element, and pulse-forming networks in place of a simple storage capacitor.

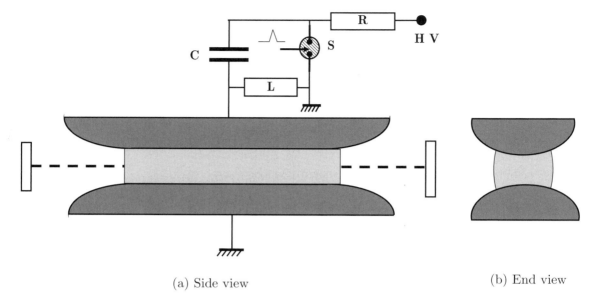

(a) Side view   (b) End view

Fig. 12.12: Schematic diagram of the layout and circuit of a TEA CO_2 laser showing: (a) a side view; (b) an end view. Only the transverse electrodes are inside the gas envelope (not shown). Pre-ionization of the gas volume between the electrodes is provided by a series of sparks or a corona discharge (also not shown), imposed immediately prior to application of voltage across the main discharge gap.

electrons and ions produced by photoionization of minority contaminant species—usually organic molecules—present in the gas. When voltage is applied to the main gap, the field uniformity and the presence of conducting gas at all points in the active volume ensure that the current can flow in a uniform distribution for long enough to deposit the energy and create an optically uniform gain medium for the duration of the current pulse. Since these lasers are mostly used for applications in which the power needs to be delivered in as short and intense a pulse as possible, there is usually no need to sustain the excitation for time scales longer than milliseconds. Eventually, discharge-instability effects will tend to collapse the high-pressure discharge from a uniform glow into a series of discrete arc (or proto-arc) channels. It is the time over which the discharge can be maintained as a reasonably uniform glow that determines the length of the excitation pulse, and ultimately the length of the laser pulse[26] that can be extracted.

The requirement for uniform pre-ionization of the discharge volume can be understood from the following scenario. If the surface of one of the electrodes were to have a slight departure from perfection, for example a small bump as indicated in Fig. 12.13 then, in the absence of pre-ionization of the discharge volume, when the rising voltage pulse is applied across the main discharge gap the breakdown value of E/N is reached first at the bump, where the electric field is higher than elsewhere. The gas in this region breaks down and starts to conduct. Since the electrodes are good conductors, the current flowing in this local region loads the circuit and prevents the voltage elsewhere on the electrode from increasing above the breakdown value. As more current flows into the incipient arc, the gas temperature rises locally, reducing the local gas

[26] The duration of the laser pulse can be considerably longer than that of the current pulse since transfer of energy to CO_2 from the pool of excited N_2 molecules via reaction (12.11) can continue well after the discharge current has decayed to negligible values.

Fig. 12.13: Arc formation in an unpre-ionized transverse discharge.

density and thus further increasing the local value of E/N. Given time, a high-current arc develops at the high spot, leading to irreversible damage of the electrode as an arc pit is formed. It is impossible in practice to manufacture electrodes so perfect and parallel that the field is sufficiently uniform for this problem to be avoided—even a small dust mote will cause enough departure from the degree of perfection required. Further, if no other cause of breakdown is supplied, then the ionization track of a random cosmic ray (as in the spark-chamber technology used in particle physics) can seed the localized discharge growth that leads to arc formation. Flooding the entire discharge volume with hard-UV radiation, before the voltage across the gap starts to rise, can create a substantial electron and ion density sufficient to ensure that the current is launched in a uniform way throughout the discharge region. Once the discharge is established, the processes[27] that act to disrupt its uniformity and fragment it into filaments (albeit on longer time scales than those that characterize arc formation in the unpre-ionized case) ultimately set the limit to the length of the current pulse that can be usefully applied.

Modern lasers of this type can achieve specific output energies up to $10\,\mathrm{J\,l^{-1}\,atm^{-1}}$, so that single-pulse energies of up to 10–20 J are obtainable from fairly compact devices. Moreover, because these pulsed lasers have relatively high gain on many transitions, output can be obtained on as many as 60–70 transitions in the 10.4-μm band and 40–50 individual transitions in the 9.4-μm band. The selection of particular transitions is made with a diffraction grating mounted in the Littrow[28] configuration replacing the high-reflectance cavity mirror. Diffraction gratings used for this purpose are usually 'master gratings' (i.e. ruled directly on a metal substrate) that can withstand higher incident powers than metallized 'replica gratings' cast in plastic. Moreover, by filling the TEA laser with gas mixtures at pressures of several atmospheres the pressure broadening of individual laser lines within the two bands can be made so large that the rotational structure is lost and continuous tuning within each of the two laser bands becomes possible.

12.7 Other types of CO_2 laser

12.7.1 Gas-dynamic CO_2 lasers

The very highest powers are obtained from gas-dynamic CO_2 lasers for which the basic energy input to excite the gas is simply heat. (A discharge may be used, but only for its heating effect). The basic principle of the gas-dynamic laser, as illustrated in Fig. 12.14, is to expand the gas rapidly through a supersonic nozzle to a high Mach number in order to lower the gas-mixture temperature and pressure downstream of the nozzle in a time that is short compared with the vibrational relaxation time of the upper laser level. At the same time, by addition of water[29] or helium, the lower level is caused to relax in a time comparable to, or shorter than, the expansion time.

Because of this rapid expansion, the upper laser level population cannot follow the rapid change in temperature and pressure and thus becomes 'hung up' at a population characteristic of that in the stagnation region. Because the vibrational relaxation times associated with the upper level are long,[30]

---

[27] See Section 13.8.3.

[28] In the Littrow configuration the grating is inclined to the direction of the incoming incident beam so that the diffracted order (usually first order) is returned back along the direction of incidence. The grating equation $\sin\alpha + \sin\beta = m\lambda$, which applies in the case where α (the angle of incidence) and β (the angle of diffraction) are both on the same side of the normal and are both intrinsically positive, reduces to $2\sin\alpha = m\lambda$. In order to make best use of the available resolving power of the grating, the beam may be expanded to cover the full width of the grating by a lens or combination of lenses and prisms or by using the grating at or near-grazing incidence. See Fig. 9.21 and Exercise 14.3 for applications of this technique in other laser systems.

[29] Either H_2O or He could be used to promote relaxation of molecules in the (01^10) level, but since in this type of laser the gas is used once only, the relative cost of He and water vapour dictates the choice.

[30] As discussed in Section 12.3.1, the rate of collisional depopulation of the lower laser levels is an order of magnitude faster than that of the upper laser levels because of the effect of the process of eqn (12.11) in providing an efficient path for population to leave the lower laser levels rapidly.

350 *Infra-red molecular gas lasers*

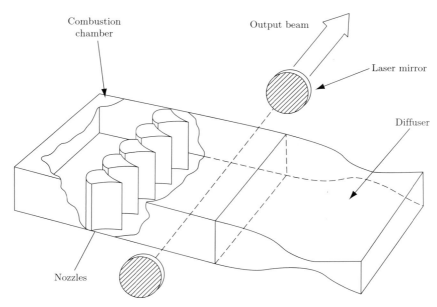

Fig. 12.14: Schematic illustration of a gas-dynamic CO_2 laser. The laser gas mixture, comprising 7.5% CO_2, 91.3% N_2, and 1.2% H_2O is confined at high pressure (17 bar) in the upstream combustion zone where it is heated to temperatures of order 1400 K before being allowed to expand via the nozzle section. At the gain region downstream of the nozzles the gas has a velocity of Mach 4.2, a pressure of 1 bar and a temperature of 350 K. The diffuser section of the gas duct is shaped to match the exhaust to the atmosphere without reflecting shocks upstream that might spoil the optical homogeneity of the gain region. (After Anderson (1976).)

the population will stay 'hung up' (a process known as **vibrational freezing**) providing a gain region for many centimetres downstream of the nozzle, as illustrated in Fig. 12.15. After the nozzle, the gas expands into a duct whose area is approximately 14 times that of the nozzle and flows downstream into the gain zone with a speed of Mach 4.2 at a pressure of about 1 bar and a kinetic temperature T_{gas} of about 350 K. These huge devices, the size of the jet engines they closely resemble, produce essentially c.w. output powers of 100 kW or more for durations of several seconds.

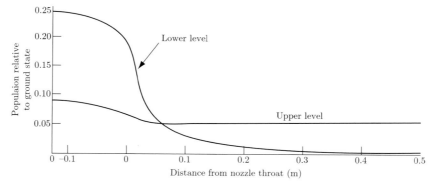

Fig. 12.15: Populations of upper and lower laser levels of CO_2 as a function of distance from the expansion nozzle throat in a gas-dynamic CO_2 laser. The gain region extends from 0.2–0.4 m downstream of the nozzles. (After Anderson (1976).)

12.7.2 Waveguide CO_2 lasers

At the opposite extreme of the power range and size, CO_2 lasers exist in miniaturized form as waveguide lasers. The discharge region of these devices comprises a tube of BeO (beryllium oxide)[31] of typical length 5 to 25 cm and bore diameters as small as 1 mm. A low-loss optical mode can propagate in such a waveguide for radiation with free-space wavelength around 10 μm. Concave mirrors at each end of the waveguide section provide feedback and form the optical cavity.[32] Because of the small bore diameter and the good thermal conductivity of BeO, the discharge is cooled efficiently, so that the gain is relatively high. Waveguide CO_2 lasers produce output powers of around 0.2 W/cm of tube length so that a laser only 20 cm long can produce an output power of 4 W.

Waveguide CO_2 lasers are compact and easily transportable. They have found applications in communications, pollution monitoring, and optical ranging systems.

[31] BeO is used because it is a ceramic material of high thermal conductivity with good resistance to erosion by ion bombardment and low optical loss as a waveguide material for radiation in the 9–11 μm region.

[32] The radius of curvature and separation of these mirrors do not have to satisfy the condition for a low-loss cavity (see eqn (6.22)), since the mode propagating in the space between them is not a free-space mode but a waveguide mode.

12.8 Applications of CO_2 lasers

Because of their multi-kilowatt power capability, CO_2 lasers have been widely adopted in industry for many tasks that used to be carried out with mechanical saws or oxy-acetylene torches. Examples are the cutting and welding of heavy-gauge metals. CO_2 lasers of 4 kW output are capable of cutting 20 mm thick mild steel at a speed of 900 mm/min, and 10 mm mild steel can be cut at speeds of 2000 mm/min. In some cutting applications a stream of oxygen gas may supplied at the focus of the beam to enhance cutting speed, or if oxidation of the material is a problem, the heated region may be shrouded in a stream of nitrogen gas. As well as metals, a wide variety of other materials are routinely shaped and cut by techniques involving CO_2 lasers, including plastics, rubber, wood, even fabrics and clothing materials. Moulding details for decorating furniture are carved from wood using CO_2 lasers guided by numerically controlled machines following patterns originally crafted by hand.

It is also possible to produce local changes in the surface properties of metals to increase their resistance to wear or corrosion via a process known as 'laser cladding'. Selected alloys in powder form are delivered via a carrier gas stream to the melt pool on the surface of the substrate metal caused by a multi-kilowatt CO_2 laser beam. As the laser beam moves across the surface a bead of molten coating material, typically 3–5 mm wide and 1 mm high, is bonded to the substrate to form a new surface layer with the desired properties.

Pulsed TEA lasers are used in industry to mark plastics and semiconductor chips with identifying numbers or date of manufacture by ablating small amounts of material or thermally inducing local changes of colour or reflectivity. Waveguide CO_2 lasers, because of their small size, have found applications in airborne lidar systems and in probing the air ahead of aircraft for the signature of dangerous wind-shear conditions.

Although CO_2 laser technology was the first to be adopted on a large scale by manufacturing industry, and represents a well-established and ma-

[33]The lack of a practical means of delivering power from the laser to the point of use via a flexible fibre-optic link has also limited the medical applications of CO_2 lasers to those where the treatment site is accessible to direct illumination from the handpiece of a delivery system made up of articulated rigid segments. In other respects the 10 μm radiation has many advantages for medical applications, since this radiation is strongly absorbed by water (the main component of tissue and blood) and therefore seals blood vessels as it cuts.

ture technology for many industrial processes, in recent years there has been strong competition from solid-state lasers for many of the same applications. Nd:YAG lasers operating at their fundamental wavelength of 1.06 μm, or in their frequency-doubled form at 532 nm (see Chapter 7), are capable of producing focal-spot diameters smaller by factors of 10 or 20, respectively than their CO_2 counterparts. Moreover, unlike radiation in the 9–11 μm wavelength region, the output from these solid-state lasers is easily delivered by fibre-optic cable from the point of generation to the point of use[33] so that several workstations can be serviced by a central laser system. With recent developments in diode laser pumping of solid-state systems and fibre lasers, output powers in the several hundred watt range have become commercially available. Further, because solid-state laser systems tend to be more compact than their CO_2 counterparts (especially in the case of diode-pumped fibre lasers – see Chapter 10), many of the industrial applications originally opened up by CO_2 lasers may in the future become the province of solid-state laser systems. Considerations of long-term reliability and cost of ownership as well as the process time required and the quality of the finished product will ultimately determine which type of system will become the system of choice for a particular industrial process.

Further reading

For a more detailed discussion of the underlying physics and operation of CO_2 lasers, the reader should consult DeMaria (1973) and Hall (2004). A review of the wide range of applications of the CO_2 laser may be found in Ready (2001) and Webb and Jones (2004).

Exercises

(12.1) Thermal conductivity of He gas

(a) In the case where the mean free path λ is short compared to vessel dimensions, use classical kinetic theory to show that the average height above a reference plane of the last collision that a molecule makes before crossing the plane is $2\lambda/3$.

[Hint: Let N be the number of molecules per unit volume, then Z the average collision frequency for molecules is given by $Z = N\sigma\overline{v}$, where $\overline{v}$ is the average velocity and σ is the molecule–molecule collision cross-section. In volume δV the number of free paths that originate per unit time is $NZ\delta V$, and of these the fraction that lie within the solid angle $\delta\Omega$ is simply $\delta\Omega/4\pi$. The number of free paths δN that start out in δV and survive to cross δA per unit time is thus $\delta N = \frac{\delta\Omega}{4\pi} NZ \exp\left(-\frac{r}{\lambda}\right)\delta V$. The average height of the last collision is given by $\int r \cos\theta \, dN / \int dN$ with $\delta\Omega$ and δV expressed in spherical polar coordinates with ($0 \leq \theta \leq \frac{\pi}{2}$), ($0 \leq \phi \leq 2\pi$), ($0 \leq r \leq \infty$).]

(b) Using these results show that the net transport rate of some local property G (per unit volume) across a plane normal to the z-direction is given by $\frac{dG}{dt} = \frac{1}{3}N\overline{v}\lambda\frac{dG}{dz}$.

[You may assume that the average number of molecules striking unit area per unit time is given by $\frac{1}{4}N\overline{v}$.]

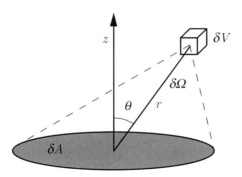

Fig. 12.16: Schematic diagram of the geometry employed in Exercise 12.1.

(c) In the case where G is the average thermal energy per molecule ($G = \frac{3}{2}k_B T$ for monatomic gases, $G = \frac{5}{2}k_B T$ for diatomic gases) use the result of the previous part to show that the thermal conductivity of a particular gas can be written as

$$\kappa = \mu \sqrt{\frac{8k_B T}{\pi M}} \frac{k_B}{\sigma} \text{ W m}^{-1}\text{ K}^{-1},$$

where M is the mass of the of the molecule, σ is the cross-section for collisions of the molecule with its own kind and $\mu = \frac{1}{2}$ for monatomic gases, $\mu = \frac{5}{6}$ for diatomic gases.

Note that this result shows that the thermal conductivity of a gas is *independent of pressure* in the case where the mean free path is small compared to the vessel dimensions.

(d) Use these results to calculate the thermal conductivity of helium gas and compare it to that of oxygen given that for He–He collisions $\sigma = 5.8 \times 10^{-20}$ m^2, and for $O_2 - O_2$ collisions $\sigma = 2.0 \times 10^{-19}$ m^2.

(12.2) Usually, the strongest transition in the 10.6 μm band output of the CO_2 laser is found to be the P(20) transition from $J' = 19$ of (00^01) to $J'' = 20$ of (10^00) at 10.591 μm. Table 12.2 lists other members of the P branch lines of this band, which are among the strongest in the entire c.w. output of the laser.

(a) Explain why all the transitions are even numbered.

(b) Using the data in the table, and assuming that B is the same for rotational levels of the 00^01 and 10^00 levels, calculate the value of the effective rotational constant $\overline{B}$ using the data for the P(14) and P(36) transitions.

(c) Estimate the carbon–oxygen bond length for these two cases, and suggest a reason for the difference.

(d) Explain why the lowest excited symmetric stretching vibrational level does not decay radiatively to the ground state but is coupled non-radiatively to the bending modes of vibration of the molecule.

(12.3) A CO_2 laser operates in the 10.6 μm band in a cavity of 1 m length with a total gas pressure of 15 Torr and an effective temperature of 400 K. The spacing of the P and R rotational transitions in this band is of order 2 cm^{-1}. By considering the longitudinal mode spacing of the cavity, and the collision- and Doppler-broadened linewidths, explain the distribution of frequencies that may be expected to be present on the laser output from such an oscillator

[The collisional width of the CO_2 vibration-rotation transitions is given by $\Delta\nu_H$ (collisions) $= 4.95 P(300/T)^{1/2}$ MHz, where P is the total gas pressure in Torr and T is the effective gas temperature in Kelvin.]

Table 12.2 Vacuum wavelengths and wave numbers of P-branch transitions of the $00^01 - 10^00$ band of the c.w. CO_2 laser.

Transition	Vacuum wavelength (μm)	Vacuum wave number (cm^{-1})	$-\Delta$ (cm^{-1})
P(12)	10.513114	951.193	0
P(14)	10.532080	949.480	1.713
P(16)	10.551387	947.743	1.737
P(18)	10.571037	945.981	1.762
P(20)	10.591035	944.195	1.786
P(22)	10.611385	942.384	1.811
P(24)	10.632090	940.549	1.835
P(26)	10.653156	938.689	1.860
P(28)	10.674586	936.804	1.885
P(30)	10.696386	934.895	1.909
P(32)	10.718560	932.961	1.934
P(34)	10.741113	931.002	1.959
P(36)	10.764052	929.018	1.984
P(38)	10.787380	927.009	2.009

(12.4) **Relative gain of P- and R-branches**

The population N_J of the rotational level J of a particular vibrational level of a linear molecule in rapid collisional equilibrium can be characterized by a single rotational temperature T such that

$$N_J = \frac{N}{Q_r} g_J \exp\left[-F(J)\frac{hc}{k_B T}\right],$$

where N is the total population of the vibrational level, $g_J = (2J+1)$, $F(J) = BJ(J+1) - DJ^2(J+1)^2$ and

$$Q_r = \sum_{J=0}^{\infty} (2J+1) \exp\left[-\frac{F(J)hc}{k_B T}\right].$$

(a) Show that if $D \ll B$, then for small B and large T we may write $Q_r \approx k_B T/hcB$.

[Hint: Within this approximation, we can represent the summation as an integral over J such that

$$Q_r \approx \int_0^{\infty} (2J+1) \exp[-BJ(J+1)hc/k_B T] dJ.]$$

(b) Hence, show that

$$N_J = N \frac{hcB}{k_B T} g_J \exp\left[-\frac{BJ(J+1)hc}{k_B T}\right].$$

(c) With the additional approximation that the value of B is the same for both upper and lower laser levels of the CO_2 molecule, show that if the population of the upper 00^01 vibrational level is *equal* to that of the 10^00 lower vibrational level then all of the P-branch transitions of the corresponding band exhibit gain, while all of the R-branch transitions show loss.

Ultraviolet molecular gas lasers

13

13.1 The UV and VUV spectral regions

The ultraviolet (UV) region of the spectrum covers the range of wavelengths from 400 nm (the short-wavelength limit of human vision) down to 200 nm, the wavelength at which atmospheric air ceases to be transparent to electromagnetic radiation. As in the visible region of the spectrum, it is conventional to quote wavelength values in the UV region as those measured in air, which has a refractive index[1] of about 1.0003. Radiation of shorter wavelengths, down to 100 nm, can only be transmitted along evacuated optical paths or paths filled with pure helium, while for even shorter wavelengths only vacuum paths can be used. For this reason the spectral region from 200 nm down to about 20 nm is called the vacuum ultraviolet (VUV) region. For radiation in this region of the spectrum the wavelength values quoted are always those measured in vacuum. The spectral region from 20 nm down to 1 nm is called the XUV region or soft X-ray region.[2]

It is no coincidence that most laser systems that generate radiation directly in the VUV region operate on transitions of atoms, ions or *diatomic* rather than polyatomic molecules. This is because polyatomic molecules tend to have such complicated rotation–vibration structure in their absorption transitions from the ground state that the absorption is virtually continuous over the VUV region. In the case of diatomic molecules the vibration–rotation bands associated with the strong electronic transitions from the ground state still offer gaps where laser transitions can show gain above the background loss.

Ordinary optical quality glass (e.g. BK7 borosilicate crown glass) does not transmit light of wavelength shorter than about 315 nm so that over much of the UV region materials such as fused silica or crystalline quartz are used for laser windows and other optics. Materials such as magnesium fluoride or calcium fluoride may be used down to wavelengths of approximately 130 nm, but below this one is restricted to lithium fluoride, which has a cut-off at 106 nm. For radiation with a wavelength below 106 nm all optical paths from the point of generation must be evacuated and windowless.

13.1 The UV and VUV spectral regions	355
13.2 Energy levels of diatomic molecules	356
13.3 Electronic transitions in diatomic molecules: The Franck–Condon principle	358
13.4 The VUV hydrogen laser	361
13.5 The UV nitrogen laser	364
13.6 Excimer molecules	364
13.7 Rare-gas excimer lasers	367
13.8 Rare-gas halide excimer lasers	370
Further reading	377
Exercises	378

[1] For wavelengths between 200 nm and 2 μm the refractive index of standard air (dry air at 15 °C and 101 325 Pa, containing 0.045% by volume of carbon dioxide) n is given by $(n-1) \times 10^8 = 8\,342.54 + \frac{2\,406\,147}{130 - 1/\lambda_{vac}^2} + \frac{15\,998}{38.9 - 1/\lambda_{vac}^2}$, where λ_{vac} is the wavelength in vacuum expressed in μm. See Birch and Downs (1994) or www.kayelaby.npl.co.uk.

[2] Going to still shorter wavelengths we reach the hard X-ray region (for which all materials of low mass density are transparent) and ultimately to the γ-ray region. The boundaries between all of these regions are somewhat arbitrary, and different authors assign different values to the wavelengths at which one region ends and another begins.

13.2 Energy levels of diatomic molecules

13.2.1 Separation of the overall wave function

The Schrödinger equation for a diatomic molecule may be written in the form

$$[T_n + T_e + V(\mathcal{R}, r)] \Psi(\mathcal{R}, r) = E \Psi(\mathcal{R}, r), \tag{13.1}$$

where: $\mathcal{R} \equiv \boldsymbol{R}_1, \boldsymbol{R}_2$ represents the coordinates of the two nuclei; $\boldsymbol{r} \equiv \boldsymbol{r}_1, \boldsymbol{r}_2, \ldots \boldsymbol{r}_N$ represents the coordinates of all the electrons; T_n is the sum of the kinetic energy operators of the nuclei; T_e is the sum of the kinetic energy operators of the electrons; and $V(\mathcal{R}, r)$ is the electrostatic potential energy, comprised of terms arising from the attraction of the electrons to the two nuclei, mutual repulsion between the electrons, and the repulsion between the two nuclei.[3]

[3] The coordinates $\mathcal{R} \equiv \boldsymbol{R}_1, \boldsymbol{R}_2$ and $\boldsymbol{r} \equiv \boldsymbol{r}_1, \boldsymbol{r}_2, \ldots \boldsymbol{r}_N$ are with respect to an origin that is fixed in the laboratory frame.

Provided the velocity with which the nuclei move relative to one another is much smaller than typical electron orbital velocities, it may be shown that the overall wave function $\Psi(\mathcal{R}, r)$ can be separated into a product of an electronic wave function $\psi_e(\mathcal{R}, r)$ and a nuclear wave function $\psi_n(\mathcal{R})$. Note that the electronic wave function depends on the coordinates of the electrons *and* the nuclei, whilst the nuclear wave function depends only on those of the nuclei. In this, the **Born–Oppenheimer approximation**,[4] the Schrödinger equation for the molecule can be separated into Schrödinger equations for the electronic and nuclear wave functions:

[4] The Born–Oppenheimer approximation relies on the fact that if the nuclear motion is very slow compared to the electronic motion then certain terms involving derivatives with respect to R can be neglected, allowing eqn (13.1) to be separated into equations for the electronic and nuclear wave functions. For details see, for example, Bransden and Joachain (2003) or Haken and Wolf (2004).

$$[T_e + V(\mathcal{R}, r)] \psi_e(\mathcal{R}, r) = U(\mathcal{R}) \psi_e(\mathcal{R}, r) \tag{13.2}$$

$$[T_n + U(\mathcal{R})] \psi_n(\mathcal{R}) = E \psi_n(\mathcal{R}). \tag{13.3}$$

These equations may in principle be solved as follows. The Schrödinger equation (13.2) for the electronic wave function $\psi_e(\mathcal{R}, r)$ can be solved for a fixed value of $\mathcal{R}$ to yield an energy eigenvalue, $U(\mathcal{R})$. Repeating this process for different values of $\mathcal{R}$ yields $U(\mathcal{R})$ that, as shown by eqn (13.3), acts as an effective potential in which the nuclei move. The energy E of the nuclei moving in this potential is the *total energy of the molecule*.

This procedure is simplified considerably if we realize that the potential energy term $V(\mathcal{R}, r)$ depends only on the separation of the two nuclei and is independent of the orientation of the vector joining the two nuclei. We therefore define $\boldsymbol{R} = \boldsymbol{R}_2 - \boldsymbol{R}_1$, and solve

$$[T_e + V(R, r)] \psi_e(R, r) = U(R) \psi_e(R, r), \tag{13.4}$$

with the magnitude of the nuclear separation, R, as a parameter. This gives the effective potential $U(R)$ in which the nuclei move.

To solve the Schrödinger equation for the nuclear wave function we replace the coordinates $(\mathcal{R}) \equiv (\boldsymbol{R}_1, \boldsymbol{R}_2)$ with $(\boldsymbol{R}_c, \boldsymbol{R})$, where $\boldsymbol{R}_c$ is the coordinate of the centre of mass. This allows a further separation of eqn (13.3) into an equation describing the motion of the centre of mass,[5] and

[5] Strictly, this equation should be solved in terms of the potential well describing the macroscopic container in which the gas is held. In practice, the energy levels of any macroscopic potential well are so closely spaced that the problem may be treated classically, in which case the energy of this part of the motion is simply the classical kinetic energy associated with the motion of the centre of mass.

$$[T'_n + U(R)] \psi_n(\boldsymbol{R}) = E \psi_n(\boldsymbol{R}), \tag{13.5}$$

where T'_n is the kinetic energy operator for a particle of mass M_r. This is simply the Schrödinger equation for a *single particle* of reduced mass[6] M_r moving in a central potential $U(R)$. The equation may be solved in a spherical coordinate system (R, θ, ϕ), where R is the internuclear separation and θ and ϕ give the angular position of the particle of mass M_r or, equivalently, the orientation of the internuclear axis. Just as in atomic physics,[7] the nuclear wave function may be written as a product of functions depending only on the radial and angular parts:[8]

$$\psi_n(\boldsymbol{R}) = \frac{1}{R}\psi_v(R)\psi_r(\theta, \phi), \tag{13.6}$$

where $\psi_v(R)$ is a vibrational eigenfunction, which depends only on $(R - R_0)$, the difference in internuclear separation R from its equilibrium value R_0. The $\psi_r(\theta, \phi)$ wave function is the rotational eigenfunction that depends only on the orientation of the internuclear axis in space.

Thus, to summarize, within the Born–Oppenheimer approximation the overall wave function $\Psi(\mathcal{R}, \boldsymbol{r})$ can be written as:

$$\Psi(\mathcal{R}, \boldsymbol{r}) = \frac{1}{R}\psi_e(R, \mathbf{r})\psi_v(R)\psi_r(\theta, \phi). \tag{13.7}$$

The potential energy $U(R)$ in which the reduced mass M_r of the two nuclei moves is a curve that rises steeply as R approaches zero—as a result of the nuclear repulsion—but tends to a horizontal asymptote as R tends to infinity. Around the region $R \approx R_0$ the curve is usually approximately quadratic in $(R - R_0)$,[9] and hence in this region the Schrödinger equation for the vibrational wave function resembles that of a pure harmonic oscillator. Deviations from a quadratic variation with $(R - R_0)$ become important for large values of $(R - R_0)$, and hence the potential shows anharmonic effects—in particular the energy eigenvalues of the oscillator are no longer equally spaced in energy.

[6] The reduced mass of a diatomic molecule with nuclei of masses M_1 and M_2 is defined by $1/M_r = 1/M_1 + 1/M_2$.

[7] See eqn (7.8).

[8] The $1/R$ factor is included explicitly by convention.

[9] For the same reasons given in Note 68 on page 172.

13.2.2 Vibrational eigenfunctions

In the parabolic approximation we can write the potential energy $U(R)$ as:

$$U(R) = \frac{1}{2}k(R - R_0)^2 = \frac{1}{2}kx^2 = \frac{1}{2}M_r\omega^2 x^2, \tag{13.8}$$

where $x = R - R_0$ is the displacement of the mass point of reduced mass M_r and in which $k = M_r\omega^2$ represents an effective 'spring constant' describing the potential well, where ω is the angular frequency of vibration. It is shown in standard texts on quantum mechanics[10] that eigenfunctions of the linear harmonic oscillator can be written as:

$$\psi_v = \left(\frac{\alpha}{\sqrt{\pi}2^v v!}\right)^{\frac{1}{2}} \exp\left[-\left(\frac{\alpha^2 x^2}{2}\right)\right] H_v(\alpha x), \tag{13.9}$$

where $v = 0, 1, 2, \ldots$ is the vibrational quantum number, $\alpha = (M_r k/\hbar^2)^{1/4}$, and $H_v(\alpha x)$ are the Hermite polynomials. The first few Hermite polynomials are given in Table 13.1.[11]

[10] See, for example, Bransden and Joachain (2003).

[11] Comprehensive tables of the Hermite polynomials may be found in Gradshteyn and Ryzhik (2000).

Table 13.1 The first few Hermite polynomials $H_v(y)$.

v	$H_v(y)$
0	1
1	$2y$
2	$4y^2 - 2$
3	$8y^3 - 12y$
4	$16y^4 - 48y^2 + 12$
$\vdots$	$\vdots$

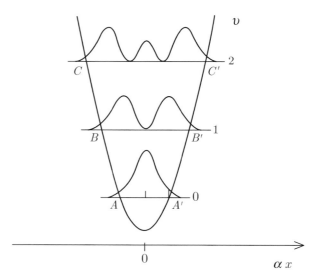

Fig. 13.1: Probability density $|\psi_v|^2$ for the harmonic oscillator for values of the vibrational quantum number $v = 0, 1, 2,$ and 10.

From eqn (13.9), and as illustrated in Fig. 13.1, the vibrational ground state $|\psi_0|^2$ is a Gaussian function that peaks at $x = 0$, indicating that in this state the harmonic oscillator spends most of its time around $R = R_0$. This is totally unlike the behaviour of a classical oscillator (e.g. a pendulum) that spends most of its time around the turning points of its motion where its speed is smallest. In Fig. 13.1 the turning points of the classical motion are indicated by A, A' for $v = 0$, and B, B' for $v = 1$, etc. As the vibrational quantum number increases, the likelihood of finding the nuclei with a separation near R_0 gets steadily smaller. Instead, it becomes more and more likely to find the nuclei with internuclear separations near the two turning points of the classical motion as indicated, for example, by D, D' for the case of $v = 10$.[12]

[12]This is an example of the correspondence principle that states that in the limit of high quantum numbers (i.e. large energies) the predictions of quantum mechanics must approach those of classical physics.

13.3 Electronic transitions in diatomic molecules: The Franck–Condon principle

13.3.1 Absorption transitions

A statement of the Franck–Condon principle to be found in many texts is that: "the electron jump in a molecule takes place so rapidly in comparison to the vibrational motion that immediately afterwards the nuclei still have very nearly the same relative position and velocity as before the jump." This implies that the electronic transitions that are the most probable are those denoted by vertical arrows in Fig. 13.2.

This statement is a good way to remember the result, but it is of no great help in calculating the relative intensities of the various transitions. The principle was put on a sound quantum-mechanical footing by Condon who showed that the matrix element of the electric dipole operator $\hat{\boldsymbol{D}}$—the square of which gives the transition probability between two states of the molecule—can be written as the product of a part $\hat{\boldsymbol{D}}_{\mathrm{el}}(R)$ (which is assumed to be only weakly dependent on R) with a factor

$$f_{v'v''} = \int_0^\infty \psi_{v'}^* \psi_{v''} \mathrm{d}R, \qquad (13.10)$$

where $\psi_{v''}$ is the vibrational eigenfunction of the lower level, and $\psi_{v'}$ is that of the upper level.

The probable transitions are therefore those for which $|f_{v'v''}|$ is largest. These are 'vertical' in the following sense. The wave function of the lowest vibrational level ($v = 0$) of an electronic level is usually approximately Gaussian and peaked near the equilibrium molecular separation R_0, as can be seen for the two electronic levels shown in Fig. 13.2. In contrast, the amplitude of excited vibrational wave functions ($v > 0$) are typically largest near the 'limit of classical motion,' i.e. where the vibrational energy equals the potential energy. Thus, for excited vibrational levels the wave functions are largest in regions away from the equilibrium molecular separation R_0. The Franck–Condon factor will be largest for transitions between vibrational levels with vibrational wave functions that have peaks[13] at similar molecular separations. For example, we would expect the largest Franck–Condon factor for transitions from the $v'' = 0$ level of the lower electronic level of Fig. 13.2 to be to that vibrational level in the upper electronic level for which the vibrational wave

[13] Since the transition rate is proportional to the square of the Franck–Condon factor, it doesn't matter if the 'peak' is a minimum.

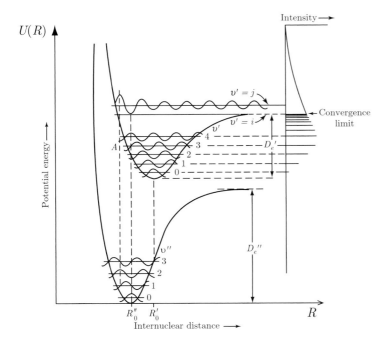

Fig. 13.2: Schematic potential-energy curves $U(R)$ for the electronic ground state and an electronically excited state of a diatomic molecule. In the case shown here the equilibrium internuclear separation R_0' of the excited electronic energy level happens to be larger than the equilibrium internuclear separation R_0'' of the ground state. Figure reproduced by kind permission of J. G. Calvert and J. N. Pitts (1966).

function has a maximum amplitude at a similar molecular separation, which in this case is $v' = 4$. In contrast, there is little overlap of the vibrational wave functions of the $v = 0$ vibrational levels of the upper and lower electronic levels, and consequently for the case shown the $(v' = 0) \leftarrow (v'' = 0)$ transition should be relatively weak. So, we see that provided R'_0 is sufficiently different from R''_0, strong transitions in absorption from the ground state are expected to states v' that have their limit of classical motion close to R''_0.

We note that transitions to states like that labelled $v' = j$ in Fig. 13.2 can also occur, but lead to dissociation of the molecule since the energy of the vibrational state is above the dissociation limit[14] for the electronic level and the absorption in this region of the spectrum will form part of a structureless continuum at short wavelengths.

[14]The quantities labelled as D''_e and D'_e in Fig. 13.2 (which refer to the energy asymptotes above the minimum in the energy curves for the electronic ground state and excited state, respectively) are not exactly the 'dissociation energies' of the two states since they do not take into account the $\frac{1}{2}\hbar\omega$ zero point energy of the corresponding vibrational energy ladder. The true dissociation energy D''_0 of the ground-state molecule represented in Fig. 13.2 is $D''_e - \frac{1}{2}\hbar\omega$.

13.3.2 The 'Franck–Condon loop'

The Franck–Condon principle applies equally to the case of emission, and leads to the concept of a 'Franck–Condon loop' for absorption and re-emission. In the case shown in Fig. 13.3, the $\psi_{v'}$ of the upper level with $v' = 2$ has the best overlap with $\psi_{v''}$ for $v'' = 0$, so the $0 \to 2$ transition is strongest in absorption. However, because the vibrational wave function for $v' = 2$ has another peak near the outer classical turning point, this particular eigenfunction also shows good overlap with $\psi_{v''}$ for $v'' = 5$. Assuming that there are no collisions that de-excite (quench) the molecule before it emits, when a sample of molecules of this type absorbs photons that take them from $v'' = 0$ in

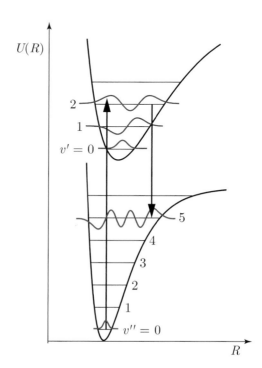

Fig. 13.3: Schematic diagram of an optically pumped 'Franck-Condon loop' laser system in a diatomic molecule.

the lower level to $v' = 2$ in the upper level, they will subsequently re-emit a set of bands with two maxima in their intensities, one corresponding to $(v' = 2) \rightarrow (v'' = 0)$ and another for $(v' = 2) \rightarrow (v'' = 5)$.

Some of the radiative transitions from the upper electronic state will terminate on high vibrational levels of the electronic ground state. These are initially empty since the spacing of neighbouring vibrational levels is typically many times larger than $k_B T$ at room temperature. Hence, optical pumping of diatomic molecules to the vibrational levels of an excited electronic state can form population inversions on transitions down to the vibrationally excited levels of the ground electronic state.

An example of a laser operating on a Franck–Condon loop of this type is the nitric oxide VUV laser.[15] In this system absorption[16] of photons from a molecular fluorine laser[17] operating at a wavelength of 157.6 nm excites NO molecules from the $(v'' = 0)$ level of the $X^2\Pi$ ground electronic state to the $(v' = 3)$ vibrational level of an upper electronic state labelled[18] as $B'^2\Delta$. As a result, a population inversion is formed on several transitions from this upper level to several vibrational states of the electronic ground state. At the low gas densities involved in this case, collisional de-excitation of excited NO does not compete with optical decay rates and laser emission occurs on the $(v' = 3) \rightarrow (v'' = 1, 6, 9, 10, 12)$ transitions of the $B'^2\Delta \rightarrow X^2\Pi$ electronic transition with wavelengths of 162, 190, 210, 218 and 234 nm, respectively.

The discussion above ignores the effects of collisions in which the vibrational energy is quenched, but the electronic energy is not. Collisions of this type can become important at higher densities, and lead to a change in the spectrum of the fluorescence from the upper electronic level. At sufficiently high densities the molecule may be quenched all the way down to the lowest vibrational state ($v' = 0$) of the upper electronic state before it emits. If this happens, say, for the case of the molecule of Fig. 13.2, the band that will be strongest in emission might be $(v' = 0) \rightarrow (v'' = 2)$. Quenching is much more likely in the liquid phase than in the gas phase since collisions are much more frequent at liquid densities.[19]

[15] Hooker and Webb (1990).

[16] In fact, the frequency of the absorption transition is shifted into exact resonance with the fixed frequency of the molecular fluorine laser by applying a magnetic field of order 1 T.

[17] See Section 13.8.5.

[18] For a full description of the notation used to specify excited electronic levels of diatomic molecules, the reader is directed to standard works on molecular spectroscopy such as that by Bransden and Joachain (2003).

[19] As discussed in Section 14.1, rapid collisional de-excitation of vibrationally excited molecules within the upper electronic level is a key step in the achievement of inversion within dye lasers. See Fig. 7.23, and the discussion relating to it, for an example of the influence of quenching on absorption and emission spectra of ions in a *solid* host.

13.4 The VUV hydrogen laser

So far, we have considered the achievement of inversions in optically pumped molecular systems. Because the cross-sections for excitation of atoms and molecules in collisions with fast electrons tend to reflect the features of the corresponding optical transition probabilities (see Section 11.1.6) the electronic transitions that have the strongest optical cross-sections for absorption transitions from the ground state also tend to have the largest cross-section for excitation by collisions with electrons. In other words, the Franck–Condon principle is also helpful in determining which vibrational levels of electronically excited levels are most likely to be excited in collisions of energetic electrons with ground-state molecules.

The molecular-hydrogen laser is an example of the operation of a laser system on a Franck–Condon loop in which the excitation is provided not by absorption of photons but by electron collisions. The ground state of the H_2

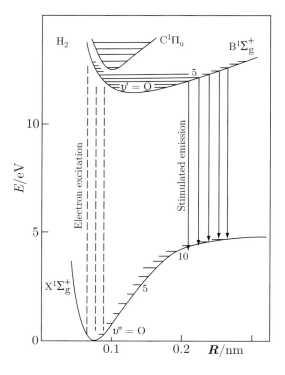

Fig. 13.4: Energy levels of molecular hydrogen involved in VUV laser action. For clarity only inversions on transitions of the $B^1\Sigma \to X^1\Sigma$ Werner band (100–120 nm) are shown. Inversion can also be produced on transitions of the $C^1\Pi \to X^1\Sigma$ Lyman band (140–165 nm). Reprinted figure with permission from R. T. Hodgson, Phys. Rev. Lett., 25, pp 494–496, (1970), http://link.aps.org/doi/10.1103/Phys. Rev. Lett. 25, 494, Copyright (1970) by the American Physical Society.

molecule is labelled $X^1\Sigma$. The two excited states shown in Fig. 13.4, the $C^1\Pi$ and the $B^1\Sigma$ are connected to the $X^1\Sigma$ electronic ground state by strongly allowed transitions with radiative lifetimes of 0.6 and 0.8 ns, respectively. For the reasons discussed below, this laser is restricted to pulsed operation and requires extremely short-duration excitation (less that 0.5 ns).

This excitation is provided by a very fast-rising discharge current sent through a low pressure of hydrogen gas. The fast electrons excite several vibrational levels of the B and C state via:

$$e_{\text{fast}} + H_2(X^1\Sigma, v'' = 0) \to e_{\text{slow}} + H_2(B^1\Sigma, v' = 3, 4, \ldots 12). \quad (13.11)$$

$$e_{\text{fast}} + H_2(X^1\Sigma, v'' = 0) \to e_{\text{slow}} + H_2(C^1\Pi, v' = 0, 1\ldots 4). \quad (13.12)$$

Provided that the excitation occurs on a very short time scale, such that the threshold excitation rate is reached in a time of order the upper-level lifetime, then inversion can occur between various vibrational levels of the B state and X (Werner-band emission 100 to 120 nm) and on the C to X transition (Lyman-band emission 140 to 165 nm). The laser is said to operate on bound–bound transitions since the molecule is stable against dissociation both in upper and in lower electronic levels.

Note that this laser spectacularly fails to satisfy the necessary but not sufficient condition for a steady-state population inversion (eqn (2.16)). As such it can only operate in a pulsed mode.

The rate of radiative decay from the lower laser levels is very slow indeed since the matrix elements are small and the ω^2 (see eqn (3.15)) factors for the infra-red vibrational transitions are also minute in comparison with those for visible or UV transitions. The lifetimes of the lower laser levels are very much longer than the time scales of interest here, to the extent that we can ignore decay from these levels.

These laser transitions are therefore self-terminating[20] since the lower laser levels cannot decay sufficiently quickly compared to the rate at which they are populated by stimulated emission from the upper laser level. As such, the excitation must establish the upper-level population before the lower levels have had time to fill, and consequently it is necessary for the excitation to be in the form of pulses with duration $\lesssim A_{21}^{-1}$. Only when the laser medium has relaxed back to its initial thermal equilibrium population can the pulse be repeated. The recovery time is determined by whatever mechanisms (wall collisions or quenching collisions) dominate the relaxation process.

The hydrogen laser can only operate if the excitation is applied extremely rapidly—either by a pulsed electron beam or directly by a high-voltage discharge run in low-pressure hydrogen gas. Electrical circuits can only generate sufficiently fast excitation pulses if they have a very low inductance (typically less than 10 nH). One solution is to use strip-line technology, such as that shown in Fig. 13.5 in which energy is stored in the distributed capacitor comprising two stainless steel or copper conductors forming a sandwich with a central insulator such as a printed-circuit board or glass plate. The discharge

[20] See Section 2.2.2.

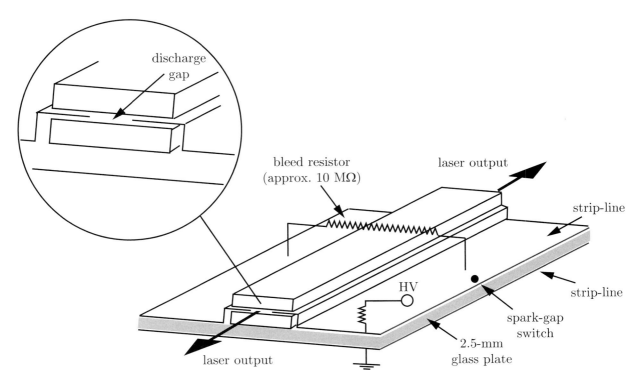

Fig. 13.5: Schematic diagram of a molecular-hydrogen laser employing fast-discharge strip-line circuitry. The discharge is contained in a channel of length 120 cm, width 1.2 cm and height 0.04 cm. (After Hodgson (1970)).

channel is formed by a gap in the conductor on the top side of the board. This gap defines the sides of the laser gas envelope, the top and bottom of which are sealed by layers of glass strip. VUV transmitting windows are sealed directly onto the ends of the box so formed. There are no mirrors since the gain-pulse duration is so short that there is no time for light to traverse the structure more than once; the output of the laser is therefore amplified spontaneous emission (ASE), as discussed in more detail in Section 18.3.

Initially the two sections of conductor on the top side are charged to the same high voltage with respect to the grounded conductor on the lower surface. A voltage wave is launched along the strip-line by suddenly connecting one of the top sections of conductor to the adjacent bottom sector via a fast-acting switch such as a spark gap. The switched side rapidly rings to invert in voltage with respect to ground. When this reverse-polarity voltage wave arrives at the break in the top conductor (whose edges form the transverse electrodes of the discharge gap) the hydrogen gas between them rapidly undergoes breakdown and conducts a high current pulse of subnanosecond rise time.

Because the volume of the H_2 laser is small, and the stored energy that can be transferred in such a short time is small, the output of such devices tends to be rather low—only a few microjoules per pulse.

13.5 The UV nitrogen laser

The energy curves for some of the low-lying states of the N_2 molecule are shown in Fig. 13.6. As in the case of H_2, laser action in the N_2 molecule occurs on a bound–bound transition forming a Franck–Condon loop, except that in this case laser action occurs on the $C^3\Pi(v'=0) \to B^3\Pi(v''=0)$ transition at 337 nm in the near-UV region. For exactly the same reasons as in the H_2 laser, the N_2 laser is a self-terminating system, but because the lifetime of the $C^3\Pi(v'=0)$ upper laser level is 40 ns as against ≈ 0.5 ns in the case of the H_2 upper laser level, the constraint on the risetime of the excitation pulse is correspondingly less severe. Consequently, it is possible to use much larger discharge volumes and discrete circuit components in the construction of N_2 lasers and, because there is more time in which to feed excitation energy into the discharge, the output-pulse energy can be a few millijoules per pulse.

Since the output wavelength of 337 nm lies in a convenient region of the spectrum for providing the excitation for dye lasers operating in the near-UV and visible regions (see Section 14.5), and because nitrogen is a non-hazardous gas, nitrogen lasers are still in use as low-cost pump lasers for low-repetition-rate pulsed dye lasers. However, for many applications in photochemistry laboratories they have been replaced by XeCl or KrF lasers (see Section 13.8.2) that are capable of much higher pulse energies (several hundred millijoules) and higher repetition rates.

13.6 Excimer molecules

The basic features of the energy-level scheme of an excimer molecule are illustrated in Fig. 13.7. An excited diatomic molecule, denoted XY^*, is formed

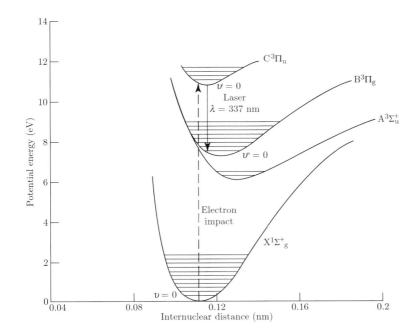

Fig. 13.6: Energy levels of molecular nitrogen involved in UV laser action at 337 nm on the $C^3\Pi \to B^3\Pi$ transition. (After Davies (2000).)

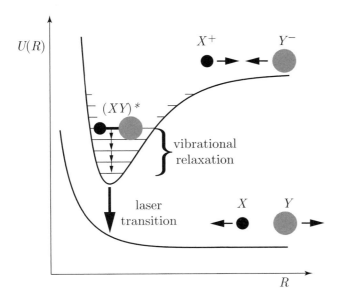

Fig. 13.7: Schematic energy level diagram of an excimer molecule.

either via an association reaction involving an excited atom X^* and another (ground-state) atom Y, or by the recombination of a positive ion X^+ and a negative ion Y^-. In either case the reactants spiral in towards one another from an initially large value of internuclear separation.

The formation reaction also involves the cooperation of a third atom Z (which could be a separate bystander atom of type X or Y or a different species) that allows the excited complex (XY)* to stabilize in a high vibrational state by removing excess energy and angular momentum. Subsequent collisions of vibrationally excited (XY)* with Z de-excite it towards its lowest vibrational level with R values close to R_0, at which the energy curve for the excited bound molecular level (XY)* has its minimum.

Excimer lasers operate on transitions from bound upper states to unstable lower states of molecules of this general type. The upper laser level (XY)* in the special case when X and Y are identical corresponds to an *excited dimer*, from which the name 'excimer' was coined. It is fast becoming a universally accepted practice to apply the same name to combinations in which X and Y are not the same (i.e. excited complexes, which would more logically be called 'exciplexes').

Apart from the fact that the population of the upper laser level is fed not by excitation of the ground state but directly from reactions that lead to the formation of the excimer already in the excited state, the other feature shared by all other excimer (exciplex) systems is that the lower laser level (the molecular ground state) is either fully repulsive or has at most only a very shallow potential well (e.g. due to van der Waals attraction), not much deeper than thermal energy values.

A molecule XY making a transition from the upper to the lower level will dissociate within one vibrational period ($\lesssim 0.1$ ps) owing to the steepness of the lower-level curve under the minimum of the upper-level curve. The short lifetime of the lower level smears out any rotational or vibrational structure that the transition might have exhibited, so that emission on these bands is typically in the form of a featureless[21] quasi-continuum some 0.3 to 30 nm broad.

[21] One of the few examples of an excimer molecule that does show some evidence of vibrational structure in its gain spectrum is the XeCl rare-gas halide molecule. The structure arises from a weakly bound van der Waals potential well (of order $300\,\text{cm}^{-1}$ well depth) in its ground electronic level.

The fact that the lower level is unstable is very advantageous in obtaining a population inversion—the excimer lasers satisfy the condition of eqn (2.16) automatically—but the associated broadness of the transition spreads the available gain over a very wide bandwidth. From eqn (4.14), the peak gain cross-section $\sigma^L_{21}(0)$ for a homogeneously broadened transition, is given by

$$\sigma^L_{21}(0) = \frac{\lambda^2}{2\pi}\frac{A_{21}}{\Delta\omega_H}. \qquad (13.13)$$

The spontaneous emission coefficient A_{21} for the excimer systems of interest is of order $10^8 - 10^9\,\text{s}^{-1}$,[22] so that typical values of $\sigma^L_{21}(0)$ for excimer systems tend to be in the range 7×10^{-18} to $5 \times 10^{-16}\,\text{cm}^2$.

[22] This is typical of the 1–10 ns lifetimes of atoms or molecules for fully allowed electronic transitions in the visible or near-UV.

As well as gain, excimer lasers show an inherent optical loss, some of which is due to transiently created molecular species whose broadband absorption overlaps the frequency band of the laser transition. Typical background losses amount to $5 \times 10^{-3}\,\text{cm}^{-1}$, and the small-signal gain must therefore exceed this value in order for oscillation to occur. Since it takes an energy of at least 10 eV to create the precursor excited atom or ion that goes into making one excimer molecule (which decays in 10 ns), these figures imply[23] a minimum power input of 3–300 kW cm^{-3} to reach threshold on a steady-state basis even if excimer formation were 100% efficient.

[23] See Exercise 13.3.

It is therefore not surprising that excimer lasers require very high specific input powers to work at all—they tend to be all-or-nothing systems, emitting pulses with peak output powers of tens of kilowatts or more once they do start oscillating.

13.7 Rare-gas excimer lasers

Although the suitability of the rare-gas dimer emission bands as potential laser transitions was pointed out as long ago as 1960 by Houtermans, it was not until 1971 that successful operation of the first Xe_2^* laser was announced by Basov and coworkers (1971) who employed electron-beam pumping of *liquid* xenon. A more practical approach was to use high pressure samples of rare *gas* as the active medium, and over the following three years laser oscillation was obtained in Xe_2^* in gas form at a wavelength of 172 nm, in Kr_2^* at 146 nm (Hoff, et al. 1973), and in Ar_2^* at 126 nm (Hughes, et al. 1974). In all cases e-beam pumping of high-pressure gas was used. Since the mechanisms of all the rare-gas excimer lasers are similar, it will suffice to describe only those of Xe_2^*; the description applies equally well to Kr_2^* and Ar_2^*.

Figure 13.8 shows some of the lower lying-states of the Xe_2 molecule. The ground-state $A^1\Sigma_g^+$ belong to levels that asymptotes at infinite R to two ground-state Xe atoms. Similarly, the curves marked $B^3\Sigma_u^+$ and $C^1\Sigma_u^+$ asymptote to combinations of one ground-state Xe atom and one excited Xe* atom of the first excited configuration $5s5p^56s$ in the $^3P_2$ or $^3P_1$ levels, respectively. The laser transition at 172 nm connects $B^3\Sigma_u^+$ with the $A^1\Sigma_g^+$ ground state.

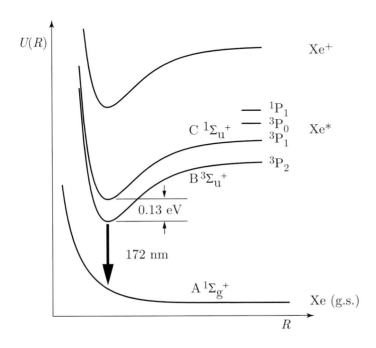

Fig. 13.8: Energy levels of the rare-gas dimer Xe_2^* relevant to laser action at 172 nm.

The corresponding transition from the $^3\Sigma$ level, has a much smaller gain cross-section, reflecting the spin change involved, and so does not exhibit laser oscillation.

When xenon gas at several atmospheres pressure is excited by an electron beam, the overwhelming effect of the incoming beam of electrons (at 200–500 keV incident energy) is to make ion–electron pairs, and in so doing losing about 25 eV on average per ion pair. Thus, each incoming beam electron leaves a trail of Xe^+ ions and slow electrons in its wake, making tens of thousands of such collisions before it is finally slowed to thermal energies.

The Xe^+ ions created in this way undergo association reactions in three-body collisions with two Xe atoms to form Xe_2^+ ions—the third body as usual being needed to stabilize the system in a bound state by removing excess energy. The Xe_2^+ ions can dissociatively recombine with slow electrons to form highly excited Xe^{**} neutrals, which can be quenched via three-body reactions involving Xe atoms or electrons to form Xe^* atoms in the $^3P_{2,1}$ levels. The three-body reaction that finally leads to Xe_2^* in the $^1\Sigma$ level is:

$$Xe^*(^3P_1) + Xe + Xe \rightarrow Xe_2^*(^1\Sigma) + Xe, \qquad (13.14)$$

with a similar three-body reaction of $Xe(^3P_2)$ forming $Xe_2^*(^3\Sigma)$.

The high gas pressure (typically several atmospheres) is necessary in order to ensure that the three-body reactions such as eqn (13.14) are sufficiently fast for the excitation rate of the upper laser level to be competitive with its radiative decay rate. The overall conversion of incoming electron beam energy to fluorescence on the excimer bands can be very efficient—of order 40% under favourable conditions.

A typical arrangement for pumping Xe_2^* and other rare-gas excimer lasers is shown in Fig. 13.9. High-pressure xenon gas is contained within a thin-walled stainless-steel cylinder that forms the grounded anode of a vacuum diode. The anode is situated along the axis of a cylindrical cathode that has sharp blade-like projections on its inner surface facing the anode. The cathode is connected, via a pulse-shaping line, to a high-voltage source (e.g. a Marx generator).

When the cathode is pulsed strongly negative (−500 kV), small plasma blobs are formed on its blade surfaces that then act as sources of electrons. The electrons are accelerated across the vacuum gap and impinge on the outer surface of the anode as a beam of particles with energy corresponding to the applied voltage. Because the anode is made of such thin material (such as titanium or stainless steel), the electrons lose very little energy (perhaps only 5–10%) in penetrating it and deposit their energy in the high-pressure gas inside the anode tube by the process described above.

Although output energies as large as 8 J have been reported from very large e-beam pumped Xe lasers, the 172-nm output versus input energy curve of the Xe_2^* laser shows a disappointing tendency to turn over and decrease if the specific input energy is increased beyond a certain point. A likely explanation involves the difference in stimulated emission cross-section between the $^3\Sigma$ and $^1\Sigma$ levels. Photons of the 172-nm laser radiation are capable of photoionizing Xe_2^* molecules in both $^3\Sigma$ and $^1\Sigma$ levels and consequently photoionization represents a distributed loss at the laser wavelength. The $^1\Sigma$ level lies

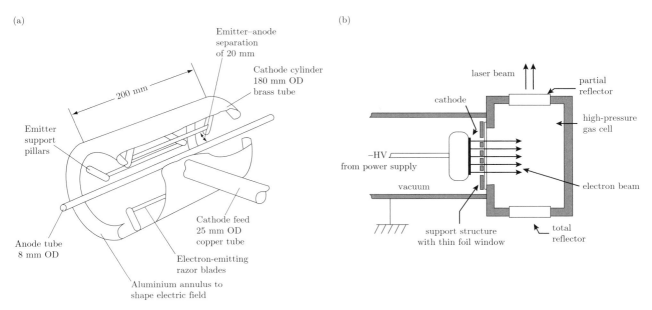

Fig. 13.9: (a) Cutaway drawing of coaxial e-beam diode for pumping rare-gas excimer lasers. The rare-gas at multi-atmosphere pressure is contained in the thin-walled cylindrical anode tube. The entire structure is contained in a chamber (not shown) maintained at high vacuum. Reprinted figure with permission from C. B. Edwards et al., Rev. Sci. Instrumen., 50, pp 1201–07, (1979), http://link.aip.org./link/?RSINAK/50/1201/1. Copyright (1979) by the American Institute of Physics. (b) Linear electron gun for single-sided pumping of an excimer laser gas cell. The thin foil separating the high vacuum of the electron gun from the high-pressure rare gas in the laser chamber is supported by a grid structure called a 'hibachi' by analogy with a type of Japanese charcoal grill.

some 0.13 eV above the $^3\Sigma$ level at its minimum, and has a stimulated emission cross-section that is larger than its cross-section for photoionization by photons of 172 nm laser radiation. In contrast, the photoionization cross-section of $^3\Sigma$ is larger than its stimulated-emission cross-section at 172 nm. Thus, a photon of 172 nm encountering an excimer in $^1\Sigma$ will on average cause amplification, while if it encounters $^3\Sigma$ the average result will be absorption. Pumping the gas mixture too strongly causes a high electron density to be produced, and with it, rapid collisional transfer between $^3\Sigma$ and $^1\Sigma$ via:

$$\text{Xe}_2^*(^3\Sigma) + e_{\text{fast}} \leftrightarrow \text{Xe}_2^*(^1\Sigma) + e_{\text{slow}}. \tag{13.15}$$

The net result of this fast electron collisional transfer is to pull the populations of $^3\Sigma$ and $^1\Sigma$ towards Boltzmann equilibrium at the electron temperature. Since $^3\Sigma$ lies lower in energy and has the higher statistical weight, the result of pumping the laser too hard is to populate the loss-causing $^3\Sigma$ state at the expense of the useful $^1\Sigma$ state, a situation that can only be harmful for laser oscillation.

Although the output wavelengths lie in a spectral region of very great intrinsic interest for photochemistry the size, cost and complexity of e-beam pumped rare-gas excimer lasers has, so far, restricted their availability to all but a very few well-equipped national laboratories.

13.8 Rare-gas halide excimer lasers

13.8.1 Spectroscopy of the rare-gas halides

In 1974 a series of new emission bands was observed when halogen gases were titrated into the flowing afterglow of a rare-gas discharge. These bands were identified by Golde and Thrush (1974) as the $B\,^2\Sigma \to X^2\Sigma$ transitions of rare-gas halide (RGH) excimer molecules comprising one rare-gas atom and one halogen atom. In 1975 laser action in XeBr was demonstrated by Searles and Hart and in XeF, XeBr, XeCl and KrF by Brau and Ewing.[24] Table 13.2 lists the known $B \to X$ transition wavelengths of the various RGH molecules.

Laser action is not possible on all of the RGH bands. Some RGH molecules are unstable against pre-dissociation in the B-state particularly those towards the bottom left-hand corner of Table 13.2.

As shown in Fig. 13.10, the KrF molecule (in common with all Kr and Ar halides) exhibits no binding at all in its $X\,^2\Sigma$ ground state, and hence in the Kr- and Ar-halide lasers gain occurs over a band some 2–3 nm wide that shows little evidence of vibrational structure. In contrast, the ground states of the Xe halides do possess a shallow (0.04 eV) potential well, and hence in XeF and XeCl the gain regions are broken into groups of vibrational bands each about 0.1 nm wide; however, even in these molecules the lower levels are thermally unstable and still retain most of the characteristics of the lower level of an excimer laser. One feature the RGH systems have in their favour compared to the pure rare-gas excimer lasers is that the problem of photoionization of the upper laser level seems to be very much less severe, indeed for many RGH molecules it is altogether absent since the laser photon energy is insufficient to cause ionization. For this reason, it has proved much easier to scale rare-gas halide lasers to high output energies than was the

[24] See Searles and Hart (1975) and Brau and Ewing (1975).

Table 13.2 Emission bands of the rare-gas halogen excimer molecules. Wavelengths (in nm) of known laser transitions are underlined.

	Rare gas			
	Ne	Ar	Kr	Xe
F	107	<u>193</u>	<u>248</u>	<u>351</u>
Cl		<u>175</u>	<u>223</u>	<u>308</u>
Br		166	206	<u>282</u>
I			185	252

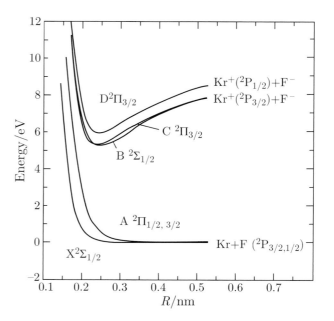

Fig. 13.10: Energy levels of the KrF molecule. (After Brau (1979).)

case for pure rare-gas excimer lasers. Progress in developing RGH lasers was extremely rapid, and within a few months of their initial discovery, single-pulse output energies of several joules were being obtained in KrF in a number of laboratories.

13.8.2 Rare-gas halide laser design

Early RGH lasers were pumped by electron beams, using both the co-axial and planar[25] electron beam geometries that had earlier been established for the rare-gas excimer lasers (see Section 13.7).

The gas mixture of an e-beam pumped KrF laser is made up of three components:

1. the rare gas (Kr) at some tens of Torr pressure,
2. a few Torr of the halogen donor (F_2),
3. two or three atmospheres of argon.

The argon is there because a high pressure of something is needed to scatter the incoming beam electrons so that they deposit their energy in the gas cell instead of sailing straight through the gas mixture. A second important function of this background gas is the provision of an abundant supply of 'third bodies' for the three-body reactions involved in the formation kinetics. Reactions such as:

$$Kr^* + F_2 \rightarrow KrF^* + F, \quad (13.16)$$

which were initially thought to play the dominant role in the formation kinetics of KrF^* in e-beam pumped systems, are now recognized to be less important than the three-body ion-recombination reaction:

$$Kr^+ + F^- + Ar \rightarrow KrF^* + Ar. \quad (13.17)$$

It was clear at an early stage in the development of RGH lasers that if a 'pure discharge' device could be developed (i.e. one not requiring an e-beam generator) it would have widespread applications. In these cases helium or neon at up to four atmospheres pressure is used as the buffer gas since it is found experimentally that these gases are less prone to arcing than the argon used in e-beam pumped devices. However, the problem of discharge stability still provided a major difficulty in developing a pure discharge technology. In order to launch a transverse discharge in a high-pressure gas in a uniform manner, it is necessary to seed the gas volume with $10^7 - 10^9$ electrons cm^{-3} before applying the high-voltage pulse to the main electrodes. This preionization makes the spatial distribution of discharge current much more immune to small perturbations that might otherwise allow a local arc to develop and grow. In other words, if the plasma is uniformly highly conducting, constriction of the plasma to form an arc does not increase the conduction of the plasma any further. As discussed in Section 12.6 in connection with the high-pressure pulsed CO_2 laser, in the absence of pre-ionization any small region undergoing breakdown and ionization growth ahead of other regions would lead to the development of an arc channel at that point.

[25] In the planar geometry the electron beam irradiates the high-pressure gas chamber via a flat window of thin titanium of stainless steel foil supported by a grid (commonly known as a hibachi) that provides mechanical support.

In early RGH lasers the pre-ionization was provided by an array of bare sparks whose XUV emission irradiated and ionized the discharge region between the transverse electrodes. However, gradual erosion of the spark electrodes introduces dust and unwanted impurities into the gas mixture, which severely limits the useful lifetime of the laser gas fill. Many modern discharge-excited RGH lasers instead rely on irradiation of the main discharge gap from both sides by corona discharges that provide a weaker but less contaminating means of pre-ionization.[26]

Even with pre-ionization, the transverse discharge in RGH laser gas mixtures is unstable: for reasons we shall describe below, the uniform glow will eventually break up into filaments and arcs. To combat this tendency it is necessary to make the discharge pulse short and intense so as to couple in most of the discharge excitation before the instability has time to develop.

Another approach towards obtaining discharge uniformity in a large-volume device is the use of X-ray pre-ionization.[27] As indicated in Fig. 13.11 the gas volume between the main electrodes is irradiated with low-energy X-rays provided by linear X-ray sources on each side of the grounded electrode. The X-ray sources each comprise an evacuated ceramic tube containing an internally heated cathode tube made from porous tungsten (impregnated with low work function material such as BaO) that runs the length of the tube opposite an anode (target) in the form of a solid rod of tungsten or tantalum. The complications of providing a high-voltage supply (70 kV peak) to power the linear X-ray sources and the necessary X-ray shielding are compensated by the highly uniform penetration of the laser gas by the pre-ionizing radiation and the absence of sparks—and the contamination they cause—in the laser gas.

For operation at repetition rates above a few pulses per second it is necessary to circulate the gas mixture by means of a fan system so as to replace the gas in

[26] The corona electrodes run close to the grounded main electrode along its full length on either side. Each comprises a narrow central conductor sheathed in a tube of ceramic insulator of high dielectric strength. The corona discharge is initiated in the gas around the surface of the ceramic tube when a pulse of several tens of kilovolts is applied to the central conductor. The corona discharge is self-extinguishing since the insulating tube blocks any possible DC current path to ground, so that current can flow only until sufficient charge is accumulated on the outer surface of the dielectric tube to reduce the field between it and the grounded main electrode to zero.

[27] See, for example, Müller-Horsche et al. (1993).

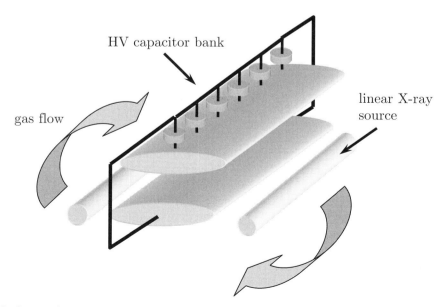

Fig. 13.11: Schematic diagram of the key components of an X-ray pre-ionized RGH laser. Note that in practice the part of the circuit connecting the high-voltage capacitor bank to the lower electrode is formed from low-inductance strip-lines placed edge-on to the gas flow.

the active volume with fresh gas between successive discharge pulses to avoid the onset of arcing via the type of thermal feedback instability described in Section 12.6. Any region of hot gas remaining in place between successive main discharge pulses provides a locally higher E/N value and hence a locally increased electron avalanching rate and thus a locally increased power deposition.

13.8.3 Pulse-length limitations of discharge-excited RGH lasers

Unlike the H_2 and N_2 lasers described in Sections 13.4 and 13.5, the energy-level scheme of the RGH laser systems does satisfy, indeed it satisfies extremely well, the necessary but not sufficient condition (eqn (2.16) for c.w. laser action.

This then raises the question why it is that c.w. laser action has never been observed in discharge-excited RGH lasers and pulse lengths reported in the literature are much shorter than 500 ns, typically only 20–50 ns FWHM. The answer lies in the instability mechanisms that limit the period for which a uniform discharge can be maintained in the gas mixture. The laser gas mixtures contain significant concentrations of strongly electron-attaching molecular species such as F_2 (in the case of KrF and ArF lasers) and HCl (in the case of XeCl and KrCl lasers). These species are necessary to act as halogen donors to provide the F^- and Cl^- ions in the RGH formation reactions such as eqn (13.17) and its analogues. However, the ability of these halogen donor species to attach electrons via processes such as

$$F_2 + e^- \rightarrow F^- + F, \qquad (13.18)$$

limits the rate of growth of electron density.[28] If for some reason the halogen density in a region is decreased slightly from its ambient value, the electron density will grow faster in that region and hence the concentration of neutral halogen donor will be further diminished locally as the result of the increased availability of electrons. This is clearly a positive feedback, and unchecked it will lead to the formation of a column low in halogen donor concentration and high in electron density, ultimately forming the site of an arc that disrupts the uniformity of the gain medium. Such an arc would lead to a fast collapse of the voltage across the main gap, terminating the laser pulse.[29] To avoid the damage to the electrodes which arcing can cause, the duration of the voltage pulse applied to the main discharge is deliberately limited by the design of the high-voltage pulse-forming network to values that are compatible with the maximum duration of the uniform glow expected for the optimum operating conditions of the laser.

[28] The electron density grows from the preionization level of order 10^8 cm^{-3} to values of order 10^{15} cm^{-3} at the peak of the pulse.

[29] This type of discharge instability, known as the halogen donor depletion instability, is much more aggressive than the instabilities that ultimately limit pulse duration in transverse discharge CO_2 lasers (see Section 12.6), and is the reason why the speed of discharge circuits in RGH lasers is much more critical than in CO_2 lasers that in other respects they superficially resemble. See, for example, Coutts and Webb (1986).

13.8.4 Cavity design and beam properties of RHG lasers

The gain region in an excimer laser (ideally) takes the form of an extended block of uniform glow some 3–30 mm in height, 5–30 mm in width, and 0.2–1 m in length. It would be impossible to couple power into the low-order resonant modes of the sort of stable concave-mirror cavity employed for a

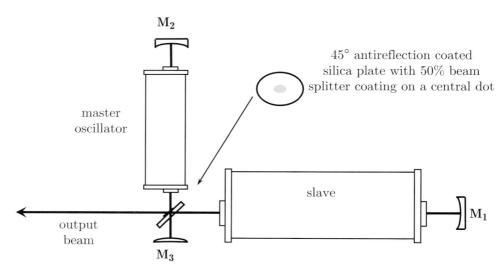

Fig. 13.12: Injection-seeded 2-J KrF laser with 3× diffraction-limited output beam (after Fletcher and Webb (2007)). The slave-laser discharge (indicated by the hatched area) was of 30 mm × 30 mm cross-section, and 860 mm in length. That of the small master oscillator was 7 mm width × 19 mm height × 220 mm length. The two confocal positive-branch cavities were coupled via a beam-splitter plate comprising an anti-reflection coated silica flat with a central 4 mm × 3 mm spot coated for 50% reflectivity. A cavity formed by high reflectance concave mirrors M_1 (focal length 1.50 m), M_2 (focal length 0.695 m) and convex mirror M_3 (focal length −60 mm) gave the lowest divergence output of 30 μrad when the peak of the pulse from the master oscillator was timed so as to occur 17 ns before the peak of the slave oscillator gain pulse.

He-Ne or argon-ion laser. Those stable cavities (see Section 6.3.2) have mode patterns of a slender 'dog-bone' shape[30] suited for extracting power from the longitudinal discharges in tubes of 0.2–1 m length and 1–3 mm diameter, but their mode overlap with the gain volume of an excimer laser would be very poor.

Fortunately, the very low loss of such stable-cavity designs is not needed for excimer lasers that typically have round-trip gain factors of 100 or more. Power can be extracted efficiently with a plane–plane mirror cavity comprising one high-reflectance plane mirror and an uncoated silica flat (reflectivity approximately 4%) as output coupler. However although the mode volume of the plane–plane cavity[31] overlaps the entire gain region, the focusability of the output beam is poor since its intrisic divergence is much higher than the diffraction-limited values associated with c.w. lasers operated in stable cavities. To understand why this is so we need to consider the fact that in the time during which the discharge has gain (typically 10–20 ns) a pulse of light can travel only 3–6 m, i.e. it can make at most 3–6 round trips of a cavity of 500 mm length. The divergence of the output beam for light that has made one round trip will be at best determined by the aperture provided by the discharge cross-sectional area as seen from a distance of two cavity lengths. After each successive round trip the divergence decreases since the effective length at which the output aperture is seen increases by a further two cavity lengths. So, with a plane–plane cavity, although the output pulse divergence decreases throughout the duration of the output pulse it never reaches values smaller than a few milliradians.

---

[30] In planes containing the cavity axis.

[31] In Fig. 6.9, the plane–plane resonator lies on the boundary between stable and unstable resonators at the point (1, 1).

A way out of this difficulty is provided by the use of confocal unstable resonators. For the present application, the arrangement illustrated in Fig. 6.13(a) is unsuitable since the power density reached at the common internal focus would be sufficient to cause air breakdown, with a disastrous effect on the optical quality of the cavity. The positive-branch design shown in Fig. 6.13(b) avoids this problem, but still suffers from the small number of cavity round trips that can be accomplished during the gain period. From the discussion in Section 6.3.3 it is clear that the divergence of the beam is reduced on every successive round trip of the cavity, and that only the very last part of the output pulse is of very low divergence.

A solution to this problem is to use the injection-seeded coupled-cavity arrangement shown in Fig. 13.12, in which two confocal unstable cavities are coupled via a small 50% beam splitter common to both. By adjusting the relative inductances of the two halves of the common high-voltage cicuitry it is possible to introduce a pre-determined delay between the excitation pulse applied to the large slave oscillator relative to that of the smaller master oscillator. This enables the low-divergence output of the small master oscillator (achieved towards the end of its output pulse) to be injected into the cavity of the slave oscillator so that its radiation field is built up from a low-divergence seed. With this arrangement beam divergences of a few times the diffraction limit can be obtained. The plasma formed when the beam from this laser is focused on a tungsten-wire target yields sufficient soft X-rays that, in single laser shot exposures, excellent contact images of live tissue samples could be obtained in the water-window (2–4 nm) region of the spectrum.[32]

[32] Fletcher and Webb (2007), Fletcher (1993) and Cotton and Webb (1993).

13.8.5 Performance and applications of RGH excimer laser

The fundamental parameters and performance characteristics of typical commercial discharge-excited RGH lasers are shown in Table 13.3.

The intrinsic efficiency (ratio of energy output to the energy deposited in the gas) of e-beam pumped KrF is relatively high, of order 8–10%. Very large single pulse energies can be obtained from e-beam pumped RGH lasers—the Titania KrF laser at the Rutherford Laboratory produced 100 J in a single pulse of 20 ps duration,[33] and devices capable of more than 5 kJ per pulse have been demonstrated at the Naval Research Laboratory in the USA.

Such large-scale e-beam excited RGH lasers, used in conjunction with optical pulse-compression techniques, have been proposed as potential drivers for laser-induced thermonuclear fusion schemes. The potential advantages of using KrF lasers as drivers for practical laser fusion reactors in which pellets containing a mixture of deuterium and tritium are compressed and heated by multiple laser beams so that a thermonuclear burn is initiated, are twofold. First, the wavelength of the laser is generated directly in the required UV range without the need for non-linear conversion from the infrared in two steps. Secondly, because the working medium of the laser is a gas, it can be circulated around a closed loop in which heat exchangers

[33] See M. H. Key, Edward Teller Medal Lecture: High Intensity Lasers and the Road to Ignition, June 1997. Reprints of this lecture are available at http://www.osti.gov/energycitations/servlets/purl/641284-BIFeOP/webviewable/.

Table 13.3 Fundamental parameters and typical operating characteristics of compact discharge-excited RGH lasers. (Data on operating characteristics of rare-gas fluoride and fluorine lasers from Görtler and Strowitski (2001).)

		XeCl	KrF	ArF	F_2
Wavelength	(nm)	308	249	193	157.6
Upper-level lifetime	(ns)	11	6.7	4.2	3.7
Gain cross-section	(10^{-16} cm^2)	4.5	2.2	2.4	6.8
Broadening type		homogeneous[a]	homogeneous	homogeneous	homogeneous
Typical pulse duration	(ns)	20	12	12	15[b]
Beam dimensions	(h [mm] × w [mm])	10 × 24	10 × 24	10 × 24	10 × 24
Single output pulse energy	(mJ)	500	800	450	50
Repetition rate	(pulses s^{-1})	100	100	100	50
Average output power	(W)	50	80	45	2.5
Beam divergence[c]	(mrad)	1 × 3	1 × 3	1 × 3	1 × 3

[a] Although the gain curve of XeCl shows evidence of distinct vibrational structure, the broadening is pre-dominantly homogeneous and the inversion can be accessed at all wavelengths across the gain band.
[b] Output pulses as long as 70 ns have been observed in specially designed discharge-pumped F_2 lasers; with electron beam pumping, output pulses as long as 160 ns have been observed.
[c] With plane–plane resonator.

cool the gas continuously. Heat removal from the glass discs of large solid-state amplifier modules is, in comparison, a much more difficult problem. To explore the possibility of making a repetitively pulsed KrF laser system the Electra laser shown in Fig. 13.13 has been constructed and is undergoing evaluation.[34]

[34] See Hegeler et al. (2008).

Discharge-excited RGH lasers have overall 'wall-plug' efficiencies of 1–2% in practical systems. Large numbers of small discharge-excited RGH excimer lasers are currently in use in chemistry laboratories throughout the world for fundamental research in photochemistry. The range of operating wavelengths (193–351 nm) available from a (relatively) small and inexpensive discharge-excited RGH laser (just by changing the gas mixture) covers a region of great interest for photochemists. It is even possible to operate the same kind of

Fig. 13.13: Large-scale e-beam pumped KrF Electra laser at the Naval Research Laboratory, Washington, DC, USA. The laser amplifier volume is pumped from opposite sides by 500 kV, 110 kA, 149 ns electron beams. The pre-amplifiers are seeded with beams from commercial (Lambda Physik) KrF discharge-excited lasers, and the overall system can produce output energies up to 700 J per shot. The main amplifier is provided with a gas-circulation system (the accordion-like structure visible at the top of the main amplifier) and can be operated at repetition rates of 5 Hz. Photo courtesy Dr. John Sethian, Program Manager, NRL, KrF Laser Fusion Program and Dr F Hegeler, Contractor.

device on a transition of F_2 to obtain output pulse energies of order 200 mJ in the vacuum ultraviolet at 157 nm.[35]

Excimer lasers are widely used to provide the pump radiation for pulsed dye laser systems especially those operating in the near-ultraviolet region of the spectrum. A single excimer laser unit can be used for photochemistry experiments to provide radiation for photoexcitation or dissociation and, with suitable time delay, the tunable radiation to probe the products.

The RGH lasers have found many industrial applications involving photochemical processing of materials, and in particular the fabrication of integrated circuits by lithographic techniques. For many years the density of components that could be written was limited by diffraction and consequently there was a steady progression towards employing shorter-wavelength sources. The ArF laser, operating at 193 nm, is the shortest-wavelength laser that can (just) be operated without the additional inconvenience of working in the vacuum or extreme ultraviolet spectral regions, and is now routinely used in lithography.[36] Plans to use the F_2 laser (operating at 157 nm) as the radiation source for production of future generations of chips with feature sizes significantly smaller than 200 nm have recently been reported to have been put on hold since the introduction of new techniques which allow the production of feature sizes considerably smaller than the source wavelength.[37] The convenience of using ArF lasers in air, rather than shorter wavelength lasers in vacuum, is sufficiently attractive to justify the extra complication of the optical system.

The fact that the short-wavelength radiation from KrF and ArF lasers are very strongly absorbed by organic materials, such as plastics or living tissue, coupled with the high peak-power capability of lasers of even modest size, means that these lasers can be used for micromachining applications exploiting the direct photoablation of material from the workpiece. Intricate patterns of components on the scale of 10 μm or even smaller can be made by projecting a reduced image of an appropriate mask on to the plastic surface, relying on the explosive action of the laser radiation to ablate material directly, and to dissociate the heavier molecular fragments in the ablation plume into easily evaporated lighter species.[38] The same direct photoablative action of ArF laser radiation is also exploited in the various schemes that are available for the correction of refractive problems of human eyesight. Sections of the cornea (the layer of tissue at the front of the eyeball) are reshaped by projecting high-intensity ArF laser radiation via masks whose shape and/or transmission is carefully engineered so as to remove tissue in a controlled way and thus improve the focusing ability of the lens system of the eye.[39]

[35] Strictly speaking, the F_2 laser is not an excimer laser since the ground state of the molecule is stable. It is more like the N_2 laser in that it operates between on a bound–bound transition between excited states of a molecule with a stable ground state. However, the binding energy of the lower level of the F_2 laser is only 0.15 eV; under the operating conditions of the laser the effective lifetime of this level is only 1 ns, which is sufficiently short to prevent build-up of population in the lower level. Since it can be made to operate in the same discharge-excited devices as the RGH lasers by simply changing the gas mixture, it is often included in the category of excimer lasers.

[36] Okazaki (2004).

[37] The new techniques involve the use of high ($n \approx 1.7$) refractive index matching fluids between the fluid immersion objective lens and photoresist, the development of special resist materials, and the use of computer-generated mask features tailored to produce diffraction patterns at the required size on the chip. Currently feature sizes of 65 nm can be produced with 193 nm laser sources. See Jones-Bey (2006) and Owa et al. (2004).

[38] Cennamo and Forte (2004).

[39] Gower (2004).

Further reading

Modern accounts of molecular structure and spectroscopy may be found in Bransden and Joachain (2003) and Haken and Wolf (2004). An encyclopedic treatment of molecular spectroscopy, which goes beyond the requirements of the present text, is to be found in Herzberg (1989).

Exercises

(13.1) **The correspondence principle**

(a) A classical harmonic oscillator comprises a mass point on a spring or a pendulum. Show that classically the most likely position to find the mass point is at the extreme limits of its motion.

(b) Compare this result to the predictions of quantum mechanics for the ground state of a harmonic oscillator and for a state of high vibrational quantum number as exemplified by the diagram for $v = 10$ in Fig. 13.1.

(c) By considering the motion of a pendulum comprising a 10 g mass suspended by an inextensible string of 1 m length, find the energy of the quantum state with $v = 1$.

(d) Find the amplitude of the motion described by the quantum state with $v = 1\,000\,000$ and compare the energy of this state to the kinetic energy of a monatomic molecule at room temperature.

(13.2) **Franck–Condon loop laser**

In this problem we consider a laser operating on the Franck–Condon loop illustrated in Fig. 13.3

(a) The interatomic potential of a diatomic molecule can be described by a Morse potential of the form

$$U(R) = D\left[1 - e^{-\beta(R-R_0)}\right]^2.$$

Explain the physical significance of the constants D and R_0. Find an expression for the vibrational frequency of this molecule for low vibrational levels.

(b) In the NO laser an intense pulse of radiation at 157.6 nm from an F_2 laser pumps NO molecules from the ground state to a vibrational level of an excited electronic state. Laser output is observed at wavelengths of 163.1, 211.0, 218.2, 226.1 and 234.5 nm. Sketch the potential curves for the ground and excited electronic states and explain why population inversion occurs between one particular vibrational level of the excited state and certain vibrational levels of the electronic ground state.

(c) Assuming that any change in the rotational state of the molecule is negligible for all absorption and emission transitions, and given that the 163.1-nm transition terminates on the $v'' = 1$ level, while that at 234.5 nm terminates on $v'' = 12$, estimate the spacing of the vibrational levels of the ground state and make reasoned suggestions for identifications for other lower levels of the laser transitions.

(d) What is the thermal population of the $v'' = 1$ level at room temperature?

(13.3) **Excimer lasers: threshold specific power input**

(a) Use eqn (13.13) to show that, if the lineshape is considered to be Lorentzian, the peak gain cross-section of an excimer laser is given by

$$\sigma_{21}^H = \frac{\lambda^4 A_{21}}{4\pi^2 c \Delta\lambda},$$

where $\Delta\lambda$ is the width of the excimer band.[40]

(b) In the case of the KrF laser the minimum energy required to create one ion pair is of order 15 eV, $\tau_2 = A_{21}^{-1} = 6.7$ ns, and $\Delta\lambda \approx 2$ nm. Estimate the minimum input power required to achieve threshold if the internal loss is 0.5% cm^{-1}.

(c) Repeat the calculation for the Xe$_2$ excimer system with the same internal loss. Take $\tau_2 = A_{21}^{-1} = 10$ ns and $\Delta\lambda \approx 15$ nm.

(13.4) **Excimer lasers: beam divergence**

(a) The cavity of a KrF laser comprises two planar mirrors of diameter 3 cm, spaced by 1 m. The gain medium may be taken to fill the entire length of the cavity, and to provide uniform gain over a diameter of 2 cm but to be zero outside this region. Assuming that the duration of the gain is of order 20 ns, estimate the divergence that the output laser beam achieves towards the end of the laser pulse.

(b) Figure 13.14 shows a positive-branch, unstable, confocal resonator for a KrF laser. It comprises a high-reflectance mirror M_2 of 200 cm radius of curvature separated by 95 cm from M_1, a convex

---

[40] Some authors prefer to model the shape of the gain band as a Gaussian, although it should be stressed that the broadening in an excimer laser is dominated by homogeneous broadening and that the Doppler broadening may be neglected. If a Gaussian lineshape is assumed an extra factor of $\sqrt{\pi \ln 2} \approx 1.44$ appears on the RHS of eqn (13.13).

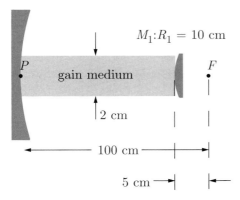

Fig. 13.14: The positive-branch unstable resonator considered in Exercise 13.4.

high-reflectance mirror of small diameter and 10 cm radius of curvature so that the mirrors share a common focal point F. A point P on the axis of the gain medium 95 cm from the surface of the convex mirror (i.e. at the far end of the gain medium away from the convex mirror) may be considered to be the kernel of spontaneous emission for initiating laser oscillation, so that the final laser output is dominated by the initial emission from this source. Use geometric optics to calculate the position of I_1, the image of the kernel in M_1, and that of I_2, the image of I_1 in M_2. Iterate this procedure to find the position of I_3, the image of I_2 in M_1 and finally the position I_4 of the image of I_3 in M_2. Hence, estimate the divergence of the annular output beam after two round trips of the cavity given that the diameter of the gain region is 2 cm.

14

Dye lasers

14.1 Introduction

The earliest lasers, such as the ruby laser, the helium-neon laser, and the Nd:YAG laser all operated on transitions between well-defined energy levels of the gain medium and therefore produced output at one or more fixed wavelengths. For many applications this was entirely acceptable—indeed for some applications (e.g. interferometry and precision metrology) it was a distinct benefit. However, many applications in atomic and molecular spectroscopy require a laser whose output can be to tuned into resonance with an absorption feature of the atomic or molecular species in question, and it was not until the 1966 demonstration by Sorokin and Lankard[1] of broad-bandwidth stimulated emission from an organic compound in solution that one path to development of truly tunable lasers became clear. Dye lasers, in which the active medium comprises a solution of a particular type of organic molecule in a solvent (such as ethanol, methanol or glycol) are the first class of tunable laser in which the operating wavelength can be adjusted over wide ranges using tuning elements such as prisms, gratings or birefringent filters within its cavity. For detailed descriptions of the theory of dye laser operation and construction the reader is referred to the texts by Schäfer (1990) or Titterton (2004a, 2004b).

The 1990s saw rapid development of other tunable sources based on semiconductor diode lasers, Ti:sapphire, and optical parametric oscillators. To a large extent these have taken over many of the applications that were previously the sole province of dye lasers. However, because of the extremely large range of wavelengths that a single dye laser can produce (with the appropriate changes of laser dye material and solvent) dye lasers seem likely to remain an important tool for spectroscopists and photochemists for many years to come.

Dye lasers do not create their radiative output by direct conversion of electrical energy, rather they absorb radiation from a fixed-wavelength 'pump' laser or a high-intensity flashlamp and generate tunable output at a longer wavelength whose precise value is determined by the cavity.

14.2 Dye molecules

The feature that characterizes all organic dye molecules is that their backbone is composed of a chain of carbon atoms in which C–C single bonds alternate with double bonds—such double bonds are said to be 'conjugated'.

[1] See Sorokin and Lankard (1966).

14.1	Introduction	380
14.2	Dye molecules	380
14.3	Energy levels and spectra of dye molecules in solution	382
14.4	Rate-equation models of dye laser kinetics	387
14.5	Pulsed dye lasers	388
14.6	Continuous-wave dye lasers	391
14.7	Solid-state dye lasers	395
14.8	Applications of dye lasers	396
Further reading		398
Exercises		398

(a)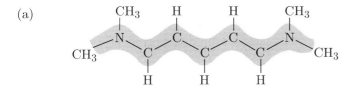

(b)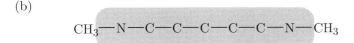

(c)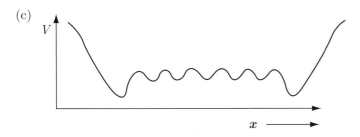

(d)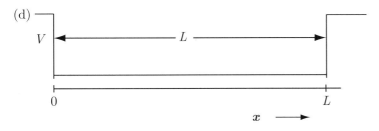

Fig. 14.1: The shaded area represents the spatial distribution of π-electrons in a cyanine dye molecule: (a) seen from above the plane containing the C–H bonds; (b) from the side. (c) the potential energy V experienced by one of the delocalized π-electrons as a function of distance along the molecular backbone. (d) a square-well approximation to the potential shown in (c). (After Schäfer (1990).)

As indicated in Fig. 14.1, the π-orbitals of the C atoms overlap to form a region above and below the plane of C–H bonds in which the electrons are delocalized. To a simple approximation, this region forms a deep square potential well with a length L equal to the length of the carbon backbone. The ground electronic configuration of all such molecules is a spin singlet in which the total electronic spin is zero.

The pairs of electrons of opposite spins occupy the available energy states of the square well from the lowest state E_1 to the highest occupied state E_n, where $n = N/2$ and N is the number of π-electrons. For a deep square well (see Fig. 14.2) the energy of the nth state is given by

$$E_n = \left(\frac{\pi^2 \hbar^2}{2m_e L^2}\right) n^2. \tag{14.1}$$

The energy gap $\Delta E_{n+1,n}$ between adjacent states is

$$\Delta E_{n+1,n} = (2n+1)\frac{h^2}{8m_e L^2}, \tag{14.2}$$

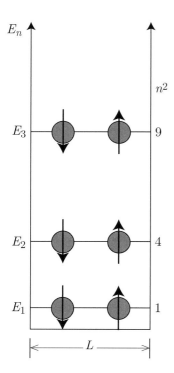

Fig. 14.2: Energy levels of the deep square-well potential, which approximate those of the delocalized π-electrons of the C atoms of a dye molecule.

[2] The symbols S_0, S_1, S_2, etc. simply denote the various singlet electronic levels (i.e. levels having zero electronic spin) numbered in sequence from the ground state upwards. Similarly, T_1, T_2, T_3, etc. denote the triplet electronic levels (i.e. levels having a total spin of 1) numbered from the lowest triplet upwards. These symbols are nothing to do with the conventional spectroscopic notation S, P, D for atoms.

[3] The Pauli principle forbids the existence of T_0.

[4] For example, in He, $1s2s\ ^3S_1$ lies lower than $1s2s\ ^1S_0$.

corresponding to an electronic transition of wavelength λ_0 where

$$\lambda_0 = \frac{8m_e c}{h} \frac{L^2}{2n+1}. \quad (14.3)$$

This simple model provides some indication of the wavelength region λ_0 in which the principal electronic absorption feature of the dye molecule will lie, and the scaling of λ_0 with the size of the molecule. Since the C–C bond has a certain characteristic length (0.140 nm in the case of the cyanines) the value of L must be proportional to N. So

$$\lambda_0 \propto \frac{N^2}{N+1} \approx N, \quad (14.4)$$

where the approximation holds for large N. Thus, choosing a dye molecule with a long backbone ensures that its principle absorption maximum is shifted to long wavelengths and vice versa. By choosing which type of dye molecule to use in the solution forming the active medium of the laser, it is therefore possible to select the approximate spectral region in which the dye laser will operate.

Dyes are available covering the region from the near-UV at 300 nm to the near-infra-red at 1 μm. Each dye species typically has a useful operating range of 20–100 nm.

14.3 Energy levels and spectra of dye molecules in solution

14.3.1 Energy-level scheme

All dye molecules have an electronic ground state comprising a pair of electrons of oppositely directed spins, i.e. they are always singlet levels. The ground state is conventionally denoted[2] by S_0. The excited electronic states can either be singlets S_1, S_2 ..., etc. or triplets T_1, T_2 ..., etc.[3] Just as in the atomic case,[4] the lowest triplet state T_1 always lies below the corresponding singlet S_1, and radiative electronic transitions in which $\Delta S \neq 0$ tend to be relatively forbidden, while those for which $\Delta S = 0$ are allowed unless forbidden for some other reason. As for diatomic molecules, the electronic states are typically separated by a few eV and have vibrational and rotational structures on scales of 0.1 to 0.3 eV and 0.001 to 0.01 eV, respectively. Remembering that 300 K is equivalent to 2.6×10^{-2} eV, it is evident that at room temperature nearly all of the molecules will be in the lowest vibrational state ($v'' = 0$) of the electronic ground state.

14.3.2 Singlet–singlet absorption

The shape of the absorption spectrum of dye molecules is governed the Franck–Condon principle (see Section 13.3), which determines the relative probabilities of exciting the molecule to the various vibrational states of the upper electronic state.

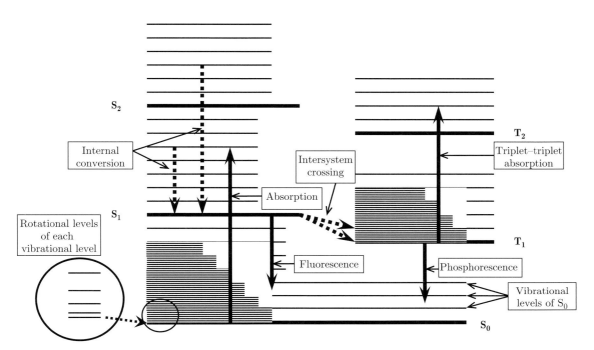

Fig. 14.3: Schematic diagram of the energy levels of a typical dye molecule. For clarity, only a few of the closely spaced rotational levels of the lowest vibrational levels are indicated; rotational levels belonging to higher vibrational levels are omitted altogether.

In the case of a diatomic molecule the electronic energy curves are functions of a single variable—the internuclear separation R. However, for a complex polyatomic molecule the electronic energy of a given electronic configuration will be a function of the separation of all the nuclei involved, and the electronic energy 'curve' must now be a surface plotted in a multidimensional manifold. Clearly it is impossible to render this information in a 3-dimensional diagram, let alone a 2-dimensional curve. The best we can do is to make a 2-dimensional cut through the surface corresponding to the variation of electronic energy with one particular representative configuration coordinate[5]—the length of one of the C–H bonds, for example. Such a representation is shown in Fig. 14.4 and if, as shown in Fig. 14.4(a), the energy minimum of S_1 happens to occur at the same value of configuration coordinate as that of S_0, then the Franck–Condon principle dictates that the $(v'' = 0) \rightarrow (v' = 0)$ transition at λ_{00} is strongest, with $(v'' = 0) \rightarrow (v' = 1)$ at λ_{01} less strong and so on. If, on the other hand, the minimum in the energy curve for S_1 lies at a greater value of R than that for S_0, then the strongest absorption will occur for a transition to a higher vibrational level of S_1; in the particular case illustrated in Fig. 14.4(b) this transition is λ_{03}, corresponding to $(v'' = 0) \rightarrow (v' = 3)$.

If the dye solution were cooled well below room temperature then the individual vibrational transitions would be just discernible from one another, but the rotational structure would be entirely smeared out. The rotational structure is completely lost because individual rotational states are so strongly broadened that they merge into one another completely. The two effects giving rise to this broadening are:

[5] A configuration coordinate representation was also used on page 172 to describe the energy levels of ions doped in a crystal.

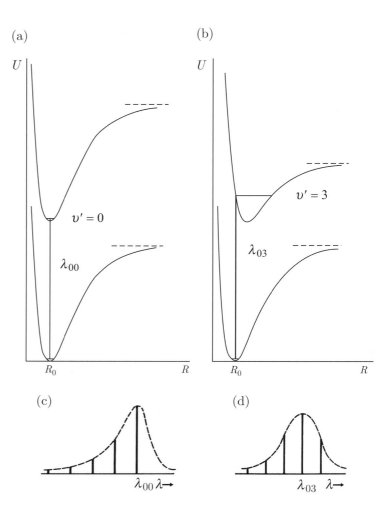

Fig. 14.4: The influence of the form of the electronic potential curves in determining the wavelength of the strongest feature in singlet–singlet absorption. In (a) the minimum in the potential curves for both S_0 and S_1 lie at the same value of R and consequently the strongest absorption transition is at λ_{00} corresponding to the transition $(v'' = 0) \rightarrow (v' = 0)$. In contrast, in (b) the minimum in the potential curve for S_1 lies at a greater value of R than that of S_0 and hence in this case the strongest absorption occurs for λ_{03} corresponding to the transition $(v'' = 0) \rightarrow (v' = 3)$. The corresponding absorption spectra are indicated in (c) and (d), respectively. At very low temperatures, where collisional transfer between the neighbouring vibrational levels is slow, the individual vibrational transitions (indicated by the vertical bars) can be resolved to some extent. However, at room temperature the transfer between vibrational levels is so rapid that the vibrational transitions merge into the featureless quasi-continua indicated by the dashed line defining the envelope.

1. **Collisions with solvent molecules:** In a liquid the molecules are in close contact with one another and 'collisions' occur on time scales of picoseconds or less.
2. **Internal conversion:** Because the energy surfaces corresponding to the various rotational states of different vibrational states of the electronic states intersect one another in the multidimensional configuration coordinate space, the point representing the instantaneous electronic energy of the molecule can 'slide' down the surfaces from one to another until it reaches the lowest energy (the lowest vibrational state of a given electronic state) from which it can go no further without making an electronic transition. This process is called internal conversion and can occur on time scales of a few tens of femtoseconds. From the uncertainty principle, the energy broadening associated with τ (the lifetime against such collisions) is of order $\hbar/\tau$. In this case, the energy broadening is of order 0.1 eV, which is larger than the relative separation of the rotational

states themselves, and nearly as large as the separation between adjacent vibrational states.

At room temperature, the combined effects of these two broadening mechanisms usually leads to a complete smoothing of the envelope of the absorption feature with hardly any vibrational structure still discernible.

Although these rapid transfer mechanisms lead to thermalization of the populations at the temperature of the solvent in subpicosecond time scales, this is not true of the transfer across the large energy gap between S_1 and S_0. Here, the collisional mechanism is ineffective in converting such a large amount of electronic excitation energy into kinetic energy of the solvent molecules (and/or high vibrational energy of the dye molecule) in a single collision and the internal conversion from S_1 to S_0 does not occur because the relevant energy surfaces do not intersect.

14.3.3 Singlet–singlet emission spectra

As indicated in Fig. 14.5, dye molecules in solution can absorb photons from a broadband source to arrive initially in various excited vibrational states of S_1 (or even those of S_2 if the light contains sufficiently short-wavelength components). Whatever their initial vibrational excitation, they promptly relax

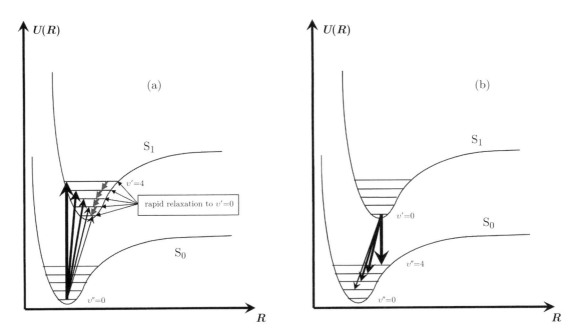

Fig. 14.5: (a) The relative strengths of the absorption transitions from the ground state of a typical dye molecule are represented by the width of the arrow connecting the lowest vibrational level ($v'' = 0$) of S_0 to the various vibrational levels of S_1. Rapid internal conversion and collisional processes lead to thermalization of the population among the vibrational levels of S_1, with the result that almost the entire population is quickly transferred to the lowest vibrational state ($v' = 0$) of S_1. (b) From this vibrational level the population decays radiatively to the various vibrational levels of S_0. The relative strengths of the emission features are indicated by the width of the arrows, with those closest to vertical corresponding to the strongest transitions.

to the lowest vibrational state of S_1 by internal conversion and collisional transfer—even those originally excited to S_2 will be internally converted to S_1 and on downwards through its vibrational manifold. Emission on $S_1 \rightarrow S_0$ transitions is known as fluorescence (see Fig. 14.3) and occurs with radiative lifetimes typical of fully allowed visible emission of order 1–10 ns. The shape of the quasi-continuum emission spectrum corresponding to the $S_1 \rightarrow S_0$ transition is a 'mirror image' version of the absorption spectrum shifted to longer wavelength, since the Franck–Condon principle dictates that the vibrational states of S_0 that are reached from the lowest vibrational level ($v' = 0$) of S_1 will have similar relative strengths[6] to those of the various states of which are reached in absorption from the ground vibrational state ($v'' = 0$) of S_0.[7]

Figure 14.6 illustrates the 'mirror image' relationship between the shapes of the $S_0 \rightarrow S_1$ absorption cross-section $\sigma_a(\lambda)$, and the corresponding emission (gain) $S_1 \rightarrow S_0$ cross-section $\sigma_e(\lambda)$. It is evident from the 100× magnified version of the $\sigma_a(\lambda)$ curve that there is a long-wavelength limit beyond which any absorption in the $S_0 \rightarrow S_1$ transition is negligible. Moreover, there is a region beyond this limiting wavelength towards longer wavelengths throughout which the gain cross-section $\sigma_e(\lambda)$ is appreciable, while the absorption cross-

[6] If the potential wells of the upper and lower electronic levels have the same shape, i.e. they are merely displaced, then the vibrational wave functions in the two electronic levels will be the same. It then follows that the Franck–Condon factor for the $v' \leftarrow v''$ transition is equal to that for the $v' \rightarrow v''$ transition.

[7] The form of the absorption and emission spectra is similar to that of the vibronic transitions observed in some crystals (see page 172), reflecting the fact that similar physical processes are responsible.

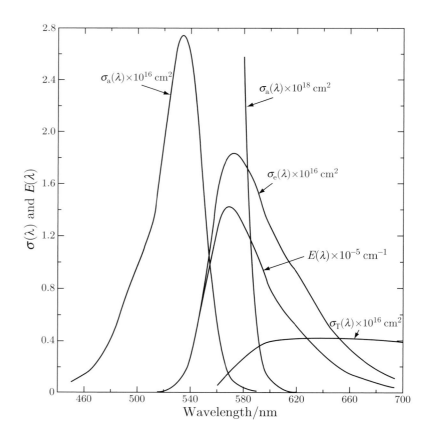

Fig. 14.6: Spectrophotometric data for a $10^{-4}\%$ molar solution of the dye rhodamine 6G in water showing the variation with wavelength λ of the singlet absorption cross-section $\sigma_a(\lambda)$, singlet emission cross-section $\sigma_e(\lambda)$, triplet absorption cross-section $\sigma_T(\lambda)$, and fluorescent emission intensity $E(\lambda)$ normalized so that $\int E(\lambda) \, d\lambda = 0.92$, the measured quantum yield for fluorescence. (After Drexhage (1990).)

section $\sigma_a(\lambda)$ is essentially zero. The possibility therefore exists of obtaining laser action in this wavelength range.

14.3.4 Triplet–triplet absorption

If there were no other competing effects, then obtaining laser action on excited dye solutions would be relatively simple. Unfortunately, there are competing absorption processes that complicate matters considerably. Molecules initially excited to S_1 can be transferred to T_1 (the lowest triplet state, see Fig. 14.3) by the process of internal conversion. Although the process of conversion to T_1 occurs at a very much slower rate than that of internal conversion within the singlet state itself, with typical values of the rate constant for singlet to triplet conversion k_{ST} of $\approx 3 \times 10^6$ s^{-1} of molecules in S_1 occurs on a time scale of order 300 ns. This is to be compared to the 3 ns lifetime of molecules in S_1 against making a radiative transition to one of the vibrational levels of S_0. Unless something is done to prevent it, population in T_1 builds up[8] and can act as a source of absorption of photons of the fluorescence band $S_1 \rightarrow S_0$ on which the laser action may occur. Although $\sigma_T(\lambda)$ the cross-section for absorption on the fully allowed $T_1 \rightarrow T_2$ band is smaller than the gain cross-section $\sigma_e(\lambda)$, as indicated in Fig. 14.6, it covers the entire region for which $\sigma_e(\lambda)$ is large and $\sigma_a(\lambda)$ is effectively zero. If nothing is done about it, the population of T_1 can rise to values large enough that absorption on the $T_1 \rightarrow T_2$ band can overcome gain on the $S_1 \rightarrow S_0$ band and prevent laser action occurring.

[8] The build-up of population in T_1 occurs because emission on the triplet–singlet $T_1 \rightarrow S_0$ transition is relatively forbidden by the $\Delta S = 0$ selection rule. Such emission corresponds to the slow decay mode of excited dye molecules known as the phosphorescent emission (see Fig. 14.3). Lifetimes against phosphorescent decay can be extremely long—ranging from several microseconds to milliseconds. For the same reason, absorption on the $S_0 \rightarrow T_1$ is negligible compared to that on the $S_0 \rightarrow S_1$ fully allowed transition.

14.4 Rate-equation models of dye laser kinetics

The first and most important simplification we can make in setting up rate-equation models for dye laser kinetics is that within each electronic level the rates of internal conversion and collisional coupling are so fast that no

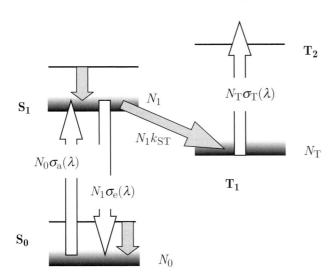

Fig. 14.7: Simplified model of dye laser population kinetics. Rates of transfer induced by absorption or emission of radiation are indicated by open arrows. Rates of transfer by fast internal conversion process are indicated by grey shaded arrows. The distribution over states is indicated schematically by the shaded continuum, the density of shading representing the relative density of population.

[9]Because of the fast internal conversion and collisional transfer processes, we can consider the populations of all rotational and vibrational states in each electronic level as completely coupled and populated relative to the lowest vibrational level according to a Boltzmann distribution at the solvent temperature.

other rate, not even stimulated emission, can compete with them. As indicated in Fig. 14.7 this means that even in the presence of lasing we can treat the entire population of each electronic state as being distributed over the rotational levels of the lowest[9] vibrational state with the thermal distribution appropriate to the solvent temperature. This distribution over rotational states constitutes a homogeneously broadened frequency distribution given by the spectral dependence of the corresponding cross-section. Thus, we can simply write the net gain at a wavelength λ as

$$\alpha(\lambda) = N_1 \sigma_e(\lambda) - N_0 \sigma_a(\lambda) - N_T \sigma_T(\lambda), \quad (14.5)$$

in which N_0 is the population of the ground state S_0, N_1 is the population of S_1 and N_T is the population of the triplet T_1 state.

In the case illustrated in Fig. 14.6 at a wavelength of 580 nm all three terms on the right-hand side eqn 14.5 have non-zero values, but at wavelengths longer than 620 nm the term involving $\sigma_a(\lambda)$ is essentially zero.

14.5 Pulsed dye lasers

14.5.1 Flashlamp-pumped systems

The construction of the earliest dye lasers was similar to that of ruby lasers, with a tube containing the dye solution taking the place of the ruby rod and excitation supplied by a linear flash tube running the length of the opposing focal line in an elliptical cylinder reflector housing.[10] There is no way to quench the population of T_1 in time scales of interest within a single pulse, so the build up of population in T_1 limits the maximum pulse duration that can obtained from such a laser even if the pump radiation could be turned on to a constant high power with arbitrarily fast rise time.

[10]See Fig. 7.11.

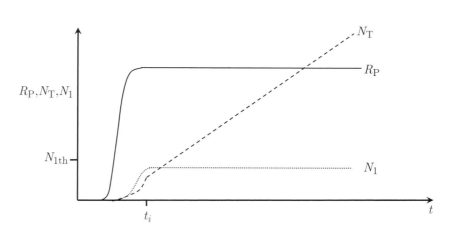

Fig. 14.8: Population kinetics of a pulsed dye laser. Here, R_P represents the rate of pumping of the S_1 population via absorption of flashlamp photons on the $S_0 \rightarrow S_1$ transition. In this highly simplified model, the pump radiation intensity is assumed to rise in a very short time, and to stay constant thereafter. By time t_i the population of S_1 has risen to $N_{1\text{th}}$, the threshold for laser oscillation, and is then clamped at the value for all subsequent time. The population N_T of the triplet state T_1 rises linearly with time after t_i.

As indicated in Fig. 14.8, once the laser starts to oscillate the value of the upper level population N_1 is clamped at the threshold value $N_{1\text{th}}$ for laser oscillation.[11] During laser oscillation, the rate at which the triplet state is fed depends on k_{ST}, the rate constant for singlet to triplet conversion, and the value of $N_{1\text{th}}$:

$$\frac{dN_T}{dt} = N_{1\text{th}}k_{\text{ST}} - \frac{N_T}{\tau_T} \quad (14.6)$$

$$\simeq N_{1\text{th}}k_{\text{ST}}, \quad (14.7)$$

where the approximation follows because the time scale for triplet decay τ_T is very long compared to other time scales within a single pulse, and hence we can neglect[12] the spontaneous decay of T_1 molecules. Integrating eqn (14.7) we reach

$$N_T(t) \approx N_{1\text{th}}k_{\text{ST}} \cdot (t - t_i) + N_T(t_i). \quad (14.8)$$

where t_i is the time at which N_1 reaches the lasing threshold value $N_{1\text{th}}$. Now, from eqn (14.5), lasing at wavelength λ_L will extinguish at time t_f such that[13]

$$N_T(t_f)\sigma_T(\lambda_L) = [N_{1\text{th}}k_{\text{ST}} \cdot (t_f - t_i) + N_T(t_i)]\sigma_T(\lambda_L) \geq N_{1\text{th}}\sigma_e(\lambda_L). \quad (14.9)$$

Lasing will therefore extinguish when

$$N_{1\text{th}}k_{\text{ST}} \cdot (t_f - t_i) + N_T(t_i) \approx QN_{1\text{th}}, \quad (14.10)$$

where the ratio $\sigma_e(\lambda_L)/\sigma_T(\lambda_L) = Q$ is characteristic of the dye. Since, in practical cases, the value of $N_T(t_i)$ is quite negligible compared to $N_{1\text{th}}k_{\text{ST}} \cdot (t_f - t_i)$, within the spirit of our approximation the laser pulse duration $(t_f - t_i)$ is given from eqn (14.10) by

$$(t_f - t_i) \lesssim \frac{Q}{k_{\text{ST}}}. \quad (14.11)$$

In the particular case of the dye Rhodamine 6G, $Q \approx 10$ (see Fig. 14.6) and $k_{\text{ST}} \approx 1.6 \times 10^7 \text{ s}^{-1}$ so from eqn (14.11) we expect a pulse duration of this typical dye laser to be of order 600 ns. This implies that if the flashlamp takes more than a few µs to reach full power, most of its energy will arrive too late to be useful for laser pumping and will only heat up the dye solution. Such short risetimes were a challenging specification for the high-voltage circuits, switches and flashlamp designs in the early days of dye lasers and spurred the search for alternative pump sources. However, flashlamp-pumped dye lasers are still in use today—typical commercial devices can produce pulses up to 10 J and single-pulse devices producing over 100 J per pulse have been constructed.

[11] See Section 6.6.1.

[12] In this simplified model we also neglect any depletion of population in T_1 due to absorption of laser photons on the $T_1 \rightarrow T_2$ transition since any molecules thus promoted to T_2 can also absorb at the laser wavelength. The symbol N_T represents the sum of population in all triplet levels.

[13] In this simplified model of the population kinetics of a flashlamp-pumped dye laser, we assume that there is no means of removing molecules in the T_1 level on the relevant time scales. It is the build up of population in T_1 and the consequent absorption of laser radiation on the $T_1 \rightarrow T_2$ transition that brings laser oscillation to an end. It is assumed that the switch from net gain to net loss on the laser transition occurs so rapidly that the difference between the time at which this occurs and the time when net gain falls below the cavity round-trip loss is negligible. In order to maintain c.w. laser action it would be necessary to depopulate T_1. This can be achieved by providing a chemical additive that collisionally quenches T_1 molecules, or by rapidly flowing the dye through the active medium to physically remove T_1 molecules. Lasers that employ such techniques will be described in Section 14.6.

14.5.2 Dye lasers pumped by pulsed lasers

As an alternative to flashlamp excitation, the radiation provided by a visible or near-UV pulsed laser can be used to pump dye lasers. Several types of laser that can deliver pulses with a peak power of several 100 kW, or even a few megawatts, in a few nanoseconds and therefore can easily satisfy the risetime constraints outlined above. Specifically, the nitrogen laser (operating at 337 nm), the CVL (copper vapour laser with outputs at 511 and 578 nm), frequency-doubled Nd:YAG (532 nm) and the excimer lasers such as XeCl (308 nm) and KrF (249 nm) have all been successfully employed for this application. Using one laser to pump another one may not at first seem very sensible, but it must be borne in mind that the pump lasers are single-wavelength (or are at most microtunable) devices, whereas dye lasers can be made to oscillate at a wavelength determined by the user. Clearly, there is an optimum pump wavelength for each type of dye molecule since the pump radiation wavelength must lie in a region of reasonably good absorption by the dye, not too far away from the fluorescence region, although its exact wavelength is not very critical.

Because the peak intensity of the pump beam provided by any of these pulsed laser sources is extremely high, the gain of the dye medium in the dye laser is typically well over a factor of 10 per pass, so that the inclusion inside the dye laser cavity of dispersive optics having quite appreciable losses can be tolerated without unacceptably impairing the efficiency of the dye laser.

The layout of a dye laser suitable for pumping by a pulsed laser source is shown in Fig. 14.9. The wavelength-tuning element is a reflection grating. In order to achieve good wavelength narrowing it is necessary to expand the beam from the value of 1 mm, or less, it typically has in the optically pumped region of the dye cell, since the resolving power of a diffraction grating is proportional to the number of lines illuminated. In early designs a beam-expanding telescope was included inside the dye laser cavity between the dye cell and the grating, but nowadays it is more usual to employ one or more asymmetric prisms for this purpose. Such prisms expand the beam in one dimension because a parallel beam incident on the plane boundary between

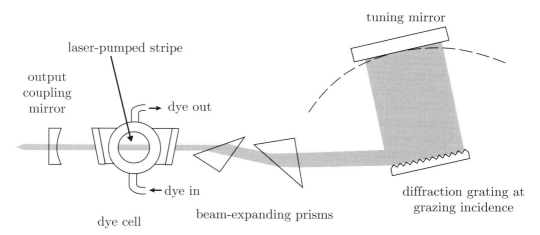

Fig. 14.9: Schematic diagram of a dye laser pumped by pulsed laser source. The layout of all components except the dye cell is shown in plan view. For clarity the dye cell is shown in side view. The flowing dye is excited by focusing the beam of a pulsed laser (usually of a horizontal 'letterbox' cross-section) to a narrow stripe within the dye medium by means of a cylindrical lens.

air and glass at a non-zero angle of incidence experiences a change in width as it is slewed by refraction.

In the laser design of Fig. 14.9 the diffraction grating is used at near-grazing incidence, and one of the diffracted beams (usually the first order) is reflected back on its own path by the high-reflectance mirror. Tuning is accomplished by moving the tuning mirror so that its surface is always tangent to a circle centred on an axis parallel to the grating rulings and passing through the centre of the grating surface.[14] To reduce the lasing linewidth below a few GHz it may be necessary to include a tilted etalon within the cavity.

Dye lasers pumped by high repetition rate pulsed lasers (up to 10 kHz in the case of CVLs) require the dye solution to be flowed through the dye cell at a sufficient rate for the solution in the pumped stripe to be replaced in the interval between successive pump pulses. This may require careful attention to the design of the dye solution circulation system. No special measures need be taken to quench the triplet molecules since the build-up of T_1 population in the short duration of the pump laser pulse is not sufficient to cause problems.

14.6 Continuous-wave dye lasers

14.6.1 Population kinetics

The problem of triplet population dominates the design of laser systems that operate in a truly c.w. fashion. Since the pump power is provided by a c.w. laser system, such as the argon-ion laser or the frequency-doubled Nd:YAG or Nd:YLF laser, the maximum available pump power is only of order 10 W–far less than the 100 kW to 1 MW peak powers available from pulsed laser systems discussed in Section 14.5. As a consequence the gain of a c.w. dye laser is very low indeed—at most a few per cent per pass. The cavity must therefore be designed to keep optical losses to a minimum. For this reason gratings cannot be used as a tuning element and combinations of a birefringent filter[15] with one or more tilted etalons must be used instead. Again to minimize losses, the gain region of such lasers is in the form of a parallel-sided, thin stream of dye solution forced through a slit-like nozzle as a laminar-flow jet by a high-pressure circulating pump.

Because the excitation is provided by a c.w. laser of high beam quality (TEM_{00} mode), it can be focused into the dye jet at a common waist shared by both the pump laser beam and the dye laser cavity. Since the diameter of such a waist is typically 15 μm and dye flow velocity in the jet is of order $10\,\mathrm{m\,s^{-1}}$, the residence time of a dye molecule in the pumped region is only 1.5 μs, so that triplet molecules are flushed away in time scales of this order. From the calculation of Section 14.5.1, it is clear that the inequality eqn (14.11) is satisfied and the triplet population can never build up to values where it would represent a serious problem for c.w. operation.[16]

Ignoring absorption from the excited singlet level S_1 and the triplet level T_1, the dynamics of the laser can be represented by the simple 4-level scheme shown in Fig. 14.10, in which relaxation within S_1 and S_0 is assumed to occur on an extremely rapid time scale.

[14] With this type of near-grazing incidence arrangement, the incident beam may not require expansion by a prism if the angle of incidence is chosen to be very close to 90°. However, gratings used at such high incidence angles are very lossy indeed since the highest part of the profile of each groove tends to shadow its neighbour and prevent the beam from reaching fully into the depth of the groove profile. A practical compromise is to use a slightly less steep angle of incidence and one or two stages of prism beam expansion.

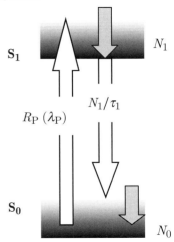

Fig. 14.10: Rate-equation model of a c.w. dye laser. The rate of excitation of dye molecules from S_0 to S_1 via the absorption of photons of pump radiation at wavelength λ_P is denoted by $R_P(\lambda_P)$. For pump intensities below laser threshold the rate of decay from S_1 back to S_0 is governed by spontaneous emission at wavelength λ_L characterized by the fluorescence lifetime τ_1.

[15] See page 183.

[16] Additional means to quench the triplet population can be employed if necessary. For example, simply leaving the jet dye stream open to air entrains oxygen in the solution, which is an effective quencher. Alternatively, special additives such as COT (1, 3, 5, 7, -cyclooctatetraene) that are effective quenchers of triplet dye molecules can be employed.

If N is the total number density of dye molecules we have,

$$N = N_0 + N_1. \tag{14.12}$$

In the absence of stimulated emission on the dye laser transition, the rate equation for the population of the upper laser level is

$$\frac{dN_1}{dt} = R_P - \frac{N_1}{\tau_1}, \tag{14.13}$$

and hence in steady state

$$N_1 = R_P \tau_1. \tag{14.14}$$

Note that in the above equation, and throughout the remainder of this section, N_1 and N_0 are steady-state values. Thus, by analogy[17] with eqn (4.25) we can write the rate of excitation from S_0 to S_1 by absorption of photons of energy $\hbar\omega_P$ as:

$$R_P = \frac{I_P}{\hbar\omega_P}\sigma_a(\lambda_P)N_0 = \beta N_0 = \frac{N_1}{\tau_1}, \tag{14.15}$$

where $\beta = \frac{I_P}{\hbar\omega_P}\sigma_a(\lambda_P)$ and I_P is the intensity of the pump beam.

From eqns (14.12) and (14.15)

$$N_1 = \frac{\beta\tau_1}{1+\beta\tau_1}N. \tag{14.16}$$

So, the round-trip gain at laser threshold, ignoring triplet and singlet absorption, is

$$\alpha_{th}(\lambda_L) = N_1^{th}\sigma_e(\lambda_L) = \frac{T}{2\ell_g}, \tag{14.17}$$

since the round-trip cavity loss is simply equal to T, the useful output mirror transmission, in the low-loss approximation (see Section 6.7.1). The length of gain medium ℓ_g is simply the dye jet thickness in this case.

From eqns (14.16) and (14.17) it follows that at threshold:

$$\beta_{th}\tau_1 = \frac{T}{2N\sigma_e(\lambda_L)\ell_g - T}. \tag{14.18}$$

Whence, substituting for β_{th}, the threshold value of pump intensity I_P^{th} is given by:

$$I_P^{th} = \frac{\hbar\omega_P}{\sigma_a(\lambda_P)\tau_1}\frac{T}{2N\sigma_e(\lambda_L)\ell_g - T}. \tag{14.19}$$

Substituting typical values into eqn (14.19), i.e. $\hbar\omega_P \approx 4.4 \times 10^{-19}$ J, $\sigma_a(\lambda_P) \approx \sigma_e(\lambda_L) \approx 1.0 \times 10^{-16}$ cm^2, $\tau_1 \approx 3.0 \times 10^{-9}$ s, $\ell_g \approx 0.02$ cm, $T \approx 0.10$, and $N = 7.5 \times 10^{17}$ cm^{-3}, we find $I_P^{th} \approx 4.5 \times 10^4$ W cm^{-2}.

Only lasers[18] can supply such high values of focal intensity. If the beam waist is 15 μm in diameter, it implies that the threshold pump power of 80 mW is needed, which is well within the capability of c.w. Ar$^+$ or Kr$^+$ lasers or their solid-state analogues.

---

[17] Equation (4.25) applies to the case where the spectral width of the incident radiation is considerably smaller than the homogenous width of the absorption feature. For the moment we shall consider the pump radiation to be supplied by a laser and to be much narrower in spectral width than the $S_0 \to S_1$ absorption band. In this case eqn (4.25) applies, and the pump beam intensity I_P is the total intensity of the beam. We will later (in note 18) consider the differences implied by using a broadband pump source such as a flashlamp or arc lamp.

[18] This is not strictly true, with a truly mammoth array of arc lamps supplying several kW of light output, excitation of a c.w. laser has been demonstrated. However, this represents something of a technological 'tour de force' and is not a practical approach. In comparing the focal intensity of pump radiation required to bring the laser to threshold in the case of a broadband source such as a flashlamp or an arc lamp, it must be borne in mind that perhaps only 20% or less of the pump radiation lies within the spectral range of the $S_0 \to S_1$ absorption band. Thus, the total intensity in the case of an unfiltered broadband source would need to be 5 or more times larger than that given by eqn (14.19) for a laser source operating at the centre of the $S_0 \to S_1$ absorption band. This is the caveat we alluded to in note 17.

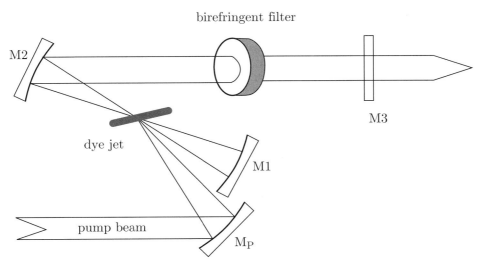

Fig. 14.11: A c.w. dye laser employing a simple two-way cavity comprising high-reflectance mirrors M1 and M2 and output coupling mirror M3. The gain region comprises a thin jet of dye solution at Brewster's angle to the cavity axis. The jet, shaped by a finely finished nozzle to ensure optical flatness, is excited by the beam of a c.w. argon- or krypton-ion laser brought to a focus at the beam waist of the three-mirror cavity by the mirror M_P. An intra-cavity birefringent filter (also at Brewster's angle to the cavity axis) is used to tune the cavity within the wavelength tuning range of the dye. (After Yarborough (1974).)

14.6.2 Continuous waves dye laser design

Figure 14.11 shows a simple two-way cavity. For tuning, the cavity contains one or more frequency-selective elements such as a birefringent filter to achieve coarse control of the oscillating wavelength. If finer control is required a set of tilted etalons can be included in the cavity (as discussed in Section 16.2.2) to restrict the band of oscillation even further, right down to a single axial mode of the cavity. However, as discussed on page 114, spatial hole-burning allows other longitudinal modes to oscillate and consequently it is difficult to ensure that the dye laser will oscillate on a single axial mode at all times as its frequency is swept across the gain curve.

The ring-cavity design shown in Fig. 14.12 avoids the problem of spatial hole-burning associated with standing-wave patterns in the cavity by employing a ring cavity, which does not require the nodes of a standing-wave electric field to be pinned to the mirror positions, since its resonance condition is simply that one round trip of the cavity must correspond to an integer number of half-wavelengths without reference to the absolute positions of the mirrors. In order to avoid standing waves altogether a Faraday isolator[19] is inserted into the cavity, which restricts laser oscillation to one direction round the ring.

Figure 14.13 shows the range of wavelengths available from argon-ion and krypton-ion laser-pumped c.w. dye lasers for various dye species. The shortest wavelength available as fundamental output from the Ar^+ laser is 351 nm.[20] Dye laser oscillation in the violet region of the spectrum requires pumping with this UV line, while over the yellow and red regions of the spectrum pumping by the strong blue-green 488-nm and green 514-nm transitions provides

[19] See Section 16.2.3.

[20] Actually this is a transition of doubly ionized argon, ArIII, see Section 11.3.2.

394   *Dye lasers*

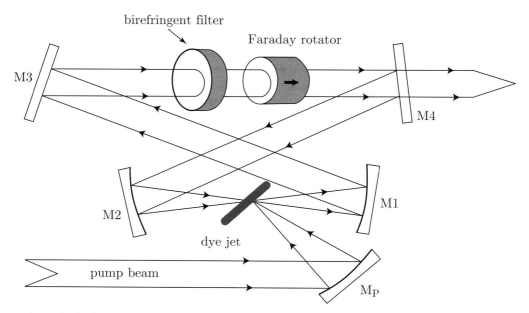

Fig. 14.12: c.w. ring cavity dye laser. The four-mirror cavity of this laser includes a unidirectional device employing the Faraday effect to allow high transmission of travelling waves in the left-to-right direction as shown, but has high loss for light travelling in the opposite direction. (After Jarret and Young (1979).)

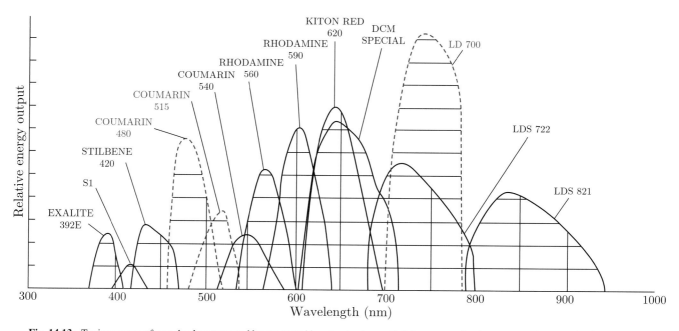

Fig. 14.13: Tuning ranges of c.w. dye laser pumped by argon and krypton ion lasers. Solid curves are for dyes pumped by argon-ion lasers; curves with dashed outlines are for dyes pumped by krypton-ion lasers. Data courtesy Exciton Ltd.

the strongest output. To cover the near-infra-red region out to about 1 μm requires the red output of a krypton-ion laser.

Using the largest argon-ion lasers available, providing about 25 W of output on the strong blue-green lines, c.w. dye laser output powers up to a maximum of around 3 W are possible for the most favourable dye region around 500–600 nm.

14.7 Solid-state dye lasers

The dye laser systems we have so far considered have all involved gain media in the form of solutions of the molecular dye species in solvents such as ethanol or methanol. There are several beneficial side-effects that arise from the fact that the gain medium is a liquid. First, the dye solution can be circulated around a closed loop with only a small fraction of the total volume exposed to the pump laser or flashlamp radiation at any time. Not only does this help minimize the build-up of triplet population and thermal distortion effects on the optical quality in the gain region, it also increases the working life of the dye solution.[21] Further, the fast circulation of dye solution allows heat[22] to be removed from the gain region: the dye solution may be be further cooled by heat exchangers if necessary. This enables very high average powers (in the range of several kilowatts) to be extracted from a gain region of only a few cm^3 volume.

However, there are disadvantages to the use of dye laser media in liquid form. Making up the solutions is a messy business, and the carcinogenic and flammable nature of many of the species employed means that it can only be undertaken in properly equipped laboratories. For this reason, there has been a demand for solid-state versions of the dye laser in which the active medium is in the form of a glass or plastic rod or disc containing the dye that (once made) is safe and easy to handle and cheap to replace. To this end, a number of host materials have been investigated. Directly incorporating the dye material into glass is impossible, because to melt the glass the constituent materials must be heated to temperatures in excess of 1000 °C, thereby destroying the laser dye. Consequently, alternative host materials and techniques for fabrication at lower temperatures have been sought.

There are two major approaches: synthesizing polymer or glassy materials at low temperature or producing a glass with a porous structure. The former technique allows a laser dye to be added to the constituent chemicals and is known as pre-doping. A porous structure allows a dye solution to be introduced into the structure to impregnate it by diffusion, and is termed post-doping of the solid matrix.

Pre-doped hosts usually produce solid materials with good optical quality and uniform dye concentration. These dye-doped solids can have optical properties approaching those of glass but tend to be somewhat softer and less scratch-resistant. In contrast, post-doped materials are usually more glass-like, with many of the excellent bulk properties of glass, but they do suffer from losses caused by scattering from the pores in the host matrix.

---

[21] Many dye molecules, especially those that operate in the red or near-infra-red region, are complex and relatively fragile. They are subject to photo dissociation by absorption of energetic photons in single-step processes, or by visible photons in multiple-step absorption processes. The aging process is accelerated at high temperatures. The efficiency of the laser degrades as the dye solution ages in use, and once its performance has dropped below an acceptable level, the dye solution must be replaced.

[22] The waste heat comprises virtually all of the energy input to the gain region less the energy emitted as laser output. It must be removed if the optical uniformity of the dye solution is to be maintained and the rate of thermal decomposition of the active species is to be minimized.

Because the host materials used for solid-state dye lasers tend to have rather poor thermal conductivity compared to those of conventional solid-state lasers, it is not possible to operate them at sustained high mean power levels with a static gain medium since the gain medium would melt after a fraction of a second. In one attempt[23] to overcome this problem, a gain medium comprising MMA:HEMA (a copolymer of methyl methacrylate and 2-hydroxethyl methacrylate) doped with the dye rhodamine 6G, the material was cast into discs of 50 mm diameter and 2 mm thickness. The discs were spun at 30 revolutions per second and slowly translated across the position of the focal spot of the pump laser—a frequency-doubled Nd:YLF laser operating at 527 nm, producing 20-ns pulses at 10 kHz repetition rate with a mean output power of 3.5 W. The pumped spots on the disc traced out a spiral pattern on the surface of the disc so that each pulse of the pump laser accessed a different spot from the previous pulse. It was found that the mean output power of the dye laser at 565 nm dropped to half its initial value of 563 mW after about 6.6 min, i.e. after about 4×10^6 shots. After the disc had been in use for 40 min its output power had dropped to 10% of its initial value, and it could not be used again.

The situation is much better for low repetition rate devices. A recent study[24] used gain media in the form of static rods of 10 mm length and 10 mm diameter with a flat surface of 4×10 mm ground parallel to the axis of the cylinder to form the transversely pumped face. The rods were fabricated from MMA copolymerized with a variety of linear and cross-linked polymers and doped with a special dye (based on pyrromethene 567). The pump laser provided pulses of 5.5 mJ energy and 6 ns duration at 534 nm. With the cross-linked polymers, initial lasing efficiencies of up to 40% were obtained that dropped by only 15% after 2.8 h of operation at 10 Hz pulse repetition rate.

[23] Abedin et al. (2003).

[24] Alvarez et al. (2005).

14.8 Applications of dye lasers

The introduction of dye laser technology in the late 1960s had a profound effect upon the development of techniques of high-resolution spectroscopy of atoms and molecules—indeed it can be said to have brought about a renaissance in experimental spectroscopy. With the ability to tune laser radiation to the exact frequency of atomic transitions, and the availability of beams of sufficient spectral brightness to perturb the populations of ground and excited states, techniques such as saturated absorption spectroscopy and two-photon absorption spectroscopy became possible and allowed sub-Doppler resolution to be achieved at room temperature. Continuous-wave dye lasers pumped by argon-ion lasers played a significant role in these developments. More recent developments have enabled the trapping and cooling of atoms to the extent that they form Bose–Einstein condensates, and have demonstrated the manipulation of beams of ultracold atoms by laser light so that it is now possible to speak of atom optics.[25]

[25] For a detailed discussion of these techniques the reader is referred to the texts by Foot (2005) and Letokhov (2007).

The technique of resonant ionization spectrometry (RIMS) uses pulsed dye lasers tuned to the frequencies of a sequence of transitions that ionize ground-state atoms via one or more intermediate levels. Atoms of the bulk materials in the vaporized sample are not resonant to the laser frequencies and hence will not be ionized. The ions of the target species created by this highly selective process are presented to the entrance port of a mass spectrometer, which further separates the atomic species present in the sample. The technique allows the quantitative detection of trace amounts of a particular atom present in a sample containing much larger quantities of other species and has been used, for example, to measure the amount of plutonium in samples of house dust.[26]

[26] Ruster et al. (1989).

High-power dye lasers pumped by high repetition rate pulsed copper-vapour lasers are also key elements of the proposed scheme for enriching the content of the fissile isotope ^{235}U in natural uranium—perhaps the only example to date of an industrial-scale photochemical process to rely on laser technology, but one that has still to reach commercial viability. The process relies on the fact that many of the visible transitions of ^{238}U display a considerable frequency shift (many times the Doppler width) from the corresponding transitions of ^{235}U. It is thus possible, by irradiating a stream of atomic uranium vapour entrained in a rare-gas flow, to ionize selectively the ^{235}U atoms by stepwise absorption of radiation from three different dye laser beams tuned to frequencies that lead to excitation and finally ionization. The ions (exclusively ^{235}U if no mixing collisions occur) are then removed from the gas stream by deflecting them onto the surface of a metal collector plate under the influence of an electric field. Dye lasers developed for this application at the Lawrence Livermore National Laboratory in California USA were reported[27] in 1988 to have reached average powers of over 600 W in a single beam tuned to a single isotope transition.

[27] Paisner et al. (1988).

The technique of photodynamic therapy (PDT), which is a promising treatment for cancer,[28] uses the output from a c.w. or high repetition rate dye laser operating in the region of 630 nm. In this technique a photosensitive compound, which binds to the collagen associated with a tumour, is introduced into the bloodstream of the patient. The cancerous area is illuminated by laser radiation (but not at intensities to cause burning), which activates the photosensitive drug and causes it to attack all tissue in is vicinity. The tissue in the illuminated region (some of which will have been non-cancerous) dies and forms a wound that heals in the normal way.

[28] Wilson and Bown (2004).

Flashlamp-pumped dye lasers have also been used in a variety of medical applications. For example, the in-situ pulverization of kidney stones or gall stones by the shock wave created when a plasma is formed at the surface of the stone by a pulse of several Joules of dye laser light delivered by an optical fibre. Perhaps the most widespread medical application of flashlamp-pumped dye lasers (operating broad band in the yellow region of the spectrum) is to the treatment of skin blemishes such as port-wine stains[29] by sealing off the subcutaneous blood vessels causing the skin colouration by absorption of yellow light that passes easily through normal tissue but is strongly absorbed by the haemoglobin component of blood.

[29] Lanigan (2004).

Further reading

Further information on the design and operation of dye lasers may be found in the books by Schäfer (1990) and Duarte (1991), and in the article by Titterton (2004a).

Exercises

(14.1) **Conversion from $\varepsilon(\lambda)$ to $\sigma_a(\lambda)$**

In describing the absorption of light of wavelength λ by dye molecules in solution, many authors prefer to quote a value for the extinction coefficient $\varepsilon(\lambda)$ instead of the absorption cross-section $\sigma_a(\lambda)$, where $\varepsilon(\lambda)$ is defined in terms of the decadic version of Beer's law:

$$I(\lambda, x) = I(\lambda, 0) 10^{-\varepsilon(\lambda) c' x},$$

and c' is the concentration of (ground-state) dye molecules in moles per litre and x is measured in centimetres.

(a) Show that $\varepsilon(\lambda) = 2.16 \times 10^{20} \sigma_a(\lambda)$ if $\sigma_a(\lambda)$ is given in units of cm^2. Note that the units of $\varepsilon(\lambda)$ are $1\,\text{moles}^{-1}\,\text{cm}^{-1}$.

(14.2) **Flashlamp-pumped dye lasers: Significance of triplet quenching**

Here, we illustrate the importance in flashlamp-pumped dye lasers of quenching the population of the triplet state. We do this by solving the rate-equation model of Section 14.5.1 without neglecting the loss term N_T/τ_T on the RHS of eqn (14.6).

(a) Use this approach to show that the triplet population approaches a constant value $N_T(\infty)$ as $t \to \infty$.

(b) Assuming a constant pump intensity, show that the population in the triplet level will be too small to terminate laser oscillation provided that

$$\tau_T < \frac{\sigma_e(\lambda_L)}{\sigma_T(\lambda_L)} \frac{1}{k_{ST}}.$$

[In the case of the dye Rh6G this implies that sufficient quenching additive is needed to ensure $\tau_{T_1} < 600\,\text{ns}$.]

(14.3) **High-gain pulsed dye laser: tuning mechanism**

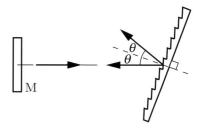

Fig. 14.14: The optical arrangements for wavelength tuning a dye laser considered in Exercise 14.3.

The right-hand side of Fig. 14.14 shows a diffraction grating used in reflection, in the 'Littrow configuration' such that one of the first-order diffracted beams propagates back along the incident direction. The separation of the grating elements is d and the width of each element (a facet tilted at angle θ_F to the plane of the grating blank) can be taken to be equal to d.

(a) Find the wavelength λ_0 such that the first diffracted order is reflected antiparallel to the incident direction.

(b) Show that if d is chosen to be very slightly less than λ_0 the grating does not produce any diffracted orders other than the zero order and retro-reflected first order for any angle of incidence.

(c) Light of wavelength λ_0 is used with a grating of spacing d very slightly less than λ_0. A blaze angle θ_{blaze} is chosen so that the zero-order beam and the retro-reflected beam have the same power. How is the '1st-order blaze wavelength' related to λ_0? (The 1st-order blaze wavelength is defined to be that wavelength for which the intensity of the 1st-order diffracted beam is maximized when the grating is used at normal incidence.)

(d) The plane mirror M is then introduced and illuminated normally such that the transmitted light propagates towards the grating. The mirror and the 1st-order reflection from the grating thus form an optical cavity for light of wavelength λ_0. Given that the mirror has reflectivity 98%, and that the grating absorbs 20% of the light incident upon it, what is the finesse of the Fabry–Perot cavity thus formed for light of wavelength λ_0? Why would the finesse be expected to be reduced when light of other wavelengths is used? Why can't such an arrangement be used for tuning c.w. dye lasers?

(14.4) **Prism beam expander**
Many designs of dye laser tuning optics employ one or more beam-expanding prisms in the beam incident on the grating, as illustrated in Fig. 14.15.

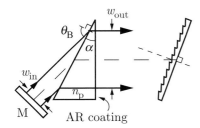

Fig. 14.15: The prism beam expander considered in Exercise 14.4.

(a) What is the advantage of expanding the beam in this way?

(b) Calculate the value of the apex angle α required for a beam entering the right-angled prism (of refractive index n_p) at Brewster's angle, θ_B, to leave the prism normal to the antireflection coated face.

(c) What is the corresponding lateral magnification $w_{\text{out}}/w_{\text{in}}$?

(d) If four such prisms are used in series, what beam-expansion factor might be achieved for fused silica prisms with $n_p \approx 1.46$?

(e) Estimate the round-trip loss for a multiple prism beam expander comprising a series of 4 fused silica prisms whose right-hand normal-incidence faces are each coated with a single quarter-wave layer of MgF_2 of refractive index $n_{MgF_2} \approx 1.38$.
[Hint: the reflectivity of a quarter-wave layer of a material of refractive index n_{layer} coated on a substrate of refractive index n_s is $R = \left[(n_s - n_{\text{layer}}^2)/(n_s + n_{\text{layer}}^2)\right]^2$, assuming normal incidence from a medium of refractive index $n = 1$.]

(14.5) For a particular c.w. dye laser the dye concentration is 10^{-4} molar. The dye jet is 1 mm thick and absorbs 70% of the pumping laser light at 514 nm (close to the wavelength of maximum absorption). The lifetime of the upper level is $\tau_2 = 3$ ns.

(a) Estimate the value of the saturation intensity, stating clearly all your assumptions and approximations.

(b) If the pump laser can be focused to a spot of diameter 10 μm in the dye jet, estimate the speed of the dye stream necessary to ensure that significant build-up of triplet population does not occur.

15 Non-linear frequency conversion

15.1 Introduction 400
15.2 Linear optics of crystals 400
15.3 Basics of non-linear optics 405
15.4 Phase-matching techniques 409
15.5 SHG: practical aspects 420
15.6 Three-wave mixing and third-harmonic generation (THG) 421
15.7 Optical parametric oscillators (OPOs) 424
Further reading 428
Exercises 428

15.1 Introduction

The application of non-linear optical techniques has greatly extended the wavelength range over which laser radiation can be obtained. In this chapter we investigate three of the most important examples: second-harmonic generation (SHG), third-harmonic generation (THG), and optical parametric generation.

Probably the most widespread and familiar application of non-linear optics is the generation of output at 532 and 355 nm by the intra-cavity frequency doubling and tripling of the fundamental radiation of Nd:YAG lasers at 1064 nm. Even though these are non-tunable devices, the availability of powerful visible and near-UV output from compact solid-state laser sources has opened up a wide vista of applications, ranging from their use as pump lasers for tunable sources such as dye lasers, titanium-doped sapphire lasers and optical parametric oscillators (OPOs), as well as applications to the cutting and drilling of metals and ceramics. Further, the wavelength range covered by such widely tunable sources such as the Ti:sapphire laser (whose fundamental tunes from approximately 660 to 1180 nm) has been significantly increased by SHG to provide tunable output over selected regions of the near-UV and visible spectrum.

The generation of second-harmonic radiation is only a special case of the more general process of sum-frequency generation. In SHG the two input beams are derived from the same source. Using separate sources it is possible to generate outputs at the frequency sum and difference, using the same processes that underlie the operation of optical parametric amplifiers and oscillators. Indeed, there are a vast number of ways in which non-linear optical techniques can be applied to extend the capabilities of lasers. By way of introduction to the practicalities and physical principles of these techniques we shall consider the process of second-harmonic generation in anisotropic crystals, but first it is helpful to review some aspects of the propagation of light in anisotropic media in the regime of linear optics.

15.2 Linear optics of crystals

15.2.1 Classes of anisotropic crystals

At low intensity the polarization P induced in optical materials is directly proportional to the electric field E of the electromagnetic signal. However, even in this linear regime, crystalline materials show a variety of behaviours that reflect the symmetry of their structure. One such manifestation is the phenomenon of **double refraction** or **birefringence**[1] in which a beam of light incident on an anisotropic crystal is, in general, transmitted as two orthogonally polarized beams. In the most familiar case of uniaxial crystals, one of the beams is refracted at the interface according to Snell's law and is called the **ordinary wave**. The other wave (polarized orthogonally to the ordinary wave) is refracted differently and is known as the **extraordinary wave**.

[1] The terms 'birefringence' and 'double refraction' apply equally to the refractive properties of both uniaxial and biaxial materials. It is the fact that they separate a single incident beam into two refracted beams that is important, rather than the number of optical axes.

For a detailed account of the optics of anisotropic crystals the reader is referred to standard works such as Born and Wolf (1999); for the purposes of the present chapter we shall draw heavily upon the treatment set out in Chapter 15 of Brooker (2003). Following his approach, we define an arbitrary set of rectangular Cartesian coordinates within the crystal, in terms of which the most general linear relationship fields between the fields $\mathbf{D}$ and $\mathbf{E}$ can be expressed as:

$$\frac{D_1}{\epsilon_0} = \epsilon'_{11} E_1 + \epsilon'_{12} E_2 + \epsilon'_{13} E_3 \tag{15.1}$$

$$\frac{D_2}{\epsilon_0} = \epsilon'_{21} E_1 + \epsilon'_{22} E_2 + \epsilon'_{23} E_3 \tag{15.2}$$

$$\frac{D_3}{\epsilon_0} = \epsilon'_{31} E_1 + \epsilon'_{32} E_2 + \epsilon'_{33} E_3, \tag{15.3}$$

where for the moment we have used $E_1 \equiv E_x$, $E_2 \equiv E_y$, $E_3 \equiv E_z$ in order to liberate numerical subscripts for use in the equations below, and to develop a more compact notation. For example, we can write the equations above in the shortened form:

$$\frac{D_i}{\epsilon_0} = \sum_{j=1}^{3} \epsilon'_{ij} E_j, \tag{15.4}$$

or, using the 'summation convention' in which it is understood that repeated subscripts are to be summed over:

$$\frac{D_i}{\epsilon_0} = \epsilon'_{ij} E_j. \tag{15.5}$$

Clearly, eqns (15.1)–(15.3) can be written in matrix form using a 3×3 matrix—or, equivalently, a second-rank tensor—whose elements are ϵ'_{ij}. In the case of a material that is loss-free and doubly refracting in the usual way,[2] the matrix ϵ'_{ij} is real and symmetric. It is a property of such real symmetric matrices that they can always be written in diagonal form by a suitable rotation of axes. From hereon it will be assumed that such a rotation of axes has been performed so that referred to these new **principal axes** eqns (15.1)–(15.3) reduce to:

[2] For a discussion of what is implied by 'the usual way' see Brooker (2003), Section 15.4.

$$D_x = \epsilon_0 \epsilon_{11} E_x, \quad D_y = \epsilon_0 \epsilon_{22} E_y, \quad D_z = \epsilon_0 \epsilon_{33} E_z, \tag{15.6}$$

where we have written the ϵ_{ij} symbols without primes to emphasize that they are different from the ϵ'_{ij} referred to the original axes.

In terms of their optical properties, crystalline materials are classified in terms of the three classes described in Table 15.1.

In the case of uniaxial crystals the z-axis is always chosen so that ϵ_{33} is the 'different' element.

Table 15.1 Classes of crystalline material.

Crystal type	Properties
isotropic[a]	$\epsilon_{11} = \epsilon_{22} = \epsilon_{33}$
uniaxial[b]	$\epsilon_{11} = \epsilon_{22} \neq \epsilon_{33}$
biaxial[c]	$\epsilon_{11} \neq \epsilon_{22} \neq \epsilon_{33}$

[a] Cubic crystals are optically isotropic. Their linear optical properties are equivalent to those of amorphous materials.

[b] Comprises crystals belonging to the trigonal, tetragonal and hexagonal systems with three-, four- and six-fold symmetry, respectively.

[c] Includes crystals belonging to the orthorhombic, monoclinic and triclinic systems.

15.2.2 Vectors

For plane electromagnetic waves propagating in isotropic media, the wave normals, the ray direction, and the Poynting vector are parallel. However, since in anisotropic media the refractive index depends on the polarization and direction of propagation, it is necessary to distinguish carefully between the directions of these vectors.

Consider a monochromatic plane wave:

$$\boldsymbol{D} = \boldsymbol{D}_0 \exp[i(\boldsymbol{k} \cdot \boldsymbol{r} - \omega t)] = \boldsymbol{D}_0 \exp(i\varphi). \quad (15.7)$$

We will define the direction of polarization to be that of $\boldsymbol{D}$. The wavefront is the plane where the phase ϕ is everywhere the same, and is clearly normal to the direction of $\boldsymbol{k}$. Since $\boldsymbol{D}$, $\boldsymbol{E}$, and $\boldsymbol{H}$ share a common dependence on $\exp[i(\boldsymbol{k} \cdot \boldsymbol{r} - \omega t)]$, we can replace the operations: $\frac{\partial}{\partial t}$ with multiplication by $-i\omega$; $\nabla\cdot$ by $i\boldsymbol{k}\cdot$; and $\nabla\times$ by $i\boldsymbol{k}\times$.

Thus, Maxwell's equations for an uncharged, non-conducting medium:

$$\nabla \cdot \boldsymbol{D} = 0 \quad (15.8)$$

$$\nabla \cdot \boldsymbol{B} = 0 \quad (15.9)$$

$$\nabla \times \boldsymbol{E} = -\frac{\partial \boldsymbol{B}}{\partial t} \quad (15.10)$$

$$\nabla \times \boldsymbol{H} = \frac{\partial \boldsymbol{D}}{\partial t} \quad (15.11)$$

may be written as[3]

$$\boldsymbol{k} \cdot \boldsymbol{D} = 0 \quad (15.12)$$

$$\boldsymbol{k} \cdot \boldsymbol{B} = 0 \quad (15.13)$$

$$\boldsymbol{k} \times \boldsymbol{E} = \mu\omega \boldsymbol{H} \quad (15.14)$$

$$\boldsymbol{k} \times \boldsymbol{H} = -\omega \boldsymbol{D}, \quad (15.15)$$

where we have used $\boldsymbol{B} = \mu_0\mu_r\boldsymbol{H} = \mu\boldsymbol{H}$.

From eqns (15.12) and (15.13) we see immediately that $\boldsymbol{D}$ and $\boldsymbol{B}$ are perpendicular to $\boldsymbol{k}$ and therefore lie in the plane of the wavefront. Further, from eqns (15.14) and (15.15), $\boldsymbol{H}$ is normal to the three vectors $\boldsymbol{D}$, $\boldsymbol{E}$ and $\boldsymbol{k}$, which must therefore be coplanar.

These vector relations hold for any uncharged, non-conducting, magnetically isotropic[4] medium. If the medium is also electrically isotropic, so that $\boldsymbol{D} \parallel \boldsymbol{E}$, eqns (15.14) or (15.14) show that the direction of energy flow—given by the Poynting vector $\boldsymbol{S} = \boldsymbol{E} \times \boldsymbol{H}$—is parallel to $\boldsymbol{k}$. However, as discussed

[3] See Exercise 15.1.

[4] That is, one for which $\boldsymbol{B} = \mu_0\mu_r\boldsymbol{H} = \mu\boldsymbol{H}$.

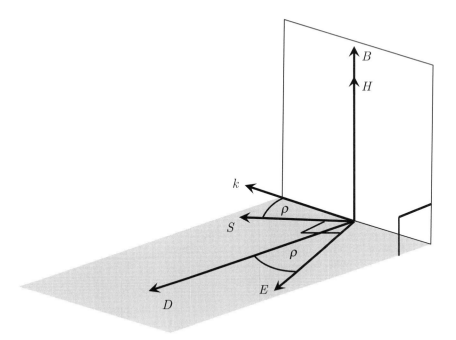

Fig. 15.1: Orientation of electric and magnetic field vectors in non-magnetic anisotropic media.

above, for electrically non-isotropic media $D \nparallel E$, and in this case there is an angle ρ between the D and E vectors given by

$$\rho = \cos^{-1}\left(\frac{E \cdot D}{|E||D|}\right). \tag{15.16}$$

Since E and D are both perpendicular to H, it is clear that ρ is also the angle between S and k.[5] The relative orientations of the field vectors is illustrated in Fig. 15.1.

15.2.3 Field directions for o- and e-rays in a uniaxial crystal

Referring to the coordinate system shown in Fig. 15.2, the reader can readily verify[6] that if the E-field of a travelling wave in a uniaxial crystal lies in the (x, y) plane (the optic axis being the z-axis), then for this restricted case $D = \epsilon_0 \epsilon_{11} E$. When substituted into Maxwell's equations this yields a wave propagating with speed

$$v_o = \frac{c}{\sqrt{\epsilon_{11}}} = \frac{c}{n_o} \quad \text{(ordinary wave).} \tag{15.17}$$

By the arguments of Section 15.2.2 it follows that such a wave travels in a direction k lying in a plane perpendicular to E and that no directions of k are excluded—for example, k can lie in the (x, y) plane or the (x, z) plane or any other plane just so long as $k \cdot E = 0$. The wave therefore has a spherical

[5] Imagine first that E and D are parallel such that S and k are parallel. Now, rotate D through an angle ρ about H. It should be clear that the vector $S = E \times H$ will also rotate by an angle ρ. See Exercise 15.2 for a more mathematical proof.

[6] The treatment presented here is based entirely on Chapter 15 of Brooker (2003), in particular Exercises 15.1 and 15.2 of that text. The reader would be well advised to work through those problems.

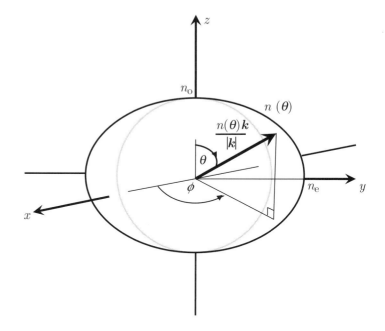

Fig. 15.2: Refractive indices n_o for ordinary wave (inner sphere) and $n(\theta)$ extraordinary wave (outer spheroid) in the case of a *positive* ($n_e > n_o$) uniaxial crystal. The optic axis is the z-axis. The value of $n(\theta)$ is given by the length of a radius vector from the origin to the surface of the spheroid in the direction of the wavevector $\boldsymbol{k}$. The maximum value of $n(\theta)$ is n_e and is reached at $\theta = 90°$, i.e. in the (x, y) plane. Note that the figure is rotationally symmetric about the z-axis.

[7] There are several different types of diagram used to represent the propagation of o- and e-waves in uniaxial crystals and, confusingly, they have a superficial resemblance to one another. Figure 15.2 is a polar plot of the refractive indices n_o and $n(\theta)$, and is known as the *index ellipsoid*. This must be carefully distinguished from other representations such as the wave-normal surface and ray-surface ellipsoid (see, for example, Born and Wolf (1999)). For a uniaxial crystal the Huygens wave surface (i.e. the surface of constant phase) for the o-wave is a sphere, and that for the e-wave is a spheroid but for a positive crystal the spheroid of the e-wave lies *inside* the sphere corresponding to the o-wave.

[8] See Exercise 15.3.

[9] The wave with $E_y \neq 0$ is the ordinary wave.

[10] See Exercise 15.3.

Huygens' wave surface[7] and must represent the ordinary wave. For this reason $n_o \equiv \sqrt{\epsilon_{11}}$ is called the ordinary refractive index. The ordinary wave *always* has its $\boldsymbol{E}$-field perpendicular to the optic axis.

Now, consider a wave whose electric field $\boldsymbol{E}$ is parallel to the z-axis. For such a wave $D_z = \epsilon_0 \epsilon_{33} E_z$ that, for this special case, can be written as $\boldsymbol{D} = \epsilon_0 \epsilon_{33} \boldsymbol{E}$. When substituted into Maxwell's equations this yields the equation of a wave that propagates with speed

$$v_e = \frac{c}{\sqrt{\epsilon_{33}}} = \frac{c}{n_e} \quad \text{(extraordinary wave, special case).} \quad (15.18)$$

This particular choice of direction for $\boldsymbol{E}$ implies that the wave vector $\boldsymbol{k}$ lies in the (x, y) plane and consequently the wave is one example of an extraordinary wave. Other directions of $\boldsymbol{E}$ and $\boldsymbol{k}$ (not lying in the (x, y) plane) can correspond to extraordinary waves, but this particular one corresponds to an extremum value of extraordinary-wave speed (either a maximum or a minimum depending on the type of crystal). The quantity $n_e = \sqrt{\epsilon_{33}}$ corresponding to the extremum value is called the 'extraordinary refractive index'.

We now turn to the case of an extraordinary wave propagating in a uniaxial crystal.[8] We suppose that the wave propagates in the (x, z) plane, and propagates with its $\boldsymbol{k}$-vector making an angle of θ with the z-axis (the optic axis). In this case the electric and magnetic fields of the extraordinary wave vary as

$$E_x, E_z \propto \exp\left[i\left(kz\cos\theta + kx\sin\theta - \omega t\right)\right], \quad (15.19)$$

and $E_y = 0$.[9] From eqns (15.14) and (15.15) it may be shown that[10]

$$\frac{\omega^2}{c^2 k^2} = \frac{\cos^2\theta}{\epsilon_{11}} + \frac{\sin^2\theta}{\epsilon_{33}}. \quad (15.20)$$

For the case of a wave travelling with $\theta = \pi/2$, eqn (15.20) reduces to $ck/\omega = \sqrt{\epsilon_{33}} = n_\mathrm{e}$; for a wave travelling along the z-axis we have $ck/\omega = \sqrt{\epsilon_{11}} = n_\mathrm{o}$.

It is convenient to define a refractive index $n(\theta)$ to describe the phase *speed* $c/n(\theta)$ of an extraordinary wave whose phase *velocity* makes an angle θ to the z-axis.[11] From eqn (15.20), recognizing that $n(\theta) = ck/\omega$,

$$\frac{1}{n(\theta)^2} = \frac{\cos^2\theta}{n_\mathrm{o}^2} + \frac{\sin^2\theta}{n_\mathrm{e}^2}, \tag{15.21}$$

which is the equation of the ellipsoidal surface of Fig. 15.2. The case illustrated in Fig. 15.2 is that of a positive uniaxial crystal, i.e. one for which $n_\mathrm{e} > n_\mathrm{o}$. In this case the ellipsoid representing the refractive index $n(\theta)$ is an oblate spheroid lying outside the sphere representing the ordinary refractive index n_o. In the opposite case, that of a negative uniaxial crystal ($n_\mathrm{e} < n_\mathrm{o}$), the ellipsoid representing the extraordinary refractive index is a prolate spheroid lying inside the sphere representing the ordinary refractive index.

It should be borne in mind that the surfaces represented in diagrams such as Fig. 15.2 refer to a particular wavelength of light, a fact that is of particular relevance for the index-matching techniques to be described in Section 15.4.

[11] Note that for uniaxial crystals the symbol n_e, written with no argument θ, denotes the extremum value of the refractive index experienced by an extraordinary wave propagating perpendicular to the optic axis. The symbol $n(\theta)$ (i.e. written without the subscript e) always denotes the refractive index experienced by an extraordinary wave propagating at an angle of θ to the optic axis. The symbol n_o denotes the refractive index experienced by an ordinary wave propagating in any direction.

15.3 Basics of non-linear optics

15.3.1 Maxwell's equations for non-linear media

The polarization $\boldsymbol{P}$ (dipole moment per unit volume) produced in a medium by an intense applied electric field $\boldsymbol{E}$ is of the form:[12]

$$\boldsymbol{P} = \epsilon_0 \left[\chi^{(1)} \boldsymbol{E} + \chi^{(2)} \boldsymbol{E}^2 + \chi^{(3)} \boldsymbol{E}^3 + \cdots \right], \tag{15.22}$$

in which the term involving $\chi^{(1)}$ describes ordinary linear optics, and the term in $\chi^{(2)}$ describes the second-order non-linear effects responsible for second-harmonic generation (SHG) and sum- and difference-frequency generation.[13]

We can therefore express $\boldsymbol{P}$ in general as

$$\boldsymbol{P} = \epsilon_0 \chi^{(1)} \boldsymbol{E} + \boldsymbol{P}^\mathrm{NL}. \tag{15.23}$$

Now, since $\boldsymbol{D} \equiv \epsilon_0 \boldsymbol{E} + \boldsymbol{P} = \epsilon_0 \left(1 + \chi^{(1)}\right) \boldsymbol{E} + \boldsymbol{P}^\mathrm{NL} = \epsilon_0 \epsilon_\mathrm{r} \boldsymbol{E} + \boldsymbol{P}^\mathrm{NL}$, it follows that for a non-linear medium Maxwell's $\nabla \times \boldsymbol{H}$ equation can be written as

$$\nabla \times \boldsymbol{H} = \boldsymbol{J} + \epsilon_0 \epsilon_\mathrm{r} \frac{\partial \boldsymbol{E}}{\partial t} + \frac{\partial \boldsymbol{P}^\mathrm{NL}}{\partial t}. \tag{15.24}$$

[12] There are several texts that deal in detail with the subject of non-linear optics. Two that employ SI units, and are used as the basis of the summary presented here, are Binks (2004) and Butcher and Cotter (1990).

[13] Only crystalline systems lacking a centre of inversion symmetry can have a non-zero $\chi^{(2)}$, and so only they can be used for SHG. Note also that for the moment we have ignored the tensor nature of $\chi^{(2)}$. The effects arising from this tensor we shall examine the next section.

For the cases of interest here, where most of the crystalline media are to a good approximation loss free at all of the wavelengths involved, we can set $\boldsymbol{J} = 0$ and treat them as pure dielectrics with a real ϵ_r. Further, since all media of interest here are non-magnetic ($\mu_\mathrm{r} = 1$) in the optical range, Maxwell's $\nabla \times \boldsymbol{E}$ equation can be written:

$$\nabla \times \boldsymbol{E} = -\mu_0 \frac{\partial \boldsymbol{H}}{\partial t}. \tag{15.25}$$

Taking the curl of both sides of eqn (15.14)

$$\nabla \times (\nabla \times E) = -\mu_0 \frac{\partial}{\partial t}(\nabla \times H)$$

$$= -\mu_0 \frac{\partial}{\partial t}\left(\epsilon_0 \epsilon_r \frac{\partial E}{\partial t} + \frac{\partial P^{\text{NL}}}{\partial t}\right)$$

$$= -\mu_0 \epsilon_0 \epsilon_r \frac{\partial^2 E}{\partial t^2} - \mu_0 \frac{\partial^2 P^{\text{NL}}}{\partial t^2}. \quad (15.26)$$

But, by a standard vector–operator relationship,

$$\nabla \times (\nabla \times E) = \nabla(\nabla \cdot E) - \nabla^2 E. \quad (15.27)$$

Now, for an *isotropic* medium with no free charges $\nabla \cdot D = \nabla \cdot E = 0$. If we apply this result to eqn (15.26) we reach:

$$\nabla^2 E = \mu_0 \epsilon_0 \epsilon_r \frac{\partial^2 E}{\partial t^2} + \mu_0 \frac{\partial^2 P^{\text{NL}}}{\partial t^2}, \quad (15.28)$$

which is the basic wave equation of loss-free non-linear optics.

In the following sections we shall, for the sake of simplicity, apply eqn (15.28) to the case of uncharged *anisotropic* media. For these media $\nabla \cdot D = 0$, but $\nabla \cdot E \neq 0$. However, the results that we shall derive are still a good approximation to reality in most cases of practical interest, and where it is necessary to take explicit account of the consequences of this simplification we shall draw attention to them.

15.3.2 Second-harmonic generation in anisotropic crystals

In this section we shall be mainly concerned with the effects arising from the second-order term of eqn (15.22) in *anisotropic* media and for the moment shall ignore higher-order terms. Because we are dealing with *anisotropic* media we shall have to take account of the tensor nature of $\chi^{(2)}$.

We consider first the situation when two intense monochromatic beams at angular frequencies ω_1 and ω_2 propagate in the form of plane waves in a uniaxial crystal along directions given by their wave vectors k_1 and k_2, respectively. The (real) electric fields can be represented as:[14]

$$E_1(r, t) = \frac{1}{2}[\widehat{E}_1 \exp[i(k_1 \cdot r - \omega_1 t)] + \text{c.c.}, \quad (15.29)$$

with corresponding expressions for $E_2(r, t)$ and for the field $E_3(r, t)$ that is, as we shall see, generated as a result of the polarization $P_3^{\text{NL}}(r, t)$ induced at the sum frequency:

$$\omega_3 = \omega_1 + \omega_2 \quad (15.30)$$

by the second-order interaction of E_1 and E_2.[15] The non-linear polarization $P^{\text{NL}}(r, t)$ that is the source term giving rise to the generated signal at angular frequency ω_3 is given by:[16,17]

$$P^{\text{NL}}(r, t) = \epsilon_0 \chi^{(2)}(-\omega_3, \omega_1, \omega_2)E_1(r, t)E_2(r, t), \quad (15.31)$$

[14] The electric-field amplitude $\widehat{E}_1(r, t)$ and non-linear polarization $\widehat{P}^{\text{NL}}(r, t)$ introduced here are both complex signals defined so as to include implicitly any phase differences. Also, note that by choosing to factor out explicitly the wave-like part $\exp(ik \cdot r - \omega t)$ from $E_1(r, t)$, the remaining 'amplitude' $\widehat{E}_1(r, t)$ represents the wave envelope that is a much less rapidly varying function of r and t than $E_1(r, t)$ itself.

[15] The subscript 1 on $E_1(r, t)$ and the corresponding complex amplitude $\widehat{E}_1(r, t)$ denotes that these signals are at angular frequency ω_1, etc.

[16] See Exercise 15.4

[17] Here, we have used the standard notation in non-linear optics: the frequency of the polarization generated by the non-linear interaction (ω_3) is written first, with a minus sign in front of it; those of the driving fields ($\omega_{1,2}$) are then listed (without minus signs). Equation (15.30) then requires that the terms in the bracket must sum to zero.

or, in terms of the Cartesian components of the vectors involved

$$\widehat{P}_i^{NL}(\boldsymbol{r}, t) \exp[i(\boldsymbol{k}_3 \cdot \boldsymbol{r} - \omega_3 t)]$$
$$= \frac{\epsilon_0}{4} \sum_{jk} \chi_{ijk}^{(2)}(-\omega_3, \omega_1, \omega_2) \widehat{E}_{1j} \widehat{E}_{2k}$$
$$\times \exp[i(\boldsymbol{k}_1 + \boldsymbol{k}_2) \cdot \boldsymbol{r} - (\omega_1 + \omega_2)t] \quad (15.32)$$

where any of i, j, k can take the values x, y, z, of the coordinates referred to the principal axes.

A third-rank tensor such as $\chi_{ijk}^{(2)}$ would in general have 27 independent components, but because there is no physical significance to the order in which products of interaction fields are formed (i.e. $\widehat{E}_{1j}\widehat{E}_{2k} = \widehat{E}_{2k}\widehat{E}_{1j}$) the number here can be reduced to 18. In addition, in applications to SHG, many authors[18] prefer to use instead of the tensor elements $\chi_{ijk}^{(2)}$ the tensor elements d_{ijk} defined as:[19]

$$\chi_{ijk}^{(2)}(-\omega_3, \omega_1, \omega_2) = 2d_{ijk}(-\omega_3, \omega_1, \omega_2). \quad (15.33)$$

When the expression for the Cartesian components of $P_i^{NL}(\boldsymbol{r}, t)$ from eqn (15.32) and for the corresponding Cartesian components of $\boldsymbol{E}_{1,2}(\boldsymbol{r}, t)$ from eqn (15.29) are inserted in the wave equation (15.28), and the **slowly varying amplitude approximation** is applied,[20] we reach several equations, including the following one that describes the growth of $\widehat{E}_3$ with respect to l where l is a vector in the $\boldsymbol{k}_3$-direction:

$$\frac{d\widehat{E}_3}{dl} = i\frac{\omega_3^2}{2k_3 c^2} d_{\text{eff}} \widehat{E}_1 \widehat{E}_2 \exp[i(\boldsymbol{k}_1 + \boldsymbol{k}_2 - \boldsymbol{k}_3) \cdot \boldsymbol{r}] = i\kappa_3 \widehat{E}_1 \widehat{E}_2 \exp(i\Delta k l), \quad (15.34)$$

where $d_{\text{eff}} \widehat{E}_1 \widehat{E}_2$ represents the sum $\sum_{jk} d_{3jk} \widehat{E}_{1j} \widehat{E}_{2k}$ and

$$\kappa_3 = \frac{\omega_3^2}{2k_3 c^2} d_{\text{eff}}, \quad (15.35)$$

$$\Delta \boldsymbol{k} = \boldsymbol{k}_1 + \boldsymbol{k}_2 - \boldsymbol{k}_3. \quad (15.36)$$

But since for SHG all three beams must propagate along the same direction this implies

$$\Delta k = k_1 + k_2 - k_3. \quad (15.37)$$

Now,

$$k_3 = \omega_3 \sqrt{\mu_0 \epsilon_0 \epsilon_3} = \frac{\omega_3 n_3}{c}, \quad (15.38)$$

where n_3 is the refractive index of the crystal at angular frequency ω_3 for light travelling in the $\boldsymbol{k}_3$-direction and polarized in the $\boldsymbol{E}_3(\boldsymbol{r}, t)$ direction. The effective tensor element d_{eff} introduced in eqns (15.34) and (15.35) is a compact way[21] of expressing the effect of summing over the repeated indices j and k involved in eqn (15.32) as we shall see later in specific examples.

[18] The reader should be aware that a variety of units is employed even within the SI system. We have chosen to define the polarization in the form $\boldsymbol{P} = \epsilon_0 \chi^{(2)} \boldsymbol{E}^2 = 2\epsilon_0 d\boldsymbol{E}^2$. The polarization $\boldsymbol{P}$ is the dipole moment (C m) per unit volume, so in the SI system the units of $\boldsymbol{P}$ are C m^{-2}. Since $\boldsymbol{E}$ is measured in V m^{-1} and the units of ϵ_0 are F m^{-1} = C V^{-1} m^{-1} it follows that the units of $\chi^{(2)}$ and d used here are both m V^{-1}. However, some authors (e.g. Yariv (1989)) do not include the factor of ϵ_0 on the RHS of eqn (15.31) relating $\boldsymbol{P}$ to $\boldsymbol{E}^2$ and consequently the units of d in their equations are F m^{-1}.

[19] The components $\chi_{ijk}^{(2)}$ (or equivalently d_{ijk}) are conventionally denoted by writing the ω_1, ω_2 and ω_3 argument of the function in such a way that the three frequencies sum to zero, emphasizing the fact that energy is conserved by the interaction. Two symmetry conditions apply to the coefficients of $\chi_{ijk}^{(2)}$ and d_{ijk}: the first is that the coefficients remain the same under a permutation of indices provided the frequencies are also permuted, i.e. $d_{ijk}(-\omega_3, \omega_1, \omega_2) = d_{ikj}(-\omega_3, \omega_2, \omega_1) = d_{kij}(\omega_2, -\omega_3, \omega_1)$, and the second is Kleinman's conjecture (see Appendix C).

[20] This approximation assumes that $\widehat{P}$ and $\widehat{E}$ change slowly on the length scale of the wavelength and on the time scale of a period of an optical cycle, so that terms such as $d_3^2 \widehat{E}/dr^2$ are neglected in comparison with $2k_3 d\widehat{E}_{3i}/dr$, terms such as $\partial^2 \widehat{P}/\partial t^2$ and $\omega \partial \widehat{P}/\partial t$ are neglected in comparison with $\omega^2 \widehat{P}$ and $\partial \widehat{E}/\partial t$ is neglected in comparison with $\omega \widehat{E}$.

[21] See Appendix C.

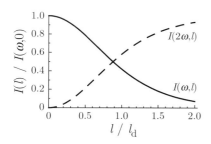

Fig. 15.3: Depletion of fundamental intensity $I(\omega)$ and growth of second-harmonic intensity $I(2\omega)$ with distance l along the crystal, calculated under conditions of perfect phase matching ($\Delta k = 0$).

In the specific case of SHG, $\omega_2 = \omega_1$, and $\omega_3 = 2\omega_1$

$$\widehat{E}_1(r,t) = \widehat{E}_2(r,t). \tag{15.39}$$

Note that eqn (15.32) does *not* imply that $E_1(r,t)$ and $E_2(r,t)$ are necessarily perpendicular to one another—they could both be in the same direction; the important point is that $\widehat{E}_1$ and $\widehat{E}_2$ are those components of E_1 and E_2 that give non-zero products with the d_{eff} appropriate to generating a plane wave propagating in the k_3 direction.

If we ignore the depletion of the signals at the fundamental frequency we can treat $\widehat{E}_1(r,t)$ as a constant and integrate eqn (15.34) for the variation of $\widehat{E}_3$ over a crystal of length L in the k_3 direction to reach:

$$\widehat{E}_3(L) = \mathrm{i}\kappa_3|\widehat{E}_1|^2 \frac{e^{\mathrm{i}\Delta kL}-1}{\mathrm{i}\Delta k} = \mathrm{i}\kappa_3|\widehat{E}_1|^2 e^{\mathrm{i}(\Delta kL/2)} \operatorname{sinc}(\Delta kL/2). \tag{15.40}$$

Now the intensities of the fundamental and second harmonic beams are given by:

$$I(2\omega_1) = |\widehat{E}_3(L)|^2 \frac{n_3}{2Z_0}, \tag{15.41}$$

$$I(\omega_1) = |E_1|^2 \frac{n_1}{2Z_0}, \tag{15.42}$$

where Z_0 is the intrinsic impedance of free space. Hence, from eqns (15.40) and (15.41) we finally reach

$$I(2\omega_1) = \frac{2\omega_1^2}{n_3 n_1^2 c^3 \epsilon_0} I^2(\omega_1) d_{\text{eff}}^2 L^2 \operatorname{sinc}^2(\Delta kL/2). \tag{15.43}$$

Thus, the intensity of second-harmonic radiation is, as expected, proportional to the square of the intensity at the fundamental. In the idealized case of plane waves with no depletion[22] of the fundamental power (i.e. very little conversion to second-harmonic) we would expect from eqn (15.43) the second harmonic intensity to increase as the product of the square of the crystal length and the $\operatorname{sinc}^2(\Delta kL/2)$ factor. It is the influence of this latter factor that we shall examine in the next section.

[22]When the conversion to second-harmonic becomes an appreciable fraction of the intensity initially available at the fundamental, the small-signal approximations of eqns (15.40) and (15.41) are no longer valid. In that case, eqn (15.34) for the growth of $\widehat{E}_{3k}$ and the corresponding expression for the decrease of $\widehat{E}_{1i}$ can be solved exactly as coupled equations (see, for example, Boyd et al. (1965)) to give:

$$\frac{I(\omega_1,l)}{I(\omega_1,0)} = \operatorname{sech}^2(l/l_\mathrm{d}) = 1 - \frac{I(2\omega_1,l)}{I(2\omega_1,0)},$$

where l_d is a characteristic length for the interaction given by:

$$l_\mathrm{d} = \frac{nc}{\omega d_{\text{eff}}}\sqrt{\frac{n}{2I(\omega,0)Z_0}},$$

where $n = n_1 = n_2 = n_3$ is the common refractive index under phase-matched conditions and Z_0 is the impedance of free space. Figure 15.3 plots these functions.

[23]For a finite value of Δk the intensity of the second-harmonic radiation oscillates with distance along its axis of propagation with a period

$$l_\mathrm{c} = \frac{\pi}{\Delta k},$$

known as the coherence length. See Section 15.4.7.

15.3.3 The requirement for phase matching

Figure 15.4 plots $\operatorname{sinc}^2(\Delta kL/2)$ as a function of $\Delta kL/2$ and illustrates the controlling influence the phase-mismatch parameter $\Delta kL/2$ has on the efficiency with which second-harmonic radiation can be generated according to eqn (15.43). Only when $|\Delta kL/2| < \pi$ is it possible to generate appreciable second-harmonic intensities, and clearly generation is most efficient if it can be arranged that $\Delta k = 0$.[23]

For maximum conversion we require perfect phase matching such that

$$\Delta k = k_3 - k_2 - k_1 = k_{2\omega_1} - 2k_{\omega_1} = 0. \tag{15.44}$$

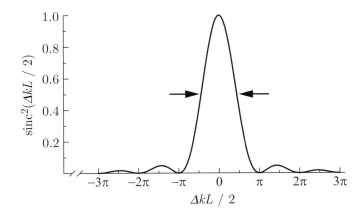

Fig. 15.4: The function $\text{sinc}^2(\Delta kL/2)$ as a function of the phase-mismatch parameter $\Delta kL/2$. The double-ended arrow indicates the FWHM of the curve.

Since $k_3 \equiv k_{2\omega_1} = n_3 2\omega_1/c$ and $k_1 \equiv k_{\omega_1} = n_1\omega_1/c$, eqn (15.44) is equivalent to requiring that

$$2n_3 = n_1 + n_2, \quad \textbf{SHG phase-matching condition} \quad (15.45)$$

for perfect phase matching.

In the case of Type I phase matching (see Section 15.4.1), the physical significance of eqn (15.45) is that unless the fundamental wave propagates with the same phase velocity as that at the second harmonic,[24] wavelets of the second harmonic generated further down the length of the crystal will not be in phase with those generated earlier, leading to destructive interference. As a consequence, the amplitude of the second-harmonic wave alternately grows and diminishes as second-harmonic radiation is converted from fundamental and back again with a period of $2l_c$.

For the generation of any significant intensity of second harmonic in anisotropic bulk materials[25] it is therefore necessary to achieve *perfect* phase matching by arranging that, for some chosen direction of propagation, the phase velocity of the second harmonic and fundamental are equal. Because in general the refractive indices of optical materials for light polarized in one particular plane show considerable dependence on frequency, it is necessary to achieve this equality of refractive index by careful exploitation of the polarization properties of anisotropic materials.

[24] This is strictly true only for the case of Type I phase matching. For the case of Type II phase matching, the criterion is that the propagation of $\boldsymbol{P}^{\text{NL}}$ through the crystal, which is proportional to $\exp[\mathrm{i}(\boldsymbol{k}_1 + \boldsymbol{k}_2)\cdot\boldsymbol{r} - (\omega_1 + \omega_2)t]$, must have the phase velocity of the second harmonic.

[25] As we shall see in Section 15.4.7 it is possible to use periodically layered materials rather than bulk materials. The layer thickness is engineered to be approximately l_c, so that the harmonic radiation generated in one layer does not backconvert to radiation at the fundamental frequency in the succeeding layer, but continues to generate new second-harmonic radiation in that and subsequent layers.

15.4 Phase-matching techniques

15.4.1 Birefringent phase matching in uniaxial crystals

To illustrate the technique of phase matching in birefringent uniaxial crystals, we shall describe the use of BBO (β-barium borate) a crystal widely used for harmonic generation. Over its transparency range (approximately $0.2 - 2.6\,\mu\text{m}$) the ordinary and extraordinary refractive indices of BBO are given by:

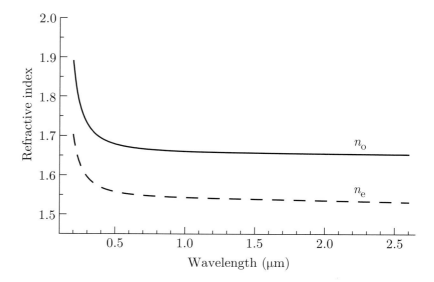

Fig. 15.5: Ordinary and extraordinary refractive indices of β-barium borate as a function of wavelength, calculated from eqns (15.46) and (15.47).

$$n_o^2 = 2.7405 + \frac{0.0184}{\lambda^2 - 0.0179} - 0.00155\lambda^2, \quad (15.46)$$

$$n_e^2 = 2.3730 + \frac{0.0128}{\lambda^2 - 0.0156} - 0.0044\lambda^2, \quad (15.47)$$

respectively, where λ is in units of μm; these variations are plotted in Fig. 15.5.

It is clear from eqns (15.46) and (15.47) and Fig. 15.5 that the refractive index for the second harmonic and fundamental of the same polarization will be different. To achieve exact phase matching the polarization of the wave $\boldsymbol{E}_3$ at the second-harmonic frequency $\omega_3 = 2\omega_1$ must be different from that of at least one of the two fundamental waves $\boldsymbol{E}_1, \boldsymbol{E}_2$ at frequency $\omega_1 = \omega_2$. Whether the second-harmonic ray should be ordinary or extraordinary to achieve phase matching depends on whether the crystal is positive ($n_e > n_o$) or negative ($n_e < n_o$) uniaxial.[26]

Two[27] types of phase matching can be used for second-harmonic generation: Type I, in which both fundamental waves have the same polarization; and Type II, for which the two fundamental waves are orthogonally polarized one to another.

We now consider each of these cases in turn.

Type I phase matching in a positive crystal

As indicated in Fig. 15.6(a) the condition for phase matching at angle θ_m is:

$$n^\omega(\theta_m) = n_o^{2\omega}. \quad \textbf{Type I, positive crystal} \quad (15.48)$$

Combining eqn (15.21) with eqn. (15.48) leads to the following expression for $\sin \theta_m$:[28]

$$\sin^2 \theta_m = \frac{(n_o^{2\omega})^{-2} - (n_o^\omega)^{-2}}{(n_e^\omega)^{-2} - (n_o^\omega)^{-2}}. \quad (15.49)$$

---

[26] Strictly it also depends on the sign of the dispersion. We have assumed in the following paragraphs that both n_e and n_o increase with frequency since all solid media transparent in the visible and near UV show positive dispersion. Negative dispersion, where n_e and n_o decrease with frequency, occurs only in the neighbourhood of strong absorption features.

[27] Some authors, e.g. Binks (2004) distinguish a third type of phase matching, Type III in which the order of the fundamental waves is reversed from that of the Type II entries in Table 15.1. This has relevance to the generation of sum and difference frequencies in which fundamental waves of two unequal frequencies are involved, but in the case of second harmonic generation this distinction is lost.

[28] See Exercise 15.6.

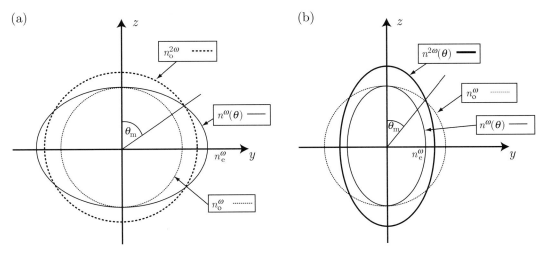

Fig. 15.6: (a) *Type 1* phase matching in a *positive* uniaxial crystal: sections in the (z, y) plane of the refractive-index ellipsoids for fundamental $n^\omega(\theta), n_o^\omega$ and second harmonic $n_o^{2\omega}$. (b) *Type 1* phase-matching in a *negative* uniaxial crystal: sections in the (z, y)-plane of the refractive-index ellipsoids for fundamental $n^\omega(\theta), n_o^\omega$ and extraordinary wave at second harmonic $n^{2\omega}(\theta)$. (The ellipse corresponding to $n^{2\omega}(\theta)$ is omitted from (a) and the circle corresponding to $n_o^{2\omega}$ is omitted from (b) to avoid unnecessary clutter—neither of them is involved in phase-matching considerations.) Note that both figures are sections through three-dimensional ellipsoids that are rotationally symmetric about the z-axis.

Type I phase matching in a negative crystal

As indicated in Fig. 15.6(b) the condition for phase-matching is:

$$n^{2\omega}(\theta_m) = n_o^\omega. \quad \textbf{Type I, negative crystal} \qquad (15.50)$$

It is left as an exercise for the reader to derive an expression for $\sin \theta_m$ in this case.

Type II phase matching in a positive crystal

For Type II phase matching one of the fundamental waves is an o-ray and the other is an e-ray. The appropriate fundamental refractive index is the arithmetic mean of n_o^ω and $n^\omega(\theta)$ that, for phase matching, must be equal to the ordinary refractive index at 2ω, since as indicated in Table 15.2, the second harmonic is a o-ray. Thus, for phase matching we require:

$$n_o^{2\omega} = \frac{1}{2}\left[n_o^\omega + n^\omega(\theta_m)\right]. \quad \textbf{Type II, positive crystal} \qquad (15.51)$$

Solving this for $\sin \theta_m$ using eqn (15.21) we reach

$$\sin \theta_m = \frac{2n_e^\omega}{(2n_o^{2\omega} - n_o^\omega)} \left[\frac{n_o^{2\omega}(n_o^{2\omega} - n_o^\omega)}{(n_e^\omega)^2 - (n_o^\omega)^2}\right]^{\frac{1}{2}}. \qquad (15.52)$$

Table 15.2 Phase-matching types in uniaxial crystals.[a]

Type	Positive crystal	Negative crystal
I	$e(\omega)e(\omega) \to o(2\omega)$	$o(\omega)o(\omega) \to e(2\omega)$
II	$e(\omega)o(\omega) \to o(2\omega)$	$o(\omega)e(\omega) \to e(2\omega)$

[a] $o(\omega)$ and $e(\omega)$ denote, respectively, ordinary and extraordinary waves of frequency ω.

Type II phase matching in a negative crystal

As noted in Table 15.2, the second harmonic in this case corresponds to a e-ray at 2ω, so the phase-matching condition in this case becomes:

$$n^{2\omega}(\theta) = \frac{1}{2}\left[n_o^\omega + n^\omega(\theta_m)\right]. \qquad \textbf{Type II, negative crystal} \qquad (15.53)$$

Once again it is left as an exercise for the reader to derive an expression for $\sin\theta_m$.

It can be seen from the above results that θ_m for Type II phase matching is *always* greater than θ_m for Type I and that if the Type II process is possible in a given crystal, then so is Type I.[29]

15.4.2 Critical and non-critical phase matching

We now turn to the question of how closely the angle θ must approach the angle of exact matching θ_m in order that good phase matching still occurs. From Fig. 15.4 it is evident that $|\Delta k L/2|$ must be less than approximately π for efficient SHG. Since the refractive index of the extraordinary wave(s) depends on the direction of propagation, θ, changes in the propagation direction will change the wave vectors of the extraordinary wave(s) and hence change Δk. As a consequence, efficient harmonic generation can only occur within a small range of angles known as the **angular acceptance**. Using eqn (15.21) it may be shown that the angular acceptance $\Delta\theta$ is given by[30]

$$\Delta\theta < \frac{1}{2\beta} \frac{\lambda_0}{L|n_e - n_o|\sin 2\theta_m}, \quad \beta = \begin{cases} 1 & \text{Type I} \\ \frac{1}{2} & \text{Type II,} \end{cases}$$

Angular acceptance (critical) (15.54)

where λ_0 is the wavelength of the fundamental radiation, and the expression applies for both positive and negative birefringence.

The angular acceptance in practical cases can be so small as to cause practical difficulties. For example, for Type I SHG in KDP, $\theta_m \approx 42°$ for fundamental radiation at 1064 nm. If the coherence length[31] l_c is to be of order 10 mm then the angular acceptance $\Delta\theta$ is no greater than 1 mrad. Not only does this make stringent demands on the precision with which the crystal must be cut and its housing made and adjusted, it is also restrictive on the divergence of the beam presented to the crystal. Phase matching under these unfavourable

[29] See Hobden (1967).

[30] See Hobden (1967) and Exercise 15.7.

[31] See Section 15.4.7 and note 23.

conditions ($\theta_m \neq 90°$), where the angle, divergence, and crystal temperature must all be maintained at their optimum values within very tight tolerances, is called **critical phase matching**.

If, by variation of a parameter such as temperature or chemical composition, the refractive indices can be adjusted so that $\theta_m = 90°$ then eqn (15.54) suggests that the acceptance angle is infinite. Clearly this is unphysical, the reason being that the linear dependence of $\Delta\theta$ on $1/\sin 2\theta_m$ of eqn (15.54) is not valid in this limit. In fact, the FWHM angular acceptance is given by:[32]

$$\Delta\theta < \sqrt{\frac{1}{4\beta}\frac{\lambda_0}{L\,|n_e - n_o|}}, \quad \beta = \begin{cases} 1 & \text{Type I} \\ \dfrac{1}{2} & \text{Type II.} \end{cases}$$

Angular acceptance (non-critical) (15.55)

[32] Hint: to find the correct limit multiply each side of eqn (15.54) by $\sin 2\theta_m$, and take the limit as $2\theta_m \to 180° \pm \Delta\theta_m$. Thus, $\lim_{2\theta \to 2\Delta\theta}(\sin 2\theta_m \Delta\theta) \to 2(\Delta\theta)^2$, leading to eqn (15.55).

The case of $\theta_m \approx 90°$ is known as **non-critical phase matching (NCPM)**, and is illustrated in Fig. 15.7.

An example of this situation occurs for Type I phase-matched SHG from $1.065-\mu$m radiation in LiNbO$_3$ at a temperature of $\approx 47\,°$C. The allowable

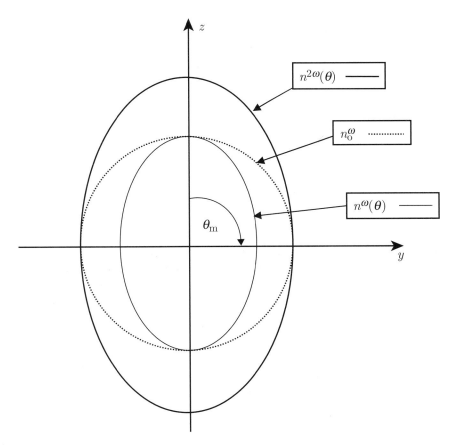

Fig. 15.7: Noncritical (90°) Type I phase-matching in a negative crystal.

[33] See Dimitrev et al (1991).

[34] The plane-wave analysis we have used hitherto shows that the intensity of the second-harmonic beam is proportional to the square of intensity of the fundamental beam. One might therefore expect that greater conversion efficiency could be attained from a Gaussian beam that is tightly focused into the non-linear crystal. However, two factors operate to offset the potential advantage of focusing the beam very tightly: (i) the reduced Rayleigh range Z_R (See Section 6.3.2; and (ii) the increased divergence of the beam that may exceed the angular acceptance of the crystal discussed in Section 15.4.2. If the Rayleigh range of the fundamental beam is much shorter than the interaction length (see Note 22) then despite the fact that the peak intensity near the beam waist may be increased by the tight focusing, the average intensity over the interaction length is reduced and the overall conversion efficiency is reduced. In a well-designed scheme one would choose the crystal length L to be a few times the interaction (depletion) length l_D. Boyd and Kleinman (1968) have provided a detailed analysis of the interaction of focused Gaussian beams in non-linear crystals and have shown that the optimum conversion occurs when $L = 2.84 Z_R$. They also showed that under these conditions the generated second-harmonic intensity grows linearly with L, rather than as L^2 as in the plane-wave analysis of eqn (15.43).

[35] See Section 6.3.2.

divergence from the direction of exact phase matching is greater than 10 mrad for a coherence length of 10 mm. This situation is clearly preferred for device applications. It also has the advantage of avoiding the effects of 'walk-off' (see below) that could otherwise limit the length of crystal usable for SHG.

15.4.3 Poynting vector walk-off in birefringent phase matching

As discussed in Section 15.2.2, the Poynting vector S of an e-wave is not, in general, parallel to the wave vector k. As shown in Fig. 15.8, for an extraordinary ray propagating through an anisotropic crystal such that its k-vector makes an angle of θ with the optic axis then the S-vector will be at an angle ρ to the k-vector, where

$$\rho = \pm \tan^{-1} \left[\left(\frac{n_o}{n_e} \right)^2 \tan \theta \right] \mp \theta, \quad (15.56)$$

and the upper (lower) signs correspond to a negative (positive) crystal.[33]

A fundamental beam with uniform intensity over a diameter $2a$ will no longer overlap the SHG beam, and therefore no longer contribute to SHG, after a distance $l_a = a/\rho$ from the entrance face of the crystal. The distance l_a is known as the aperture length. In the case of a Gaussian beam[34] of spot size[35] w the corresponding aperture length is given by

$$l_a = w\sqrt{\pi}/\rho. \quad (15.57)$$

In practice, ρ might be ≈ 30 mrad, and $w \approx 50$ μm, yielding $l_a \approx 2.5$ mm. Clearly, in most practical cases there is no point in using a crystal longer than a few mm.

15.4.4 Other factors affecting SHG conversion efficiency

While it is true that no SHG can occur unless the angle θ (the angle between the wave vector k and the optic axis) is chosen to be very close to optimum

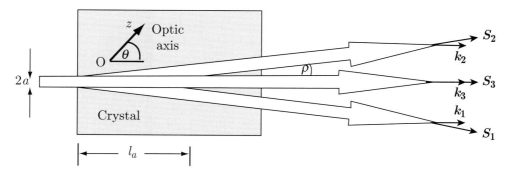

Fig. 15.8: Poynting-vector walk-off for Type I phase matching in a positively birefringent crystal. The collinear wave-vector direction of k_1 (e-wave at ω), k_2 (e-wave at ω) and k_3 (o-wave at 2ω) makes an angle of θ with the optic axis Oz. The Poynting vectors S_1 and S_2, giving the directions of energy flow in the two fundamental waves, make angles of $\pm\rho$ with the direction of $k_{1,2,3}$ and S_3.

for phase matching, we have so far ignored any influence on the efficiency of SHG of the angle ϕ the angle between the projection of $\boldsymbol{k}$ on the (x, y)-plane and the x-axis.[36] The dependence on ϕ arises because, from eqn (15.34), the intensity of second-harmonic radiation $I(2\omega_1)$ generated depends on d_{eff}^2 and the value of d_{eff} is, in general, a function of ϕ.[37] For example, in the case of BBO (β-barium borate) the value of d_{eff} is given by

[36]See As shown in Fig. 15.2.

[37]See Exercise 15.5.

$$|d_{\text{eff}}| = d_{22} \cos\theta \sin 3\phi - d_{31} \sin\theta. \tag{15.58}$$

The value of θ is constrained by phase-matching considerations to be $21°$, but we are free to choose ϕ in such a way as to maximize $|d_{\text{eff}}|$; in this case clearly $\phi_{\text{optimum}} = 90°$.

Crystal manufacturers will usually supply a non-linear crystal cut from the boule of bulk material in the form of a rod of rectangular cross-section with the optic axis very close to the correct matching angle to the long axis of the rod. Further, the crystal may be cut so that, with one of the sides of the end face in a vertical position, vertically polarized radiation is presented at close to ϕ_{optimum} to the crystal axes. Although the crystal mount needs to provide for very precise adjustment of the θ angle, this is not so necessary in the case of ϕ.

15.4.5 Phase-matched SHG in biaxial crystals

Calculation of the refractive index experienced by a wave transmitted though a biaxial crystal at an arbitrary angle to the principal axes is more complex than for the uniaxial case, because it is not generally possible to resolve the wave into an e- and an o-wave. If, however, the ray is propagating parallel to one of the three principal planes then the calculation of refractive indices for two orthogonally polarized waves reduces to that for the uniaxial case. For example, if the propagation is restricted to the (x, y) plane then the wave polarized normal to this plane (i.e. with $\boldsymbol{E}$ parallel to the z-direction) experiences the same refractive index in all directions in this plane and becomes the effective o-ray with $n_o = n_z = \sqrt{\epsilon_{33}}$. On the other hand, the refractive index experienced by the wave polarized in the plane depends on the angle between $\boldsymbol{k}$ and the x-axis, and can be regarded as the effective extraordinary wave for this plane with one extremum value of $n_e = n_x$ and the other $n_e = n_y$.[38] Well-behaved biaxial crystals are those in which the principal axes x, y, z as defined in Section 15.2.1 do not change direction with frequency and for which

[38]The subscripts on the *principal refractive indices* n_x, n_y and n_z refer to the directions in which the $\boldsymbol{E}$ field is polarized, not the direction of propagation.

$$n_i(2\omega) > n_i(\omega) \quad i = x, y, z \tag{15.59}$$

$$n_z(\omega') > n_y(\omega') > n_x(\omega') \quad \omega' = \omega, 2\omega. \tag{15.60}$$

Sections of the index surfaces for a single frequency in the (x, y), (x, z), and (y, z) planes are shown for crystals of this type in Fig. 15.9.

The conditions for phase matching are the same as those for uniaxial media, i.e. for Type I phase matching we have to find a direction in the crystal for which the refractive index at the fundamental is exactly equal to that at the second harmonic for the wave of opposite polarization. These form a cone around

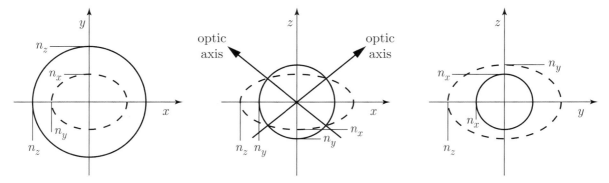

Fig. 15.9: Sections through the index ellipsoid surfaces of a biaxial crystal for one frequency showing in particular the intersection in the (z, x) plane of the circle representing the o-wave (with $\mathbf{E}$ parallel to the y-axis) with the ellipse representing the e-wave ($\mathbf{E}$ in the (z, x) plane) giving rise to two optic axes equally inclined to the z-axis. (Solid curves are circles, dashed curves are ellipses.)

each of the optic axes. In the more complicated case of Type II phase-matching with orthogonally polarized fundamental waves, Hobden (1967) shows that the matching directions define the surface of an irregular cone encompassing both an optic axis and the z-axis.

15.4.6 Birefringent materials for SHG

One of the most widespread applications of SHG is converting the fundamental output of Nd:YAG lasers at 1064 nm to visible (green) radiation at 532 nm. For that reason, in summarizing the properties of some of the materials that can be used for SHG, we have listed their properties for a fundamental wavelength of 1064 nm; but of course they can be used at many other wavelengths within their transparency range. Table 15.3 summarizes the properties of some common non-linear materials.[39]

[39] For more detailed information the reader is referred to the text by Binks (2004) upon which this section is based.

KDP, KD*P

Good optical-quality crystals of KDP potassium dihydrogen phosphate KH_2PO_4 and its deuterated analogue KD*P can be grown from aqueous solution with diameters up to 27 cm. However these crystals must be kept below 100 °C and heated and cooled only slowly. The non-linear coefficients of KDP are rather low compared to other materials but, because they were among the

Table 15.3 Properties of some common nonlinear materials. Data apply to 1064 nm fundamental radiation.[a]

Material	Uniaxial + ve or - ve	Matching type	Ph. matching angle θ_m	d_{eff} pm V^{-1}	Transparency range (μm)	Angular bw[b] $\Delta\theta$ (mrad)	Temperature ΔT (°C)	Wavelength b.w. $\Delta\lambda$ nm
KD*P	−ve	II	53.5°	0.38	0.20–2.0	5.0	13	6.5
KTP	Biaxial	II	25°	5.27	0.35–4.5	15–68	25	0.56
LiNbO$_3$	−ve	I	78°	5.89	0.33–5.5	3.1	1.1	0.3
BBO	−ve	I	21°	2.01	0.20–2.6	1.8	50	0.66
LBO	Biaxial	I	90°	1.17	0.16–2.6	70	5.8	4.37

[a] Data taken from Binks (2004) and Dimitrev et al. (1991).
[b] Angular bw = angular acceptance bandwidth for $l_c = 10$ mm.

first to be used for non-linear optics, coefficients of newer materials are usually stated as values relative to those of KDP for which $d_{36} = 0.44\,\mathrm{pm\,V^{-1}}$, and KD*P for which $d_{36} = 0.38\,\mathrm{pm\,V^{-1}}$.

KTP and KTA

Potassium titanyl phosphate, $KTiOPO_4$ (KTP), has good overall properties as a birefringent material. It has large nonlinear coefficients ($d_{31} = 6.5\,\mathrm{pm\,V^{-1}}$ and $d_{32} \approx 5\,\mathrm{pm\,V^{-1}}$), may be phase matched over a wide wavelength range, and has broad angular and temperature bandwidths. However, it is a difficult material to grow and is relatively expensive. In addition, it needs to be held at an elevated temperature ($> 65\,°\mathrm{C}$) to avoid photochemical degradation (grey tracking). Potassium titanyl arsenate KTA is an isomorph of KTP with broadly similar properties. Unlike KTP it cannot be used for frequency doubling 1064 nm radiation, but works well at 1053 nm (the fundamental wavelength of Nd:YLF lasers) for which it has an effective nonlinear coefficient some 1.6 times higher. It is also less absorbing in the $3-5\,\mu\mathrm{m}$ region, and has found use for optical parametric oscillators in that region (see Section 15.7.3).

Lithium niobate

Lithium niobate $LiNbO_3$ has good chemical and mechanical properties—it is non-hygroscopic and is straightforward to cut and polish. It has relatively large non-linear coefficients ($d_{31} = -5.4\,\mathrm{pm\,V^{-1}}$ and $d_{22} = 2.8\,\mathrm{pm\,V^{-1}}$), and can be temperature tuned when non-critically phase matched. However, it has a relatively low damage threshold and must be subjected to high temperature ($> 170\,°\mathrm{C}$) to reverse the effects of photodegradation. Doping with 5% MgO improves the damage threshold. It was the first material to be successfully periodically poled for use in the quasi-phase-matching technique described in Section 15.4.7.

BBO

Barium borate BaB_2O_4 in the β-phase is non-centrosymmetric with somewhat smaller non-linear coefficients ($d_{22} = 2.2\,\mathrm{pm\,V^{-1}}$ and $d_{31} = 0.16\,\mathrm{pm\,V^{-1}}$) than many other non-linear materials. However its wide temperature bandwidth, low optical absorption ($0.01\,\mathrm{cm^{-1}}$ at 532 nm), and a transparency range that extends all the way down to 189 nm at the long-wavelength edge of the VUV, as well as its high damage threshold, more than compensate for this slight disadvantage.

LBO

Lithium triborate LiB_3O_5 has a transparency range that extends even further into the UV (down to 155 nm) than BBO. It is a hard, chemically stable, non-hygroscopic material that is easy to handle and polish. It can be non-critically phase matched and temperature tuned. However, it has non-linear coefficients ($d_{31} = 1.1\,\mathrm{pm\,V^{-1}}$ and $d_{32} = 1.2\,\mathrm{pm\,V^{-1}}$) that are even smaller than those of BBO, but it does not suffer from the same narrow angular bandwidth.

15.4.7 Quasi-phase matching techniques

For a finite wave vector mismatch Δk, eqn (15.43) gives the dependence of the second-harmonic intensity as

$$I(2\omega_1) \propto L^2 \operatorname{sinc}^2(\Delta k L/2) = \left(\frac{2}{\Delta k}\right)^2 \sin^2(\Delta k L/2). \tag{15.61}$$

It is clear that the intensity of the second harmonic will oscillate with the length L of the crystal, as illustrated in Fig. 15.10. The second-harmonic output first reaches a peak when $L = l_\text{c}$, where the **coherence length** l_c is defined by $(\Delta k L/2) = \pi/2$, i.e.

$$l_\text{c} = \frac{\pi}{\Delta k}. \tag{15.62}$$

The second-harmonic intensity therefore oscillates in L with a period of $2l_\text{c}$. The reason for this behaviour is as follows. For short distances into the crystal all the generated second-harmonic radiation is in phase, leading to a growth of the second-harmonic intensity. However, owing to dispersion, the second-harmonic radiation generated further into the crystal will be out of phase with that generated earlier. As a consequence, second-harmonic radiation is converted back to the fundamental frequency, causing the net intensity of the generated harmonic radiation to reduce as L is increased beyond l_c—becoming zero for $L = 2l_\text{c}$. It is clear that this oscillation of the second-harmonic intensity will continue for larger values of L.

As indicated in Fig. 15.10, if we could arrange that after a distance l_c the value of d_eff is reversed to become $-d_\text{eff}$ then the intensity of second harmonic would be enhanced—instead of backconverting to fundamental radiation—as the waves traverse the second l_c. This concept, called **quasi-phase matching**, was suggested by Armstrong et al. as long ago as 1962, but practical means of implementing it were not realized until the 1990s.

Many of the non-linear materials used for SHG and similar applications are ferroelectric i.e. they can maintain a permanent electric dipole moment

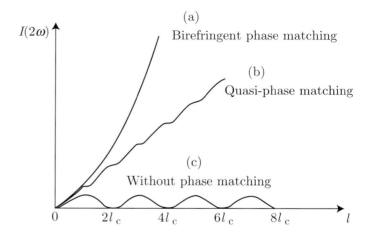

Fig. 15.10: Intensity of second-harmonic radiation (ignoring depletion of the fundamental) as a function of distance l from the entrance face of a birefringent crystal in the cases of (a) perfect phase-matching, (b) quasi phase matching and (c) no phase matching.

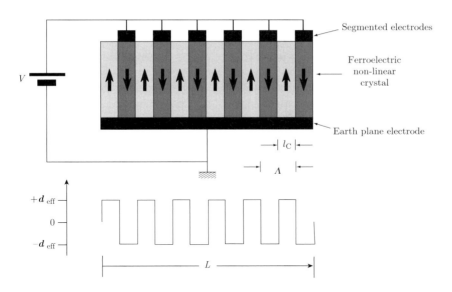

Fig. 15.11: Periodic poling of ferroelectric material for quasi-phase matching.

provided the temperature is below the Curie temperature. The moment (and thus the value of d_{eff}) can be reversed by applying an opposing electric field larger than the coercive field strength—a process known as *'poling'*. Figure 15.11 shows how a sample of such material (e.g. lithium niobate) can poled with alternating slabs of material polarized in directions opposite to one another, each of thickness l_c to form a repeating structure of period $\Lambda = 2l_c$. The resulting artificial material is known as PPLN (periodically poled lithium niobate) and referred to as 'piplin'.

A real advantage of periodically poled materials is that the directions of propagation and polarization of the light beams can be chosen so as to maximize the value of d_{eff}. This is to be compared with the situation in birefringent phase-matching materials for which the directions of propagation and polarization are both severely constrained and in general do not correspond with the maximum d_{eff}. Further, for periodically poled materials the Poynting vector of all the waves is along the direction of propagation, so there is no beam walk-off. Another advantage is that by appropriate choice of the spatial period Λ the structure can be manufactured so as to provide quasi-phase matching at any desired wavelength within the transparency range of the crystal. This contrasts strongly with the situation for birefringent phase-matching where we have to rely on finding a specific angle where phase-matching is possible over a limited wavelength range. The quasi-phase matching approach has another significant merit—it can be applied to isotropic[40] as well as anisotropic materials.

The main disadvantage of periodically poled non-linear materials is their relatively low damage threshold. For this reason, birefringent phase matching in hard crystals such as BBO and LBO is still preferred for frequency doubling high-power pulsed lasers.

[40] As pointed out by Armstrong et al. (1962) crystals of the cubic ZnS structure are isotropic but lack a centre of inversion symmetry (i.e. they have an ϵ_{ij} that is isotropic but their d_{ijk} are not). Thus, although such crystals can generate second-harmonic radiation, they do not show the double refraction necessary for birefringent phase matching.

15.5 SHG: practical aspects

While it is, of course, possible to generate SHG by simply focusing a laser beam into a non-linear crystal, for the reasons discussed in Note 34 there is a limit to the payoff that can be gained by tight focusing. While this approach can be used for pulsed lasers where the peak output powers are high, it is not very useful for c.w. lasers where the power is typically a few watts or less.[41]

[41] The practical implementation of the schemes to allow worthwhile SHG intensities to be achieved are discussed in detail in the texts by Binks (2004) and Koechner (2006).

One way round this problem is to place the non-linear crystal inside the cavity, when the intensity is higher than in the output beam by a factor of approximately T^{-1}, where T is the intensity transmission of the output coupler at the fundamental wavelength. In the case of the laser described by Geusic et al. (1968), shown in Fig. 15.12, the dimensions of the Nd:YAG rod and its position in the cavity are chosen so as to limit the transverse mode to TEM_{00}. The cavity is provided with mirrors that both have high reflectivity (as near 100% as possible) for the fundamental, and the output coupler has high transmission at the second-harmonic wavelength. The doubling crystal is placed in the position of the beam waist in the near-hemispherical cavity formed by the mirrors of 0.5 m and 10 m radius of curvature.

Because the second harmonic at 532 nm is within the absorption band of Nd:YAG, the radiation at this wavelength initially travelling towards the left-hand mirror in Fig. 15.12 does not contribute appreciably to the 532 nm output—only the radiation travelling to the right can contribute—so not all of the available SHG efficiency can be utilized.

[42] See Hitz and Falk (1971).

A scheme[42] that overcomes this difficulty is illustrated in Fig. 15.13. Here, the non-linear crystal is situated at a beam waist in the part of a three-mirror cavity that does not contain the laser rod. This enables second-harmonic radiation emitted in both directions within the cavity to contribute to the emergent beam.

However there is one drawback (the so-called 'green problem') that is common to all schemes that employ a cavity shared by the gain medium and the doubling crystal. This arises from beating between the randomly phased longitudinal modes that produces fluctuating power at the

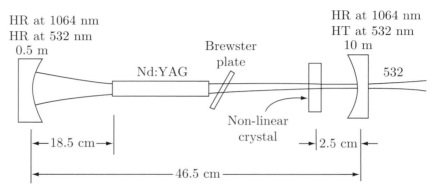

Fig. 15.12: Intra-cavity SHG with single-ended output; HR = highly reflecting, HT = highly transmitting. Reprinted figure with permission from J. E. Geusic et al., Appl. Phys. Lett., 12, pp 306–08, (1968), http://link.aip.org/link/?APPLAB/12/306/1, ©1968 by the American Institute of Physics.

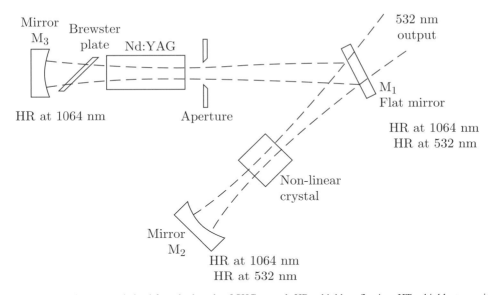

Fig. 15.13: Intra-cavity SHG with output derived from both ends of SHG crystal; HR = highly reflecting, HT = highly transmitting. (After Hitz and Falk (1971).)

fundamental—an effect that is reflected even more strongly at the second harmonic and fed back to the circulating power in the fundamental via their non-linear coupling.

To avoid this problem it is necessary either: (a) to make the cavity very long so that the longitudinal modes are very closely spaced, ensuring that oscillation occurs on many modes and hence that fluctuations cancel out statistically; or (b) to include inside the cavity frequency selective elements such as one or more tilted Fabry–Perot etalons to ensure that oscillation is restricted to a single longitudinal mode.[43]

Another approach, which avoids strong fluctuations in the output power caused by feedback between the gain medium and the non-linear process, is to place the doubling crystal in its own resonant cavity external to the laser cavity. Feedback from the external cavity into the laser is avoided using the three-mirror travelling-wave cavity illustrated in Fig. 15.14, since there is no backwards wave travelling along the laser cavity axis. However, this approach brings its own problem of the need to stabilize the external cavity resonance frequency to match that of the laser.

[43] See Section 16.2.2.

15.6 Three-wave mixing and third-harmonic generation (THG)

15.6.1 Three-wave mixing processes in general

SHG is a special case of a more general phenomenon—three-wave mixing. In SHG the waves at ω_1 and ω_2 are derived from the same source, but this is not the case for three-wave mixing in general where $\omega_1 \neq \omega_2$ and the interaction

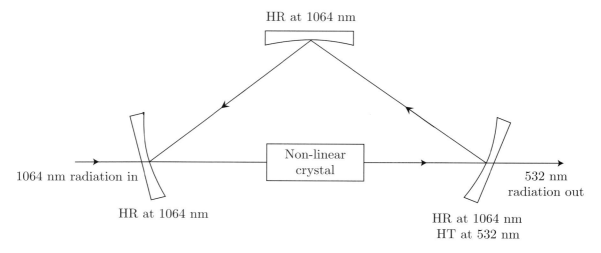

Fig. 15.14: SHG generated in an external resonant cavity; HR = highly reflecting, HT = highly transmitting.

of the waves at ω_1 and ω_2 produces a polarization and consequenlty a wave at ω_3. In the case where ω_3 represents the sum frequency, energy conservation dictates that

$$\omega_3 = \omega_1 + \omega_2 \tag{15.63}$$

$$\Rightarrow \frac{1}{\lambda_3} = \frac{1}{\lambda_1} + \frac{1}{\lambda_2} \quad \text{sum-frequency generation,} \tag{15.64}$$

while momentum conservation (see eqn (15.36)) requires that $\boldsymbol{k}_3 = \boldsymbol{k}_1 + \boldsymbol{k}_2$. This relationship allows for the possibility of non-collinear waves, but for the usual case of collinear propagation, and sum-frequency generation, we have

$$\frac{n_3}{\lambda_3} = \frac{n_1}{\lambda_1} + \frac{n_2}{\lambda_2}. \tag{15.65}$$

However, two waves interacting in a non-linear medium can in general also produce a wave at the difference frequency.[44] To see how this arises consider the polarization amplitude P_3 produced by the interaction of two waves with real amplitudes E_1 and E_2. According to eqn (15.32)

$$(2/\epsilon_0) P_3 \cos(\boldsymbol{k}_3 \cdot \boldsymbol{r} - \omega_3 t) = d_{\text{eff}} E_1 \cos(\boldsymbol{k}_1 \cdot \boldsymbol{r} - \omega_1 t) E_2 \cos(\boldsymbol{k}_2 \cdot \boldsymbol{r} - \omega_2 t)$$
$$= \frac{d_{\text{eff}}}{2} E_1 E_2 \{\cos[(\boldsymbol{k}_1 + \boldsymbol{k}_2) \cdot \boldsymbol{r} - (\omega_1 + \omega_2)t] + \cos[(\boldsymbol{k}_1 - \boldsymbol{k}_2) \cdot \boldsymbol{r} - (\omega_1 - \omega_2)t]\}. \tag{15.66}$$

[44] See Exercise 15.4.

Equating the time-varying parts of both sides of eqn (15.66) for difference-frequency mixing yields:

$$\omega_3 = \omega_1 - \omega_2 \tag{15.67}$$

$$\Rightarrow \frac{1}{\lambda_3} = \frac{1}{\lambda_1} - \frac{1}{\lambda_2} \quad \text{(difference-frequency generation)} \tag{15.68}$$

and equating the space-varying parts yields

$$k_3 = k_1 - k_2, \quad \text{difference-frequency generation} \quad (15.69)$$

which for collinear propagation implies

$$\frac{n_3}{\lambda_3} = \frac{n_1}{\lambda_1} - \frac{n_2}{\lambda_2}, \quad (15.70)$$

where we have assumed $\omega_1 > \omega_2 > \omega_3$, and $k_1 > k_2 > k_3$.

In the usual situation, the signals at ω_1 and ω_2 are provided by monochromatic sources so there are only two possibilities for ω_3—either the sum or the difference frequency. Whichever of the two (if either) satisfies one of the phase-matching conditions (15.65) or (15.70) will determine whether it is the sum or difference frequency that is generated.

15.6.2 Third-harmonic generation (THG)

It might seem possible to generate the third harmonic of a given frequency by processes exploiting the $\chi^{(3)}$ term in eqn (15.16); however, the magnitudes of $\chi^{(3)}$ encountered in practical materials are so small that no useful amounts of power can be generated by this route. The solution is to employ sum-frequency generation mixing a beam of second harmonic at $2\omega_1$ with residual fundamental at ω_1 as indicated in Fig. 15.15.

By judicious choice[45] of the polarization angle in the frequency-doubling crystal, it is possible to arrange that 2/3 of the original N photons at the fundamental frequency ω_1 are converted to $N/3$ photons at the second harmonic $2\omega_1$, leaving an equal number of photons of residual fundamental and ensuring optimum photon balance in the two beams prior to the final frequency-summing step.

Clearly, this technique of sequential frequency doubling (and sum generation if needed) can be used to generate fourth, fifth and higher harmonics of the fundamental laser frequency, albeit with diminishing returns as the final-harmonic wavelength approaches the short-wavelength transparency limit[46] of suitable non-linear materials.

[45] With the use of a waveplate to rotate polarizations to the optimum angle between stages it is theoretically possible to achieve 100% conversion to third harmonic, and in practice 80% conversion efficiency has been reported. However, for simplicity, schemes that obviate the need for polarization rotation between stages are generally preferred in practice.

[46] See Table 15.1.

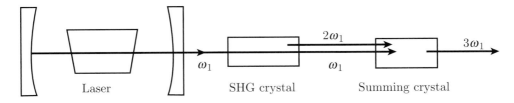

Fig. 15.15: Third-harmonic generation.

15.7 Optical parametric oscillators (OPOs)

15.7.1 Parametric interactions

One of the most interesting phenomena arising from the non-linear polarization field due to $\chi^{(2)}$ is that of the **parametric amplification** of a small signal at frequency ω_s in the presence of a strong optical wave, called the *pump*, at higher frequency ω_p. Via the $\chi^{(2)}$ interaction the pump and signal produce a third wave (called the idler wave) at the difference frequency ω_i. The interaction of the pump wave and the non-linear medium gives rise to gain at both signal and idler frequencies—gain that can be used as the basis of amplifiers and oscillators. For historical reasons such amplifiers and oscillators are known as 'parametric' devices.[47]

Both the waves at signal and idler frequency[48] arise from amplified random noise, and both[49] grow in amplitude as a result of the frequency-difference mixing interaction illustrated in Fig. 15.16. For the interaction to occur with any worthwhile strength, the waves must be phase matched.

As illustrated schematically Fig. 15.16(b), an optical parametric amplifier (OPA) merely comprises a non-linear material pumped by a strong monochromatic signal. The precise frequencies of the generated pair of beams at ω_s and ω_i are determined by the requirement that the phase-matching condition of eqn (15.70):

$$\frac{n_s}{\lambda_s} = \frac{n_p}{\lambda_p} - \frac{n_i}{\lambda_i}, \qquad (15.71)$$

and energy conservation:

$$\frac{1}{\lambda_s} = \frac{1}{\lambda_p} - \frac{1}{\lambda_i}, \qquad (15.72)$$

are simultaneously satisfied.

An expression for G_s, the gain factor of an OPA for a beam at the signal frequency after traversing a length L of non-linear medium can be obtained[50] by solving a set of coupled equations of the form of eqn (15.34), with $\omega_p = \omega_3$, $\omega_s = \omega_1$, and $\omega_i = \omega_2$. This yields

$$G_s = \frac{I_s(l = L)}{I_s(l = 0)} = \cosh^2 \Gamma L, \qquad (15.73)$$

[47] See, for example, Butcher and Cotter (1990).

[48] For historical reasons it is usual to call the higher of the two signal and idler frequencies the signal, but the choice is arbitrary. In the case where $\omega_s = \omega_i$ the process is said to be degenerate.

[49] This is in contrast to the situation for the frequency sum interaction where power is transferred from two strong input waves and the third, at the sum frequency, is the only one to grow from random noise.

[50] This assumes perfect phase matching, no depletion of the pump beam, and no external input at the idler frequency. In the case where phase matching is not perfect, and $\Delta k = k_3 - (k_1 - k_2) \neq 0$, then

$$G_s = \frac{I_s(z = L)}{I_s(L = 0)}$$

$$= 1 + \Gamma^2 L^2 \frac{\sinh^2[\Gamma^2 L^2 - (\Delta k L/2)^2]^{\frac{1}{2}}}{[\Gamma^2 L^2 - (\Delta k L/2)^2]}$$

$$\approx 1 + \Gamma^2 L^2 \left[\frac{\sin(\Delta k L/2)}{(\Delta k L/2)}\right]^2,$$

where the approximation holds in the limit of small gain, i.e. when $\Gamma^2 L^2 \ll (\Delta k L/2)^2$.

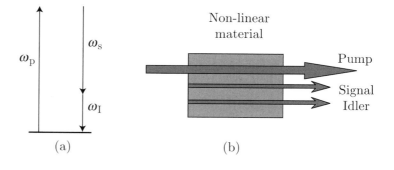

Fig. 15.16: Parametric interactions. (a) Frequency-difference mixing interaction, where ω_p, ω_s, and ω_i are the pump, signal, and frequencies, respectively. (b) Schematic diagram of an optical parametric amplifier.

in which

$$\Gamma^2 = \frac{8\pi d_{\text{eff}}^2}{\epsilon_0 c n_s n_i n_p \lambda_s \lambda_i} I_p. \qquad (15.74)$$

15.7.2 Optical parametric oscillators (OPOs)

As with a conventional laser, the way to turn an amplifier into an oscillator is to apply optical feedback by providing the non-linear crystal with a cavity resonant to the signal wavelength. Unlike a conventional laser, however, there is no inversion of population—the gain medium is the non-linear crystal pumped by a laser beam.[51]

In the case of a singly resonant oscillator (SRO) shown schematically in Fig. 15.17(a) the cavity is resonant at λ_s, as is the 'pump-enhanced' (PE-SRO) version shown in Fig. 15.17(b). This has the additional feature of returning some of the pump radiation not absorbed in the first pass back into the crystal for a second chance of conversion. In the doubly resonant oscillator (DRO) shown in Fig. 15.17(c) the cavity is simultaneously resonant at both signal and idler wavelengths. The advantage of the SRO system is that it has simple tuning characteristics, but the pump radiation must reach a relatively high threshold intensity before any output can be generated. The DRO scheme has a relatively low threshold, but the difficulty of maintaining resonance at both wavelengths as the device is tuned (by slight rotation of the non-linear crystal) makes for very complex tuning arrangements, especially if smooth tuning over an extended wavelength range is to be achieved whilst avoiding mode hopping. In certain limited circumstances it may be possible to achieve operation of triply resonant oscillator (TRO) in which radiation the pump wavelength also resonated, but the possible tuning range is rather limited.

[51] The wavelength of the pump laser has to be sufficiently short that eqn (15.72) can be satisfied.

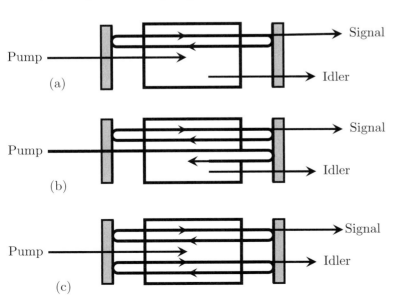

Fig. 15.17: OPO cavities. (a) SRO: singly resonant oscillator; (b) PE-SRO: singly resonant oscillator with pump beam enhanced by reflection; (c) DRO: doubly resonant OPO cavity. The left-hand mirror is transparent at λ_p but has high reflectivity at λ_s and λ_i. The right-hand mirror allows some transmission as output coupling at λ_s and λ_i but may also have a high reflectance at λ_p. Reproduced with permission from S. Wu, G. A. Blake, Z. Sun, and J. Ling, Appl. Opt., 36, 5898–5901 (1997). ©OSA (1997)

15.7.3 Practical parametric devices

It is clear from Table 15.4 that OPOs operating in the picosecond (or subpicosecond) regime with modelocked pump lasers have huge single-pass gain, despite the short lengths of non-linear medium dictated by considerations of phase matching and spatial overlap in the face of group-velocity dispersion.[52] In the c.w. regime the gain per pass is extremely small, and consequently only low-loss frequency-selective cavity elements (such as tilted Fabry–Perot etalons) can be used to reduce the linewidth.

Ebrahimzadeh (2004) has reviewed the performance characteristics of OPO systems, and the reader is directed to that review for details of device performance over very diverse temporal and wavelength regimes. A particular advantage of OPOs, as pointed out in Section 14.1, is their ability to provide widely tunable coverage of the near- and mid-infra-red regions where other tunable sources (such as dye lasers) are lacking. To give just a few examples from Ebrahizadeh's list, the introduction of PPLN as a non-linear material has enabled c.w. OPO coverage of the range from 3.25 μm to 3.95 μm with output powers of order 3.6 W for 13.5 W of pump power provided by a diode-pumped Nd:YAG laser. Pump radiation from a tunable c.w. Ti:sapphire laser has enabled mid-infra-red output from a PE-SRO, tunable from 4.07 μm to 5.25 μm.

In the picosecond regime, PPLN-based devices have shown average output powers exceeding 10 W and potential wavelength coverage of 1.3 μm to 5.0 μm for output signal or idler pulses of 50 ps at 75 MHz repetition rate.[53] LBO-based devices have demonstrated average output powers of 650 mW and tuning ranges 0.6 μm to 2.8 μm for 22-ps pulses at 75 MHz repetition frequency. However, without elaborate measures to produce line narrowing the linewidths of such devices tend to be too large for high-resolution molecular spectroscopy. Even the transform-limited linewidths of femtosecond devices (1.6 THz for a 100 fs pulse), are too large to resolve individual rotational lines of electronic transitions, which are of order 1–10 GHz due to Doppler and pressure broadening.

For the demands of molecular spectroscopy, OPOs pumped by nanosecond-pulsed lasers, such as the device shown schematically in Fig. 15.18(a), are

[52] See Section 17.1.3.

[53] See Ebrahimzadeh (2004).

Table 15.4 Parametric gain factor ΓL and fractional single-pass intensity gain G_s at the signal wavelength of OPOs for different pumping regimes (After Ebrahimzadeh (2004).)

	c.w.	Q-switched	Modelocked	Modelocked amplified
Pump pulse energy	–	10 mJ	15 nJ	10 μJ
Pump pulse duration	–	10 ns	100 fs	200 fs
Peak pump power	5 W	1 MW	150 kW	50 MW
Focused waist radius	20 μm	1 mm	15 μm	15 μm
Peak intensity I_3	400 kW cm^{-2}	30 MW cm^{-2}	20 GW cm^{-2}	7 TW cm^{-2}
Crystal length L	10 mm	10 mm	1 mm	1 mm
ΓL	0.09	0.77	1.99	37
$G_s(L)$	1.008	1.72	13.88	3.4×10^{31}

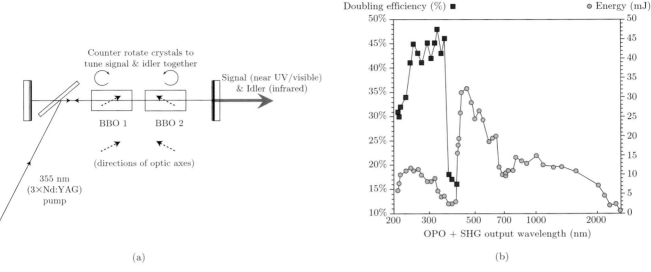

Fig. 15.18: Practical SRO pumped by frequency-tripled nanosecond-pulsed Nd:YAG laser. (a) Schematic diagram of cavity; (b) tuning characteristics. From 420–710 nm the output (circular data points) is the signal wave, from 710–2250 nm it is the idler wave. For wavelengths shorter than 420 nm the curve shows the result of frequency doubling in an extra-cavity crystal, with the efficiency of doubling denoted by square data points. Reproduced with permission from S. Wu, G. A. Blake, Z. Sun, J. and Ling, Appl. Opt., 36, 5898–5901 (1997). ©OSA (1997).

much more promising. This OPO uses two BBO crystals with Type II phase matching.[54] It has good conversion efficiency (30% at the peak of the OPO tuning curve) and is tunable from 440 to 2250 nm, with a linewidth of order 20–45 GHz without intra-cavity frequency-selective elements. Tuning is accomplished by counter-rotating the BBO crystals about axes normal to the plane of the diagram, an arrangement that cancels the effect of Poynting vector walk-off. The 355-nm pump laser provides 2.5-ns pulses of 150 mJ energy at repetition rates up to 100 Hz.

[54] Phase-matching curves for parametric generation in BBO are given in Fig. 15.19.

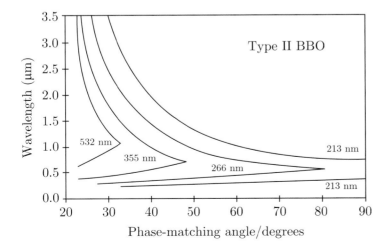

Fig. 15.19: Tuning curves at various pump wavelengths for parametric generation in BBO with Type II phase matching. For a given pump wavelength the signal and idler wavelengths at each phase-matching angle θ are given by the intersection of the appropriate curve with a vertical line passing through the abscissa at θ. (Data courtesy of Casix Inc.)

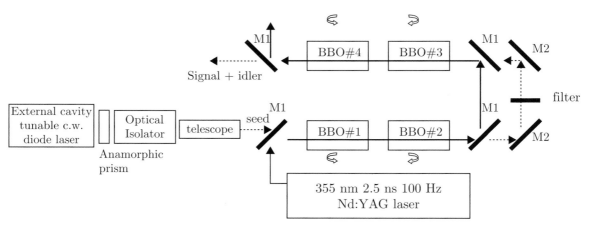

Fig. 15.20: Diode laser injection-seeded nanosecond optical parametric generator (OPG)/OPA combination. The pump steering mirrors M1 have high reflectivity at 355 nm and good transmission at signal and idler wavelengths. The filter selectively absorbs radiation at the signal wavelength, leading to seeding of the OPA stage by the idler output of the OPG. (Reprinted from S. Wu, V. A. Kapinus, and G. Blake, Optics Communications, 159, pp 74–79, (1999) with permission from Elsevier.)

[55] See Wu et al. (1999).
[56] See Fig. 9.21.

To reduce the linewidth of nanosecond pulsed parametric devices even further, the same group[55] has used seeding from a 6-mW c.w. external-cavity tunable diode laser[56] to control the linewidth of a parametric generator/amplifier combination based on the OPO design of Fig. 15.18(a). With this arrangement, shown in Fig. 15.20, they obtained output linewidths of 650 ± 150 MHz over the region near 815 nm where water vapour shows absorption on the third O–H stretch overtone of the H_2O molecule.

Further reading

The following articles from 'The Handbook of Laser Technology and Applications,' edited by C. E. Webb and J. Jones, are recommended further reading: Binks (2004), Ebrahimzadeh (2004), and Boyd (2004).

A more advanced treatment of non-linear optics is provided by Butcher and Cotter (1990).

Exercises

(15.1) **Maxwell's equations for plane waves**

  (a) Show that for plane waves of the form $D_0 \exp[i(\mathbf{k} \cdot \mathbf{r} - \omega t)]$:
   (i) $\frac{\partial}{\partial t} \mathbf{D} = -i\omega \mathbf{D}$.
   (ii) $\nabla \cdot \mathbf{D} = i\mathbf{k} \cdot \mathbf{D}$.
   (iii) $\nabla \times \mathbf{D} = i\mathbf{k} \times \mathbf{D}$.

  (b) Hence, derive eqns (15.12) to (15.15) from Maxwell's equations for a magnetically isotropic medium.

(15.2) **Field orientations in non-isotropic media**

  (a) Consider first the case of a uniform *isotropic* medium. Use eqns (15.12) to (15.15) to show that in this case $S = E \times H$ and $D \times H$ are both parallel to k. Sketch the relative orientations of all the vectors.

(b) Now consider a non-isotropic medium in which $D \not\parallel E$ and $B = \mu H$. Show that:
  (i) $D \times H \parallel k$.
  (ii) $E \times H \not\parallel k$.

(c) Use eqn (15.15) to show that:
  (i) $\omega |D| = |k| |H|$.
  (ii) $\omega E \cdot D = S \cdot k$.

(d) Hence, show that the angle between E and D is equal to that between S and k.

(e) Sketch the relative orientations of the vectors in this case.

(15.3) **Refractive index of the extraordinary wave**

We consider an extraordinary wave propagating in a uniaxial crystal such that its wave vector k lies in the (x, z)-plane and makes an angle of θ with the z-axis (the optic axis). In this case the electric and magnetic fields vary as

$$\exp[\mathrm{i}(kz\cos\theta + kx\sin\theta - \omega t)],$$

and the extraordinary wave has electric-field components in the x- and z-directions only (the wave polarized with its electric-field in the y-direction is the ordinary wave).

(a) Show that in this case the Maxwell 'curl' eqn (15.10) yields,

$$0 = H_x \quad (15.75)$$
$$k\sin\theta E_z - k\cos\theta E_x = -\mu_0 \omega H_y \quad (15.76)$$
$$0 = H_z. \quad (15.77)$$

(b) Similarly, show that the Maxwell 'curl' eqn (15.11) yields,

$$k\cos\theta H_y = \omega\epsilon_0\epsilon_{11} E_x \quad (15.78)$$
$$k\sin\theta H_y = -\omega\epsilon_0\epsilon_{33} E_z. \quad (15.79)$$

(c) By eliminating the fields E_x, E_z, and H_y, show that

$$\frac{k^2\sin^2\theta}{\epsilon_{33}} + \frac{k^2\cos^2\theta}{\epsilon_{11}} = \mu_0\epsilon_0\omega^2, \quad (15.80)$$

and hence that

$$\frac{1}{n(\theta)^2} = \frac{\sin^2\theta}{n_\mathrm{e}^2} + \frac{\cos^2\theta}{n_\mathrm{o}^2}. \quad (15.81)$$

(d) Explain the values of $n(\theta)$ that arise when $\theta = 0$ and $\theta = \pi/2$, and sketch $n(\theta)$ for a positive birefringent crystal.

(15.4) **Calculation of non-linear polarization terms**

As an example of a calculation of the non-linear polarization P_NL, here we consider the terms arising from the second-order susceptibility $\chi^{(2)}$. For the moment we shall ignore the tensor nature of $\chi^{(2)}$ that non-linear media typically display.

Let us suppose that the medium is subjected to two waves, $E_1(r, t)$ and $E_2(r, t)$, each of which may be written in the form,

$$E_i(r, t) = \frac{1}{2}[\widehat{E}_i \exp[\mathrm{i}(k_i \cdot r - \omega_i t)] + \mathrm{c.c.} \quad (i = 1, 2).$$

(a) Show that the non-linear polarization $P_\mathrm{NL} = \epsilon_0 \chi^{(2)} E^2$, where $E = E_1(r, t) + E_2(r, t)$, may be written in the form,

$$\frac{2}{\epsilon_0 \chi^{(2)}} P_\mathrm{NL} = |\widehat{E}_1|^2 + |\widehat{E}_2|^2$$
$$+ \frac{1}{2}\widehat{E}_1^2 \exp[2\mathrm{i}(k_1 \cdot r - \omega_1 t)] + \mathrm{c.c.}$$
$$+ \frac{1}{2}\widehat{E}_2^2 \exp[2\mathrm{i}(k_2 \cdot r - \omega_2 t)] + \mathrm{c.c.}$$
$$+ \widehat{E}_1 \cdot \widehat{E}_2 \exp\{[\mathrm{i}(k_1 + k_2) \cdot r - (\omega_1 + \omega_2)t]\} + \mathrm{c.c.}$$
$$+ \widehat{E}_1 \cdot \widehat{E}_2^* \exp\{[\mathrm{i}(k_1 - k_2) \cdot r - (\omega_1 - \omega_2)t]\} + \mathrm{c.c.}$$

(b) Describe the effects arising from each term on the right-hand side of the above equation.

(15.5) **The effective tensor element d_eff**

In terms of the coordinate system defined in Fig. 15.2, an electric field of amplitude $E(\omega)$ at the fundamental frequency ω is applied to a BBO (negative uniaxial) crystal with the optic axis parallel to the z-axis. The fundamental corresponds to an o-wave and the second harmonic to an e-wave. The components of the fundamental wave E are:

$$E_x = E\sin(\phi) \qquad E_y = -E\cos(\phi) \qquad E_z = 0.$$

From the Kleinman conjecture (See Appendix C) the following relations apply:

$$d_{21} = d_{16}, \quad d_{24} = d_{32}, \quad d_{31} = d_{15}, \quad d_{13} = d_{35},$$
$$d_{14} = d_{36} = d_{25}, \quad d_{12} = d_{26}, \quad d_{32} = d_{24}.$$

In addition, for the particular case of BBO:

$$d_{22} = -d_{21} = -d_{16}, \quad d_{31} = d_{32}, \quad d_{15} = d_{24},$$

with all other components zero except d_{35}.

(a) Using the matrix equation relating the components of P_NL to those of E, as given in Appendix C, show that

$$\frac{2}{\epsilon_0} P_x^{\text{NL}} = -2d_{16} \sin\phi \cos\phi E^2 = d_{22} E^2 \sin(2\phi) \quad (15.82)$$

$$\frac{2}{\epsilon_0} P_y^{\text{NL}} = \left[d_{21} \sin^2(\phi) + d_{22} \cos^2(\phi)\right] E^2$$
$$= d_{22} E^2 \cos(2\phi) \quad (15.83)$$

$$\frac{2}{\epsilon_0} P_z^{\text{NL}} = \left[d_{31} \sin^2(\phi) + d_{32} \cos^2(\phi)\right] E^2$$
$$= d_{32} E^2. \quad (15.84)$$

(b) Show also that if (see Binks (2004)) we write P^{NL} as:

$$P^{\text{NL}}(2\omega) = \sin(\theta + \pi/2)$$
$$\left[P_x^{\text{NL}} \cos\phi + P_y^{\text{NL}} \sin\phi\right]$$
$$+ P_z^{\text{NL}} \cos(\theta + \pi/2), \quad (15.85)$$

then,

$$\frac{2}{\epsilon_0} P^{\text{NL}}(2\omega) = \left[d_{22} \cos\theta \sin(3\phi) - d_{31} \sin\theta\right] E^2. \quad (15.86)$$

(c) Hence, show that in this case the effective tensor element is given by,

$$|d_{\text{eff}}| = |d_{22} \cos\theta \sin(3\phi) - d_{31} \sin\theta|. \quad (15.87)$$

(15.6) Derivation of phase-matching angle

Use eqn (15.21) and the phase-matching conditions of Section 15.4.1 to derive expressions for the phase-matching angle θ_m for:

(a) Type I phase matching in a material with positive birefringence ($n_e > n_o$);
(b) Type I phase matching in a material with negative birefringence ($n_e < n_o$);
(c) Type II phase matching in a material with positive birefringence ($n_e > n_o$);
(d) Type II phase matching in a material with negative birefringence ($n_e < n_o$). [Hint: for this case there is no analytical solution; try using a binomial expansion to find an approximate expression for $\sin\theta_m$.]

(15.7) Angular acceptance for SHG

In this problem we derive expressions for the angular acceptance for SHG. We first find the change in the wave vector mismatch $\Delta(\Delta k)$ arising from small changes in the propagation angle of the extraordinary ray.

(a) Use eqn (15.44) to show that for Type I phase matching the change in wave vector mismatch $\Delta(\Delta k)$ arising from a change $\Delta n(\theta)$ of the refractive index of the extraordinary ray is

$$\Delta(\Delta k) = 2\frac{\omega}{c}\Delta n(\theta), \quad (15.88)$$

for *both* positive and negative birefringence, where ω is the angular frequency of the fundamental wave.

(b) Find the equivalent expression for Type II phase matching in materials with positive birefringence ($n_e > n_o$).

(c) Show that you get the same expression for $\Delta(\Delta k)$ for Type II phase matching in materials with negative birefringence if dispersion is small.

(d) Hence, show that for both Type I and Type II phase matching, and for both positive and negative birefringent materials,

$$\Delta(\Delta k) \approx 2\beta\frac{\omega}{c}\Delta n(\theta), \quad (15.89)$$

where $\beta = 1$ for Type I phase matching and $\beta = 1/2$ for Type II phase matching.

(e) From eqn (15.21), show that

$$\Delta n(\theta) = \frac{1}{2} n(\theta)^3 \left(\frac{1}{n_o^2} - \frac{1}{n_e^2}\right) \sin(2\theta)\Delta\theta. \quad (15.90)$$

(f) Given that for significant second-harmonic generation we require $\Delta k L/2 < \pi$, show that the angular acceptance is given by

$$\Delta\theta < \frac{1}{\beta}\frac{\lambda_0}{L \sin 2\theta_m}\frac{1}{n(\theta)^3}\left(\frac{1}{n_o^2} - \frac{1}{n_e^2}\right)^{-1}, \quad (15.91)$$

where λ_0 is the vacuum wavelength of the fundamental radiation, and n_o, n_e, and $n(\theta)$ are evaluated at the frequency of the extraordinary wave.[57]

(g) Show that this result may be simplified to eqn (15.54).

---

[57] For the case of Type II phase matching in materials with negative birefringence we have already assumed that the refractive indices at ω and 2ω are similar.

Precision frequency control of lasers[†]

16

- 16.1 Frequency pulling — 431
- 16.2 Single longitudinal mode operation — 433
- 16.3 Output linewidth — 440
- 16.4 Frequency locking — 448
- 16.5 Frequency combs — 453
- Further reading — 456
- Exercises — 456

Since the earliest days of the laser it has been known that lasers can, at least in principle, generate radiation of very narrow linewidths. This feature gives rise to many important applications that fall under the broad categories of spectroscopy and interferometry. Examples of the former category include the establishment and transfer of frequency standards and measurement of the fundamental constants; examples of the later are the accurate measurement of lengths and detection of gravitational waves.

In this chapter we describe some techniques for achieving narrow-bandwidth operation, stabilizing the output frequency of a laser, and measuring the output frequency in terms of the primary frequency standard.[1]

16.1 Frequency pulling

As discussed in Section 6.5.2, a laser can only oscillate at frequencies that: (i) correspond to a mode of the cavity; (ii) experience sufficient gain to overcome the cavity losses.

The modes of a cavity are determined by the condition (6.7):

$$\phi_{\mathrm{rt}}(\omega_p) = 2\pi p \quad p = 0, 1, 2, 3, \ldots, \quad (16.1)$$

where here we have explicitly indicated the frequency dependence of ϕ_{rt}.

The term **frequency pulling** refers to the fact that the frequencies of the cavity modes are changed, or 'pulled', when a population inversion is produced, and hence the mode frequencies of the 'cold' cavity—i.e. one in which there is no population inversion—differ slightly from those of a 'hot' cavity, in which the population inversion is non-zero.

Consider a laser comprised of a gain medium of length ℓ_{g} located in a linear cavity of length L_{c}. In the absence of a population inversion, eqn (16.1) becomes for this cavity

$$\frac{2\omega_p}{c}L_{\mathrm{c}} + \frac{2\omega_p}{c}(n_{\mathrm{g}} - 1)\ell_{\mathrm{g}} = 2\pi p,$$

where ω_p is the angular frequency of the cold cavity mode, and n_{g} is the refractive index of the gain medium in the absence of a population inversion. This result may be written more conveniently as

[1] The underlying theory of many of the sections of this chapter are often most conveniently developed in terms of the angular frequency (units of rad s^{-1}), whereas the application of these results or comparison with real systems is most conveniently made in terms of the frequency (units of Hz). Both kinds of frequency will be used, as should be clear from the context. However, to avoid doubt, in this chapter all angular frequencies and linewidths will be denoted by ω or Ω.

$$\omega_p T_c = 2\pi p, \qquad (16.2)$$

where T_c is the time taken to complete one round trip of the cold cavity.[2]

[2] Note that here we are ignoring the difference between the phase and group velocity. It should be clear that the velocity determining the cavity modes is the phase velocity. The role of the group and phase velocities in the behaviour of modelocking is explored in Exercise 17.2.

Now, suppose that a population inversion is created in the gain medium. This will change the real part of the refractive index of the gain medium by an amount Δn_g^r, whereupon the frequencies of the cavity modes will become,

$$\omega_p' T_c + 2\frac{\omega_p'}{c}\Delta n_g^r \ell_g = 2\pi p. \qquad (16.3)$$

In order to progress beyond this point it is necessary to make some assumptions about the gain medium. It is shown in Exercise 16.5 that for a homogeneously broadened transition of width $\Delta\omega_H$ the real and imaginary parts of the change in refractive index arising from the population inversion are related by,

$$\frac{\Delta n_g^i}{\Delta n_g^r} = \frac{1}{2}\frac{\Delta\omega_H}{\omega_0 - \omega}.$$

Further, the gain coefficient α_I is related to the imaginary part of the change in refractive index by,[3]

[3] Both the change in refractive index Δn_g^r and the gain coefficient depend on the populations of the two laser levels. We require the value of Δn_g^r under the operating conditions of the laser, and hence we must use the *saturated* gain coefficient.

$$\frac{\alpha_I}{2} = -\frac{\omega}{c}\Delta n_g^i.$$

Hence, we may rewrite eqn (16.3) as,

$$\omega_p' T_c + 2\frac{\omega_p' - \omega_0}{\Delta\omega_H}\alpha_I \ell_g = 2\pi p.$$

Eliminating $2\pi p$ with the help of eqn (16.3), we find the new mode frequencies to be:

$$\omega_p' = \frac{\left(\frac{T_c}{2\alpha_I \ell_g}\right)\omega_p + \left(\frac{1}{\Delta\omega_H}\right)\omega_0}{\left(\frac{T_c}{2\alpha_I \ell_g}\right) + \left(\frac{1}{\Delta\omega_H}\right)}. \qquad (16.4)$$

We see that the frequency of a mode of the hot cavity is a weighted average of the mode frequency of the cold cavity and the centre frequency of the transition, and hence *the frequencies of the hot cavity are pulled towards the centre frequency ω_0*.

We may develop a more useful expression for the weighting factors in eqn (16.4) for the case of a low-gain laser. In such systems the intra-cavity intensity, and hence the saturated gain coefficient, is approximately constant, and hence in one round trip the beam is amplified by a factor of approximately $\exp(2\alpha_I \ell_g)$. In the absence of cavity losses other than those that occur at the mirrors, the condition for steady-state laser oscillation may then be written[4]

[4] This result is analogous to eqn (6.65), with $\kappa = 0$. The difference is simply that the unsaturated gain coefficient α_0, which should be used when calculating the *threshold* condition, has been replaced by the saturated gain coefficient, α_I, which should be used when considering the laser under *operating* conditions.

$$R_1 R_2 \exp(2\alpha_I \ell_g) = 1,$$

and hence we may write

$$\frac{2\alpha_I \ell_g}{T_c} = -\frac{1}{T_c}\ln R_1 R_2 \approx \frac{1}{\tau_c},$$

where τ_c is the cavity lifetime.[5] Now, from eqn (6.58) $\Delta\omega_c \approx 1/\tau_c$, where $\Delta\omega_c$ is the linewidth of the cold cavity. Hence, the pulled frequencies of the hot cavity may be written in the alternative form,

$$\omega'_p \approx \frac{\left(\frac{1}{\Delta\omega_c}\right)\omega_p + \left(\frac{1}{\Delta\omega_H}\right)\omega_0}{\left(\frac{1}{\Delta\omega_c}\right) + \left(\frac{1}{\Delta\omega_H}\right)}. \quad (16.5)$$

[5] See Section 6.5.1. Here we have assumed that the mirror reflectivitites are close to unity so that we may write $\ln R_1 R_2 \approx -(1 - R_1 R_2)$.

We see that the weighting factors are simply the reciprocals of the linewidths of the cold cavity and the laser transition. Since typically $\Delta\omega_c \ll \Delta\omega_H$ the shift in the mode frequency from that of the cold cavity is very small. In the limit of $\Delta\omega_c/\Delta\omega_H \ll 1$ we may write,

$$\omega'_p \approx \omega_p + \frac{\Delta\omega_c}{\Delta\omega_H}(\omega_0 - \omega_p). \quad (16.6)$$

Example: As an example we consider the effect of frequency pulling in a Nd:YAG laser operating at 1064 nm. For a cavity of length $L_c = 0.5$ m with $R_1 R_2 = 0.9$ we find $\tau_c = 32$ ns, $\Delta\nu_c = 4.9$ MHz, and $\Delta\nu_{p,p-1} = 300$ MHz. The homogeneous linewidth of the transition is 160 GHz and hence $\Delta\nu_H/\Delta\nu_c \approx 3 \times 10^4$. Since $|\nu_p - \nu_0|$ will be of order $\Delta\nu_{p,p-1}$, we see that the frequency shift arising from this effect will be less than 10 kHz.

16.2 Single longitudinal mode operation

In order to obtain the narrowest possible bandwidth it is necessary to ensure that the laser oscillates on a single mode of the cavity. Restricting oscillation to a single transverse mode—almost always the lowest—may be achieved by placing a suitable aperture within the cavity, as discussed on page 96. The measures required to restrict oscillation to a single longitudinal mode depend on whether the laser transition is homogeneously or inhomogeneously broadened. We note that the term **single longitudinal mode (SLM)** operation usually refers to operation on a single longitudinal *and* single transverse mode of the cavity.

As discussed in Section 6.6.2, in principle a homogeneously broadened laser transition will only oscillate on the cavity mode nearest to the peak of the gain curve. However, in practice spatial hole-burning can allow other longitudinal modes to reach the threshold for laser oscillation by feeding off regions of unsaturated gain near the nodes of an oscillating mode.[6] In such cases it is necessary to take additional measures to ensure SLM operation, such as those described below.

[6] See page 114.

For an inhomogeneously broadened transition, as discussed in Section 6.6.3, laser oscillation is possible on any cavity mode for which the small-signal round-trip gain exceeds the round-trip loss. For inhomogeneously broadened laser transitions it is therefore almost always necessary to take special steps to ensure oscillation on a single mode.

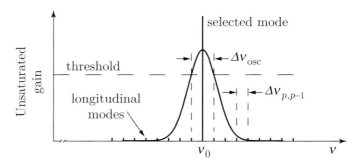

Fig. 16.1: Selection of a single longitudinal mode by using a sufficiently short laser cavity. It has been assumed that the selected mode has a frequency equal to the centre frequency of the transition.

Some techniques for achieving SLM operation on both homogeneously and inhomogeneously broadened laser transitions are described below.

16.2.1 Short cavity

The simplest way to ensure oscillation on a single longitudinal mode is to make the cavity very short, so that the longitudinal mode spacing $\Delta\nu_{p,p-1} \approx c/2L_c$ is comparable to or greater than the linewidth of the laser transition. More quantitatively, as illustrated in Fig. 16.1, if the range of frequencies above the threshold for oscillation is $\Delta\nu_{osc}$, single-mode oscillation may occur provided $\Delta\nu_{p,p-1} > \frac{1}{2}\Delta\nu_{osc}$, or[7]

$$L_c < \frac{c}{\Delta\nu_{osc}}. \qquad (16.7)$$

Of course, it is also necessary for the small-signal round-trip gain to exceed the cavity losses and hence the cavity cannot be made too short. It is therefore easiest to achieve single mode oscillation with this technique for laser transitions with a narrow bandwidth. For example, for the He-Ne laser $\Delta\nu_D \approx \Delta\nu_{osc} = 1.7$ GHz. From eqn (16.7) we then expect single longitudinal mode operation to be possible for cavity lengths shorter than approximately 175 mm.

The broader linewidths (100s of GHz) of solid-state lasers means that achieving SLM operation with this approach requires the cavity length to be no longer than about 1 mm. This can be realized using the 'microchip' geometry illustrated in Fig. 7.19.

If a short laser cavity is used it is often desirable to be able to tune the frequency of the oscillating mode to the centre frequency of the laser transition. This can be achieved by mounting one of the cavity mirrors on piezoelectric transducers. It is also worth emphasizing that the cavity length required to achieve SLM operation depends on the rate of pumping of the laser transition, since increasing the pump rate increases $\Delta\nu_{osc}$, allowing other longitudinal modes to oscillate.

---

[7] We have assumed that the selected mode is maintained at or near the centre of the gain curve by some external means (e.g. by one of the techniques described in Section 16.4). In the absence of frequency locking to line centre, the oscillating modes are free to drift relative to the gain curve, in which case in order to ensure that only one of them can oscillate it would be necessary to reduce the length of the cavity so that $\Delta\nu_{p,p-1} > \Delta\nu_{osc}$.

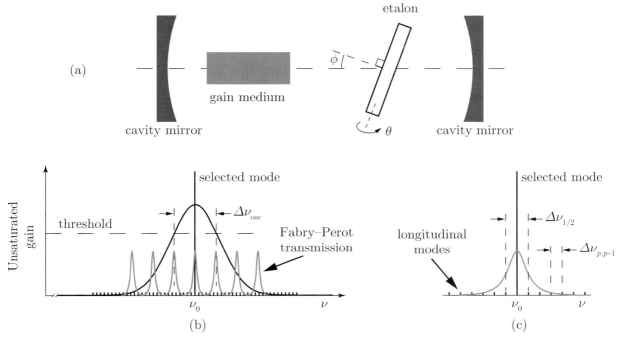

Fig. 16.2: (a) Use of an intra-cavity etalon to select a single longitudinal mode. Note that the angle of tilt ϕ is exaggerated. (b) Relation between the peaks of the etalon transmission and the range of frequencies above threshold when eqn (16.9) is satisfied. (c) A close-up showing the relation between the full width at half-maximum of a peak in the etalon transmission, $\Delta\nu_{\frac{1}{2}}$, and the longitudinal mode spacing $\Delta\nu_{p,p-1}$ when eqn (16.8) is satisfied and the frequency of the selected cavity mode coincides with the etalon peak.

16.2.2 Intra-cavity etalons

Figure 16.2 illustrates how an intra-cavity etalon can restrict oscillation to a single longitudinal mode by increasing the round-trip losses for other modes.[8]

In order to prevent oscillation on other longitudinal modes the etalon must satisfy two conditions: (i) the width of the etalon peaks must be sufficiently narrow to discriminate between nearby longitudinal modes; (ii) the separation of the etalon peaks must be sufficiently large to ensure that the cavity modes within adjacent etalon peaks lie outside the range at which oscillation can occur.

We may express condition (i) as:

$$\frac{\Delta\nu_{\frac{1}{2}}}{2} < \Delta\nu_{p,p-1}, \tag{16.8}$$

where $\Delta\nu_{\frac{1}{2}}$ is the full width at half-maximum frequency width of one of the etalon peaks.[9]

[8] For further information on Fabry–Perot etalons see Section 6.5.1 and standard textbooks on optics, such as Brooker (2003).

[9] It should be apparent that there is a degree of arbitrariness in this condition. If the frequency of the selected longitudinal mode coincides with a peak in the transmission of the etalon, then satisfying eqn (16.8) would mean that the intensity transmission of adjacent longitudinal modes would be 50% of that of the selected mode for each pass through the etalon. The difference in transmission that is actually required to discriminate between the selected mode and a nearby mode depends in detail on the properties of the laser transition and the laser cavity. Equation (16.8) should therefore be regarded as a reasonable estimate of the requirement on $\Delta\nu_{\frac{1}{2}}$.

Condition (ii) may be written as,

$$\Delta \nu_{\text{FSR}} > \frac{\Delta \nu_{\text{osc}}}{2}, \quad (16.9)$$

where $\Delta \nu_{\text{FSR}}$ is the free spectral range of the etalon.

For an etalon of thickness d we have from eqns (6.52), (6.53) and (6.54):

$$\Delta \nu_{\frac{1}{2}} = \frac{\Delta \nu_{\text{FSR}}}{\mathcal{F}}; \quad \Delta \nu_{\text{FSR}} = \frac{1}{2nd}; \quad \mathcal{F} = \frac{\pi \sqrt{R}}{1 - R},$$

where $\mathcal{F}$ is the finesse of the etalon, n is the refractive index between the reflecting surfaces of the etalon, and $R = \sqrt{R_1 R_2}$, where R_1 and R_2 are the reflectivity of the two surfaces.

Equations (16.8) and (16.9) yield a condition on the thickness of the etalon:

$$\frac{\Delta \nu_{\text{osc}}}{2} < \frac{c}{2nd} < \mathcal{F} \frac{c}{L_c}. \quad (16.10)$$

Comparing the extreme left and extreme right of eqn (16.10), we conclude,

$$L_c < (2\mathcal{F}) \frac{c}{\Delta \nu_{\text{osc}}}. \quad (16.11)$$

and hence we see that the etalon can increase the maximum length of the cavity for which SLM operation is possible by a factor of $2\mathcal{F}$ (compare eqn (16.7)). This is a substantial advantage since values of the finesse up to about 30 may be achieved relatively straightforwardly.

It is worth pointing out several practical aspects. The etalons employed in laser cavities are usually solid etalons comprising a block of transparent material, such as fused silica, with high-reflectivity coatings on each surface. Solid etalons are more robust than air-spaced etalons since the two reflecting surfaces cannot become misaligned. In order to avoid the etalon acting as a cavity mirror, the etalon is rotated by a small angle ϕ, as illustrated in Fig. 16.2(a).[10]

[10] This angle must be kept as small as possible whilst ensuring that reflections from the etalon do not couple radiation back into the laser cavity. The reason for this is that in order for the etalon to promote interference it is necessary that the forward- and backward-propagating beams within the etalon overlap; if they do not, interference does not occur and the etalon does not provide frequency selection.

In order to ensure that the selected longitudinal mode coincides with a peak in the etalon transmission it is necessary either to adjust the length of the cavity, or to adjust the angle of the etalon. For an etalon illuminated at non-normal incidence the frequencies transmitted with peak intensity are given by,

$$\nu_m = \frac{c}{2nd \cos \theta'} m \quad m = 1, 2, 3, \ldots, \quad (16.12)$$

where θ' is the angle the rays make to the normal of the etalon *within* the etalon.

In order to prevent the required angular adjustment from becoming unmanageably small,[11] the angle θ must be close to zero, which appears to contradict the requirement that ϕ is large enough to prevent feedback into the cavity. The solution is to adjust the frequencies of the etalon peaks by rotating the etalon about the axis orthogonal to that which adjusts ϕ. This small rotation can be adjusted and controlled by mounting the etalon on a galvanometer mechanism or by using a piezoelectric crystal.

[11] See Exercise 16.2.

> **Example:** As an example we consider single longitudinal mode operation of an argon-ion laser operating at 488.0 nm. Taking $\Delta \nu_{\text{osc}} \approx \Delta \nu_D = 4.1$ GHz, the maximum length the cavity could be to ensure SLM operation in the absence of an intra-cavity etalon is $c/\Delta \nu_{\text{osc}} \approx 73$ mm. Using an intra-cavity etalon with a finesse $\mathcal{F} = 30$, the cavity length could be increased by a factor of $\mathcal{F}$ to 2.2 m.
>
> Let us design the etalon in more detail. In order to satisfy the first inequality of eqn (16.10) we require the etalon spacing $d < c/n\Delta \nu_{\text{osc}} = 49$ mm, where we have taken $n = 1.5$. Choosing d too small will require unreasonably high values of the finesse, or short cavity lengths. Hence, we choose $d = 25$ mm that incorporates a 'safety factor' of 2. The second inequality of eqn (16.10) then gives $L_c/\mathcal{F} < 2nd = 75$ mm. Taking the length of the cavity to be $L_c = 1.5$ m, single longitudinal mode operation would require the finesse of the intra-cavity to be greater than 20, which can be achieved relatively easily.

Multiple etalons

For laser transitions with large linewidths it may not be possible to satisfy eqn (16.10) with realistic cavity lengths L_c and reasonable values of the finesse $\mathcal{F}$. In these circumstances, multiple etalons may be used to achieve SLM operation.

The way this works is straightforward. For a given finesse, the linewidth of an etalon (etalon 1) can be made as small as desired by increasing the etalon thickness. However, doing so decreases the free spectral range of the etalon so that the adjacent etalon peaks lie within $\Delta \nu_{\text{osc}}$. This problem may be overcome by introducing a second etalon (etalon 2) that satisfies two conditions: (i) the width of the etalon peaks must be sufficiently narrow to discriminate between adjacent transmission peaks of etalon 1; (ii) the separation of the etalon peaks must be sufficiently large to ensure that its adjacent transmission peaks lie outside the range at which oscillation can occur. In other words, the parameters of the second etalon are chosen according to eqns (16.8) and (16.9), but with the longitudinal mode spacing $\Delta \nu_{p,p-1}$ replaced by the free spectral range of etalon 1. If (i) and (ii) cannot be simultaneously satisfied by the second etalon, additional etalons may be employed.

It is shown in Exercise 16.3 that for two etalons of finesse $\mathcal{F}_1$ and $\mathcal{F}_2$ the maximum cavity length for which SLM operation can be achieved is given by

$$L_c < (2\mathcal{F}_1)(2\mathcal{F}_2)\frac{c}{\Delta \nu_{\text{osc}}}, \tag{16.13}$$

i.e. increased by a further factor of $2\mathcal{F}_2$ from the single-etalon result.

16.2.3 Ring resonators

Although a simple analysis suggests that homogeneously broadened laser transitions should always oscillate on only a single cavity mode, in practice spatial hole-burning can allow more than one mode to oscillate.[12] Hence, one way of ensuring SLM operation with a homogeneously broadened gain

[12] See page 114.

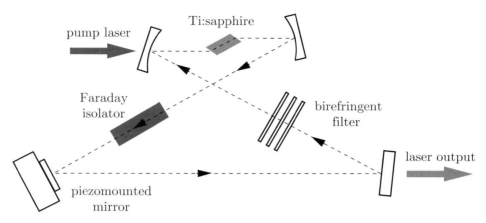

Fig. 16.3: Single longitudinal mode tunable Ti:sapphire laser employing a unidirectional ring cavity. Wavelength tuning is achieved by rotating the birefringent filter; one of the cavity mirrors is mounted on a piezoelectric crystal so that the length of the cavity may be adjusted as the wavelength of the laser is tuned.

medium is to avoid spatial hole-burning by preventing the formation of a standing-wave pattern in the cavity. This may be done by employing a ring cavity, such as that illustrated in Fig. 16.3, and *ensuring that laser radiation can only propagate round the ring in one direction*. A further advantage of a unidirectional ring cavity is that power is extracted uniformly from the length of the gain medium—not just near the antinodes—which increases the efficiency with which energy is extracted from the gain medium.

In order to ensure that the laser mode propagates in one direction only, it is necessary to incorporate some sort of 'optical diode' within the cavity that transmits light in one direction with low loss, but introduces high loss for light propagating the opposite way. Figure. 16.4 illustrates how this may be achieved using the **Faraday effect**.

The Faraday effect[13] occurs in any transparent material in which an applied magnetic field has a component B along the direction of propagation of light. To understand how the Faraday effect can be used to make an optical diode, imagine that linearly polarized light propagates through a material in which a magnetic field is applied parallel to the direction of the light propagation. As a result of the Faraday effect the plane of polarization of the light will rotate through some angle β as observed by an observer looking towards the source of light. Imagine now that the light is reflected by a mirror. In propagating back through the material the plane of polarization of the light will continue to rotate in the same direction *with respect to the original observer*, so that it leaves the material with the plane of polarization at an angle 2β with respect to the original plane of polarization. In other words, in propagating back through the material the plane of polarization *does not* 'unwrap' itself and restore the original direction of polarization.

Figure 16.4(c) shows how the Faraday effect can be used to make a **Faraday isolator**. Light incident from the left passes through a linear polarizer and a glass rod in which a longitudinal magnetic field is applied by an array of permanent magnets. After leaving the glass rod the light passes through a half-

[13]Named after its discoverer, Michael Faraday (1791–1867). Further information on the Faraday effect may be found in Hecht (2002).

16.2 *Single longitudinal mode operation* 439

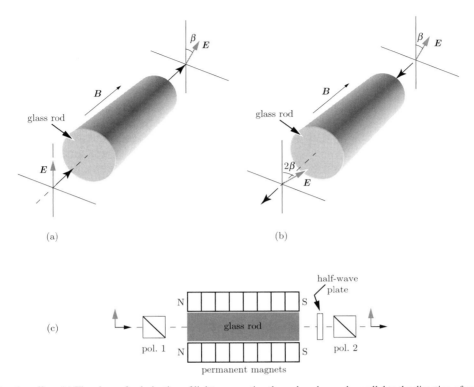

Fig. 16.4: The Faraday effect. (a) The plane of polarization of light propagating through a glass rod, parallel to the direction of an applied magnetic field $\boldsymbol{B}$, is rotated through an angle β. (b) If the transmitted light is returned through the glass rod the plane of polarization is rotated by a further angle β so that the plane of polarization of the double-passed light makes an angle 2β with respect to that of the incident radiation. (c) An optical diode—known as a Faraday isolator—can be constructed by placing a rod of transparent material in a longitudinal magnetic field formed by an array of permanent magnets. Either side of the rod are placed linear polarizers with their transmission axes parallel. A half-wave plate is placed between the exit end of the rod and polarizer 2, with its fast axis oriented so that (in the case shown) light propagating from left to right reaches the polarizer 2 with its plane of polarization parallel to the transmission axis of that polarizer. Light propagating in the reverse direction reaches polarizer 1 with its plane of polarization rotated by an angle 2β with respect to the transmission axis of polarizer 1, and hence experiences significant loss.

wave plate and a second linear polarizer oriented parallel to the first. After leaving the glass rod the plane of polarization of the light will be rotated with respect to its original direction by the Faraday effect; the half-wave plate is oriented so as to return the plane of polarization to its original direction so that it passes through the second polarizer without loss. For the reasons discussed above, if the plane of polarization of light propagating through the glass rod from left to right is rotated through an angle β, light propagating through the Faraday isolator from right to left will be incident on the leftmost polarizer with its plane of polarization at an angle of 2β to the transmission axis of the polarizer. As such, light propagating through the isolator from left to right suffers no – or at least low—loss; light propagating in the other direction experiences high losses. If $\beta = 45°$ light propagating in the 'wrong' direction would be completely blocked; in practice, however, β does not need to be as large as this, and a rotation of a few degrees is sufficient to introduce enough loss to suppress the counter-propagating mode.

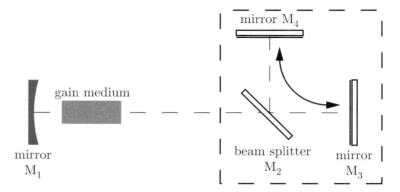

Fig. 16.5: Schematic diagram of a laser cavity employing a Fox–Smith interferometer. The optical components within the dashed box provide frequency-selective feedback into the main laser cavity.

16.2.4 Other techniques

A variety of other techniques have been developed for achieving single longitudinal mode operation. Figure 16.5 shows one of these, the Fox–Smith interferometer. In this, one of the cavity mirrors is replaced by a beam splitter and a pair of mirrors. The amplitude of the waves reflected from this combination is sharply peaked about certain frequencies determined by the optical path between mirrors M_3 and M_4, providing highly frequency-selective feedback into the main laser cavity.[14] By adjusting the relative separation of the two mirrors of the interferomter it is possible to control which frequencies are reflected back into the laser cavity. A wide variety of other multiple mirror laser cavities can be employed to achieve the same objective.[15]

Many other techniques have been employed to achieve SLM operation, including:

- Replacing one of the cavity mirrors with an etalon. This acts as a frequency-dependent mirror in a similar way to the Fox–Smith interferometer.
- Placing a thin absorber, such as a metal film coated onto a plate, at a node of the selected mode. Other modes will not have their nodes at this position and hence will experience greater losses and so be suppressed.

[14] In many respects the beam splitter and mirror pair behave like a Fabry–Perot etalon. The difference is that the intensity of waves reflected from an etalon is generally high apart from sharp dips at those frequencies, that are transmitted by the etalon. See Exercise 16.4.

[15] For a comprehensive review of these see Smith (1972).

16.3 Output linewidth

Suppose that the techniques described above are used to restrict laser oscillation to a single mode of the cavity. What would be the linewidth of the output radiation?

We know that the linewidth must be less than that of the laser transition, $\Delta\nu_H$ or $\Delta\nu_D$, since the laser cavity provides additional restrictions on the frequencies that can oscillate. Given that we have restricted oscillation to a single cavity mode, we might expect the output linewidth to be equal to the linewidth $\Delta\nu_c$ of the cold cavity or, equivalently, the frequency width of a transmission peak if the cavity were used as a Fabry–Perot etalon. This is not the case, however. The reason that the modes of the cold cavity have a finite frequency width is that the amplitude of radiation oscillating within the cavity decays on each round trip because of the finite reflectivity of the cavity mirrors and other intra-cavity losses. In other words, radiation is only 'stored' within the cavity for a time of order the cavity lifetime τ_c, and hence we expect the modes to have a width of order $1/\tau_c$ (see eqn (6.58)). However, in a laser oscillator operating under steady-state conditions the cavity losses are exactly compensated for by optical gain, corresponding to $\tau_c \to 0$ and hence $\Delta \nu_c \to 0$! This cannot be the case; something else must determine the linewidth of the laser output.

16.3.1 The Schawlow–Townes limit

The fundamental limit to the linewidth of a laser operating continuously on a single cavity mode is determined by the role of spontaneous emission, as first recognized by Schawlow and Townes in their pioneering paper of 1958.

There will always be some spontaneous emission into the same cavity mode as the oscillating mode of the laser; and, since these spontaneously emitted photons cannot be distinguished from those resulting from stimulated emission, they form an intrinsic part of the laser output. As we show below, the effect of the spontaneous emission is a slow diffusion of the phase of the laser output that is associated with a finite frequency width.

Suppose that the oscillating mode within the laser cavity contains an average of $\bar{n}$ photons. The energy of the mode will be proportional to $\bar{n}$ or, equivalently, to the square of the amplitude of the electric field E_0 of the mode. Thus, we have $E_0 \propto \sqrt{\bar{n}}$.

If the cavity lifetime is τ_c, the photons in the mode are lost from the cavity at a rate $\bar{n}/\tau_c$: these form the output beam. Under steady-state conditions the lost photons are replaced at exactly the rate they are lost by stimulated and spontaneous emission into the oscillating mode. The photons arising from stimulated emission are perfectly coherent with the existing photons in the mode; consequently, if there were no spontaneous emission the phase of the electric field at any point in the cavity would be constant, and hence the electric field could be written as $E(t) = E_0 \exp(-\mathrm{i}\,[\omega_L t - \phi])$, where ω_L is the angular frequency of the oscillating mode, and ϕ is an arbitrary constant phase. The amplitude and phase of this field could be represented by a constant vector in the complex plane, as illustrated in Fig. 16.6(a).

Occasionally, however, a photon will be emitted into the mode spontaneously. This spontaneous photon will have a random phase relative to that of the stimulated photons, leading to a small shift in the phase (and amplitude) of the oscillating mode. This process is illustrated schematically in Fig. 16.6(b) that shows schematically the change in the amplitude and phase of the mode after several photons have been emitted spontaneously into the mode. The change in amplitude of the mode is very small, and not relevant to our present purpose. However, the change in phase of the mode is important: spontaneous

Fig. 16.6: Origin of the Schawlow–Townes linewidth. (a) In the absence of spontaneous emission, if a laser oscillates on a single cavity mode the electric field may be written as $E(t) = E_0 \exp(-\mathrm{i}[\omega_L t - \phi])$, where ϕ is an arbitrary constant phase. This may be represented by a constant vector in an Argand diagram, as shown, where the axes correspond to $\Re\left[E(t)\exp(\mathrm{i}\omega_L t)\right]$ and $\Im\left[E(t)\exp(\mathrm{i}\omega_L t)\right]$. The effect of spontaneous emission is shown in (b). After each spontaneously emitted photon is emitted into the oscillating mode, the tip of the electric-field vector moves in a random direction by an amount E_{sp}. The dots shown in (b) show the positions occupied by the tip of the electric-field vector after a very large number of photons have been emitted spontaneously. The inset in (b) shows the effect on the electric field vector of the emission of 8 spontaneously emitted photons, and in particular the net displacement, ΔE_θ, along a direction, $\hat{\boldsymbol{\theta}}$, orthogonal to the original electric-field vector. Note that the electric-field vector of each spontaneously emitted photon has a different component along $\hat{\boldsymbol{\theta}}$, and consequently on average the displacement ΔE_θ is given by the result of a one-dimensional random walk (along $\hat{\boldsymbol{\theta}}$) of varying step size. The resultant displacement causes the orientation of the electric field vector to rotate through an angle θ, as shown.

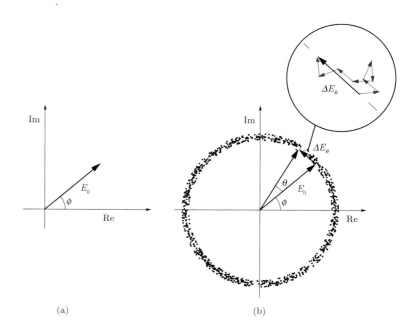

[16] Remember that a purely monochromatic wave must have a constant phase and amplitude over an infinite interval.

[17] See, for example, Reif (1985).

emission causes the phase of the mode to change randomly by small amounts. In fact the phase of the oscillating mode diffuses with a characteristic time, τ_{ST}, so that the relative phase of two points on the wave separated by a delay of order τ_{ST} is uncertain. This corresponds to a finite frequency width $\Delta\omega_{\mathrm{ST}} = 1/\tau_{\mathrm{ST}}$.[16]

We now estimate the magnitude of the linewidth arising from the effect of spontaneous emission. A spontaneously emitted photon will contribute an amplitude of magnitude E_{sp} to the electric field of the mode. Now, since $E_0 \propto \sqrt{n}$ we may write

$$\frac{E_{\mathrm{sp}}}{E_0} = \frac{\sqrt{1}}{\sqrt{n}}. \tag{16.14}$$

Let us suppose that N photons have been emitted spontaneously into the mode. From Fig. 16.6(b) we see that the change θ of the phase of the mode is given by

$$\theta = \frac{\Delta E_\theta}{E_0}, \tag{16.15}$$

where ΔE_θ is the component of the vector sum of the electric fields of the N photons along the $\hat{\boldsymbol{\theta}}$ direction.

The value of ΔE_θ depends on the phases of the spontaneously emitted photons. Since these are uncorrelated, we may use the results of the theory of a one-dimensional random walk of variable step size.[17] These state that for N

steps in the random walk—i.e. after the emission of N spontaneous photons—the mean amplitude and mean square amplitude along $\hat{\boldsymbol{\theta}}$ are given by:

$$\overline{\Delta E_\theta} = N\overline{E_{\theta\,\text{sp}}} \tag{16.16}$$

$$\overline{\Delta E_\theta^2} = N\overline{E_{\theta\,\text{sp}}^2}, \tag{16.17}$$

where $\overline{E_{\theta\,\text{sp}}}$ and $\overline{E_{\theta\,\text{sp}}^2}$ are the mean amplitude and mean square amplitude along $\hat{\boldsymbol{\theta}}$ of a single spontaneously emitted photon.[18] Since for spontaneously emitted photons E_{sp} is the same, but their phase can take any value with equal probability, it is easy to show that

$$\overline{E_{\theta\,\text{sp}}} = 0 \tag{16.18}$$

$$\overline{E_{\theta\,\text{sp}}^2} = \frac{E_{\text{sp}}^2}{2}, \tag{16.19}$$

and hence that after N photons have been emitted into the mode:

$$\overline{\theta} = 0 \tag{16.20}$$

$$\overline{\theta^2} = \frac{\overline{\Delta E_\theta^2}}{E_0^2} = \frac{N}{2}\left(\frac{E_{\text{sp}}}{E_0}\right)^2. \tag{16.21}$$

If photons are spontaneously emitted into the mode at a rate R_{sp}, then after a time τ the root mean square value of θ will be

$$\theta_{\text{rms}} = \sqrt{\overline{\theta^2}} = \sqrt{\frac{R_{\text{sp}}\tau}{2\bar{n}}}, \tag{16.22}$$

where we have used eqn (16.14).

We may calculate the rate at which θ evolves if we know the rate at which photons are emitted spontaneously into the mode. It was shown in Exercise 2.2 that the rate of spontaneous emission into a mode is equal to $1/\bar{n}$ of the rate of stimulated emission from the upper laser level. However, under steady-state conditions the rate of stimulated emission from the upper level must equal $\bar{n}/\tau_c$, the rate at which photons leave the cavity.[19] Hence, we may write:

$$R_{\text{sp}} = \frac{1}{\bar{n}}\frac{\bar{n}}{\tau_c} = \frac{1}{\tau_c}. \tag{16.23}$$

Estimate of linewidth

We may obtain an estimate of the Schawlow–Townes linewidth in a straightforward way if we can estimate the time τ_{ST} for the uncertainty in the phase of the mode to become significant. Clearly, the phase is uncertain by a 'significant' amount once $\theta_{\text{rms}} \approx \pi/2$ and hence we may write,

$$\sqrt{\frac{R_{\text{sp}}\tau_{\text{ST}}}{2\bar{n}}} \approx \frac{\pi}{2},$$

from which we deduce

$$\Delta\omega_{\text{ST}} = \frac{1}{\tau_{\text{ST}}} \approx \left(\frac{2}{\pi}\right)^2 \frac{R_{\text{sp}}}{2\bar{n}}. \tag{16.24}$$

[18] To be more precise, the averages on the left-hand sides of eqns (16.16) and (16.17) are *ensemble* averages. To calculate an ensemble average one imagines that a large number of systems are prepared identically, and N photons are emitted into the mode spontaneously. The ensemble average $\overline{\Delta E_\theta}$ is the value of ΔE_θ averaged over all the members of the ensemble. The averages on the right-hand sides of eqns (16.16) and (16.17) are averages over the distribution of possible values of a given step. For further details see, for example, Reif (1985).

[19] This is not quite correct, as explored in Exercise 16.6.

Full result

In Exercise 16.7 we show how the approach above may be used to show that the lineshape of the emission of the laser is Lorentzian with a full width at half-maximum of:

$$\Delta\omega_{ST} = \frac{R_{sp}}{2\bar{n}}, \qquad (16.25)$$

and hence our estimate above is within a factor of $(2/\pi)^2$ of the correct result.

It is useful to write $\bar{n}$ in terms of the output power P of the laser. Since the rate of loss of photons from the cavity is $\bar{n}/\tau_c$ we have $P = (\bar{n}/\tau_c)\hbar\omega_L$. Using eqn (16.23) we may then write the linewidth in frequency units as:

$$\Delta\nu_{ST} = \left(\frac{N_2}{N^*}\right)_{th} \frac{\pi h \nu_L (\Delta\nu_c)^2}{P}. \quad \textbf{Schawlow–Townes limit}$$

(16.26)

[20]See Exercise 16.6.

where the additional factor of $(N_2/N^*)_{th}$ arises from a more careful derivation of R_{sp}.[20] This factor, the ratio of the density of the upper laser level to the population inversion density—both evaluated under the operating conditions of the laser—is typically close to unity.

> **Example:** As an example we calculate the Schawlow–Townes linewidth for the Nd:YAG laser considered in the Example on page 433. There we found that for a cavity of length $L_c = 0.5$ m with $R_1 R_2 = 0.9$: $\tau_c = 32$ ns and $\Delta\nu_c = 4.9$ MHz. Since Nd:YAG is a four-level laser we may take $(N_2/N^*)_{th} \approx 1$. Assuming a modest output power of 1 W we find $\Delta\nu_{ST} = 1.4 \times 10^{-5}$ Hz, or a relative bandwidth of $\Delta\nu_{ST}/\nu_L = 5 \times 10^{-20}$!

16.3.2 Practical limitations

The linewidth of the output of a real free-running laser is always greater—and usually very much greater—than the Schawlow–Townes limit since any disturbance of the laser cavity will affect the cavity mode.

The frequency of a cavity mode will be affected by anything that changes the round-trip phase of the mode, processes that may be conveniently divided into: rapid fluctuations, which determine the linewidth of the output; and slow changes, which cause the mean frequency of the output to drift.

Changes in cavity length

To provide more quantitative information we consider the effect of a change in length of a linear cavity. For a given longitudinal mode p, changing the length of the cavity by ΔL_c changes the frequency of the mode by an amount $\Delta\nu_p$ such that

$$\frac{\Delta\nu_p}{\nu_p} = -\frac{\Delta L_c}{L_c}, \qquad (16.27)$$

where L_c is the original length of the cavity. Substitution into eqn (16.26) shows that in order for this frequency shift to be smaller than $|\Delta\nu_{ST}|$ we require,

$$\frac{\Delta L_c}{L_c} < \frac{1}{2}\left(\frac{N_2}{N^*}\right)_{th} \frac{h\nu_L}{P\tau_c} \frac{1}{Q}.$$

We note that the term in brackets is of order unity, and the remaining terms are equal to the reciprocal of the cavity Q times the number of photons emitted in one cavity lifetime τ_c. For example, to achieve the Schawlow–Townes limit in a 1-mW laser operating at 500 nm in a 1-m long cavity with $\tau_c = 100$ ns requires $\Delta L_c \lesssim 5 \times 10^{-3}$ fm. This is much smaller than the size of an atomic nucleus!

Thermal changes

The change in length, ΔL, of a rod of length L in response to a temperature change T is given by

$$\frac{\Delta L}{L} = \alpha_T \Delta T,$$

where α_T is the coefficient of thermal expansion of the rod. Hence, from eqn (16.27), thermal expansion of the material on which the cavity mirrors are mounted will cause the frequency of a cavity mode to change by

$$\frac{\Delta\nu_p}{\nu_p} = -\alpha_T \Delta T. \quad (16.28)$$

Once again, it is found that in order to keep such frequency shifts below the Schawlow–Townes linewidth, the required temperature stability is impracticably high.

Example: As an example we consider the effect of thermal expansion in the invar spacer rods separating the cavity mirrors of a laser operating at a wavelength of 500 nm. Invar is a steel alloy with a very low coefficient of thermal expansion $\alpha_T \approx 10^{-6}$. Suppose we wish to keep shifts in the frequency of the cavity to less than 1 MHz. This corresponds to $\Delta\nu_p/\nu_p < 1.7 \times 10^{-9}$, and from eqn (16.28) we see that this requires $\Delta T < 1.7$ mK.

Pressure changes

If the medium within a linear cavity of length L_c has a refractive index n, the change in frequency of a cavity mode resulting from a change in refractive index Δn is given by[21]

$$\frac{\Delta\nu_p}{\nu_p} = -\frac{\Delta n}{n}. \quad (16.29)$$

[21] For simplicity, we have assumed that the medium undergoing the change in pressure fills the cavity completely.

The refractive index of an ideal gas varies with pressure P according to

$$n - 1 = \beta P,$$

where β is a constant. Hence, a change in pressure of ΔP causes a change in the frequency of the cavity mode given by,

$$\frac{\Delta \nu_p}{\nu_p} = -\frac{\beta}{n}\Delta P. \tag{16.30}$$

Example: Consider the effect of pressure changes in the cavity of a laser operating at a wavelength of 500 nm. Once again, let us suppose we wish to keep shifts in the frequency of the cavity to less than 1 MHz so that $\Delta \nu_p/\nu_p < 1.7 \times 10^{-9}$. For standard air (dry air at 15 °C at a pressure of 101 325 Pa) the refractive index at 500 nm is 1.00028. Hence, $\beta = 0.00028/101\,325 = 2.8 \times 10^{-9}\,\text{Pa}^{-1}$. From eqn (16.30) we then see that a shift in frequency of 1 MHz corresponds to a pressure change of 0.6 Pa or 6×10^{-3} mbar. We note that changes in weather conditions cause changes in atmospheric pressure of order 10 mbar.

From the examples above we see that in order to minimize long-term drifts in the output frequency of a laser it is crucial that the pressure and temperature of the environment of the laser are stabilized.

16.3.3 Intensity noise

In addition to shifts and broadening of the output frequency, changes in environmental conditions also cause fluctuations or drifts in the *intensity* of the laser output known, respectively, as **intensity noise** and **intensity drift**. Some sources of intensity noise for different types of lasers are listed in Table 16.1.

Relative intensity noise

The intensity noise of a laser is usually described in terms of the **relative intensity noise (RIN)**. This is calculated as follows.

Table 16.1 Sources of intensity noise in different types of laser.

Type of laser	Sources of intensity noise
Gas	Fluctuations in the current from the power supply
	Discharge instabilities
Dye	Density fluctuations
	Air bubbles
Solid state	Fluctuations in pumping
Semiconductor	Fluctuations in driving current
	Electron–hole recombination noise[a]

[a] In a semiconductor laser changes in the number of electrons or holes change the refractive index of the material. Hence, fluctuations in the carrier densities caused by spontaneous emission can lead to frequency broadening and intensity noise. See Note on page 255 and Henry (1982).

Suppose that the average output power of a laser is $\bar{P}$ and that the fluctuations in the output power are described by $\Delta P(t)$. We may then define an intensity autocorrelation function by

$$C_{pp}(\tau) = \frac{\overline{\Delta P(t)\Delta P(t+\tau)}}{\bar{P}^2}. \tag{16.31}$$

The RIN is defined[22] as the Fourier transform of the intensity autocorrelation:

$$\text{RIN}(\omega) = \int_{-\infty}^{\infty} C_{pp}(\tau) \exp(-i\omega\tau)\,d\tau \tag{16.32}$$

$$C_{pp}(\tau) = \frac{1}{2\pi}\int_{-\infty}^{\infty} \text{RIN}(\omega) \exp(i\omega\tau)\,d\omega, \tag{16.33}$$

where the second relation follows from the fact that $\text{RIN}(\omega)$ is a Fourier transform of $C_{pp}(\tau)$.

The relative intensity noise of a laser may often be reduced by employing a so-called **noise eater**, which measures the output power of the laser and applies negative feedback to either the laser pumping or to an electro-optic modulator located prior to the final output of the laser. As an example, Fig. 16.7 shows the measured RIN spectrum of a diode-pumped Yb:YAG ring laser with and without negative feedback applied to the drive current of the diode laser.[23] Note that the RIN of the free-running laser has a pronounced peak for frequencies of approximately 500 kHz, corresponding to the frequency of the relaxation oscillations of the laser.[24] The noise eater reduces the RIN at low frequencies by as much as 40 dB. At very high frequencies the noise is dominated by internal noise processes and hence cannot be reduced by feedback control of the pump power.

[22] The relative intensity noise may also be expressed in frequency units. Since the RIN is a spectral quantity, i.e. it has units of (frequency)$^{-1}$, we have $\text{RIN}(\nu) = 2\pi\text{RIN}(\omega)$.

[23] The RIN is given in units of dB Hz^{-1}, this is simply $\text{RIN}\,[\text{dB Hz}] = 10\log[\text{RIN}(\nu) \times 1\,\text{Hz}]$.

[24] For a discussion of relaxation oscillations, see Section 8.1.2. For an estimate of the frequency of the relaxation oscillations in this system, see Exercise 8.2.

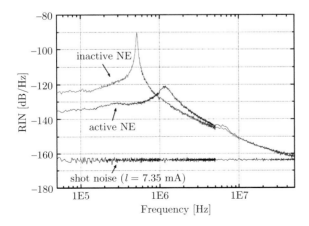

Fig. 16.7: Measured relative intensity noise of a diode-pumped Yb:YAG laser with an active and inactive noise eater. (After Burdack et al. (2006).)

> **Example:** It is worth highlighting the connection between the RIN and the magnitude of the intensity fluctuations. From eqn. (16.33), $C_{pp}(0) = \frac{1}{2\pi}\int_{-\infty}^{\infty} \text{RIN}(\omega)d\omega \approx \frac{1}{2\pi}\text{RIN}_{\text{peak}}(\omega)\Delta\omega = \text{RIN}_{\text{peak}}(\nu)2\pi\Delta\nu$. For the case shown in Fig. 16.7, $\text{RIN}_{\text{peak}}(\nu) \approx -90\,\text{dB}\,\text{Hz}^{-1}$ and $\Delta\nu \approx 50\,\text{kHz}$. Hence, $C_{pp}(0) = \overline{\Delta P^2}/\bar{P}^2 = 3 \times 10^{-4}$, corresponding to intensity fluctuations of about 2%.

16.4 Frequency locking

From the sections above it is clear that even if a laser operates on a single cavity mode, interactions with the environment can cause the frequency of the output of the laser to fluctuate and drift, which increases the linewidth of the laser output significantly above the limit set by the Schawlow–Townes linewidth. In order to reduce the linewidth, and to prevent drift, it is necessary to lock the frequency to a stable reference frequency. In principle, this is straightforward:

1. The frequency ν of the laser to be locked is compared with some fixed frequency ν_{ref}.
2. Any difference in frequency is converted to an error signal, which increases with the magnitude of the frequency error and differs in sign according to the sign of $(\nu - \nu_{\text{ref}})$.
3. The frequency of the oscillating mode is adjusted—usually by changing the length of the laser cavity—so as to minimize the magnitude of the error signal, thereby minimizing the frequency error.

In order to increase the signal-to-noise level of the error signal it is advantageous to modulate the frequency of the laser to be locked and employ phase-sensitive detection (PSD). In many implementations the frequency of the laser is modulated by mounting one of the cavity mirrors on a piezoelectric stage and modulating the cavity length by a small amount. However, there are disadvantages to this approach. The frequencies of many perturbations (such as acoustic noise) are in the kHz range, and in order to minimize their effect it is necessary to modulate the mirror at frequencies above this, which is difficult to achieve in practice. Further, if the cavity length is modulated then so is the output frequency of the laser, which may be undesirable. Hence, it is often better to modulate a portion of the beam external to the laser cavity by passing it through an electro-optic phase modulator, as shown in Fig. 16.8. With this arrangement it is possible to modulate a portion of the beam at frequencies of a few MHz, whilst leaving the beam sent to the experiment unmodulated.

Form of error signal

We now outline the form of the error signal for the general case of a frequency-modulated wave passing through (or reflecting from) a reference system with a frequency-dependent *amplitude* response $a(\omega)$ that is strongly peaked in the region of the reference frequency ω_{ref}. From eqn (8.61), the amplitude

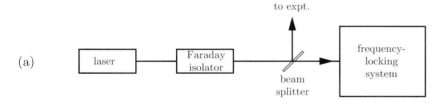

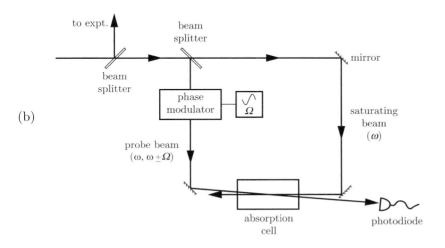

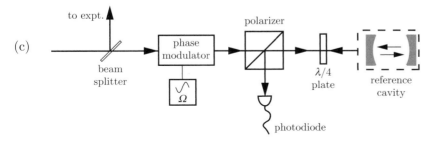

Fig. 16.8: (a) Schematic diagram of the arrangement used to frequency lock a laser. (b) Frequency locking to a Doppler-free absorption line. (c) Frequency locking to a reference cavity.

of the wave leaving a frequency modulator with a time-dependent amplitude transmission $\exp(i\beta \sin \Omega t)$ may be written as,

$$E(t) = E_0 \exp(-i\omega t) \exp(i\phi_0) \sum_{m=-\infty}^{\infty} J_m(\beta) \exp(im\Omega t), \qquad (16.34)$$

where β is proportional to the phase modulation introduced by the modulator. If the amplitudes of the sidebands other than $m = \pm 1$ may be ignored, we have

$$\begin{aligned} E(t) = E_0 \exp(i\phi_0) \{ \\ - J_1(\beta) \exp[-i(\omega + \Omega)t] \\ + J_0(\beta) \exp(-i\omega t) \\ + J_1(\beta) \exp[-i(\omega - \Omega)t] \}, \end{aligned}$$

where we have used the fact that $J_{-1}(\beta) = -J_1(\beta)$. After passing through the reference system the amplitude of the radiation is

$$E(t) = E_0 \exp(i\phi_0) \{$$
$$- J_1(\beta) a(\omega + \Omega) \exp[-i(\omega + \Omega)t]$$
$$+ J_0(\beta) a(\omega) \exp(-i\omega t)$$
$$+ J_1(\beta) a(\omega - \Omega) \exp[-i(\omega - \Omega)t] \}.$$

In Exercise 16.9 it is shown that the signal at the modulation frequency Ω is given by,

$$S_\Omega(t) \propto \Omega |a(\omega)| \left. \frac{\partial |a(\omega)|}{\partial \omega} \right|_\omega \cos \Omega t. \qquad (16.35)$$

The amplitude of this signal—which is known as the error signal—is proportional to the derivative of $|a(\omega)|$, and hence the sign of the error signal depends on whether ω is greater or less than the reference frequency ω_{ref}. Figure 16.9 shows the form of the error signal for the simplified case of a reference system with a Lorentzian amplitude response function centred at $\omega = \omega_{\text{ref}}$.

A wide variety of techniques for frequency-locking lasers has been developed.[25] Two of these are described below.

[25] For a review of these, see, for example, Barwood and Gill (2004).

16.4.1 Locking to atomic or molecular transitions

It might be thought that the frequencies of transitions in atoms or molecules could provide an absolute frequency reference able to be reproduced in any laboratory. However, this is not the case since the frequencies of atomic transitions depend to some extent on environmental factors such as temperature, pressure and the concentration of impurity species. Further, the details of the frequency-locking technique, such as the depth of the frequency modulation, can affect the frequency of the locked laser. For this reason an agreed set of reference transitions, with associated uncertainties, and a so-called *mise en pratique*, which specifies the conditions under which the laser and reference should be operated in order to obtain the specified reproducibility, is overseen by the Comité International des Poids et Mesures (CIPM).[26]

[26] See Quinn (2003).

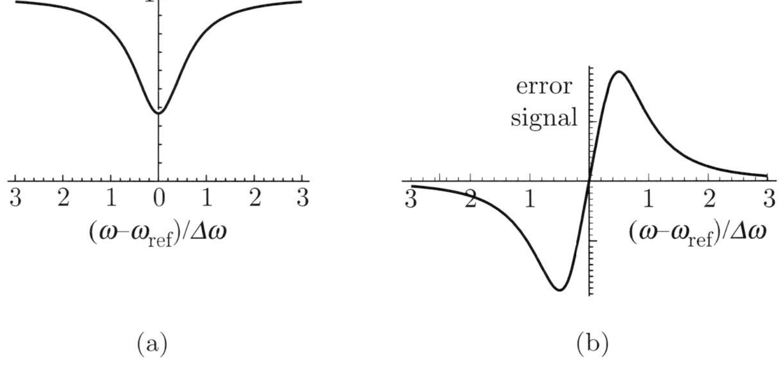

Fig. 16.9: Frequency locking to a Lorentzian response function centred at ω_{ref} with a full width at half-maximum $\Delta\omega$. (a) the amplitude response function $a(\omega)$; (b) the amplitude of the error signal $S_\Omega(t)$ as a function of the output frequency ω of the laser.

Notwithstanding these comments, locking to atomic or molecular transitions provides the best long-term stability and gives highly reproducible results: the frequency reproducibility is typically found to be 1 part in 10^{11}.

The most convenient transitions suitable as reference frequencies occur in atoms or molecules in the gaseous state, and consequently it is necessary to use **Doppler-free** techniques to eliminate Doppler broadening of the reference transition,[27] as illustrated in Fig. 16.8(b). In this approach the laser beam is divided into a strong 'saturating' beam and a weak 'probe' beam. The saturating and probe beams pass through the absorption cell in an exact, or near, counter-propagating geometry, and the intensity of the transmitted probe beam is measured by a photodiode. This is the same arrangement used in saturation absorption spectroscopy.[28]

Prior to passing through the absorption cell, the probe beam is passed through a phase modulator, operating at a frequency Ω, which generates sidebands at $\omega \pm \Omega$ and allows an error signal to be produced by phase-sensitive detection.[29] Following the discussion above, in order to determine the form of the error signal we must find the frequency response $a(\omega)$ of the reference system, which in this case is the frequency dependence of the amplitude transmission of the probe beam. Let us first consider the amplitude transmission of the probe as a function of frequency *in the absence of the saturating beam*. In this case, the transmitted amplitude will be large for frequencies ω such that $|\omega - \omega_0| \gg \Delta\omega_D$, where ω_0 and $\Delta\omega_D$ are the frequency and Doppler width, respectively, of the absorption transition. In contrast, for detunings $|\omega - \omega_0|$ comparable to or small compared to $\Delta\omega_D$ the transmission of the probe beam will be small. This behaviour is shown in Fig. 16.10.

The intensity of the probe beam is chosen to be small compared to the saturation intensity of the absorption transition. In contrast, that of the saturating beam is comparable to, or larger than, the saturation intensity. Thus, the saturating beam reduces the ground-state population of those atoms with which it interacts. Now, since the probe and saturating beams have the same frequency ω but propagate in opposite directions, they in general interact with

[27] For further details of a range of Doppler-free techniques used in spectroscopy see Foot (2005)

[28] See, for example, Foot (2005).

[29] When saturated absorption spectroscopy is undertaken in this way it is known as FM spectroscopy.

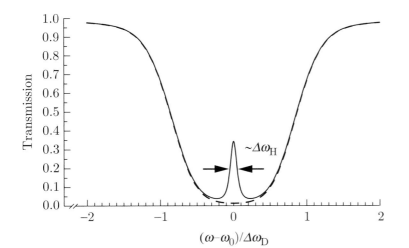

Fig. 16.10: Calculated frequency dependence of the transmission of a weak probe through a sample of Doppler-broadened atoms with (solid line) and without (dashed line) a counter-propagating saturating beam with the same frequency as that of the probe beam. Note that the main absorption has a frequency width of order the Doppler width $\Delta\omega_D$, but the spike of reduced absorption has a width approximately equal to the homogeneous width of the atoms, $\Delta\omega_H$. For the purposes of illustration, in this calculation $\Delta\omega_H$ was set equal to $0.1\Delta\omega_D$; in practice the ratio of the homogeneous to inhomogeneous linewidths would usually be very much smaller.

atoms with different velocities v_z. If we take the probe beam to propagate along the z-axis, an atom moving with velocity v_z appears to have a centre frequency $(1 - v_z/c)\omega_0$ to the probe beam, but $(1 + v_z/c)\omega_0$ to the saturating beam. Thus, the two beams interact with atoms of equal and opposite velocity:[30]

[30] Also see Section 6.6.3.

$$\text{Probe:} \quad v_z = \left(1 - \frac{\omega}{\omega_0}\right)c \qquad (16.36)$$

$$\text{Saturating:} \quad v'_z = -\left(1 - \frac{\omega}{\omega_0}\right)c. \qquad (16.37)$$

For most values of ω, $v_z \neq v'_z$ and consequently the ground-state population of those atoms with which the probe interacts is not saturated by the saturating beam, and hence the transmission of the probe beam is the same as when the saturating beam is switched off. However, if $\omega \approx \omega_0$ then $v_z \approx v'_z \approx 0$ and the two beams interact with the same atoms. In this case, the transmission of the probe beam is increased above that obtained in the absence of the saturating beam since the intense saturating beam reduces the ground-state population of the same atoms with which the probe interacts. Since atoms with a given velocity v_z are homogeneously broadened with a linewidth $\Delta\omega_H$, the frequency width of the increase in probe transmission near $\omega = \omega_0$ is of order $\Delta\omega_H$, as illustrated in Fig. 16.10. We see that this technique of saturated absorption spectroscopy has circumvented the effects of Doppler broadening and generated a spectral feature with a width of order the homogeneous linewidth of the transition, which will usually be very much narrower than the Doppler width.[31]

[31] For more detailed considerations see Exercise 16.10.

16.4.2 Locking to an external cavity

Locking to the resonant frequency of an external cavity provides better short-term stability than locking to an atomic transition since the error signals are large and hence have a high signal-to-noise ratio. There are many ways in which this can be done, but the most widely used is the **Pound–Drever–Hall technique** illustrated schematically in Fig. 16.8(c).

The Pound-Drever-Hall technique is similar to the FM spectroscopy technique mentioned above with the transmission of a frequency-modulated beam through an absorbing medium replaced by reflection from a high-finesse cavity.[32] The advantage of using the signal *reflected* from the cavity is that the error signal has a much wider 'capture' range than that which would be obtained by monitoring the transmitted beam. To see this we note that the cavity acts as a Fabry–Perot etalon, and hence the transmission of the cavity is only large for frequencies within of order the linewidth, $\Delta\omega_c$, of the cavity. The linewidth of the cavity must be small if the frequency of the laser is to be locked tightly to one of the resonant frequencies of the cavity. However, for a reference system with a response function comprising a single peak of width $\Delta\omega$, the error signal is only significant for frequencies ω within approximately $\Delta\omega$, as illustrated in Fig. 16.9(b). Consequently, in transmission a reference cavity can only achieve and maintain locking if the laser frequency is within approximately $\Delta\omega_c$ of one of the resonance frequencies of the cavity. This behaviour may also be considered in the time domain; a narrow linewidth

[32] In this application cavity finesses up to 2×10^5 may be obtained by using 'super-polished' mirrors.

cavity corresponds to a long cavity lifetime τ_c and consequently the transmitted beam does not respond to fluctuations in the beam that occur in a time significantly shorter than τ_c, or, equivalently, to frequency shifts greater than of order $1/\tau_c \approx \Delta\omega_c$.

The beam reflected from a reference cavity behaves quite differently since it is comprised of a component reflected directly from the first mirror plus leakage of radiation circulating within the cavity. The response function is more complex, and leads to an error signal that has a very wide capture range, yet a narrow central feature that can be used to lock the frequency to a tight tolerance.

Figure 16.11(a) shows as a function of frequency the amplitude and phase of the signal reflected from a reference cavity. It may be seen that the reflected amplitude is approximately unity apart from sharp dips in the region of the resonance frequencies of the cavity; this behaviour is to be expected from our understanding of the *transmission* of a Fabry–Perot etalon. Figure 16.11(c) shows the error signal derived from a beam modulated at a frequency Ω. The error signal is large, and has the correct sign, for frequencies within approximately $\pm\Omega$ of the resonance frequency. At the same time the central feature has a much smaller width, of order $\Delta\omega_{1/2} = \Delta\omega_{\rm FSR}/\mathcal{F}$, where $\mathcal{F}$ is the finesse of the reference cavity, allowing narrow frequency-locked linewidths to be obtained. This approach can be used to achieve extremely narrow linewidths if the reference cavity is isolated against mechanical and thermal fluctuations and drift. For example, a dye laser operating at 563 nm has been stabilized to a linewidth of only 0.2 Hz over an averaging time of 32 s, with a frequency stability of 3×10^{-16} at 1 s.[33]

[33] See Young et al. (1999).

16.5 Frequency combs

In many applications of stabilized, narrow-linewidth laser radiation it is also necessary to know the frequency of the laser to high accuracy. Until quite recently relating the frequency of a laser to the primary frequency standard[34] was a difficult task, involving complex chains of klystrons and lasers linking the laser frequency to the primary frequency standard.

In recent years this approach has been largely superseded by optical frequency combs based on the output of a femtosecond laser oscillator.[35] The output of a modelocked laser comprises a series of pulses separated by the cavity round-trip time T_c. As first recognized by Theodor Hänsch, straightforward Fourier-transform considerations tell us that the frequency spectrum of the pulse train must comprise a series of discrete frequencies given by

$$\omega_p = \omega_{\rm ce} + \frac{2\pi}{T_c} p \tag{16.38}$$

$$\text{or} \quad \nu_p = \nu_{\rm ce} + \frac{1}{T_c} p \tag{16.39}$$

$$= \nu_{\rm ce} + p f_{\rm rep} \quad p = 0, 1, 2, 3, \ldots, \tag{16.40}$$

[34] The primary frequency standard is presently based on a hyperfine transition in caesium, which has a defined frequency of 9 192 631 770 Hz. This transition defines the second. The metre is defined as the length of the path travelled by light in vacuum during a time interval of $1/299\,792\,458$ s, which fixes the speed of light in vacuo at $c = 299\,792\,458\,{\rm m\,s^{-1}}$.

[35] This achievement was recognized by the award of half the 2005 Nobel Prize in Physics to John Hall and Theodor Hänsch "for their contributions to the development of laser-based precision spectroscopy, including the optical frequency comb technique." The other half of the prize was awarded to Roy Glauber "for his contribution to the quantum theory of optical coherence."

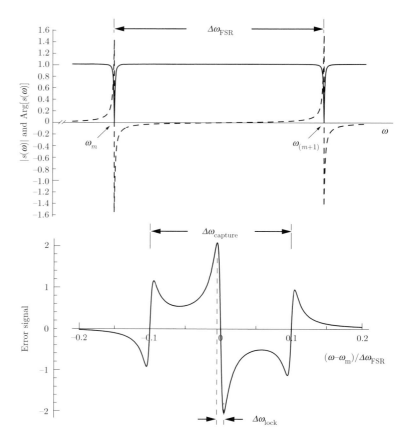

Fig. 16.11: The Pound–Drever–Hall technique. (a) shows the amplitude (solid line) and phase (dashed line) of the beam reflected from a cavity as a function of frequency, where ω_m and $\omega_{(m+1)}$ are two adjacent resonant frequencies of the cavity. (b) shows the error signal as a function of frequency near one of the resonant frequencies. Note that the 'capture' range of the error signal is a significant fraction of the free spectral range, $\Delta\omega_{\text{FSR}}$, of the cavity, but that the error signal has a slope of very large magnitude near the resonant frequency; the frequency range of this feature is of the order of $\Delta\omega_{\text{FSR}}/\mathcal{F}$, where $\mathcal{F}$ is the finesse of the reference cavity. This combination allows the frequency of the laser to be locked to that of the cavity easily, whilst ensuring that the linewidth of the locked laser is narrow.

[36] See Section 8.3.2 and Exercise 8.3.

[37] It is straightforward to measure the beat frequency by illuminating a non-linear detector with both sources simultaneously.

where ν_{ce} is known as the 'off-set frequency' and $f_{\text{rep}} = 1/T_c$ is the pulse repetition rate of the modelocked laser. We note that if $\nu_{\text{ce}} = 0$ then the pulses in the mode-locked train are identical in amplitude and phase. If $\nu_{\text{ce}} \neq 0$ then the carrier-envelope phase of subsequent pulses will slip by a constant amount ϕ_{slip}.[36]

The spectrum of discrete frequencies output by a modelocked laser is known as an **optical frequency comb**, and it forms a convenient ruler against which an unknown frequency ν can be measured. If ν_{ce} and f_{rep} are known accurately, then the unknown frequency can be determined by measuring the lowest beat frequency, δ, between the modelocked train and the unknown frequency.[37] The unknown frequency is then given by

$$\nu = \nu_{\text{ce}} + p f_{\text{rep}} \pm \delta. \tag{16.41}$$

It is assumed that ν has been established by lower-resolution methods sufficiently well that the integer p and the sign of $\pm\delta$ may be determined unambiguously.

The pulse repetition rate f_{rep} may be determined by measuring the beat frequency between a high harmonic of f_{rep} and a known frequency generated by a high-stability radio-frequency synthesiser; the laser repetition rate may then be stabilized by a feedback loop that adjusts the laser cavity so as to keep this beat frequency constant.

The solution to the problem of measuring ν_{ce} is to use the **self-referencing technique** illustrated in Fig. 16.12. In this method, frequencies near the red end of the spectrum of the laser output are frequency doubled in a non-linear crystal such that each frequency ν_p in the modelocked train generates a frequency $2\nu_p = 2\nu_{ce} + 2pf_{rep}$. The frequency-doubled beam is combined on a non-linear photodetector with frequencies near the blue end of the original spectrum. The lowest frequency beat generated is that between $2\nu_p$ and ν_{2p} and is given by,

$$f_{beat} = 2(\nu_{ce} + pf_{rep}) - (\nu_{ce} + 2pf_{rep}) = \nu_{ce}. \quad (16.42)$$

Thus, the beat frequency gives ν_{ce} directly, which allows it to be stabilized by feedback to the cavity of the modelocked laser.[38,39]

In order for the self-referencing technique to work it is necessary for the spectrum of the modelocked laser to contain frequencies at ν_p and $2\nu_p$, i.e. the spectrum must span an octave. The broadest bandwidth modelocked laser available today is the Ti:sapphire laser discussed in Sections 7.3.5 and 17.3.1. However, even these very broad bandwidth sources typically have an output spectrum covering only 30 THz compared to a centre frequency of 370 THz. To allow self-referencing it is therefore necessary to broaden the spectrum of pulses from the modelocked laser. This is achieved by propagating the pulses through an optical fibre that causes substantial broadening by self-phase-modulation so that, typically, the spectrum of the transmitted pulses extends from 500 nm to 1100 nm.[40] Since the new frequencies are generated by sum- and difference-mixing of the frequencies input to the fibre, the frequency spectrum of this 'white-light continuum' generated in the fibre is also given by eqn (16.40).

[38] If ν_{ce} is stabilized then the carrier-envelope off-set changes in successive pulses by $\phi_{slip} = \omega_{ce} T_c$ (see Exercise 8.3). Setting $\phi_{slip} = 2\pi/n$ ensures that every nth mode-locked pulse has the same CEO phase. Systems in which this is achieved are said to have 'stabilized CEO phase'; see, for example Jones et al. (2000) and Baltuska et al. (2003).

[39] The reader may have noticed that two parameters of the train of modelocked pulses are stabilized by adjusting the laser cavity: the pulse repetition rate f_{rep} and the off-set frequency ν_{ce}. The pulse repetition rate depends on the cavity length and the group velocity of the pulse within the cavity; the off-set frequency depends on the cavity length and the difference between the group and phase velocities (see Exercise 17.2). Control of f_{rep} and ν_{ce} therefore requires control of the cavity length and the cavity dispersion. The cavity length may be adjusted by moving one of the cavity mirrors with a piezoelectric transducer; the dispersion of the cavity may be controlled by changing the length of dispersive material in the cavity or by adjusting the pump power, which alters the dispersion through the Kerr effect. Although these adjustments are not usually orthogonal, they affect the round-trip group and phase delays differently, which is sufficient to control both f_{rep} and ν_{ce}. For further details see Reichert et al. (1999) and Jones et al. (2000).

[40] For efficient broadening of the spectrum it is necessary for the fibre to have zero, or close to zero, dispersion at the frequencies of interest. This is usually achieved by using so-called photonic crystal fibre, an air-silica microstructure that may be engineered to control the dispersion properties. See, for example, Russell (2003).

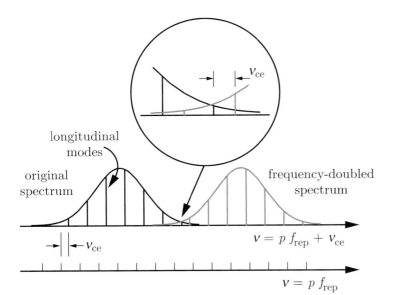

Fig. 16.12: Self-referencing technique for determining the off-set frequency ν_{ce} in pulses output by a modelocked laser oscillator. The beat frequency between modes in the blue-wing of the original spectrum and those in the red wing of the frequency-doubled light is equal to ν_{ce}. For this method to work the spectrum of the original pulse must span at least one octave. For clarity we have used an unrealistically large mode spacing; typically of order 10^5 longitudinal modes span the laser linewidth.

Using optical frequency combs, measuring frequencies in the optical region is reduced to measuring three, much lower frequencies ($f_{\rm rep}$, $\nu_{\rm ce}$, and δ) that may be compared directly with a microwave frequency standard. The frequency comb technique has been demonstrated[41] to generate a uniform comb of frequencies spaced by the pulse repetition frequency—i.e. obeying eqn (16.40)—to 1 part in 10^{19}, and can, for example, be used to measure frequency differences in the visible part of the spectrum with a fractional uncertainty of 2×10^{-15} in an averaging time of only 10 s. The technique allows frequencies to be transferred between the microwave and optical spectral regions, and hence allows frequencies within this range to be compared directly with the Cs primary standard. However, the accuracy with which this can be done (in a reasonable time) is limited by the relatively low frequency, 10^{10} Hz of the primary standard. There is therefore considerable interest in developing a stable and reproducible *optical* frequency standard that would operate at $10^{14} - 10^{15}$ Hz and could be transferred by optical frequency combs to lower frequencies, including the microwave region, with high precision.

[41] See Udem et al. (1999) and Ma et al (2004).

Further reading

An excellent recent review of optical frequency and wavelength references has been provided by Hollberg et al. (2005). Details of the techniques to control the carrier-envelope phase of a mode-locked laser may be found in Jones et al. (2000). Further details of the development of laser-based precision spectroscopy may be found in the Nobel lectures given by the winners of the 2005 prize, which are available at http://nobelprize.org/nobel_prizes/physics/laureates/2005/.

Exercises

(16.1) Practical limitations to the linewidth

Consider a laser oscillating on a single longitudinal mode with a wavelength close to 500 nm. Take the length of the cavity to be 1 m, and assume that most of the space between the cavity mirrors is open to the atmosphere. Calculate the shift in the frequency of the mode caused by:

(a) A change in the length of the cavity by 1 mm.

(b) A change in the temperature of the invar spacer rods of $0.1°$.

(c) A change in atmospheric pressure of 1 mbar.

(16.2) Intra-cavity etalon

We consider SLM operation of a Rhodamine 590 dye laser in a linear cavity of the type illustrated in Fig. 14.9. We suppose that the grating restricts oscillation to a bandwidth $\Delta \nu_{\rm osc} = 10$ GHz near the centre wavelength of the dye laser transition at 600 nm.

(a) What would be the maximum cavity length in which SLM operation could be achieved?

(b) Find suitable values of the spacing d and finesse $\mathcal{F}$ of an etalon made of material with a refractive index equal to 1.5 that could restrict oscillation to a single longitudinal mode in a cavity of length 0.5 m.

(c) In practice, the normal of the etalon is tilted by an angle θ relative to the axis of the cavity. Show that if the tilt of the etalon is changed by a small amount, the frequency ν_m of the mth order peak transmitted by the etalon changes by

$$\Delta \nu_m = -\frac{m\, c \sin \theta'}{2nd \cos^2 \theta'} \Delta \theta' = -\nu_m \tan \theta' \Delta \theta', \quad (16.43)$$

where θ' is the angle of the rays to the normal *within* the etalon.

(d) Calculate the change in angle $\Delta \theta'$ required to scan the frequency transmitted by the etalon by the longitudinal mode spacing of the laser cavity for:
 (i) $\theta' = 5°$;
 (ii) $\theta' = 0.5°$;
and comment on these results.

(16.3) **Multiple etalons**
Here, we consider the design required to achieve SLM operation with two intra-cavity etalons.

(a) We consider first the design of a (thick) etalon, 1, which selects the desired longitudinal mode. Show that this requires that

$$\frac{\Delta \nu_{\text{FSR}}^{(1)}}{2\mathcal{F}_1} < \Delta \nu_{p,p-1}, \quad (16.44)$$

where $\Delta \nu_{\text{FSR}}^{(1)}$ and $\mathcal{F}_1$ are the free spectral range and finesse, respectively, of etalon 1.

(b) The second etalon must discriminate between adjacent transmission peaks of etalon 1. Show that this requires

$$\frac{\Delta \nu_{\text{FSR}}^{(2)}}{2\mathcal{F}_2} < \Delta \nu_{\text{FSR}}^{(1)}, \quad (16.45)$$

where $\Delta \nu_{\text{FSR}}^{(2)}$ and $\mathcal{F}_2$ are the free spectral range and finesse, respectively, of etalon 2.

(c) The adjacent transmission peaks of the second etalon must also lie outside the range of possible oscillation frequencies. Show that this condition requires,

$$\Delta \nu_{\text{FSR}}^{(2)} < \frac{\Delta \nu_{\text{osc}}}{2}. \quad (16.46)$$

(d) Hence, show that the thickness d_2 of etalon 2 must satisfy

$$\frac{\Delta \nu_{\text{osc}}}{2} < \frac{c}{2n_2 d_2} < (2\mathcal{F}_1)(2\mathcal{F}_2)\, \Delta \nu_{p,p-1}, \quad (16.47)$$

where n_2 is the refractive index of etalon 2, and that the maximum length of the cavity for which SLM operation may be achieved is increased to

$$L_c < (2\mathcal{F}_1)(2\mathcal{F}_2) \frac{c}{\Delta \nu_{\text{osc}}}. \quad (16.48)$$

(e) A c.w. Ti:sapphire laser oscillates on multiple longitudinal modes owing to spatial hole-burning. The oscillating modes cover a frequency range of 10 THz.
 (i) Estimate the longest cavity that could be used, in the absence of an intra-cavity etalon, to restrict oscillation to a single longitudinal mode.
 (ii) Estimate the minimum etalon finesse required to restrict oscillation to a single longitudinal mode in a cavity of length 0.5 m.
 (iii) What etalon finesses would be required if two etalons were used?
 In each case, comment on your results.

(16.4) †**Fox–Smith interferometer**
Here, we calculate the amplitude and intensity reflection coefficient of a Fox–Smith interferometer using the geometry illustrated in Fig. 16.13. We will assume that mirrors M_3 and M_4 have reflectivities of 100%, and will let the amplitudes of the waves within the interferometer be described by

$$u_\pm = \alpha_\pm \exp(i[\pm kx - \omega t]) \quad (16.49)$$
$$v_\pm = \beta_\pm \exp(i[\pm ky - \omega t]). \quad (16.50)$$

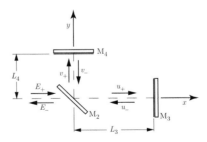

Fig. 16.13: Geometry of the Fox–Smith interferometer investigated in Exercise 16.4.

Further, we write the waves incident on, and reflected from, the beam splitter M_2 as

$$E_+ = a_i \exp(i[+kx - \omega t]) \quad (16.51)$$
$$E_- = a_r \exp(i[-kx - \omega t]). \quad (16.52)$$

Note that the origin of the xy coordinate system is taken to be at the surface of M_2.

(a) By considering the waves at M_3, show that

$$\alpha_- = \alpha_+ \exp\left[i(2kL_3 + \phi_3)\right], \quad (16.53)$$

where ϕ_3 is the phase shift introduced by reflection from M_3.

(b) Similarly, by considering the waves at M_4, show that

$$\beta_- = \beta_+ \exp\left[i(2kL_4 + \phi_4)\right], \quad (16.54)$$

where ϕ_4 is the phase shift introduced by reflection from M_4.

(c) By considering the waves at M_2, show that

$$\beta_+ = \alpha_- r_2 \exp(i\phi_2) \quad (16.55)$$

$$\alpha_+ = a_i t_2 + \beta_+ \exp\left[i(2kL_4 + \phi_2 + \phi_4)\right], \quad (16.56)$$

where t_2 is the amplitude transmission coefficient of the beam splitter for waves propagating along the x-axis towards positive x.

(d) Hence, show that

$$\alpha_+ = \frac{a_i t_2}{1 - R_2 \exp(i\delta)}, \quad (16.57)$$

where $R_2 = r_2^2$ and $\delta = 2k(L_3 + L_4) + 2\phi_2 + \phi_3 + \phi_4$ is the total phase accumulated in one trip round the cavity formed by M_3, M_2, and M_4.

(e) Find the amplitude a_r, and hence show

$$\frac{a_r}{a_i} = T_2 \frac{\exp(2ikL_3)}{1 - R_2 \exp(i\delta)}, \quad (16.58)$$

where T_2 is the intensity transmission coefficient of M_2.

(f) Hence, show that the intensity reflection coefficient R of the Fox–Smith interferometer is given by

$$R = \left|\frac{a_r}{a_i}\right|^2 = \left(\frac{T_2}{1 - R_2}\right)^2 \frac{1}{1 + F \sin^2\left(\frac{\delta}{2}\right)}, \quad (16.59)$$

where

$$F = \frac{4R_2}{(1 - R_2)^2}. \quad (16.60)$$

(g) Sketch R as a function of δ and compare the behaviour of the Fox–Smith interferometer with that of a Fabry–Perot etalon.

(16.5) †Frequency pulling in detail

Here, we derive the relationship between the real and imaginary parts of the refractive index of a homogeneously broadened atom. We first note that the quantum-mechanically correct expression for the refractive index is:

$$n^2 - 1 = \sum_{ij} \frac{N_i f_{ij} e^2}{m_e \epsilon_0} \frac{1}{\omega_{ji}^2 - \omega^2 - i\omega \Delta\omega_{ji}}, \quad (16.61)$$

where N_i is the density of atoms in level i, f_{ij} is the oscillator strength, ω_{ji} the angular frequency, and $\Delta\omega_{ji}$ the linewidth of the $i \to j$ transition.

We are interested in the *change* in refractive index $\Delta n_g = \Delta n_g^r + i \Delta n_g^i$ of a gain medium when a population inversion density N^* is created. This change will usually be very small compared to the refractive index of the medium when $N^* = 0$ and the contribution of the right-hand side of eqn (16.61) will therefore be small.

(a) Write down an expression for the refractive index in the absence of a population inversion. Assume that all the atoms are in the ground state (which is denoted level 0).

(b) Show that for frequencies close to that of the $2 \to 1$ transition, the change in refractive index caused by populating levels 1 and 2 of an atom is given by

$$2n_0 \Delta n_g \approx \frac{e^2}{m_e \epsilon_0} \frac{N_1 f_{12} + N_2 f_{21}}{\omega_0^2 - \omega^2 - i\omega \Delta\omega_H}, \quad (16.62)$$

where n_0 is the refractive index of the material with $N_1 = N_2 = 0$, and $\omega_0 \equiv \omega_{21}$.

(c) Using the relation between the oscillator strengths, $g_1 f_{12} = -g_2 f_{21}$, show that this result may be written as

$$2n_0 \Delta n_g \approx -\frac{N^* e^2 |f_{21}|}{m_e \epsilon_0} \frac{1}{\omega_0^2 - \omega^2 - i\omega \Delta\omega_H}, \quad (16.63)$$

where $N^* = N_2 - (g_2/g_1)N_1$ is the population inversion.

(d) Hence, show that near resonance ($\omega \approx \omega_0$)

$$2n_0 \Delta n_g^{\text{r}} \approx -\frac{N^* e^2 |f_{21}|}{2m_e \epsilon_0} \frac{\omega_0 - \omega}{(\omega_0 - \omega)^2 + \left(\frac{\Delta\omega_H}{2}\right)^2} \tag{16.64}$$

$$2n_0 \Delta n_g^{\text{i}} \approx -\frac{N^* e^2 |f_{21}|}{2m_e \epsilon_0 \omega_0^2} \frac{\frac{\Delta\omega_H}{2}}{(\omega_0 - \omega)^2 + \left(\frac{\Delta\omega_H}{2}\right)^2}. \tag{16.65}$$

(Hint: use the approximation employed in Exercise 3.1.)

(e) Hence show that

$$\frac{\Delta n_g^{\text{i}}}{\Delta n_g^{\text{r}}} = \frac{1}{2} \frac{\Delta\omega_H}{\omega_0 - \omega}. \tag{16.66}$$

(f) By considering the propagation of a plane wave through a medium with a complex refractive index $n = n_\text{r} + i n_\text{i}$, show that the wave is absorbed with an absorption coefficient κ (or, equivalently a gain coefficient $-\alpha$) given by,

$$\kappa = -\alpha = \frac{2\omega}{c} n_\text{i}.$$

(16.6) Schawlow–Townes linewidth: Correct calculation of R_{sp}

Here, we revisit the derivation of R_{sp}, the rate of spontaneous emission into a single cavity mode.

(a) Consider a laser exhibiting c.w. oscillation on a single cavity mode. Explain why the total rate of spontaneous and stimulated emission from the upper laser level into that mode may be written in the form

$$k N_2 (\bar{n} + 1),$$

where k is a constant and N_2 is the population density of the upper laser level under the operating conditions of the laser. [Hint: See Exercise 2.4.3.]

(b) Use this result to write down the *net* rate of stimulated emission into the cavity mode in terms of k and the population-inversion density N^*.

(c) The rate of loss of photons in the mode from the cavity is equal to $\bar{n}/\tau_\text{c}$, where τ_c is the cavity lifetime. Using the fact that this rate of loss is balanced by the total net rate of stimulated and spontaneous emission into the cavity mode, show that

$$R_{\text{sp}} = \left(\frac{N_2}{N^*}\right)_{\text{th}} \frac{\bar{n}}{\bar{n}+1} \frac{1}{\tau_\text{c}}$$

$$\approx \left(\frac{N_2}{N^*}\right)_{\text{th}} \frac{1}{\tau_\text{c}}. \tag{16.67}$$

(d) Explain why the population densities appearing in this result are the *threshold* values for laser oscillation.

(16.7) †Schawlow–Townes linewidth: Derivation of the lineshape

Here, we develop in detail the approach of Section 16.3.1 and derive the lineshape of the broadening caused by spontaneous emission into a single oscillating mode.

In order to derive the lineshape we need to establish the *distribution* of θ, the change of phase arising from the spontaneous emission of N photons into the oscillating mode.

The theory of one-dimensional random walks tells us that for N steps, where N is large, the probability $p(x)$ of a net displacement between x and $x + \delta x$ is

$$p(x) = \sqrt{\frac{1}{2\pi \overline{x^2}}} \exp\left[-\frac{(x-\bar{x})^2}{2\overline{x^2}}\right],$$

where $\bar{x} = N\bar{s}$, $\overline{x^2} = N\overline{s^2}$, and $\bar{s}$ and $\overline{s^2}$ are, respectively, the mean length and mean square length of a single step. Throughout the problem we will refer to Fig. 16.6.

(a) By considering the spontaneous emission of N photons, show that

$$\overline{\Delta E_\theta} = 0 \tag{16.68}$$

$$\overline{\Delta E_\theta^2} = N \frac{E_{\text{sp}}^2}{2}. \tag{16.69}$$

(b) Hence, show that after the spontaneous emission of N photons

$$\bar{\theta} = 0 \tag{16.70}$$

$$\overline{\theta^2} = \frac{N}{2}\left(\frac{E_{\text{sp}}}{E_0}\right)^2 = \frac{N}{2\bar{n}}, \tag{16.71}$$

where $\bar{n}$ is the mean number of photons in the oscillating mode.

(c) Use these results to show that the distribution of the values of θ is given by

$$p(\theta) = \sqrt{\frac{1}{2\pi} \frac{1}{\theta_{\text{rms}}}} \exp\left[-\left(\frac{\theta}{\sqrt{2}\theta_{\text{rms}}}\right)^2\right].$$

(d) In order to find the spectrum produced by this variation in phase we calculate the autocorrelation of the field: $c(\tau) = \langle E(t)E(t+\tau)\rangle$. Here, $E(t)$ is the electric field at time t, $E(t+\tau)$ the field a time τ later, and the brackets indicate an average over time t. We will write the field as

$$E(t) = E_0 \cos\left[\omega_0 t + \phi(t)\right].$$

Show that

$$c(\tau) = \frac{E_0^2}{2} \langle \cos\left[2\omega_0 t + \omega_0 \tau + \phi(t) + \phi(t+\tau)\right]$$
$$+ \cos\left[\omega_0 \tau + \theta(\tau)\right]\rangle$$
$$= \frac{E_0^2}{2} \langle \cos\left[\omega_0 \tau + \theta(\tau)\right]\rangle,$$

where $\theta(\tau)$ is the change in phase that occurs over a delay τ.

(e) Hence, show that

$$c(\tau) = \frac{E_0^2}{2} \int_{-\infty}^{\infty} \cos\left[\omega_0 \tau + \theta(\tau)\right] p(\theta) d\theta.$$

(f) Given that for $\Re(\beta) > 1$:

$$\int_{-\infty}^{\infty} \exp(ix) \exp\left(-\beta x^2\right) dx = \sqrt{\frac{\pi}{\beta}} \exp\left(-\frac{1}{4\beta}\right),$$

show that

$$c(\tau) = \frac{E_0^2}{2} \exp\left(-\frac{\theta_{\text{rms}}^2}{2}\right) \cos(\omega_0 \tau).$$

(g) Hence, show that

$$c(\tau) = \exp\left(-\frac{|\tau|}{2\tau_{\text{ST}}}\right) \cos(\omega_0 \tau),$$

where

$$\tau_{\text{ST}} = \frac{2\bar{n}}{R_{\text{sp}}}. \tag{16.72}$$

(h) Having calculated the autocorrelation of the field we may find the spectrum by using the **Wiener–Khintchine theorem**.[42] This states that the power spectrum of a field is proportional to the Fourier transform of its autocorrelation:

$$g(\omega) \propto \int_{-\infty}^{\infty} c(\tau) e^{i\omega\tau} d\tau.$$

Explain why in our case the imaginary part of the integral is zero, and hence that we may write

$$g(\omega) \propto 2 \int_0^{\infty} \exp\left(-\frac{\tau}{2\tau_{\text{ST}}}\right) \cos(\omega_0 \tau) \cos(\omega \tau) d\tau.$$

(i) Hence, show that

$$g(\omega) \propto \frac{\frac{1}{\tau_{\text{ST}}}}{(\omega - \omega_0)^2 + \left(\frac{1}{2\tau_{\text{ST}}}\right)^2}$$
$$+ \frac{\frac{1}{\tau_{\text{ST}}}}{(\omega + \omega_0)^2 + \left(\frac{1}{2\tau_{\text{ST}}}\right)^2}$$
$$\approx \frac{\frac{1}{\tau_{\text{ST}}}}{(\omega - \omega_0)^2 + \left(\frac{1}{2\tau_{\text{ST}}}\right)^2}.$$

(j) Describe the shape of the spectrum, find its full width at half-maximum, and hence derive eqn (16.26).

(16.8) **Schawlow–Townes linewidth of a He-Ne laser**
Here, we evaluate the key parameters involved in determining the Schawlow–Townes linewidth in a He-Ne laser. Suppose that the laser operates on a single longitudinal mode, that it delivers 1 mW on the 632.8 nm line, and that the cavity is 30 cm long with $R_1 = 1$, $R_2 = 0.98$. Find:

(a) the cavity lifetime τ_c;

(b) the cavity linewidth $\Delta \nu_c$;

(c) the Schawlow-Townes linewidth $\Delta \nu_{\text{ST}}$.

(d) the coherence time τ_{ST} against dephasing by spontaneous emission into the oscillating mode;

(e) the mean number of photons $\bar{n}$ in the oscillating mode.

(16.9) **Derivation of the error signal**
Here we derive the form of the error signal obtained by passing a frequency-modulated beam through a system with a frequency-dependent amplitude transmission

[42] See, for example, Brooker (2003).

$a(\omega)$. We suppose that a beam of angular frequency ω is passed through a phase modulator operating at an angular frequency Ω.

(a) Show that if the only frequency components with appreciable amplitude are ω and $\omega \pm \Omega$, the amplitude of the beam transmitted by the reference system is given by,

$$E(t) = E_0 \exp(i\phi_0) \{$$
$$- J_1(\beta) a(\omega + \Omega) \exp[-i(\omega + \Omega)t]$$
$$+ J_0(\beta) a(\omega) \exp(-i\omega t)$$
$$+ J_1(\beta) a(\omega - \Omega) \exp[-i(\omega - \Omega)t] \}.$$

(b) Hence, show that the signal detected by a square-law detector at an angular frequency Ω is given by,

$$S_\Omega(t) = J_0(\beta) J_1(\beta) [$$
$$a(\omega) a^*(\omega - \Omega) e^{-i\Omega t} + a^*(\omega) a(\omega - \Omega) e^{i\Omega t}$$
$$- a^*(\omega) a(\omega + \Omega) e^{-i\Omega t} - a(\omega) a^*(\omega + \Omega) e^{i\Omega t}]$$

(c) By writing $a(\omega \pm \Omega) \approx a(\omega) \pm \Omega \left. \frac{\partial a}{\partial \omega} \right|_\omega$, show that

$$S_\Omega(t) = -4 J_0(\beta) J_1(\beta) \Omega \Re \left(a^*(\omega) \left. \frac{\partial a}{\partial \omega} \right|_\omega \right) \cos \Omega t.$$

(d) By writing the amplitude transmission of the reference in the form $a(\omega) = |a(\omega)| \, e^{i\phi(\omega)}$, show that the error signal ϵ_Ω is given by,

$$\epsilon_\Omega \propto -\Omega \, |a(\omega)| \left. \frac{\partial |a|}{\partial \omega} \right|_\omega.$$

Sketch how the error signal depends on the frequency detuning $\omega - \omega_0$.

(e) What would be the error signal if the reference system changes the phase, but not the amplitude of the transmitted frequencies?

(16.10) **Saturated absorption**
Here, we derive an expression for the absorption coefficient experienced by a weak probe beam in the presence of an intense (i.e. saturating) counter-propagating beam of the same frequency ω. Since we have already established how the optical gain saturates, it will be easiest to do this by first considering saturation of the gain coefficient.

(a) We consider the contribution of those atoms with a velocity v_z along the direction of the *probe* beam.

Show that for the probe beam the apparent centre frequency of these atoms is ω_c, whereas for the saturating beam it is ω'_c, where

$$\omega_c = \left(1 - \frac{v_z}{c}\right) \omega_0$$
$$\omega'_c = \left(1 + \frac{v_z}{c}\right) \omega_0.$$

(b) Show that

$$\omega'_c = 2\omega_0 - \omega_c.$$

(c) Explain why the population-inversion density of this class of atoms is saturated by the saturating beam according to

$$\Delta N^*(v_z) = \frac{\Delta N_0^*(v_z)}{1 + I/I_s(\omega - \omega'_c)},$$

where $\Delta N_0^*(v_z)$ is the inversion density of this class of atoms in the absence of the saturating beam, I is the intensity of the saturating beam, and $I_s(\omega - \omega_c)$ is the saturation intensity of the transition for atoms with centre frequency ω_c.

(d) Hence, show that the optical gain coefficient experienced by the *probe* beam is given by

$$\alpha(\omega) = N^* \int \frac{g_D(\omega_c - \omega_0) \sigma_{21}(\omega - \omega_c)}{1 + I/I_s(\omega + \omega_c - 2\omega_0)} d\omega_c.$$

where $g_D(\omega_c - \omega_0)$ is the distribution of centre frequencies experienced by the probe and N^* is the total population inversion.

(e) Show that the absorption coefficient experienced by the probe is given by,

$$\kappa(\omega) = -N^* \int \frac{g_D(\omega_c - \omega_0) \sigma_{21}(\omega - \omega_c)}{1 + I/I_s(\omega + \omega_c - 2\omega_0)} d\omega_c,$$
(16.73)

(f) Use this result to explain qualitatively why the absorption coefficient is unaffected by the saturating beam for frequencies such that $|\omega - \omega_0| \gg \Delta \omega_H$. Sketch the variation of the absorption coefficient as a function of frequency.

(g) (optional) Write a computer program to calculate the intensity transmission $a(\omega) = \exp[-\kappa(\omega) \ell]$ of the probe, where ℓ is the length of the absorption cell for the case when the homogeneous lineshape is Lorentzian.

17 Ultrafast lasers

17.1 Propagation of ultrafast laser pulses in dispersive media 462
17.2 Dispersion control 474
17.3 Sources of ultrafast optical pulses 482
17.4 Measurement of ultrafast pulses 489
Further reading 495
Exercises 495

In this chapter we discuss the issues associated with the generation and control of ultrafast laser pulses. Of course, 'ultrafast' is a relative term—we will take it to refer to optical pulses shorter than 1 ps. As we will see, laser pulses with a duration of a few tens of femtoseconds (1 fs $\equiv 10^{-15}$ s) may now be generated routinely, and pulses shorter than 10 fs with a little more effort. As mentioned in Section 18.7.1, the current frontier in the generation of ultrafast laser pulses is the attosecond regime (1 as $\equiv 10^{-18}$ s).

It is worth recalling how these time scales relate to various physical phenomena. The period of: the lattice vibrations in a solid is of order 1 ps; the vibrational motion of the atoms in an H_2 molecule is 7.5 fs; and the orbit of the electron in the ground state of a hydrogen atom is 150 as. These time scales are pertinent to one application of low-energy (fraction of a millijoule) ultrafast pulses: 'pump-probe' experiments in which an event—such as the vibration of a crystal lattice—is triggered by a short laser pulse, and the subsequent evolution is studied using a second, short probe laser pulse. The ability to control the phase structure of short laser pulses is opening up the field of 'femtochemistry' in which, for example, the products of photoinitiated chemical reactions are selected by controlling the precise variation of the electric field within the laser pulse.[1]

At somewhat higher pulse energies, focusing the laser pulse to a beam diameter of a few tens of micrometers results in a peak laser intensity of above 10^{15} W cm^{-2}. At such intensities the electric field of the electromagnetic wave is comparable, or stronger than, the Coulomb field binding the valence electrons in to an atom. A number of applications arise: high-harmonic generation (Section 18.7.1); short-wavelength lasers (Section 18.4.3); plasma accelerators;[2] inertial-confinement fusion;[3] and 'laboratory astrophysics'.[4]

17.1 Propagation of ultrafast laser pulses in dispersive media

17.1.1 The time–bandwidth product

It may be shown that the root-mean-square (RMS) angular frequency bandwidth, $\Delta\omega_{\mathrm{RMS}}$, of the *intensity* of the spectrum of a pulse is related to the RMS duration, Δt_{RMS}, of the temporal profile of the *intensity* of the pulse by

$$\Delta\omega_{\mathrm{RMS}} \Delta t_{\mathrm{RMS}} \geq \pi. \tag{17.1}$$

---

[1] The 1999 Nobel prize for chemistry was awarded to Ahmed Zewail for his work on femtochemistry. His Nobel lecture may be found at www.nobelprize.org or in a published collection of Nobel lectures Zewail (2003). A review article based on this work is also available Zewail (2000).

[2] See the short review articles by Joshi (2006) and Krushelnick and Najmudin (2006).

[3] See the review article by Lindl et al. (2004).

[4] See the article by Rose (2004).

In general, the value of the **time-bandwidth product** $\Delta\omega_{\text{RMS}}\Delta t_{\text{RMS}}$ depends on:

- the exact shape of the pulse envelope (i.e. Gaussian, square, sech^2 etc.);
- the amount of frequency chirp or phase structure within the pulse.

In addition, of course, different definitions of $\Delta\omega$ and Δt—such as the full width at half-maximum—will yield different time–bandwidth products, although any reasonable definition of these quantities should give a result similar to eqn (17.1). For example, as explored in Exercise 17.1, in the absence of phase structure within the pulse, laser pulses with Gaussian, top-hat, and sech^2 temporal profiles have FWHM time–bandwidth products of 2.77, 5.56, and 1.98, respectively.

Equation (17.1) tells us that short laser pulses are associated with a large bandwidth. For propagation *in vacuo*, this presents no difficulties, and the pulse will propagate over arbitrary distances without distortion. However, optical materials are generally dispersive—that is, their refractive index varies with frequency—which can lead to distortion of the pulse as it propagates. It is no use generating a short laser pulse if it is stretched and distorted beyond recognition when it reaches the interaction region in an experiment. Further, as discussed in Section 17.3.1, in order to achieve the shortest possible pulses from a modelocked laser it is necessary to control the phases of the oscillating modes. An understanding of the effect of dispersion, and how it may be controlled, is therefore crucial for generating and using ultrafast laser pulses.

17.1.2 General considerations

Here, we develop a general description of the propagation of an optical pulse through a linear optical system. Let us suppose that the pulse propagates along the z-axis, towards positive z, and that in the plane $z = 0$ the electric field of the pulse is given by[5]

$$E_{\text{in}}(t) = f(t)e^{-i\omega_0 t}, \quad (17.2)$$

where the function $f(t)$ describes the **envelope** of the pulse, and ω_0 is the angular frequency of the **carrier wave**. The advantage of dividing the pulse into two parts, the envelope and carrier wave, arise from the fact that they usually have very different time scales: the carrier wave oscillates with a period equal to the optical period, while the pulse envelope changes on the time scale of the duration of the optical pulse. Since short pulses contain a range of frequencies there is some choice in the value of ω_0. The value chosen is not critical, provided it lies within the bandwidth of the pulse; it is usual to take ω_0 as the mean angular frequency, or the frequency with the greatest amplitude.[6]

The electric field $E_{\text{in}}(t)$ may be written as a superposition of harmonic waves:

$$E_{\text{in}}(t) = f(t)e^{-i\omega_0 t} = \frac{1}{\sqrt{2\pi}}\int_{-\infty}^{\infty} a(\omega)e^{-i\omega t}\,d\omega. \quad (17.3)$$

[5] It should be noted that we have ignored all transverse structure of the pulse. As a consequence, we have ignored the possibility that different frequencies are imaged differently in the system or are displaced transversely relative to each other. Pulses that have transverse structure of this type are not uncommon, and are said to exhibit 'spatial chirp.'

[6] It is straightforward to see that if the carrier frequency is changed from ω_0 to ω_0' the pulse envelope is changed from $f(t)$ to $f(t)\exp[-i(\omega_0 - \omega_0')t]$.

The amplitude per unit frequency interval is given by the inverse Fourier transform:

$$a(\omega) = \frac{1}{\sqrt{2\pi}} \int_{-\infty}^{\infty} \left[f(t) e^{-i\omega_0 t} \right] e^{i\omega t} dt, \quad (17.4)$$

which may more conveniently be written as

$$a(\omega_0 + \omega') \equiv a'(\omega') = \frac{1}{\sqrt{2\pi}} \int_{-\infty}^{\infty} f(t) e^{i\omega' t} dt, \quad (17.5)$$

where $\omega' = \omega - \omega_0$. This last result emphasizes that the shape and width of the distribution over frequency is determined by the shape of the pulse envelope; the location of this distribution on the frequency axis is determined by ω_0.

In propagating through an optical system each frequency ω will accumulate a total phase shift of $\phi(\omega)$. The form of the optical pulse in the output plane of the optical system is given by the superposition of the amplitudes of the phase-shifted frequencies; i.e. an integral of the form of eqn (17.3):

$$E_{\text{out}}(t) = \frac{1}{\sqrt{2\pi}} \int_{-\infty}^{\infty} \left[a(\omega) e^{i\phi(\omega)} \right] e^{-i\omega t} d\omega. \quad (17.6)$$

This last result may be written in the alternative form:

$$E_{\text{out}}(t) = \underbrace{e^{-i\omega_0 t}}_{\text{carrier wave}} \overbrace{\frac{1}{\sqrt{2\pi}} \int_{-\infty}^{\infty} \left[a'(\omega') e^{i\phi(\omega_0 + \omega')} \right] e^{-i\omega' t} d\omega'}^{\text{pulse envelope}}, \quad (17.7)$$

which shows clearly that the output pulse comprises a carrier wave of angular frequency ω_0 and a pulse envelope that in general will be modified—possibly significantly—from that of the input pulse.

Propagation through a uniform medium

As a simple example, consider the case of propagation of a pulse through a uniform medium located between the planes $z = 0$ and $z = z$, as illustrated schematically in Fig. 17.1. The phase accumulated at each frequency component is given by

$$\phi(\omega) = k(\omega)z = \frac{\omega}{c} n(\omega) z, \quad (17.8)$$

where $n(\omega)$ is the refractive index of the medium, which will in general be a function of the frequency of the light.[7]

We can expand the wave vector about its value at the carrier frequency ω_0:

$$k(\omega) = k(\omega_0) + \left.\frac{\partial k}{\partial \omega}\right|_{\omega_0} (\omega - \omega_0) + \frac{1}{2} \left.\frac{\partial^2 k}{\partial \omega^2}\right|_{\omega_0} (\omega - \omega_0)^2 + \ldots. \quad (17.9)$$

---

[7] The frequency dependence of many optical materials may conveniently be summarized by a Sellmeier equation of the form

$$n^2(\lambda) = 1 + \sum_{i=1}^{i_{\max}} \frac{B_i \lambda^2}{\lambda^2 - C_i},$$

where λ is the wavelength in the medium and B_i and C_i are Sellmeier coefficients. Typically, $i_{\max} = 3$ is sufficient to describe the variation of the refractive index over the wavelength range of interest.

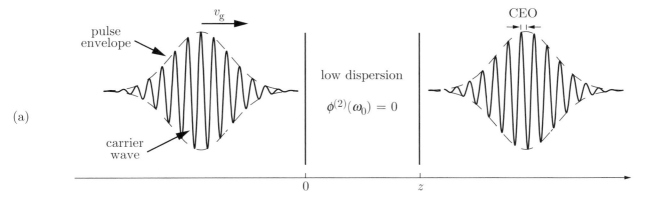

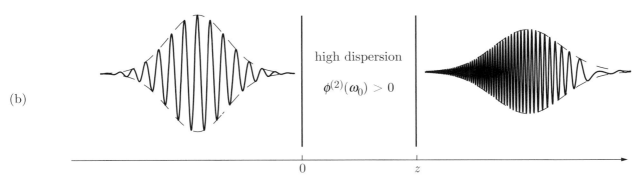

Fig. 17.1: Propagation of optical pulses through low- and high-dispersion media. (a) Propagation through a low-dispersion medium for which the group delay (and higher-order) dispersion is zero. In this case, the envelope of the pulse is undistorted and propagates through the medium at the group velocity v_g. However, since, in general, the group and phase velocities are different, the phase of the carrier wave shifts relative to the pulse envelope as the pulse propagates through the medium. This causes the carrier-envelope offset (CEO) to change, as indicated schematically. (b) Distortion of an initially unchirped optical pulse by propagation through a dispersive medium with positive group-delay dispersion. Now, the transmitted pulse is stretched in time, and (since $\phi^{(2)} > 0$) develops a positive frequency chirp.

If we retain just the first two terms in this expansion, substitution into eqn (17.7) yields,

$$E_{\text{out}}(t) = \underbrace{e^{-i[k(\omega_0)-\omega_0 t]}}_{\text{carrier wave}} \overbrace{\frac{1}{\sqrt{2\pi}} \int_{-\infty}^{\infty} a(\omega') e^{-i\omega'(t-z/v_g)} d\omega'}^{f(t-z/v_g)}. \qquad (17.10)$$

We see that if only the first two terms of eqn (17.9) are retained the pulse envelope maintains a constant shape $f(t - z/v_g)$ that moves at the **group velocity**

$$v_g = \left.\frac{\partial \omega}{\partial k}\right|_{\omega_0}, \qquad (17.11)$$

Table 17.1 Terms in the expansion of the optical phase $\phi(\omega)$.

$\phi(\omega_0)$	$\phi^{(0)}$		Determines CEO
$\left.\frac{\partial \phi}{\partial \omega}\right|_{\omega_0}$	$\phi^{(1)}$		Group delay. Determines CEO
$\left.\frac{\partial^2 \phi}{\partial \omega^2}\right|_{\omega_0}$	$\phi^{(2)}$	GDD	Group delay dispersion or quadratic phase
$\left.\frac{\partial^3 \phi}{\partial \omega^3}\right|_{\omega_0}$	$\phi^{(2)}$	TOD	Third-order dispersion or cubic phase
$\left.\frac{\partial^n \phi}{\partial \omega^n}\right|_{\omega_0}$	$\phi^{(n)}$		Quartic, quintic... phase.

whilst the carrier wave propagates at the **phase velocity**

$$v_p = \frac{\omega_0}{k(\omega_0)}. \tag{17.12}$$

Note that in the absence of dispersion (i.e. the refractive index of the medium is independent of frequency), $v_g = v_p$ and the pulses entering and leaving the medium are indistinguishable. However, in general, $v_g \neq v_p$, which causes the position of the carrier wave relative to the peak of the pulse envelope—known as the **carrier-envelope offset (CEO)** or **carrier-envelope phase (CEP)**—to change as the pulse propagates. For long pulses, changes in the carrier-envelope phase have little effect on the pulse. However, for short pulses the carrier-envelope phase has a significant effect on the shape and peak amplitude of the pulse (see Exercise 17.3).

17.1.3 Propagation through a dispersive system

Calculation of the propagation of an optical pulse through a dispersive system follows the procedure established above. Provided we can find the total phase accumulated at each frequency, the form of the optical pulse at the exit of the system can be calculated from eqn (17.7).

Just as we did in eqn (17.9), it is useful to expand the optical phase about the value at the carrier frequency:

$$\phi(\omega) = \phi(\omega_0) + \left.\frac{\partial \phi}{\partial \omega}\right|_{\omega_0}(\omega - \omega_0) + \frac{1}{2}\left.\frac{\partial^2 \phi}{\partial \omega^2}\right|_{\omega_0}(\omega - \omega_0)^2 + \dots . \tag{17.13}$$

It is worth considering each of the terms in the expansion of $\phi(\omega)$:

- $\phi(\omega_0) \equiv \phi^{(0)}$. The total phase accumulated at ω_0. Together with the group delay it determines where the peaks of the carrier wave are located with respect to the envelope of the pulse. The phase difference between the peak of the pulse and a peak of the carrier wave is known as the carrier-envelope offset (CEO). It is easy to see that this is given by $\phi_{\text{CEO}} = \phi^{(0)} - \omega_0 \phi^{(1)}$. The carrier-envelope offset becomes important for pulses that comprising only a few optical cycles.

- $\left.\frac{\partial \phi}{\partial \omega}\right|_{\omega_0} \equiv \phi^{(1)}$. This term is known as the **group delay**. To see this, we note that for propagation through a single medium of length L, $\phi^{(1)} = L/v_g$, and is therefore equal to the time taken for the envelope to propagate between the two planes.

- $\left.\frac{\partial^2 \phi}{\partial \omega^2}\right|_{\omega_0} \equiv \phi^{(2)}$. This term is known as the **group delay dispersion (GDD)** or the **quadratic phase**. It is the lowest term in the expansion responsible for distortion of the pulse envelope. **Positive dispersion** is defined as $\phi^{(2)}(\omega_0) > 0$. The GDD is generally expressed in units of fs^2. A related quantity is the **group velocity dispersion (GVD)**; this is defined to be $\left.\frac{\partial}{\partial \omega}\right|_{\omega_0}\left(\frac{1}{v_g}\right)$, and hence in a uniform medium is equal to $\phi^{(2)}/L$.

- $\left.\frac{\partial^n \phi}{\partial \omega^n}\right|_{\omega_0} \equiv \phi^{(n)}$.[8] There are of course higher terms in the expansion: the cubic, quartic, quintic... phase. For the production and propagation of sufficiently short pulses, these will also be important.

[8] To reduce clutter from hereon we take it as understood that $\phi^{(n)}$ is evaluated at the carrier frequency.

Effect of group delay dispersion

It is useful to imagine changing the carrier frequency from ω_0 to ω and to calculate the time for a pulse to propagate through the system as a function of ω. This time, $T(\omega)$, will be given by

$$T(\omega) = \frac{\partial \phi}{\partial \omega}. \tag{17.14}$$

For narrowband pulses propagating through a low-dispersion system $T(\omega) \approx T(\omega_0)$ for all frequencies within the bandwidth of the pulse. In such cases the pulse will not change shape appreciably, as in Fig. 17.1(a). However, in general $T(\omega)$ will be a function of frequency so that we may once again expand about the mean frequency:

$$T(\omega) = T(\omega_0) + \left.\frac{\partial T}{\partial \omega}\right|_{\omega_0}(\omega - \omega_0) + \ldots \tag{17.15}$$

$$= \phi^{(1)} + \phi^{(2)}(\omega - \omega_0) + \ldots. \tag{17.16}$$

We see that the term in $\phi^{(2)}$ causes the transit time to increase linearly with frequency. Hence, for positive values of the GDD, the high-frequency components of the pulse will move to the rear of the pulse; the low-frequency components will lead. This will cause the pulse to develop a **positive frequency chirp**, that is for a stationary observer the frequency of the radiation increases as the pulse passes.

The variation in the group delay for different frequencies will cause the duration of the pulse to change. For an initially unchirped pulse the duration will increase, and we can estimate by how much by considering the range of values taken by the second term on the right-hand side of eqn (17.16). Hence, the estimated increase in pulse duration is

$$\Delta T_{\text{GDD}} \approx \phi^{(2)}(\omega_0)\Delta \omega, \tag{17.17}$$

where $\Delta \omega$ is the bandwidth of the pulse. These processes are illustrated schematically in Fig. 17.1(b).

Worked example:
To provide an example, in Table 17.2 are tabulated representative values of the GDD for various materials. Let us estimate the increase in duration

Table 17.2 Values of the non-linear refractive index (see Section 17.1.5 n_2 and dispersion at a wavelength of 800 nm for various optical materials. The dispersion terms have been calculated by Backus et al. (1998) for 10-mm thick samples and a wavelength of 800 nm. Non-linear refractive indices are from the data of Adair et al. (1989), Milam et al. (1976), and Pennington et al. (1989) for a wavelength of 1 μm. Materials denoted by † are birefringent, in which case average values are given.

Material	n_2 (10^{-16} cm^2 W^{-1})	$\phi^{(2)}$ (fs^2)	$\phi^{(3)}$ (fs^3)	$\phi^{(4)}$ (fs^4)
Fused silica	2.5	361.626	274.979	−114.35
BK7	3.4	445.484	323.554	298.718
KD*P†	2.1	290.22	443.342	−376.178
Calcite†	2.6	780.96	541.697	−118.24
Sapphire†	3.0	581.179	421.756	−155.594
Air	0.005	0.0217	0.0092	2.3×10^{-11}

of a pulse of duration $\tau_p = 10$ fs caused by propagating through a 10-mm thick fused silica window. From eqn (17.1) the bandwidth of the pulse must be at least $\pi/\tau_p = 3.1 \times 10^{14}$ s^{-1}. Multiplying this by the GDD for 10-mm fused silica of 362 fs^2 we find $\Delta T_{GDD} \approx 110$ fs. It is clear from this that very short pulses are rather delicate and can be distorted severely by propagating though dispersive materials.

Focusing short optical pulses

The pulse envelope of a short optical pulse may also be affected when the pulse is focused by a lens, owing to the fact that parts of the beam at different distances from the axis of propagation pass through different thicknesses of the lens material. The focusing property of a lens arises from the variation with transverse position of the optical path through the lens;[9] in other words $\phi^{(0)}$ will depend on the transverse coordinates of the beam. However, if the lens is dispersive, higher-order terms—and in particular the group delay, $\phi^{(1)}$—will vary with distance from the optical axis. As a consequence, after the lens the loci of constant group delay will not in general be spherical or centred on the focal point of the lens. In simple terms, rays passing through the thicker central part of the lens are delayed relative to those passing through the edge of the lens, as illustrated schematically in Fig. 17.2.

The duration of the pulse at the focus will therefore be longer than the incident pulse. It may be shown[10] that the time taken for a ray to reach the focus increases with the initial distance r of the ray from the axis of the lens by an amount

$$\Delta \tau_p(r) = -\frac{r^2}{2cf^2} \lambda \frac{df}{d\lambda}. \quad (17.18)$$

As an example, the difference in propagation time between a ray initially 10 mm from the propagation axis and the axial ray amounts to approximately 50 fs for a BK7 lens of 100 mm focal length. The degree of stretching may

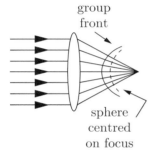

Fig. 17.2: Schematic diagram illustrating how a short optical pulse is stretched when it is focused by a lens.

[9]To be clear, here we refer to the optical path through the lens only. By Fermat's principle we know that in a perfect imaging system the *total* optical path from the source to the image must be independent of the route taken—i.e. independent of the transverse position at the lens through which the light passes. Although the total phase shift experienced by light propagating from the source to the image is independent of the route taken, the magnitude of this phase shift will in general vary with the wavelength of the light. Dispersion within the imaging system causes chromatic aberration (dependence of the position of the image with wavelength) and distortion of the pulse envelope, as discussed in the text.

[10]See Walmsley et al. (2001) and references therein.

be reduced somewhat by using an achromatic lens, for which $\mathrm{d}f/\mathrm{d}\lambda \approx 0$, although it must be remembered that achromats are designed for a specific wavelength; lenses that are achromatic at a wavelength of 500 nm will not be achromatic at 800 nm! For these, and other, reasons very short optical pulses are usually brought to a focus by concave mirrors, for which it is rigorously true that $\mathrm{d}f/\mathrm{d}\lambda = 0$.

17.1.4 Propagation of Gaussian pulses

Considerable insight, and several useful analytical results, may be obtained by applying the general methods described in Section 17.1.2 to the special case of an optical pulse with a Gaussian temporal profile. The general conclusions we draw will apply to pulses with arbitrary temporal profile.[11]

Properties of a Gaussian pulse

We consider a generalized Gaussian pulse of the following form,[12]

$$E_{\mathrm{in}}(t) = \mathrm{e}^{-\Gamma t^2} \mathrm{e}^{-\mathrm{i}\omega_0 t}, \quad (17.19)$$

where the complex Gaussian parameter describing the pulse is given by,

$$\Gamma \equiv a + \mathrm{i}b. \quad (17.20)$$

Let us first establish the main properties of this pulse.[13] The full width at half-maximum duration of the temporal profile of the pulse *intensity* is given by,

$$\tau_{\mathrm{p}} = \sqrt{\frac{2\ln 2}{\Re(\Gamma)}} = \sqrt{\frac{2\ln 2}{a}}, \quad (17.21)$$

from which we note that the duration of the pulse depends only on $\Re(\Gamma) \equiv a$.

If $b \neq 0$ the instantaneous frequency changes within the pulse. To see this we note that the phase of the electric field is given by $\psi(t) = -\omega_0 t - bt^2$. The instantaneous frequency is given by,[14]

$$\omega(t) \equiv -\frac{\mathrm{d}\psi(t)}{\mathrm{d}t} = \omega_0 + 2bt. \quad (17.22)$$

Thus, in general the frequency of the pulse changes linearly with time about the mean value of ω_0; this frequency chirp will be positive if b is positive.

To investigate how such a pulse propagates through an optical system it is necessary to find the frequency spectrum of the pulse from eqn (17.5). As explored in Exercise 17.5, upon doing this we find,

$$a(\omega') = \frac{1}{\sqrt{2\Gamma}} \mathrm{e}^{-\omega'^2/4\Gamma} \quad (17.23)$$

$$= \frac{1}{\sqrt{2\Gamma}} \exp\left[-\Re\left(\frac{1}{4\Gamma}\right)(\omega - \omega_0)^2\right]$$

$$\times \exp\left[-\mathrm{i}\Im\left(\frac{1}{4\Gamma}\right)(\omega - \omega_0)^2\right]. \quad (17.24)$$

[11] Here, we follow the approach of Siegman (1986). Further considerations and examples may be found therein.

[12] Note that Siegman writes the pulse in the form $\exp(-\Gamma t^2)\exp(\mathrm{j}\omega_0 t)$ with $\Gamma \equiv a - \mathrm{j}b$. Consequently, equations in Siegman's book may be converted to our notation by making the transformation $\mathrm{j} \to -\mathrm{i}$, and vice versa.

[13] These are further explored in Exercise 17.5.

[14] We note that eqn (17.19) may be written as $E_{\mathrm{in}}(t) = \exp(-at^2)\exp\left[-\mathrm{i}(\omega_0 + bt)t\right]$. However, it is *not* correct to deduce from this that the effective frequency is $\omega_0 + bt$.

It is clear that, in general, the phase $\psi(\omega)$ of the amplitudes of the frequency components varies quadratically with $(\omega - \omega_0)$ with a curvature determined by $\Im(1/\Gamma)$. The *power* spectrum, which is proportional to $|a(\omega)|^2$, has a full width at half-maximum of,

$$\Delta\omega_{\rm p} = 2\sqrt{\frac{2\ln 2}{\Re(1/\Gamma)}} \tag{17.25}$$

$$= 2\sqrt{2\ln 2}\sqrt{a[1 + (b/a)^2]}. \tag{17.26}$$

In terms of the full widths at half-maxima, the time–bandwidth product of the Gaussian pulse is found from eqns (17.21) and (17.26) to be

$$\Delta\omega_{\rm p}\tau_{\rm p} = 4\ln 2\sqrt{1 + (b/a)^2}. \tag{17.27}$$

The time–bandwidth product clearly takes the smallest possible value if $b = 0$; that is if the pulse has no frequency chirp or, equivalently, if the phases of the frequency components are all the same.[15]

[15] If the widths in eqn (17.27) are replaced with root-mean-square values the factor of $4\ln 2$ is replaced by π. We then see that the case $b = 0$ reduces to the minimum possible value according to eqn (17.1).

Figure 17.3 summarizes the properties of Gaussian optical pulses of this type.

Propagation of a Gaussian pulse

We now consider the propagation of a Gaussian pulse through an optical system for which the linear and quadratic dispersion are finite, but the higher-order dispersion may be ignored. Hence, we assume that the optical phase accumulated in propagating through the system may be written as

$$\phi(\omega) = \phi^{(0)} + \phi^{(1)}(\omega - \omega_0) + \frac{1}{2}\phi^{(2)}(\omega - \omega_0)^2. \tag{17.28}$$

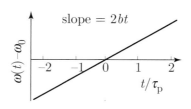
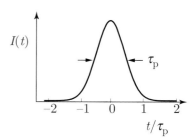
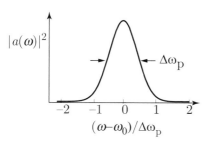
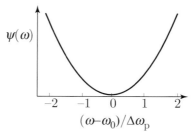

Fig. 17.3: Properties of a Gaussian optical pulse with a pulse parameter $\Gamma = a + ib$ for the case $b > 0$. Shown are the electric field $E(t)$, the instantaneous frequency $\omega(t)$, the temporal profile of the intensity $I(t)$ the power spectrum $|a(\omega)|^2$ and phase $\psi(\omega)$.

To calculate the form of the output pulse we apply eqn (17.7), i.e. we multiply the frequency spectrum of the input pulse, eqn (17.24), by $\exp[i\phi(\omega)]$ and take the inverse Fourier transform. As explored in Exercise 17.5 this turns out to be relatively straightforward, and we find

$$E_{\text{out}}(t) = \sqrt{\frac{\Gamma_{\text{out}}}{\Gamma_{\text{in}}}} \exp[i(\phi^{(0)} - \omega_0 t)] \exp\left\{-\Gamma_{\text{out}}\left[t - \phi^{(1)}\right]^2\right\}, \quad (17.29)$$

where

$$\frac{1}{\Gamma_{\text{out}}} = \frac{1}{\Gamma_{\text{in}}} - 2i\phi^{(2)}. \quad (17.30)$$

We see that the transmitted pulse still has a Gaussian temporal profile, but that the duration of the pulse envelope and the frequency chirp will, in general, be different from the input pulse since the parameter Γ will be changed. There will also be a relative shift between the peak of the Gaussian pulse envelope—which occurs at $t = \phi^{(1)}$, i.e. it is shifted in time by the group delay—and the carrier wave that is shifted in phase by $\phi^{(0)}$. Since, in general, $\phi^{(0)}/\omega_0 \neq \phi^{(1)}$ the carrier-envelope offset of the pulse will be altered by the optical system.[16]

These points are made more obvious by considering the case of a uniform medium. Equation (17.29) may then be written as

$$E_{\text{out}}(t) = \underbrace{\exp[-i\omega_0(t - z/v_{\text{p}})]}_{\text{carrier wave}} \overbrace{\sqrt{\frac{\Gamma_{\text{out}}}{\Gamma_{\text{in}}}} \exp[-\Gamma_{\text{out}}(t - z/v_{\text{g}})^2]}^{\text{pulse envelope}}, \quad (17.31)$$

since $\phi^{(0)} = k(\omega_0)z$ and the group delay $\phi^{(1)} = z/v_{\text{g}}$.

In general, the duration, time–bandwidth product, and frequency chirp of the optical pulse are altered by transmission through the optical system since the Gaussian parameter Γ is changed. Note, however, that the frequency width of the pulse is *not* altered by the propagation, as must be the case for a linear system. This may be seen from eqn (17.26); the width of the spectrum is determined by $\Re(1/\Gamma)$ and from eqn (17.30) $\Re(1/\Gamma_{\text{out}}) = \Re(1/\Gamma_{\text{in}})$.

Whether the optical pulse is compressed or stretched in time, and whether the pulse becomes more or less chirped, depends on the relative sign of the initial frequency chirp and the GDD of the optical system: if $b\phi^{(2)} \geq 0$ the pulse duration will increase throughout the propagation; if $b\phi^{(2)} < 0$ the pulse duration will decrease for at least part of the propagation (although continued propagation may increase the duration of the pulse, and reverse the frequency chirp). These points are explored in Exercise 17.6 and illustrated in Fig. 17.4.

We conclude this section by quoting a useful result. If the Gaussian pulse entering the optical system has no chirp (i.e. it is transform-limited) and of duration $\tau_{\text{p}0}$, the duration of the pulse leaving the optical system is given by

[16] There are strong parallels between eqn (17.29) describing the propagation of a Gaussian optical pulse and eqn (6.8) describing the propagation of a lowest-order Gaussian beam. Comparison of these two equations reveals that the parameter $1/\Gamma$ plays an equivalent role to that of the complex radius of curvature q, and the dimension of time (shifted by $-\phi^{(1)}$) in (17.29) is equivalent to the transverse coordinate r in eqn (6.8). Thus, compression (stretching) of a Gaussian pulse are analogous to focusing (defocusing) of a Gaussian beam. Note too that compression of a chirped Gaussian pulse requires the introduction of a medium with quadratic phase variation in frequency; whereas focusing a Gaussian beam requires a lens that imposes a phase shift that varies quadratically with the transverse coordinates of the beam. For further details see Siegman (1986, Chapter 9) and references therein.

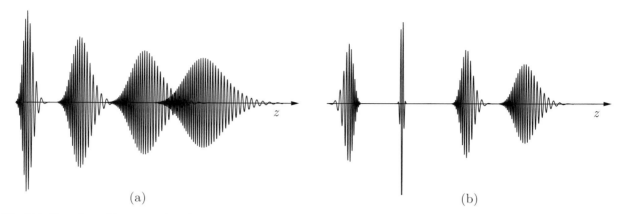

Fig. 17.4: Snap shots of the propagation of a Gaussian optical pulse through a dispersive medium when: (a) the dispersion and initial chirp have the same sign ($b_0 \phi^{(2)} \geq 0$) causing the pulse to broaden monotonically as it propagates through the medium; (b) the dispersion and initial chirp have opposite signs ($b_0 \phi^{(2)} < 0$), causing the pulse to be compressed before it rebroadens. The parameter b_0 is the imaginary part of the Gaussian parameter Γ_{in} when the pulse enters the medium.

$$\tau_p = \tau_{p0} \sqrt{1 + \left(\frac{4 \ln 2 \phi^{(2)}}{\tau_{p0}^2}\right)^2} \quad (17.32)$$

$$\approx 4 \ln 2 \frac{\phi^{(2)}}{\tau_{p0}} = \phi^{(2)} \Delta \omega_p, \quad (17.33)$$

where the approximation holds when the magnitude of the quadratic phase is large.

17.1.5 Non-linear effects: self-phase modulation and the B-integral

At high intensities the refractive index of a material becomes nonlinear with the intensity:[17]

$$n = n_0 + n_2 I. \quad (17.34)$$

In propagating a distance L a wave of vacuum wavelength λ and constant intensity I therefore accumulates a total phase

$$\phi = \frac{2\pi}{\lambda} \int_0^L n I(z) \mathrm{d}z \quad (17.35)$$

$$= \frac{2\pi}{\lambda} \int_0^L n_0 I(z) \mathrm{d}z + \frac{2\pi}{\lambda} \int_0^L n_2 I(z) \mathrm{d}z. \quad (17.36)$$

The second term gives the intensity-dependent part of the phase, ϕ_I, or the phase accumulated above that which would be accumulated by a beam of vanishingly low intensity. It is known as the **B-integral**:

---

[17] The non-linear refractive index arises from the third-order non-linear susceptibility $\chi^{(3)}$ (see Section 15.3). To see this we note that, from eqn (15.22), a wave with electric-field amplitude $\boldsymbol{E}_1(\boldsymbol{r}, t) = \frac{1}{2}\widehat{\boldsymbol{E}}_1 \exp[i(\boldsymbol{k}_1 \cdot \boldsymbol{r} - \omega_1 t)] + \text{c.c.}$ will generate through $\chi^{(3)}$ several non-linear terms, including $\boldsymbol{P}_{\text{NL}} = \frac{3\epsilon_0}{8}\chi^{(3)}|\widehat{\boldsymbol{E}}_1|^2 \widehat{\boldsymbol{E}}_1 \exp[i(\boldsymbol{k}_1 \cdot \boldsymbol{r} - \omega_1 t)] + \text{c.c.}$ This may be rewritten as $\boldsymbol{P}_{\text{NL}} = \frac{3\epsilon_0}{4}\chi^{(3)} I_1 \boldsymbol{E}_1(\boldsymbol{r}, t)$, where I_1 is the intensity of the wave. We know that the linear refractive index arises from the *linear* polarization produced by the wave: $\boldsymbol{P} = \epsilon_0 \chi^{(1)} \boldsymbol{E}_1(\boldsymbol{r}, t)$. We can therefore see that $\boldsymbol{P}_{\text{NL}}$ will contribute a non-linear refractive index that is proportional to the product of $\chi^{(3)}$ and the intensity of the wave.

Values of the nonlinear refractive index n_2 are provided for some materials in Table 17.2.

$$\phi_I \equiv B = \frac{2\pi}{\lambda} \int_0^L n_2 I(z) \mathrm{d}z. \quad \text{B-integral} \qquad (17.37)$$

The intensity of a beam of radiation will generally vary with transverse position within the beam, and usually so that the beam is more intense near the axis of propagation than in the wings. If this is the case the centre of the beam will accumulate a greater phase than the wings; in other words each slice of medium will appear to act as a positive lens, leading to self-focusing.

For a *pulse* of light the intensity-dependent phase will also vary with time during the pulse. The instantaneous phase of the wave a distance L into the medium is given by,

$$\psi(t) = \frac{2\pi}{\lambda} \left[n_0 + n_2 I(t) \right] L - \omega t. \qquad (17.38)$$

At any point in the pulse, the instantaneous frequency, $\omega'(t) = -\partial \psi / \partial t$ and hence

$$\omega'(t) = \omega - \frac{2\pi}{\lambda_0} n_2 L \frac{\partial I}{\partial t}. \qquad (17.39)$$

We see that the non-linear variation of the phase of the wave arising from the non-linear refractive index causes the frequency to vary within the pulse. This process is known as **self-phase-modulation**.

Figure 17.5 shows this process schematically. The leading edge of a laser pulse will have $\partial I / \partial t > 0$, and hence self-phase-modulation will cause a decrease in frequency, or red shift, for materials with $n_2 > 0$. Similarly, the trailing edge of the pulse will be blue shifted, and as a consequence the pulse will develop a positive chirp. The rate of change of frequency with time is approximately linear over the central part of the pulse, as shown in Fig. 17.5.[18]

Self-phase-modulation and self-focusing are nearly always undesirable in a laser system, in which case they must be kept to a minimum. A rule of thumb often employed by system designers is that the total B-integral should be kept below 1 radian.[19]

[18] We note that in combination with dispersion, self-modulation can lead to additional stretching of an intense laser pulse: self-modulation tends to increase the bandwidth of the pulse; and the increased bandwidth increases the degree of pulse stretching caused by the quadratic phase (see eqn (17.17).)

[19] See Exercise 17.4.

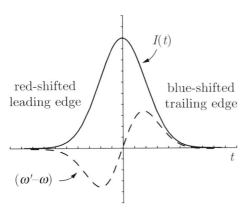

Fig. 17.5: Generation of a frequency chirp by self-phase-modulation of an intense optical pulse with initial carrier frequency ω. The solid line shows the temporal profile of the pulse; the dotted line is the frequency shift arising from self-modulation.

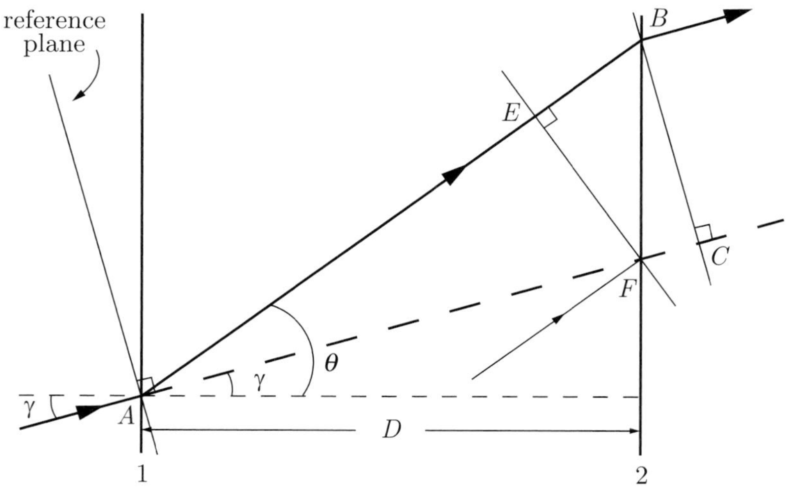

Fig. 17.6: Schematic diagram of a general optical system exhibiting geometric dispersion.

17.2 Dispersion control

The linear and non-linear dispersion discussed above distorts short laser pulses as they propagate within a system. In order to overcome this—or at least alleviate it—it is necessary to employ techniques in which additional, controllable dispersion can be introduced. Further, as discussed in Section 17.3.1 it is also important for generating short optical pulses by modelocking.

17.2.1 Geometric dispersion control

A simple way of introducing a controllable amount of dispersion is to pass the pulse through a series of optical components for which the physical (and hence optical) path is different for different wavelengths. In this way, controllable, wavelength-dependent phases can be introduced. The dispersion introduced in this way is known as **geometric dispersion**.

Figure 17.6 shows a general system exhibiting geometric dispersion. Upon reaching the plane 1, light is refracted or diffracted through a wavelength-dependent angle θ. After propagating some distance it is refracted/diffracted again so as to return the direction of propagation parallel to the initial direction. Since θ depends on the wavelength, the optical path, and hence the group delay, will depend on wavelength; the system therefore introduces dispersion.

In calculating the phase accumulated by a given wavelength, we need to define two planes (an input plane and an output plane) through which all waves pass. There is no difficulty in identifying the input plane: it is the plane passing through A, perpendicular to the *incident* ray. However, the output plane needs to be defined more carefully. For example, the plane BC is not a suitable candidate since this plane is different for each wavelength.[20] There is still a lot of flexibility: the output plane could be any fixed plane perpendicular to the rays *leaving* the system.[21] One candidate would be such a plane passing through F.

---

[20] Because the point B moves if the wavelength is changed.

[21] The calculated accumulated phase depends on where the input and output planes are defined to be. However, the calculated GDD (and higher orders) do not depend on these locations; all that will change is the total accumulated phase and the group delay for propagating between the two reference planes, just as we would expect.

However, it is convenient to move the reference output plane back along the undeflected ray to A i.e. the same plane as the input plane! We also note that EF is a wavefront of the deflected ray, and that in leaving the system this wavefront is diffracted/refracted to the wavefront BC. Using our chosen reference plane the additional *optical* path travelled by the deflected ray is equal to the geometric distance $AE - AF$.[22]

Now,

$$AF = \frac{D}{\cos \gamma} \quad (17.40)$$

$$AE = AF \cos(\theta - \gamma). \quad (17.41)$$

Hence, the optical path difference is given by,

$$OP = -\frac{D}{\cos \gamma}[1 - \cos(\theta - \gamma)], \quad (17.42)$$

and hence the phase introduced is,

$$\phi(\omega) = -\frac{\omega}{c}\frac{D}{\cos \gamma}[1 - \cos(\theta - \gamma)]. \quad (17.43)$$

It is a simple matter (in principle) to calculate the GDD, etc. by differentiating eqn (17.43), providing the relation between θ and ω is known. In fact, as considered in Exercise 17.10, provided the angle $(\theta - \gamma)$ is not too large the GDD arising from geometric dispersion is always negative, irrespective of the sign of $\partial \theta / \partial \omega$.

The grating pair

Figure 17.7 shows a ray propagating via a pair of parallel reflection gratings. Light of wavelength λ will be diffracted at an angle θ from the first grating according to,

[22] Alternatively, we can imagine fixing the output reference plane through F. The optical path of the deflected ray to this plane is $AB - FC$. But, $FC \equiv EB$ and consequently the optical path of the deflected ray to this alternative plane is equal to AE. If we now move the output reference plane to A, as in the text, we must subtract a distance AF.

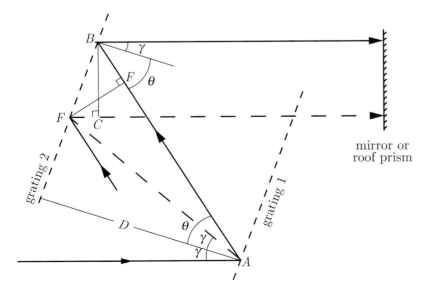

Fig. 17.7: Schematic diagram of a grating pair used to introduce negative GDD.

476 Ultrafast lasers

$$m\lambda = d[\sin\theta - \sin\gamma] \quad m = 0, \pm 1, \pm 2, \pm 3, \ldots, \qquad (17.44)$$

where γ is the incident angle and d the grating spacing. At the second grating the incident angle will now be θ and consequently the second grating will diffract the light at an angle γ if the order of diffraction is m for the first grating and $-m$ for the second. The diffracted ray therefore leaves parallel to the second.

It is easy to deduce the sign of the GDD for a grating pair. Consider the time taken for light to propagate from A to (say) the position of the mirror or roof prism, which we take as a convenient reference plane. We know that the time taken for the undeflected light (dashed line) to reach this plane is a minimum, since the path it takes is the same as if the gratings were replaced by mirrors. All other paths must take a longer time, and since the diffraction angle θ increases with wavelength, we conclude that the group delay is smaller for shorter wavelengths (higher frequencies). Hence, the group delay decreases with frequency, and the GDD must be negative.[23] It should also be clear that the GDD will increase linearly with the grating separation D, and hence the grating pair can provide controllable negative GDD.

[23]Note that the group delay is *not* equal to the optical path difference of eqn (17.42) divided by the speed of light. The reason for this is that the optical path of eqn (17.42) is merely a convenient measure of the path difference used to calculate a phase difference; it is not the physical distance travelled by rays propagating through the system.

Clearly the different wavelengths leave the grating pair at different heights above the incident ray. Consequently, in the exiting beam there is a correlation between position and wavelength, a situation known as **spatial chirp**. In general, this is undesirable, and the spatial chirp is undone by reflecting the beam back through the grating pair to form a beam propagating antiparallel to the incident beam. This counter-propagating beam may be selected by allowing the incident beam to enter the system at a small angle to the plane of the page; the exiting beam therefore leaves above or below the incident beam and may be picked off by a small mirror. Alternatively, the beam may be returned by reflection from a roof mirror that returns the light parallel to the incident rays, but at a different height. Double-passing the beam through the grating pair will introduce twice the GDD generated from a single pass.

Quantitative expressions for the GDD, and higher-order terms in the phase expansion, are obtained by differentiating eqns (17.43) and (17.44). For example, for the GDD we find, after some labour, that for a *single pass* of the grating pair

$$\phi^{(2)} = -\frac{4cD\pi^2 m^2}{d^2\omega^3 \cos^3\theta} = -\frac{Dm^2\lambda^3}{2\pi c^2 d^2 \cos^3\theta}, \qquad (17.45)$$

where θ is given by eqn (17.44). We see that the GDD is negative and proportional to D, as deduced in the discussion above.

The prism pair

Dispersion control may also be achieved with the arrangement of prisms shown in Fig. 17.8.[24] It is clear that the optical paths are different for different wavelengths, and consequently the arrangement will introduce geometric dispersion. Note that the plane AB is a plane of symmetry and consequently the total GDD is simply twice that introduced by the first prism pair.

[24]For more detailed consideration of the dispersion introduced by prism pairs see Fork et al. (1984), Sherriff (1998) and Osvay et al. (2002).

Provided that the distance the light propagates in the material of the prism is sufficiently small compared to the separation l of the apex of the prisms, the

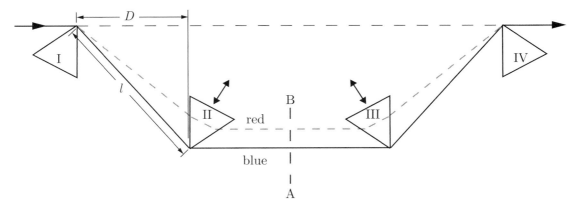

Fig. 17.8: A prism pair used to introduce negative GDD. Note that the plane AB is a plane of symmetry; it would be possible to place a mirror here, reflecting rays back through the first two prisms to yield a counter-propagating beam with the same GDD as produced by the four prisms shown. One advantage of the arrangement of four prisms is that the emerging beam propagates along the same axis as the incident beam. This means that the optical system (in particular a laser cavity) can be aligned *without* the prisms in place, and once this is done the prisms can be introduced without changing the alignment of other components. The ray labelled as 'blue' is the extreme ray that just passes through the apexes of the prisms; longer wavelength rays will pass through more prism material, as indicated by the ray labelled 'red'.

GDD of the prism pair is dominated by geometric dispersion and is therefore negative.[25] The material of the prisms introduces positive GDD, and consequently the GDD may be controlled by moving prisms II and III perpendicular to their bases, as indicated in Fig. 17.8. This reduces l and introduces more prism material into the path, and therefore increases the GDD.

[25] See Exercise 17.10.

The prisms are usually oriented at Brewster's angle so as to minimize the insertion loss; and the apex angles of the prisms are chosen so that they operate at minimum deviation (i.e. the incident and exit angles are equal). The GDD introduced by prism pairs is typically much smaller than that provided by grating pairs, but prisms have the advantage that the insertion loss can be significantly lower. For these reasons prism pairs are often used to provide dispersion control in modelocked laser oscillators.

Introduction of positive GDD

The prism and grating pairs discussed above both introduce negative GDD. Clearly, we would also like to be able to introduce positive GDD, but this would appear to be difficult owing to the fact that geometric dispersion nearly always gives rise to negative GDD for a positive separation D.

It was realized by Martinez et al. (1984) that the solution to this problem is to introduce an imaging system between the two dispersive elements that can project the first element in front, behind, or at the same position as the second. In this way the effective separation of the elements may be made to be positive, zero, or negative, which allows the GDD to be negative, zero, or positive.

In practice, this system is usually implemented with diffraction gratings, as shown in Fig. 17.9. The two lenses form an image, with unit magnification, of grating $G1$. By adjusting the position of $G1$, its image, $G1'$, may be positioned either side of the second grating $G2$. Notice that the longitudinal magnification

478 *Ultrafast lasers*

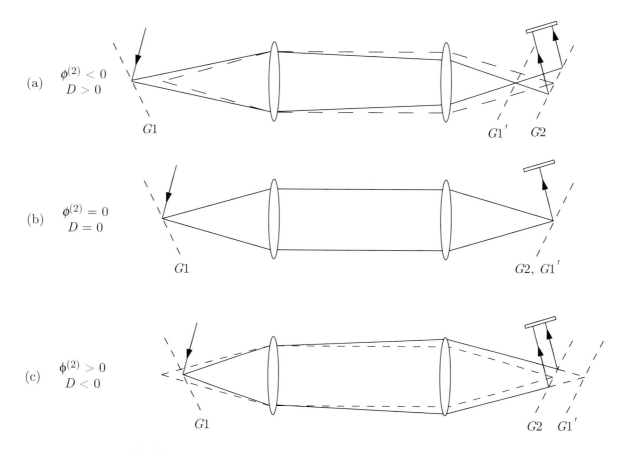

Fig. 17.9: Grating configurations giving (a) negative, (b) zero, and (c) positive GDD.

is equal to -1, and hence the imaging system changes the orientation of $G1'$ with respect to $G1$.

For controlling the dispersion of large-bandwidth pulses it is advantageous to avoid the chromatic aberrations and material dispersion arising from lenses and instead use all-reflective telescope. Figure 17.10 shows a common design known as an Öffner triplet stretcher. In this design the imaging system comprises two spherical mirrors with a common centre of curvature C. This arrangement yields a perfectly aberration-free image for objects located at C, and only weak aberrations for objects located close to C. In the design shown in Fig. 17.10 only a single grating is employed, which avoids the difficulties of ensuring that the image of the first grating G' is perfectly parallel to the second grating $G2$.

17.2.2 Chirped mirrors

A rather different way of controlling GDD (and higher-order phase terms) is reflection from a **chirped mirror**. A standard 'dielectric mirror' comprises

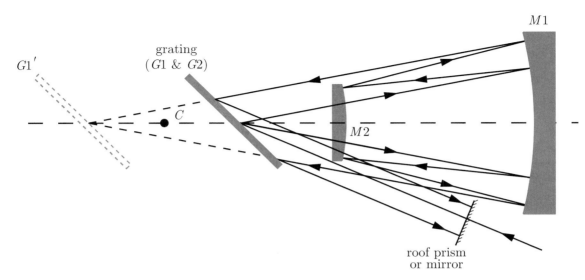

Fig. 17.10: Schematic design of an Öffner triplet stretcher employing only a single grating. If the grating were located at the common centre of curvature C of the two mirrors it would be imaged at C with a magnification of -1. In order to introduce non-zero GDD the grating is moved away from C as shown. After striking the grating, the rays of the diffracted input beam are reflected by a concave mirror $M1$ (of radius of curvature R) and a convex mirror $M2$ (of radius $R/2$). The rays would form an image of the grating $G1'$, but are intercepted by the grating and diffracted to form a parallel beam. Having undertaken a single pass through the stretcher, the beam is returned through the system at a different height by a roof prism, or a mirror slightly tilted in the vertical plane. The double-passed beam (not shown) leaves the stretcher propagating antiparallel to the input beam and can be directed to the rest of the laser system via a pick-off mirror located above or below the input beam. (After Cheriaux et al. (1996).)

a stack of alternating quarter-wave layers of high and low refractive index material, and more generally a Bragg mirror may be formed by a series of such layers in which the relative phase shift between light reflected from the first surface of the first layer, and the last surface of the second layer, is a multiple of 2π.[26] In a chirped mirror the thickness of the layers is varied within the stack. Each wavelength is reflected most strongly at the point in the stack where the Bragg condition is satisfied, and consequently a chirped–mirror structure leads to different wavelengths penetrating the stack to different depths and thereby experiencing a different group delay. The operation of a chirped mirror is illustrated schematically in Fig. 17.11.

[26] See Section 9.10.4, and especially Note 53 on page 254.

Interference between the weak reflection from the front surface of the mirror and reflection within the stack can lead to unwanted structure in the wavelength dependence of the phase of the reflected light. One way to overcome this is to vary the *relative* thickness of the high- and low-index layers as well as the Bragg period; such mirrors are known as 'double-chirped'.[27] If the GDD produced by one mirror proves insufficient, it may be increased by multiple reflections from the mirror.

[27] See, for example, Szipocs et al. (1994) and Kartner et al. (1997).

Chirped mirrors have the advantage that they are simple to align and set up, and introduce a minimum of optical elements into the system. A disadvantage is that they must be designed specifically for each application, and are therefore less flexible.

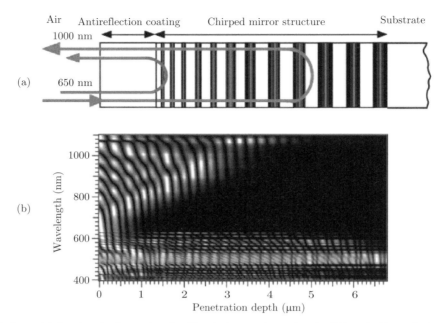

Fig. 17.11: Chirped mirrors. (a) the principle of operation of a chirped mirror; each wavelength within an incident pulse is reflected most strongly at the point in the stack where the Bragg condition is satisfied. (b) the calculated standing-wave pattern in a chirped mirror as a function of wavelength. For this mirror the penetration depth increases with wavelength, giving negative GDD. In this example, the large penetration depth for wavelengths close to 500 nm is used to transmit the pump laser radiation. (From Steinmeyer et al. (1999). Reprinted with permission from AAAS.)

17.2.3 Pulse shaping

Perhaps the ultimate example of dispersion control is shaping of the temporal profile and phase of femtosecond and picosecond pulses, which has a number of applications falling under the heading of 'coherent control' as well as being useful in controlling the amplification of short laser pulses to high peak powers (see Section 17.3.3).

Figure 17.12 shows a generic technique for pulse shaping based on a zero-dispersion stretcher, i.e. the optical arrangement of Fig. 17.9(b). In this application the telescope comprises a pair of identical lenses (or mirrors) of focal length f; these are placed a distance f from the nearest grating and separated by $2f$. The effective separation of the image of the first grating ($G1'$) and the second grating ($G2$) is zero. Mid-way between the lenses is located a plane in which the spectrum of the light pulse is dispersed,[28] which allows each frequency component to be accessed individually. Control of the output optical pulse can then be achieved by introducing amplitude and phase masks into this plane.

The most flexible type of mask is a liquid crystal spatial light modulator, which comprises a liquid crystal sandwiched between two square grids of electrodes. The birefringence of the liquid crystal located between each pair of electrodes may be controlled by adjusting the voltage between them, which enables the phase and amplitude of the transmitted light to be varied. The great advantage of this approach is that the transmission

[28] This plane is often referred to as the 'Fourier plane' since with this so-called $4f$-geometry the amplitude of the light in this plane is proportional to the Fourier transform of the amplitude of light on the grating surface. For completeness, we note that if the grating (or object) plane is some other distance from the lens, the amplitude of the light in the back focal plane is instead equal to the product of the Fourier transform of the amplitude in the object plane and a phase term that varies with transverse position.

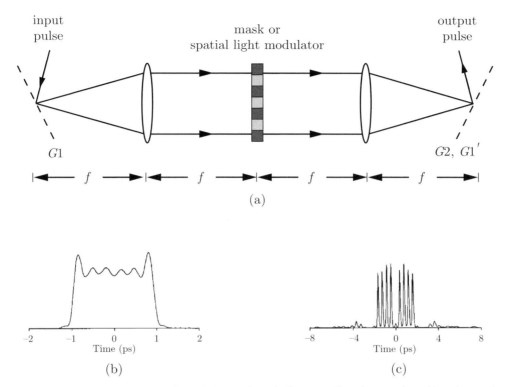

Fig. 17.12: Pulse shaping in a zero-dispersion stretcher. (a) shows schematically a zero-dispersion stretcher with a phase and amplitude mask located in the back(front) focal plane of the first(second) lens. The spectrum of the light is dispersed in this plane allowing the amplitude and phase of each frequency component of the light to be modified by a mask. (b) An approximately square pulse formed by a fixed phase and amplitude mask corresponding to a truncated sinc function; (c) a pulse train generated by a phase mask. ((b) and (c) from Weiner et al. (1998) © 1998 IEEE.)

properties of the mask can be adjusted continuously to give the desired output pulse.[29]

An alternative approach to pulse shaping employs an **acousto-optic programmable dispersive filter (AOPDF)**,[30] the operation of which is illustrated schematically in Fig. 17.13. As discussed on p. 196, an acoustic wave of wavelength λ_a propagating through an acousto-optic crystal will partially diffract an optical wave of wavelength λ if the Bragg condition is satisfied: $m\lambda = 2\lambda_a \sin\theta$, where m is the diffraction order and θ is the angle the ray makes with the wavefronts of the acoustic wave.

In an AODPF a frequency-chirped acoustic wave and the laser pulse to be controlled are launched coaxially through an acousto-optic crystal. Since the phase velocity of the acoustic wave is very much slower than that of the optical wave we may consider the acoustic wave to be stationary on the time scales of interest. Each frequency component of the laser pulse will propagate until it reaches the point in the crystal where the wavelengths of the acoustic and optical waves satisfy the Bragg condition, whereupon the optical wave may be partially diffracted into a different mode; this second mode forms the output pulse.[31] In an AOPDF the crystal is anisotropic so

[29] In some applications this adjustment is achieved with the aid of a genetic algorithm that adjusts the electrode voltages to optimize one or more output parameters of the experiment or process being investigated.

[30] See, for example, Verluise et al. (2000).

[31] For details see, for example, Yariv (1989).

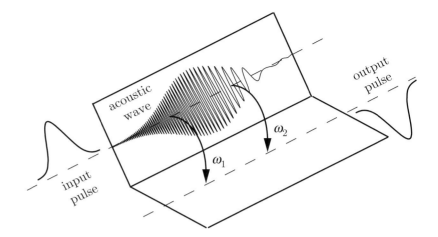

Fig. 17.13: Pulse shaping in an acousto-optic programmable dispersive filter (AOPDF). The two frequencies, ω_1 and ω_2, of the incident pulse are diffracted into the output optical mode only at the points in the crystal that the wavelength of the acoustic wave satisfies the Bragg condition for that optical frequency.

that the diffracted optical mode experiences a different refractive index from the original mode. As a consequence, the total phase accumulated by each frequency depends upon the distance into the crystal at which the frequency is diffracted into the output mode; the amplitude of each frequency is determined by the amplitude of the acoustic wave at the point at which the frequency is diffracted.

Acoustic-optic programmable dispersive filters can typically diffract up to 30% of the incident pulse into the output mode. They are very widely used in laser systems employing chirped-pulse amplification (see Section 17.3.3) for precise control of the phase structure, and hence pulse shape, of short laser pulses; the AOPDF is usually placed early in the amplifier chain so that the losses incurred in the AOPDF may be made up for in the subsequent amplifiers, and the amplitude and phase of the acoustic wave are adjusted so that the optical pulse at the output of the entire system has the desired properties.

17.3 Sources of ultrafast optical pulses

17.3.1 Modelocked lasers

Importance of dispersion control

Dispersion control is particularly important for modelocked lasers producing very short pulses.[32] To see this we note that, as discussed in Section 16.5, the frequency spectrum of a modelocked pulse is given by eqn (6.63):

$$\omega_p = \omega_{ce} + \frac{2\pi}{T_c} p \qquad p = 0, 1, 2, 3, \ldots, \tag{17.46}$$

where ω_{ce} is the off-set frequency and T_c is the interval between pulses, which corresponds to the time for one round trip of the cavity.

However, the angular frequencies ω_n of the longitudinal modes of an optical cavity are given by the condition that the round-trip accumulated phase is an integer multiple of 2π, i.e. by

[32] For further details, see Exercise 17.2.

$$\phi(\omega_p) = 2\pi p \qquad p = 0, 1, 2, 3, \ldots. \qquad (17.47)$$

In the absence of dispersion the frequencies solving eqn (17.47) have a constant spacing $\Delta\omega = 2\pi/T_c$ and consequently eqns (17.46) and (17.47) are both satisfied. However, in general, an optical cavity will contain elements that to some extent are dispersive—at the very least, for example, the gain medium will exhibit some dispersion. In this case the mode frequencies are not evenly spaced and do not obey eqn (17.46). In such cases it is important to control the dispersion within the cavity so that, as far as possible, the frequencies of the modes are evenly spaced. In practice, the dispersion cannot be compensated perfectly. However, since the spectrum of the modelocked pulse *must* obey eqn (17.46), the modes contributing to the pulse are pulled by dispersion and the modelocking mechanism into agreement with eqn (17.46). Residual dispersion will limit the bandwidth over which this can be achieved, and hence increase the duration of the modelocked pulse above that which would be possible in the absence of dispersion.

The problem of dispersion becomes more acute as the desired duration of the modelocked pulses is reduced, and the associated bandwidth is increased. In practice, it is found that dispersion control is necessary to generate pulses shorter than approximately 150 fs. Most materials exhibit positive dispersion, and hence a source of controllable negative dispersion must be introduced into the cavity. It is also helpful to keep the intrinsic dispersion of the cavity as low as possible; for example, by using a short gain medium.

17.3.2 Oscillators

The most common source of ultrafast optical pulses is the Kerr-lens modelocked Ti:sapphire laser, described in Section 7.3.5 and illustrated in Fig. 7.29(b). These routinely generate trains of modelocked pulses with durations as short as 30 fs at pulse repetition rates of about 80 MHz.

In principle, the bandwidth of Ti:sapphire can support pulses as short as approximately 2 fs. However, in order to generate pulses below 10 fs the GDD must be kept as small as possible—by using short, highly-doped laser rods—and carefully controlled with prism pairs and chirped mirrors. To date, the shortest pulses that have been generated directly[33] from a modelocked laser oscillator are of approximately 5 fs duration, corresponding to fewer than two cycles of the carrier frequency![34]

17.3.3 Chirped-pulse amplification (CPA)

The mean power output by a modelocked Ti:sapphire oscillator is of the order of 1 W, and consequently for a typical pulse repetition rate of 80 MHz the energy of each pulse in the train is of order 10 nJ. Many applications require pulses of higher energy and peak power.

Efficient amplification of a pulse requires that the fluence of the pulse prior to the amplifier is of the order of the saturation fluence $\Gamma_s = \hbar\omega_L/\beta\sigma_{21}$, as discussed in Section 5.2. For very short pulses, however, a pulse with a fluence equal to Γ_s may have a peak intensity above the damage thresholds

[33] See Morgner et al. (1999) and Sutter et al. (1999) and Ell et al. (2001).

[34] Shorter pulses may be generated by increasing the bandwidth of a short pulse through self-phase-modulation and subsequent recompression. The shortest pulses generated with this technique are below 4 fs duration, see, for example, Schenkel et al. (2003).

484 *Ultrafast lasers*

of the amplifier rod or other optical components. For example, for Ti:sapphire $\Gamma_s \approx 1\,\mathrm{J\,cm^{-2}}$, so that a 30-fs pulse with a fluence of Γ_s would have a peak intensity of approximately $3 \times 10^{13}\,\mathrm{W\,cm^{-2}}$; this is high enough to damage the laser rod and other optical components. Further, in practice, the intensity to which the pulses can be amplified is even lower than the limit set by optical damage; at lower intensities self-phase-modulation leads to self-focusing and growth in localized non-uniformities (known as 'hot spots') in the intensity profile of the beam. For example, the B-integral corresponding to propagation of a 30-fs pulse with a fluence equal to Γ_s through 10 mm of Ti:sapphire is[35] nearly 10^3!

[35] See Exercise 17.12.

The solution to this problem is **chirped-pulse amplification (CPA)** in which low-energy pulses are amplified to high-energy, high peak power pulses in three stages:

1. The pulses are stretched in time by passing them through a pulse stretcher with a large GDD. Stretch factors of order 10^3 to 10^5 are typical.
2. The stretched pulses are then amplified in a chain of one or more amplifiers. Since the duration of the stretched pulses is many orders of magnitude longer than the original pulses, they may be amplified to higher energies than in the absence of stretching by a factor approximately equal to the stretch factor.[36]
3. The amplified pulse is then recompressed to (approximately) the duration of the original pulse by passing the amplified pulse through a pulse compressor with the opposite GDD to that of the stretcher.

[36] The dynamic behaviour of the population inversion in the amplifier will be unchanged provided that the duration of the stretched pulse remains short compared to the recovery time of the laser transition. For the case of Ti:sapphire $\tau_R \approx \tau_2 = 3.8\,\mu\mathrm{s}$, and consequently this condition is easily met.

The CPA process is illustrated schematically in Fig. 17.14.

The pulse stretcher usually comprises the grating pair illustrated in Fig. 17.9(c), set to produce positive dispersion. One reason for this choice is

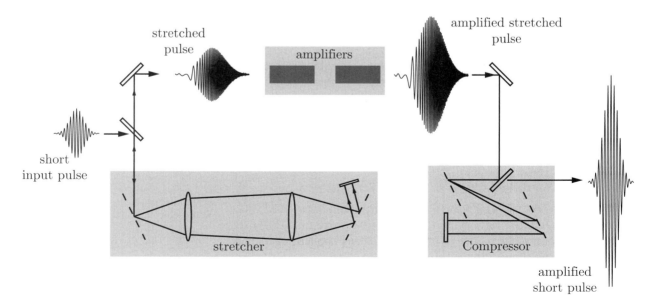

Fig. 17.14: Schematic diagram of chirped-pulse amplification.

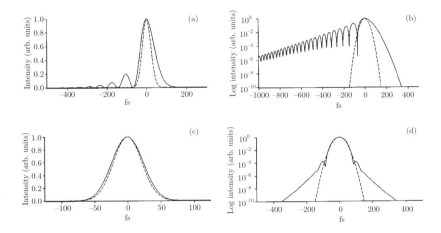

Fig. 17.15: Calculated temporal intensity profiles of re-compressed pulses in a CPA system. A transform-limited 50-fs pulse (dashed line) is stretched by a pair of 1200 lines/mm gratings, then amplified by 12 round trips in a regenerative amplifier, and finally compressed by a second pair of 1200 lines/mm gratings (solid line). When the grating separation in the compressor is adjusted to compensate for the total GDD introduced by the stretcher and material of the amplifier, distortions due to the remaining third-order dispersion are significant, as shown on a linear (a) and logarithmic (b) scale. Significant improvements are made if both the grating separation and angle of incidence are adjusted to minimize the duration of the compressed pulse; the remaining pulse distortions are due to residual higher order phase, as can be seen on a linear (c) and logarithmic (d) scale. (Reprinted with permission from Walmsley et al. (2001). Copyright 2001, American Institute of Physics.)

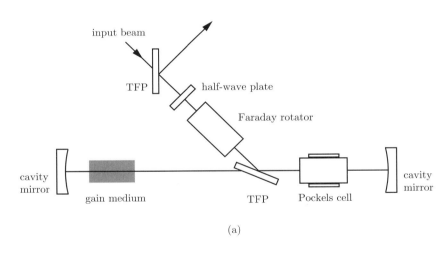

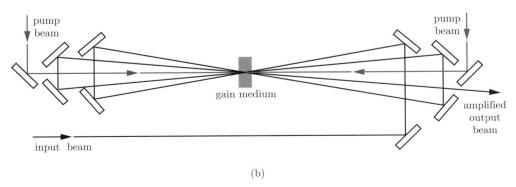

Fig. 17.16: Schematic diagram of: (a) a regenerative; and (b) a multipass amplifier.

that the material of the following chain will tend to introduce more positive dispersion, and therefore giving the pulses a positive initial chirp ensures that there is no danger of partially recompressing the laser pulses as they are amplified (See Fig. 17.4(b)).

It might be supposed that the pulse compressor should be designed to exactly negate the GDD and higher-order phase terms introduced by the stretcher. However, this does not normally lead to good pulse compression. The reason is the material of the amplifiers introduces large amounts of GDD and higher phase terms; the total path of amplifier material through which the pulse propagates can be of order 1 m. Adjusting the spacing D of the compressor to eliminate the total GDD introduced by the stretcher plus the amplifier material also does not lead to the shortest pulses, as illustrated in Fig. 17.15. The reason for this is that adjusting the compressor to eliminate the total GDD of the stretcher *and* amplifier material results in non-zero values for the total third- and higher-order phase, i.e. the stretcher and compressor will introduce a net third- and higher-order phase that is not in general cancelled by that of the amplifier material. Better results can be achieved by adjusting the grating separation and the angle of incidence γ in the compressor to minimize the duration of the compressed pulse, as shown in Fig. 17.15. Even this does not lead to perfect recompression, and consequently a variety of schemes for achieving perfect compression up to and including the fourth-order phase have been investigated.[37]

Amplifiers

In a CPA laser system there will typically be several stages of amplification. Immediately after the pulse has been stretched, there will be one or more pre-amplifiers in which most of the *gain* occurs (perhaps a factor of 10^7). After the pre-amplifiers the energy of the stretched pulses will be sufficiently high that they can extract energy efficiently from one or more power amplifiers. Most of the *energy* in the final pulse is added by the power amplifiers.

There are two main classes of amplifier used in CPA systems, as illustrated schematically in Fig. 17.16.

Regenerative amplifiers

In a regenerative amplifier the amplification takes place in an optical cavity. Laser pulses are injected into the cavity by applying a voltage pulse to a Pockels cell, as shown schematically in Fig. 17.16(a). In the absence of any voltage on the Pockels cell the incident laser pulses will, after reflection from the cavity end mirror, be reflected out of the cavity by the thin-film polarizer (TFP). However, if a quarter-wave voltage is applied to the Pockels cell, the pulses will return to the intra-cavity polarizer with the orthogonal polarization and so will be trapped in the cavity and hence amplified on each round trip. After amplification the pulses are switched out by applying a second quarter-wave pulse to the Pockels cell.[38]

Since the gain medium is located within an optical cavity, the pumping must not be too strong, or lasing—or at least strong amplification of spontaneous emission—will occur even in the absence of an injected pulse. As such, in

[37] For details see Walmsley et al. (2001) and references therein.

[38] In this design the Faraday rotator is included in the arrangement in order to prevent pulses being reflected back into earlier parts of the laser chain (such as the laser oscillator), since such reflections can cause damage or instabilities. Its inclusion complicates a description of the operation of the regenerative amplifier somewhat, although the essential features are included in the discussion of the main text. Let us suppose that the input pulses are vertically polarized. The half-wave plate is oriented at $45°$ to the vertical so as to rotate the plane of polarization to the horizontal plane. On passing through the Faraday rotator the plane of polarization is rotated by (say) $+45°$. In the absence of any voltage on the Pockels cell, the pulses will be rejected by the cavity and their polarization further rotated by the Faraday rotator in the *same sense*, i.e. another $+45°$, to give vertically polarized light. The half-wave plate then flips the polarization to the horizontal plane, and consequently the pulses are reflected by the external polarizer. Trapping laser pulses within the cavity operates as discussed above, i.e. applying a voltage to the Pockels cell rotates the polarization of the laser pulses so that they are transmitted by the intra-cavity polarizer. The pulses are rejected from the cavity by applying a second pulse to the Pockels cell, whereupon they are reflected by the external polarizer in the manner discussed in this note.

order to extract energy from the gain medium it is necessary for the pulse to undergo many cavity round trips, typically about 20.

The advantages of regenerative amplifiers are that the optical cavity imposes a good transverse mode structure on the amplified pulses, leading to good beam quality. There are several disadvantages, however. Leakage through the polarizers and Pockels cell can give rise to a series of lower-energy pulses propagating ahead of the main pulse. Pre-pulses of this type can be problematic in many applications. Further, the many round trips required to extract the energy stored in the population inversion introduces a long length of material into the path of the pulse and thereby introducing large amounts of GDD and higher-order phase terms that must be removed in the compressor. Nevertheless, regenerative amplifiers can be used to generate high-power pulses with pulse durations as short as 30 fs.

Multipass amplifiers

In a multipass amplifier the pulse is passed through the gain medium several times by reflection from a series of mirrors – sometimes referred to as angular multiplexing. Since the gain medium is not located within a cavity the single-pass gain may be much higher than in a regenerative amplifier. Multipass amplifiers therefore have the advantage of introducing less material into the path of the pulse, as well as not introducing pre-pulses. The disadvantages are that the overlap between the pump laser beam and the amplified laser pulse is not as good as in a regenerative amplifier, and consequently the efficiency will not be as high as in a regenerative amplifier. A typical efficiency for a multipass pre-amplifier is 15%; although multipass power amplifiers may reach 30%, which is comparable to that achieved in a regenerative amplifier.

OPCPA

A rather different type of amplifier that is being used increasingly in CPA systems is the optical parametric amplifier discussed in Section 15.7; this approach is known as **optical parametric chirped-pulse amplification (OPCPA)**.

The advantage of optical parametric amplification is that all of the energy of the pump photons is converted to light, in the form of the signal and idler waves. As a consequence, essentially no thermal energy is deposited in the gain medium, which eliminates problems caused by thermal lensing in the amplifier rod and avoids the necessity of cooling.[39]

A second advantage of OPCPA is that the bandwidth of parametric amplification can be very large and the gain of the amplifier does not saturate. Hence, at least to the extent that the pump laser pulses have a constant intensity, the amplification experienced by the front and back of the chirped pulse is constant. OPCPA can therefore amplify very large bandwidth pulses without distortion. Finally, optical parametric amplification does not give rise to amplified spontaneous emission at the output wavelengths; this ensures that the contrast of the final, compressed pulses is high.[40]

The amplifier in an OPCPA system therefore merely comprises a rod of material with a non-zero $\chi^{(2)}$, such as BBO or KDP, pumped by a laser of frequency ω_p. To date OPCPA systems have been used to generate pulses with a peak power as high as 200 TW.[41]

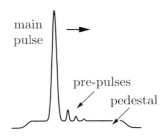

Fig. 17.17: Schematic diagram of the output typically provided by a high-power laser system, showing how the main laser pulse is heralded by one or more pre-pulses and sits on a pedestal of low-intensity radiation.

[39] This is in distinct contrast to a Ti:sapphire amplifier, for which the quantum efficiency is approximately 65%. Thus, around 35% of the pump energy remains in the rod, which must be removed by flowing coolant around it. For high-energy Ti:sapphire amplifiers the coolant is often liquid nitrogen since the thermal conductivity of sapphire increases very rapidly as the temperature is decreased.

[40] The importance of pulse contrast depends on the application. Applications involving the interaction of high-intensity pulses with solid surfaces require the total energy contained in the low-intensity pedestal of the pulse (see Fig. 17.17) to be sufficiently low that the target is not heated appreciably before the arrival of the peak of the laser pulse. In other applications, such as OFI lasers (see Sections 18.5.4 and 18.6.2), the intensity of any pre-pulses must be below the threshold for ionization of the target gas.

[41] For a review of OPCPA systems see Dubietis et al. (2006).

488 *Ultrafast lasers*

TW laser systems

Chirped-pulse amplification has made the generation of laser pulses with peak powers of up to 10 TW relatively routine; indeed systems of this type are commercially available.[42] Figure 17.18 shows schematically the layout of a small-scale TW laser system.

PW laser systems

Large-scale laser systems based upon Ti.Sapphire with peak powers of order 500 TW have recently been constructed. These systems are able to generate such high peak powers by ensuring that the duration of the pulse is only a few tens of femtoseconds.[43]

In order to reach very high pulse energies (100s of Joules) it is necessary to increase the beam diameter to 100 mm or more. At present, it is not possible to grow large-diameter crystals of Ti:sapphire with sufficient optical quality

[42] It is perhaps worth remembering that the total electrical-generating capacity of all the power stations on Earth is less than 5 TW! (For further information see http://www.eia.doe.gov/.)

[43] For example, the Astra-Gemini laser system at Rutherford Appleton Laboratory is designed to generate pulses of energy 15 J and duration 30 fs, corresponding to a peak power of 0.5 PW. The laser is designed to deliver pulses every 20 s. Further details may be found at www.clf.rl.ac.uk.

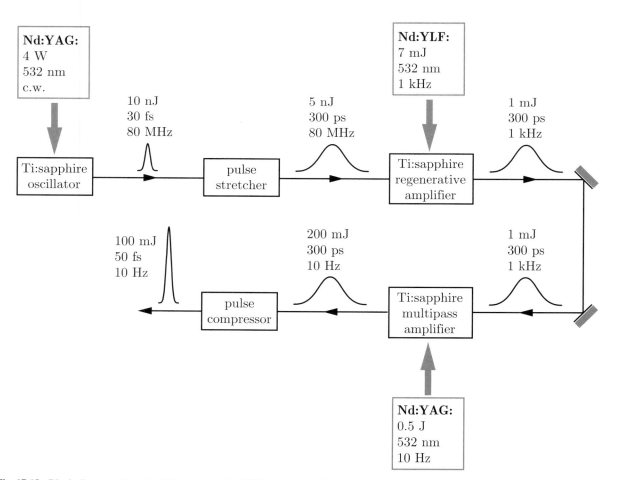

Fig. 17.18: Block diagram of a typical laboratory-scale TW laser system. The main parameters of the Ti:sapphire pulse at various points in the system are given, and the parameters of the pump lasers are summarized in the grey boxes. The output pulses have a peak power of 2 TW and a duration close to those from the oscillator.

for this purpose; high-energy laser systems therefore use as the active medium Nd:Glass, since the technology for producing optical-quality rods and disks of this material is well developed.

The laser transition in Nd:Glass is inhomogeneously broadened with a linewidth of approximately 50 THz, and hence can produce modelocked laser pulses with durations as short as approximately 100 fs. Modern high-energy 'glass' laser systems amplify pulses from a Ti:sapphire oscillator and pre-amplifier system, the Ti:sapphire being tuned in wavelength to match the peak of the gain of the Nd:Glass laser. Typically, a series of multipass rod amplifiers is used for amplification to pulse energies of several tens of Joules. Amplification to higher energies requires the use of larger beam diameters, and hence large-area discs of Nd:Glass are employed.

A good example of a PW class laser (there are not many!) is the Vulcan laser at the Rutherford Appleton Laboratory, which has the following key features:

- Pulses with an energy of 5 nJ, a duration of 120 fs, and a wavelength of 1055 nm are generated by a modelocked Ti:Sapphire oscillator.
- These pulses are stretched by a factor of 4×10^4 to a duration of 5 ns, and then amplified to an energy of order 10 mJ using the OPCPA technique.
- The pre-amplified pulses are passed to a chain of flashlamp-pumped rod and disk Nd:Glass amplifiers, which amplifies the pulse energy to 670 J.
- Two gold-coated holographic diffraction gratings (940 mm diameter!) with a groove spacing of 480 lines mm^{-1} recompress the pulse to 650 fs; somewhat longer than the original pulses, but not impressively short considering that they have been amplified by a factor of 10^{11}!
- After losses caused by reflection at the compressor gratings and beam-steering optics, the energy of the pulses available to the target chamber is approximately 500 J, i.e. a peak power of 0.8 PW. The laser operates at a pulse repetition rate of one shot every 20 min.[44]

[44] In principle, these pulses can be focused to a spot size of order 7 μm diameter, corresponding to a peak focused intensity of order 10^{21} W cm^{-2}. Note that at this intensity the radiation pressure is 11 orders of magnitude greater than atmospheric pressure!

17.4 Measurement of ultrafast pulses

In most applications of short optical pulses it is necessary to know, at least approximately, how long the pulse is. However, the temporal profile of pulses shorter than a few nanoseconds are too fast to be recorded directly by photodiodes. More complex devices, known as streak cameras,[45] are able to measure pulses as short as a few hundred femtosecond but for shorter pulses it is necessary to use an altogether different approach.

[45] In a streak camera the temporal variation of the pulse is converted into a spatial variation. For example, this may be done by irradiating a photcathode to generate a pulse of electrons that are then accelerated onto a phosphor screen. A rapidly rising electric field is applied transverse to the electron beam such that the deflection of the beam on the screen varies linearly in time.

17.4.1 Autocorrelators

The simplest method for measuring the duration of ultrafast laser pulses is to use a copy of the pulse as a probe. Figure 17.19 illustrates how this may be done with a Michelson interferometer. In this approach the pulse to be measured is split into two copies at the beam splitter and recombined after a suitable temporal delay τ has been introduced between them. The two pulses

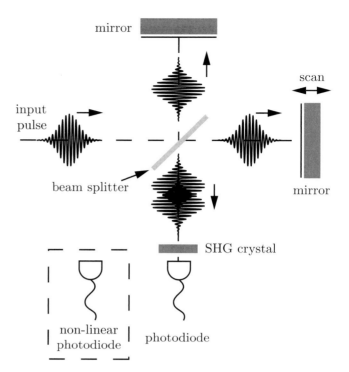

Fig. 17.19: A scanning autocorrelator used to measure the duration of ultrafast laser pulses. The autocorrelation signal may be measured by recording the second harmonic generated in a non-linear crystal, or by using a non-linear photodiode.

are then directed to a non-linear detector that produces a signal proportional to the square of the incident intensity. For example, the detector may comprise a second-harmonic crystal and a filtered photodiode that detects the generated second-harmonic light. Alternatively, the detector could be a photodiode with a band gap greater than the photon energy of the laser pulse; in such cases electrons in the valence band of the material of the photodiode are only excited across the band gap if two photons are absorbed simultaneously, and consequently the photocurrent is proportional to the square of the incident intensity.

The autocorrelation signal $S(\tau)$ is therefore given by,

$$S(\tau) \propto \int_{-\infty}^{\infty} \left(|E(t) + E(t+\tau)|^2 \right)^2 dt \qquad (17.48)$$

$$\propto \int_{-\infty}^{\infty} \left(|E(t)|^2 + |E(t+\tau)|^2 + \underbrace{2\Re\left[E(t)E(t+\tau)^*\right]}_{\text{interference}} \right)^2 dt, \qquad (17.49)$$

where $E(t)$ is the electric field of the laser pulse.

As discussed in Exercise 17.13, the form of the autocorrelation signal as a function of the delay τ depends on whether the bandwidth of the detector is able to resolve the interference terms present in eqn (17.49). If these terms *are*

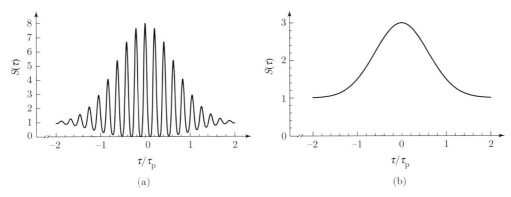

Fig. 17.20: Autocorrelations of an ultrashort laser pulse with a Gaussian temporal intensity profile of FWHM τ_p: (a) the interferometric autocorrelation; (b) the intensity autocorrelation. The autocorrelations have been normalized so that the autocorrelation signal at large delays τ is unity.

resolved the autocorrelation is said to be an **interferometric autocorrelation**, and the signal resembles that shown in Fig. 17.20(a). This comprises interference fringes of spacing $\Delta\tau = 2\pi/\omega_0$, and an amplitude that decays with $|\tau|$, superimposed on a constant background signal. As shown in Exercise 17.13, the ratio of the peak signal to that of the constant background is 8:1.

If the interference terms are not resolved the autocorrelation is an **intensity autocorrelation**, which may be written as

$$S_i(\tau) \propto \int_{-\infty}^{\infty} I(t)^2 dt + 2\int_{-\infty}^{\infty} I(t)I(t+\tau) dt. \qquad (17.50)$$

As shown in Fig. 17.20(b), an intensity autocorrelation does not exhibit interference fringes and instead the autocorrelation signal decays monotonically with $|\tau|$ to a constant value. It is shown in Exercise 17.13, that for an intensity autocorrelation the ratio of the peak signal to that of the constant background is 8:1.

For both types of autocorrelation the signal is symmetric about $\tau = 0$, and consequently it is evident that the detailed shape of the pulse envelope cannot be deduced from the autocorrelation alone.[46] However, it is clear that the autocorrelation will decrease to the constant level for delays greater than the order of the duration of the pulse. In order to make a more quantitative estimate of the duration of the pulse it is necessary to *assume* the shape of the laser pulse envelope, whereupon the width of the laser pulse may be deduced from that of the autocorrelation. Table 17.3 gives the relation between the FWHM τ_{ac} of the intensity autocorrelation above the constant background for various temporal intensity profiles of FWHM τ_p.

[46] For example, a pulse with a fast leading edge and a slow trailing edge cannot be distinguished from its time-reversed counterpart.

Table 17.3 Deconvolution factors for intensity autocorrelation of laser pulses with an intensity temporal profile $I(t)$ of FWHM duration τ_p. The FWHM of the autocorrelation above the background level is τ_{ac}.

$I(t)$	τ_{ac}/τ_p
top hat	1
$\exp\left(-\ln 2 \frac{t}{\tau_p}\right)$	2
$\exp\left[-4\ln 2 \left(\frac{t}{\tau_p}\right)^2\right]$	$\sqrt{2}$
$\mathrm{sech}^2\left(1.76\frac{t}{\tau_p}\right)$	1.54

Single-shot autocorrelators

The scanning autocorrelators described above require the autocorrelation signal to be recorded as one of the mirrors of the interferometer is scanned. This procedure becomes inconvenient for laser systems operating at pulse repetition rates below around 1 kHz, and instead so-called single-shot autocorrelators are used. In these devices two copies of the pulse are brought together with

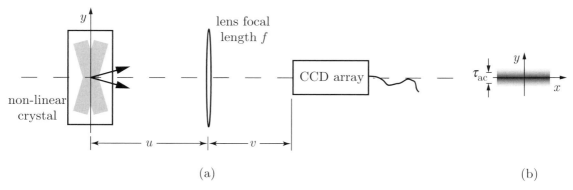

Fig. 17.21: A single-shot autocorrelator showing: (a) the arrangement employed; (b) the form of the autocorrelation signal recorded by the camera. Note that the lens images the plane in the non-linear crystal in which the two copies of the laser pulse intersect, so that $1/u + 1/v = 1/f$.

a small angle between their axes of propagation; in practice, this may be achieved by tilting one of the mirrors of the interferometer of Fig. 17.19 slightly, or by passing the beam to be measured through a Fresnel biprism to form two beams propagating at a small angle. As shown in Fig. 17.21, the relative delay between the two beams varies linearly[47] with the distance y. If a non-linear crystal is placed at the intersection of the two beams, the variation of the generated second-harmonic signal with y will be proportional to the autocorrelation.

The single-shot autocorrelation is recorded by imaging the plane of intersection onto a detector such as a CCD array. In practice, the interference terms are not resolved in single-shot autocorrelators, and the autocorrelation recorded is an intensity autocorrelation.[48]

17.4.2 Methods for exact reconstruction of the pulse

In recent years, several techniques for measuring the amplitude and phase of ultrashort pulses have been developed. These are discussed briefly below.

FROG

The first method developed to measure amplitude and phase of an optical pulse is the **frequency-resolved optical gating (FROG)** technique developed by Kane and Trebino.[49] This is essentially a spectrally resolved autocorrelation measurement.

To understand this in more detail, imagine that a section of the pulse to be measured in the region $t = \tau$ is isolated and its spectrum measured. Mathematically, this can be achieved by multiplying the amplitude $E(t)$ of the pulse by a gate function $g(t - \tau)$ that might, for example, be a top-hat function. The intensity of the measured spectrum would then be given by,

$$I_s(\omega, \tau) \propto \left| \int_{-\infty}^{\infty} E(t) g(t - \tau) e^{-i\omega t} dt \right|^2. \qquad (17.51)$$

[47] It is straightforward to show that the optical delay is given by $\tau = 2y \sin\theta$, where 2θ is the angle between the direction of propagation of the two beams. The y-axis of the autocorrelator may be calibrated in terms of delay τ by measuring the change in the y position of the peak of the autocorrelation signal when one of the beams is delayed by a known amount (with the interferometer arrangement of Fig. 17.19 this may be achieved by translating one of the mirrors through a known distance).

[48] It should be apparent that this method relies on the 'pulse front' of the beam—i.e. the locus of the peak of the pulse—being perpendicular to the beam, and the duration of the pulse being the same across the beam. These assumptions may not be valid in practice. In particular, tilting of the pulse front can lead to significant errors in the measurement of the pulse duration by single-shot autocorrelators, as well as unexpected behaviour in applications of the short pulses. For further details see Pretzler et al. (2000) and Raghuramaiah et al. (2003).

[49] See Kane and Trebino (1993). A detailed account of the FROG technique may be found in the book edited by Trebino (2002).

17.4 Measurement of ultrafast pulses

It is easy to see that as the position of the gate is varied, by adjusting τ, the spectrum of the pulse as a function of time τ can be deduced.[50,51] The function $I_s(\omega, \tau)$ is known as a **spectrogram (or sonogram)**. The key to the FROG technique is the fact that the optical pulse is essentially uniquely determined once the spectrogram is determined.[52]

Particular varieties of the FROG technique may introduce other ambiguities. For example, in SHG-FROG the direction in time is not determined since the FROG trace is symmetric in delay τ. This ambiguity may be removed by prior information or additional measurements; for example, passing the pulse through a glass plate introduces chirp of a known sign. Since there is a unique solution for $E(t)$, given $I_s(\omega, \tau)$, the pulse amplitude and phase may be found by an iterative retrieval algorithm.

In practice, the gate pulse is provided by the pulse to be measured. For example, in the single-shot autocorrelator shown in Fig. 17.21 the amplitude of the second-harmonic signal is proportional to $E(t)E(t-\tau)$, corresponding to a gate pulse $g(t-\tau) = E(t-\tau)$. In order to record the spectrogram it is necessary to replace the CCD in Fig. 17.21 by a spectrograph with its entrance slit oriented in the plane of the paper so that the delay τ varies along the slit and the spectrum of the signal is dispersed in the plane normal to the paper.

The FROG technique can be used to measure the laser pulse in a single laser shot; and the retrieval algorithm can reach convergence in a time of order 1 s, allowing real-time implementation. Examples of spectrograms, or 'FROG traces', are given in Fig. 17.22.

[50] Of course, the spectrum is averaged over the duration of the gate function.

[51] It might be thought that the duration of the gate should be as short as possible. However, a very short gate would result in the spectrogram only reproducing the intensity profile of the pulse, meaning that the phase variation within the pulse could not be determined. Similarly, a very wide gate pulse would cause the spectrogram, for all delays τ, to be equal to the spectrum of the entire pulse.

[52] The pulse is said to be 'essentially uniquely' determined since the terms $\phi^{(0)}$ and $\phi^{(1)}$ in the spectral phase expansion of eqn (17.13) and Table 17.1 cannot be determined, a restriction that also applies to the SPIDER technique (see below). As a consequence, neither the arrival time of the pulse nor the carrier-envelope offset (CEO) can be determined from these measurements.

SPIDER

An alternative method for measuring the amplitude and phase of an ultrashort laser pulse is known as **spectral phase interferometry for direct electric-field**

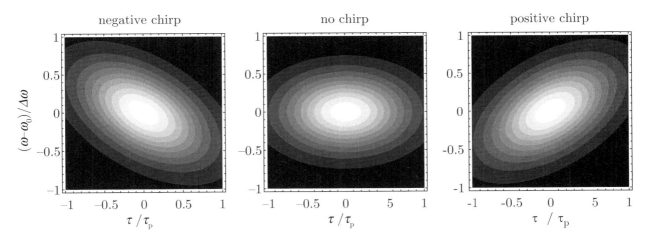

Fig. 17.22: Calculated FROG traces for optical pulses with negative, zero, and positive frequency chirps. The traces are calculated for the case of polarization gating for which the gate function is given by $g(t-\tau) = |E(t-\tau)|^2$. In polarization-gated FROG the pulse to be measured and the gate pulse (a copy of the pulse to be measured) are polarized orthogonally and passed through a non-linear medium with a finite value of $\chi^{(3)}$. The non-linear interaction can generate a wave with a polarization that is orthogonal to that of the pulse to be measured, and with an amplitude proportional to $E(t)|E(t-\tau)|^2$. This wave forms the polarization-gated FROG signal. It is worth noting that for the second-harmonic gate discussed in the text there is an ambiguity in the direction of time such that the measured FROG traces for positive and negative chirp are identical.

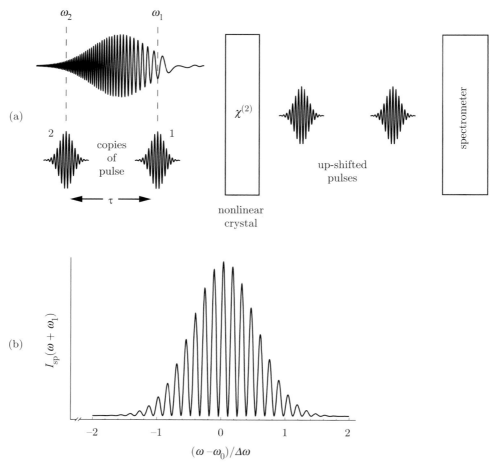

Fig. 17.23: The SPIDER technique for measuring the amplitude and phase of ultrafast laser pulses. (a) Two copies of the pulse to be measured are up-shifted by frequencies ω_1 and ω_2 by sum-frequency mixing them with a strongly chirped pulse (derived from the pulse to be measured). The up-shifted pulses interfere in the output plane of a spectrometer to give the SPIDER spectrum $I_{\rm sp}(\omega+\omega_1)$ shown in (b).

reconstruction (SPIDER). As illustrated in Fig. 17.23, this technique is a form of spectral shearing interferometry in which two copies of the pulse to be measured are shifted in frequency with respect to each other and allowed to interfere in a spectrometer.

Two copies of the pulse, with a separation τ of a few picoseconds, may easily be formed from the reflections from the two surfaces of a thin plate. The required spectral shear is achieved by sum-frequency mixing each pulse with a strongly chirped copy of the original pulse; the stretched pulse can be generated from a copy of the pulse to be measured using the grating stretcher discussed on page 475.

It is shown in Exercise 17.14 that the intensity of the SPIDER spectrum may be written as

$$I_{\rm sp}(\omega+\omega_1) \propto a(\omega)^2 + a(\omega-\Omega)^2 + 2a(\omega)a(\omega-\Omega)\cos\left[\phi_{\rm sp}(\omega)\right],$$

(17.52)

where $a(\omega)$ is the real part of the amplitude per unit frequency interval of the spectrum of the pulse, the pulses 1 and 2 are up-shifted by frequencies ω_1 and ω_2, respectively, $\Omega = \omega_2 - \omega_1$, and

$$\phi_{\text{sp}}(\omega) = \phi(\omega) - \phi(\omega - \Omega) - (\omega + \omega_1)\tau. \qquad (17.53)$$

The SPIDER spectrum comprises an up-shifted spectrum modulated by the interference fringes described by $\cos[\phi_{\text{sp}}(\omega)]$; there are standard techniques[53] for finding the phase $\phi_{\text{sp}}(\omega)$ from this fringe pattern. The pulse separation τ may be measured straightforwardly using the method described in Exercise 17.14, and hence from the measured $\phi_{\text{sp}}(\omega)$ it is possible to deduce the phase differences $\phi(\omega) - \phi(\omega - \Omega)$ to within the constant $\omega_1\tau$, and thus the phase $\phi(\omega)$ to within a constant. The amplitude per unit frequency interval of the spectrum, $a(\omega)$, is determined directly by measuring the spectrum of the pulse to be measured.

The main advantage of the SPIDER technique is that it is fast, since the phase of the pulse may be recovered directly from the measured signal without resorting to an iterative procedure.

[53] See, for example, Takeda et al. (1982).

Further reading

Comprehensive reviews of the role of dispersion in ultrafast optics and techniques for shaping these pulses may be found in the articles by Walmsley et al. (2001) and Weiner (1995), respectively. Useful reviews of the generation of ultrafast laser pulses have been provided by Steinmeyer et al. (1999) and Keller (2003), and a very readable account of high-power ultrafast lasers has been provided by Backus et al. (1998). A recent review of OPCPA systems is provided by Dubietis et al. (2006), and an article in Scientific American by Mourou and Umstadter (2002) covers the generation and applications of high-intensity laser pulses.

The textbook by Siegman (1986) provides a comprehensive description of the propagation of optical pulses in linear, non-linear, and amplifying media.

A comprehensive review of techniques for measuring ultrafast laser pulses has been provided by Walmsley and Dorrer (2009); the book by Trebino (2002) provides a complete discussion of the FROG technique.

Exercises

(17.1) Orders of magnitude

(a) Calculate the FWHM duration τ_p of the temporal profile of the pulse intensity; the FWHM width $\Delta\omega$ of the power spectrum; and the time–bandwidth product $\Delta\omega\tau_p$ for pulses with the following profiles of the pulse *amplitude*:

(i) $f(t) = e^{-at^2}$;

(ii) $f(t) = \begin{cases} e^{-at} & t > 0 \\ 0 & t < 0 \end{cases}$;

(iii) $f(t) = \begin{cases} 1 & 0 < t < \tau_p \\ 0 & t < 0, t > \tau_p \end{cases}$;

(iv) $f(t) = \text{sech}(at)$,

where a is a real and positive constant. Use standard integrals where necessary.

(b) Taking as an example the Gaussian pulse of part (a), calculate the minimum FWHM bandwidth, $\Delta\lambda$, and minimum relative bandwidth, $\Delta\lambda/\lambda$, required to generate pulses of duration 1 ps, 100 fs, and 10 fs for light of wavelength 500 nm.

(c) Using the data of Table 17.2, estimate the thicknesses of fused silica, sapphire, and air that would double the duration of a pulse of mean wavelength 800 nm and of duration:

(i) 100 fs;

(ii) 10 fs.

(17.2) **Modes in a dispersive cavity**

Here, we investigate in more detail the connections between dispersion in a laser cavity, the frequency spectrum of modelocked pulses generated in the cavity, and the phase slip ϕ_{slip}. We suppose, for simplicity, that the optical cavity is filled uniformly with a dispersive medium. The frequencies of the cavity modes are given by the condition:

$$k(\omega_p)2L = 2\pi p, \quad (17.54)$$

where $k(\omega)$ is the wave vector in the medium at an angular frequency ω, and L is the cavity length. However, we also expect the frequencies of the modelocked pulses to be given by eqn (17.46). In this problem we investigate whether these two conditions can be satisfied simultaneously.

(a) By expanding $k(\omega)$ about a carrier frequency ω_0 (not necessarily equal to the frequency of one of the cavity modes), show that the condition for a mode (17.54) may be written as

$$k(\omega_0) + \frac{\omega_p - \omega_0}{v_g}$$

$$+ \sum_{n\geq 2} \frac{1}{n!} \left.\frac{\partial^n k}{\partial \omega^n}\right|_{\omega_0} (\omega_p - \omega_0)^n = \frac{\pi}{L} p. \quad (17.55)$$

Solution of this equation leads to a series of mode frequencies ω_p, which in general will not be evenly spaced.

(b) Now let us suppose that the modes *are* evenly spaced in frequency, i.e. that they are given by eqn (17.46). Show that for this to hold we must have

$$k(\omega_0) + \left(\frac{\Delta\omega}{v_g} - \frac{\pi}{L}\right) p - \frac{\omega_0 - \omega_{\text{ce}}}{v_g}$$

$$+ \sum_{n\geq 2} \frac{1}{n!} \left.\frac{\partial^n k}{\partial \omega^n}\right|_{\omega_0} (\omega_p - \omega_0)^n = 0. \quad (17.56)$$

(c) Explain why this condition cannot be satisfied for all the oscillating modes p unless:

$$T_c = \frac{2L}{v_g} \quad (17.57)$$

$$\left.\frac{\partial^n k}{\partial \omega^n}\right|_{\omega_0} = 0 \quad \text{for } n \geq 2. \quad (17.58)$$

Interpret these conditions.

(d) Show that when eqn (17.58) is satisfied $\phi_{\text{slip}} = \omega_{\text{ce}} T_c$ is given by[54]

$$\phi_{\text{slip}} = \omega_0 \left(\frac{2L}{v_g} - \frac{2L}{v_p}\right),$$

where $v_p = \omega_0/k(\omega_0)$. Interpret this result.

(17.3) **Carrier-envelope offset (CEO)**

As a simple example we consider the effect of CEO on a pulse with an amplitude given by

$$E(t) = \begin{cases} \cos\left(\frac{\pi}{T}t\right)\cos(\omega_0 t + \phi) & |t| \leq \frac{T}{2} \\ 0 & |t| > \frac{T}{2} \end{cases},$$

where T is a constant.

Sketch the temporal variation of the amplitude of the field for the following cases, and comment on your results:

(a) $\omega_0 \gg \frac{2\pi}{T}$ and $\phi = 0$;

(b) $\omega_0 = \frac{4\pi}{T}$ and $\phi = 0$;

(c) $\omega_0 = \frac{4\pi}{T}$ and $\phi = \frac{\pi}{2}$.

(17.4) **Simple examples of B-integral calculations**

In a Ti:sapphire CPA system the laser pulse leaves a multipass amplifier with a peak energy of 200 mJ, a stretched pulse duration of 300 ps, and a diameter of 8 mm. The beam is then expanded to a diameter of 50 mm and passed through a compressor, which reduces the duration of the

---

[54]See Exercise 8.3.

pulse to 50 fs. The energy transmission of the compressor is 50%.

Calculate the peak intensity and B-integrals associated with the following stages of the system:

(a) The final pass through the 15-mm long laser rod.

(b) A BK7 window placed immediately after the compressor.

(c) A 1-m long air path located immediately after the compressor.

In each case comment on the values you calculate.

(17.5) **Basic properties of Gaussian optical pulses**

Here, we derive the basic properties of the Gaussian optical pulse described by eqns (17.19) and (17.20).

(a) Show that the full width at half-maximum duration of the *intensity* profile of the pulse is given by

$$\tau_p = \sqrt{\frac{2 \ln 2}{\Re(\Gamma)}} = \sqrt{\frac{2 \ln 2}{a}}.$$

(b) Given the standard integral,

$$\int_{-\infty}^{\infty} e^{-\beta t^2} e^{i\omega t} dt = \sqrt{\frac{\pi}{\beta}} e^{-\omega^2/4\beta}$$

if $\Re(\beta) > 0$, (17.59)

calculate the Fourier transform of eqn (17.5) to show that the amplitude per unit frequency interval is given by

$$a(\omega') = \sqrt{\frac{1}{2\Gamma}} e^{-\omega'^2/4\Gamma}, \quad (17.60)$$

where $\omega' = \omega - \omega_0$.

(c) Show that the full width at half-maximum width of the *power* spectrum is given by

$$\Delta \omega_p = 2\sqrt{2 \ln 2}\sqrt{a[1 + (b/a)^2]},$$

and hence show that the time–bandwidth product of the Gaussian pulse is equal to

$$\Delta \omega_p \tau_p = 4 \ln 2 \sqrt{1 + (b/a)^2}.$$

(17.6) †**Propagation of Gaussian pulses in dispersive media**

We now consider the propagation of the Gaussian optical pulse through a dispersive medium in which the accumu-

lated phase is given by

$$\phi(\omega) = \phi^{(0)} + \phi^{(1)}(\omega - \omega_0) + \frac{1}{2}\phi^{(2)}(\omega - \omega_0)^2.$$

(a) By considering the propagation of each frequency component, show that the amplitude of the electric field at the end of the medium is given by,

$$E_{\text{out}}(t) = \sqrt{\frac{1}{4\pi \Gamma_{\text{in}}}} e^{i[\phi^{(0)} - \omega_0 t]}$$

$$\times \int_{-\infty}^{\infty} \exp\left[-\left(\frac{1}{4\Gamma_{\text{in}}} - i\frac{\phi^{(2)}}{2}\right)\omega'^2\right]$$

$$\times \exp\left\{-i\omega'\left[t - \phi^{(1)}\right]\right\} d\omega'.$$

(b) Using eqn (17.59), or by comparing the above integral to eqn (17.60), show that the amplitude of the transmitted pulse is given by eqns (17.29) and (17.30).

(c) We now consider the form of the Gaussian pulse as a function of the distance z it propagates through a quadratically dispersive medium. Use eqn (17.30) to show that the real and imaginary parts of the Gaussian parameter $\Gamma(z) = a(z) + ib(x)$ are given by

$$a(z) = \frac{a_0}{[1 + 2b_0\phi^{(2)}]^2 + [2a_0\phi^{(2)}]^2} \quad (17.61)$$

$$b(z) = \frac{b_0 + 2\phi^{(2)}(a_0^2 + b_0^2)}{[1 + 2b_0\phi^{(2)}]^2 + [2a_0\phi^{(2)}]^2}, \quad (17.62)$$

where a_0 and b_0 are the real and imaginary parts of Γ at $z = 0$.

(d) Show that the duration of the transmitted pulse is given by

$$\tau_p(z) = \tau_p(0)\sqrt{\left[1 + 2b_0\phi^{(2)}\right]^2 + \left[2a_0\phi^{(2)}\right]^2}.$$

(17.63)

(e) Equation (17.63) shows that the duration of the pulse always increases with z if $b_0\phi^{(2)} \geq 0$. Explain qualitatively why this occurs.

(f) Sketch the behaviour of $\tau_p(z)$ for the case $b_0\phi^{(2)} > 0$ and $b_0\phi^{(2)} < 0$.

(g) Discuss how the bandwidth of the pulse varies as it propagates through the medium.

(17.7) **Stretching of Gaussian pulses in dispersive media**

Here, we use the results of Exercise 17.6 to determine how a pulse stretches when $b_0\phi^{(2)} \geq 0$.

(a) We consider first the case of an initially un-chirped input pulse. Show that in this case the duration of the pulse leaving the optical system is given by eqn (17.32).

(b) Show that in the limit of large $|\phi^{(2)}|$, the duration of the transmitted pulse is given by eqn (17.17):

$$\tau_p \approx \phi^{(2)} \Delta\omega_p.$$

(c) We now consider the opposite limit: that of a large initial chirp. Use eqn (17.26) to show that in this case the bandwidth of the pulse is given by

$$\Delta\omega_p \approx 2\frac{\sqrt{2\ln 2}}{a_0}b_0.$$

(d) Using eqn (17.61), show that for large values of the GDD

$$a(z) \to \frac{a_0}{4\phi^{(2)}b_0^2}.$$

(e) Hence, show that the duration of the pulse is also given by eqn (17.17).

(17.8) Compression of Gaussian pulses in dispersive media
Here, we consider in more detail how a Gaussian pulse may be compressed in a quadratically dispersive medium when $b_0\phi^{(2)} < 0$.

(a) Using eqn (17.61), show that the duration of the transmitted pulse is minimized if

$$2b_0\phi^{(2)} = -\frac{b_0^2}{a_0^2 + b_0^2}. \quad (17.64)$$

(b) Hence, use eqn (17.63) to show that the minimum possible duration of the compressed pulse is given by

$$\tau_p^{\min} = \tau_p(0)\sqrt{\frac{1}{1 + (b_0/a_0)^2}}.$$

(c) Show that the time–bandwidth product of the compressed pulse has the minimum possible value for a Gaussian pulse, i.e. $\Delta\omega_p \tau_p^{\min} = 4\ln 2$.

(17.9) Grating stretcher

(a) Use eqn (17.45) to calculate the GDD in units of fs^2 at a wavelength of 800 nm arising from a single pass of a grating stretcher comprised of a pair of 1200 lines/mm gratings separated by 50 cm. Assume that $\theta \approx \gamma$, where the angle of incidence γ is close to satisfying the Littrow condition ($m\lambda = 2\sin\gamma$), and that the diffraction is first order ($m = 1$).

(b) Hence, calculate the duration to which a transform-limited 50-fs pulse would be stretched following a double pass through the stretcher.

(17.10) Geometric dispersion
Here, we follow the approach of Martinez (1984) and show that the group delay dispersion arising from geometric dispersion is always negative.

(a) Using the geometry of Fig. 17.6, show that the group delay dispersion is given by,

$$\frac{\partial^2 \phi}{\partial\omega^2} = -\frac{D}{c\cos\gamma}\left\{\omega\cos\beta\left(\frac{\partial\beta}{\partial\omega}\right)^2 \right.$$
$$\left. + \sin\beta\left[\omega\frac{\partial^2\beta}{\partial\omega^2} + 2\frac{\partial\beta}{\partial\omega}\right]\right\},$$

where $\beta = \theta - \gamma$.

(b) Give a simple argument (i.e. without further calculus) to explain why the GDD is always negative, provided that the angle of diffraction or refraction is small compared to that of the reference ray.

(17.11) Here, we determine the effect of optical gain and dispersion on a Gaussian pulse, and use this to estimate when dispersion control is necessary in a modelocked laser. We suppose that the gain medium is homogeneously broadened with a Lorentzian lineshape, so that the (saturated) gain coefficient can be written in the form,

$$\alpha(\omega - \omega_0) = \alpha_p\left[1 + \left(\frac{\omega - \omega_0}{\Delta\omega_H/2}\right)^2\right]^{-1},$$

where α_p is the peak gain coefficient and $\Delta\omega_H$ is the homogeneous linewidth.

We also suppose that immediately before the gain medium the pulse is unchirped, so that the amplitude per unit frequency interval is given by

$$a(\omega, 0) = \sqrt{\frac{1}{2\Gamma_{in}}}e^{-(\omega-\omega_0)^2/4\Gamma_{in}},$$

and the initial Gaussian parameter Γ_{in} is real.

(a) Show that the *amplitude* per unit frequency interval after passing through a length ℓ_g of the gain medium is given by,

$$a(\omega, \ell_g) = \sqrt{\frac{1}{2\Gamma_{\text{in}}}} e^{-(\omega-\omega_0)^2/4\Gamma_{\text{in}}}$$

$$\times \exp\left\{\frac{\alpha_p \ell_g}{2}\left[1 + \left(\frac{\omega - \omega_0}{\Delta\omega_H/2}\right)^2\right]^{-1}\right\}.$$

(b) Hence, show that provided $|\omega - \omega_0| \ll \Delta\omega_H$, i.e. that the bandwidth of the pulse is small compared to the homogeneous linewidth, the amplified pulse has a Gaussian frequency spectrum with a Gaussian parameter given by,

$$\frac{1}{\Gamma'} = \frac{1}{\Gamma_{\text{in}}} + \frac{8\alpha_p \ell_g}{\Delta\omega_H^2}.$$

(c) Use the results of Section 17.1.4 to show that if in propagating once through the cavity (i.e. half of one round trip) the pulse experiences a total GDD of $\phi^{(2)}$, the Gaussian parameter of the pulse after a single pass through the cavity will be

$$\frac{1}{\Gamma'} = \frac{1}{\Gamma_{\text{in}}} + \frac{8\alpha_p \ell_g}{\Delta\omega_H^2} - 2i\phi^{(2)}. \quad (17.65)$$

(d) We first consider the effect of the amplification on the bandwidth of the pulse. Remembering that the bandwidth depends on $\Re(1/\Gamma')$, show that the bandwidth is given by

$$\Delta\omega_p = \Delta\omega_{p0}\left[1 + \frac{\alpha_p \ell_g}{\ln 2}\left(\frac{\Delta\omega_{p0}}{\Delta\omega_H}\right)^2\right]^{-1/2},$$

where $\Delta\omega_{p0}$ is the bandwidth of the pulse before the half-round trip. Is the bandwidth of the pulse increased or decreased?

(e) Show that

$$\frac{1}{\Re(\Gamma')} = \frac{1}{\Gamma_{\text{in}}}\left[1 + \alpha'\Gamma_{\text{in}} + \frac{(\phi'\Gamma_{\text{in}})^2}{1 + \alpha'\Gamma_{\text{in}}}\right],$$

where $\alpha' = 8\alpha_p \ell_g/\Delta\omega_H^2$ and $\phi' = 2\phi^{(2)}$. Comment on whether the duration of the pulse will be increased or decreased from its initial value.

(f) Show that provided the change in duration of the pulse is small, the duration of the pulse after one pass through the cavity is given by,

$$\tau_p \approx \tau_{p0}\left\{1 + 8\ln 2\left[\frac{\alpha_p \ell_g}{(\Delta\omega_H \tau_{p0})^2}\right.\right.$$

$$\left.\left. + \ln 2\left(\frac{\phi^{(2)}}{\tau_{p0}^2}\right)^2\right]\right\}.$$

(g) By how much must the action of the modelocking mechanism reduce the duration of the pulse in a single pass through the cavity?

(h) Show that for the effects of dispersion to be negligible compared to those of the optical gain,

$$\tau_{p0} \gg \sqrt{\frac{\ln 2}{\alpha_p \ell_g}}\left|\phi^{(2)}\right|\Delta\omega_H.$$

(i) As an example, estimate the shortest pulse duration that can be obtained in a modelocked Ti:sapphire laser without dispersion control. Take the homogeneous linewidth to be 100 THz, assume $\alpha_p \ell_g = 0.1$, and suppose that $\phi^{(2)} = 120\,\text{fs}^2$ (i.e. equal to the GDD arising from propagation through 2 mm of sapphire).

(17.12) Required duration of the stretched pulse in a CPA system

(a) For efficient operation the beam input to a Ti:sapphire amplifier must have a fluence greater than or equal to the saturation fluence $\Gamma_s = \hbar\omega_L/\beta\sigma_{21}$. Find the saturation fluence for Ti:sapphire using the data of Table 7.3, assuming $\beta = 1$.

(b) Calculate the peak intensities of pulses with a fluence equal to Γ_s and durations equal to 1 ps, 100 fs, and 10 fs. Comment on your results.

(c) Given that for sapphire $n_2 = 3 \times 10^{-16}\,\text{cm}^2\,\text{W}^{-1}$, calculate the B-integral corresponding to propagation of a 30-fs pulse with a fluence equal to Γ_s through 10 mm of Ti:sapphire.

(d) Given that the damage threshold of typical mirror coatings is approximately given by $5\sqrt{\tau_p}\,\text{J}\,\text{cm}^{-2}$, where τ_p is the duration of the laser pulse in nanoseconds, find the minimum duration to which the pulse should be stretched if mirror damage is to be avoided.

(e) Comment on any disadvantages there might be in stretching the pulse to very long durations.

(17.13) Autocorrelation of an ultrafast laser pulse

Here, we consider in detail the measurement of an ultrafast laser pulse by the autocorrelation method of Section 17.4.1. We assume that the electric field of the pulse may be written in the form $E(t) = f(t)e^{-i\omega_0 t} = g(t)e^{i\phi(t)}e^{-i\omega_0 t}$, where ω_0 is the angular carrier frequency, $g(t) = |f(t)|$, and $\phi(t) = \arg[f(t)]$ is the phase of the field amplitude.

(a) Show that

$$|E(t) + E(t+\tau)|^2 \propto I(t) + I(t+\tau)$$
$$+ 2\sqrt{I(t)I(t+\tau)}\cos\left[\omega_0\tau - \delta\phi(t,\tau)\right],$$

where the intensity of the pulse $I(t) \propto |E(t)|^2$ and $\delta\phi(t,\tau) = \phi(t+\tau) - \phi(t)$.

(b) Hence, show that the autocorrelation signal of eqn (17.49) may be written as

$$S_c(\tau) \propto \int_{-\infty}^{\infty} I(t)^2 dt + 2\int_{-\infty}^{\infty} I(t)I(t+\tau) dt$$
$$+ \int_{-\infty}^{\infty} I(t)I(t+\tau)\cos\{2[\omega_0\tau - \delta\phi(t,\tau)]\} dt$$
$$+ 2\int_{-\infty}^{\infty} \cos[\omega_0\tau - \delta\phi(t,\tau)]$$
$$\times \left[I(t)\sqrt{I(t)I(t+\tau)}\right.$$
$$\left. + I(t+\tau)\sqrt{I(t)I(t+\tau)}\right] dt. \quad (17.66)$$

This is the form of the *interferometric autocorrelation* of the pulse.

(c) Show that for a pulse of finite duration the ratio of the autocorrelation signal at zero delay to that at large delay is equal to $S_c(0)/S_c(\infty) = 8$, and hence sketch the form of $S_c(\tau)$.

(d) Show that if the detection system is not able to resolve the interference terms in eqn (17.66), the *intensity autocorrelation* is given by eqn (17.50).

(e) Show that for a pulse of finite duration the ratio of the intensity autocorrelation signal at zero delay to that at large delay is equal to $S_i(0)/S_i(\infty) = 3$, and hence sketch the form of $S_i(\tau)$.

(f) As an example of the calculation of the deconvolution factor, let us calculate this factor for a pulse with a Gaussian temporal profile with $f(t) = e^{-at^2}$. Show that,

$$S_i(\tau) \propto \left(1 + 2e^{-a\tau^2}\right)\int_{-\infty}^{\infty} e^{-4at^2} dt,$$

i.e. the autocorrelation comprises a constant background signal plus a component that depends on the delay τ.

(g) Hence, show that the FWHM of the autocorrelation signal *above the background level* is given by,

$$\tau_{ac} = \sqrt{2}\tau_p,$$

where τ_p is given by eqn (17.21).

(h) †It is interesting to investigate whether the envelope of the *interferometric* autocorrelation has the same form as the intensity autocorrelation. It is easiest to do this by plotting numerically an example, which we will take to be a Gaussian pulse of FWHM duration τ_p and carrier frequency such that $\omega_0\tau_p = 50$. Plot the interferometric and intensity autocorrelations on a shifted and normalized scale such that $S_{i,c}(0) = 1$ and $S_{i,c}(\infty) = 0$. Comment on whether the shifted and normalized intensity autocorrelation coincides with the envelope of the shifted and normalized interferometric autocorrelation. Explain your observations in the light of eqn (17.66).

(17.14) Analysis of the SPIDER technique
Here, we derive an expression for the intensity of the signal in the SPIDER technique for measuring ultrafast laser pulses that is discussed in Section 17.4.2.

(a) Let the spectrum of the pulse to be measured be described by $a(\omega)e^{i\phi(\omega)}$, where $a(\omega)$ is real. Suppose that the frequency of that part of the chirped pulse that is overlapped with pulse 1 as it passes through the non-linear crystal is ω_1. Explain why the amplitude of the light up-shifted from pulse 1 to frequency $\omega + \omega_1$ is given by,

$$\tilde{E}_1(\omega + \omega_1) \propto a(\omega)e^{i\phi(\omega)},$$

which may be written

$$\tilde{E}_1(\omega) \propto a(\omega - \omega_1)e^{i\phi(\omega - \omega_1)}.$$

(b) Similarly, show that if the frequency of that part of the chirped pulse overlapped with pulse 2 as it passes through the non-linear crystal is ω_2, the amplitude of the up-shifted frequency may be written as

$$\tilde{E}_2(\omega) \propto a(\omega - \omega_2)e^{i\phi(\omega - \omega_2)}e^{i\omega\tau},$$

where τ is the separation in time of the two copies of the pulse to be measured.

(c) Hence show that the intensity of the signal recorded by the spectrometer in a SPIDER apparatus is given by,

$$I_{\text{sp}}(\omega) \propto a(\omega - \omega_1)^2 + a(\omega - \omega_2)^2$$
$$+ 2a(\omega - \omega_1)a(\omega - \omega_2)$$
$$\times \cos\left[\phi(\omega - \omega_1) - \phi(\omega - \omega_2) - \omega\tau\right].$$

(d) The SPIDER spectrum is up-shifted from the frequency ω_0 of the pulse to be measured by approximately $\omega_1 \approx \omega_2 \approx \omega_0$, i.e. it is at approximately twice the frequency of the original pulse. It is convenient, therefore, to let $\omega \to \omega + \omega_1$ and to define $\Omega = \omega_2 - \omega_1$. Show that with these definitions the signal may be written as,

$$I_{\text{sp}}(\omega + \omega_1) \propto a(\omega)^2 + a(\omega - \Omega)^2$$
$$+ 2a(\omega)a(\omega - \Omega)\cos\left[\phi_{\text{sp}}(\omega)\right],$$

where

$$\phi_{\text{sp}}(\omega) = \phi(\omega) - \phi(\omega - \Omega) - (\omega + \omega_1)\tau.$$

(e) Sketch the SPIDER spectrum $I_{\text{sp}}(\omega + \omega_1)$. Do this qualitatively, or calculate the spectrum numerically for reasonable values of the parameters. What is the spacing of the interference fringes produced in the case when $\phi(\omega)$ is constant?

(f) As discussed in Section 17.4.2, in order to deduce the phase $\phi(\omega)$ of the pulse the parameter τ must be found. This is achieved by blocking the chirped pulse, removing the non-linear crystal, and recording the spectrum of pulses 1 and 2 alone. Show that the spectrum produced in this case can be described by,

$$I_{\text{calib}} \propto a(\omega)^2 \left[1 + \cos(\omega\tau)\right].$$

(g) Hence, summarize the steps required to deduce from the SPIDER spectrum the amplitude per unit frequency interval $a(\omega)$ and phase $\phi(\omega)$ of the pulse to be measured.

(h) In general, the phase and amplitude of those parts of the *chirped* pulse that are mixed with pulse 1 and 2 will be different. How would such differences manifest themselves in the recorded spectrum. Would the analysis of the spectrum need to be modified?

18 Short-wavelength lasers

18.1 Definition of wavelength ranges 503
18.2 Difficulties in achieving optical gain at short wavelengths 503
18.3 General properties of short-wavelength lasers 505
18.4 Laser-generated plasmas[†] 510
18.5 Collisionally excited lasers 517
18.6 Recombination lasers 530
18.7 Other sources 535
Further reading 541
Exercises 541

[1] In principle, short-wavelength lasers could also be used to decrease the feature size—and hence increase the density—of components on computer chips. However, to be economically viable it is necessary to achieve a throughput of the order of 100 wafers per hour, and it may be shown that this requires an *average* source power of order 1 kW. This is many orders of magnitude greater than can presently be reached by short-wavelength lasers or high-harmonic generation, and consequently sources for XUV and soft X-ray lithography are likely to be incoherent sources based on laser- or discharge-produced plasmas. See Banine (2004) and other papers in the same issue.

[2] The K-edges of carbon and oxygen are at 284 eV and 543 eV, corresponding to radiation of wavelengths 4.4 nm and 2.3 nm, respectively.

Lasers operating at short wavelengths have many potential applications; these arise from the short wavelength itself, or from the associated high photon energy. For example, many emerging technologies require the production or control of structures with nanometre-scale dimensions. Sources of short-wavelength radiation could be used to create such structures through nanoscale patterning,[1] and to probe them through microscopy, spectroscopy, and interferometry with spatial resolution of order 10 nm.

Short wavelengths are also useful for probing biological materials. In particular, the so-called 'water-window' between 2.3 nm and 4.4 nm is of great interest since this wavelength region lies between the K-edges of carbon and oxygen.[2] Hence, radiation in the water-window is strongly absorbed by proteins, but only weakly by water, enabling biological systems to be studied in their natural, watery environments.

Some of these applications are already realized using the incoherent radiation generated by synchrotrons. However, these are very large and expensive machines; more compact—so-called 'table-top'—sources of short-wavelength radiation are desirable since they would bring these techniques into ordinary laboratories, enabling them to be used more widely.

In order for the transitions between valence electrons to lie in the extreme ultraviolet or soft X-ray regions it is necessary for the atom to be ionized, and consequently all short-wavelength lasers operate on transitions in ions. Two generic methods for achieving population inversions have been investigated to date: lasers driven by **electron collisions** and those driven by **electron–ion recombination**. These are discussed in detail in Sections 18.5 and 18.6.

In this chapter we discuss the physics underlying the operation of short-wavelength lasers. As explained below, it is difficult to achieve a population inversion on short-wavelength transitions. As a consequence, there are far fewer lasers operating in this spectral region than in the visible, infra-red, or ultraviolet, and much remains to be done before this spectral region is as well served with sources of coherent radiation as are the visible and infra-red regions. For this reason we also describe briefly some non-laser sources of short-wavelength radiation.

18.1 Definition of wavelength ranges

The naming conventions for the wavelength regions below the ultraviolet are not rigidly established. However, the following definitions are in accord with usage in laser physics:

- **Vacuum ultraviolet (VUV): 200–100 nm.** Below approximately 200 nm radiation is absorbed very strongly by oxygen, and consequently experiments in this spectral region must be performed in vacuum. The lower limit of the VUV is determined by the shortest cut-off wavelength of materials suitable as optical windows: 106 nm, corresponding to the short-wavelength cut-off of lithium fluoride.
- **Extreme ultraviolet (XUV or EUV): 100–30 nm.** Work in this region (and at shorter wavelengths) is hampered by the lack of suitable materials with which to make windows. Instead, radiation can only be coupled between experimental regions at different ambient pressures through small holes, or arrays of holes. The difference in pressure is maintained by continuous gas flow and/or operation of vacuum pumps in an arrangement known as 'differential pumping.'
- **Soft X-ray: 30–0.2 nm.** The onset of the X-ray region is often taken to be at 30 nm, and is divided between soft X-rays and the onset of hard X-radiation at 0.2 nm.

18.2 Difficulties in achieving optical gain at short wavelengths

18.2.1 Pump-power scaling

The main reason that there are relatively few lasers operating at short wavelengths is that very much higher pump powers are required to generate a population inversion on a short-wavelength laser transition. To show this, we investigate how the pump power density P needed to reach a certain gain per unit length varies as the wavelength of the transition is reduced towards the X-ray region. We therefore imagine scaling the same transition to shorter wavelengths by increasing the charge on the nucleus whilst keeping the number of electrons in the ion the same. As an example, we could consider scaling to shorter wavelengths the well-known $4p \rightarrow 4s$ transition at 488 nm of the argon-ion (Ar$^+$) laser[3] by operating the laser in K^{2+}, Ca^{3+}, Sc^{4+} .... Along an **an isoelectronic sequence** of this type we would expect the oscillator strength f_{21} of the transition to remain approximately the same. From eqn (3.15) we may write the Einstein A coefficient in terms of the oscillator strength[4] as,

$$A_{21} = \frac{e^2 \omega_0^2}{2\pi \epsilon_0 m_e c^3}(-f_{21}), \quad (18.1)$$

and hence $A_{21} \propto \omega_0^2$, where ω_0 is the centre frequency of the transition.

[3]See Section 11.3.

[4]Remember, that for an emission transition the oscillator strength is negative.

The optical gain cross-section of the laser transition can be written as:

$$\sigma_{21}(\omega - \omega_0) = \frac{\pi^2 c^2}{\omega_0^2} A_{21} g(\omega - \omega_0) \tag{18.2}$$

$$= \frac{\pi e^2}{2\epsilon_0 m_e c}(-f_{21}) g(\omega - \omega_0), \tag{18.3}$$

where $g(\omega - \omega_0)$ is the lineshape of the transition.

In scaling to shorter wavelengths we will fix the peak small-signal gain coefficient $\alpha(0) = N^* \sigma_{21}(0)$ to some constant value. We will also assume that the population of the lower laser level is small compared to that of the upper laser level so that $N^* \approx N_2$, and hence $\alpha(0) \approx N_2 \sigma_{21}(0)$.

Now, atoms in the upper laser level must decay at a rate at *least* as fast as $N_2 A_{21}$. Further, since the energy required to excite an ion to the upper laser level is at least $\hbar \omega_0$, the pump power density required to generate an upper-level population density N_2 satisfies the inequality:

$$P > N_2 A_{21} \hbar \omega_0 \tag{18.4}$$

$$> N_2 \sigma_{21}(0) \frac{A_{21}}{\sigma_{21}(0)} \hbar \omega_0 \tag{18.5}$$

$$> \alpha(0) \frac{\hbar \omega_0^3}{\pi^2 c^2 g(0)}. \tag{18.6}$$

At the centre frequency of the transition, irrespective of the lineshape, the value of the lineshape is inversely proportional to the linewidth of the transition $\Delta \omega$. Hence, we deduce the following scaling for the pump-power density:

$$P \propto \omega_0^3 \Delta \omega. \tag{18.7}$$

To proceed further with this analysis we need to know the nature of the broadening mechanisms affecting the laser transition. For example, if the width of the laser transition is determined by natural broadening of the upper laser level, and the upper laser level decays pre-dominantly on the laser transition, we have $\Delta \omega = 1/\tau_2^{\text{rad}} \approx A_{21}$. Since $A_{21} \propto \omega_0^2$ we find $P \propto \omega_0^5$. Alternatively, for Doppler broadening $\Delta \omega \propto \omega_0$, and hence $P \propto \omega_0^4$. The point is that irrespective of the detailed nature of the broadening mechanisms affecting the laser transition, the required pump-power density increases very rapidly with the frequency of the laser transition.

We can illustrate this strong scaling by comparing the pump-power density used in the Ar^+ laser operating at 488 nm and the Ar^{8+} laser operating at 46.9 nm. As discussed in Section 11.3, in a small-frame Ar^+ laser the discharge current is typically 20–50 A at a voltage of about 400 V. For a 1-m long discharge tube of 3 mm diameter, this corresponds to a pump power density of approximately 2 kW cm^{-3}. This may be compared to the peak current of 40 kA at approximately 400 kV used to excite the Ar^{8+} laser discussed in Section 18.5.3. For a typical capillary of length 12 mm and diameter 4 mm, this corresponds to a (pulsed) pump power density of 100 GW cm^{-3}! The increase in pump-power density by nearly 6 orders of

18.3 General properties of short-wavelength lasers

Lasers operating in these spectral region are significantly different from those operating at longer wavelengths owing to the fact that it is not usually possible to construct a laser cavity. There are two reasons for this:

- The duration of the optical gain is short, typically less than a nanosecond, owing to the difficulties in maintaining the high pump power that is required. Even if the pump power could be maintained for longer than this, the optical gain is often terminated by the onset of bottlenecking in the lower level, which occurs on a time scale of order the lifetime of the upper laser level. The radiative lifetimes of the upper levels of short-wavelength transitions vary, from eqn (18.1), approximately as ω_0^{-2} and are typically 1–100 ps. In 100-ps light travels only 30 mm and consequently there is not usually sufficient time for the generated radiation to make many cavity round trips in the interval for which optical gain persists.
- It is difficult to manufacture mirrors in this spectral region—especially output couplers that, of course, must have a finite transmission.

Lasers operating in these spectral regions therefore do not in the main employ an optical cavity, and instead their output comprises **amplified spontaneous emission (ASE)** from a relatively long and narrow column of the gain medium. As illustrated in Fig. 18.1, the output of an ASE laser diverges with a solid angle approximately equal to that subtended at one end of the gain medium by the cross-sectional area of the other end.

18.3.1 Travelling-wave pumping

The short duration, τ_g, of the optical gain means that if the population inversion is created at all points simultaneously, radiation from any given point will be amplified over a distance of $c\tau_g$ only; the population inversion in regions of the gain medium lying beyond this will have decayed to zero by the time that

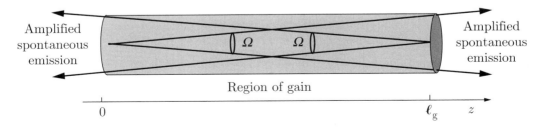

Fig. 18.1: A short-wavelength laser operating by amplified spontaneous emission.

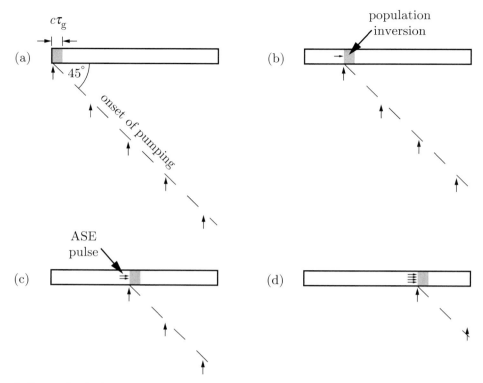

Fig. 18.2: Schematic diagram showing how travelling-wave pumping enables amplification over a length longer than $c\tau_g$, where τ_g is the duration of the optical gain, by pumping each region just as the ASE pulse arrives. Here, we have assumed that both the generated radiation and the pumping propagate at c so that the line delineating the onset of pumping makes an angle of 45° with the axis of the gain region. In this diagram time increases from (a) to (d).

the radiation reaches it. The output power of an ASE laser of length $\ell_g \gg c\tau_g$ would then be approximately $\ell_g/c\tau_g$ times that of a laser $c\tau_g$ long.

In order to extend the length over which amplification is possible—and hence enable exponential amplification to continue until saturation is reached—it is necessary to stagger the pumping along the length of the gain region, so that the population inversion is established only when ASE from one end of the laser medium reaches it. In other words, if at $z = 0$ the pumping of the upper laser level can be described by $R_2(t)$, at a distance z along the axis of the gain medium the pumping should be described by the function $R_2(t - z/v_g)$, where v_g is the group velocity of the generated laser radiation. This so-called **travelling-wave** pumping is illustrated schematically in Fig. 18.2.

If travelling-wave pumping is employed, the ASE output occurs predominantly from one end of the laser; if it is not, the output is generally from both ends.

18.3.2 Threshold and saturation behaviour in an ASE laser

As the gain in a *laser oscillator* is increased the output power exhibits an unambiguous threshold when the round-trip small-signal gain just equals the round-trip loss; below this threshold the output power is essentially zero,[5] above it the output power increases approximately linearly with the small-signal gain.

It is not possible unambiguously to define a threshold condition for an ASE laser, and in general the onset of lasing is somewhat 'softer' than observed in a laser oscillator. Figure 18.3(a) shows the calculated variation of the output intensity of an ASE laser as a function of the single-pass small-signal gain $\alpha_0 \ell_g$. The variation of output intensity may be roughly divided into three regions:[6]

- $0 < \alpha_0 \ell_g \lesssim 3$: The output increases linearly with $\alpha_0 \ell_g$. Here, the output of the ASE laser is dominated by spontaneous emission, and the intensity of this increases linearly as the gain—and hence the population density of the upper laser level—is increased. For these low levels of $\alpha_0 \ell_g$ the system is not really a laser at all, but simply an elongated source of spontaneous emission. In the absence of self-trapping,[7] the intensity of the spontaneous emission detected will increase linearly with the length of the emitting region.
- $3 \lesssim \alpha_0 \ell_g \lesssim 15$: The output increases approximately exponentially with $\alpha_0 \ell_g$. In this region the stimulated emission dominates spontaneous emission. As $\alpha_0 \ell_g$ increases the output becomes ever more dominated by that spontaneous emission originating near the far end of the gain region, since this radiation is amplified over the longest length.
- $\alpha_0 \ell_g \gtrsim 15$: Once again the output increases linearly with $\alpha_0 \ell_g$. The output is still dominated by stimulated emission, but here the amplification is strongly saturated since for these very large values of $\alpha_0 \ell_g$ spontaneous emission originating at one end of the ASE laser is amplified to

[5] In fact, below threshold the output of the laser will comprise spontaneous emission and weak stimulated emission leaked from the cavity. However, the intensity of this radiation will be very low compared to that above threshold.

[6] This behaviour is analysed in more detail in Exercises 18.1 and 18.2.

[7] See Section 4.5.2.

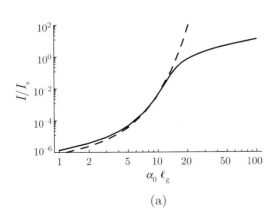
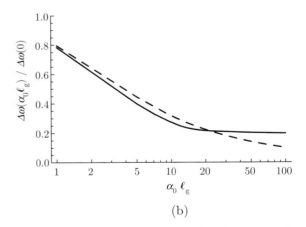

Fig. 18.3: Calculated output of an ASE laser operating on a homogeneously broadened transition, showing: (a) the ratio of the total output intensity to the saturation intensity I_s (note the log-log scale) and (b) the relative spectral width as a function of the single-pass small-signal gain $\alpha_0 \ell_g$ (note the lin-log scale). The solid lines show plots of the analytical formulae derived by Pert (1994) for the idealized case of a homogeneously broadened ASE laser with unidirectional output operating under steady-state conditions. The dashed lines are plots of the Linford formulae, derived in Exercise 18.2, which assume the gain is not saturated. Note that the Linford formulae are in good agreement with the more general expressions until the onset of saturation at $\alpha_0 \ell_g \approx 15$. Finally, we note that the detailed forms of these curves depend on the parameters of the laser transition and the geometry of the gain region. For further details see Pert (1994), references therein, and Exercise 18.2.

an intensity of order of the saturation intensity after propagating only a small fraction of ℓ_g. Under these conditions energy is transferred to the ASE beam at the maximum possible rate (determined by the pump rates and lifetimes of the laser levels).[8] For a homogeneously broadened laser transition the intensity of the ASE radiation then grows according to eqn (5.21), i.e. linearly with $\alpha_0 \ell_g$.

[8] See Hooker and Spence (2006). It is worth noting that whilst several short-wavelength lasers have been demonstrated to reach saturation, operation in this regime of very strong saturation has not yet been demonstrated.

We emphasize that these regions are not defined precisely. Demonstration of optical gain on a particular short-wavelength transition usually requires $\alpha_0 \ell_g$ to be at least 5 so that exponential growth of the output with the length of the gain medium is clear. The onset of saturation in an ASE laser is also not well defined, but is usually considered to occur for values of $\alpha_0 \ell_g$ greater than 15.[9]

[9] See Exercise 18.1.

18.3.3 Spectral width of the output

Figure 18.3(b) shows the calculated relative width of the output from a homogeneously broadened ASE laser as a function of the small-signal single-pass gain $\alpha_0 \ell_g$. As we would expect, the spectral width of the output decreases as the single-pass gain is increased, owing to the fact that radiation emitted at frequencies close to the line centre will experience a larger gain than radiation in the wings of the line profile. This behaviour is an example of the gain narrowing mentioned in Section 4.1.4.

Note that the rate of spectral narrowing with $\alpha_0 \ell_g$ is reduced considerably once the degree of saturation becomes significant, that is for $\alpha_0 \ell_g$ greater than approximately 15. The reason for this is simply that the rate of growth of each frequency is determined by the *saturated* gain coefficient, which becomes significantly smaller than the small-signal gain coefficient α_0 once saturation occurs. It is worth emphasizing that for homogeneously broadened transitions the frequency dependence of the optical gain is the same at all points in the ASE laser since it is determined by the lineshape of the transition. In general, one finds that gain narrowing can reduce the spectral width of the ASE down to about 0.2 of the homogeneous linewidth; as illustrated in Fig. 18.3(b) further reduction in the linewidth occurs only very slowly with increasing $\alpha_0 \ell_g$.

The behaviour of the spectral width of an ASE laser operating on an inhomogeneously broadened laser transition is quite different. As $\alpha_0 \ell_g$ is increased, where α_0 is calculated using the inhomogeneous lineshape, gain narrowing initially causes the spectral width to decrease just as for the homogeneously broadened case. However, since for an inhomogeneously broadened transition each frequency component interacts with a separate class of atom, frequencies near the line centre experience saturation before those in the wings of the inhomogeneous lineshape. As a result, the spectral width of the ASE output can re-broaden,[10] as illustrated in Fig. 18.4(b), and for large values of $\alpha_0 \ell_g$ is approximately equal to the initial linewidth $\Delta \omega_D$.

[10] In the case of mixed broadening, that is when the homogeneous and inhomogeneous linewidths are similar, re-broadening of the ASE spectrum after the onset of saturation is suppressed, reflecting the fact that frequencies near the line centre can access less saturated classes of atoms via the tail of their homogeneous lineshape. See Pert (1994).

The linewidths of several types of short-wavelength laser have been measured, usually through measurements of the longitudinal coherence. The relative linewidths $\Delta \lambda / \lambda$ are typically found to be of order 10^{-5}.

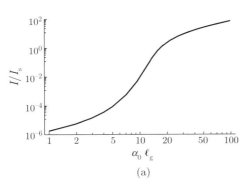 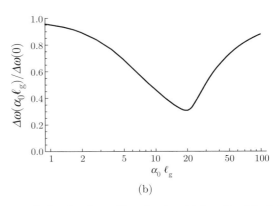

Fig. 18.4: Calculated output of an ASE laser operating on an inhomogeneously broadened transition, showing: (a) the ratio of the total output intensity to the saturation intensity I_s (note the logarithmic scale) and (b) the relative spectral width as a function of the single-pass small-signal gain $\alpha_0 \ell_g$. Note that the spectral width of the ASE output rebroadens for large values of the single-pass small-signal gain, once the intensity of the ASE becomes large enough for strong saturation of the transition. The gain coefficient α_0 and the saturation intensity I_s are calculated in terms of the peak *inhomogeneous* gain cross-section. This calculation follows the approach developed in Exercise 18.4 for the idealized case of a steady-state, unidirectional ASE laser operating on a transition exhibiting extreme inhomogeneous broadening. The detailed forms of these curves depend on the parameters of the laser transition and the geometry of the gain region.

18.3.4 Coherence properties of ASE lasers

The longitudinal and transverse coherence of an ASE laser cannot be as good as in a conventional laser since no laser cavity is employed. In a laser oscillator the optical cavity can be configured to restrict laser oscillation to a single longitudinal and transverse mode; resulting in essentially perfect transverse coherence, and a linewidth that can be much narrower than that of the laser transition.[11]

[11] See Chapter 16.

Longitudinal coherence

The longitudinal coherence length $L_c = c\tau_c$, where τ_c is the coherence time, of any source is related to its spectral width through the relation $\Delta\omega\tau_c \approx 2\pi$.[12] Thus, that a narrow linewidth is associated with a long coherence time that, for example, allows interferometry with a wider range of path differences.

The longitudinal coherence lengths of several short-wavelength lasers have been determined.[13] These show that the coherence length of soft X-ray lasers are typically in the range $L_c = 100\text{--}400$ µm, corresponding to coherence times of $0.3\text{--}1.3$ ps, and relative linewidths of $\Delta\lambda/\lambda = 2 - 5 \times 10^{-5}$. In some cases, the pulses output by short-wavelength lasers can be close to being bandwidth limited (see Section 17.1.1), that is the duration of the laser pulse is only a few times longer than the minimum set by the spectral width of the laser pulse.[14]

[12] The precise value of $\Delta\omega\tau_c$ depends on the spectral lineshape and our definition of $\Delta\omega$ and τ_c. For our present purpose the correct order of magnitude will suffice.

[13] For examples, see Celliers et al (1995), Smith et al. and Klisnick et al. (2006).

[14] Klisnick et al. (2006).

Transverse coherence

In the absence of gain and self-trapping, the spontaneous emission from a uniform cylinder of length ℓ_g and radius a could be described as a superposition of approximately F^2 transverse modes, where the Fresnel number is given by $F = (2\pi a^2)/(\lambda \ell_g)$. The transverse coherence length of such a source is approximately a/F. Thus, in principle, if the length of the cylinder is very long compared to its radius, the emission can appear to consist of only a few transverse modes and the transverse coherence can be high.[15]

Optical gain, gain saturation, and refraction of the ASE out of the gain medium by strong transverse density gradients all influence the transverse coherence of an ASE laser, and hence the theoretical treatments of the transverse coherence of ASE laser are complex.[16]

In practice, the transverse coherence of short-wavelength lasers is typically rather poor. Even in the best cases the transverse coherence length is only of the order of 10 to 20% of the beam diameter.[17]

18.4 Laser-generated plasmas[†]

It is difficult to achieve a population inversion between two levels if there are many other atomic levels lying between them, since in that case the upper level can decay by many possible radiative and non-radiative routes other than that to the lower level. Hence, in order to generate a population inversion on a short-wavelength transition it is necessary for the adjacent levels to have large energy separations, which in turn necessitates the use of highly ionized species.

In practice this means that the active medium of all short-wavelength lasers is a plasma, and in most cases these are generated by the interaction of intense laser radiation with solid or gaseous targets. In this section we shall give only a brief outline of those aspects of laser-generated plasmas required for a basic understanding of short-wavelength lasers operating in these media.[18]

18.4.1 Inverse bremsstrahlung heating

Consider the motion of an electron in a laser field in an electric field of the form $E(t) = E_0 \cos \omega_\mathrm{L} t$. The electric field of the laser radiation will cause the electron to oscillate at the frequency of the driving laser radiation, ω_L. This oscillatory motion is known as the **quiver motion** of the electron, and it is easy to show (see Exercise 18.6) that the time-averaged energy of the quiver motion—known as the **ponderomotive energy**—is given by,

$$U_\mathrm{p} = \frac{\mu_0 e^2 c}{2 m_\mathrm{e} \omega_\mathrm{L}^2} I_0 = \frac{r_\mathrm{e}}{2\pi c} I_0 \lambda_\mathrm{L}^2, \qquad (18.8)$$

where I_0 is the intensity of the driving laser radiation, and $r_\mathrm{e} = e^2/(4\pi m_\mathrm{e} c^2) \approx 2.818 \times 10^{-15}$ fm is known as the classical electron radius. The quiver energy can be significant. For example, for laser radiation of wavelength 1 μm the ponderomotive energy is approximately 1 eV for an intensity of 10^{13} W cm^{-2}, and nearly 1 keV for an intensity of 10^{16} W cm^{-2}.

The driven velocity of an electron quivering in a laser field is proportional to the electric field of the laser radiation, and consequently a *pulse* of radiation will cause the quiver energy of the electron to increase during the rising edge of the laser pulse and to decrease on its falling edge. It might appear, therefore,

---

[15]This can be understood with reference to the higher-order Hermite–Gaussian modes discussed in Section 6.4. We can, in principle, describe the radiation field at any point within the gain region of the ASE laser as a superposition of such modes. For reasonably high-order modes, the half-width of the nth-order mode is approximately $\sqrt{n} w$, where w is the spot size of the lowest-order mode. The highest-order mode with appreciable amplitude in the output beam will be that mode that roughly fills the width of the gain region *at both ends*. Suppose the waist is located at one end of the gain region; for mode n_max to fill this region we require $\sqrt{n} w(0) \approx a$. Now, in propagating to the other end of the gain region the modes will diffract, so that the lowest-order mode has a spot size or order $(\lambda/w(0))\ell_\mathrm{g}$; hence the condition for the mode n_max to fill the other end of the gain region is $\sqrt{n_\mathrm{max}}(\lambda/w(0))\ell_\mathrm{g} \approx a$. These two conditions may be solved to give $w(0) \approx \sqrt{\lambda \ell_\mathrm{g}}$ and $n_\mathrm{max} \approx a^2/\lambda \ell_\mathrm{g} \approx F$, where F is the Fresnel number F. Given that the radiation field will be a superposition of approximately n_max modes, we can expect that the beam will be reasonably coherent for transverse dimensions of the order of the size of the nodes of the highest-order mode with appreciable amplitude, i.e. a width of $2a/n_\mathrm{max}$.

[16]See London et al. (1990), Feit and Fleck (1991), and Amendt et al. (1996).

[17]For examples of measurements of the transverse coherence of short-wavelength lasers see Trebes et al. (1992), Burge et al. (1998), Marconi et al. (1997), and Larotonda et al. (2004).

[18]For more details the interested reader should consult specialist texts such as that by Kruer Kruer (1988).

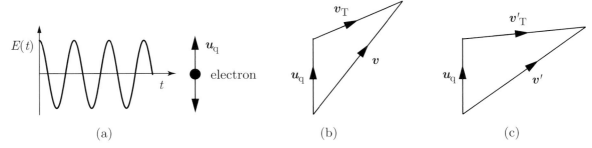

Fig. 18.5: Inverse bremsstrahlung heating of electrons in a plasma. (a) shows schematically the quiver motion of an electron in the electric field of incident laser radiation. In (b) and (c) an electron with initial velocity $v = u_q + v_T$ experiences a collision, after which it has a velocity $v' = u_q + v'_T$. On average, collisions of this type increase the magnitude of the thermal velocity v_T of the electron.

that no net energy could be transferred to the electron from a pulse of laser radiation since the quiver velocity starts and finishes at zero.

We have, however, neglected the role of collisions. When a quivering electron undergoes a collision with a plasma ion, its velocity will be deflected from some initial value v to v', as illustrated in Fig. 18.5.

It is helpful to resolve the velocity of the electron into two components: $v = u_q + v_T$, where u_q is the quiver velocity that the electron would have if it had not undergone any collisions. This quiver velocity is always parallel to the electric field of the laser, and its value is determined only by the instantaneous value of the electric field. In contrast, the second component, v_T, can be in any direction and its value depends on the history of collisions the electron has undergone. We see that this random velocity, v_T, is the velocity associated with the thermal motion of the electron. After the laser pulse has passed, the quiver velocity u_q will be zero, leaving the electron with a velocity v_T. Collisions tend to cause the velocity of the electron to differ from the quiver velocity by ever-increasing amounts, and therefore to increase the thermal velocity and hence thermal energy of the electron; in other words, collisions cause energy to be transferred from the 'coherent' quiver motion to random, thermal motion. This process is known as **inverse bremsstrahlung (IB)** heating.[19] It should also be noted that only collisions with plasma ions or atoms lead to a net heating of the electrons; electron–electron collisions simply redistribute thermal energy between the two electrons and there is no net gain or loss of the thermal component of the energy.

[19] So called because it is the reverse process of bremsstrahlung radiation—the *emission* of radiation by a charged particle undergoing a collision.

As discussed in Exercise 18.8, for moderate laser intensities, of order $10^{13}\,\text{W}\,\text{cm}^{-2}$, such that the quiver velocity is less than the electron thermal velocity, the rate of IB heating is approximately proportional to both the laser intensity and the plasma density. In this regime IB heating rates can be $1 - 10\,\text{eV}\,\text{ps}^{-1}$, allowing rapid heating and further ionization by laser pulses with durations of 100 ps–1 ns.

18.4.2 Generation of highly ionized plasmas from laser–solid interactions

In several types of short-wavelength laser the gain medium is a highly ionized plasma generated by the interaction of a laser pulse with a peak intensity in

512 *Short-wavelength lasers*

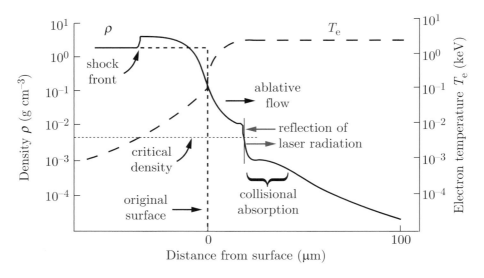

Fig. 18.6: Schematic plot of the density and temperature profiles in a laser-ablated solid.

[20] A comprehensive description of laser–solid interactions may be found in Gibbon (2005).

the range $10^{13} - 10^{14}$ W cm^{-2} incident from vacuum onto a solid surface. It is therefore worth outlining some of the key processes that lead to the formation of the gain medium; these are summarized in Fig. 18.6.[20]

A small fraction of the energy contained in the leading edge of the pulse is absorbed at the surface forming a very dense plasma. This plasma absorbs further laser energy, and becomes more highly ionized in a positive feedback loop between IB heating—which increases the plasma temperature, and hence the rate of collisional ionization—and collisional ionization—which increases the electron density, and hence the rate of IB heating. Since one edge of this hot plasma is unconstrained, the plasma streams out into the surrounding vacuum with a velocity close to that of acoustic waves in the plasma, forming a low-density region—known as the corona—with a density that decreases with distance from the solid surface.

There is a limit to the maximum plasma density that the laser radiation an reach. The refractive index of a plasma may be written as,[21]

[21] See, for example, Bleaney and Bleaney (1989) or Lorrain and Corson (1988).

$$n = \sqrt{1 - \frac{N_e e^2}{m_e \epsilon_0 \omega_L^2}} = \sqrt{1 - \left(\frac{\omega_p}{\omega_L}\right)^2}, \quad (18.9)$$

where $\omega_p = (N_e e^2/m_e \epsilon_0)^{1/2}$ is known as the **plasma frequency**. At low plasma densities $\omega_L > \omega_p$, and hence n is real (and less than unity). However, at high densities $\omega_L < \omega_p$ and n is imaginary. Electromagnetic waves cannot propagate in a medium with a wholly imaginary refractive index, and in fact the incident laser radiation is reflected at the point in the plasma for which $\omega_p = \omega_L$, i.e. at the so-called **critical density** $N_c = m_e \epsilon_0 \omega_L^2 / e^2$, which defines a surface known as the **critical surface**. Since the rate of IB heating increases with density, the power per unit volume deposited in the plasma is higher in denser regions, and hence highest at the critical surface. It follows that the plasma temperature is also highest at this point. From the critical surface there is a temperature gradient to the colder solid. Heat from the plasma is conducted down this gradient to the solid surface, which ablates more material and keeps the whole process going.

Distribution over ionization states

For a given plasma density and temperature the plasma will in general consist of a mixture of ionization stages. If the plasma is in thermal equilibrium the population densities of the various ion stages are related by the **Saha equation**:

$$\frac{N_e N_{Z+1}}{N_Z} = 2 \left(\frac{2\pi m_e k_B T_e}{h^2} \right)^{3/2} \frac{g_{Z+1}}{g_Z} \exp\left(-\frac{\Delta E_{Z+1,Z}}{k_B T_e} \right), \quad (18.10)$$

where N_Z is the density and g_Z the ground-state degeneracy of ions of charge Ze, and $\Delta E_{Z+1,Z}$ is the energy difference between the ground states of the two ions.[22] To form a closed solution the Saha equation must be supplemented by the equations relating the total ion density N_T and the electron density to the densities of the various species:

$$N_T = \sum_{i=1}^{i=Z_{max}} N_i \quad (18.11)$$

$$N_e = \sum_{i=1}^{i=Z_{max}} i N_i, \quad (18.12)$$

where Z_{max} is the maximum possible ion charge.[23]

Figure 18.7 shows an example of a calculation of the relative abundance of different ion stages as a function of the plasma temperature. It can be seen that typically two or three adjacent ion stages dominate the distribution over ionization stages. However, for certain ranges of density and temperature the plasma can be dominated by a single ion stage. For example, for the plasma of Fig. 18.7 the Ne-like (10 electrons) ion has an abundance above

[22] It has been assumed here that the populations of the excited levels of the ions are small compared to the ground-state populations. Of course, this is not strictly the case, but it can be shown that the relative populations of the excited levels is small and hence to a good approximation can be ignored. For further details, and a derivation of the Saha equation, see Hutchinson (2002).

[23] The maximum possible ion charge is equal to the atomic number of the plasma species, although frequently the summation can safely be terminated at lower values of Z than Z_{max} since the density of such highly ionized species is low.

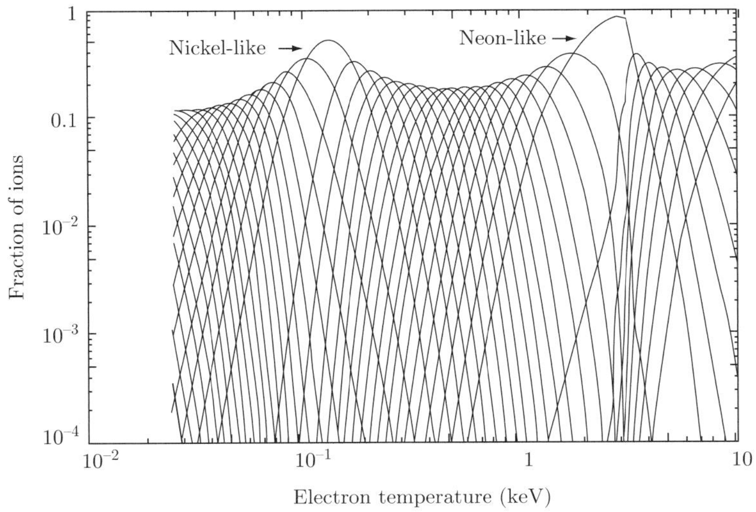

Fig. 18.7: Calculated relative abundance of different ion stages as a function of the plasma temperature for a tin (Sn) plasma with a total ion density of 10^{20} cm^{-3}. (After Daido (2002).)

50% for plasma temperatures between 2 and 3 eV, and can reach a relative abundance as high as 90%. At lower temperatures the Ni-like (28 electrons) ion can also dominate the plasma. These ions can dominate the plasma, under appropriate conditions, since they have closed shells and hence high ionization energies relative to adjacent ion states. For this reason closed-shell ions play an important role in short-wavelength lasers; for example, almost all collisionally excited lasers are based on transitions within ions of this type.

Ultimate stability against further ionization is achieved if the ion is completely stripped of electrons. Such ions cannot be used in a collisionally excited scheme, but they are the precursor ion for several examples of recombination laser, as discussed in Section 18.6.

18.4.3 Optical field ionization

For laser intensities of approximately 10^{16} W cm^{-2}, or above, the electric field of the electromagnetic wave is comparable to that binding the valence electrons in atoms or ions. As a consequence, as shown in Fig. 18.8, the binding potential of the electrons is severely distorted, creating a potential barrier through which the bound electrons can quantum mechanically tunnel to freedom.

As might be expected, the rate of this so-called **tunnelling ionization** depends very strongly on the height and thickness of the barrier through which the electron must tunnel. The ionization rate therefore shows a threshold behaviour with the strength of the electric field—and hence with the intensity of the radiation—and depends strongly on the ionization energy of the atom or ion undergoing ionization. For very large values of the incident field, as illustrated in Fig. 18.8(c), the barrier can be completely removed. In such cases the electrons will ionize as soon as they have 'found' the hole in the potential well created by the applied field. In this extreme limit the ionization is referred to as **over-the-barrier ionization** or **barrier-suppression ionization**; both tunnelling and over-the-barrier ionization can be said to be different regimes of **optical field ionization (OFI)**.

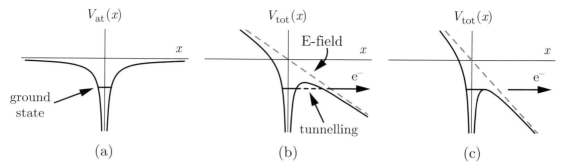

Fig. 18.8: Optical field ionization (OFI). (a) shows the potential energy $V_{at}(x)$ of a valence electron in the absence of an applied laser field. (b) When the atom is exposed to intense laser radiation, the electric field of the electromagnetic wave distorts the total potential, $V_{tot}(x)$, creating a potential barrier through which the outermost valence electron can quantum mechanically tunnel. This is the regime of tunnelling ionization. (c) If the intensity of the laser radiation is increased sufficiently, the potential barrier is removed and the electron may undergo rapid 'over-the-barrier' ionization. Note that in reality the laser field, and hence the potential energy of the electron in this field, will oscillate at ω_L. However, as long as the time taken for the electron to be ionized is short compared to the period of the laser radiation the quasi-static picture presented above is valid.

An estimate of the threshold intensity for rapid optical field ionization can be found by considering the case shown in Fig. 18.8(b) in which the applied field moves the top of the potential barrier down to the energy of the unperturbed ground state of the atom or ion. As shown in Exercise 18.5, if we consider the initial potential well to be a Coulomb potential, the threshold intensities for OFI by linearly and circularly polarized radiation can be estimated to be[24]

$$I_{\text{th}}^{\text{lin}} = \frac{1}{2}\frac{\pi^2 \epsilon_0^2}{e^6 \mu_0 c} \frac{E_I^4}{Z_{\text{ion}}^2} \approx 4.00 \times 10^9 \frac{(E_I\,[\text{eV}])^4}{Z_{\text{ion}}^2} \quad \text{W cm}^{-2} \quad (18.13)$$

$$I_{\text{th}}^{\text{circ}} = \frac{\pi^2 \epsilon_0^2}{e^6 \mu_0 c} \frac{E_I^4}{Z_{\text{ion}}^2} \approx 8.00 \times 10^9 \frac{(E_I\,[\text{eV}])^4}{Z_{\text{ion}}^2} \quad \text{W cm}^{-2}, \quad (18.14)$$

where E_I is the ionization energy of the ion undergoing ionization, and $Z_{\text{ion}} e$ is the net charge of the ion *after* ionization.

[24] In the practical units of the right-hand sides of eqns (18.13) and (18.14) the ionization energy is in units of electron volts and the threshold intensities in units of W cm^{-2}.

Distribution over ionization states

The strong dependence of the ionization rate on the ionization energy is reflected in the strong dependence on E_I in eqns (18.13) and (18.14). This property turns out to be useful in the OFI lasers discussed in Sections 18.5.4 and 18.6.2 since it allows the degree of ionization of a plasma formed by OFI to be controlled by adjusting the intensity of the incident laser radiation. For example, we see from eqns (18.13) and (18.14) that for circularly polarized radiation the threshold intensity for ionizing Xe^{7+} ($E_I = 103.8$ eV, $Z_{\text{ion}} = 8$) is approximately 1.5×10^{16} W cm^{-2}, whereas to ionize Xe^{8+} ($E_I = 178.7$ eV, $Z_{\text{ion}} = 8$) requires an intensity nearly an order of magnitude higher: about 1.0×10^{17} W cm^{-2}.

Electrons ionized by OFI are 'born' essentially at rest into the laser field, and are then driven by it. As explored in Exercise 18.6, the motion of the ionized electron then comprises an oscillation at the laser frequency—this is the quiver motion, which has a mean (ponderomotive) energy given by eqn (18.8)—plus a constant drift velocity. The drift velocity is determined by the condition that the electron is born at rest, and consequently depends on the value of the amplitude of the laser field at the moment of ionization.

In Exercise 18.6 it is shown that for linearly polarized radiation the drift velocity of the electron is given by,

$$v_{\text{drift}} = \frac{eE_0}{m_e \omega_L} \sin \phi_0, \quad (18.15)$$

where ϕ_0 is the phase of the laser field at the moment the electron is ionized. As indicated schematically in Fig. 18.9(a) for linearly polarized radiation the electrons are much more likely to be ionized near the peaks of the laser field, and hence we expect $\phi_0 \approx 0$.

If we now imagine that the incident radiation is not a continuous wave, but is pulsed, we see that after the laser pulse the amplitude of the driven (quiver) motion will be zero and the electron will retain only the drift motion, with an energy of[25]

[25] See Exercise 18.7 and note 71.

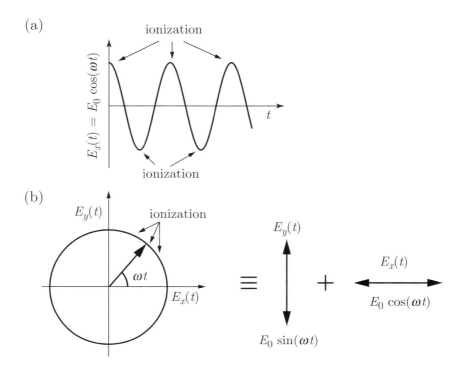

Fig. 18.9: Optical field ionization in (a) linearly polarized and (b) circularly polarized laser fields. Since the rate of OFI depends very strongly on the magnitude of the electric field, for linearly polarized radiation ionization only occurs near the peaks of the electric field. With circularly polarized radiation the magnitude of the electric field remains constant, but the direction of the field rotates with an angular frequency ω about the axis of propagation. As a consequence, OFI is equally probably at any point in the optical cycle. Circularly polarized radiation can be considered to be a superposition of two orthogonal linearly polarized waves with a phase difference of $\pm\pi/2$.

$$E_{\text{drift}}^{\text{lin}} = \frac{1}{2} \frac{e^2 E_0^2}{m_e \omega_L^2} \sin^2 \phi_0 = 2 U_{\text{p}0} \sin^2 \phi_0, \qquad (18.16)$$

where $U_{\text{p}0}$ is the ponderomotive energy of the laser field at the moment the electron is ionized. Since ϕ_0 will be small, the energy retained by the electrons after optical field ionization with a linearly polarized laser pulse will be much smaller than the ponderomotive energy.

As illustrated in Fig. 18.9(b), circularly polarized laser radiation can be regarded as two orthogonal linearly polarized fields with a $\pi/2$ phase difference between them. Since the amplitude of the electric field is constant throughout the optical cycle, with circular polarization OFI can occur with equal probability at any time. For convenience, let us suppose that at the moment of ionization the electric field of the laser radiation points in the x-direction. For this component of the field $\phi_0 = 0$; in contrast, for the y-component $\phi_0 = \pi/2$. Thus, the only contribution to the energy retained by the electrons after the laser pulse arises from the motion in the y-direction. The energy of this drift motion is given by eqn (18.16) with $\phi_0 = \pi/2$ and $U_{\text{p}0} \to U_{\text{p}0}/2$, since only half of the ponderomotive energy is associated with motion in the y-direction.[26] Thus, for circularly polarized radiation the electrons retain an energy of

---

[26] Alternatively, linearly polarized radiation with an electric-field amplitude E_0 has the same intensity as circularly polarized radiation in which the amplitudes of the x- and y-components of the electric field are $E_0/\sqrt{2}$.

$$E_{\text{drift}}^{\text{circ}} = U_{\text{p0}}. \qquad (18.17)$$

If an atom undergoes several stages of ionization it will do so at different points in the laser pulse, corresponding to the moments when the laser intensity reaches the threshold intensity of each ion. The electrons produced from each ion will therefore have different energies, and consequently the electron-energy distribution will comprise Z_{ion} peaks where the final net charge of the ion is $Z_{\text{ion}}e$.

We see that OFI plasmas can be very different from the essentially thermal plasmas formed by the interaction of relatively low intensity laser pulses with solid targets. The distribution over ion stages can be much more sharply peaked around a single stage, and the energy distribution of the plasma electrons is generally highly non-Maxwellian and depends on the polarization of the laser radiation.[27]

18.5 Collisionally excited lasers

The first collisionally excited X-ray laser was demonstrated in 1985 at Lawrence Livermore National Laboratory in the United States by Matthews et al. following theoretical work by Zherikhin et al. and Vinagradov et al.[28] The active species in that first work was a Ne-like ion, Se^{24+}, lasing occurring on two $2p^53p \rightarrow 2p^53s$ transitions with wavelengths of 20.63 and 20.96 nm.

Figure 18.10 shows schematically the main energy levels relevant to collisonally excited lasers operating in Ne-like and Ni-like ions. In both cases electron collisions excite the upper laser level directly by strong, monopole (i.e. $\Delta l = 0$) transitions,[29] as well as cascade from higher-lying levels. Collisional excitation of the lower-lying levels occurs at a slower rate, and consequently these systems exhibit some selective excitation.

For Ne-like ions the upper laser levels belong to the $2p^53p$ configuration. These can only decay to the levels of the $2p^53s$ configuration, their decay by electric-dipole transitions to the $2p^6$ ground state being forbidden since the levels have the same parity. In contrast, the lower laser levels belong to the $2p^53s$ configuration and so do decay by strong electric-dipole transitions to the ground state. Since the energy gap between the $2p^53p$ and $2p^53s$ configurations is substantially smaller than that between the $2p^53s$ configuration and the ground state, radiative decay from the lower laser levels is much faster than that of the upper laser levels, and consequently this system of energy levels therefore has a very favourable lifetime ratio. In fact, as discussed below, provided the plasma is optically thin for radiative decay from the lower laser levels, the laser levels satisfy the necessary but not sufficient condition for continuous-wave lasing (eqn (2.16))! Of course, c.w. operation of these—and any other—X-ray lasers cannot be achieved in practice owing to the extremely high pump power that needs to be applied. Lasing has been observed in Ne-like ions from silicon ($Z = 14$) to silver ($Z = 47$), corresponding to output wavelengths between 87 and 8 nm.

The 4d $\rightarrow$ 4p laser transitions in Ni-like ions are analogous to the 3p $\rightarrow$ 3s in Ne-like ions, but offer the advantage that the energy of the laser transition

[27] It should, however, be emphasized that we have ignored collisions. These comprise electron–electron collisions, which act to thermalize the electron-energy distribution, and electron-ion collisions that can lead to heating through inverse bremsstrahlung heating. Both types of collisions occur during and after the incident laser pulse. The thermalization of the electron energy distribution can proceed extremely rapidly, to the extent that the energy features discussed in this section are washed out on the time scales of interest. However, it is still the case that the mean energy of thermalized OFI electron distributions depend strongly on the polarization of the laser radiation. Further discussion and calculations of these effects may be found in Pert (2006) and David and Hooker (2003).

[28] Matthews et al. (1985), Zherikhin et al. (1976) and Vinogradov et al. (1983)

[29] Note that the electron collisional cross-sections for excitation of *ions* obey different rules from those identified for neutral atoms in Section 11.1.6. For neutral atoms it is expected that the collisional cross-sections reflect the strength of the corresponding *optical* transition, which, for a monopole transition is zero.

The principle reason that the cross-sections for collisional excitation of ions by electrons behave quite differently is that the colliding electron is now 'sucked in' as it approaches the ion. This means, for example, that an electron can gain sufficient energy as it approaches the ion for it to excite transitions with energies above the initial energy of the electron (i.e. the energy it had at a large distance from the ion).

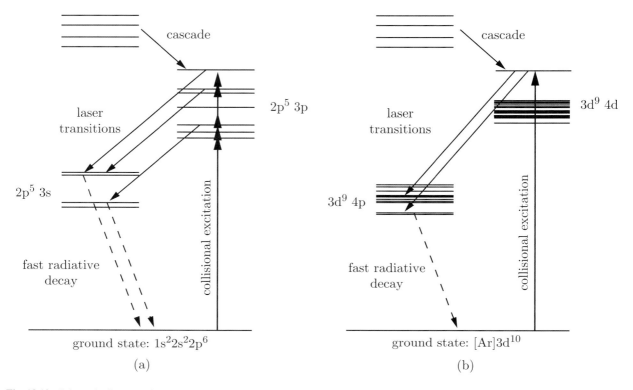

Fig. 18.10: Schematic diagram of the main energy levels relevant to: (a) Ne-like; (b) Ni-like collisional lasers.

is a higher fraction of the energy required to generate the ion species. Consequently, shorter wavelengths may be achieved for the Ni-like ions for a given available pump power. The upper laser level is pre-dominantly pumped by monopole electron collisions with ground-state ions, the rate of cascade from higher-lying levels being only about 10% of the collisional excitation rate. As for the Ne-like ions, the population inversion is maintained by a very favourable lifetime ratio arising from the significantly smaller energy difference between the laser levels than between the lower laser level and the ground state. Lasing has been obtained in Ni-like ions with atomic number ranging from 36 (krypton) to 79 (gold), yielding output wavelengths from 32.8 down to 3.56 nm in the important 'water-window' region.

18.5.1 Ne-like ions[†]

In these highly ionized ions the spin-orbit interaction is not small compared to the residual electrostatic interaction, and consequently the angular momentum is not coupled according to the LS coupling scheme.[30] Instead, the coupling is closer to so-called 'jj-coupling' in which the spin-orbit interaction couples the orbital and spin angular momentum of each electron to give the total angular momentum of that electron $j = \ell + s$. The smaller residual electrostatic interaction may then be treated as a perturbation that couples the j_i of the electrons to form a total angular momentum $J = \sum_i j_i$. The configurations of the excited

[30] See Fig. 7.3 and associated text. Further details may be found in texts on atomic physics, such as Foot (2005).

levels of interest are all of the form $2p^5 3\ell$. The $2p^5$ subshell is equivalent to a closed subshell plus an electron 'hole' with principal quantum number $n = 2$ and angular momentum $\ell = 1$. It is therefore convenient to denote the excited configurations by $\overline{2p}3\ell$, where the bar indicates an electron hole. The levels of each configuration are then denoted $(\overline{2p}_{j_1} 3\ell_{j_2})_J$, where j_1 is the total angular momentum of the 2p hole, j_2 that of the outer 3ℓ electron, and J is the total angular momentum of the level.

The original theoretical work suggested that the strongest laser transition would be the $(\overline{2p}_{1/2} 3p_{1/2})_0 \rightarrow (\overline{2p}_{1/2} 3s_{1/2})_1$ transition. In much of the early work, however, the strongest observed laser lines were the two $J = 2 \rightarrow 1$ transitions $(\overline{2p}_{1/2} 3p_{3/2})_2 \rightarrow (\overline{2p}_{1/2} 3s_{1/2})_1$ and $(\overline{2p}_{3/2} 3p_{3/2})_2 \rightarrow (\overline{2p}_{3/2} 3s_{1/2})_1$. This anomaly was later resolved, as discussed in Section 18.5.3.

As an example, Fig. 18.11 shows the main energy levels of interest for the Ne-like ion Ge^{22+}, and indicates the $J = 0 \rightarrow 1$ transition at 19.6 nm and the

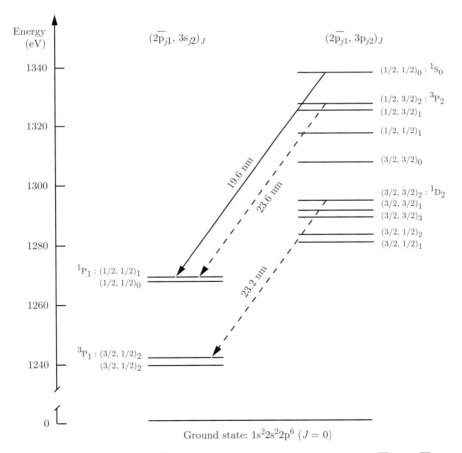

Fig. 18.11: Energy-level diagram of the Ne-like ion Ge^{22+}, showing the ground state and the levels of the $\overline{2p}3p$ and $\overline{2p}3s$ configurations, labelled by $(j_1, j_2)_J$. Also shown are the three most important $3p \rightarrow 3s$ laser transitions: the $J = 0 \rightarrow 1$ transition at 19.6 nm (solid line), and the two $J = 2 \rightarrow 1$ transitions at 23.2 and 23.6 nm (dashed lines). For these transitions the levels are also labelled in LS notation. Note that the energies of the levels are only approximate.

two $J = 2 \to 1$ transitions at 23.2 and 23.6 nm. This ion is formed in plasmas with a temperature of order 800 eV and electron density of $2 - 5 \times 10^{20}$ cm^{-3}. The radiative lifetimes of the upper laser levels are of the order 100 ps; much longer than those of the lower levels, which are of order 0.5 ps. Under the operating conditions of the laser, rapid electron collisions reduce the lifetime of the upper levels to around 10 ps (depending on the plasma density), but the lifetime ratio remains very favourable for creating and maintaining a population inversion.

The very short lifetime of the lower laser levels is a significant contribution to the homogeneous linewidth of the laser transition. Alone, this contributes a homogeneous width of order 300 GHz; electron collisions typically increase the homogeneous width by a factor of about two. The Doppler width of the laser transitions is approximately 1.8 THz, i.e. approximately three times the homogenous linewidth. The broadening of the laser transitions is therefore mixed, the lineshape of the transitions being given by the Voigt profile discussed in Section 3.4.

18.5.2 Ni-like ions[†]

The ground-state configuration of Ni-like ions comprises full $n = 1$, 2, and 3 shells, which may conveniently be written as [Ar]3d^{10}, where the symbol [Ar] represents the ground configuration of the Ar atom. The lowest-lying excited configurations arise from excitation of one of the 3d electrons and so may be labelled $\overline{3d}n\ell$, where $\overline{3d}$ indicates a hole in the 3d subshell. As for the Ne-like ions, the residual electrostatic interaction is not small compared to the spin-orbit interaction and the coupling of angular momentum in the ions is close to jj-coupling.

As an example, Fig. 18.12 shows the lowest-lying energy levels of Ni-like Ag, which shows strong laser action on the 4d $\to$ 4p laser transitions: the $J = 0 \to 1$ transition at 13.89 nm.

For Ni-like ions lasing is possible on several of the 4d $\to$ 4p transitions. The strongest are two $J = 0 \to 1$ transitions: for low-Z ions the strongest transition is usually the $(\overline{3d}_{3/2}4d_{3/2})_0 \to (\overline{3d}_{5/2}4p_{3/2})_1$ transition; for higher-Z ions the shorter-wavelength $(\overline{3d}_{3/2}4d_{3/2})_0 \to (\overline{3d}_{3/2}4p_{1/2})_1$ transition dominates.

18.5.3 Methods of pumping

Quasi-steady-state operation

The earliest collisionally excited X-ray lasers operated in a quasi-steady state. The lasant plasma was created by laser heating a solid foil of the appropriate species with a laser pulse focused by cylindrical optics to a line with a height of approximately 100 μm and length of 1–2 cm at the foil surface. The lasers used in this work were very large scale laser systems delivering infra-red or visible radiation in pulses of order 1 ns duration with energies of approximately 1 kJ. Figure 18.13 shows the typical experimental setup used in this early work.

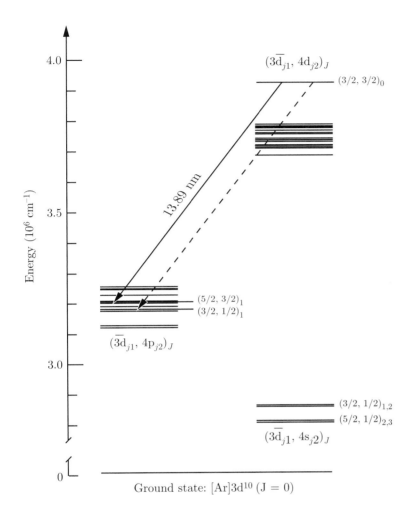

Fig. 18.12: Energy-level diagram of the Ni-like ion Ag^{19+}, showing the ground state and the levels of the $\overline{3d}4s$, $\overline{3d}4p$ and $\overline{3d}4d$ configurations, labelled by $(j_1, j_2)_J$. The dominant laser transition in Ni-like Ag is the $4d \rightarrow 4p$ $J = 0 \rightarrow 1$ transition at 13.89 nm (solid line). In higher-Z Ni-like ions the stronger laser transition can be the shorter-wavelength $4d \rightarrow 4p$ $J = 0 \rightarrow 1$ transition indicated by the dashed arrow. Note that for low-Z Ni-like ions the LS coupling approximation is quite good, and consequently the laser transition can be designated $4d\,^1S_0 \rightarrow 4p\,^1P_1$. (Energies of levels from Rahman and Rocca (2004).)

In these experiments the duration of the output short-wavelength laser pulses could be relatively long, up to 2 ns. Output energies as high as 1 mJ were achieved, corresponding to a conversion of pump laser energy to output short-wavelength energy of approximately 10^{-7} per laser line.

The very large driving lasers used in early work on collisional X-ray lasers were extremely expensive to construct, to the extent that only a few such systems existed worldwide. Further, high-energy laser systems of this type require a long time for the amplifier rods to cool down between laser pulses, and consequently can fire at most a handful of shots per hour. It was quickly recognized that if X-ray lasers were to be widely adopted in applications it would be necessary to reduce the size, complexity and cost of the driving laser, whilst increasing its pulse repetition rate.

Very significant progress in this direction has been made over the last few years by employing one or more pre-pulses and by operating in a transient regime.

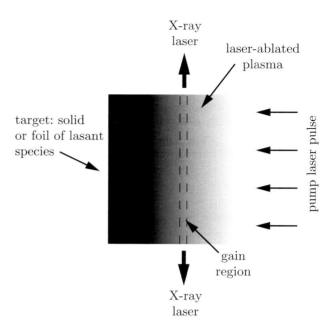

Fig. 18.13: Typical experimental arrangement used in early work on collisionally excited short-wavelength lasers.

Pre-pulses

In the 1990s it was found that the efficiency of the short-wavelength laser could be increased very significantly by illuminating the solid target with one or more laser 'pre-pulses'. The role of the pre-pulse(s) is to form a target plasma by laser ablation, as discussed in Section 18.4.2. The temperature and degree of ionization of the plasma formed by the pre-pulse is usually much lower than those established by the main driving pulse. The optimum energy of the pre-pulse is found to be a few per cent of that of the main driving laser pulse, and the optimum time of arrival of the target is a few nanoseconds prior to the arrival of the main pulse. Pre-pulses improve the efficiency of the laser for several reasons. In the absence of a pre-pulse, much of the incident laser energy is wasted in driving the expansion of the plasma away from the solid target; this waste is reduced by employing a low-energy prepulse. The plasma formed by the pre-pulse expands significantly before the arrival of the main pulse, and consequently the main driving pulse interacts with a much more uniform plasma than would be the case if it had generated the plasma itself. This has two significant benefits. First, absorption of the pump radiation occurs throughout the plasma, not just near the critical surface, and is more complete. Secondly, refraction of the generated X-ray radiation away from the target surface is reduced very considerably; this allows the short-wavelength laser radiation to be amplified over longer lengths before it is refracted out of the gain region.[31]

The use of pre-pulses also increases the relative intensity of the $J = 0 \rightarrow 1$ relative to the $J = 2 \rightarrow 1$ transitions, to the extent that the $J = 0 \rightarrow 1$ transition can dominate the output, as originally expected from theoretical modelling of Ne-like lasers. It is thought that this behaviour reflects the improved

---

[31] With solid targets, the electron density of the plasma will usually decrease with distance from the target surface. This density gradient causes the refractive index to increase with distance from the surface, and consequently radiation initially propagating parallel to the target is refracted away from the target. In the case of the short-wavelength laser radiation, this refraction will eventually cause the generated beam to refract out of regions of high gain. This limits the effective length of the gain region, although there has been some success in using curved targets, or two targets with opposite density gradients, to overcome this effect (for examples see Kodama et al. (1994) and Daido et al. (1995)). Pump laser radiation propagating at an angle to the target normal is also refracted by this density gradient, as is ingeniously utilized in the GRIP laser scheme described on page 525.

conditions in the plasma, and in particular the strong reduction in refraction of the short-wavelength output, which affects the two types of transition differently since their gain is optimized at different plasma densities.

The use of a pre-pulse enabled the total energy of the driving lasers to be reduced by approximately an order of order of magnitude to below 100 J, as well as reducing the divergence of the beam owing to reduced refraction.

Transient operation

The Ne-like and Ni-like laser systems described above are capable, in principle, of continuous operation, provided that the correct plasma conditions can be maintained. However, for short periods, the gain coefficient of the laser transition can be increased by one or two orders of magnitude by heating the plasma faster than the relaxation rate of the upper laser levels. This type of *transient* operation allows the total pulse energy of the driving lasers to be reduced very considerably from the kilojoules of early, quasi-steady-state lasers to only a few Joules. Transient operation has been achieved in both Ne-like and Ni-like ions.

Transient operation is achieved as follows. A long pulse, with an energy of a few Joules and a duration of order 1 ns, is used to generate a plasma from a solid target as described in Section 18.4.2. The intensity of this long pulse is adjusted so that the temperature of plasma is high enough that the plasma contains a high proportion of the lasant species, but low enough that most of these are in the ground state. A second pump pulse, typically of picosecond duration and with an energy of a few Joules, is then applied to the plasma that rapidly raises the electron temperature, resulting in very fast excitation of the upper laser level. The very high transient gain is terminated by bottlenecking of the lower laser level,[32] or further ionization of the plasma.

Figure 18.14 shows a diagram of the COMET laser system developed at Lawrence Livermore National Laboratory for transient excitation of collisional

[32] A further advantage of transient operation is that the density of the lasant plasma does not have to be sufficiently low that the plasma is optically thin on the radiative decay transitions from the lower laser levels. Operating at higher densities than this can increase the gain.

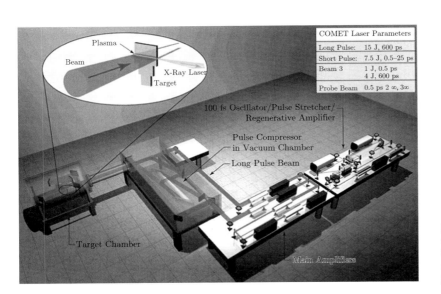

Fig. 18.14: The COMET laser system developed at Lawrence Livermore National Laboratory for pumping transient collisionally excited soft X-ray lasers. (Reprinted with permission.)

524 Short-wavelength lasers

X-ray lasers. The pump laser occupies two 1.2 m × 3.6 m optical tables and comprises a Ti:sapphire oscillator running at 1050 nm and a chain of Nd phosphate-glass amplifiers. The system utilizes the CPA technique described in Section 17.3.3, but splits the amplified beam to produce two beams: a long pulse beam with a pulse energy of 15 J and a pulse duration of 600 ps; a short pulse beam providing 7.5 J pulses with a duration of 500 fs. This laser system can operate at a pulse repetition rate of one shot every four minutes.

Since the duration of the optical gain in a transiently pumped short-wavelength laser is typically only a few tens of picoseconds, it is necessary for the pumping pulse to be applied in a travelling-wave configuration. Figure 18.15 shows how a stepped mirror may be used to divide the incident pump laser beam into several segments, each segment being delayed by an additional $2h/c$. The beam is then focused by a cylindrical lens (or mirror) to form a line focus on the surface of the target.[33] There are several alternative ways in which travelling-wave pumping can be invoked. For example, if the laser pump pulses are produced by a CPA system (see Section 17.3.3), it is possible to tilt

[33] For the pumping to travel along the surface of the target at c, the step angle θ of the mirror is given by $\tan\theta = 1/2$.

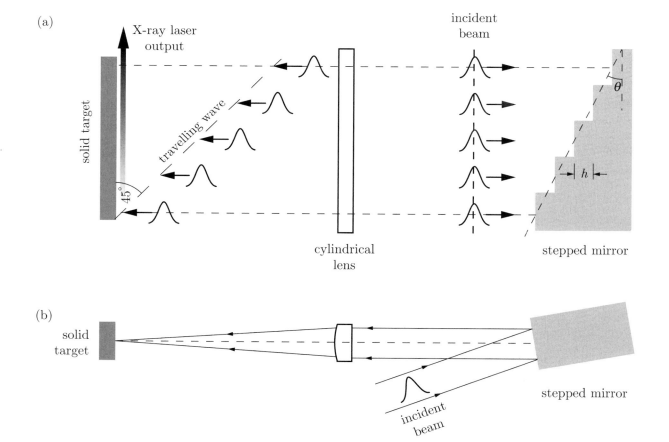

Fig. 18.15: Schematic diagram of the use of a stepped mirror for producing a travelling-wave pump pulse at the surface of a solid target shown: (a) in plan; (b) in elevation. Note that the pre-pulse used to generate the initial plasma is not shown.

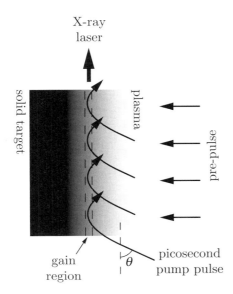

Fig. 18.16: Schematic diagram of grazing-incidence pumping (GRIP) of a short-wavelength laser.

the pulse front relative to the direction of propagation of the beam (just as in Fig. 18.15) by tilting and translating one of the compressor gratings.[34]

[34] See Chanteloup et al. (2000).

Grazing-incidence pumping (GRIP)

Recently, the energy of the pump laser in collisionally excited soft X-ray lasers has been decreased still further by employing so-called grazing-incidence pumping (GRIP). As illustrated in Fig. 18.16, in this geometry a 100 ps–1 ns pre-pulse is used to form a plasma from a solid target, as in the transiently-pumped lasers discussed above. After a suitable delay, typically 100–700 ps, the pump pulse is focused into the plasma at angle θ relative to the target surface. Typically θ is between 10 and 30 degrees.

The pump laser beam is refracted away from the target surface by the gradient in the density of the plasma formed by the pre-pulse. The electron density $N_e^{\max}$ at the turning point of the pump beam is related to the angle of incidence by

$$\theta = \sqrt{\frac{N_e^{\max}}{N_c}}, \qquad (18.18)$$

where N_c is the critical electron density for radiation at the frequency of the pump laser. The maximum electron density reached by the pump beam therefore depends only on the angle of incidence and its wavelength (see Exercise 18.9).

Grazing-incidence pumping offers several advantages over pumping at normal incidence:

- The curved path of the pump laser increases the interaction length of the pump beam with the plasma, thereby increasing the efficiency with which it is absorbed.

- Most of the energy of the pump beam is deposited in the region near the turning point of the beam. By adjusting the angle of incidence θ it is possible to control the plasma density at which this occurs.
- The geometry is close to travelling-wave pumping.

For a given laser transition, and pump and pre-pulse laser energies, there is an optimum angle of incidence. Large values of θ cause most of the pump laser energy to be deposited close to the target, where the electron density and density gradients are both high. As we have seen, the gain coefficient of the short-wavelength laser transition can be high in such regions, but radiation generated is refracted away from the target after propagating only a short distance so that the total gain experienced is low. Decreasing θ deposits more pump energy in regions of lower density gradients, where the gain coefficient of the short-wavelength laser is lower but refraction of the generated radiation is reduced.

A further restriction on the angle of incidence is mismatching of the velocity of the travelling wave. For a target of length ℓ_g the delay arising from the velocity mismatch is given by,

$$\Delta \tau \approx \frac{\ell_g}{c}(1 - \cos\theta). \tag{18.19}$$

This delay can become significant for large values of θ (clearly for $\theta = 90°$ there is no travelling-wave pumping at all!). For example, for $\ell_g = 10$ mm and $\theta = 30°$, the delay is 9 ps, which is comparable to the duration of the optical gain in these transiently pumped systems.

The GRIP geometry has proved to be very successful,[35] and has allowed saturated gain to be achieved on several laser transitions in Ne-like and Ni-like ions with output wavelengths as short as 13.9 nm (Ni-like Ag). Since the total required pump energy is reduced to less than 2 J the systems may be operated at pulse repetition rates of 5–10 Hz. The high pulse repetition rates give rise to relatively high mean output powers for the short-wavelength lasers; the typical output energy is of order 0.5 µJ, yielding a mean output power of 2.5 µW for operation at 5 Hz.

[35] For a summary of recent work, see Luther (2006).

Discharge pumping

Remarkably, the very high pump powers required to generate a population inversion at short wavelengths have been achieved in electrical discharges using an approach developed by Jorge Rocca and his group at Colorado State University in the United States.[36] The key to reaching the high power required is the generation of current pulses with high peak currents and very short risetimes, coupled with magnetohydrodynamic compression of the discharge plasma to small volumes.

The gain medium in these lasers is formed within a polyacetyl or ceramic capillary with a diameter of 3 to 4 mm and a length up to 350 mm. The capillary is filled with a gas, most frequently Ar, to an initial pressure of about 0.5 mbar. Application of a current pulse with a peak of around 25 kA and a risetime of only 40 ns causes the gas to ionize and form a plasma. Since the rate of rise of current is so high, the skin effect causes the current to flow pre-dominantly

[36] The first report of short-wavelength lasing using this approach was by Rocca et al. (1994). A summary of the development of these lasers may be found in Rocca (1999) and (2004).

near the capillary wall. During the early phase of the discharge, therefore, most of the current flows in an annulus near the capillary wall. Just as in the case of a current-carrying wire, associated with this current is an azimuthal magnetic field; this generates an inwardly directed force on the current-carrying electrons via the Lorentz ($v \times B$) force, which causes the plasma to constrict towards the axis of the capillary.[37] For sufficiently fast-rising current pulses a shock wave is driven towards the axis of the capillary, compressing the plasma to a diameter of order 300 μm. As the plasma column collapses the current density increases dramatically and the plasma is strongly heated—eventually reaching densities of $0.3 - 1.0 \times 10^{19}$ cm^{-3} and temperatures of 60–80 eV. These conditions are sufficient to ionize Ar to the Ne-like stage and to generate a population inversion by electron collisions.

[37] The skin and pinch effects are discussed in, for example, Bleaney and Bleaney (1989) or Lorrain and Corson (1988).

Figure 18.17 illustrates the discharge circuit employed. The capillary is mounted coaxially with a liquid-filled parallel-plate capacitor of typically 3 nF capacitance. This is pulse-charged to a voltage of several hundred kV via a Marx generator, and then discharged through the capillary by closing a spark-gap switch[38] that is also mounted coaxially to the capillary. The very low inductance of this arrangement allows the discharge current to rise in only a few tens of nanoseconds.

[38] A spark-gap switch comprises two electrodes mounted in a chamber containing high-pressure gas. Closing of the switch is achieved by applying a high-voltage trigger pulse that initiates electrical breakdown between the electrodes.

This approach has been used to achieve lasing in Ne-like Ar (at 46.9 nm), Ne-like Cl (52.9 nm) and Ne-like S (60.8 nm). The most highly developed of these lasers is the Ne-like Ar system, which can generate pulses with energies

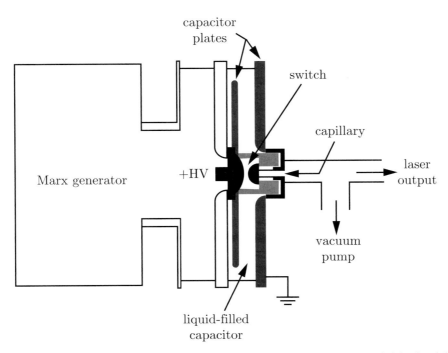

Fig. 18.17: Schematic diagram of a discharge-pumped Ne-like Ar laser. Note that when the spark-gap switch is closed the positively charged capacitor plate is connected to the left end of the capillary. The full charging voltage of the capacitor is now connected across the capillary, which causes the gas within it to break down, allowing current to flow from one capacitor plate to the other.

of up to 0.9 mJ at a pulse repetition frequency of 4 Hz, corresponding to a mean output power of 3.5 mW. This is by far the highest mean power of any laser system operating at extreme ultraviolet wavelengths. The duration of the output pulses of the Ne-like Ar laser is approximately 1 ns, and the divergence is of order 4.5 mrad. The spatially coherent output per unit bandwidth of these lasers has time-averaged values comparable to that from a third-generation synchrotron, and peak values that are several orders of magnitude larger.

18.5.4 Collisionally excited OFI lasers

As discussed in Section 18.4.3, OFI is attractive for creating the ionized gain medium of a short-wavelength laser since the degree of ionization can be controlled through the intensity of the driving laser and the electron temperature can be controlled—at least to some extent—through the laser polarization. Further, as discussed in Section 17.3.3, the high laser powers required to reach the intensities necessary for OFI can now be produced by relatively small-scale visible laser systems delivering femtosecond laser pulses at a pulse repetition rate of 10 Hz or more. OFI-driven lasers therefore offer one route to so-called 'table-top' short-wavelength lasers. As first pointed out by Corkum and Burnet,[39] two types of OFI laser may be driven: circularly polarized driving radiation generates hot electrons, which can be used to drive collisionally excited lasers; linearly polarized radiation generates cold electrons, which can be used to promote rapid recombination in a recombination laser.

The first OFI laser operated[40] on the $5d^1S_0 \rightarrow 5p^1P_1$ transition at 41.8 nm in the Pd-like ion Xe^{8+}. The ground state of Xe^{8+} is $[Kr]4d^{10}$, and consequently the mechanisms responsible for forming the population inversion in the Xe^{8+} laser are analogous to those of the Ni-like lasers discussed in Section 18.5.2.

The threshold intensity for generating Xe^{8+} with circularly polarized radiation is approximately 3×10^{16} W cm^{-2}. Figure 18.18(a) shows the calculated electron-energy distribution produced by OFI of Xe atoms with 40-fs laser pulses with a peak intensity of 3×10^{16} W cm^{-2}. The 8 peaks in the energy distribution correspond to removing 8 electrons from each atom: each successive electron has a higher ionization energy and hence is ionized later in the pulse, when the intensity of the driving laser reaches the threshold intensity given by eqn (18.14); after the laser pulse the electron retains an energy equal to the ponderomotive energy of the laser field at the moment of ionization. Figure 18.18(b) shows the most important levels of the Xe^{8+} ion. It can be seen that most of the electrons produced by OFI have sufficient energy to excite the upper laser level through collisions.

Figure 18.19 illustrates schematically the geometry employed in the first collisional OFI laser. The circularly polarized driving radiation was provided by a Ti:sapphire laser delivering 70 mJ, 40 fs laser pulses at a pulse repetition rate of 10 Hz. This beam was focused by a spherical mirror into a cell containing Xe gas at a pressure of about 15 mbar. The driving radiation entered and exited the Xe cell through two small pinholes and generated Xe^{8+} over the

[39]Corkum and Burnett (1988).

[40]Lemoff et al.(1995).

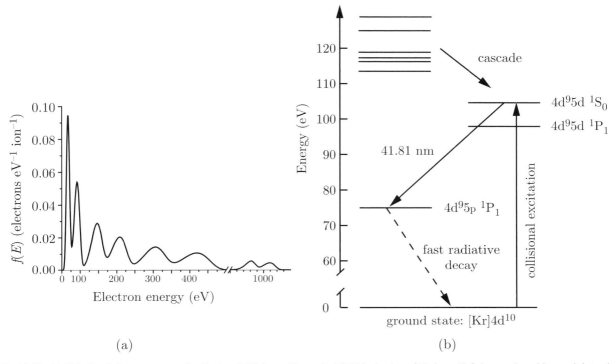

Fig. 18.18: (a) Calculated electron-energy distribution $f(E)$ formed by optical field ionization of Xe by a 50-fs laser pulse with a peak intensity of 5×10^{16} W cm^{-2}. (b) Simplified energy-level diagram of Xe^{8+}.

8 mm length of the gas cell. Note that this pumping geometry is intrinsically travelling-wave.

A major difference between collisionally excited OFI lasers and Ne-like and Li-like lasers driven in plasmas generated by laser–solid interactions is that the density of the OFI plasma is some three orders of magnitude lower and the ions are much colder. As a consequence, the linewidth of the laser transition is expected to be much narrower. For example, the natural linewidth of the $5d\,^1S_0 \rightarrow 5p\,^1P_1$ transition in Xe^{8+} is approximately 20 GHz, but electron collisions and Stark broadening increase the homogeneous linewidth to approximately 300 GHz. The Doppler width of the laser transition is estimated to be only 60 GHz, and consequently the laser transition is expected to be predominantly homogeneously broadened.[41]

[41] See Lemoff et al. (1994).

Although they were first demonstrated more than a decade ago, collisionally-excited OFI lasers have not yet realized their promise. One reason for this is strong refractive defocusing of the pump laser radiation by the large transverse gradients in the electron density of the OFI plasma; since the driving radiation is most intense along the axis of the gain region, the electron density will be highest on axis and consequently the driving radiation is refracted away from the axis (see Exercise 18.9). The length of the gain region that can therefore be generated is typically only a few millimetres, and consequently the gain coefficient of the short-wavelength laser transition must be very large—

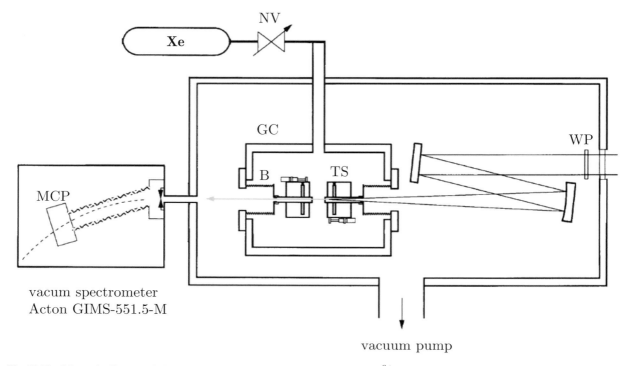

Fig. 18.19: Schematic diagram of the experimental arrangement used to drive the Xe^{8+} OFI laser. The quarter-wave plate (WP) converts the linearly polarized driving radiation to circular polarization, which is then focused by a spherical mirror into a gas cell differentially pumped through pinholes drilled in situ by the incident laser. In this arrangement the pinholes are sealed to flexible bellows (B) that connect to the inner wall of the gas cell, allowing the distance between them—and hence the length of the gain region—to be varied by moving the translation stages (TS). Radiation leaving the gas cell is dispersed by a vacuum spectrometer and detected by a microchannel plate detector (MCP). (After Lemoff et al. (1996).)

at least 10 cm^{-1}—for ASE lasing to occur. A solution to this problem is to channel or guide the pump laser radiation. This is not a simple matter at the very high intensities required for OFI, but some success has been achieved with so-called plasma waveguides[42] and hollow capillary waveguides.[43]

18.6 Recombination lasers

The potential for achieving a population inversion transitions following electron–ion recombination was recognized by Gudzenko and Shelepin[44] only a few years after the first demonstration of the laser. Recombination lasers operate via so-called **three-body** recombination in which an ion of charge $(Z+1)e$ recombines with an electron to form an ion of charge Ze:

$$A^{(Z+1)+} + e^- + X \to A^{Z+} + X, \qquad (18.20)$$

where X is a third body that takes up the spare momentum and energy. Since the precursor ion, $A^{(Z+1)+}$, is positively charged the third body is most likely to be another plasma electron.

---

[42] A plasma waveguide comprises a cylinder of plasma in which the electron density increases with radial distance from the axis. The refractive index therefore decreases with distance from the axis, and consequently the cylinder of plasma acts just like the gradient refractive index silica fibres discussed in Section 10.1.2. Of course, a grin fibre made from plasma can withstand much higher laser intensities than one formed from a solid!

[43] See Butler et al. (2003), Mocek et al. (2005) and Chou et al. (2007).

[44] Gudzenko and Shelepin (1965).

Three-body recombination preferentially populates the high-lying levels of the ion A^{Z+}.[45] Since some of the energy given up by the recombining electron is taken up by the third body, three-body recombination is associated with an overall heating of the plasma electrons known as **recombination heating**.[46]

As illustrated schematically in Fig. 18.20, recombination to the high-lying energy levels of A^{Z+} is followed by a rapid electron collisional de-excitation to lower levels. The rate of collisional de-excitation varies as $\Delta E^{-1/2}$, where ΔE is the energy gap between the two levels, and consequently the rate of the collisional cascade slows as the ion reaches the lower, more widely spaced levels. At some point the rate of collisional de-excitation becomes smaller than the rate of de-excitation by radiative transitions—which varies approximately as ΔE^2 (see Section 18.2.1)—whereupon a bottleneck is formed and population builds up in one of the low-lying levels.

The energy-level structure of hydrogen-like ions is particularly favourable for generating a population inversion through three-body recombination. In particular, if the plasma conditions can be arranged so that the bottleneck in the collisional cascade occurs at the level $n = 3$, then a population inversion

[45] The rate of three-body recombination may be calculated by an argument based on detailed balance, similar to that employed in Exercise 11.3. In thermal equilibrium the rate of collisional ionization of an atom in level n of ion A^{Z+} is balanced by the rate of three-body recombination of ions $A^{(Z+1)+}$ (in any energy level) back to level n of A^{Z+}. Thus, we may write $S_{nI} N_Z^n N_e = \alpha_3^n N_{Z+1} N_e^2$, where S_{nI} is the ionization coefficient (see Section 11.1.5 and eqn (11.17)), α_3^n is the three-body recombination rate, N_Z^n is the density of population of level n in ion A^{Z+}, and N_Z is the total density of population of ions in A^{Z+}. In thermal equilibrium the ratio N_Z^n/N_Z may be calculated from the Boltzmann distribution, and the ratio N_Z/N_{Z+1} from the Saha distribution (eqn (18.10)). Thus, the three-body recombination rate coefficient α_3^n may be calculated once the ionization coefficient S_{nI} is known. Applying the procedure to hydrogenic ions shows that $\alpha_3^n \propto n^4$, showing that the rate of recombination to high-lying levels is very much greater than that to lower-lying levels.

[46] Since the mass of the ion is so much greater than that of the electrons, the ions cannot take up the energy of recombination without also greatly changing their linear momentum.

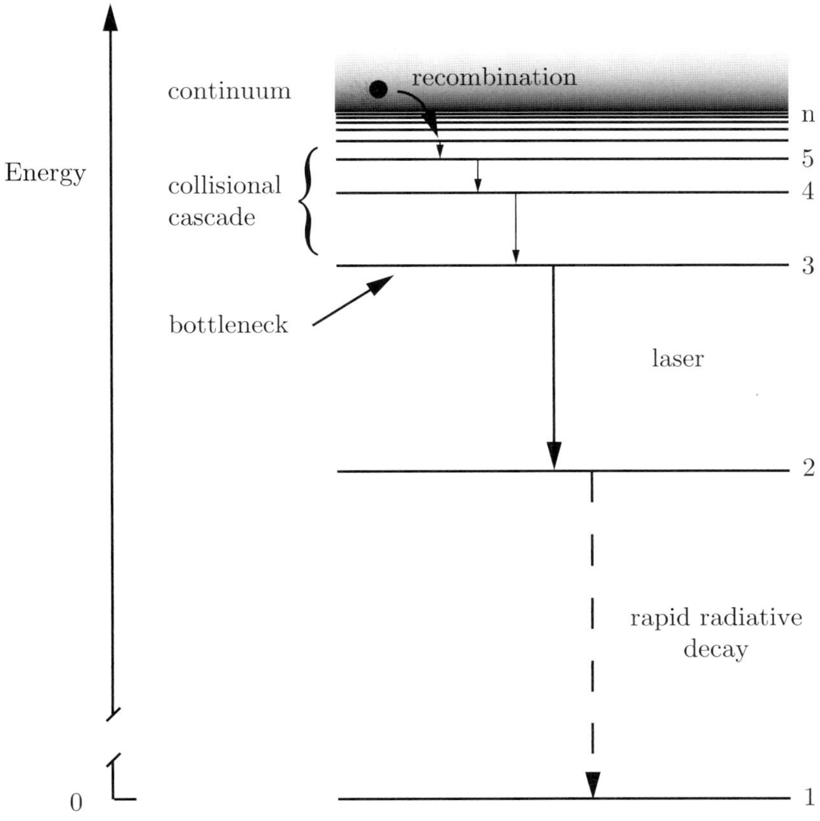

Fig. 18.20: Recombination in a H-like ion, showing the formation of a population inversion on the $n = 3 \rightarrow 2$ transition.

may be formed on the $n = 3 \rightarrow 2$ transition since there is a favourable lifetime ratio arising from the fact that the energy gap between the $n = 3$ and $n = 2$ levels is 3.6 times smaller than that between $n = 2$ and the ground state. Short-wavelength recombination lasers have been operated in several H-like ions, most notably in C^{5+} as discussed below. Recombination lasing has also been observed in lithium-like and sodium ions; both of these ions have very stable precursor ions (helium-like and neon-like, respectively), and as such can form a high proportion of the ions in a thermal plasma.

The rate of three-body recombination is proportional to the square of the plasma (and therefore electron) density, and varies with the electron temperature as $T_e^{-9/2}$. This very strong temperature dependence is the cause of one of the main difficulties of the recombination scheme, in that it requires two apparently conflicting requirements: (i) the generation of a highly ionized plasma, which favours high electron temperatures; (ii) rapid recombination, which requires very low electron temperatures. Overcoming this conflict requires a two-step approach: creating the precursor ions in a hot dense plasma; then rapidly cooling the plasma by adiabatic expansion, heat conduction, or radiation of energy by high-Z ions doped into the plasma.

The main advantage of the recombination scheme is that the laser transition involves a change in principal quantum number, and therefore the energies of the output laser photons are a larger fraction of the ionization energy of the lasant ion than is the case for collisionally excited lasers, all of which operate on $\Delta n = 0$ transitions.[47]

18.6.1 H-like carbon

The first short-wavelength recombination laser was reported by Suckewer et al. in 1985. This laser operated on the $n = 3 \rightarrow 2$ transition at 18.2 nm of H-like C (C^{5+}) formed by three-body recombination of electrons with fully stripped carbon atoms. As illustrated schematically in Fig. 18.21, the initial plasma was formed by focusing 300 J, 75 ns pulses from a CO_2 laser[48] to a peak intensity of 5×10^{12} W cm^{-2} at the surface of a solid carbon target. The plasma so-formed was confined by a magnetic field of 9 T provided by a solenoid oriented with its axis coaxial to the axis of the short-wavelength laser. A 20 mm long carbon blade was mounted parallel to the axis to help form a uniform plasma and to provide cooling by heat conduction. Short-wavelength laser pulses with an energy of about 3 mJ were generated.

Just as for the collisionally pumped lasers described above, there has been a considerable effort to reduce the size and complexity of the pump lasers required in recombination lasers. For example, Zhang et al. (1995) have demonstrated lasing on the 18.2 nm transition H-like C using picosecond-duration pump pulses of only 20 J. In that work, the pump laser pulses were focused to a peak intensity of 6×10^{15} W cm^{-2} along a line 7 mm long at the surface of a 7 μm diameter carbon fibre. A gain coefficient of 12.5 cm^{-1} was measured, corresponding to a small-signal gain–length product of $\alpha_0 \ell_g \approx 8$.

Notwithstanding these advances, recombination lasers have not been developed to the same extent as collisionally excited short-wavelength lasers. This is primarily because of the difficulty in meeting the stringent requirements

---

[47] For transitions involving a change of principal quantum number, the energy of the transition varies approximately as Z^2, where Ze is the charge on the ion. Consequently, for these transitions the ratio of the output photon energy to the ionization energy (which is a measure of how much energy is required to create the ion) of the ion is constant. In contrast, for transitions involving no change in n the energy of the transition varies approximately as Z, and hence as one moves along an isoelectronic sequence to higher Z, and hence shorter output wavelength, the energy of the transition becomes a smaller fraction of the ionization energy of the lasant ion.

[48] See Section 12.6.

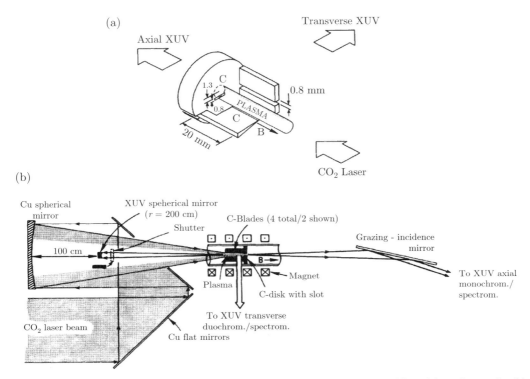

Fig. 18.21: The first short-wavelength recombination laser. (a) shows the solid carbon target, the position of the carbon cooling blade, and the orientation of the longitudinal and transverse XUV spectrometers used to diagnose the plasma. In (b) is shown the arrangement employed for focusing the CO_2 pump radiation onto the solid target and applying the longitudinal magnetic field for confining the plasma. Also shown is an optional XUV mirror for double-passing left-going ASE radiation back through the gain region. Reprinted figure with permission from S. Suckewer et al., *Physical Review Letters* **55** 1753 (1985). ©(1985) by the American Physical Society http://link.aps.org/doi/10.1103/PhysRevLett.55.1753.

on the plasma temperature for rapid recombination in the presence of competing processes such as inverse bremsstrahlung and recombination heating. The extreme sensitivity to the plasma conditions makes it difficult to produce gain over long lengths of plasma, to the extent that at the time of writing no recombination laser has been demonstrated to reach saturation.

18.6.2 OFI recombination lasers

Plasmas created by optical field ionization with linearly polarized radiation offer, at least in principle, almost ideal conditions for lasing by recombination. The key advantage that OFI brings is the ability to create a highly ionized plasma, whilst the electrons remain relatively cold. These desirable features led Burnett and Corkum[49] to propose that it would even be possible to achieve a population inversion on transitions to the ground state in an OFI plasma. This would be extremely difficult to achieve in a plasma produced by laser–solid interactions since such plasmas contain a distribution over several ion stages, and therefore some fraction of the plasma ions will be in the ground state of the recombined ion (which is the lower level of the laser transition). Further, for a population inversion to be formed on a transition to the ground state the rate of

[49] Burnett and Corkum (1989).

recombination must exceed the rate of radiative decay of the upper laser level. These lifetimes are very short for levels with strong transitions to the ground state: for example, the lifetimes of the upper level of the $n = 2 \to 1$ transitions in H-like Li and C are only 26 ps and 1.6 ps, respectively. The electrons must be very cold if such rapid radiative decay is to be overcome.

The first report of recombination lasing in an OFI plasma was by Nagata et al.[50] In that work, a plasma of singly ionized lithium was formed by using a cylindrical lens to focus 200 mJ, 20 ns pulses from a KrF laser to a line at the surface of a solid Li target. After a delay of 700 ns the plasma so-formed was fully ionized by linearly polarized 50 mJ, 500 fs pulses (from a second KrF laser system) propagating along an axis parallel to the surface with an intensity of 1×10^{17} W cm^{-2}, as shown schematically in Fig. 18.22. Radiation on the $n = 2 \to 1$ transition at 13.5 nm of H-like Li was amplified along the direction of propagation of the subpicosecond KrF laser pulses. The small-signal gain coefficient was determined to be 20 cm^{-1} by recording the energy of the 13.5 nm radiation as a function of the length of the initial plasma formed by the nanosecond KrF laser. However, since the maximum length of the target was only 2 mm, the maximum single-pass small-signal gain product was only $\alpha_0 \ell_g \approx 4$, and consequently the laser operated well below saturation.

There have been several other reports of recombination lasing in OFI plasmas, but in all cases the single-pass small-signal gain product has been less than around 6.5. There are several reasons why this approach has not been more successful. First, as for collisionally excited OFI lasers, refraction and diffraction of the pump laser pulses limits the gain length that can be generated. Secondly, processes such as inverse bremsstrahlung heating raise the electron temperature well above that expected from OFI alone. In order to reduce the impact of IB heating it is necessary to keep the density of the plasma low, but doing so reduces the rate of three-body recombination and hence reduces the optical gain.[51] At the time of writing these problems have not been overcome and hence the full promise of OFI recombination lasers has yet to be realized.

[50] Nagata et al. (1993).

[51] There have been several interesting proposals for overcoming this problem. For example, Grout et al. (1997) have suggested adding hydrogen to the target material. The temperature of the electrons released by OFI of hydrogen is much lower than that of electrons ionized from species with higher ionization energies. Rapid electron–electron collisions can then significantly reduce the *average* electron temperature significantly. This approach has been shown theoretically to increase the gain of OFI recombination lasers significantly, but it has yet to be demonstrated.

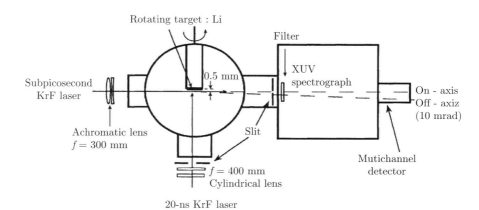

Fig. 18.22: Schematic diagram of the experimental arrangement used to produce gain on the $n = 2 \to 1$ transition at 13.5 nm in H-like Li. Reprinted figure with permission from Y. Nagata et al. *Physical Review Letters* **71** 3774 (1993). ©(1993) by the American Physical Society http://link.aps.org/doi/10.1103/PhysRevLett.71.3774.

18.7 Other sources

18.7.1 High-harmonic generation

An increasingly important source of coherent radiation at short wavelengths is so-called **high-harmonic generation (HHG)**, in which very high harmonics of a longer-wavelength laser are generated. Although the radiation generated in this way is not laser radiation, in that it does not arise directly from stimulated emission, it is nevertheless coherent and is produced in a well-defined beam.

Figure 18.23(a) shows one experimental arrangement employed. Laser radiation is focused to an intensity of $10^{14} - 10^{15}$ W cm^{-2} into a gaseous target, typically a rare gas such as argon or neon.[52] The gas may be delivered in a jet (as shown) or contained in a capillary, which, as discussed below, can increase the energy of the harmonic radiation. Of the radiation leaving the gas jet, by far the most intense component is that at the driving laser frequency ω_L. However, it is also found that radiation with frequencies equal to the odd harmonics of ω_L is also generated, i.e. at angular frequencies given by $\omega_q = q\omega_L$, where $q = 1, 3, 5 \ldots$.

Figure 18.23(b) shows an idealized high-harmonic spectrum, the important features of which are:

[52]Gaseous targets are by far the most common, although solid and liquid targets have been employed. A significant difference between gaseous targets and liquid or solid targets is that with gaseous targets only the odd harmonics are produced; solid or liquid targets generate both even and odd harmonics (see note 55).

- The intensity of the very lowest-order harmonics decreases very rapidly with q.
- There is a range of harmonics, known as the **plateau** in which the intensity of the harmonics is almost constant.
- There is a noticeable cut-off, beyond which the intensities of the harmonics decreases very rapidly.

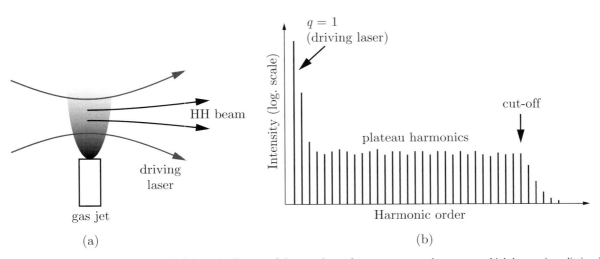

Fig. 18.23: High-harmonic generation. (a) Schematic diagram of the experimental arrangement used to generate high-harmonic radiation in a gas-jet target. (Note that the gas jet is mounted within a vacuum chamber.) (b) Idealized high-harmonic spectrum showing the intensity of each harmonic (on a logarithmic scale) as a function of the harmonic order q.

Mechanism of generation

We briefly outline here a very simple model of the generation of high harmonics originally due to Corkum.[53] As shown in Fig. 18.24, at the intensities under which significant HHG occurs, the potential in which the valence electrons of an atom of ion move can be significantly distorted by the electric field of the incident electromagnetic wave. As discussed in Section 18.4.3, a valence electron can tunnel through the potential barrier so-formed and be accelerated away from its parent ion by the electric field of the driving laser radiation. After the electric field of the laser changes sign the ionized electron can[54] return to the atom with a kinetic energy that depends on the phase of the field of the driving laser at the moment the electron was ionized. The returning electron can recombine with its parent ion, thereby converting its kinetic energy into radiation. In Exercise 18.10 it is shown that the maximum kinetic energy of the returning electron is approximately $3.17\,U_{p0}$, where U_{p0} is the ponderomotive energy of the driving laser field at the moment the electron is ionized. As a consequence, the maximum energy of the photon emitted following recombination is given by,

$$\hbar\omega_{\text{cut-off}} \approx E_{\text{I}} + 3.17\,U_{p0}, \tag{18.21}$$

where E_{I} is the ionization energy of the parent ion. This expression for the cut-off frequency derived from this simple model is found to be in good agreement with experimental results.

Note that if the ionized electron does not recombine with the parent ion on the first collision, it will have further opportunities to do so every $T_{\text{L}}/2$, where $T_{\text{L}} = 2\pi/\omega_{\text{L}}$ is the laser period. The radiation emitted from an ensemble of atoms will therefore comprise bursts of emission separated in time by $T_{\text{L}}/2$. The spectrum of this emission, given by the Fourier transform of the time-dependent emission, must therefore comprise a series of frequency spikes separated by $2\pi/(T_{\text{L}}/2) = 2\omega_{\text{L}}$. Since one of the frequency components must be the fundamental driving frequency (i.e. $q = 1$), the harmonic spectrum will comprise odd[55] harmonics of ω_{L}. The simple model also predicts—as

[53] Corkum (1993).

[54] In Exercise 18.10 it is shown that half the ionized electrons return to their parent ion.

[55] An alternative explanation for the fact that only the odd harmonics are generated is as follows. The atom driven by the laser field of frequency ω_{L} develops a dipole moment that, since the atom is grossly distorted, depends non-linearly on the electric field E_{L} of the driving laser. As discussed in Section 15.3, the polarization of the gas as a whole can therefore be written in the form $\boldsymbol{P} = \epsilon_0 \sum_n \chi^{(n)} E_{\text{L}}^n$, where $\chi^{(n)}$ is the non-linear susceptibility. The polarization of a gas must change sign if the electric field is reversed, and consequently $\chi^{(n)}$ is only non-zero for n odd. The polarization of the gas therefore contains only odd harmonics of the frequency of the driving laser, and consequently only odd harmonics are produced. With solid or liquid targets the harmonics are generated at the surface; the aforementioned symmetry is therefore broken and both even and odd harmonics can be generated.

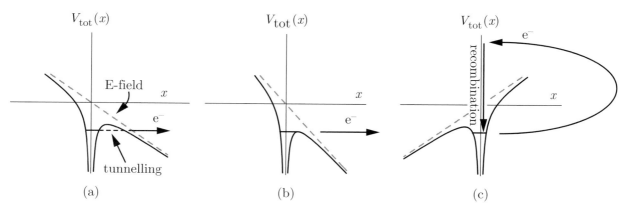

Fig. 18.24: Simple model of high-harmonic generation. (a) Electrons undergo tunnelling ionization in an intense laser field; (b) the ionized electrons are accelerated in the laser field; (c) once the laser field reverses sign the ionized electron can be accelerated back to the parent ion, whereupon it may recombine and emit a high-energy photon.

is observed—that significant high-harmonic generation will only occur for linearly polarized driving radiation since only for this polarization will the electron be returned to the parent ion.

Typical output

For a simple gas-jet target the conversion efficiency[56] for harmonics in the plateau region is typically 10^{-7}. One reason for this low efficiency is that the phase velocity of the generated radiation is, in general, different from that of the driving laser. Consequently, radiation generated some distance into the target gas will be out of phase with that generated at the beginning of the gas. To overcome this problem it is necessary to phase match the harmonic generation, just as for the second- and third-harmonic generation discussed in Section 15.3.2. However, the techniques adopted there are not possible for HHG in gases since the non-linear medium (a gas) is not birefringent. Instead, phase matching has been achieved by driving the harmonic generation in gases held within a capillary waveguide; this allows the dispersion of the gas to be balanced by the dispersion of the waveguide modes.[57] This approach has been used to increase the conversion efficiency[58] of moderately high harmonics ($q < 100$) to better than 10^{-5}.

The laser used most commonly to drive HHG is the Ti:sapphire laser, since this is a convenient source of laser pulses with high peak power. A typical driving laser system comprises an oscillator-regenerative amplifier system delivering 800 nm laser pulses of about 10 fs duration with energies of a few millijoules. Laser systems of this type can operate at a pulse repetition rate of a few kHz allowing, under optimized conditions, average powers of up to 1 mW per harmonic to be generated.

The beams generated by HHG can have very high spatial coherence. The fractional linewidth of each harmonic is relatively broad, typically 1%. The duration of an individual harmonic pulse is typically somewhat less than that of the driving laser, which reflects the fact that the harmonics are pre-dominantly generated on the rising edge of the driving laser pulse. The total bandwidth of the high-harmonic spectrum is sufficient to support pulses with durations in the attosecond regime (1 as $= 10^{-18}$ s). The generation of attosecond pulses by phaselocking several high harmonics is presently a very active area of research.[59]

[56] Although the efficiency is low, it is much higher than would be expected from extrapolating the rapid decrease of intensity with harmonic order found for the very lowest-order harmonics.

[57] See Rundquist et al. (1998).

[58] This approach cannot be used for the very highest harmonics since at the driving intensities required to produce these there is also significant ionization of the gas. The dispersion of a partially-ionized gas is large and has the same sign as the waveguide modes and so the two cannot be balanced.

[59] See, for example, Agostini and DiMauro (2004) and a corrigendum to this article (Agostini and DiMauro, 2004).

18.7.2 Free-electron lasers

Free-electron lasers (FELs) are able to deliver coherent radiation with high average power, and are generally broadly wavelength tunable. They are, however, complex and very expensive and as such are restricted to applications requiring high power and/or wavelength tunability, or to spectral regions that are difficult to access by other means. As we have seen, the latter consideration certainly applies to the X-ray spectral region and this, and the wide range of novel science that could be done, provide strong motivations for applying this technique to the X-ray spectral region. As discussed below, significant efforts in this direction are underway, and it is likely that over the next decade

X-ray FELs will become an unrivalled source of extremely bright coherent X-radiation.

Free-electron lasers are very different from the other lasers discussed in this book since they do not operate on transitions between bound levels. Instead, transitions occur between the continuum of energy levels available to a free electron propagating in a vacuum. Figure 18.25 shows schematically the typical arrangement of a FEL. An electron beam from a linear or ring accelerator is switched into an **undulator** or **wiggler**[60] by beam-steering coils or permanent magnets. In its simplest form an undulator is a periodic array of pairs of permanent magnets arranged with alternating polarity. As the electron beam propagates through this structure it is deflected transversely in alternate directions so that it executes an approximately sinusoidal motion, and therefore radiates. Once the electron beam reaches the end of the undulator it is switched out of the undulator.

The undulator may be located within an optical cavity that couples radiation emitted by the electrons back into the undulator, where it may be further amplified just as in a conventional laser. Of more interest to the short-wavelength region, however, are so-called self-amplified spontaneous emission (SASE) FELs in which no optical cavity is employed. These are directly analogous with the ASE lasers discussed in Section 18.3.

[60] Although they look similar, there is a technical difference between undulators and wigglers. In an undulator the maximum deflection angle of the oscillating electron is small compared to the cone angle of the radiation emitted by the electron; a wiggler corresponds to the other extreme. For further details see Attwood (1999).

Physics of operation

Suppose that an electron moves along the axis of the undulator with a velocity v_z. In a frame moving with velocity v_z with respect to the undulator—i.e.

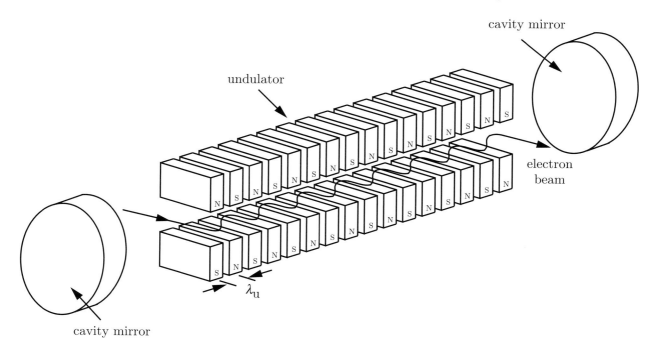

Fig. 18.25: Schematic diagram of a free-electron laser.

the electron's frame—the period of the undulator will be Lorentz contracted to $\lambda' = \lambda_u/\gamma$, where λ_u is the undulator period measured in the laboratory frame and $\gamma = (1 - \beta^2)^{-1/2}$ in which $\beta = v_z/c$. Thus, in the electron's frame it will undergo oscillations in a plane normal to the magnetic field of the undulator, with a frequency $\nu' = c/\lambda' = \gamma c/\lambda_u$. This oscillatory motion will cause the electron to radiate; in the electron's frame the frequency and spatial distribution of the emitted radiation will be that of an electric dipole oscillating at a frequency ν'.[61]

[61] See, for example, Lorrain and Corson (1988).

In transforming back to the laboratory frame the radiation will be Doppler-shifted and strongly peaked in the forward direction. In the laboratory frame the frequency of the radiation measured in the forward direction at an angle θ to the axis is given by[62]

[62] See standard texts on special relativity such as French (2001).

$$\nu = \frac{\nu'}{\gamma(1 - \beta \cos\theta)}. \quad (18.22)$$

In the case of interest the beams will be highly relativistic (i.e. $\beta \approx 1$), so that $1 - \beta \approx 1/2\gamma^2$ and the radiation will be confined close to the axis to that we may write $\cos\theta \approx 1 - \theta^2/2$. We then find

$$\nu \approx \frac{\nu'}{\gamma(1-\beta)} \frac{1}{1 + [\beta/(1-\beta)]\theta^2/2} \approx \frac{2\gamma\nu'}{1 + \gamma^2\theta^2}. \quad (18.23)$$

Substituting for λ', we find the wavelength of the emitted radiation to be

$$\lambda \approx \frac{\lambda_u}{2\gamma^2}\left(1 + \gamma^2\theta^2\right). \quad \textbf{Undulator radiation} \quad (18.24)$$

We see that the wavelength of the undulator radiation is roughly $\lambda_u/2\gamma^2$. Since γ may be of the order of 1000, wavelengths in the soft X-ray spectral region may be generated in undulators with periods of a few centimetres. More detailed calculations[63] allow for the fact that the motion of the electron in the z-direction is not constant, which introduces correction factors to eqn (18.24). Further, since v_z is not constant, the motion of the electron deviates from a pure sinusoidal oscillation and hence the electron also radiates at harmonics of the wavelength given by eqn (18.24).

[63] See, for example, Attwood (1999).

In traversing the undulator each electron will undergo N_u oscillations, where N_u is the number of undulator periods, and hence in any given direction (such as $\theta = 0$) the radiation will comprise N_u periods. It is straightforward to show that the full width at half-maximum of the power spectrum of a truncated sinusoidal wave of N_u periods is approximately

$$\frac{\delta\lambda}{\lambda} = \frac{1}{N_u}. \quad (18.25)$$

In the laboratory frame the radiation is peaked in the forward direction in a cone with a half-angle

$$\theta \approx \frac{1}{2\gamma}. \quad (18.26)$$

Note that, from eqn (18.24), the wavelength of the radiation varies by approximately 25% over this range of angles. It is often desirable to restrict the range of angles θ so that the angular contribution to the bandwidth matches the inherent bandwidth of eqn (18.25). This requires that the radiation is passed through an aperture that presents a half-angle θ_{cen} to the end of the aperture, where

$$\theta_{\text{cen}} = \frac{1}{\gamma\sqrt{N_{\text{u}}}}. \qquad (18.27)$$

Example: Consider a 500 MeV ($\gamma \approx 1000$) electron beam passing through an undulator of period 10 mm. The generated undulator radiation will have a wavelength of approximately 5 nm, and will be emitted in a cone of half-angle 0.5 mrad. For an undulator with $N = 100$, $\Delta\lambda/\lambda = 1\%$ and $\theta_{\text{cen}} = 100$ mrad.

So far, our discussion has dealt only with *spontaneous* emission of radiation in the undulator.[64] In this regime: the electrons in the beam have a random longitudinal distribution, and consequently the radiation emitted by each electron is not coherent with that emitted by other electrons in the beam; the power of the radiation increases linearly with the undulator length. As the electron beam propagates along the undulator, or if the emitted radiation is coupled back into the undulator, interactions between the electron beam and the generated radiation cause the electrons to group in so-called microbunches that are spaced at the wavelength of the emitted radiation. The radiation emitted by each microbunch will now be in phase with that emitted by neighbouring microbunches, increasing the intensity of the generated radiation. In turn, this will increase the degree of microbunching, leading to further increases in the radiation intensity. This positive feedback causes the intensity of the radiation to grow exponentially with the length of the undulator—at least until saturation processes become significant—and is the characteristic behaviour of a free-electron laser.[65] The brightness of the FEL radiation can be ten orders of magnitude greater than that generated by state-of-the-art synchrotrons.

Parameters of an X-ray FEL

Free-electron lasers operating at visible or infra-red or ultraviolet wavelengths can use pulsed or continuous electron beams with energies of a few MeV to a few tens of MeV. For pulsed electron beams the duration of each pulse is typically 1–10 ps, each pulse (known as an 'electron bunch') containing 10^8–10^{10} electrons. A typical undulator used in this region has a period of a few centimetres and a magnetic flux density of 0.1–0.5 T. The average output power of FELs operating in these spectral regions can be as high as a few Watts, for pulsed systems the peak power can reach 1 GW.

Of particular interest in the present context are FELs operating at X-ray wavelengths. At the time of writing two such systems are being constructed, the 'linac coherent light source (LCLS)' at the Stanford Linear Accelerator Center (SLAC) in the United States, and the XFEL facility at Hamburg in Germany.[66] These devices will employ very short electron bunches of high energy, together with very long undulators, to generate tunable, coherent radiation with wavelengths as short as 0.15 nm.

[64] This spontaneous emission from an undulator can be a very useful source of tunable radiation, and is one of the sources of radiation provided at synchrotron facilities.

[65] For further details, see for example, Pellegrini (2001).

[66] Further information on these extraordinary machines may be found at: www-ssrl.slac.stanford.edu/lcls and www.xfelinfo.desy.de

As an example the LCLS at SLAC will use the last 1 km of the existing two-mile long linear accelerator at SLAC to accelerate electron to energies of up to 14 GeV in bunches of a few hundred femtoseconds duration and with a peak current of 3.4 kA. These will be passed through a 112-m long undulator. The output of the FEL will be tunable between 0.15 nm and 1.5 nm with a peak power of 20 GW and an *average* power of 1.5 W. Huge machines like this are expensive. The total cost of the LCLS project is estimated to be over $350 million, but this is relatively cheap since an estimated $300 million is saved by using the existing linear accelerator at SLAC. By comparison the total cost of the XFEL project is estimated to be over €900 million.

Further reading

A great deal of detailed information on the production, detection, and applications of short-wavelength radiation may be found in the book by Attwood (1999). Many more details of the physics of short-wavelength lasers are given in the book by Elton (1990).

A thorough review of the early work at Lawrence Livermore National Laboratory on X-ray lasers can be found in the review by MacGowan (1992); more recent reviews are provided by Daido (2002), Rocca (1999) and (2004), and Suckewer and Jaegle (2009).

Recent reviews of high-harmonic generation have been provided by Eden (2004) and Winterfeldt et al. (2008).

The textbook by Gibbon (2005) on the interaction of intense laser pulses with matter provides an excellent description of much of the physics underlying the operation of short-wavelength lasers.

Exercises

(18.1) **Simple treatment of a homogeneously broadened ASE laser**

In this problem we derive simple expressions for the spectral intensity output by the ASE laser shown schematically in Fig. 18.26. We assume that the transition is broadened homogeneously, that the laser is pumped continuously, and that the pumping is uniform throughout the gain region. We will also ignore saturation of the gain.

(a) Consider spontaneous emission emitted from the section of the gain region lying between z' and $z' + \delta z'$. Show that the rate at which photons with angular frequencies lying between ω and $\omega + \delta \omega$ are emitted into a solid angle Ω is given by,

$$N_2(A\delta z')A_{21}\frac{\Omega}{4\pi}g(\omega - \omega_0)\delta\omega,$$

where N_2 is the population density of the upper laser level.

(b) Now suppose a beam of spectral intensity $\mathcal{I}(z, \omega)$, and lying within the solid angle Ω, is incident on the element. By considering the conservation of energy, show that *in the absence of optical gain* the spectral

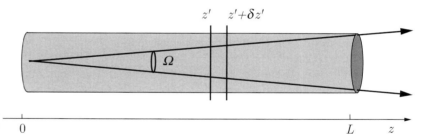

Fig. 18.26: Geometry required for the calculation in Exercise 18.1 of the output of an ASE laser.

intensity would grow according to:

$$\frac{\partial \mathcal{I}}{\partial z} = N_2 A_{21} \hbar \omega \frac{\Omega}{4\pi} g(\omega - \omega_0).$$

(c) Hence, show that in the presence of optical gain the equation describing the growth of spectral intensity is:

$$\frac{\partial \mathcal{I}}{\partial z} = N^* \sigma_{21}(\omega - \omega_0) \mathcal{I}(z, \omega)$$
$$+ N_2 A_{21} \hbar \omega \frac{\Omega}{4\pi} g(\omega - \omega_0). \quad (18.28)$$

Comment on the form of this equation, and discuss how it differs from the case of a normal laser amplifier.

(d) Show that eqn (18.28) may be written in the form,

$$\frac{\partial \mathcal{I}}{\partial z} = \alpha_0 \left[\mathcal{I}(z, \omega) + S_0 \right] \phi(\omega - \omega_0), \quad (18.29)$$

where the peak small-signal gain coefficient is given by $\alpha_0 = N^* \sigma_{21}(0)$, the constant S_0 is given by

$$S_0 = \frac{N_2}{N^*} I_s A_{21} \tau_R \frac{\Omega}{4\pi} g(0), \quad (18.30)$$

where we have assumed $\omega \approx \omega_0$, and in which $I_s = \hbar\omega_0/\sigma_{21}(0)\tau_R$ is the saturation intensity of the transition at line centre. We have introduced a new lineshape function scaled so as to have a peak value of unity: $\phi(\omega - \omega_0) = g(\omega - \omega_0)/g(0)$.

(e) Show that the constant S_0 may be written as,

$$S_0 = \frac{N_2}{N^*} \frac{\hbar \omega_0^3}{\pi^2 c^2} \frac{\Omega}{4\pi}, \quad (18.31)$$

which emphasizes that S_0 depends only on the steady-state populations of the laser levels, the geometry of the gain region, and the frequency of the transition.

(f) By integrating eqn (18.29), show that at a distance z along the ASE laser the spectral intensity is given by,

$$\mathcal{I}(z, \omega) = S_0 \left\{ \exp\left[\alpha_0 z \phi(\omega - \omega_0)\right] - 1 \right\}. \quad (18.32)$$

(g) Use eqn (18.32) to show that for small values of $\alpha_0 z$ the intensity of the ASE grows linearly with $\alpha_0 z$ and that the ASE spectrum has the same spectral lineshape as the spontaneous emission.

(h) Sketch how the spectral intensities of radiation at the centre frequency and at $\omega = \omega_0 + \Delta\omega_H/2$, where ω_H is the homogeneous linewidth, vary with $\alpha_0 z$. Hence explain in qualitative terms how the spectral width of the ASE will vary with $\alpha_0 z$.

(i) The expression we have derived ignores saturation of the gain. However, we may estimate when saturation starts to occur as follows. If we ignore spectral narrowing, the *total* intensity of the ASE beam is approximately given by $\mathcal{I}(z, \omega_0)\Delta\omega_H$. We can estimate that saturation will occur when this total intensity is greater than or equal to the saturation intensity I_s. Show that this condition gives the following condition for saturation to be reached:

$$[\alpha_0 L]_{\text{sat}} \geq \ln\left[1 + \frac{1}{\Delta\omega_H g(0)} \frac{N^*}{N_2} \frac{1}{A_{21}\tau_R} \frac{4\pi}{\Omega}\right].$$

(j) Show that if the homogeneous lineshape has a Lorentzian profile, and if $N^* \approx N_2$, this condition becomes:

$$[\alpha_0 L]_{\text{sat}} \geq \ln\left[1 + \frac{2\pi^2}{\Omega} \frac{1}{A_{21}\tau_R}\right].$$

(k) Estimate the single-pass, small-signal gain-length product $\alpha_0 L$ required to reach saturation in an ASE laser with a cylindrical gain region 20 mm long and of 100 μm diameter for a transition with $A_{21}\tau_R \approx 1$. Comment on the value you obtain.

(18.2) †More advanced treatment of an ASE laser: The Linford formula

We now extend the treatment presented in Exercise 18.1 and derive approximate expressions for the total intensity and spectral width of the ASE output.[67] The total intensity output by the ASE laser is given by integrating eqn (18.32) over all frequencies. However, this is not possible to do analytically. A way forward is to assume that for frequencies close to the centre frequency the spectrum of the ASE may be approximated by a Gaussian function of some width $\Delta\omega_{\rm ASE}$. This enables the integration over frequency to be achieved straightforwardly once $\Delta\omega_{\rm ASE}$ is known. The width of the Gaussian lineshape is found by finding the best fit to the true ASE spectrum for frequencies close to ω_0.

(a) Let us suppose that at a point z the spectrum of the ASE is proportional to $\exp(-kx^2)$, where $x = 2(\omega - \omega_0)/\Delta\omega_{\rm H}$ is a normalized frequency offset and the constant k is to be determined. Show that we may then write:

$$\mathcal{I}(z,\omega) = S_0 \left\{ \exp\left[\alpha_0 z \phi(\omega - \omega_0)\right] - 1 \right\}$$
$$= S_0 \left\{ \exp\left[\alpha_0 z\right] - 1 \right\} \exp(-kx^2). \quad (18.33)$$

(b) Hence, show that k is given by,

$$kx^2 = \ln\{\exp[\alpha_0 z] - 1\}$$
$$- \ln\{\exp[\alpha_0 z \phi(\omega - \omega_0)] - 1\}. \quad (18.34)$$

(c) Let the right-hand side of eqn (18.34) be denoted $f(x)$. We may then expand this function as a Taylor series $f(x) = f(0) + f^{(1)}(0)x + (1/2)f^{(2)}(0)x^2 + \ldots$, where $f^{(n)}(0) \equiv d^n f/dx^n|_{x=0}$. Show that $f(0) = f^{(1)}(0) = 0$ and hence that $k \approx (1/2)f^{(2)}(0)$.

(d) Assuming that the homogeneous lineshape is Lorentzian, so that $\phi(\omega - \omega_0) = (x^2 + 1)^{-1}$, show that,

$$k = \alpha_0 z \frac{\exp(\alpha_0 z)}{\exp(\alpha_0 z) - 1}. \quad (18.35)$$

(e) It is now straightforward to find the total intensity $I_{\rm T}(z)$ by integrating eqn (18.33) over all frequency. Show that this gives,

$$I_{\rm T}(z) = \frac{\sqrt{\pi}}{2} S_0 \Delta\omega_{\rm H} \frac{\left[\exp(\alpha_0 z) - 1\right]^{3/2}}{\left[(\alpha_0 z)\exp(\alpha_0 z)\right]^{1/2}}. \quad (18.36)$$

This result is known as the Linford formula;[68] the small-signal gain, α_0, may be found by fitting it to measurements of the total ASE intensity as a function of the length of the gain region.

(f) Show that the full width at half-maximum of the ASE is given by,

$$\Delta\omega_{\rm ASE} = \Delta\omega_{\rm H} \sqrt{\frac{\ln 2}{k}},$$

and hence that we may write

$$\Delta\omega_{\rm ASE}(\alpha_0 z) = \frac{\Delta\omega_{\rm ASE}(0)}{\sqrt{\alpha_0 z}} \sqrt{\frac{\exp(\alpha_0 z) - 1}{\exp(\alpha_0 z)}}$$
$$\approx \frac{\Delta\omega_{\rm ASE}(0)}{\sqrt{\alpha_0 z}}, \quad (18.37)$$

where the approximation holds for reasonably large $\alpha_0 z$.[69]

(18.3) †Saturation in a homogeneously broadened ASE laser

We may use the results of Exercise 18.1 to gain some insight into how saturation affects the growth of the beam intensity in a homogeneously broadened ASE laser. To do this we make the simplifying assumption that the lower laser level is always empty, so that we may write $N^* = N_2$. We also assume that only the ASE travelling towards positive z has a significant intensity, i.e. we ignore backward-propagating ASE. With these assumptions, eqn (18.28) may be written as

$$\frac{\partial \mathcal{I}}{\partial z} = N^*(I)\sigma_{21}(\omega - \omega_0)\mathcal{I}(z,\omega)$$
$$+ N^*(I)A_{21}\hbar\omega\frac{\Omega}{4\pi}g(\omega - \omega_0),$$

where now we have allowed for the population inversion density to be saturated.

---

[67]This treatment is based on that presented by Svelto et al. (1998).
[68]See Linford et al. (1974) and Svelto et al. (1998).
[69]We have defined the width of the ASE spectrum at $\alpha_0 z = 0$ to be $\Delta\omega_{\rm ASE}(0)$. This is *not* equal to $\Delta\omega_{\rm H}$ since we have approximated the ASE spectrum by a Gaussian function even though for small values of $\Delta\omega_{\rm ASE}(0)$ the spectrum is close to the homogeneous lineshape, which has been assumed to be Lorentzian. It is easy to show that $\Delta\omega_{\rm ASE}(0) = \Delta\omega_{\rm H}\sqrt{\ln 2}$, but this is relatively unimportant; the main purpose of the calculation is to deduce how the spectral width of the ASE should vary with $\alpha_0 z$.

(a) Explain why the population inversion density, allowing for saturation, may be written in the form,

$$N^*(I) = \frac{N^*(0)}{1 + \frac{1}{I_{s0}} \int_0^\infty \phi(\omega - \omega_0) \mathcal{I}(z, \omega) d\omega},$$

where $I_{s0} \equiv I_s(0)$ is the saturation intensity at $\omega = \omega_0$ of the homogeneously broadened transition, and $\phi(\omega - \omega_0) = g(\omega - \omega_0)/g(0)$ is the scaled homogeneous lineshape.

(b) Hence, show that, allowing for saturation, the growth of the spectral intensity is described by,

$$\frac{\partial \mathcal{I}}{\partial z} = \frac{\alpha_0}{f(z)} \phi(\omega - \omega_0) [\mathcal{I}(z, \omega) + S_0], \quad (18.38)$$

where S_0 is given by eqn (18.31) for the case $N^* = N_2$, and,

$$f(z) = 1 + \frac{1}{I_{s0}} \int_0^\infty \phi(\omega - \omega_0) \mathcal{I}(z, \omega) d\omega.$$

(c) Explain qualitatively why saturation slows the rate of growth with z of the beam intensity and the rate of decrease of the spectral width of the ASE once the total intensity reaches values of the order of the saturation intensity I_{s0}.

(d) As an optional exercise you might like to integrate eqn (18.38) numerically to deduce $\mathcal{I}(z, \omega)$, and hence $I_T(z)$ and $\Delta\omega_{ASE}(z)$. Compare your results with Fig. 18.3, which was calculated for a Lorentzian lineshape for the case $N^* = N_2$, $A_{21}\tau_R = 1$, and $4\pi\Omega = 10^{-6}$.

(18.4) †Simple treatment of an inhomogeneously broadened ASE laser

Surprisingly it is not that difficult to analyse the case of an inhomogeneously broadened ASE laser, provided that we may make the assumption that the inhomogeneous linewidth $\Delta\omega_D$ is much greater than the homogeneous linewidth $\Delta\omega_H$. To do this we first make the same assumption as in Exercise 18.3, i.e. that $N^* = N_2$. We also take the population inversion density for atoms with centre frequencies lying in the range ω_c to $\omega_c + \delta\omega_c$ to be given by $N^* g_D(\omega_c - \omega_0) \delta\omega_c$.

(a) Explain why the growth of radiation of spectral intensity $\mathcal{I}(z, \omega)$ arising from atoms with centre frequencies ω_c and $\omega_c + \delta\omega_c$ is described by,

$$\frac{\partial \mathcal{I}}{\partial z} = \frac{N^* g_D(\omega_c - \omega_0) \delta\omega_c}{1 + \int_0^\infty \frac{\mathcal{I}(z, \omega)}{I_s(\omega - \omega_c)} d\omega}$$

$$\left[\sigma_{21}(\omega - \omega_c) \mathcal{I}(z, \omega) + A_{21} \hbar \omega \frac{\Omega}{4\pi} g(\omega - \omega_c) \right],$$

and hence that, including the contribution from all atoms, the rate of growth of spectral intensity is given by:

$$\frac{\partial \mathcal{I}}{\partial z} = \int_0^\infty \left\{ \frac{N^* g_D(\omega_c - \omega_0)}{1 + \int_0^\infty \frac{\mathcal{I}(z,\omega)}{I_s(\omega-\omega_c)} d\omega} \right.$$
$$\times \left[\sigma_{21}(\omega - \omega_c) \mathcal{I}(z, \omega) \right.$$
$$\left. \left. + A_{21} \hbar \omega \frac{\Omega}{4\pi} g(\omega - \omega_c) \right] \right\} d\omega_c.$$

(b) Consider first the term

$$1 + \int_0^\infty \frac{\mathcal{I}(z, \omega)}{I_s(\omega - \omega_c)} d\omega.$$

Provided that $\Delta\omega_D \gg \Delta\omega_H$, the spectral width of the ASE will be much greater than the homogeneous linewidth. Show that in this limit the above may be written as,

$$1 + \frac{\mathcal{I}(z, \omega_c) \Delta\omega_D^{eff}}{I_{s0}^D},$$

where $\Delta\omega_D^{eff} \equiv 1/g_D(0)$ and I_{s0}^D is the saturation intensity at line centre calculated with the inhomogeneous optical gain cross-section.

(c) Using a similar argument, show that

$$\frac{\partial \mathcal{I}}{\partial z} = \frac{\alpha_0^D}{1 + \frac{\mathcal{I}(z,\omega) \Delta\omega_D^{eff}}{I_{s0}^D}} \phi_D(\omega - \omega_0)$$
$$\times [\mathcal{I}(z, \omega) + S_0], \quad (18.39)$$

where α_0^D is the unsaturated inhomogeneous optical gain coefficient evaluated at the centre of the inhomogeneous distribution, and $\phi_D(\omega - \omega_0) = g_D(\omega - \omega_0)/g_D(0)$.

(d) Compare this expression with the equivalent result for homogeneously broadened transitions (eqn (18.38).

(e) Use eqn (18.39) to show that at $\alpha_0^D z \ll 1$ and $\alpha_0^D z \gg 1$ the lineshape of the ASE is approximately the same as the inhomogeneous lineshape (and hence that $\Delta\omega_{ASE} \approx \Delta\omega_D$), and to explain qualitatively why gain narrowing occurs for intermediate values of $\alpha_0^D z$.

(f) As an optional exercise you might like to integrate eqn (18.39) numerically to deduce $\mathcal{I}(z, \omega)$ and hence $I_T(z)$ and $\Delta\omega_{ASE}(z)$. Compare your results with Fig. 18.4, which was calculated for a Gaussian inhomogeneous lineshape for the case $N^* = N_2$, $A_{21}\tau_R = 1$, and $4\pi\Omega = 10^{-6}$.

(18.5) Threshold for optical field ionization
In this problem we use a simple model of the atomic potential to estimate the threshold intensity for ionizing an atom or ion by OFI: we assume that the binding potential of the valence electron is well approximated by the Coulomb potential of a charge $Z_{ion}e$ equal to the net charge of the nucleus and the electrons remaining *after* the ionization.[70]

(a) Write down the total potential energy V_{tot} of an electron in the Coulomb field described above and a static electric field of strength E_0, and sketch this potential for small and large values of E_0.

(b) Show that the total potential has a local maximum at a distance x_0 from the nucleus where,

$$x_0 = \sqrt{\frac{Z_{ion}e}{4\pi\epsilon_0 E_0}},$$

at which point the total potential has a value of

$$V_{tot}(x_0) = -2\sqrt{\frac{Z_{ion}eE_0}{4\pi\epsilon_0}}.$$

(c) We now argue that the threshold for very rapid ionization is reached when the potential energy of this local maximum just equals the energy, $-E_I$, of the electron in the ground state of the unperturbed ion. Show that with this criterion the threshold electric field is given by

$$E_{th} = \frac{\pi\epsilon_0}{e^3}\frac{E_I^2}{Z_{ion}}.$$

(d) Given that, for propagation in vacuum, the intensity of the laser radiation is related to its electric field by $I = \mu_0 c \overline{E^2}$, show that the threshold laser intensities for OFI are given by eqns (18.13) and (18.14).

(e) For the case of linearly polarized light, calculate the threshold intensities for OFI of H, He$^+$, Ar^{7+} and Ar^{8+}, which have ionization energies of 13.6 eV, 54.4 eV, 143.4 eV and 422.5 eV, respectively. Comment on your results.

(18.6) Ponderomotive energy
In this problem we calculate the quiver energy of an electron oscillating in a laser field.

(a) Consider the motion of an electron born at $t = t_0$ into a linearly polarized laser field of the form $\boldsymbol{E}(t) = E_0\boldsymbol{i}\cos\omega_L t$. Show that if the electron is born at rest the velocity of the electron in the laser field is given by

$$v_x(t) = -\frac{eE_0}{m_e\omega_L}\sin\omega_L t + v_{drift},$$

where the drift velocity is given by,

$$v_{drift} = \frac{eE_0}{m_e\omega_L}\sin\phi_0,$$

where $\phi_0 = \omega_L t_0$.

(b) The quiver motion of the electron is that part of the electron's motion that is driven by the laser field. Show that the ponderomotive energy, the cycle-averaged quiver energy, is given by eqn (18.8):

$$U_p = \frac{\mu_0 e^2 c}{2m_e\omega_L^2}I_0 = \frac{r_e}{2\pi c}I_0\lambda_L^2,$$

where I_0 is the intensity of the laser radiation and $r_e = e^2/(4\pi m_e c^2)$ is the classical electron radius.

(c) Evaluate the ponderomotive energy of electrons oscillating in laser radiation with a wavelength of 1 μm for intensities of:
  (i) 10^{13} W cm^{-2};
  (ii) 10^{16} W cm^{-2}.

(d) Our treatment is non-relativistic. Estimate intensity of laser radiation of wavelength $\lambda_L = 1$ μm for the motion of the electron starts to become relativistic.

(e) Show that if the electron is born into a circularly polarized laser field the ponderomotive energy is also given by eqn (18.8).

[70]This approach follows that of Bethe and Salpeter (1977).

(18.7) **ATI energy**

In this problem we calculate the energy retained by the electrons after optical field ionization by a pulse of laser radiation. This energy is often called the above-threshold ionization (ATI) energy of the ionized electrons.[71]

(a) Use the results of Exercise 18.6 to show that if the electron is optically field ionized by a pulse of linearly polarized laser radiation the electron retains a kinetic energy given by eqn (18.16):

$$E_{\text{drift}}^{\text{lin}} = \frac{1}{2} \frac{e^2 E_0^2}{m_e \omega_L^2} \sin^2 \phi_0 = 2 U_{p0} \sin^2 \phi_0,$$

where U_{p0} is the ponderomotive energy of the laser field when the electron is ionized.

(b) Discuss whether this energy is likely to be large or small compared to U_{p0}.

(c) Repeat the calculation for the case of circularly polarized radiation, and show that in this case the electron retains a kinetic energy given by eqn (18.17):

$$E_{\text{drift}}^{\text{circ}} = U_{p0},$$

and comment on the magnitude of the energy in this case.

(18.8) **Inverse bremsstrahlung heating**

At low laser intensities the rate of heating, per electron, in a plasma of electrons and ions of charge Ze is given by:

$$R_{\text{IB}} = U_p \nu_{\text{ei}},$$

where U_p is the ponderomotive energy of the electron and the electron–ion collision frequency ν_{ei} is given by,

$$\nu_{\text{ei}} = \frac{4\sqrt{2\pi}}{3} \frac{Z N_e e^4 \ln \Lambda}{(4\pi \epsilon_0)^2 m_e^{1/2} (k_B T_e)^{3/2}},$$

in which T_e is the initial temperature of the electrons, and the so-called Coulomb logarithm is given by,

$$\ln \Lambda = \ln \left[\frac{1}{2Ze^3} \sqrt{\frac{(4\pi \epsilon_0 k_B T_e)^3}{\pi N_e (1+Z)}} \right].$$

(a) Use these results to calculate the rate of inverse bremsstrahlung heating in units of $\text{eV}\,\text{ps}^{-1}$ of a Ge^{21+} plasma with an electron density of $5 \times 10^{20}\,\text{cm}^{-3}$ and a temperature of 800 eV, for laser radiation with a wavelength of 1 μm and a laser intensity of $1.5 \times 10^{13}\,\text{W}\,\text{cm}^{-2}$.

(b) Hence, estimate how short the laser pulse would have to be if the plasma is *not* to be heated significantly by inverse bremsstrahlung.

(c) The rate of inverse bremsstrahlung heating is modified once the electron quiver velocity becomes greater than the thermal velocity. Estimate the laser intensity at which the rate of inverse bremsstrahlung heating will become modified for a plasma with an electron temperature of 10 eV.

(18.9) **Refraction in plasmas**

Here, we analyse a simple model of the effects of the refraction on both the pump and the generated short-wavelength radiation in a plasma with a gradient in density, using the geometry of Fig. 18.27.

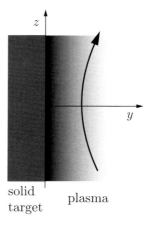

Fig. 18.27: Geometry required for the calculation of refraction effects in the plasma formed from a solid target.

We consider the simplest possible case: a plasma with a density that decreases linearly with distance from the

---

[71] If an atom or ion is ionized by a single photon the energy of the ionized electron is equal to $\hbar \omega_L - E_I$. Highly ionized atoms typically have ionization energies that are much greater than the photon energy (typically a few eV) of the incident laser radiation, and consequently require absorption of at least $N_{\min}$ photons, where $N_{\min} \hbar \omega_L \geq E_I$. If this minimum number of photons is absorbed the energy of the ionized electron is simply $E_{\min} = N_{\min} \hbar \omega_L - E_I$. Under optical field ionization, particularly with circularly polarized radiation, the ionized electrons are typically left with energies that are much greater than $E_{\min}$, corresponding to the absorption of more than $N_{\min}$ photons. The energy of the electrons above $E_{\min}$ is known as the ATI energy of the electron.

target according to

$$N_e(y) = N_{e0}\left(1 - \frac{y}{L_s}\right),$$

where the characteristic length L_s is known as the **scale-length** of the plasma.

In the paraxial approximation the equation describing the ray trajectories may be written as,[72]

$$n\frac{d^2\mathbf{r}}{dz^2} = \nabla n,$$

where $\mathbf{r}$ is a vector from the origin to a point on the trajectory.

(a) Show that for our model plasma the trajectory of the ray is given by,

$$\frac{d^2 y}{dz^2} \approx \frac{1}{2}\frac{\omega_{p0}^2}{\omega^2}\frac{1}{L_s}, \quad (18.40)$$

where ω_{p0} is the plasma frequency at the maximum electron density N_{e0} and we have taken n to be close to unity.

(b) Let us arrange the coordinate system so that $dy/dz = 0$ at $z = 0$. Show that the ray trajectory is then given by,

$$y = \frac{1}{4}\frac{\omega_{p0}^2}{\omega^2}\frac{z^2}{L_s} + c_1, \quad (18.41)$$

where c_1 is a constant.

(c) We may use this result to calculate the refraction of an X-ray generated in the plasma. Suppose that the X-ray initially propagates in a direction parallel to the target surface.
  (i) Show that the ray refracts a distance Δy away from the surface of the target after propagating a distance L along it, where:

$$L = 2\frac{\omega}{\omega_{p0}}\sqrt{\Delta y L_s}.$$

  (ii) Calculate the distance a ray of 20 nm radiation propagates before deflecting by $\Delta y = 10\,\mu\text{m}$ for a plasma with a peak electron density $N_{e0} = 10^{21}\,\text{cm}^{-3}$ and a scale length of 100 μm.

(d) The same results can also be used to analyse the propagation of the pump beam in a GRIP laser system.

In our model the boundary between the plasma and the surrounding vacuum occurs at $y = L_s$, since here the electron density is zero. At this point the incident rays of the pump laser make an angle θ to a line parallel to the z-axis, i.e. at this point $dy/dz = -\tan\theta$.
  (i) Use this information to show that the constant c_1 in eqn (18.41) is given by,

$$c_1 = L_s\left(1 - \frac{\omega^2}{\omega_{p0}^2}\tan^2\theta\right).$$

  (ii) Hence, show that at the turning point of the ray the electron density $N_e^{\max}$ is given by,

$$N_e^{\max} = N_c \tan^2\theta,$$

where N_c is the critical density at the wavelength of the pump laser. Compare this result with eqn (18.18).

(18.10) **High-harmonic generation**
In this problem we outline the argument of Corkum (1993) to derive an expression for the cut-off frequency in HHG.

(a) Use the approach of Exercises 18.6 and 18.7 to derive an expression for the position of an electron at time t moving in a linear polarized field, assuming it was ionized at time t_0.

(b) By considering the trajectories of the ionized electrons for different values of $\phi_0 = \omega_L t_0$, show that only half of the electrons that are ionized will re-collide with the atom (that is that they pass $x = 0$ again).

(c) Deduce an expression for the energy of the electron at the return time. Use a numerical method to find the value of ϕ_0 for which the return energy is a maximum. Show that this maximum energy is $3.17 U_{p0}$, where U_{p0} is the ponderomotive energy of the driving laser field at the moment that the electron was ionized, and hence show that the maximum photon energy in the harmonic spectrum is given by eqn (18.21).

(d) What does this imply for the minimum harmonic wavelength that can be generated by this interaction in He, using laser pulses of intensity $10^{15}\,\text{W cm}^{-2}$ at a wavelength of 800 nm? Approximately, what is the harmonic order q at the harmonic cut-off? [The ionization energy of helium is 24.58 eV.]

---

[72] See, for example, Born and Wolf (1999).

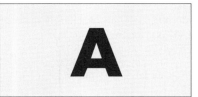

The semi-classical theory of the interaction of radiation and matter

A.1 The amplitude equations 548
A.2 Calculation of the Einstein B coefficient 551
A.3 Relations between the Einstein coefficients 555
A.4 Validity of rate equations 555

In Section 2.3 we presented the main results of the semi-classical model of the interaction of radiation and matter. Here, we discuss this model in more detail.

A.1 The amplitude equations

A.1.1 Derivation of the amplitude equations

We consider a simplified atom consisting of just two non-degenerate levels 1 and 2. In the absence of the radiation field, the energy eigenvalues E_1 and E_2, and wave functions Ψ_1 and Ψ_2 of the two levels are given by solutions of the time-dependent Schrödinger equation:

$$H_0 \Psi_i = i\hbar \frac{\partial \Psi_i}{\partial t}, \tag{A.1}$$

where H_0 is the Hamiltonian of the unperturbed atom and

$$\Psi_2(\boldsymbol{r}, t) = \phi_2(\boldsymbol{r}) e^{-iE_2 t/\hbar} \tag{A.2}$$

$$\Psi_1(\boldsymbol{r}, t) = \phi_1(\boldsymbol{r}) e^{-iE_1 t/\hbar}, \tag{A.3}$$

as discussed in Section 2.3.

Let us now suppose that the atom interacts with a plane harmonic electromagnetic wave, the electric field of which is given by

$$\boldsymbol{E}(\boldsymbol{r}, t) = \boldsymbol{E}_0 \cos(\boldsymbol{k} \cdot \boldsymbol{r} - \omega t), \tag{A.4}$$

where $\boldsymbol{E}_0$ is the peak electric field vector, ω the angular frequency, and the wave vector $\boldsymbol{k}$ has magnitude $k = \omega/c$ and points in the direction of propagation.

The electric field will interact with the instantaneous electric dipole moment $\boldsymbol{p}$ of the optically active electrons of the atom with an energy H' given by eqn (2.19). With the introduction of this interaction the time-dependent Schrödinger equation becomes,

$$\left(H_0 + H'\right)\Psi = i\hbar \frac{\partial \Psi}{\partial t}, \tag{A.5}$$

which will have new, time-dependent wave functions $\Psi(\mathbf{r}, t)$. For the simple two-level atom considered here, the wave functions $\Psi_1(\mathbf{r}, t)$ and $\Psi_2(\mathbf{r}, t)$ form a complete set of basis functions, in terms of which the new wave functions may be expressed as a superposition:[1]

$$\Psi(\mathbf{r}, t) = c_1(t)\Psi_1(\mathbf{r}, t) + c_2(t)\Psi_2(\mathbf{r}, t). \tag{A.6}$$

Note that the amplitudes $c_i(t)$ are also time dependent, but they must at all times satisfy the normalization condition,

$$|c_1(t)|^2 + |c_2(t)|^2 = 1. \tag{A.7}$$

At any moment the probability of finding the atom in level 1 is given by $|c_1(t)|^2$, whilst the probability that the atom is in level 2 is $|c_2(t)|^2$. Hence in order to deduce how population is transferred between the atomic levels, we must calculate $|c_1(t)|^2$ and $|c_2(t)|^2$.

To this end, substitution of eqn (A.6) into eqn (A.5) yields,

$$c_1 H'\phi_1 e^{-iE_1 t/\hbar} + c_2 H'\phi_2 e^{-iE_2 t/\hbar} = i\hbar \frac{dc_1}{dt}\phi_1 e^{-iE_1 t/\hbar} + i\hbar \frac{dc_2}{dt}\phi_2 e^{-iE_2 t/\hbar}, \tag{A.8}$$

where we have used eqn (A.1) and have dropped the time and spatial dependencies of all quantities to avoid clutter. Remembering that ϕ_1 and ϕ_2 are orthonormal, we can extract the term in dc_2/dt on the right-hand side of eqn (A.8) by multiplying by $\phi_2^* \exp(iE_2 t/\hbar)$ and integrating over all space to yield,

$$c_1 e^{i(E_2 - E_1)t/\hbar} \int \phi_2^* H' \phi_1 d\tau = i\hbar \frac{dc_2}{dt} \tag{A.9}$$

$$\Rightarrow c_1 e^{i(E_2 - E_1)t/\hbar} \langle 2| H' |1\rangle = i\hbar \frac{dc_2}{dt}, \tag{A.10}$$

where we have used Dirac's bra-ket notation $\langle 1| H' |2\rangle$ for the integral, together with the fact that $\langle 2| H' |2\rangle = 0$, as discussed in Section 2.3.2. Our result may be written more conveniently as,

$$i\hbar \frac{dc_2}{dt} = c_1 H_{21} e^{i(E_2 - E_1)t/\hbar} \cos \omega t, \tag{A.11}$$

where,

$$H_{21} = -\mathbf{D}_{21} \cdot \boldsymbol{\epsilon} E_0,$$
$$= -\mathbf{D}_{21} \cdot \hat{\boldsymbol{\epsilon}} E_0 \tag{A.12}$$

describes the strength of the interaction with the radiation field, and the electric dipole matrix element is given by,

$$\mathbf{D}_{21} = \langle 2| - e \sum_j \mathbf{r}_j |1\rangle. \tag{A.13}$$

[1] The wave functions ϕ_i depend on the coordinates of *all* the electrons and so should be written $\psi_i(\mathbf{r}_1, \mathbf{r}_2, \mathbf{r}_3 \ldots, \mathbf{r}_N, t)$, etc., where N is the number of electrons in the atom. As in Section 2.3, to avoid clutter we will simply let $\mathbf{r} \equiv \mathbf{r}_1, \mathbf{r}_2, \mathbf{r}_3 \ldots, \mathbf{r}_N$.

Returning to eqn (A.11) and writing the cosine in terms of a complex exponential we find,

$$i\hbar \frac{dc_2}{dt} = \frac{1}{2} c_1 H_{21} e^{i(\omega_{21}+\omega)t} + \frac{1}{2} c_1 H_{21} e^{i(\omega_{21}-\omega)t}, \quad (A.14)$$

where $\hbar\omega_{21} = (E_2 - E_1)$ is the energy spacing of the two levels.

If the radiation is reasonably close to resonance, i.e. $\omega \approx \omega_{21}$, the first term on the right-hand side oscillates extremely rapidly, whilst the second term oscillates at a frequency $\Delta\omega = \omega_{21} - \omega$ equal to the frequency detuning of the electromagnetic wave from the transition frequency. If we imagine integrating eqn (A.14) over time to find $c_2(t)$, we see that the first term will sum to approximately zero. Hence, providing that we are close to resonance we may make the so-called **rotating-wave approximation** and retain only the second term:

$$i\hbar \frac{dc_2}{dt} = \frac{1}{2} c_1 H_{21} e^{-i\Delta\omega t}. \quad (A.15)$$

In a similar way we can use eqn (A.8) to find the equivalent equation for $c_1(t)$:

$$i\hbar \frac{dc_1}{dt} = \frac{1}{2} c_2 H_{21}^* e^{i\Delta\omega t}. \quad (A.16)$$

Equations (A.15) and (A.16) describe, within the rotating-wave approximation, the interaction of radiation with the two-level atom.

A.1.2 Solution of the amplitude equations

We may find solutions to eqns. (A.15) and (A.16) straightforwardly by eliminating $c_1(t)$ to give the differential equation describing the evolution of the amplitude of the upper level:

$$\frac{d^2 c_2}{dt^2} + i\Delta\omega \frac{dc_2}{dt} + \frac{|H_{21}|^2}{4\hbar^2} c_2 = 0. \quad (A.17)$$

Substituting trial solutions of the form $c_2(t) \sim e^{st}$ we find, trivially,

$$s = -i\frac{\Delta\omega}{2} \pm i\frac{\Omega}{2} \quad (A.18)$$

where,

$$\Omega = \sqrt{\Delta\omega^2 + \omega_R^2}, \quad (A.19)$$

and the **Rabi frequency** ω_R is defined by,

$$\hbar\omega_R = |H_{21}|. \quad (A.20)$$

The Rabi frequency is an important parameter describing the interaction of atoms with radiation. It is a measure of the strength of the atom-field interaction, and the magnitude of the Rabi frequency relative to the frequency bandwidth of the radiation and the linewidth of the atomic transition is crucial in determining the manner in which the atom responds to the radiation field.

The solution of eqn (A.17) may therefore be written as,

$$c_2(t) = e^{i\Delta\omega t/2} \left(A e^{i\Omega t/2} + B e^{-i\Omega t/2} \right), \quad (A.21)$$

where the constants A and B are to be determined from the boundary conditions. If we consider the case when the atom is initially in the lower level, we have $c_2(0) = 0$, $c_1(0) = 1$, and, from eqn (A.15), $i\hbar \dot{c}_2(0) = H_{21}/2$. With these boundary conditions the population of the upper level is given by,

$$|c_2(t)|^2 = \frac{\omega_R^2}{\Omega^2} \sin^2\left(\frac{1}{2}\Omega t\right). \quad (A.22)$$

If the radiation field is exactly resonant with the transition frequency, the solution becomes

$$|c_2(t)|^2 = \sin^2\left(\frac{1}{2}\omega_R t\right). \quad (A.23)$$

This approach may be extended to describe the effect of finite decay rates and collisions on the dynamics of the atomic populations, yielding a generalized set of equations known as the optical Bloch equations.[2] An example of a solution to the Bloch equations is given in Fig. 2.7.

[2] Further details may be found in the textbook by Loudon (2000).

A.2 Calculation of the Einstein B coefficient

A.2.1 Polarized atoms and radiation

As discussed in Section 2.4.2, the Einstein description of the interaction of radiation and matter is valid when the intensity of the radiation is weak. By examining the above solutions in this limit we can therefore calculate an expression of the Einstein B coefficient in terms of the electric-dipole matrix element of the transition.

Within the semi-classical picture the interaction with the radiation is weak when

$$\omega_R \ll \Delta\omega, \quad (A.24)$$

in which case,

$$\Omega \approx \Delta\omega. \quad (A.25)$$

In this limit the solution for $|c_2(t)|^2$ becomes,

$$|c_2(t)|^2 = \frac{\omega_R^2}{\Delta\omega^2} \sin^2\left(\frac{1}{2}\Delta\omega t\right). \quad (A.26)$$

We want to extract the explicit dependence of $|c_2(t)|^2$ on strength of the radiation field, and so rewrite eqn (A.26) in the form,

$$|c_2(t)|^2 = \frac{|\mathbf{D}_{21} \cdot \widehat{\boldsymbol{\epsilon}}|^2}{4\hbar^2} E_0^2 t^2 \operatorname{sinc}^2\left(\frac{1}{2}\Delta\omega t\right). \quad (A.27)$$

The amplitude E_0 of the electric field is related to the mean energy density U of the radiation by,

$$U = \frac{1}{2}\epsilon_0 E_0^2, \tag{A.28}$$

and hence we have,

$$|c_2(t)|^2 = \frac{|\boldsymbol{D}_{21} \cdot \widehat{\boldsymbol{\epsilon}}|^2}{2\epsilon_0 \hbar^2} U t^2 \operatorname{sinc}^2\left(\frac{1}{2}\Delta\omega t\right). \tag{A.29}$$

A real beam of radiation will have a finite frequency bandwidth described by a *spectral* energy density $\varrho(\omega)$ such that the energy density of radiation with frequencies lying between ω and $\omega + \delta\omega$ is equal to $\varrho(\omega)\delta\omega$. Each frequency component will, independently, cause transitions to the upper level, and hence we must integrate the right-hand side of eqn (A.29) over all frequencies present in the beam:

$$|c_2(t)|^2 = \frac{|\boldsymbol{D}_{21} \cdot \widehat{\boldsymbol{\epsilon}}|^2}{2\epsilon_0 \hbar^2} t^2 \int_0^\infty \varrho(\omega) \operatorname{sinc}^2\left(\frac{1}{2}[\omega_{21} - \omega]t\right) d\omega. \tag{A.30}$$

Now, *provided that t is sufficiently large*, the function $\operatorname{sinc}^2\left[\frac{1}{2}(\omega_{21} - \omega)t\right]$ is strongly peaked around the transition frequency ω_{21}. For the moment, we take this to be the case, and also assume that the radiation field is sufficiently broadband that $\varrho(\omega)$ varies slowly compared to the sinc^2 function. We may then take $\varrho(\omega)$ outside the integral to find,

$$|c_2(t)|^2 = \frac{|\boldsymbol{D}_{21} \cdot \widehat{\boldsymbol{\epsilon}}|^2}{2\epsilon_0 \hbar^2} \varrho(\omega_{21}) t^2 \int_0^\infty \operatorname{sinc}^2\left(\frac{1}{2}[\omega_{21} - \omega]t\right) d\omega. \tag{A.31}$$

Making the substitution $x = \frac{1}{2}(\omega_{21} - \omega)t$, we find,

$$|c_2(t)|^2 = \frac{|\boldsymbol{D}_{21} \cdot \widehat{\boldsymbol{\epsilon}}|^2}{\epsilon_0 \hbar^2} \varrho(\omega_{21}) t \int_{-\infty}^{\omega_{21}t/2} \operatorname{sinc}^2 x \, dx \tag{A.32}$$

$$= \frac{\pi |\boldsymbol{D}_{21} \cdot \widehat{\boldsymbol{\epsilon}}|^2}{\epsilon_0 \hbar^2} \varrho(\omega_{21}) t, \tag{A.33}$$

where we have set the upper limit to ∞, and used the standard integral,

$$\int_{-\infty}^\infty \operatorname{sinc}^2(x) dx = \pi. \tag{A.34}$$

If we now consider an ensemble of such atoms, all initially in the lower level, the rate of excitation of the atoms is simply,[3]

$$N_1 \frac{|c_2(t)|^2}{t} = N_1 \frac{\pi |\boldsymbol{D}_{21} \cdot \widehat{\boldsymbol{\epsilon}}|^2}{\epsilon_0 \hbar^2} \varrho(\omega_{21}) \tag{A.35}$$

$$\equiv N_1 B_{12} \varrho(\omega_{21}). \tag{A.36}$$

---

[3] It is only after integrating eqn (A.29) over the bandwidth of the source that the probability of the atom being found in the upper level is proportional to time t, and it becomes possible to define a (time-independent) rate of excitation of the atom, and hence an Einstein B coefficient. In other words, the Einstein treatment necessarily assumes that the bandwidth of the radiation, or, equivalently, the atoms, is sufficiently large.

Hence, we have the following expression for the Einstein B coefficient:[4]

$$B_{12} = \frac{\pi |\boldsymbol{D}_{12} \cdot \hat{\boldsymbol{\epsilon}}|^2}{\epsilon_0 \hbar^2}. \qquad \text{Polarized} \qquad (A.37)$$

[4]We have written $|\boldsymbol{D}_{21} \cdot \hat{\boldsymbol{\epsilon}}|^2 = |\boldsymbol{D}_{12} \cdot \hat{\boldsymbol{\epsilon}}|^2$, which follows from the fact that $\boldsymbol{D}_{21} = \boldsymbol{D}_{12}^*$.

The value of the B coefficient depends on the relative orientation of the atomic dipole to the electric field of the electromagnetic wave through the term $\boldsymbol{D}_{21} \cdot \hat{\boldsymbol{\epsilon}}$, and consequently this result is only valid for the case when the relative orientations of the electric field and the axes of the atom are fixed.

To clarify this, suppose that the electric field of the incident radiation is parallel to the x-axis of the atomic coordinate system, so that $\hat{\boldsymbol{\epsilon}}$ is a unit vector along the x-direction. Then, in eqn (A.37) we have

$$\boldsymbol{D}_{12} \cdot \hat{\boldsymbol{\epsilon}} = X_{12} = \langle 1| - e \sum_j x_j |2\rangle. \qquad (A.38)$$

The value of this matrix element will, in general, be different to, say, Y_{12}, which would be the matrix determining the transition rate if the radiation were polarized parallel to the y-axis.

This scenario is important in some cases of importance in laser physics. For example, the Ti^{3+} ions in a Ti:sapphire laser rod (which are spatially aligned by virtue of the internal field of the crystal) have different values of the Einstein B coefficients depending on the orientation of the polarization of the radiation relative to the crystal axis.

A.2.2 Unpolarized atoms and/or radiation

We now consider the case where either or both the atoms or radiation are unpolarized.

Let us again suppose that the electric field of the radiation is polarized along the x-axis. Then,

$$X_{12} = D_{12} \cos \theta, \qquad (A.39)$$

where θ is the angle between the direction of $\boldsymbol{D}_{12}$ and the x-axis.

For an ensemble of randomly oriented atoms[5] we must average eqn (A.37) over θ. Hence,

$$B_{12} \rightarrow \frac{\pi}{\epsilon_0 \hbar^2} D_{12}^2 \overline{\cos^2 \theta},$$

where the bar indicates an average. Now, $\overline{\cos^2 \theta} = 1/3$, and hence we have,

$$B_{12} = \frac{\pi D_{12}^2}{3\epsilon_0^2}. \qquad \text{Unpolarized} \qquad (A.40)$$

It should be clear that eqn (A.40) also holds if the radiation is unpolarized, irrespective of whether or not the axes of the atoms are aligned.[6]

[5]For example, the atoms of a gas that is not subject to external magnetic or electric fields.

[6]The key quantum-mechanical quantity appearing in the expression for the Einstein B coefficient is the square modulus of the electric dipole moment:

$$D_{12}^2 = X_{12}^2 + Y_{12}^2 + Z_{12}^2.$$

Very frequently this result is written as:

$$D_{12}^2 = |\langle 1| - e \sum_j \boldsymbol{r}_j |2\rangle|^2.$$

A.2.3 Treatment of degeneracy

We now consider how our results must be modified when either or both of the upper or lower levels is degenerate, i.e. the level is composed of several *states* of the same energy. We suppose that for each level i the degenerate states are labelled by an additional quantum number m_i.[7] We take the wave function of the state to be written as $\phi_i^{m_i}$ or, in Dirac notation, as $|im_i\rangle$.

[7] For example, m_i might represent the z-component of the total angular momentum of the level.

Then, the key quantity appearing in the expression for the Einstein B coefficient becomes

$$\left|\boldsymbol{D}_{12}\cdot\hat{\boldsymbol{\epsilon}}\right|^2 = |\langle 1m_1| - e\sum_j \boldsymbol{r}_j |2m_2\rangle \cdot \hat{\boldsymbol{\epsilon}}|^2. \qquad (A.41)$$

The total transition rate from a given state, m_1, of level 1 to all possible states of level 2, is found by summing eqn (A.41) over all the states of level 2, i.e. the rate is proportional to

$$\sum_{m_2}\left|\langle 1m_1| - e\sum_j \boldsymbol{r}_j |2m_2\rangle \cdot \hat{\boldsymbol{\epsilon}}\right|^2. \qquad (A.42)$$

Unless the atoms have been specially prepared in some way, atoms in level 1 have an equal probability of being in any particular state $|1m_1\rangle$. In other words, the probability that an atom in level 1 is in the particular state $|1m_1\rangle$ is $\frac{1}{g_1}$, where g_1 is equal to the number of degenerate states in level 1. Hence, the total transition rate from level 1 to level 2 is given by summing eqn (A.42) over the states of level 1 and multiplying by $\frac{1}{g_1}$, i.e. a rate proportional to

$$\frac{1}{g_1}\sum_{m_1}\sum_{m_2}\left|\langle 1m_1| - e\sum_j \boldsymbol{r}_j |m_2\rangle \cdot \hat{\boldsymbol{\epsilon}}\right|^2.$$

This expression should replace the quantity $|\boldsymbol{D}_{12}\cdot\hat{\boldsymbol{\epsilon}}|^2$ appearing in the previous expressions found for non-degenerate atoms, yielding:

$$g_1 B_{12} = \frac{\pi}{\epsilon_0 \hbar^2}\sum_{m_1}\sum_{m_2}\left|\langle 1m_1| - e\sum_j \boldsymbol{r}_j |2m_2\rangle \cdot \hat{\boldsymbol{\epsilon}}\right|^2$$

Polarized (A.43)

$$g_1 B_{12} = \frac{1}{3}\frac{\pi}{\epsilon_0 \hbar^2}\sum_{m_1}\sum_{m_2}\left|\langle 1m_1| - e\sum_j \boldsymbol{r}_j |2m_2\rangle\right|^2.$$

Unpolarized (A.44)

A.3 Relations between the Einstein coefficients

We can obtain an equivalent expressions for B_{21} immediately by simply swapping the labels $2 \longleftrightarrow 1$. Given that $\boldsymbol{D}_{21} = \boldsymbol{D}_{12}^*$ we see immediately that

$$g_1 B_{12} = g_2 B_{21}. \tag{A.45}$$

This last result—which must hold whether or not the atoms are degenerate, and whatever the polarization of the atoms or radiation— agrees with eqn. (2.7), obtained using the thermodynamic argument of Einstein.

In Section 2.1 we deduced a relation between the Einstein A and B coefficients by considering an atom in thermal equilibrium with blackbody radiation. Blackbody radiation is weak, broadband, and unpolarized and hence the correct quantum-mechanical expression for the Einstein B coefficient in this case is eqn (A.44). We can now use eqn (2.6) to deduce the semi-classical expression for the Einstein A coefficient:

$$g_2 A_{21} = \frac{\omega_{21}^3}{3\pi \epsilon_0 \hbar c^3} \sum_{m_1} \sum_{m_2} \left| \langle 2m_2 | -e \sum_j \boldsymbol{r}_j | 1m_1 \rangle \right|^2. \tag{A.46}$$

A.4 Validity of rate equations

We briefly explore whether the condition, given in Section 2.4.3, for the rate-equation approach to be valid is likely to be satisfied under the operating conditions of a laser. For simplicity we will ignore degeneracy, i.e. assume $g_1 = g_2$, and consider the case of narrowband radiation interacting with collisionally broadened atoms.

From eqn (A.37) we have

$$B_{12} = \frac{\pi \left| \boldsymbol{D}_{12} \cdot \hat{\boldsymbol{\epsilon}} \right|^2}{\epsilon_0 \hbar^2}, \tag{A.47}$$

and hence we may write

$$B_{12} E_0^2 = \frac{\pi \left| \boldsymbol{D}_{12} \cdot \boldsymbol{E_0} \right|^2}{\epsilon_0 \hbar^2} = \frac{\pi}{\epsilon_0 \hbar^2} (\hbar \omega_R)^2. \tag{A.48}$$

Thus, we have,

$$\omega_R^2 = \frac{2 B_{12} I}{\pi c}, \tag{A.49}$$

where we have used $I = (1/2)\epsilon_0 c E_0^2$. The condition (2.32) for the validity of the rate-equation approach may therefore be written

$$\frac{2 B_{12} I}{\pi c} \ll \Delta \omega'^2, \tag{A.50}$$

where, for the case we are considering, $\Delta\omega'$ is the linewidth arising from collisions. This may be rearranged to give

$$\frac{2}{\pi g(0)} \frac{\sigma_{21}(0) I}{\hbar\omega} \ll \Delta\omega'^2, \quad (A.51)$$

where $\sigma_{21}(0)$ and $g(0)$ are, respectively, the gain cross-section and normalized lineshape evaluated at resonance ($\omega = \omega_0$). For a homogeneously broadened transition $g(0) = 2/\pi\Delta\omega_\text{H}$, and the rate equations will be valid provided

$$I \ll I_\text{s}(0) \left(\Delta\omega_\text{H}\tau_\text{R}\right), \quad (A.52)$$

where we have assumed $\Delta\omega_\text{H} \approx \Delta\omega'$, and $I_\text{s}(0) = \hbar\omega/\sigma_{21}(0)\tau_\text{R}$ is the saturation intensity evaluated at line centre, and τ_R is the recovery time.

The recovery time (eqn (5.8)) will be of the order of the fluorescence lifetimes of the upper and lower laser levels. In the absence of collisions the transition would be purely lifetime broadened, in which case $\Delta\omega_\text{H} = \Delta\omega_\text{N} \equiv 1/\tau_2 + 1/\tau_1$. In such circumstances $\Delta\omega_\text{H}\tau_\text{R} \approx 1$, and the rate equations would become invalid for radiation intensities of the order of the saturation intensity. *However*, under the typical operating conditions of a laser collisions with other atoms, with electrons, or with phonons ensure that $\Delta\omega_\text{H} \gg \Delta\omega_\text{N}$; in this case the rate equations are valid even for intensities significantly larger than the saturation intensity.

The spectral Einstein coefficients

B

For homogeneously broadened atoms, all atoms in a given level interact with radiation of angular frequency ω with the same strength. However, the strength of this interaction will depend on the extent of the frequency detuning from the central frequency of the transition ω_0. Hence, in modifying eqn (3.38) to account for homogeneous broadening, the population densities N_1 and N_2 should not be changed, but, instead the Einstein B coefficients should somehow account for the detuning from ω_0. We can accomplish this by introducing the **spectral Einstein B coefficients** a_{21}, b_{21}, and b_{12}, which can be defined by:

1. The rate at which an atom in the upper level decays to the lower level by spontaneous emission of photons with angular frequencies lying in the range ω to $\omega + \delta\omega$ is equal to $N_2 a_{21}(\omega - \omega_0)\delta\omega$, where N_2 is the density of atoms in the upper level, and ω_0 the centre frequency of the transition.
2. The rate at which atoms in the lower level are excited to the upper level by the absorption of photons with angular frequencies lying in the range ω to $\omega + \delta\omega$ is equal to $N_1 b_{12}(\omega - \omega_0)\varrho(\omega)\delta\omega$, where N_1 is the density of atoms in the lower level, and $\varrho(\omega)$ is the energy density of the radiation.
3. The rate at which atoms in the upper level decay to the lower level by the stimulated emission of photons with angular frequencies lying in the range ω to $\omega + \delta\omega$ is equal to $N_2 b_{21}(\omega - \omega_0)\varrho(\omega)\delta\omega$.

We can establish relations between the spectral Einstein coefficients in much the same way as we did for the ordinary coefficients. We imagine our homogeneously broadened atoms in thermal equilibrium with a blackbody radiation field of temperature T. Then, for each transition $2 \rightarrow 1$, the rate at which the atoms absorb radiation with frequencies in the range ω to $\omega + \delta\omega$ must be balanced by the rate at which the atoms emit such frequencies. Hence, we may write,

$$N_1 b_{12}(\omega - \omega_0)\varrho_B(\omega)\delta\omega = N_2 b_{21}(\omega - \omega_0)\varrho_B(\omega)\delta\omega + N_2 a_{21}(\omega - \omega_0)\delta\omega, \quad \text{(B.1)}$$

where $\varrho_B(\omega)$ is the energy density of the blackbody radiation field. Using the approach employed in Section 2.1.1, it is straightforward to show that,

$$a_{21}(\omega - \omega_0) = \frac{\hbar\omega^3}{\pi^2 c^3} b_{21}(\omega - \omega_0) \tag{B.2}$$

$$g_2 b_{21}(\omega - \omega_0) = g_1 b_{12}(\omega - \omega_0), \tag{B.3}$$

where g_2 and g_1 are the degeneracies of the upper and lower levels, respectively. The relations between the spectral Einstein coefficients are clearly analogous to those for the ordinary coefficients.

Determining the frequency dependence

As yet we have merely defined a new, spectral version of the Einstein coefficients, but we have not established the frequency dependence of these, or how that frequency dependence is related to the 'lineshape' of homogeneously broadened transitions that was discussed in Section 3.1. In order to make progress we first relate the spectral Einstein coefficients to the ordinary ones.

The total rate of spontaneous emission over all frequencies must be equal to the ordinary Einstein A coefficient:

$$\int_0^\infty a_{21}(\omega - \omega_0) d\omega = A_{21}. \tag{B.4}$$

Hence we can write,

$$a_{21}(\omega - \omega_0) = A_{21} g_a(\omega - \omega_0), \tag{B.5}$$

where the lineshape $g_a(\omega - \omega_0)$ contains the frequency dependence of a_{21} and is normalized so that $\int_0^\infty g_a(\omega - \omega_0) d\omega = 1$.

Similarly, if we imagine placing the atoms in a very broadband radiation field of constant spectral energy density ϱ, then the total rate of stimulated emission from atoms in the upper level must equal $N_2 B_{21} \varrho$:

$$\int_0^\infty b_{21}(\omega - \omega_0) \varrho \, d\omega = B_{21} \varrho. \tag{B.6}$$

Hence, we may write,

$$b_{21}(\omega - \omega_0) = B_{21} g_b(\omega - \omega_0), \tag{B.7}$$

where $g_b(\omega - \omega_0)$ is the normalized lineshape of b_{21}.

We can relate the two lineshapes with the aid of eqn (B.2):

$$A_{21} g_a(\omega - \omega_0) = \frac{\hbar\omega^3}{\pi^2 c^3} B_{21} g_b(\omega - \omega_0). \tag{B.8}$$

We see from eqn (B.3) that b_{12} and b_{21} have the same frequency dependence, but eqn (B.8) tells us that this is not (strictly) the same as that of a_{21} owing to the ω^3 term.

In passing, we note that by integrating eqn (B.8) over all frequencies we find,

$$A_{21} = \frac{\overline{\hbar\omega^3}}{\pi^2 c^3} B_{21}, \tag{B.9}$$

where $\overline{\omega^3}$ is the mean cube angular frequency averaged over the lineshape of b_{21}. Clearly for all but extremely broad transitions $\overline{\omega^3} \approx \omega_0^3$, and our result reduces to that obtained earlier.

Having rather carefully established that the lineshapes of the a and b coefficients are not quite the same, we note that for virtually all cases of interest, the homogeneous linewidth is small compared to ω_0 and hence the differences in the frequency dependence of the a and b coefficients is negligible. Henceforth, we shall talk only of the 'homogeneous lineshape' $g_H(\omega - \omega_0)$, which is simply the normalized lineshape of the measured emission or absorption spectrum.

C Kleinman's conjecture

The components of the non-linear polarization $P_i^{\mathrm{NL}}(\mathbf{r}, t)$ are, from eqns (15.32) and (15.33),

$$\begin{pmatrix} P_x^{\mathrm{NL}} \\ P_y^{\mathrm{NL}} \\ P_z^{\mathrm{NL}} \end{pmatrix} = \epsilon_0 \begin{pmatrix} d_{11} & d_{12} & d_{13} & d_{14} & d_{15} & d_{16} \\ d_{21} & d_{22} & d_{23} & d_{24} & d_{25} & d_{26} \\ d_{31} & d_{32} & d_{33} & d_{34} & d_{35} & d_{36} \end{pmatrix} \begin{pmatrix} E_x^2 \\ E_y^2 \\ E_z^2 \\ 2E_y E_z \\ 2E_x E_z \\ 2E_x E_y \end{pmatrix}, \quad (\mathrm{C}.1)$$

where we have compacted the notation by omitting the superscripts, and by substituting x and xx by 1, y and yy by 2, z and zz by 3, yz by 4, zx by 5, and xy by 6 for the second and third subscripts of d_{ijk}.

Further, if the polarization induced by the electric field arises entirely due to the displacement of electrons from their equilibrium positions rather than the displacement of ions, and if the crystal shows negligible optical loss at all frequencies concerned, then Kleinman's conjecture (Boyd *et al.* (1965)) is valid and the following symmetries apply:

$$d_{21} = d_{16}, \qquad d_{24} = d_{32}, \qquad d_{31} = d_{15},$$
$$d_{13} = d_{35}, \qquad d_{14} = d_{36} = d_{25}, \qquad d_{12} = d_{26},$$
$$d_{32} = d_{24}.$$

Thus, there are at most only 10 independent coefficients:

$$\begin{pmatrix} d_{11} & d_{12} & d_{13} & d_{14} & d_{15} & d_{16} \\ d_{16} & d_{22} & d_{23} & d_{24} & d_{14} & d_{12} \\ d_{15} & d_{24} & d_{33} & d_{23} & d_{13} & d_{14} \end{pmatrix}. \quad (\mathrm{C}.2)$$

In the 3×6 matrix for the crystals of interest for non-linear optics most of the elements are zero. The matrices have been tabulated for all non-centrosymmetric crystals.

It should also be noted that while E_x, E_y and E_z (and hence $\widehat{E}_{1j}$, $\widehat{E}_{2k}$, etc.) refer to the components of $\mathbf{E}$ with respect to the principal axes of the crystal, the electric-field components of interest in applying eqn (15.34) are referred to the direction of $\mathbf{k}_3$ and $\mathbf{l}$. If, for example, the fundamental beams are both ordinary waves with $\boldsymbol{E}$ perpendicular to the optic axis z, then in terms of θ and ϕ defined in Fig. 15.2:

$$E_x = E \cos\left(\phi - \frac{\pi}{2}\right) = E \sin\phi; \tag{C.3}$$

$$E_y = E \sin\left(\phi - \frac{\pi}{2}\right) = -E \cos\phi; \tag{C.4}$$

$$E_z = 0. \tag{C.5}$$

Bibliography

Abedin, K. M., Alvarez, M., Costela, A., Garcia-Moreno, I., Garcia, O., Sastre, R., Coutts, D. W., and Webb, C. E. (2003). 10 kHz repetition rate solid-state dye laser pumped by diode-pumped solid-state laser. *Optics Communications*, **218**(4–6), 359–363.

Adair, R., Chase, L. L., and Payne, S. A. (1989). Nonlinear refractive-index of optical-crystals. *Physical Review B*, **39**(5), 3337–3350.

Agostini, P. and DiMauro, L. F. (2004). The physics of attosecond light pulses. *Reports on Progress in Physics*, **67**(8), 1563.

Agostini, P. and DiMauro, L. F. (2004). The physics of attosecond light pulses. *Reports on Progress in Physics*, **67**(6), 813.

Allen, C. W. (1963). *Astrophysical quantities* (2nd edn). Athlone Press, London.

Alvarez, M., Amat-Guerri, F., Costela, A., Garcia-Moreno, I., Gomez, C., Liras, M., and Sastre, R. (2005). Linear and cross-linked polymeric solid-state dye lasers based on 8-substituted alkyl analogues of pyrromethene 567. *Applied Physics B-Lasers and Optics*, **80**(8), 993–1006.

Alvarez-Chavez, J. A., Offerhaus, H. L., Nilsson, J., Turner, P. W., Clarkson, W. A., and Richardson, D. J. (2000). High-energy, high-power ytterbium-doped q-switched fiber laser. *Optics Letters*, **25**(1), 37–39.

Amendt, P., Strauss, M., and London, R. A. (1996). Plasma fluctuations and X-ray laser transverse coherence. *Physical Review A*, **53**(1), R23–R26.

Andalkar, A. and Warrington, R. B. (2002). High-resolution measurement of the pressure broadening and shift of the Cs D1 and D2 lines by N_2 and He buffer gases. *Physical Review A*, **65**(3), 032708.

Anderson, J. D., Jr (1976). *Gas dynamic lasers—An introduction*. Academic Press, New York.

Armstrong, J. A., Bloembergen, N., Ducuing, J., and Pershan, P. S. (1962). Interactions between light waves in a nonlinear dielectric. *Physical Review*, **127**(6), 1918–1939.

Ashcroft, N. W. and Mermin, N. D. (1976). *Solid state physics* (1st edn). Saunders College, Philadelphia.

Attwood, D. (1999). *Soft X-rays and extreme ultraviolet radiation: Principles and applications* (1st edn). Cambridge University Press, Cambridge.

Auzel, F., Bonfigli, F., Gagliari, S., and Baldacchini, G. (2001). The interplay of self-trapping and self-quenching for resonant transitions in solids; role of a cavity. *Journal of Luminescence*, **94**, 293–297.

Backus, S., Durfee, C. G., Murnane, M. M., and Kapteyn, H. C. (1998). High power ultrafast lasers. *Review of Scientific Instruments*, **69**(3), 1207–1223.

Bagley, M., Wyatt, R., Elton, D. J., Wickers, H. J., Spurdens, P. C., Seltzer, C. P., Cooper, D. M., and Devlin, W. J. (1990). 242 nm continuous tuning from a GRIN-SC-MQW-BH InGaAsP laser in an extended cavity. *Electronics Letters*, **26**, 267–269.

Baltuska, A., Uiberacker, M., Goulielmakis, E., Kienberger, R., Yakovlev, V. S., Udem, T., Hansch, T. W., and Krausz, F. (2003). Phase-controlled amplification of few-cycle laser pulses. *IEEE Journal of Selected Topics in Quantum Electronics*, **9**(4), 972–989.

Banine, V. and Moors, R. (2004). Plasma sources for EUV lithography exposure tools. *Journal of Physics D: Applied Physics*, **37**(23), 3207–3212.

Bart van Zeghbroek, J. B. (1996). *Principles of electron devices*. University of Colorado, Boulder, Colorado.

Barwood, G. P. and Gill, P. (2004). C3.3 Laser stabilization for precision measurements. In *The handbook of laser technology and applications* (ed. C. E. Webb and J. Jones). Institute of Physics Publishing, Bristol.

Basiev, T. T. and Powell, Richard C. (2004). B1.7 Solid state Raman lasers. In *The handbook of laser technology and applications* (ed. C. E. Webb and J. Jones). Institute of Physics, Bristol.

Basov, N. G., Danilychev, V. A., and Popov, Yu. M. (1971). Stimulated emission in the vacuum ultraviolet region. *Soviet Journal of Quantum Electronics*, **1**, 18–22.

Basov, N. G., Vul, B. M., and Yu, M. Popov (1959). Quantum-mechanical semiconductor generators and amplifiers of electromagnetic oscillations. *Soviet JETP Pis'ma*, **37**, 587.

Beaulieu, A. J. (1970). Transversely excited atmospheric pressure co_2 layers. *Applied Physics Letters*, **16**, 504–5.

Bennett, W. R., Knutson, J. W., Mercer, G. N., and Detch, J. L. (1964). Super-radiance, excitation mechanisms, and quasi-c.w. oscillation in the visible Ar^+ laser. *Applied Physics Letters*, **4**, 180.

Bernard, M. G. A. and Duraffourg, G. (1961). Laser conditions in semiconductors. *Physica Status Solidi*, **1**(7), 699–703.

Bethe, Hans A. and Salpeter, Edwin E. (1977). *Quantum mechanics of one- and two-electron atoms*. Plenum, New York.

Bigot, L., Choblet, S., Jurdyc, A. M., Jacquier, B., and Adam, J. L. (2004). Transient spectral hole burning in erbium-doped fluoride glasses. *Journal of the Optical Society of America B-Optical Physics*, **21**(2), 307–312.

Binks, D. J. (2004). C3.1 Harmonic generation—materials and methods. In *The handbook of laser technology and applications* (ed. C. E. Webb and J. Jones). Institute of Physics, Bristol.

Birch, K. P. and Downs, M. J. (1994). Correction to the updated Edlén equation for the refractive index of air. *Metrologia*, **31**, 315–316.

Bleaney, B. I. and Bleaney, B. (1989). *Electricity and Magnetism* (3rd edn). Oxford University Press, Oxford.

Born, M., Wolf, E., and Bhatia, A. B. (1999). *Principles of optics: electromagnetic theory of propagation, interference and diffraction of light* (7th (expanded) edn). Cambridge University Press, Cambridge.

Boyd, G. D., Ashkin, A., Dziedzic, J. M., and Kleinman, D. A. (1965). Second-harmonic generation of light with double refraction. *Physical Review*, **137**(4A), 1305.

Boyd, G. D. and Gordon, J. P. (1961). Confocal multimode resonator for millimeter through optical wavelength masers. *Bell System Technical Journal*, **40**, 489–508.

Boyd, G. D. and Kleinman, D. A. (1968). Parametric interaction of focused Gaussian light beams, *Journal of Applied Physics*, **39**, 3597–3639.

Boyd, R. (2004). A4 Nonlinear optics. In *The handbook of laser technology and applications* (ed. C. E. Webb and J. Jones). Institute of Physics, Bristol.

Bransden, B. H. and Joachain, C. J. (2003). *Physics of atoms and molecules* (2nd ed. edn). Prentice Hall, Harlow.

Brau, Charles A. (1979). Rare gas halogen excimers. In *Topics in applied physics: Excimer lasers* (ed. C. K. Rhodes), Volume 30, pp. 87–133. Springer-Verlag, New York.

Brau, C. A. and Ewing, J. J. (1975). 354-nm laser action on XeF. *Applied Physics Letters*, **27**(8), 435–437.

Bridges, W. B. (1964). Laser oscillation in singly ionized argon in the visible spectrum. *Applied Physics Letters*, **4**, 128.

Brooker, Geoffrey (2003). *Modern classical optics*. Oxford master series in physics. Oxford University Press, Oxford.

Brovelli, L. R. and Keller, U. (1995). Design and operation of antiresonant Fabry–Perot saturable semiconductor absorbers for mode-locked solid-state lasers. *Journal of the Optical Society of America B*, **12**, 311–322.

Burdack, P., Fox, T., Bode, M., and Freitag, I. (2006). 1 W of stable single-frequency output at 1.03 μm from a novel, monolithic, non-planar Yb:YAG ring laser operating at room temperature. *Optics Express*, **14**(10), 4363–4367.

Burge, R. E., Slark, G. E., Browne, M. T., Yuan, X. C., Charalambous, P., Cheng, X. H., Lewis, C. L. S., Cairns, G. F., MacPhee, A. G., and Neely, D. (1998). Dependence of spatial coherence of 23.2-23.6-nm radiation on the geometry of a multielement germanium X-ray laser target. *Journal of the Optical Society of America B-Optical Physics*, **15**(10), 2515–2523.

Burgess, A. and Tully, J. A. (1978). Bethe approximation. *Journal of Physics B-Atomic Molecular and Optical Physics*, **11**(24), 4271–4282.

Burnett, N. H. and Corkum, P. B. (1989). Cold-plasma production for recombination extreme-ultraviolet lasers by optical-field-induced ionization. *Journal of the Optical Society of America B-Optical Physics*, **6**(6), 1195–1199.

Burns, Gerald (1990). *Solid state physics*. Academic Press, Boston; London.

Butcher, Paul N. and Cotter, David (1990). *The elements of nonlinear optics*. Cambridge studies in modern optics. Cambridge University Press, Cambridge.

Butler, A., Gonsalves, A. J., McKenna, C. M., Spence, D. J., Hooker, S. M., Sebban, S., Mocek, T., Bettaibi, I., and Cros, B. (2003). Demonstration of a collisionally excited optical-field-ionization XUV laser driven in a plasma waveguide. *Physical Review Letters*, **91**(20), 205001.

Calvert, Jack G. and Pitts, James N. (1966). *Photochemistry*. Wiley, New York.

Celliers, P., Weber, F., Dasilva, L. B., Barbee, T. W., Cauble, R., Wan, A. S., and Moreno, J. C. (1995). Fringe formation and coherence of a soft-X-ray laser-beam illuminating a Mach–Zehnder interferometer. *Optics Letters*, **20**(18), 1907–1909.

Cennamo, G. and Forte, R. (2004). D3.2.2 Therapeutic applications: Refractive surgery. In *The handbook of laser technology and applications* (ed. C. E. Webb and J. Jones). Institute of Physics Publishing, Bristol.

Cerullo, G., Desilvestri, S., and Magni, V. (1994). Self-starting Kerr-lens mode-locking of a Ti-sapphire laser. *Optics Letters*, **19**(14), 1040–1042.

Chanteloup, J. C., Salmon, E., Sauteret, C., Migus, A., Zeitoun, P., Klisnick, A., Carillon, A., Hubert, S., Ros, D., Nickles, P., and Kalachnikov, M. (2000). Pulse-front control of 15-TW pulses with a tilted compressor, and application to the subpicosecond traveling-wave pumping of a soft-X-ray laser. *Journal of the Optical Society of America B-Optical Physics*, **17**(1), 151–157.

Cheriaux, G., Rousseau, P., Salin, F., Chambaret, J. P., Walker, B., and Dimauro, L. F. (1996). Aberration-free stretcher design for ultrashort-pulse amplification. *Optics Letters*, **21**(6), 414–416.

Chester, A. N. (1968). Experimental measurements of gas pumping in an argon discharge. *Physical Review*, **169**, 184–193.

Chou, M. C., Lin, P. H., Lin, C. A., Lin, J. Y., Wang, J., and Chen, S. Y. (2007). Dramatic enhancement of optical-field-ionization collisional-excitation X-ray lasing by an optically preformed plasma waveguide. *Physical Review Letters*, **99**(6), 063904.

Coaton, J. R. and Marsden, A. M. (1997). *Lamps and lighting* (4th edn). Arnold, London.

Coldren, L. A. and Corzine, S. W. (1995). *Diode lasers and photonic integrated circuits*. Wiley series in microwave and optical engineering. Wiley, New York; Chichester.

Coleman, J. J. (2000). Strained-layer InGaAs quantum-well heterostructure lasers. *IEEE Journal of Selected Topics in Quantum Electronics*, **6**(6), 1008–1013.

Convert, G., Armand, M., Lagarde, P. Martinot, and Bridges, W. B. (1964). Transitions laser visibles dans l'argon ionise. *C. R. Acad. Sci.*, **258**, 4467.

Cordina, Kevin (2004). B4.5 Erbium and other doped fibre amplifiers. In *The handbook of laser technology and applications* (ed. C. E. Webb and J. Jones). Institute of Physics, Bristol.

Corkum, P.B. and Burnett, N. H. (1988). Multiphoton ionization for the production of X-ray laser plasmas. In *OSA Proceedings on Short Wavelength Coherent Radiation: Generation and Applications* (ed. R. W. Falcone and J. Kirz), Volume 2, pp. 225. Optical Society of America.

Corkum, P. B. (1993). Plasma perspective on strong-field multiphoton ionization. *Physical Review Letters*, **71**(13), 1994–1997.

Cornacchia, F., Di Lieto, A., Tonelli, M., Richter, A., Heumann, E., and Huber, G. (2008). Efficient visible laser emission of GaN laser diode pumped PR-doped fluoride scheelite crystals. *Optics Express*, **16**(20), 15932–15941.

Corney, Alan (1978). *Atomic and laser spectroscopy*. Clarendon Press, Oxford.

Cotton, R. A. and Webb, C. E. (1993). 'Soft X-ray contact microscopy using laser generated plasma sources. In *SPIE Conference on Soft X-ray microscopy*, Volume 1741, 142–153.

Coutts, J. and Webb, C. E. (1986). Stability of transverse self-sustained discharge-excited long-pulse XeCl lasers. *Journal of Applied Physics*, **59**(3), 704–710.

Cronin, A., McAtamney, C., Sherlock, R., O'Connor, G. M., and J., Glynn T. (2005). Laser-based workstation for the manufacture of fused biconical tapered coupler devices. *Proceedings SPIE*, **5287**, 505–514.

Daido, H. (2002). Review of soft X-ray laser researches and developments. *Reports on Progress in Physics*, **65**(10), 1513–1576.

Daido, H., Kodama, R., Murai, K., Yuan, G., Takagi, M., Kato, Y., Choi, I. W., and Nam, C. H. (1995). Significant improvement in the efficiency and brightness of the $j = 0 - 1$ 19.6 nm line of the germanium laser by use of double-pulse pumping. *Optics Letters*, **20**(1), 61–63.

David, N. and Hooker, S. M. (2003). Molecular-dynamic calculation of the relaxation of the electron energy distribution function in a plasma. *Physical Review E*, **68**(5), 056401.

Davis, Christopher C. (2000). *Lasers and electro-optics: fundamentals and engineering*. Cambridge University Press, Cambridge.

De Silvestri, S., Laporta, P., Magni, V., and Svelto, O. (1988). Solid-state laser unstable resonators with tapered reflectivity mirrors—the super-Gaussian approach. *IEEE Journal of Quantum Electronics*, **24**(6), 1172–1177.

DeMaria, A. J. (1973). High power CO_2 lasers. *Proceedings of the IEEE*, **61**, 731–748.

Desurvire, E., Simpson, J. R., and Becker, P. C. (1987). High-gain erbium-doped traveling-wave fiber amplifier. *Optics Letters*, **12**(11), 888–890.

Dexter, D. L. (1953). A theory of sensitized luminescence in solids. *Journal of Chemical Physics*, **21**, 836–850.

Dimitrev, V. G., Gurzadyan, G. G., and Nikogoyan, D. N. (1991). *Handbook of nonlinear optical crystals*. Springer, Berlin.

Drexhage, K.H. (1990). Structure and properties of laser dyes. In *Dye lasers* (3rd edn) (ed. F. P. Schäfer), Volume 1 of *Topics in applied physics*. Springer-Verlag, Berlin; New York.

Du Puis, R. D. (2004). The diode laser: The first 30 days 40 years ago. *Optics and Photonics News*, **15**, 30–35.

Duarte, F. J. (1991). *High-power dye lasers*. Springer series in optical sciences; v. 65. Springer-Verlag, Berlin; London.

Dubietis, A., Butkus, R., and Piskarskas, A. P. (2006). Trends in chirped pulse optical parametric amplification. *IEEE Journal of Selected Topics in Quantum Electronics*, **12**(2), 163–172.

Dumke, W. P. (1962). Interband transitions and maser action. *Physical Review*, **127**, 1559–1563.

Dunn, M. H. and Guttierrez, A. (2004). B3.5 Argon and krypton ion lasers. In *The handbook of laser technology and applications* (ed. C. E. Webb and J. Jones). Institute of Physics, Bristol.

Dunn, M. H. and Ross, J. N. (1976). The argon ion laser. *Progress in Quantum Electronics*, **4**, 233–270.

Dutta, N. K. (2004). B2.1 Semiconductor lasers. In *The handbook of laser technology and applications* (ed. C. E. Webb and J. Jones). Institute of Physics, Bristol.

Ebrahimzadeh, M. (2004). C3.2 Optical parametric devices. In *The handbook of laser technology and applications* (ed. C. E. Webb and J. Jones). Institute of Physics, Bristol.

Eden, J. G. (2004). High-order harmonic generation and other intense optical field-matter interactions: review of recent experimental and theoretical advances. *Progress in Quantum Electronics*, **28**(3-4), 197–246.

Edwards, C. B., Hutchinson, M. H. R., Bradley, D. J., and Hutchinson, M. D. (1979). Repetitive vacuum ultraviolet xenon excimer laser. *Review of Scientific Instruments*, **50**(10), 1201–1207.

Einstein, A. (1917). On the quantum theory of radiation. *Phys. Z.*, **18**, 121. This paper is reproduced in English in Knight and Allen (1983).

Ell, R., Morgner, U., Kartner, F. X., Fujimoto, J. G., Ippen, E. P., Scheuer, V., Angelow, G., Tschudi, T., Lederer, M. J., Boiko, A., and Luther-Davies, B. (2001). Generation of 5-fs pulses and octave-spanning spectra directly from a Ti:sapphire laser. *Optics Letters*, **26**(6), 373–375.

Elton, Raymond C. (1990). *X-ray lasers*. Academic Press, Boston; London.

Ernst, G. J. (1984). Uniform-field electrodes with minimum width. *Optics Communications*, **49**, 275–277.

Faist, Jerome, Capasso, Federico, Sivco, Deborah L., Sirtori, Carlo, Hutchinson, Albert L., and Cho, Alfred Y. (1994). Quantum cascade laser. *Science*, **264**(5158), 553–556.

Feit, M. D. and Fleck, J. A. (1991). Spatial coherence of laboratory soft-X-ray lasers. *Optics Letters*, **16**(2), 76–78.

Fletcher, J. H. (1993). *Soft X-ray contact microscopy using laser generated plasma sources*. Doctoral thesis, University of Oxford.

Fletcher, J. H. and Webb, C. E. (2007). Optimization of coupled confocal cavities for an injection-seeded, discharge-excited KrF laser system. *Applied Physics B-Lasers and Optics*, **88**(2), 211–219.

Foot, Christopher (2005). *Atomic physics*. Oxford master series in physics. Oxford University Press, Oxford.

Fork, R. L., Martinez, O. E., and Gordon, J. P. (1984). Negative dispersion using pairs of prisms. *Optics Letters*, **9**(5), 150–152.

Förster, T. (1949). Experimentalle und theoretische untersuchung des zwischenmolerkularen ubergangs von elektronenregungsenergie. *Naturfurshung*, **4**, 321–327.

Fox, A. G. and Li, T. (1961). Resonant modes in a maser interferometer. *Bell System Technical Journal*, **40**, 453–488.

Fox, Mark (2001). *Optical properties of solids*. Oxford master series in condensed matter physics. Oxford University Press, Oxford.

Fox, T and Li, T. (1963). Modes in a maser interferometer with curved and tilted mirrors. *Proceedings of the Institute of Electrical and Electronics Engineers*, **51**, 80–89.

Frantz, L. M. and Nodvik, J. S. (1963). Theory of pulse propagation in a laser amplifier. *Journal of Applied Physics*, **34**(8), 2346–2349.

French, A. P. (2001). *Special relativity*. M.I.T. introductory physics series. Nelson Thornes, Cheltenham.

Geusic, J. E., Levinste, H. J., Singh, S., Smith, R. G., and Vanuiter, L. G. (1968). Continuous 0.532- μm solid-state source using $Ba_2NaNb_5O_{15}$. *Applied Physics Letters*, **12**(9), 306–308.

Gibbon, Paul (2005). *Short pulse laser interactions with matter: An introduction*. Imperial College Press, London.

Gill, P. and Webb, C. E. (1977). Electron energy distributions in the negative glow and their relevance to hollow cathode lasers. *J. Phys. D: Applied Physics*, **10**, 299–311.

Gloge, D. (1971). Weakly guiding fibers. *Applied Optics*, **10**(10), 2252–2258.

Golde, M. F. and Thrush, B. A. (1974). Vacuum UV emission from reactions of metastable inert gas atoms: Chemiluminescence of ArO and ArCl. *Chemical Physics Letters*, **29**(4), 486–490.

Golla, D., Knoke, S., Schone, W., Ernst, G., Bode, M., Tunnermann, A., and Welling, H. (1995). 300-W c.w. diode-laser side-pumped Nd-YAG rod laser. *Optics Letters*, **20**(10), 1148–1150.

Gordon, E. I., Labuda, E. F., and Bridges, W. B. (1964). Continuous visible laser action in singly ionized argon, krypton, and xenon. *Applied Physics Letters*, **4**, 178.

Gordon, J. P., Zeiger, H. J., and Townes, C. H. (1954). Molecular microwave oscillator and new hyperfine structure in the microwave spectrum of NH_3. *Physical Review*, **95**, 282–284.

Gordon, J. P., Zeiger, H. J., and Townes, C. H. (1955). The maser—new type of microwave amplifier, frequency standard, and spectrometer. *Physical Review*, **99**, 1264–1274.

Görtler and Strowitski (2001). Mini-excimer lasers for industrial applications. In *Lasers in manufacturing*, Munich, Germany, 64–67.

Gower, M. C. (2004). D1.6 Micromachining. In *The handbook of laser technology and applications* (ed. C. E. Webb and J. Jones). Institute of Physics Publishing, Bristol.

Gradshteyn, I. S., Ryzhik, I. M., and Jeffrey, Alan (2000). *Table of integrals, series, and products* (6th edn). Academic Press, San Diego; London.

Graydon, O. (2006, February). Kilowatt fibre lasers drive IPG sales boom. *Optics and Lasers Europe*, 19–20.

Green, J. M. and Webb, C. E. (1975). Second-kind collisions of electrons with excited Cd^+, Ca^+, Ga^+, Tl^+ and Pb^+ ions. *Journal of Physics B-Atomic Molecular and Optical Physics*, **8**(9), 1484–1500.

Grout, M. J., Janulewicz, K. A., Healy, S. B., and Pert, G. J. (1997). Optical-field induced gas mixture breakdown for recombination X-ray lasers. *Optics Communications*, **141**(3-4), 213–220.

Gudzenko, L. I. and Shelepin, L. A. (1965). Radiation enhancement in a recombining plasma. *Soviet Physics Dokl.*, **10**, 147.

Haken, H. and Wolf, H. C. (2004). *Molecular physics and elements of quantum chemistry: introduction to experiments and theory* (2nd enl. edn). Advanced texts in physics. Springer, Berlin; London.

Hall, D. R. (2004). B3.1 Carbon dioxide lasers. In *The handbook of laser technology and applications* (ed. C. E. Webb and J. Jones). Institute of Physics, Bristol.

Hall, R. N., Fenner, G. E., Kingley, J. D., Soltys, T. J., and Carlton, R. O. (1962). Coherent light emission from GaAs junctions. *Physical Review Letters*, **9**, 366–368.

Hanna, D. (2004). B4.1 Fibre lasers. In *The handbook of laser technology and applications* (ed. C. E. Webb and J. Jones). Institute of Physics, Bristol.

Hayes, W. and Loudon, Rodney (1978). *Scattering of light by crystals*. Wiley-Interscience Publication. Wiley, New York; Chichester.

Headley, C. (2004). B4.3 Cascaded Raman fibre lasers. In *The handbook of laser technology and applications* (ed. C. E. Webb and J. Jones). Institute of Physics, Bristol.

Hecht, Eugene (2002). *Optics* (4th edn). Addison Wesley, San Francisco; London.

Hecht, Jeff (2003). *City of light: The story of fiber optics*. Sloan technology series. Oxford University Press, Oxford.

Hecht, Jeff (2005a). *Beam: The race to make the laser*. Oxford University Press, Oxford.

Hecht, J. (2005b). Beam: The race to make the laser. *Optics and Photonics News*, **July**, 24–29.

Hecht, J. (2007). The breakthrough birth of the diode laser. *Optics and Photonics News*, **18**, 38–48.

Hegeler, F., Myers, M. C., Wofford, M. F., Sethian, J. D., Burns, P., Friedman, M., Giuliani, J. L., Jaynes, R., Albert, T., and Parish, J. (2008). The Electra KrF laser system. *Journal of Physics Conference Series*, **112**, 032035.

Henry, C. H. (1982). Theory of the linewidth of semiconductor lasers. *IEEE J Quantum Electronics*, **QE-18**, 259–264.

Herzberg, Gerhard (1989). *Molecular spectra and molecular structure* (Reprint edn), Volume 1. Spectra of diatomic molecules. Krieger Publishing, Malabar, Fl.

Herzberg, Gerhard (1990). *Infrared and Raman spectra*. Krieger Publishing, Malabar, Fl. USA.

Herzig, H. P. (2004). D9.1 Holography: Holographic optical elements—computer generated holography—diffractive optics. In *The handbook of laser technology and applications* (ed. C. E. Webb and J. Jones). Institute of Physics, Bristol.

Hitz, C. B. and Falk, J. (1971). Frequency doubled neodymium laser. *US Airforce Avionics Lab*, **AFAL-TR-12**.

Hobden, M. V. (1967). Phase-matched 2nd-harmonic generation in biaxial crystals. *Journal of Applied Physics*, **38**(11), 4365–4372.

Hodgkinson, I. J. and Vukusic, J. I. (1978). Birefringent filters for tuning flashlamp-pumped dye lasers: simplified theory and design. *Applied Optics*, **17**(12), 1944–1948.

Hodgson, R. T. (1970). Vacuum-ultraviolet laser action observed in Lyman bands of molecular hydrogen. *Physical Review Letters*, **25**(8), 494–497.

Hoff, P. W., Swingle, J. C., and Rhodes, C. K. (1973). Observations of stimulated emission from high-pressure krypton and argon-xenon mixtures. *Applied Physics Letters*, **23**(5), 245–246.

Hollberg, L., Oates, C. W., Wilpers, G., Hoyt, C. W., Barber, Z. W., Diddams, S. A., Oskay, W. H., and Bergquist, J. C. (2005). Optical frequency/wavelength references. *Journal of Physics B-Atomic Molecular and Optical Physics*, **38**(9), S469–S495.

Holonyak, N., Kolbas, R. M., Dupuis, R. D., and Dapkus, P. D. (1980). Quantum-well heterostructure lasers. *IEEE Journal of Quantum Electronics*, **16**(2), 170–186.

Holstein, T. (1951). Imprisonment of resonance radiation in gases II. *Physical Review*, **83**(6), 1159–1168.

Hooker, S. M. and Spence, D. J. (2006). Energy extraction from pulsed amplified stimulated emission lasers operating under conditions of strong saturation. *Journal of the Optical Society of America B-Optical Physics*, **23**(6), 1057–1067.

Hooker, S. M. and Webb, C. E. (1990). F$_2$-pumped NO: Laser oscillation at 218 nm and prospects for new laser transitions in the 160–250 nm region. *IEEE Journal of Quantum Electronics*, **26**(9), 1529–1535.

Houghton, John Theodore and Smith, S. D. (1966). *Infra-red physics*. Clarendon Press, Oxford.

Hughes, W. M., Shannon, J., and Hunter, R. (1974). 126.1-nm molecular argon laser. *Applied Physics Letters*, **24**(10), 488–490.

Hutchinson, I. H. (2002). *Principles of plasma diagnostics* (2nd edn). Cambridge University Press, Cambridge.

Inatsugu, S. and Holmes, J. R. (1973). Transition probabilities for $5s'[1-2]_1 - 3p$ transitions of Ne I. *Physical Review A*, **8**(4), 1678–1687.

Isaev, A. A., Kazaryan, M. A., Petrash, G. G., and Sergei, G. Rautian (1974). Converging beams in unstable telescopic resonators. *Soviet Journal of Quantum Electronics*, **4**(6), 761–766.

Isaev, A. A., Kazaryan, M. A., Petrash, G. G., Sergei, G. Rautian, and Shalagin, A. M. (1975). Evolution of Gaussian beams and pulse stimulated emission from lasers with unstable resonators. *Soviet Journal of Quantum Electronics*, **5**(6), 607–614.

Isaev, A. A., Kazaryan, M. A., Petrash, G. G., Sergei, G. Rautian, and Shalagin, A. M. (1977). Shaping of the output beam in a pulsed gas laser with an unstable resonator. *Soviet Journal of Quantum Electronics*, **7**(6), 746–752.

Jackson, D. A. and Webb, D. J. (2004). D2.5 Optical fibre grating Bragg grating sensors for strain measurement. In *The handbook of laser technology and applications* (ed. C. E. Webb and J. Jones). Institute of Physics, Bristol.

Jackson, John David (1999). *Classical electrodynamics* (3rd edn). Wiley, New York; Chichester.

Jarrett, S. M. and Young, J. F. (1979). High-efficiency single-frequency c.w. ring dye-laser. *Optics Letters*, **4**(6), 176–178.

Javan, A., Bennett, W. R., Junior, and Herriott, D. R. (1961). Population inversion and continuous optical maser oscillation in a gas discharge containing a He-Ne mixture. *Physical Review Letters*, **6**, 106–110.

Jones, D. J., Diddams, S. A., Ranka, J. K., Stentz, A., Windeler, R. S., Hall, J. L., and Cundiff, S. T. (2000). Carrier-envelope phase control of femtosecond mode-locked lasers and direct optical frequency synthesis. *Science*, **288**(5466), 635–639.

Jones, J. D. C. and Webb, C. E. (2004). Handbook of laser technology and applications, Institute of Physics, Bristol, and Taylor and Francis, London.

Jones-Bey, H. A. (2006). Photolithography—optical immersion further shrinks feature sizes. *Laser Focus World*, **42**(4), 20.

Joshi, C. (2006). Plasma accelerators. *Scientific American*, **294**(2), 40–47.

Kaminskii, A. A. (2003). Modern developments in the physics of crystalline laser materials. *Physica Status Solidi A-Applied Research*, **200**(2), 215–296.

Kane, D. J. and Trebino, R. (1993). Single-shot measurement of the intensity and phase of an arbitrary ultrashort pulse by using frequency-resolved optical gating. *Optics Letters*, **18**(10), 823–825.

Kärtner, F. X., Matuschek, N., Schibli, T., Keller, U., Haus, H. A., Heine, C., Morf, R., Scheuer, V., Tilsch, M., and Tschudi, T. (1997). Design and fabrication of double-chirped mirrors. *Optics Letters*, **22**(11), 831–833.

Keller, U. (2003). Recent developments in compact ultrafast lasers. *Nature*, **424**(6950), 831–838.

Keller, U., Weingarten, K. J., Kartner, F. X., Kopf, D., Braun, B., Jung, I. D., Fluck, R., Honninger, C., Matuschek, N., and derAu, J. A. (1996). Semiconductor saturable absorber mirrors (sesam's) for femtosecond to nanosecond pulse generation in solid-state lasers. *IEEE Journal of Selected Topics in Quantum Electronics*, **2**(3), 435–453.

Kelson, I. and Hardy, A. A. (1998). Strongly pumped fiber lasers. *IEEE Journal of Quantum Electronics*, **34**(9), 1570–1577.

Kemp, A. J., Valentine, G. J., and Burns, D. (2004). Progress towards high-power, high-brightness neodymium-based thin-disk lasers. *Progress in Quantum Electronics*, **28**(6), 305–344.

Khan, A., Starrinou, P. N., and Parry, G. (2001). Semiconductor lasers. In *Low dimensional semiconductor strucutres* (ed. K. B. Vvedensky and D. D.). Cambridge University Press, Cambridge, United Kingdom.

Kittel, Charles and McEuen, Paul (2005). *Introduction to solid state physics* (8th edn). Wiley, New York; Chichester.

Klisnick, A., Guilbaud, O., Ros, D., Cassou, K., Kazamias, S., Jamelot, G., Lagron, J. C., Joyeux, D., Phalippou, D., Lechantre, Y., Edwards, M., Mistry, P., and Tallents, G. J. (2006). Experimental study of the temporal coherence and spectral profile of the 13.9 nm transient X-ray laser. *Journal of Quantitative Spectroscopy & Radiative Transfer*, **99**(1–3), 370–380.

Knight, P. and Allen, L. (1983). *Concepts of quantum optics*. Pergamon, Oxford.

Kodama, R., Neely, D., Kato, Y., Daido, H., Murai, K., Yuan, G., Macphee, A., and Lewis, C. L. S. (1994). Generation of small-divergence soft-X-ray laser by plasma wave-guiding with a curved target. *Physical Review Letters*, **73**(24), 3215–3218.

Koechner, Walter (2006). *Solid-state laser engineering* (6th rev. and updated edn). Springer series in optical sciences; 1. Springer, New York.

Koechner, W. and Bass, M. (2003). *Solid-state lasers: A graduate Text*. Springer, New York.

Kressel, H., Effenberg, M., Wittke, J. P., and Landany, I. (1980). Laser diodes and LEDs for optical communication. In *Semiconductor devices for optical communication* (2nd edn) (ed. H. Kressel). Springer-Verlag, New York.

Kruer, William L. (1988). *The physics of laser plasma interactions*. Frontiers in physics; v. 73. Addison-Wesley, Redwood City, Calif.; Wokingham.

Krushelnick, K. and Najmudin, Z. (2006). Plasmas surf the high-energy frontier. *Physics World*, **19**(2), 24–28.

Lanigan, S. (2004). D3.2.5 Therapeutic applications: dermatology. In *The handbook of laser technology and applications* (ed. C. E. Webb and J. Jones). Institute of Physics, Bristol.

Laporta, P., Taccheo, S., Longhi, S., Svelto, O., and Svelto, C. (1999). Erbium-ytterbium microlasers: optical properties and lasing characteristics. *Optical Materials*, **11**(2–3), 269–288.

Larotonda, M. A., Luther, B. M., Wang, Y., Liu, Y., Alessi, D., Berrill, M., Dummer, A., Brizuela, F., Menoni, C. S., Marconi, M. C., Shlyaptsev, V. N., Dunn, J., and Rocca, J. J. (2004). Characteristics of a saturated 18.9 nm tabletop laser operating at 5 Hz repetition rate. *IEEE Journal of Selected Topics in Quantum Electronics*, **10**(6), 1363–1367.

Lemoff, B. E., Barty, C. P. J., and Harris, S. E. (1994). Femtosecond-pulse-driven, electron-excited XUV lasers in 8-times-ionized noble-gases. *Optics Letters*, **19**(8), 569–571.

Lemoff, B. E., Yin, G. Y., Gordon, C. L., Barty, C. P. J., and Harris, S. E. (1995). Demonstration of a 10-Hz femtosecond pulse-driven XUV laser at 41.8 nm in Xe-IX. *Physical Review Letters*, **74**(9), 1574–1577.

Lemoff, B. E., Yin, G. Y., Gordon, C. L., Barty, C. P. J., and Harris, S. E. (1996). Femtosecond-pulse-driven 10-Hz 41.8-nm laser in Xe, IX. *Journal of the Optical Society of America B-Optical Physics*, **13**(1), 180–184.

Letokhov, V. S. (2007). *Laser control of atoms and molecules*. Oxford University Press, Oxford; New York.

Liao, Y., Miller, R. J. D., and Armstrong, M. R. (1999). Pressure tuning of thermal lensing for high-power scaling. *Optics Letters*, **24**(19), 1343–1345.

Lieberman, M. A. and Lichtenberg, Allan J. (2005). *Principles of plasma discharges and materials processing* (2nd edn). Wiley-Interscience, Hoboken, N.J.

Lindl, John D., Amendt, Peter, Berger, Richard L., Glendinning, S. Gail, Glenzer, Siegfried H., Haan, Steven W., Kauffman, Robert L., Landen, Otto L., and Suter, Laurence J. (2004). The physics basis for ignition using indirect-drive targets on the National Ignition Facility. *Physics of Plasmas*, **11**(2), 339–491.

Linford, G.J., Peressini, E.R., Sooy, W.R., and Spaeth, M.L. (1974). Very long lasers. *Applied Optics*, **13**, 379.

London, R. A., Strauss, M., and Rosen, M. D. (1990). Modal-analysis of X-ray laser coherence. *Physical Review Letters*, **65**(5), 563–566.

Long, D. A. (1977). *Raman spectroscopy*. McGraw-Hill International Book Company, New York; London.

Lorrain, Paul, Corson, Dale R., and Lorrain, François (1988). *Electromagnetic fields and waves: including electric cicuits* (3rd edn). W.H. Freeman, New York.

Loudon, Rodney (2000). *The quantum theory of light* (3rd edn). Oxford science publications. Clarendon Press, Oxford.

Luther, B. M., Wang, Y., Larotonda, M. A., Alessi, D., Berrill, M., Rocca, J. J., Dunn, J., Keenan, R., and Shlyaptsev, V. N. (2006). High repetition rate collisional soft X-ray lasers based on grazing incidence pumping. *IEEE Journal of Quantum Electronics*, **42**(1-2), 4–13.

Ma, L. S., Bi, Z. Y., Bartels, A., Robertsson, L., Zucco, M., Windeler, R. S., Wilpers, G., Oates, C., Hollberg, L., and Diddams, S. A. (2004). Optical frequency synthesis and comparison with uncertainty at the 10(−19) level. *Science*, **303**(5665), 1843–1845.

Macgowan, B. J., Dasilva, L. B., Fields, D. J., Keane, C. J., Koch, J. A., London, R. A., Matthews, D. L., Maxon, S., Mrowka, S., Osterheld, A. L., Scofield, J. H., Shimkaveg, G., Trebes, J. E., and Walling, R. S. (1992). Short wavelength X-ray laser research at the Lawrence-Livermore-National-Laboratory. *Physics of Fluids B-Plasma Physics*, **4**(7), 2326–2337.

Maiman, T. H., Hoskins, R. H., D'Haenens, I. J., Asawa, C. K., and Evtuhov, V. (1961). Stimulated optical emission in fluorescent solids. II. spectroscopy and stimulated emission in ruby. *Physical Review*, **123**, 1151–1157.

Marconi, M. C., Chilla, J. L. A., Moreno, C. H., Benware, B. R., and Rocca, J. J. (1997). Measurement of the spatial coherence buildup in a discharge pumped table-top soft X-ray laser. *Physical Review Letters*, **79**(15), 2799–2802.

Martinez, O. E., Gordon, J. P., and Fork, R. L. (1984). Negative group-velocity dispersion using refraction. *Journal of the Optical Society of America a-Optics Image Science and Vision*, **1**(10), 1003–1006.

Matthews, D. L., Hagelstein, P. L., Rosen, M. D., Eckart, M. J., Ceglio, N. M., Hazi, A. U., Medecki, H., Macgowan, B. J., Trebes, J. E., Whitten, B. L., Campbell, E. M., Hatcher, C. W., Hawryluk, A. M., Kauffman, R. L., Pleasance, L. D., Rambach, G., Scofield, J. H., Stone, G., and Weaver, T. A. (1985). Demonstration of a soft-X-ray amplifier. *Physical Review Letters*, **54**(2), 110–113.

Matthews, J. W. and Blakeslee, A. E. (1974). Defects in epitaxial multilayers. 1. misfit dislocations. *Journal of Crystal Growth*, **27**, 118–125.

McClung, F. J. and Hellwarth, R. W. (1962). Giant optical pulsations from ruby. *Journal of Applied Physics*, **33**(3), 828–829.

Mears, R. J., Reekie, L., Jauncey, I. M., and Payne, D. N. (1987). Low-noise erbium-doped fiber amplifier operating at 1.54 μm. *Electronics Letters*, **23**(19), 1026–1028.

Midwinter, John E. and Guo, Y. (1992). *Optoelectronics and lightwave technology*. Wiley, Chichester.

Milam, D. and Weber, M. J. (1976). Measurement of nonlinear refractive-index coefficients using time-resolved interferometry—application to optical-materials for high-power neodymium lasers. *Journal of Applied Physics*, **47**(6), 2497–2501.

Milster, T. (2004). D5.1 Optical data storage. In *The handbook of laser technology and applications* (ed. C. E. Webb and J. Jones). Institute of Physics, Bristol.

Mocek, T., McKenna, C. M., Cros, B., Sebban, S., Spence, D. J., Maynard, G., Bettaibi, I., Vorontsov, V., Gonsavles, A. J., and Hooker, S. M. (2005). Dramatic enhancement of XUV laser output using a multimode gas-filled capillary waveguide. *Physical Review A*, **71**(1), 013804.

Morgner, U., Kartner, F. X., Cho, S. H., Chen, Y., Haus, H. A., Fujimoto, J. G., Ippen, E. P., Scheuer, V., Angelow, G., and Tschudi, T. (1999). Sub-two-cycle pulses from a Kerr-lens mode-locked Ti:sapphire laser. *Optics Letters*, **24**(6), 411–413.

Mourou, Gérard A. and Umstadter, Donald (2002). Extreme light. *Scientific American*, **286**(5), 80.

Müller-Horsche et. al. (1993). US Patent No. 5,247,535.

Nagata, Y., Midorikawa, K., Kubodera, S., Obara, M., Tashiro, H., and Toyoda, K. (1993). Soft-X-ray amplification of the Lyman-alpha transition by optical-field-induced ionization. *Physical Review Letters*, **71**(23), 3774–3777.

Nelson, D. F. and Boyle, W.S. (1962). A continuously operating ruby optical maser. *Applied Optics*, **1**(2), 181–183.

Ohtsu, Motoichi (1992). *Highly coherent semiconductor lasers*. Artech House optoelectronics library. Artech House, Boston; London.

Okai, M., Suzuki, M., Taniwatari, T., and Chinone, N. (1994). Corrugation-pitch-modulated distributed feedback lasers with ultra-narrow spectral linewidth. *Japanese Journal of Applied Physics*, **33**, 2563–2570.

Okazaki, S. (2004). D1.5 Photolithography. In *The handbook of laser technology and applications* (ed. C. E. Webb and J. Jones). Institute of Physics Publishing, Bristol.

Ostendorf, Andreas (2004). C2.2 Short pulses. In *The handbook of laser technology and applications* (ed. C. E. Webb and J. Jones). Institute of Physics, Bristol.

Osvay, K., Dombi, P., Kovacs, A. P., and Bor, Z. (2002). Fine tuning of the higher-order dispersion of a prismatic pulse compressor. *Applied Physics B-Lasers and Optics*, **75**(6-7), 649–654.

Owa, S., Nagasaki, H., Ishii, Y., Shiraishi, K., and Hirukawa, S. (2004). Full-field exposure tools for immersion lithography. In *Optical Micolithography XVIII* (ed. B. W. Smith), Volume 5754, 655–668.

Paisner, J. A. (1988). Atomic vapor laser isotope-separation. *Applied Physics B-Photophysics and Laser Chemistry*, **46**(3), 253–260.

Paschotta, R., Nilsson, J., Tropper, A. C., and Hanna, D. C. (1997). Ytterbium-doped fiber amplifiers. *IEEE Journal of Quantum Electronics*, **33**(7), 1049–1056.

Pask, H. M., Carman, R. J., Hanna, D. C., Tropper, A. C., Mackechnie, C. J., Barber, P. R., and Dawes, J. M. (1995). Ytterbium-doped silica fiber lasers—versatile sources for the 1–1.2 μm region. *IEEE Journal of Selected Topics in Quantum Electronics*, **1**(1), 2–13.

Patel, C. K. N. (1964). Selective excitation through vibrational energy transfer and optical maser action in N_2—CO_2. *Physical Review Letters*, **13**, 617–619.

Pellegrini, C. (2001). Design considerations for a safe X-ray FEL. *Nuclear Instruments & Methods in Physics Research Section a-Accelerators Spectrometers Detectors and Associated Equipment*, **475**(1-3), 1–12.

Pennington, D. M., Henesian, M. A., and Hellwarth, R. W. (1989). Nonlinear index of air at 1.053 μm. *Physical Review A*, **39**(6), 3003–3009.

Pert, G. J. (1994). Output characteristics of amplified-stimulated-emission lasers. *Journal of the Optical Society of America B-Optical Physics*, **11**(8), 1425–1435.

Pert, G. J. (2006). Recombination and population inversion in plasmas generated by tunneling ionization. *Physical Review E*, **73**(6), 066401.

Powell, Richard C. (1998). *Physics of solid-state laser materials*. AIP Press/Springer, New York.

Pretzler, G., Kasper, A., and Witte, K. J. (2000). Angular chirp and tilted light pulses in cpa lasers. *Applied Physics B-Lasers and Optics*, **70**(1), 1–9.

Quinn, T. J. (2003). Practical realization of the definition of the metre, including recommended radiations of other optical frequency standards (2001). *Metrologia*, **40**(2), 103–133.

Raghuramaiah, M., Sharma, A. K., Naik, P. A., and Gupta, P. D. (2003). Simultaneous measurement of pulse-front tilt and pulse duration of a femtosecond laser beam. *Optics Communications*, **223**(1–3), 163–168.

Rahman, A., Rocca, J. J., and Wyart, J. F. (2004). Classification of the nickel-like silver spectrum (AgXX) from a fast capillary discharge plasma. *Physica Scripta*, **70**(1), 21–25.

Ready, J. F., Farson, D. F., and Feeley, T. (2001). *The Laser Institute of America handbook of materials processing*. Springer-Verlag, Berlin.

Reichert, J., Holzwarth, R., Udem, T., and Hansch, T. W. (1999). Measuring the frequency of light with mode-locked lasers. *Optics Comunications*, **172**, 59–68.

Reid, D. T. (2004). C2.3 Ultrashort pulses. In *The handbook of laser technology and applications* (ed. C. E. Webb and J. Jones). Institute of Physics, Bristol.

Reif, F. (1985). *Fundamentals of statistical and thermal physics*. McGraw-Hill series in fundamentals of physics. McGraw-Hill, Auckland; London.

Rigrod, W. W. (1965). Saturation effects in high-gain lasers. *J. Applications. Phys.*, **36**, 2487–2490.

Riley, K. F., Hobson, M. P., and Bence, S. J. (2002). *Mathematical methods for physics and engineering: a comprehensive guide* (2nd edn). Cambridge University Press, Cambridge.

Rocca, J. J. (1999). Table-top soft X-ray lasers. *Review of Scientific Instruments*, **70**(10), 3799–3827.

Rocca, J. J. (2004). B5.2 X-ray lasers. In *The Handbook of Laser Technology and Applications* (ed. C. E. Webb and J. Jones). Institute of Physics Publishing, Bristol.

Rocca, J. J., Shlyaptsev, V., Tomasel, F. G., Cortazar, O. D., Hartshorn, D., and Chilla, J. L. A. (1994). Demonstration of a discharge pumped table-top soft-X-ray laser. *Physical Review Letters*, **73**(16), 2192–2195.

Rose, S. J. (2004). Set the controls for the heart of the sun. *Contemporary Physics*, **45**(2), 109–121.

Rosenberg, H. M. (1989). *The solid state: an introduction to the physics of solids for students of physics, materials science, and engineering* (3rd edn). Oxford physics series. Oxford University Press, Oxford.

Rundquist, A., Durfee, C. G., Chang, Z. H., Herne, C., Backus, S., Murnane, M. M., and Kapteyn, H. C. (1998). Phase-matched generation of coherent soft X-rays. *Science*, **280**(5368), 1412–1415.

Russell, P. (2003). Photonic crystal fibers. *Science*, **299**(5605), 358–362.

Ruster, W., Ames, F., Kluge, H. J., Otten, E. W., Rehklau, D., Scheerer, F., Herrmann, G., Muhleck, C., Riegel, J., Rimke, H., Sattelberger, P., and Trautmann, N. (1989). A resonance ionization mass-spectrometer as an analytical instrument for trace analysis. *Nuclear Instruments & Methods in Physics Research Section a-Accelerators Spectrometers Detectors and Associated Equipment*, **281**(3), 547–558.

Schäfer, F. P. (1990). *Dye lasers* (3rd edn). Topics in applied physics; v. 1. Springer-Verlag, Berlin; New York.

Schawlow, A. L. and Townes, C. H. (1958). Infrared and optical masers. *Physical Review*, **112**, 1940–1949.

Schenkel, B., Biegert, J., Keller, U., Vozzi, C., Nisoli, M., Sansone, G., Stagira, S., De Silvestri, S., and Svelto, O. (2003). Generation of 3.8-fs pulses from adaptive compression of a cascaded hollow fiber supercontinuum. *Optics Letters*, **28**(20), 1987–1989.

Searles, S. K. and Hart, G. A. (1975). Stimulated emission at 281.8 nm from XeBr. *Applied Physics Letters*, **27**(4), 243–245.

Sheik-Bahae, M., Said, A. A., Hagan, D. J., Soileau, M. J., and Van Stryland, E. W. (1991). Nonlinear refraction and optical limiting in thick media. *Optical Engineering*, **30**(8), 1228–1235.

Sherriff, R. E. (1998). Analytic expressions for group-delay dispersion and cubic dispersion in arbitrary prism sequences. *Journal of the Optical Society of America B-Optical Physics*, **15**(3), 1224–1230.

Shine, R. J., Alfrey, A. J., and Byer, R. L. (1995). 40-W c.w., TEM(00)-mode, diode-laser-pumped, Nd-YAG miniature-slab laser. *Optics Letters*, **20**(5), 459–461.

Shockley, W. and Read, W. T. (1952). Statistics of the recombinations of holes and electrons. *Physical Review*, **87**(5), 835–842.

Siegman, A. E. (1986). *Lasers*. University Science Books, Mill Valley, Calif. Anthony E. Siegman. Ill.; 26 cm. Includes bibliographical references and index. OUP ISBN and distribution information ("Distributed by Oxford University Press outside of North America") on labels pasted to verso of title page.

Siegman, A. E. (1998). How to (maybe) measure laser beam quality. In *OSA Trends in Optical Photonics*, Volume 17, 184–199.

Siegman, A. E. (2000*a*). Laser beams and resonators: Beyond the 1960s. *IEEE Journal of Selected Topics in Quantum Electronics*, **6**(6), 1389–1399.

Siegman, A. E. (2000*b*). Laser beams and resonators: The 1960s. *IEEE Journal of Selected Topics in Quantum Electronics*, **6**(6), 1380–1388.

Singleton, John (2001). *Band theory and electronic properties of solids*. Oxford master series in condensed matter physics. Oxford University Press, Oxford.

Smith, P. W. (1966). Linewidth and saturation parameters for 6328 A transition in a He-Ne laser. *Journal of Applied Physics*, **37**(5), 2089–2093.

Smith, P. W. (1972). Mode selection in lasers. *Proceedings of the Institute of Electrical and Electronics Engineers*, **60**(4), 422–440.

Smith, R. A. (1978*a*). Excitation of transitions between atomic or molecular energy-levels by monochromatic laser-radiation. 1. excitation under conditions of strictly homogeneous line broadening. *Proceedings of the Royal Society of London Series a-Mathematical Physical and Engineering Sciences*, **362**(1708), 1–12.

Smith, R. A. (1978*b*). Excitation of transitions between atomic or molecular energy-levels by monochromatic laser-radiation. 2. excitation under conditions of strictly inhomogeneous line broadening and mixed homogeneous and inhomogeneous broadening. *Proceedings of the Royal Society of London Series a-Mathematical Physical and Engineering Sciences*, **362**(1708), 13–25.

Smith, R. A. (1979). Excitation of transitions between atomic or molecular-energy levels by monochromatic laser-radiation .3. effect of doppler broadening when phase-destroying collisions predominate. *Proceedings of the Royal Society of London Series a-Mathematical Physical and Engineering Sciences*, **368**(1733), 163–175.

Smith, R. F., Dunn, J., Hunter, J. R., Nilsen, J., Hubert, S., Jacquemot, S., Remond, C., Marmoret, R., Fajardo, M., Zeitoun, P., Vanbostal, L., Lewis, C. L. S., Ravet, M. F., and Delmotte, F. (2003). Longitudinal coherence measurements of a transient collisional x-ray laser. *Optics Letters*, **28**(22), 2261–2263.

Snitzer, E. et al. (1989), US patent No. 4, 815. 79.

Sorokin, P. P. and Lankard, J. R. (1966). Stimulated emission observed from an organic dye chloro-aluminum phthalocyanine. *IBM Journal of Research and Development*, **10**(2), 162.

Sorokin, P. P. and Stevenson, M. J. (1960). Stimulated infrared emission from trivalent uranium. *Physical Review Letters*, **5**, 557–559.

Steinmeyer, G., Sutter, D. H., Gallmann, L., Matuschek, N., and Keller, U. (1999). Frontiers in ultrashort pulse generation: Pushing the limits in linear and nonlinear optics. *Science*, **286**(5444), 1507–1512.

Stewart, G. (2004). A.6 Optical waveguide theory. In *The handbook of laser technology and applications* (ed. C. E. Webb and J. Jones). Institute of Physics, Bristol.

Stewen, C., Contag, K., Larionov, M., Giesen, A., and Hugel, H. (2000). A 1-kW c.w. thin disc laser. *IEEE Journal of Selected Topics in Quantum Electronics*, **6**(4), 650–657.

Suckewer, S. and Jaegle, P. (2009). X-ray laser: past, present, and future. *Laser Physics Letters*, **6**(6), 411–436.

Suckewer, S., Skinner, C. H., Milchberg, H., Keane, C., and Voorhees, D. (1985). Amplification of stimulated soft-X-ray emission in a confined plasma-column. *Physical Review Letters*, **55**(17), 1753–1756.

Sutherland, J. M., French, P. M. W., Taylor, J. R., and Chai, B. H. T. (1996). Visible continuous-wave laser transitions in Pr^{3+}:YLF and femtosecond pulse generation. *Optics Letters*, **21**(11), 797–799.

Sutter, D. H., Steinmeyer, G., Gallmann, L., Matuschek, N., Morier-Genoud, F., Keller, U., Scheuer, V., Angelow, G., and Tschudi, T. (1999). Semiconductor saturable-absorber mirror-assisted Kerr-lens mode-locked Ti:sapphire laser producing pulses in the two-cycle regime. *Optics Letters*, **24**(9), 631–633.

Svelto, O., Taccheo, S., and Svelto, C. (1998). Analysis of amplified spontaneous emission: some corrections to the Linford formula. *Optics Communications*, **149**(4–6), 277–282.

Szipöcs, R., Ferencz, K., Spielmann, C., and Krausz, F. (1994). Chirped multilayer coatings for broad-band dispersion control in femtosecond lasers. *Optics Letters*, **19**(3), 201–203.

Takeda, M., Ina, H., and Kobayashi, S. (1982). Fourier-transform method of fringe-pattern analysis for computer-based topography and interferometry. *Journal of the Optical Society of America*, **72**(1), 156–160.

Taverner, D., Richardson, D. J., Dong, L., Caplen, J. E., Williams, K., and Penty, R. V. (1997). 158 μJ pulses from a single-transverse-mode, large-mode-area erbium-doped fiber amplifier. *Optics Letters*, **22**(6), 378–380.

Titterton, D. (2004a). B5.3 Liquid lasers. In *The handbook of laser technology and applications* (ed. C. E. Webb and J. Jones). Institute of Physics, Bristol.

Titterton, D. (2004b). B5.4 Solid-state dye lasers. In *The handbook of laser technology and applications* (ed. C. E. Webb and J. Jones). Institute of Physics, Bristol.

Tonks, L. and Langmuir, I. (1929). A general theory of the plasma of an arc. *Physical Review*, **34**, 876–922.

Trebes, J. E., Nugent, K. A., Mrowka, S., London, R. A., Barbee, T. W., Carter, M. R., Koch, J. A., Macgowan, B. J., Matthews, D. L., Dasilva, L. B., Stone, G. F., and Feit, M. D. (1992). Measurement of the spatial coherence of a soft-X-ray laser. *Physical Review Letters*, **68**(5), 588–591.

Trebino, R. (2002). *Frequency-resolved optical gating: The measurement of ultrashort laser pulses*. Kluwer Academic Publishers Group, Dordrecht.

Tropea, C. (2004). D2.2 Laser velocimetry. In *The handbook of laser technology and applications* (ed. C. E. Webb and J. Jones). Institute of Physics, Bristol.

Tünnermann, A. and Zellmer, H. (2004). B4.2 High power fiber lasers. In *The handbook of laser technology and applications* (ed. C. E. Webb and J. Jones). Institute of Physics, Bristol.

Udem, T., Reichert, J., Holzwarth, R., and Hansch, T. W. (1999). Accurate measurement of large optical frequency differences with a mode-locked laser. *Optics Letters*, **24**(13), 881–883.

Uemura, S. and Torizuka, K. (2003). Development of a diode-pumped kerr-lens mode-locked Cr:LiSAF laser. *IEEE Journal of Quantum Electronics*, **39**(1), 68–73.

Unger, P. (2004). B2.4 high-power laser diodes and laser diode arrays. In *The Handbook of Laser Technology and Applications* (ed. C. E. Webb and J. Jones). Institute of Physics, Bristol.

Urquhart, P. (2004). D4.2 High-capacity optical transmission systems. In *The handbook of laser technology and applications* (ed. C. E. Webb and J. Jones). Institute of Physics, Bristol.

Venkatesan, T. N. C. and McCall, S. L. (1977). C. W. ruby-laser pumped by a 5145-Å argon laser. *Review of Scientific Instruments*, **48**(5), 539–541.

Verluise, F., Laude, V., Cheng, Z., Spielmann, C., and Tournois, P. (2000). Amplitude and phase control of ultrashort pulses by use of an acousto-optic programmable dispersive filter: pulse compression and shaping. *Optics Letters*, **25**(8), 575–577.

Vinogradov, A. V. and Shlyaptsev, V. N. (1983). Gain in the range of 100–1000Å in homogeneous stationary plasma. *Soviet Journal of Quantum Electronics*, **13**, 1511.

Von Engel, A. (1965). *Ionized gases* (2nd edn). Clarendon Press, Oxford.

Voo, N. Y., Horak, P., Ibsen, M., and Loh, W. H. (2004). Linewidth and phase noise characteristics of DFB fibre lasers. In *SPIE European Symposium on Optics and Photonics in Security and Defence*, London, pp. 5620–5623.

Wachsmann-Hogiu, S., Annala, A. J., and Farkas, D. L. (2004). D3.4 Laser applications in biology and biotechnology. In *The handbook of laser technology and applications* (ed. C. E. Webb and J. Jones). Institute of Physics, Bristol.

Walling, J. C., Peterson, O. G., Jenssen, H. P., Morris, R. C., and Odell, E. W. (1980). Tunable alexandrite lasers. *IEEE Journal of Quantum Electronics*, **16**(12), 1302–1315.

Walmsley, I.A. and Dorrer, C. (2009). Characterization of ultrashort electromagnetic pulses. *Adv. Opt. Photon.*, **1**(2), 308–437.

Walmsley, I., Waxer, L., and Dorrer, C. (2001). The role of dispersion in ultrafast optics. *Review of Scientific Instruments*, **72**(1), 1–29.

Webb, C. E. (1975). The fundamental discharge physics of atomic gas lasers. In *High power gas lasers*, Volume 29 of *Institute of Physics Conference Series*, 1–28.

Webb, Colin E. and Jones, Julian D. C. (2004). *Handbook of laser technology and applications*. Institute of Physics, Bristol.

Webb, C. E. and Miller, R. C. and Tang, C. L. (1968). New radiative lifetime values for 4s levels of Ar. II. *IEEE Journal of Quantum Electronics*, **QE- 4**(5), 357.

Weber, Marvin J. (1982). *Handbook of laser science and technology*. CRC Press, Boca Raton, FL.

Weiner, A. M. (1995). Femtosecond optical pulse shaping and processing. *Progress in Quantum Electronics*, **19**(3), 161–237.

Weiner, A. M. and Kan'an, A. M. (1998). Femtosecond pulse shaping for synthesis, processing, and time-to-space conversion of ultrafast optical waveforms. *IEEE Journal of Selected Topics in Quantum Electronics*, **4**(2), 317–331.

White, A. D. and Rigden, J. D. (1962). Continuous gas maser operation in the visible. *Proceedings of the Institute of Radio Engineers*, **50**, 1697.

White, A. D. and Tsufura, L. (2004). B3.6 Helium-neon lasers. In *The handbook of laser technology and applications* (ed. C. E. Webb and J. Jones). Institute of Physics, Bristol.

Willett, Colin S. (1974). *Introduction to gas lasers: population inversion mechanisms.* Pergamon, Oxford.

Wilson, B. C. and Bown, S. G. (2004). D3.2.3 Therapeutic applications: photodynamic therapy. In *The handbook of laser technology and applications* (ed. C. E. Webb and J. Jones). Institute of Physics, Bristol.

Winterfeldt, C., Spielmann, C., and Gerber, G. (2008). Colloquium: Optimal control of high-harmonic generation. *Reviews of Modern Physics*, **80**(1), 117–140.

Wisdom, J., Digonnet, M., and Byer, R. L. (2004). Ceramic lasers: Ready for action. *Photonics Spectra*, **38**(2), 50.

Woodgate, G. K. (1980). *Elementary atomic structure* (2nd edn). Clarendon Press, Oxford. G.K. Woodgate. ILL.; 24 cm. First issued in paperback 1983. Includes bibliographical references and index. Previous ed.: Maidenhead: McGraw-Hill, 1970.

Woods, S. and Flinn, G. (2006, December /January). Marking fibre lasers enhance dynamic pulse performance. *Europhotonics*, 28–29.

Wu, S., Kapinus, V. A., and Blake, G. A. (1999). A nanosecond optical parametric generator/amplifier seeded by an external cavity diode laser. *Optics Communications*, **159**(1–3), 74–79.

Yarborough, J. M. (1974). C.W. dye laser-emission spanning visible spectrum. *Applied Physics Letters*, **24**(12), 629–630.

Yariv, A. (1989). *Quantum electronics* (3rd edn). John Wiley & Sons, New York.

Yonezu, H., Kobayash, K., and Sakuma, I. (1973). Threshold current-density and lasing transverse mode in a $GaAs-Al_xGa_{1-x}$ as double heterostructure laser. *Japanese Journal of Applied Physics*, **12**(10), 1585–1592.

Young, B. C., Cruz, F. C., Itano, W. M., and Bergquist, J. C. (1999). Visible lasers with subhertz linewidths. *Physical Review Letters*, **82**(19), 3799–3802.

Zagumennyi, A. I., Mikhailov, V. A., and Shcherbakov, I. A. (2004). B1.3 rare earth ions - Nd^{3+}. In *The handbook of laser technology and applications* (ed. C. E. Webb and J. Jones). Institute of Physics, Bristol.

Zemskov, K. I., Isaev, A. A., Kazaryan, M. A., Petrash, G. G., and Sergei, G. Rautian (1974). Use of unstable resonators in achieving the diffraction divergence of the radiation emitted from high-gain pulsed gas lasers. *Soviet Journal of Quantum Electronics*, **4**(4), 474–477.

Zewail, A. H. (2000). Femtochemistry: Atomic-scale dynamics of the chemical bond. *Journal of Physical Chemistry A*, **104**(24), 5660–5694.

Zewail, A. H. (2003). Femtochemistry: Atomic-scale dynamics of the chemical bond using ultrafast lasers. *Nobel Lectures in Chemistry 1996–2000*. (ed. Ingmar Grenthe). World Scientific Books.

Zhang, J., Key, M. H., Norreys, P. A., Tallents, G. J., Behjat, A., Danson, C., Demir, A., Dwivedi, L., Holden, M., Holden, P. B., Lewis, C. L. S., Macphee, A. G., Neely, D., Pert, G. J., Ramsden, S. A., Rose, S. J., Shao, Y. F., Thomas, O., Walsh, F., and You, Y. L. (1995). Demonstration of high-gain in a recombination XUV laser at 18.2 nm driven by a 20 J, 2 ps glass-laser. *Physical Review Letters*, **74**(8), 1335–1338.

Zherikhin, A. N., Koshelev, K. N., and Letokhov, V. S. (1976). Amplification in far ultraviolet region due to transitions in multicharged ions. *Soviet Journal of Quantum Electronics*, **6**(1), 82.

Zucker, H. (1970). Optical resonators with variable reflectivity mirrors. *Bell System Technical Journal*, **49**, 2349.

Index

Major sections are indicated in bold

M^2-parameter, **103–106**
3-level lasers, 61, 78, **142–146**, 160, 174, 176, 178, 284
3-level vs. 4-level lasers, **142–146**, 299
4-level lasers, 63, **142–146**, 160, 165, 178, 179, 181, 190, 192, 199, 201, 444

A & B coefficients, *see* Einstein coefficients
above-threshold ionization (ATI), 545–546
absorption, 12, 13, 42, **54–56**, 58, 79, 139, 382–385, 387, 398, 451, 461
    coefficient, 54
acousto-optic Q-switching, *see* Q-switching, acousto-optic
AM modelocking, *see* modelocking, AM
amplification
    in saturated homogenously broadened amplifier, 66–67
    in saturated inhomogenously broadened amplifier, 71–73
    narrow band radiation, **50–51**
amplified spontaneous emission, *see* ASE
amplifiers
    CPA, *see* ultrafast pulses, chirped-pulse amplification
    design of, **77**
    EDFA, *see* laser, EDFA
    large aperture, 153, 376
    MOPA, 258
    multipass, **487**
    OPCPA, **487**, 489
    optical parametric, 424
    Raman, *see* laser, fibre Raman amplifiers
    regenerative, **486–487**
    ultrafast, *see* ultrafast pulses, amplifiers for
anisotropic media, 400–405
AOPDF, 481
applications of lasers
    alignment, 321
    CD writing, 329
    cutting, 295, 351
    eye surgery, 377
    fibre Bragg gratings, 329
    flow cytometry, 329
    inertial confinement fusion, 375–376
    isotope enrichment, 397
    laser cladding, 351
    laser Doppler velocimetry (LDV), 329
    lithography, 377
    marking and etching, 295, 351
    photochemistry, 376–377
    photodynamic therapy (PDT), 397
    printing, 295
    resonant ionization spectroscopy (RIMS), 396–397
    short pulses, 204
    spectroscopy, 329
    telecommunications, 280–282
    welding, 295, 351
ASE, 293, 505
ASE lasers, **83–85**, 506, **506–510**, 538, 541–543
    coherence properties, 509–510
    gain narrowing in, 508
    saturation in, 506–508, 543–545
autocorrelation, 489–492, 499–500
    scanning, 490
    single-shot, 491–492
    table of deconvolution factors, 491

band structure, 228–235
beam quality, **103–106**, 156, 291, 487
beam waist, *see* Gaussian beam
Beer's law, 54
Bernard–Duraffourg condition, *see* laser, diode, condition for gain
birefringence, 146, 195, 401
blackbody, 55
blackbody radiation, **7–9**, 14, 59
    energy density of, 8–9
Bloch wave functions, 228
Born–Oppenheimer approximation, *see* molecules, Born–Oppenheimer approximation
bottlenecking, **75**, 76, 82, 143, 199
broadening
    amorphous solids, **38**
    Doppler, **36–38**
    homogeneous, **27–35**
    in dye lasers, 383
    in solid state, **142**
    inhomogeneous, **35–38**
        class of atoms, 39
    natural, **27–32**
        classical model, 27–30
        quantum theory, 31–32
    phonon, 35, **142**
    power, **65**

broadening (*cont.*)
  pressure, **32–35**, 346
    classical model, 32–34
    impact approximation, 34
    quantum theory, 34–35
    quasi-static approximation, 34–35
buried mesa design, *see* laser, diode

carrier wave, 463
carrier-envelope offset (CEO), **204**, 207, **466**, 496
carrier-envelope phase (CEP), 207, **466**
cavity, **85–103**
  Q value, 109
  closed, 85–86
  cold, 106–111, 431, 432
  confocal unstable, 98
  diffraction loss, **89**, **97**, 101, 102, 128–130
  examples of, 93–96
  finesse, **109**, 131
  free spectral range, 109, 454
  high-loss, 97–103, 126–128, 373–375, 378–379
  lifetime, 109, 110, 131, 189, 200, 203, 441
  low-loss, 89–97
  near-concentric, 95, 124
  near-hemispherical, 96, 124
  near-planar, 95, 124
  stability condition, **94**
  symmetric confocal, 95, 124
  tolerance to misalignment, 125
  unstable, 97–103, 126–128, 378–379
    negative-branch, 98
    positive-branch, 98
cavity dumping, 221–222
cavity modes, 86–89, **89–103**
  calculation of, 100–103
  effect of cavity parameters, 93–96
  frequency pulling of, 431–433
  higher-order, 91, 103–106
  longitudinal, **84**, 251–254
    effect of dispersion, 496
  lowest-order, 89–91
  selection of, 96–97, 155, 198
  transverse, **87**, 250–251
central-field approximation, **133**
ceramic hosts, *see* laser, solid-state
chirped mirrors, *see* dispersion control
chirped pulse amplification, *see* ultrafast pulses, chirped-pulse amplification
classical decay rate, **29–30**
  relation to Einstein A coefficient, 30
collision cross-section, 34
concentration quenching, **139–142**
configuration coordinate diagrams, *see* energy levels, configuration coordinate diagrams
confocal parameter, *see* Gaussian beam
Correspondence principle, 173, 358, 378
cross-section
  absorption, **54**, 215
  collisional excitation, 307–309, 324
  collisional ionization, 306–307
  homogeneous broadening, **46–49**
  inhomogeneous broadening, **52–53**
  net gain, 282
  optical gain, **46–49**
    frequency dependence, 48–49
    effective, 161–162
  orders of magnitude, 53
  Raman, 286, 287
  superelastic, 310–311
  tables of, 177, 293, 315, 321, 343, 376, 386
crystal field, 35, *see* energy levels, of ions in solid state
crystalline hosts, *see* laser, solid-state, crystalline hosts

DBR, *see* distributed feedback reflector
decibel (dB), 267
density of states
  in semiconductors, 231–232
detailed balance
  principle of, 14, 330
diatomic molecules
  Franck–Condon factor, 383
diffraction
  far field, 91, 99, 377
  Kirchoff integral, 101
  loss, *see* cavity, diffraction loss
  near field, 99
diffraction-limited beam, 105, 106
diode lasers, *see* laser, diode
diode pumping, *see* pumping, optical, diode
direct band-gap semiconductor, 228
discharge physics, **298–314**
  excitation rates, 307–309, 329–330
  excited-state populations in, 311–314, 330
  glow discharges, 299–303
    temperatures in, 300–303
  ionization rates, 306–307
  low vs. high pressure, 298–299
  negative glow, 299–300
  positive column, 300
  steady-state conditions, 303–305
  superelastic collisions, 310–311, 330
dispersion
  effect of focusing, 468–469
  effect on propagation, **466–473**
  Gaussian pulse analysis, **469–472**, 497–498
  group delay, 466–468
  group-delay dispersion (GDD), 467–468
  group-velocity dispersion (GVD), 467
  in phase matching, 408–414, 418–420, 537
  non-linear, 472–473
  table of material dispersion, 468
  waveguide, 87, 274–276, 296
dispersion control, 183, **474–482**
  chirped mirrors, 478–479
  geometric, **474–478**, 498
  grating pair, 475–476
  grating stretcher, 498
  postive GDD, 477–478
  prism pair, 476–477
  pulse shaping, 480–482

distributed feedback reflector, 254, 292
doping, 233–235
Doppler broadening, *see* broadening, Doppler
Doppler-free techniques, 449, 451
double refraction, 401, 419
dye lasers, *see* laser, dye
dye molecules, *see* molecules

EDFA, *see* laser, EDFA and laser, Er^{3+}, Glass
effective electron mass, 229
Einstein coefficients, **12–15**
    allowance for broadening, **38–40**
    calculation of B coefficient, 551–555
    definition of, 13
    homogenous broadening, **38–39**
    inhomogenous broadening, **39–40**
    relations between, **14–15**, 555
    semiclassical values, 19–21
    spectral, 38–39, **557–559**
elastic collisions, 301
electro-optic Q-switching, *see* Q-switching, electro-optic
energy levels
    of dye molecules, **380–387**
        singlet–singlet absorption, 382–385
        singlet–singlet emission, 385–387
        triplet–triplet absorption, 387
    configuration coordinate diagrams, 172–174
    electron configurations, 134
    gross structure, 134
    in quantum wells, 244–247
    in semiconductors, 228–230
    of free atoms and ions, 133–136, 332–333
    of ions in solid state, 35, 38, **132–137**, 146, 158
    of molecules, 332–333
        diatomic, **356–358**
        homonuclear, 332
        P- and R-branches, 354
        rotational, 333
        vibrational, 332–333
    residual electrostatic interaction, 133
    spin-orbit interaction, 134
    terms, 135
excimer molecules, **364–367**
    rare-gas excimers, 367–369
    rare-gas halides, **370–371**
    spectroscopy of, 370–371
extinction coefficient, 398
extraordinary wave, 401, 403–405, 429

Faraday effect, 438
Faraday isolator, 393, 438, 439
Fermi energy, 232
Fermi–Dirac distribution, 232
fibre Bragg grating, **253**, 254, **254**, 268, 278, 279, 329
fibre lasers, *see* laser, fibre
fibre Raman amplifiers, *see* laser, fibre Raman amplifiers
filters (in optical fibres), 278
finesse, *see* cavity, finesse
flashlamp pumping, *see* pumping, optical, flashlamp
fluence, **76**, 215, 483

FM modelocking, *see* modelocking, FM
four–level lasers, *see* 4-level lasers
Fox–Li method, **100–102**, 128–129
Fox–Smith interferometer, 440, 457–458
fractional round-trip loss, **118**
Franck–Condon loop, *see* molecules
Franck–Condon principle, *see* molecules
Frantz–Nodvik treatment, *see* gain saturation, pulsed amplifier
free spectral range, *see* cavity, free spectral range
frequency chirp, 239, 463, 465, 467, 471, 473, 493
frequency combs, 207, **453–456**
frequency doubling, **406–408**
    materials for, 416–417
    phase-matching, 408–409
    techniques, 420–421
frequency locking, **448–453**
    Pound–Drever–Hall technique, 452
    to atomic transition, 450–452
    to external cavity, 452–453
frequency tripling, **423**
frequency pulling, 112, **431–433**, 458–459
FROG, 492–493

gain
    at short wavelengths, 503–505
    conditions for, **16–18**, 48
    fractional round-trip, **118**
    narrowing, **49**
    necessary, but not sufficient condition, 18
gain coefficient, **49**
    saturated, **63–66**, 68–71
gain cross-section, *see* cross-section, optical gain
gain saturation, **60–77**
    comparison of homogenous and inhomogeneous broadening, 71
    homogeneous broadening, 60–67
    inhomogeneous broadening, 67–73, 115
    pulsed amplifier, 73–77
    recovery time, **62–63**, 67, 80
    saturation fluence, **76**, 483
    saturation intensity, **62**, 63
gain switching, 221
Gauss–Hermite beam, *see* Gaussian beam, higher-order
Gaussian beam, **89–97**
    beam waist, 90, 93
    confocal parameter, 89, 93
    higher-order, 91, 96, 103, 510
    lowest-order, 89–91, 96, 99, 101, 103–106
    radius of curvature, 89, 90
    Rayleigh range, 89
    spot-size, 90
glass hosts, *see* laser, solid-state, glass hosts
glow discharges, *see* discharge physics
GRIN fibre, *see* optical fibres, GRIN
group velocity, 275, 465
    dispersion of, *see* dispersion, group velocity dispersion (GVD) and dispersion, group delay dispersion (GDD)
Guoy effect, **124**

harmonic generation, *see* frequency doubling and tripling, high-harmonic generation

Helmholtz equation, 4
Hermite polynomials, 357
high-harmonic generation (HHG), **535–537**, 547
   mechanism, 536–537
hole (semiconductor), **229–230**
   current carrier, 230
   effective mass, 230
   heavy/light holes, 230
   split-off band, 230
hole-burning
   spatial, **114**, 189, 257, 393, 433, 437
   spectral, **68**, 68–69, 72, 81–82, **115–117**, 167
      image frequency, 117
homogeneous broadening, *see* broadening, homogeneous
host materials, *see* laser, solid-state, host materials

image frequency, *see* hole-burning, spectral
inelastic collisions, 302, **306–311**
   Bethe and Born approximations, 308
   excitation, 307–309
   ionization, 306–307
   Klein–Rosseland relation, 310
   superelastic, 310–311
infra-red lasers
   efficiency of, **332–335**
inhomogeneous broadening, *see* broadening, inhomogeneous
inside vapour deposition, 277
intensity noise, 446–448
intermodal dispersion, *see* optical fibres, intermodal dispersion
inverse bremsstrahlung heating, **510–511**, 517, 533, 534, 546
ionization
   barrier suppression, 514
   electron collisional, **306–307**, 309, 313–314, 512
   optical field, **514–517**, 528–530, 533–534
      threshold, 545
   over-the-barrier, 514
isoelectronic sequence, 503

joint density of states, 236, 239

Kerr effect, 194, **195**, 218–219
Kerr-lens modelocking, *see* modelocking, passive, Kerr lens
Klein–Rosseland relation, 310, 330
Kleinman's conjecture, 407, 429, **560–561**

laser, 2
   alexandrite, **177–180**
      crystal properties, 177–178
      energy levels, 178
      laser parameters, 178–179
      practical implementation, 179–180
      table of parameters, 177
   Ar^+, 219
   Ar_2, 367–369
   ArCl, 370–377
   ArF, 370–377
      applications, 377
      table of parameters, 376

   argon ion, 53, **321–328**
      applications, 329, 394
      construction, 325–327
      effect of magnetic field, 327
      energy levels, 322–325
      excitation routes, 325
      limitations to output power, 327–328
      table of parameters, 321, 326
      table of transitions, 329
   CO_2, 53, 192, 219, **338–352**
      applications, 279, 351–352, 532
      COFFEE laser, 346
      effect of He, 342–343
      effect of N_2, 341–342
      energy levels, 338–341
      gas-dynamic lasers, 349–350
      high-pressure devices, 346–349
      laser parameters, 343
      low-pressure devices, 344–346
      pre-ionization of, 347–348
      table of laser parameters, 343
      table of wavelengths, 354
      TEA laser, 347
      vibrational freezing in, 350
      waveguide lasers, 351
   Cr^{3+} ions, 180
   Cr^{3+}:LiCAF, 180
   Cr^{3+}:LiSAF, 180
   diode, 132, 219, **226–264**
      applications, 150–152, 164, 268
      beam properties, 250–257
      beam shaping, 250
      buried mesa, 241
      condition for gain, 236–241
      distributed feedback (DFB), 254
      edge-emitting, 227
      general features, 226–227
      heterostructures, 241
      homostructures, 241
      linewidth of, 254–255
      longitudinal modes of, 251–254
      materials, 243–244
      MOPA configuration, 258
      output power, 257–259
      quantum cascade, 262–264
      quantum well, 244–247
      single-mode, 253–254
      strained-layer, 261–262
      table of parameters, 244
      threshold, 247–250
      transverse modes of, 250–251
      tunable cavities, 255–257
      VCSEL lasers, 259–261
   dye, 219, **380–397**
      applications, 396–397
      c.w., 391–395
      design, 393–395
      pulsed, 388–391
      rate equations, 387–388
      solid-state, 395–396

stabilized, 453
synchronously pumped, 214
tuning of, 398
tuning range, 394
EDFA, 165, 281, **282–285**
   applications, 288
   design, 284–285
   energy levels and pumping, 282
   fabrication, 285
   gain lineshape, 282–284
Er^{3+}
   energy levels, 165–167
Er^{3+} ions, **165–168**
Er^{3+}:glass, **167–168**, **282–285**
   table of parameters, 167
Er^{3+}:YAG, **165–167**
   table of parameters, 167
excimer, see laser, rare-gas halides and laser, rare-gas excimer
F_2, 377
   table of parameters, 376
fibre, **267–295**
   applications, 295
   cladding-pumped, 290–291
   high-power, 289–295
   large mode area, 293
   linewidth of, 291–293
   oscillator–amplifier, 294–295
   Q-switched, 294
fibre Raman amplifiers, **285–289**, 297
   cross-section, 286
   fibre amplifiers, 286–287
   long-haul systems, 287–289
   table of, 287
free electron, **537–541**
   operation, 538–540
   parameters, 540–541
GRIP, **525–526**
H_2, **361–364**
H-like carbon, **532–533**
He–Ne, 53, 219
He–Ne, **314–321**
   applications, 321
   construction, 318–319
   energy levels, 316–318
   limitations to output power, 319–321
   table of parameters, 315
Kr_2, 367–369
KrCl, 370–377
KrF, 53, 370–377
   table of parameters, 376
krypton ion, 328–329
   applications, 329, 393, 394
N_2, **364**
   applications, 390
Nd^{3+} ions, **157–164**
   energy levels, 157
   linewidth, 157–158
Nd^{3+}:glass, 53, **164**, 219
   applications, 489, 524
   table of parameters, 164

Nd^{3+}:vanadate, 147, 164
   table of parameters, 164
Nd^{3+}:YAG, 53, **158–163**, 219
   applications, 390
   effective cross-section, 161–162
   energy levels, 158–161
   line broadening, 161
   practical implementation, 162–163
   table of parameters, 164, 186
Nd^{3+}:YLF, 163–164
   table of parameters, 164
OPCPA, 487
OPOs, **425–428**
   practical devices, 426–428
   table of, 428
   tuning curves, 426–427
Pr^{3+}:YLF, 169
rare-gas excimer, **367–369**
rare-gas halides, **370–377**
   applications, 375–377, 390
   beam divergence, 378
   beam properties, 373–375
   design, 371–375
   discharge pumping, 371–373
   e-beam pumping, 371
   injection seeded, 374
   output parameters, 375–377
   pulse length, 373
   table of parameters, 376
   table of wavelengths, 370
   threshold pump power, 378
ruby, 53, **174–177**, 219
   crystal properties, 175
   energy levels, 175–176
   laser parameters, 176
   practical implementation, 2, 176–177
   table of parameters, 177
single longitudinal mode, see single longitudinal mode operation
solid-state
   ceramic hosts, 149
   crystalline hosts, 147–148
   glass hosts, 148–149
   host materials, 146–149
Ti:sapphire, 53, **180–184**, 219
   applications, 483–489, 528, 537
   crystal properties, 180
   energy levels, 181
   laser parameters, 181–182
   practical implementation, 182–184, 438, 488
   table of parameters, 177
trivalent-ion group, **169–184**
   energy levels, 169–174
Xe_2, 367–369
XeBr, 370–377
XeCl, 366, 370–377
   table of parameters, 376
XeF, 370–377
Yb^{3+}:glass
   table of parameters, 293

584 *Index*

laser linewidth, **440–448**
  of diode lasers, 254–255
    practical limitations, 444–446, 456
    Schawlow–Townes limit, 255, **441–444**, 459–460
    single longitudinal mode operation, **433–440**
      by intra-cavity etalon, 435–437, 456–457
      by short cavity, 434
      ring resonators, 437–439
    single-longitudinal mode operation
      by intra-cavity etalon, 184
laser oscillation
  above threshold, **111–117**
  multimode, 114
  steady-state, 112–113, 117
  threshold condition, 110–111
laser spiking, **188–193**
  numerical analysis, 192–193
  rate-equation analysis, 190
laser–solid interactions, **511–514**
lifetime
  fluorescence, 17, 31, 139–142
  radiative, **30–31**, 44–45, 56
lifetime broadening, *see* broadening, natural
line broadening, *see* broadening
lineshape
  normalized, **28**, 37
  Voigt profile, **40–42**, 45, 520
local thermodynamic equilibrium, 313
LTE, *see* local thermodynamic equilibrium

M-squared, *see* M^2-parameter
maser, 1, 2, 86, 174
  optical, 2
material dispersion, *see* optical fibres, material dispersion
Maxwell's equations, 4, 402–403, 405–406, 428–429
Maxwellian distribution, 36, 301, 302
mirror-less laser, *see* ASE lasers
modelocking, **203–221**
  and dispersion control, 482–483, 498–499
  active, **208–214**
    output pulse duration, 214
    synchronous pumping, 214
  AM, **209–211**
  analytical treatment, 206–208, 223–224
  examples, 219–221
  FM, **211–214**, 224–225
  harmonic, **208**
  passive, **214–221**
    Kerr lens, 218–219, 225
    saturable absorbers, 215–218
    SESAMs, 216–217
  pre-lasing, 220
  pulsed, 219–221
  regenerative, 208
  synchronous pumping, 220
molecules
  Born–Oppenheimer approximation, 172, **356–357**
  diatomic, **356–361**
    energy levels of, 356–358
    transitions in, 358–361

  dye, **380–387**
  energy levels of, 332–333, **356–358**
    rotational, 333
    vibrational, 332–333, 357–358
  Franck–Condon
    factor, 359, 382
    loop, 360–361, 378
    principle, 358–361
  transitions in
    electronic, 358–361
    P- and R-branches, 335
    rotational, 335
multiquantum-well device, *see* laser, diode, quantum well

natural broadening, *see* broadening, natural
negative-branch cavity, *see* cavity, unstable negative-branch
noise eater, 447
non-radiative transitions
  effect on quantum yield, 187
  in solid state, **138–142**
  phonon de-excitation, **138–139**
non-linear optics
  crystal classes, **400–405**
    table of, 402
  parametric interactions, 424–425
  quasi-phase matching, 418–420
  table of materials, 415
  three-wave mixing, 421–423
non-linear refractive index, 472
  B-integral, 472, 496–497
  table of, 468

OPOs, *see* laser, OPOs
optical fibres, **267–282**
  bulk losses in Si, 281
  components, 280
  directional couplers, 278–280
  dispersion in, 274–276
  fabrication, 276–277
  GRIN, 296
  importance of, 267–268
  intermodal dispersion, 274, 296
  material dispersion, 275–276
  mode cut-off condition, 272
  modes, 271–274
  numerical aperture, 270, **270**, 274, 293, 296
  ray-optics description, 268–270
  single-mode, 270, **273**, 274, 276, 290, 296
  step-index, 268–270
  table of LP modes, 274
  table of parameters, 270
  telecommunications bands, 280–282
  wave-optics description, 271–274
  waveguide dispersion, 275–276
optical frequency comb, *see* frequency combs
optical gain cross-section, *see* cross-section, optical gain
optical maser, *see* maser, optical
optical nutation, *see* Rabi flopping
optical parametric oscillator, *see* laser, OPOs
optical thickness, 55, 56

ordinary wave, 401, 403–405
oscillator strength, **30**, 43–45
  absorption, 30
  emission, 30
  weighted, 30
output power, **117–123**
  high-gain lasers, 120–122
  low-gain lasers, 117–120
  optimum coupling, 119–120

phase velocity, 432, 466
phase matching, **408–420**, 430
  angular acceptance, 412, 430
  critical and non-critical, 412–414
  in biaxial crystals, 415–416
  of high harmonics, 537
  of parametric generation, 427
  periodic poling, 418–420
  quasi-phase matching, 418–420
  table of types, 412
  Types I and II, 409–412
  walk-off, 414
phonon broadening, *see* broadening, phonon
phonon de-excitation, *see* non-radiative transitions, phonon de-excitation
phonons, 35, 142
Planck's Law, **7–9**
plasma
  accelerator, 462
  critical density of, 512
  critical surface, 512
  discharge, 298–314
  frequency, 512
  laser-generated, **510–517**
  refraction in, 546–547
  scale length of, 547
Pockels
  cell, **196**, 210, 214, 486
  effect, **195**
ponderomotive energy, 510, 515, 545
population inversion, **16**
  density, **47**
  partial, **335–338**
positive-branch cavity, *see* cavity, unstable positive-branch
power broadening, *see* broadening, power
pulse envelope, 463
pumping
  discharge, 307–309
  electron beam, 367, 369
  of short-wavelength lasers, 520–528
  optical, 142, **149–156**, 361
    diode, 145, **150–152**, 154–156, 258, 261, 262
    flashlamp, **152–154**
    sources, **149–152**
  threshold, 110–111, 143–146
  travelling-wave, 84, **505–506**

Q-switching, **193–203**
  acousto-optic, 196–197
  electro-optic, 195–196
  energy-utilization factor, 201
  examples, 163, 177, 180
  numerical simulations, 203
  of fibre lasers, 294
  output-pulse duration, 201–202
  over-pumping ratio, 191, 201
  peak output power, 201–202
  pulse build-up, 202–203
  rate-equation analysis, **198–203**, 222–223
  rotating mirror, 194
  saturable absorbers, 197–198
  SESAMs, **198**
  techniques for, 194–198
QCL, *see* laser, diode, quantum cascade
quantum cascade lasers, *see* laser, diode, quantum cascade
quantum ratio, 333–335, 339
quantum-well lasers, *see* laser, diode, quantum well
quarter-wave stack, 254
quiver motion, 510, 511, 515, 545

Rabi
  flopping, **22–23**
  frequency, 23, 550
radiation modes
  closed cavity, **3–7**
  density of, **7**
  free-space, 7
radiation trapping, 42–43, **56**
radiative lifetime, *see* lifetime, radiative
radiative transitions
  electric dipole, **20–21**
  higher-order, 21
  in semiconductors, 235–236
  of ions in crystals, **137–138**
Raman amplifiers, *see* laser, fibre Raman amplifiers
Raman scattering, **285–286**
rate equations, **22–24**
  for two-level atom, 25
  inhomogeneous broadening, 39–40
  narrowband radiation, **51**
  validity of, 23–24, 555–556
recombination
  three-body, 530, 531
recombination lasers, *see* short-wavelength lasers, recombination
recovery time
  absorprion, 79, 80, **215–218**
  gain, **62–63**, 63, 67, 78–79, 484
relative intensity noise, *see* intensity noise
relaxation oscillations, **188–193**, 223, 447
  rate-equation analysis, 190–192
Rigrod analysis, *see* output power, high-gain lasers
rotating-wave approximation, 550

Saha equation, **513–514**
saturable absorbers, **197–198**, 215–219
  fast, **215**
  slow, **217**
saturation, *see* gain saturation
Schawlow–Townes limit, *see* laser linewidth
second-harmonic generation, *see* frequency doubling

selection rules, 19
　electric dipole, **20–21**, 26
　higher-order, 21
　in diatomic molecules, 335
　in molecules, 333
　in quantum well, 265
　parity, 20
self-absorption, 42–43, **55**
self-phase modulation, 455, **472–473**, 483, 484
self-terminating transition, **18**, 363
self-trapping, *see* radiation trapping
semi-classical model, **19–21**
semiconductor, **228–236**
　degenerate, 234–235
　diode laser gain, 235–241
　heavy doping, 234–235
　intrinsic, 233
　light doping, 233
　p-i-n junction, 236–238
　physics of, 228–235
semiconductor diode lasers, *see* laser, diode
semiconductor materials, **243–244**
　AlAs, 243–244
　GaAlAs, 243–244, 261–262
　GaAs, 243–244
　GaN, 243
　III-V materials, 243–244
　InGaN, 243
　lattice matching, 243
　substrates, 244
SESAMs, *see* modelocking, passive, SESAMs
SHG, *see* frequency doubling
short-wavelength lasers
　coherence properties, 509–510
　collisionally excited, 517–530
　general properties, 505–510
　Ne-like ions, 518–520
　Ni-like ions, 520
　optical field ionization, 528–530, 533–534
　pumping of, 520–528
　recombination, 530–534
　spectral width, 508
　threshold and saturation, 506–508
single longitudinal mode operation, **433–440**
slope efficiency, **119**, 163, 265
spatial chirp, 463, 476
spectral regions
　extreme ultraviolet (XUV), **503**

　infra-red, 332
　soft X-ray, **503**
　terahertz, 332
　UV and VUV, **355**
　vacuum ultraviolet (VUV), **503**
SPIDER, **493–495**, 500–501
spin-orbit interaction, *see* energy levels, spin-orbit interaction
spontaneous emission, **12**
spot-radius, *see* Gaussian beam, spot-size
spot-size, *see* Gaussian beam, spot-size
stable cavities, *see* cavity, low-loss
Stefan–Boltzmann Law, 11
stimulated emission, 1, **12**
strained-layer lasers, *see* laser, diode, strained layer
superlattice, 247

temperature
　distribution-over-states T_{exn}, 301
　electron T_e, 301
　kinetic T_{gas}, 300
thermal conductivity
　of gas, 342, 352
　of solid-state hosts, 146, 147, 149, 164, 487
THG, *see* frequency tripling
three-level lasers, *see* 3-level lasers
time–bandwidth product, 224, **462–463**, 470, 495–496
tunnelling ionization, *see* ionization, optical field

ultrafast pulses
　amplifiers for, 486–487
　chirped-pulse amplification, 482–489, 499
　measurement, **489–495**
　oscillators, 483
　sources of, **482–489**
　TW and PW systems, 488–489
undulator, 538
unstable cavities, *see* cavity, high-loss

variable-reflectivity mirrors, **102–103**
VCSEL lasers, *see* laser, diode, VCSEL lasers
vibrational freezing, *see* laser, CO_2, vibrational freezing in

waveguide dispersion, *see* dispersion, waveguide
Wiener–Khintchine theorem, 460
wiggler, 538